AF352785

Advancing Complexity Research in Earth Sciences and Geography

Advancing Complexity Research in Earth Sciences and Geography

Editor

Jianbo Gao

Basel • Beijing • Wuhan • Barcelona • Belgrade • Novi Sad • Cluj • Manchester

Editor
Jianbo Gao
Faculty of Geographical
Sciences
Beijing Normal University
Beijing, China

Editorial Office
MDPI
St. Alban-Anlage 66
4052 Basel, Switzerland

This is a reprint of articles from the Special Issue published online in the open access journal *Applied Sciences* (ISSN 2076-3417) (available at: https://www.mdpi.com/journal/applsci/special_issues/Earth_Geography).

For citation purposes, cite each article independently as indicated on the article page online and as indicated below:

Lastname, A.A.; Lastname, B.B. Article Title. *Journal Name* **Year**, *Volume Number*, Page Range.

ISBN 978-3-0365-9586-3 (Hbk)
ISBN 978-3-0365-9587-0 (PDF)
doi.org/10.3390/books978-3-0365-9587-0

Contents

Jianbo Gao
Advancing Complexity Research in Earth Sciences and Geography
Reprinted from: *Appl. Sci.* **2023**, *13*, 12275, doi:10.3390/app132212275 . . . . **1**

Xinghua Cheng and Zhilin Li
Configurational Entropy for Optimizing the Encryption of Digital Elevation Model Based on
Chaos System and Linear Prediction
Reprinted from: *Appl. Sci.* **2021**, *11*, 2402, doi:10.3390/app11052402 . . . . **7**

Muhammad Jamil, Numair Ahmed Siddiqui, Abdul Hadi Bin Abd Rahman, Noor Azahar Ibrahim, Mohd Suhaili Bin Ismail, Nisar Ahmed, et al.
Facies Heterogeneity and Lobe Facies Multiscale Analysis of Deep-Marine Sand-Shale
Complexity in the West Crocker Formation of Sabah Basin, NW Borneo
Reprinted from: *Appl. Sci.* **2021**, *11*, 5513, doi:10.3390/app11125513 . . . . **25**

Liming Gao, Lele Zhang, Yongping Shen, Yaonan Zhang, Minghao Ai and Wei Zhang
Modeling Snow Depth and Snow Water Equivalent Distribution and Variation Characteristics
in the Irtysh River Basin, China
Reprinted from: *Appl. Sci.* **2021**, *11*, 8365, doi:10.3390/app11188365 . . . . **49**

Thomas Hitziger, Luisa Näke and Karel Pavelk
Ice Elevation Change Based on GNSS Measurements along the Korth-Traverse in Southern
Greenland
Reprinted from: *Appl. Sci.* **2022**, *12*, 12066, doi:10.3390/app122312066 . . . . **65**

Luigi Cucci and Francesca R. Cinti
In Search of the 1654 Seismic Source (Central Italy): An Obscure, Strong, Damaging Earthquake
Occurred Less than 100 km from Rome and Naples
Reprinted from: *Appl. Sci.* **2022**, *12*, 1150, doi:10.3390/app12031150 . . . . **81**

Niloofar Alaei, Mehrdad Soleimani Monfared, Amin Roshandel Kahoo and Thomas Bohlen
Seismic Imaging of Complex Velocity Structures by 2D Pseudo-Viscoelastic Time-Domain
Full-Waveform Inversion
Reprinted from: *Appl. Sci.* **2022**, *12*, 7741, doi:10.3390/app12157741 . . . . **93**

Mohamed Hamdache, José A. Peláez, Dragomir Gospodinov, Jesús Henares, Jesús Galindo-Zaldívar, Carlos Sanz de Galdeano and Boyko Ranguelov
Stochastic Modeling of the Al Hoceima (Morocco) Aftershock Sequences of 1994, 2004 and 2016
Reprinted from: *Appl. Sci.* **2022**, *12*, 8744, doi:10.3390/app12178744 . . . . **109**

Paul Edigbue, Ismail Demirci, Irfan Akca, Hamdan Ali Hamdan, Panagiotis Kirmizakis, Pantelis Soupios, et al.
A Comprehensive Study of Local, Global, and Combined Optimization Methods on Synthetic
Seismic Refraction and Direct Current Resistivity Data
Reprinted from: *Appl. Sci.* **2022**, *12*, 11589, doi:10.3390/app122211589 . . . . **135**

Chiara Martinello, Claudio Mercurio, Chiara Cappadonia, Miguel Ángel Hernández Martínez, Mario Ernesto Reyes Martínez, Jacqueline Yamileth Rivera Ayala, et al.
Investigating Limits in Exploiting Assembled Landslide Inventories for Calibrating Regional
Susceptibility Models: A Test in Volcanic Areas of El Salvador
Reprinted from: *Appl. Sci.* **2022**, *12*, 6151, doi:10.3390/app12126151 . . . . **157**

Chiara Martinello, Claudio Mercurio, Chiara Cappadonia, Viviana Bellomo, Andrea Conte, Giampiero Mineo, et al.
Using Public Landslide Inventories for Landslide Susceptibility Assessment at the Basin Scale: Application to the Torto River Basin (Central-Northern Sicily, Italy)
Reprinted from: *Appl. Sci.* **2023**, *13*, 9449, doi:10.3390/app13169449 . . . . **173**

Nafiseh Haghtalab, Nathan Moore and Pouyan Nejadhashemi
Would Forest Regrowth Compensate for Climate Change in the Amazon Basin?
Reprinted from: *Appl. Sci.* **2022**, *12*, 7052, doi:10.3390/app12147052 . . . . **187**

Yushuo Zhang, Boyu Liu and Renjing Sui
Evaluation and Driving Determinants of the Coordination between Ecosystem Service Supply and Demand: A Case Study in Shanxi Province
Reprinted from: *Appl. Sci.* **2023**, *13*, 9262, doi:10.3390/app13169262 . . . . **201**

Shenglei Xu, Yunjia Wang, Meng Sun, Minghao Si and Hongji Cao
A Real-Time BLE/PDR Integrated System by Using an Improved Robust Filter for Indoor Position
Reprinted from: *Appl. Sci.* **2021**, *11*, 8170, doi:10.3390/app11178170 . . . . **225**

Cong Liao, Teqi Dai, Pengfei Zhao and Tiantian Ding
Weighted Centrality and Retail Store Locations in Beijing, China: A Temporal Perspective from Dynamic Public Transport Flow Networks
Reprinted from: *Appl. Sci.* **2021**, *11*, 9069, doi:10.3390/app11199069 . . . . **255**

Zhuolin Tao, Qi Wang and Wenchao Han
Towards Health Equality: Optimizing Hierarchical Healthcare Facilities towards Maximal Accessibility Equality in Shenzhen, China
Reprinted from: *Appl. Sci.* **2021**, *11*, 10282, doi:10.3390/app112110282 . . . . **271**

Xin Huang and Xiaojuan Liu
Incorporating a Topic Model into a Hypergraph Neural Network for Searching-Scenario Oriented Recommendations
Reprinted from: *Appl. Sci.* **2022**, *12*, 7387, doi:10.3390/app12157387 . . . . **289**

Sha Sun, Haiyue Xu, Minsong He, Yao Xiao and Huayong Niu
An Alternative Globalization Barometer for Investigating the Trend of Globalization
Reprinted from: *Appl. Sci.* **2022**, *12*, 7896, doi:10.3390/app12157896 . . . . **313**

Bin Liu and Jianbo Gao
Normality in the Distribution of Revealed Comparative Advantage Index for International Trade and Economic Complexity
Reprinted from: *Appl. Sci.* **2022**, *12*, 1125, doi:10.3390/app12031125 . . . . **333**

Giuseppe Bilotta, Annalisa Cappello and Gaetana Ganci
Formal Matters on the Topic of Risk Mitigation: A Mathematical Perspective
Reprinted from: *Appl. Sci.* **2023**, *13*, 265, doi:10.3390/app13010265 . . . . **347**

Jianbo Gao and Bo Xu
Complex Systems, Emergence, and Multiscale Analysis: A Tutorial and Brief Survey
Reprinted from: *Appl. Sci.* **2021**, *11*, 5736, doi:10.3390/app11125736 . . . . **357**

Editorial

Advancing Complexity Research in Earth Sciences and Geography

Jianbo Gao [1,2]

[1] Center for Geodata and Analysis, Faculty of Geographical Science, Beijing Normal University, Beijing 100875, China; jbgao.pmb@gmail.com
[2] Institute of Automation, Chinese Academy of Sciences, Beijing 100190, China

Citation: Gao, J. Advancing Complexity Research in Earth Sciences and Geography. *Appl. Sci.* **2023**, *13*, 12275. https://doi.org/10.3390/app132212275

Received: 8 November 2023
Accepted: 10 November 2023
Published: 13 November 2023

1. Introduction

Many complex phenomena in earth sciences and geography, including nonlinear fluid motions in the atmosphere, oceans, rivers, and lakes, coastal morpho dynamics, volcanic and seismic activities, the spatiotemporal dynamics of species, human movement trajectory, and city transportation dynamics, among many others, have played significant roles in the creation and development of complexity science, particularly chaos theory and fractal geometry [1]. With big data rapidly accumulating in almost every branch of earth sciences and geography, our increasing understanding of complex systems, and the availability of richer and more powerful methods for modeling complex systems, a golden age for the study of the complexity of the earth and our living environment has emerged. This book arises from a Special Issue of *Applied Sciences* that aimed to systematically examine the many complex phenomena that occur in earth sciences and geography, employing state-of-the-art methods for modeling complex data in order to invigorate research in earth sciences and geography, and to facilitate the further development of complexity science. Altogether, this Special Issue comprises 20 papers, contributed by researchers from all over the world and covering a range of diverse topics, including the encryption of digital elevation models [2], facies heterogeneity [3], the simulation of the snow cover process [4], the exploration of ice elevation change [5], earthquake and seismic activity [6–9], landslide susceptibility [10,11], the effect of reforestation [12], coordination between the supply and demand of ecosystem services [13], indoor positioning [14], public transport flow networks and retail store locations [15], the equality of healthcare facilities [16], recommender systems for e-retail [17], globalization [18], international trade and optimal industrial structure [19], risk analysis [20], and the quantification of political processes [21]. Below, I briefly explain the premise of each work, and when appropriate, highlight what could be further explored in future.

2. Topics Covered in the Book and Future Perspectives

The encryption of digital elevation models (DEMs) is a crucial task in geosciences. In their study, Cheng and Li [2] tackle this issue by integrating a chaos system and a linear prediction technique. While their technique is innovative and interesting, in the future it would be interesting to determine which currently available encryption scheme, including those developed by electrical engineers and computer scientists, operates the best for this purpose.

In their study, Jamil et al. [3] study facies heterogeneity in the West Crocker Formation of Sabah in northwest Borneo. By using the lithological characteristics, bed geometry, sedimentary textures and structures of individual beds, they categorize the rock units into nine sedimentary lithofacies: five sandstone lithofacies (S1–S5), one hybrid bed facies (H), two siltstone facies (Si1 and Si2), and one shale or mudstone facies (M). These facies were then grouped into four facies associations (FA1–FA4), which were further interpreted as lobe axis (FA1), lobe off-axis (FA2), lobe fringe (FA3), and distal fringe to interlope (FA4) facies associations. In future, it would be interesting to determine whether this approach may be applicable for the determination of the distribution of lobes and their sub-seismic, multiscale complexities, for the purpose of characterizing the potential hydrocarbon intervals in deep-marine sand-shale systems around the globe.

The accurate simulation of the snow cover process is of great significance to the study of climate change and the water cycle. In their study, Gao et al. [4] use the China Meteorological Forcing Dataset (CMFD) and ERA-Interim as driving data to simulate the dynamic changes in the snow depth and snow water equivalent (SWE) in the Irtysh River Basin from 2000 to 2018 using the Noah-MP land surface model; they compare the simulation results with the gridded dataset of snow depth at Chinese meteorological stations (GDSD), the long-term series of the daily snow depth dataset in China (LSD), and China's daily snow depth and snow water equivalent products (CSS). The authors find that the rainfall and snowfall (SNF) scheme mainly affects the snow accumulation process, while the surface layer drag coefficient (SFC), snow/soil temperature time (STC), and snow surface albedo (ALB) schemes mainly affect the melting process.

Hitziger et al. [5] provide a fascinating account of a series of geodetic expeditions conducted in order to explore ice elevation change based on GNSS measurements along the Korth-Traverse in Southern Greenland. The efforts made by the researchers in these expeditions are truly inspirational.

In the cluster of papers on earthquakes and seismic activity, Cucci et al. [6] make efforts to compile all of the information available regarding the M6.3 earthquake that occurred in southern Lazio (Central Italy) in 1654, the strongest seismic event to have ever occurred in the area, in order to provide reliable landmarks with which to identify its seismic source. Alaei et al. [7] propose a 2D pseudo-viscoelastic time-domain full-waveform inversion approach for the seismic imaging of complex velocity structures. Hamdache et al. [8] employ a stochastic model entitled the restricted epidemic-type aftershock sequence (RETAS) to examine the similarities/differences in the three aftershock sequences that occurred in Al Hoceima, Morocco, in May 1994 (Mw 6.0), February 2004 (Mw 6.4) and January 2016 (Mw 6.3). In addition, in their study, Edigbue et al. [9] develop a combined local and global optimization approach for jointly inverting two-dimensional direct current resistivity (DCR) and seismic refraction (SR) data for the purpose of reliably estimating the corresponding physical model parameters.

On the issue of landslide susceptibility, in their study, Martinello et al. [10] first evaluate the reliability of regional landslide susceptibility models obtained by exploiting inhomogeneously collected inventories for calibration. They find that models appearing to perform well on a large scale may actually perform very poorly on a local scale. Then, they choose the Torto River Basin (Central-Northern Sicily, Italy) as an example, and propose a technique with which to overcome the limitations of Public Landslide Inventories in order to assess landslide susceptibility more reliably [11]. The assessment of landslide susceptibility is certainly of enormous practical importance. It would be interesting to observe whether some salient patterns or regularities can be found in the measured landslide data so that the assessment of landslide susceptibility is not solely data-driven, but also has a sound theoretical foundation.

Haghtalab et al. [12] examine the impacts of potential tropical reforestation on surface energy and moisture budgets, including precipitation and temperature. Using WRF.V3.9 (weather research and forecast model), they find that forest rehabilitation across the Amazon Basin can make the atmosphere cooler, with more moisture and latent heat (LH), especially between May and November. Choosing a large watershed area with a number of counties, Zhang et al. [13] employ the coupling coordination degree model (CCDM) and examine the coordination between supply and demand in ecosystem services (ESs), including crop production, water retention, soil conservation, carbon sequestration, and outdoor recreation. Within their study area, they find that different regions could be classified into four distinct types: extreme incoordination, moderate incoordination, reluctant coordination, and moderate coordination. As one could readily expect, a mountain ecosystem belongs to the first category, where the ES supply is much greater than the demand. This study is based on data collected in 2000 and 2020. It would be interesting to observe how the degree of coordination between supply and demand in ESs continuously varies with time.

In their study, Xu et al. [14] develop a real-time Bluetooth low-energy (BLE)/pedestrian dead-reckoning (PDR) integrated system for enhanced indoor positioning. The system is based on constructing a robust vector that is responsible for changing the observation co-variance matrix of the extended Kalman filter (EKF). This is achieved by detecting the gross error at different granularities. Focusing on three weighted centrality indices in the net-works of public transport flows, namely degree, betweenness, and closeness, Liao et al. [15] find that supermarkets, convenience stores, electronics stores, and specialty stores have the highest weighted degree value. In contrast, building material stores and shopping malls have the lowest weighted closeness and weighted betweenness values, respectively. In their study, Tao et al. [16] develop a hierarchical maximal accessibility equality model to examine the equality of accessibility to healthcare services in Shenzhen, China. In addition, Huang and Liu [17] propose a more accurate personalized recommendation system for e-retailers that is also computationally more efficient. While all this research is fascinating, it would be desirable to see whether the results of these studies can be applied in practice and make a profound impact on society.

Globalization is often understood in terms of an increase in human mobility with time, an increase in the number of multinational corporations with time, as well as an increase in connectedness over time, enabled by increasingly powerful communication and information technologies. Considering this, Sun et al. [18] propose an alternative globalization index, which is a valuable addition to the globalization indices proposed previously [22–25]. One can readily see that with this kind of reasoning, globalization will generally increase with time, despite being at times disrupted by some global catastrophe, such as the COVID-19 pandemic. However, it is difficult to simultaneously understand anti-globalization with regard to this concept. In future, it would be vital to develop an approach that can simultaneously understand globalization and anti-globalization, so that superior strategies can be developed to ensure that globalization benefits more people and countries.

Analyzing massive international trade data from 1991 to 2019, Liu and Gao [19] find that deviations from normality for the distribution of revealed comparative advantage (RCA) are strongly negatively correlated with the logarithm of GDP and the Economic Complexity Index (ECI). In particular, the correlation between this deviation and GDP is stronger than that between ECI and GDP post 2008. These results suggest that this deviation may serve as an excellent new index with which to quantify the economic complexity and economic performance of a country. It would be interesting to use the entropy maximization principle to gain further insights into the approach.

With extreme weather and natural disasters occurring more frequently, risk analysis and mitigation have become increasingly crucial. Rising to this challenge, Bilotta et al. [20] provide formal mathematical expressions for hazard, the exposure of hazard, vulnerability, risk, and the mitigation of risk. It remains to be seen how these expressions can actually be computed in various scenarios of real-world importance. In future, it is perhaps even more vital to pay greater attention to insurance in countries where the insurance industry lags the development of economy, since without the proper development of the insurance industry, risk analysis cannot make a real impact. Here, of course, an important issue is to properly quantify the term "lag".

When dedicating this Special Issue of *Applied Sciences* to the study of complexity in earth sciences and geography, it is assumed that a significant fraction of researchers and students in the relevant fields understand the basics of complexity science. But however significant this fraction is, there will still be many researchers and students who require help in order to catch up with the recent developments in complexity science. This book thus includes a review article by Gao and Xu [21], which first provides a tutorial introduction to complex systems and emergence, then presents two multiscale approaches that may be useful for analyzing complex temporal dynamics in earth sciences and geography, and beyond. This article can be used as a reference for an introductory but enhanced course on complexity science for geosciences; by "enhanced", it is meant that students in the class are encouraged to perform extensive hands-on exercises, including programming, as much as possible. Solely for this purpose, instructors, as well as readers interested in the relevant computer analysis programs, are encouraged to contact the authors.

Reference [21] also briefly touches on the issue of characterizing the political evolution of various countries, utilizing news media big data. Studies in geopolitics and digital humanity may well instigate new frontiers in earth sciences and geography.

Funding: J. Gao is supported by the Fundamental Research Funds for the Central Universities in China.

Conflicts of Interest: The author declares no conflict of interest.

References

1. Gao, J.; Cao, Y.; Tung, W.; Hu, J. *Multiscale Analysis of Complex Time Series—Integration of Chaos and Random Fractal Theory, and Beyond*; Wiley: Hoboken, NJ, USA, 2007.
2. Cheng, X.; Li, Z. Configurational Entropy for Optimizing the Encryption of Digital Elevation Model Based on Chaos System and Linear Prediction. *Appl. Sci.* **2021**, *11*, 2402. [CrossRef]
3. Jamil, M.; Siddiqui, N.; Rahman, A.; Ibrahim, N.; Ismail, M.; Ahmed, N.; Usman, M.; Gul, Z.; Imran, Q. Facies Heterogeneity and Lobe Facies Multiscale Analysis of Deep-Marine Sand-Shale Complexity in the West Crocker Formation of Sabah Basin, NW Borneo. *Appl. Sci.* **2021**, *11*, 5513. [CrossRef]
4. Gao, L.; Zhang, L.; Shen, Y.; Zhang, Y.; Ai, M.; Zhang, W. Modeling Snow Depth and Snow Water Equivalent Distribution and Variation Characteristics in the Irtysh River Basin, China. *Appl. Sci.* **2021**, *11*, 8365. [CrossRef]
5. Hitziger, T.; Näke, L.; Pavelka, K. Ice Elevation Change Based on GNSS Measurements along the Korth-Traverse in Southern Greenland. *Appl. Sci.* **2022**, *12*, 12066. [CrossRef]
6. Cucci, L.; Cinti, F. In Search of the 1654 Seismic Source (Central Italy): An Obscure, Strong, Damaging Earthquake Occurred Less than 100 km from Rome and Naples. *Appl. Sci.* **2022**, *12*, 1150. [CrossRef]
7. Alaei, N.; Soleimani Monfared, M.; Roshandel Kahoo, A.; Bohlen, T. Seismic Imaging of Complex Velocity Structures by 2D Pseudo-Viscoelastic Time-Domain Full-Waveform Inversion. *Appl. Sci.* **2022**, *12*, 7741. [CrossRef]
8. Hamdache, M.; Peláez, J.; Gospodinov, D.; Henares, J.; Galindo-Zaldívar, J.; Sanz de Galdeano, C.; Ranguelov, B. Stochastic Modeling of the Al Hoceima (Morocco) Aftershock Sequences of 1994, 2004 and 2016. *Appl. Sci.* **2022**, *12*, 8744. [CrossRef]
9. Edigbue, P.; Demirci, I.; Akca, I.; Hamdan, H.; Kirmizakis, P.; Soupios, P.; Candansayar, E.; Hanafy, S.; Al-Shuhail, A. A Comprehensive Study of Local, Global, and Combined Optimization Methods on Synthetic Seismic Refraction and Direct Current Resistivity Data. *Appl. Sci.* **2022**, *12*, 11589. [CrossRef]
10. Martinello, C.; Mercurio, C.; Cappadonia, C.; Hernández Martínez, M.; Reyes Martínez, M.; Rivera Ayala, J.; Conoscenti, C.; Rotigliano, E. Investigating Limits in Exploiting Assembled Landslide Inventories for Calibrating Regional Susceptibility Models: A Test in Volcanic Areas of El Salvador. *Appl. Sci.* **2022**, *12*, 6151. [CrossRef]
11. Martinello, C.; Mercurio, C.; Cappadonia, C.; Bellomo, V.; Conte, A.; Mineo, G.; Di Frisco, G.; Azzara, G.; Bufalini, M.; Materazzi, M.; et al. Using Public Landslide Inventories for Landslide Susceptibility Assessment at the Basin Scale: Application to the Torto River Basin (Central-Northern Sicily, Italy). *Appl. Sci.* **2023**, *13*, 9449. [CrossRef]
12. Haghtalab, N.; Moore, N.; Nejadhashemi, P. Would Forest Regrowth Compensate for Climate Change in the Amazon Basin? *Appl. Sci.* **2022**, *12*, 7052. [CrossRef]
13. Zhang, Y.; Liu, B.; Sui, R. Evaluation and Driving Determinants of the Coordination between Ecosystem Service Supply and Demand: A Case Study in Shanxi Province. *Appl. Sci.* **2023**, *13*, 9262. [CrossRef]
14. Xu, S.; Wang, H.; Xing, L.; Su, J.; Zhang, Y.; Zhou, C.; Zhou, Q.; Zhou, H. Identifying the Nonlinearity of the Impact of Globalization on Carbon Dioxide Emissions: A Multiple Threshold Nonlinear Model. *Appl. Sci.* **2023**, *13*, 3811.
15. Liao, C.; Dai, T.; Zhao, P.; Ding, T. Weighted Centrality and Retail Store Locations in Beijing, China: A Temporal Perspective from Dynamic Public Transport Flow Networks. *Appl. Sci.* **2021**, *11*, 9069. [CrossRef]
16. Tao, Z.; Wang, Q.; Han, W. Towards Health Equality: Optimizing Hierarchical Healthcare Facilities towards Maximal Accessibility Equality in Shenzhen, China. *Appl. Sci.* **2021**, *11*, 10282. [CrossRef]
17. Huang, X.; Liu, X. Incorporating a Topic Model into a Hypergraph Neural Network for Searching-Scenario Oriented Recommendations. *Appl. Sci.* **2022**, *12*, 7387. [CrossRef]
18. Sun, S.; Xu, H.; He, M.; Xiao, Y.; Niu, H. An Alternative Globalization Barometer for Investigating the Trend of Globalization. *Appl. Sci.* **2022**, *12*, 7896. [CrossRef]
19. Liu, B.; Gao, J. Normality in the Distribution of Revealed Comparative Advantage Index for International Trade and Economic Complexity. *Appl. Sci.* **2022**, *12*, 1125. [CrossRef]
20. Bilotta, G.; Cappello, A.; Ganci, G. Formal Matters on the Topic of Risk Mitigation: A Mathematical Perspective. *Appl. Sci.* **2023**, *13*, 265. [CrossRef]

21. Gao, J.; Xu, B. Complex Systems, Emergence, and Multiscale Analysis: A Tutorial and Brief Survey. *Appl. Sci.* **2021**, *11*, 5736. [CrossRef]
22. Altman, S.; Bastian, P. Dhl Global Connectedness Index 2020: The State of Globalization in a Distancing World. *November* 2020.
23. Figge, L.; Martens, P. Globalisation Continues: The Maastricht Globalisation Index Revisited and Updated. *Globalizations* **2014**, *11*, 875–893. [CrossRef]
24. Gygli, S.; Haelg, F.; Potrafke, N.; Sturm, J. The KOF Globalisation Index—Revisited. *Rev. Int. Organ.* **2019**, *14*, 543–574. [CrossRef]
25. Lockwood, B.; Redoano, M. The CSGR Globalisation Index: An Introductory Guide. *Cent. Study Glob. Reg. Work. Pap.* **2005**, *155*, 185–205.

 *applied sciences*

Article

Configurational Entropy for Optimizing the Encryption of Digital Elevation Model Based on Chaos System and Linear Prediction

Xinghua Cheng [1] and Zhilin Li [1,2,*]

[1] Department of Land Surveying and Geo-Informatics, The Hong Kong Polytechnic University, Hong Kong, China; xingh.cheng@connect.polyu.hk

[2] Faculty of Geosciences and Environmental Engineering, Southwest Jiaotong University, Chengdu 611756, China

* Correspondence: lszlli@polyu.edu.hk

Abstract: A digital elevation model (DEM) digitally records information about terrain variations and has found many applications in different fields of geosciences. To protect such digital information, encryption is one technique. Numerous encryption algorithms have been developed and can be used for DEM. A good encryption algorithm should change both the compositional and configurational information of a DEM in the encryption process. However, current methods do not fully take into full consideration pixel structures when measuring the complexity of an encrypted DEM (e.g., using Shannon entropy and correlation). Therefore, this study first proposes that configurational entropy capturing both compositional and configurational information can be used to optimize encryption from the perspective of the Second Law of Thermodynamics. Subsequently, an encryption algorithm based on the integration of the chaos system and linear prediction is designed, where the one with the maximum absolute configurational entropy difference compared to the original DEM is selected. Two experimental DEMs are encrypted for 10 times. The experimental results and security analysis show that the proposed algorithm is effective and that configurational entropy can help optimize the encryption and can provide guidelines for evaluating the encrypted DEM.

Keywords: digital elevation model; information security; chaos system; configurational information; configurational entropy

Citation: Cheng, X.; Li, Z. Configurational Entropy for Optimizing the Encryption of Digital Elevation Model Based on Chaos System and Linear Prediction. *Appl. Sci.* **2021**, *11*, 2402. https://doi.org/10.3390/app11052402

Academic Editor: Yosoon Choi

Received: 2 February 2021
Accepted: 5 March 2021
Published: 8 March 2021

Publisher's Note: MDPI stays neutral with regard to jurisdictional claims in published maps and institutional affiliations.

1. Introduction

A digital elevation model (DEM) is a digital representation of terrain variations and can explicitly reveal information about the topographic complexity with computer graphics. With the development of advanced equipment for data acquisition (e.g., high-resolution satellite sensors, unmanned aerial vehicle (UAV), and LiDAR (Light Detection and Ranging)), it is becoming more and more easy to acquire DEMs. In addition, DEM transmission becomes more and more frequent due to the development of advanced computer and network communication technologies. However, due to the openness and sharing of networks, there exists a serious threat in information security and confidentiality [1,2]. Therefore, information protection is desired and hence has attracted much attention. The literature on information protection can be traced back to Shannon's paper entitled "Communication Theory of Secrecy System" [3]. By now, numerous information protection methods have been proposed, and encryption is one such solution.

An increasing number of encryption algorithms have been developed to protect information from images as much as possible, and such algorithms can be employed to protect DEMs as well. Since chaotic systems are sensitive to the initial parameters, determinacy, ergodicity, and so forth [4–7], chaotic-systems-based encryption algorithms [8–15] are popular among these methods. In general, a chaotic-system-based algorithm encrypts an image via two stages (i.e., confusion and diffusion). At the confusion stage, the positions of pixels are changed. To enhance security, the pixel values are changed at the diffusion stage. Sometimes, these two stages can be achieved simultaneously. Nevertheless, one may notice that the precision of initial parameters for generating chaotic sequences can influence the encryption performance of a chaotic system. At this point, for a given image, one may ask two questions: (i) Can we employ a metric to help optimize an encryption algorithm based on the chaos system? and (ii) What abilities should such a metric have? To answer these two questions, let us first recall the viewpoint proposed by Shannon that it is possible to break many kinds of ciphers using a statistical analysis on the histogram and the correlation of adjacent pixels in the cipher image [3]. From this viewpoint, we know that both the composition (proportions of pixels) and configurational information (spatial structures) of an image should be considered when designing an encryption algorithm and when evaluating its performance. This further suggests that we may need to find metrics for capturing both compositional and configurational information of an image.

Some metrics have been developed to evaluate the performance of encryption systems upon an image, e.g., correlation [9], NPCR (Number of Pixels Change Rate) [9,16], UACI (Unified Average Changing Intensity) [9], histogram [17], and Shannon entropy [18–20]. Theoretically speaking, these metrics are not good enough for capturing both compositional and configurational information. For example, Shannon entropy is a type of statistical entropy [21] and thus is unable to completely capture the configurational information of an image since its calculation relies on the occurrence probabilities of pixels, not the two-dimensional spatial structures. Three DEMs are shown in Figure 1, where the ones in the middle and right frames are the scrambled results of the one in the left frame. They have different spatial structures, whereas their Shannon entropy values are the same. Additionally, the information content of the multiscale representation of a DEM cannot be well-quantified by these metrics.

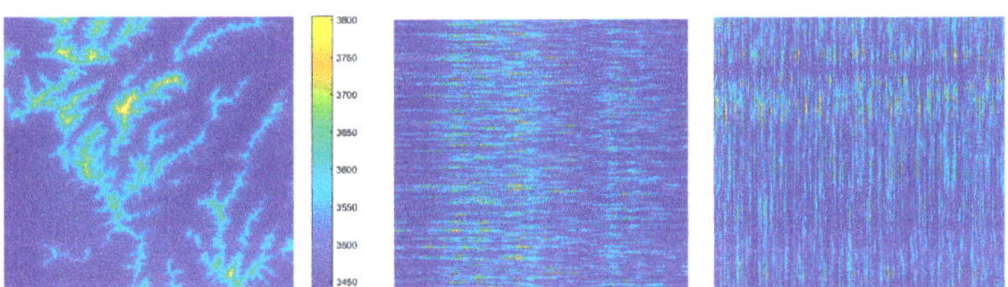

Figure 1. Three digital elevation models (DEMs) with the same histogram and, thus, same Shannon entropy values.

To bridge the gaps induced by these metrics mentioned above, this study utilizes the configurational entropy (thermodynamic entropy) to encrypt DEM. An encryption algorithm is proposed with the integration of a chaos system and linear prediction and is optimized by leveraging the configurational entropy. Apart from the Introduction section, the remainder of this study is organized as follows. The Second Law of Thermodynamics and configurational entropy are introduced first as the perspective for optimizing the DEM encryption in Section 2. Then, a novel encryption algorithm based on the leverage of configurational entropy is proposed and described in Section 3. Two DEMs are used in experiments followed by the results analysis in Section 4. Finally, a conclusion is made in Section 5.

2. The Second Law of Thermodynamics as a New Perspective for Optimizing Encryption of Numerical Raster Data

The Second Law of Thermodynamics is concerned with the direction of natural processes. This law states that an isolated and closed thermodynamic system can spontaneously evolve towards thermodynamic equilibrium, where its disorder degree (which can be measured by entropy) is at maximum [22–24]. Inspired by this law, we can assume that a DEM could be considered an isolated and closed thermodynamic system where pixels are taken as gas molecules. Different temperatures (i.e., different encryption techniques or same techniques with different initial parameters) are imposed on the same thermodynamic system (an image), and then, the gas molecules (pixels) move in different directions and finally reach one type of status. Figure 2 shows different statuses of a closed thermodynamic system under different temperatures. The disorder of the thermodynamic system represents the complexity (randomness) of an image. The gas molecules move in different directions and then form different distributions. The disorder degree of gas molecules increases from (a) to (d).

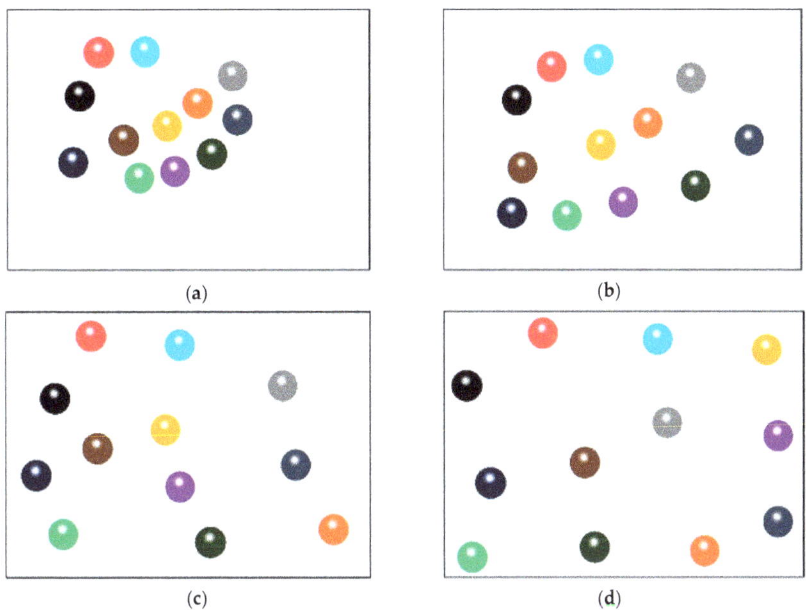

Figure 2. Four closed and isolated thermodynamic systems with the same gas molecules but different distributions.

The disorder of an isolated and closed thermodynamic system can be quantified by the thermodynamic entropy proposed by Ludwig Boltzmann [25,26]. The calculation formula for the thermodynamic entropy (configurational entropy and Boltzmann entropy) is as follows:

$$S = K \log W \tag{1}$$

where K is the Boltzmann constant (1 in the case of digital images, as suggested by [27]) and W is the number of microstates for a given macrostate. The configurational entropy of numerical raster data has been defined and computed in [28] with the assistance of the concept of multiscale representation, leading to two types of terms: relative and absolute. Concretely, the macrostate is defined as the upscaling results by an operation with a 2×2 sliding window; the microstates are all possible downscaling results, which can be seen in Figure 3. For an image, its relative configurational entropy (S_R) is the sum of configurational entropies of pixels in a sliding window of size 2×2 through the whole image. The absolute configurational entropy (S_A) is the sum of relative configurational entropies across all scales, capturing the multiscale information, which can help us enhance the analysis of the complexity of an encrypted DEM.

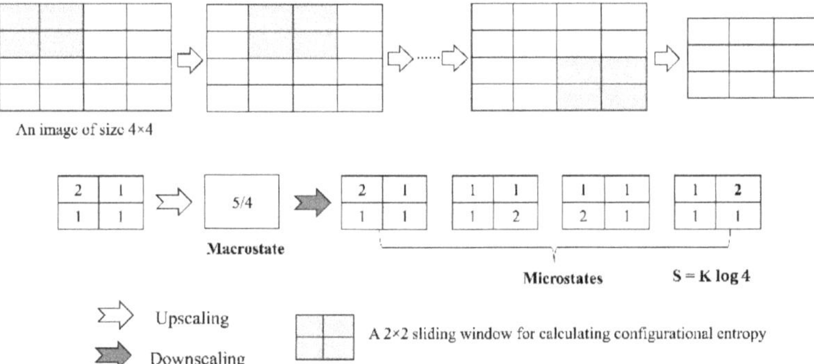

Figure 3. An example of computing the configurational entropy.

The experiments conducted in [29] demonstrates that S_R can measure the scrambling degree of grayscale images at the confusion stage. Regarding the diffusion phase included by an encryption function, the range of pixel values is modified. A good encrypted image should have various value ranges and pixel structures different from the original one. At this point, we can take the absolute configurational entropy as a metric to help choose the best one among all encrypted images. Theoretically speaking, the higher the absolute configurational entropy, the higher the complexity (and the lower the compressibility concerning lossless compression). To improve the encryption security, we should select the one with the maximum S_A value among all cases. In this study, the base of the logarithmic function in Equation (1) is set to 2 to measure the configurational information in units of bits. The configurational entropy of an image is proportional to its complexity.

3. Encryption Based on the Integration of Chaos System and Linear Prediction

Inspired by the Second Law of Thermodynamics, this section proposes an encryption algorithm consisting of two parts: (i) the encryption function and (ii) determination of the best-encrypted image with configurational entropy, which are shown in Figures 4 and 5.

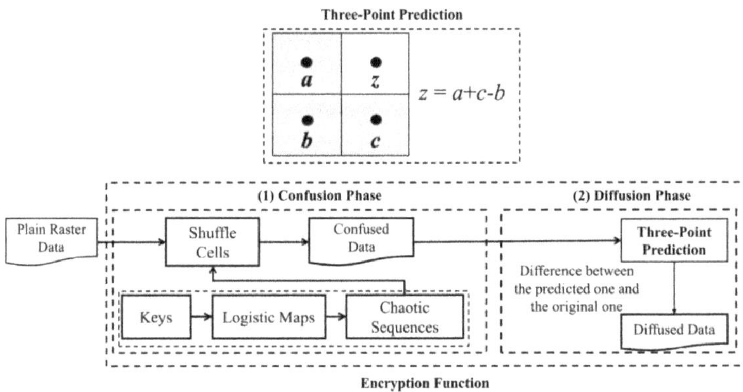

Figure 4. The proposed encryption algorithm.

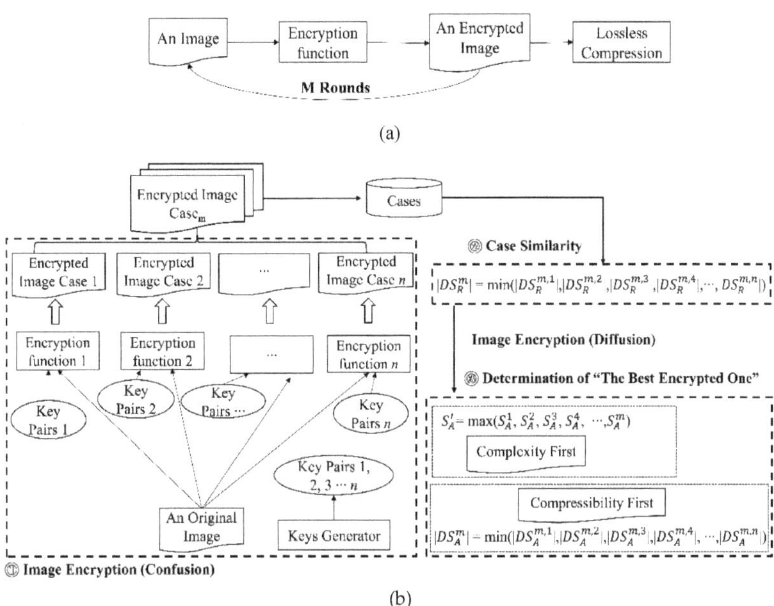

Figure 5. The schematic process of the proposed encryption function. (**a**) Encryption of a DEM (an image) for *m* rounds. (**b**) Determination of the best encrypted one with configurational entropy; *m* ($\geq$1) represents the *m* total encryption rounds; *n* ($\geq$1) represents the number of scrambled images with respect to 2*n* key pairs for generating logistic maps.

The confusion phase included under the proposed encryption function is implemented by the chaos system generated by two logistic maps with different initial parameter values. Mathematically, the logistic map [30] is written as follows

$$x_{n+1} = rx_n(1 - x_n) \tag{2}$$

where x_n is located in the interval [0,1] and $0 \leq r \leq 4$. When $r \in (3.5699456, 4)$, the sequence generated by the logistic map can show chaotic status, though there are many periodic windows in this interval. We can assume that a DEM is read as a numerical matrix of size M × N. The confusion phase scrambles the whole image, indicating that both row and column scrambling are needed. To begin this process, first, we set the initial parameter r_0 and x_0 values to iterate the chaotic system (i.e., Equation (2)) for M times and then a chaotic sequence of length M, { $x_1, x_2, x_3, x_4, x_5, ,x_m$}, is generated and referred to as S_M. Then, sorting this chaotic sequence in ascending or descending order, we get {$\bar{x}_1, \bar{x}_2, \bar{x}_3, \bar{x}_4, \bar{x}_5, ,\bar{x}_m$} named $\bar{S}_m$. Next, we need to find the position values of S_M in $\bar{S}_m$ and to record the transformation positions TP = { $tp_1, tp_2, tp_3, tp_4, tp_5, tp_m$}. When we use TP for row scrambling, we only need to move the tp_1 row to the first row and the tp_2 row to the second row until all rows are scrambled. Similarly, regarding column scrambling, new parameter r_0 and y_0 values are needed to iterate the logistic map for *N* times and then to conduct the same operation as the row scrambling.

Concerning the diffusion phase, the three-point prediction is employed. A 2 × 2 sliding window is moved pixel by pixel, which generates the predicted pixels. Regarding the edge pixels, the missing ones among pixels *a*, *b*, and *c* are automatically set to 0. Thereafter, the difference between the confused image and the predicted one is computed and then taken as the final encrypted DEM in one round. The advantages of three-point prediction are (i) reducing the correlation between pixels (increasing the complexity of an image) and (ii) changing the range of pixel values.

After introducing the encryption function, we describe how the whole encryption algorithm is optimized with the assistance of configurational entropy. As shown in Figure 5a, users can determine the total encryption rounds, m, and the number of different confusion phases, n, as illustrated in Figure 5b. The encrypted image (DEM) in the last round is taken as the input of the encryption function for the next round in the whole encryption process. Figure 5b shows how to select the best-encrypted image. An image can be scrambled by $2n$ logistic maps with $2n$ different key pairs (r_0, x_0) at the confusion phase; thus, n confused DEMs with the same histogram but different structures. Among these n confused DEMs, the one with the maximum absolute S_R difference ($|DS_R|$) compared to the original is selected as the input for the diffusion phase in which the range of pixel values is changed. The absolute configurational entropy (S_A) is finally employed to determine which one is the most suitable for transmission. From a theoretical perspective of information, the higher S_A value, the higher the complexity (lower compressibility) of a DEM, indicating higher encryption performance. Two modes are provided for users: (i) complexity first and (ii) compressibility first. For the former, the one with the maximum S_A is finally selected. Regarding the latter, the one with the minimum absolute S_A difference ($|DS_A|$) compared to the original DEM is chosen.

The encrypted image can be further processed by lossless compression techniques, such as Huffman encoding [31], free lossless image format (FLIF) [32], and multiscale compression [33], to reduce the burden on transmission and storage. To improve the encryption performance as much as possible, it is recommended that users encrypt a DEM for at least 4 times (i.e., $m \geq 4$) using the proposed algorithm.

4. Experimental Results and Analysis

4.1. Encryption Results

Two 600×600 DEMs with different complexities tabulated in Table 1 were considered experimental images. Their data formats were plain text, and their elevation values were integer. Figure 6 shows these two DEMs, showing different complexities and various ranges of pixel values.

Table 1. Two DEMs for the experiments [28]; S_R and S_A denote relative and absolute configurational entropy, respectively.

DEM	Latitude Extent	Longitude Extent	S_R	S_A	Size (KB)
A	34°27′04″ N–35°02′53″ N	100°36′21″ E–101°49′23″ E	2,502,048.3	401,204,550.0	1758
B	31°23′17″ N–32°06′40″ N	104°07′31″ E–105°06′55″ E	2,416,595.3	308,809,911.3	1459

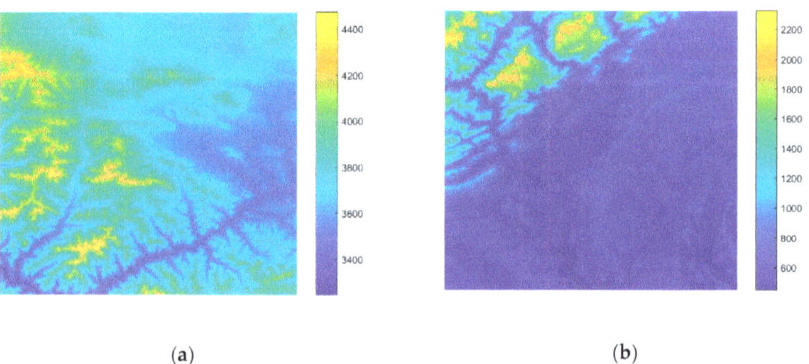

(a)   (b)

Figure 6. Two experimental DEMs with different complexities.

For convenience conducting the experiments, both m and n were set to 10 to encrypt two DEMs. The development environment was Microsoft Visual studio 2013 with .Net Framework 4.5, and the language used for programming was C#. The keys for generating chaotic sequences and corresponding $|DS_R|$ of the confused DEM A in the confusion phase of the first round are tabulated in Table 2. Figure 7 shows the scrambled images, while they have the same histogram. The fourth one was selected for the diffusion phase because its $|DS_R|$ was the maximum compared with the remaining confused images. By using the proposed encryption algorithm, we obtained 10 encrypted DEM A, which are shown in Figure 8, and the key pairs are shown in Table 3, where C_R represents the lossless compression ratio (i.e., the ratio between the bytes used for storing the original data and that for storing the compressed data) by using LZMA [34,35], which is a dictionary-based compression algorithm and takes into consideration the spatial structure of data. From Figure 8, we find that the pixel value range has been modified and the tenth one has the maximum $|DS_A|$ and S_A as shown in Table 3. Therefore, it is selected as the best one when mode (i) is activated. Regarding mode (ii), Figure 8e is considered the best one. From Figure 9, we find that the S_A values of the encrypted images increased, whereas the C_R values decreased with the increase in the total encryption rounds (i.e., m). This can be explained by the viewpoint derived from [19] that, from a theoretical perspective, the lower the redundancy (which is measured by configurational entropy here) of an image, the lower the compression ratio of the image achieved.

Table 2. Comparisons of relative configurational entropy of confused DEM A under different keys in the first round. (r_0, x_0) and (r_0, y_0) denote the keys to scramble the row and column of DEM A, respectively. $|DS_R|$ means the absolute S_R difference compared to the original one.

No.	(r_0, x_0)	(r_0, y_0)	$|DS_R|$
1	(3.6949202, 0.94)	(3.6477592, 0.16)	1,425,882.9
2	(3.7278720, 0.12)	(3.7657562, 0.64)	1,455,723.6
3	(3.6158898, 0.54)	(3.7451577, 0.34)	1,440,030.0
4	(3.6694297, 0.82)	(3.6036054, 0.56)	1,490,201.9
5	(3.7033331, 0.43)	(3.8601213, 0.80)	1,478,572.8
6	(3.5919501, 0.58)	(3.7767205, 0.53)	1,472,217.3
7	(3.7562061, 0.76)	(3.9686558, 0.74)	1,449,584.6
8	(3.7254665, 0.13)	(3.942484, 0.03)	1,455,760.0
9	(3.7873567, 0.05)	(3.6882554, 0.48)	1,443,565.6
10	(3.8638823, 0.83)	(3.6808917, 0.04)	1,453,363.9

Table 3. The best key pairs among 10 encryption times for DEM A. $|DS_A|$ means the absolute S_A difference compared to the original one. S_A is the absolute configurational entropy.

mth Round	(r_0, x_0)	(r_0, y_0)	S_A	$|DS_A|$	Size (KB)	C_R
1	(3.6694297, 0.82)	(3.6036054, 0.56)	267,144,464.1	134,060,085.9	1759	3.239
2	(3.9720618, 0.63)	(3.8118391, 0.31)	300,969,521.6	100,235,028.4	1529	2.682
3	(3.9982823, 0.61)	(3.6946723, 0.56)	329,305,875.0	71,898,675.0	1641	2.634
4	(3.6911029, 0.24)	(3.7563811, 0.84)	359,120,512.9	42,084,037.1	1764	2.602
5	(3.6761227, 0.71)	(3.9535386, 0.56)	393,026,254.4	8,178,295.6	1850	2.531
6	(3.6468789, 0.54)	(3.7460818, 0.2)	427,592,398.5	26,387,848.5	1946	2.485
7	(3.9889696, 0.77)	(3.7643806, 0.9)	464,150,752.2	62,946,202.2	2074	2.474
8	(3.7666660, 0.74)	(3.7904553, 0.39)	511,483,888.6	110,279,338.6	2172	2.435
9	(3.9793047, 0.27)	(3.7826054, 0.06)	550,795,418.0	149,590,868.0	2253	2.392
10	(3.8921841, 0.88)	(3.8734612, 0.19)	604,312,707.5	203,108,157.5	2373	2.380

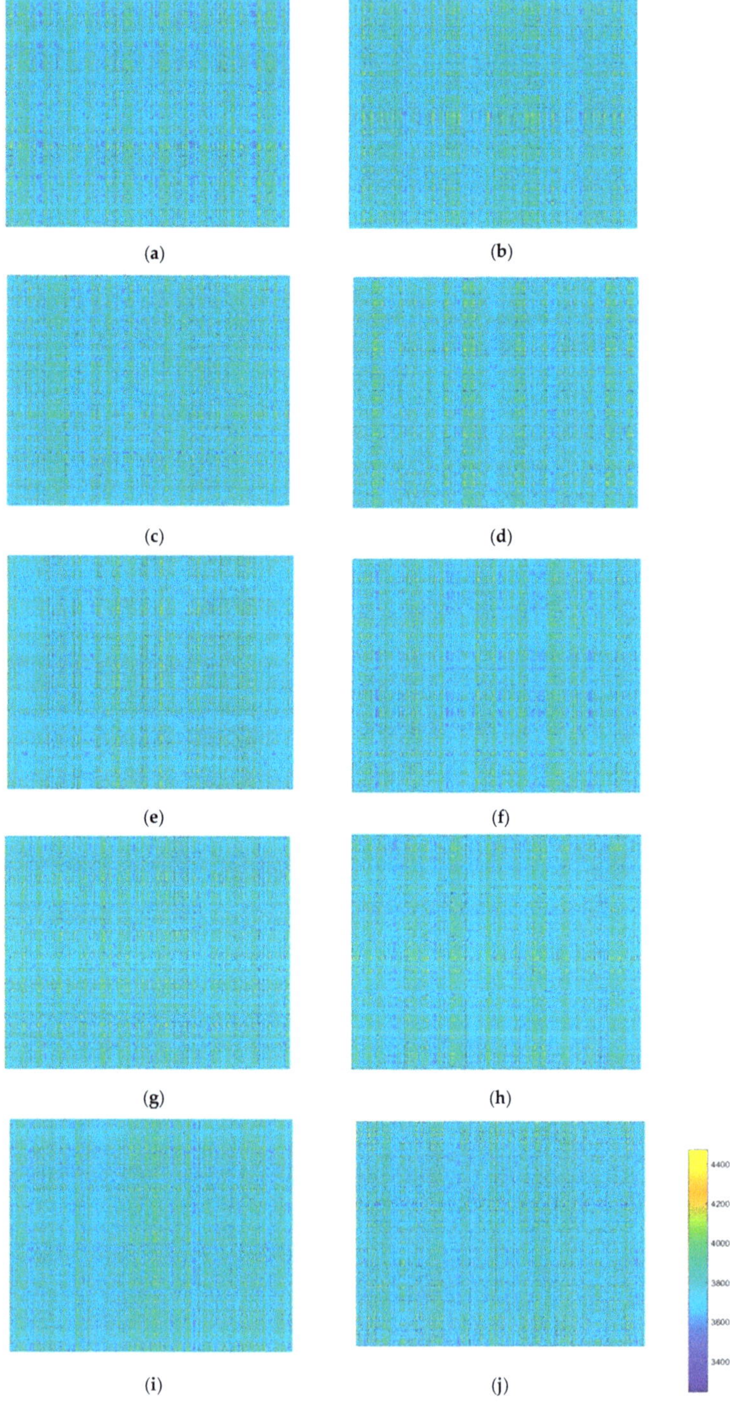

Figure 7. Ten confused DEM A.

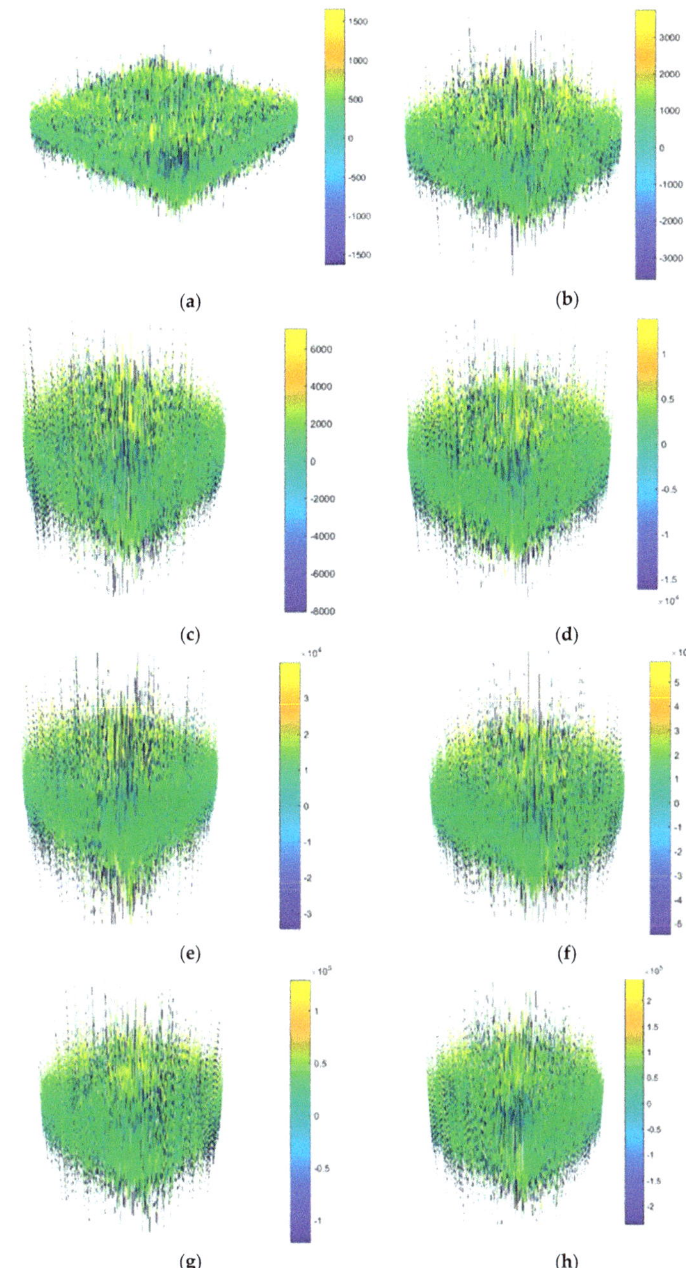

Figure 8. *Cont.*

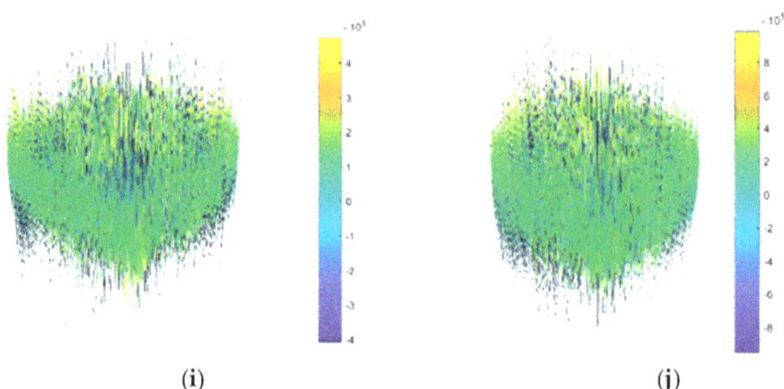

(i)   (j)

Figure 8. Three-dimensional images of confused and diffused DEM A. The numbering sequence is consistent with the encryption round.

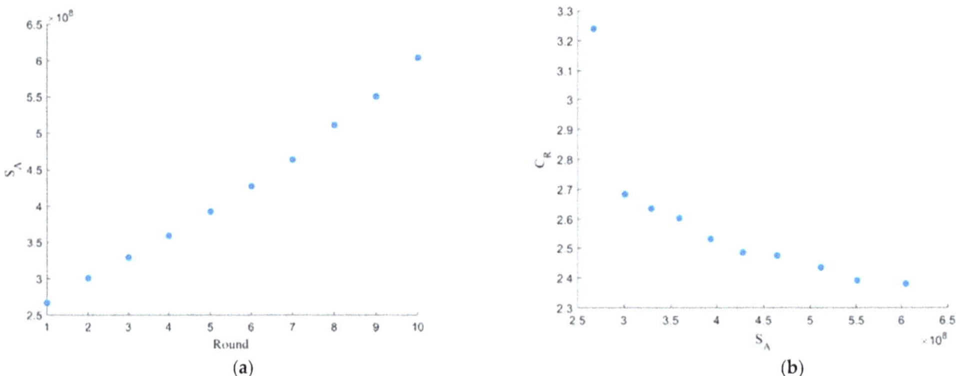

(a)   (b)

Figure 9. Scatters plot of m rounds compared to the S_A value of the encrypted DEM A and that of C_R compared to the S_A values.

Regarding DEM B, Table 4 shows the $|DS_R|$ values of 10 confused ones illustrated in Figure 10. We find the 10th one is the best in the confusion phase. Table 5 shows similar results to DEM A. Obviously, when mode (i) is employed, the 10th one is the best since it has the maximum S_A value in comparison with the others shown in Figure 11. However, the second one is selected when mode (ii) is activated. Figure 12a illustrates that the S_A values increase with the increase in encryption rounds. However, we find that the C_R values decrease in Figure 12b. These experimental results indicate that the configurational entropy is useful to optimize the proposed encrypted algorithm.

Table 4. Comparisons of the relative configurational entropy of confused DEM B under different keys in the first round.

No.	(r_0, x_0)	(r_0, y_0)	$|DS_R|$
1	(3.9127452, 0.56)	(3.9430024, 0.44)	1,366,081.4
2	(3.7803406, 0.42)	(3.7118742, 0.28)	1,371,377.5
3	(3.9446201, 0.10)	(3.6720488, 0.25)	1,401,148.1
4	(3.8410564, 0.11)	(3.7863010, 0.4)	1,201,102.3
5	(3.6867861, 0.75)	(3.5896140, 0.09)	1,373,653.2
6	(3.5921033, 0.19)	(3.9948832, 0.07)	1,381,518.3
7	(3.6462285, 0.68)	(3.5859005, 0.23)	1,375,019.6
8	(3.7970835, 0.63)	(3.8113753, 0.49)	1,377,774.8
9	(3.9591995, 0.05)	(3.9107138, 0.70)	1,382,701.9
10	(3.9429938, 0.25)	(3.7159570, 0.45)	1,388,199.0

Table 5. The best key pairs among 10 encryption rounds for DEM B.

mth Round	(r_0, x_0)	(r_0, y_0)	S_A	$|DS_A|$	Size (KB)	C_R
1	(3.9446201, 0.1)	(3.6720488, 0.25)	285,857,730.3	22,952,181.0	1491	2.696
2	(3.6978437, 0.44)	(3.9206044, 0.49)	320,402,657.6	11,592,746.3	1592	2.636
3	(3.6557073, 0.03)	(3.8864578, 0.33)	350,085,107.0	41,275,195.7	1728	2.606
4	(3.9037505, 0.02)	(3.7374396, 0.21)	383,939,349.7	75,129,438.4	1826	2.550
5	(3.7128502, 0.56)	(3.7213353, 0.35)	413,822,209.8	105,012,298.5	1911	2.492
6	(3.8007097, 0.09)	(3.9265191, 0.78)	452,157,235.0	143,347,323.7	2031	2.477
7	(3.5822104, 0.46)	(3.7138054, 0.15)	494,673,201.5	185,863,290.2	2139	2.453
8	(3.6359581, 0.57)	(3.9861503, 0.54)	532,925,355.6	224,115,444.3	2220	2.403
9	(3.9345383, 0.58)	(3.7636595, 0.61)	590,387,815.5	281,577,904.2	2325	2.382
10	(3.8741445, 0.34)	(3.900868, 0.25)	635,882,520.3	327,072,609.0	2446	2.377

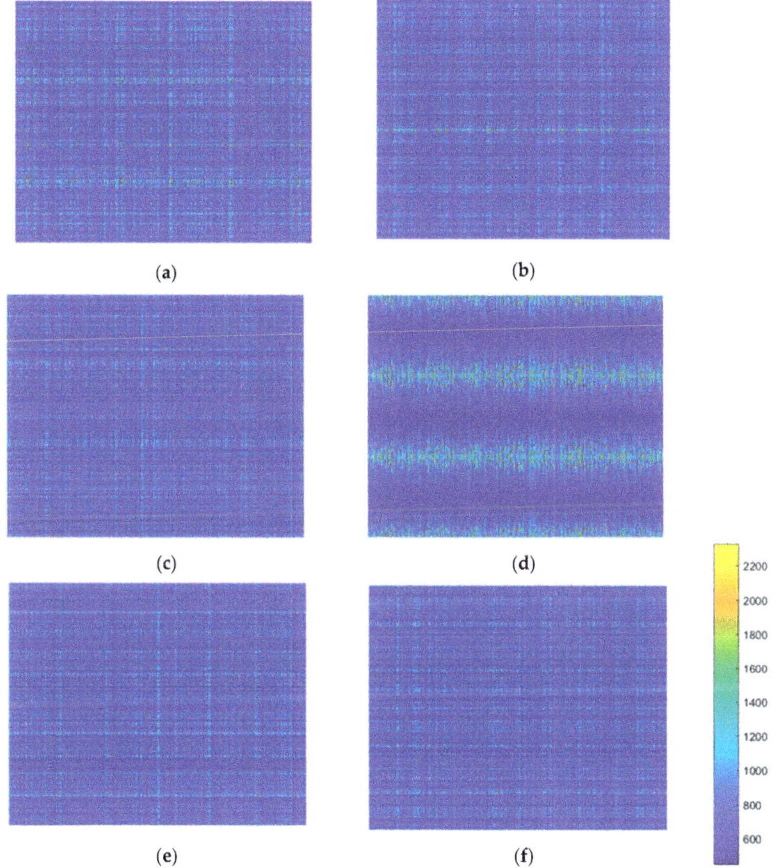

Figure 10. *Cont.*

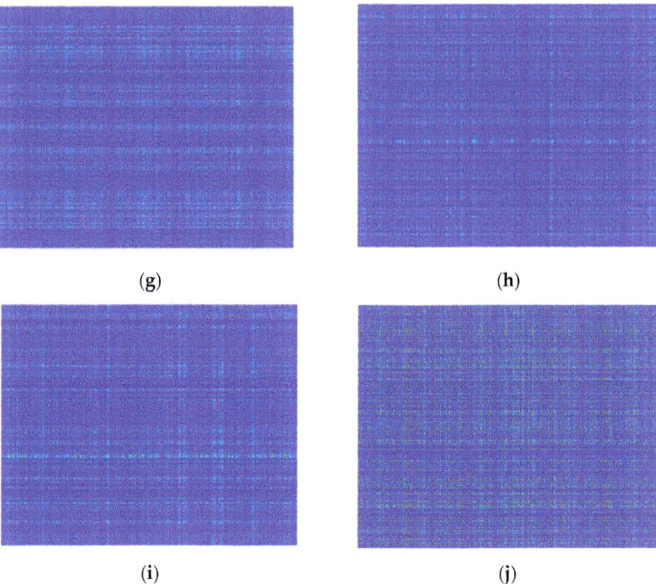

Figure 10. Ten confused DEM B. Their histograms are the same.

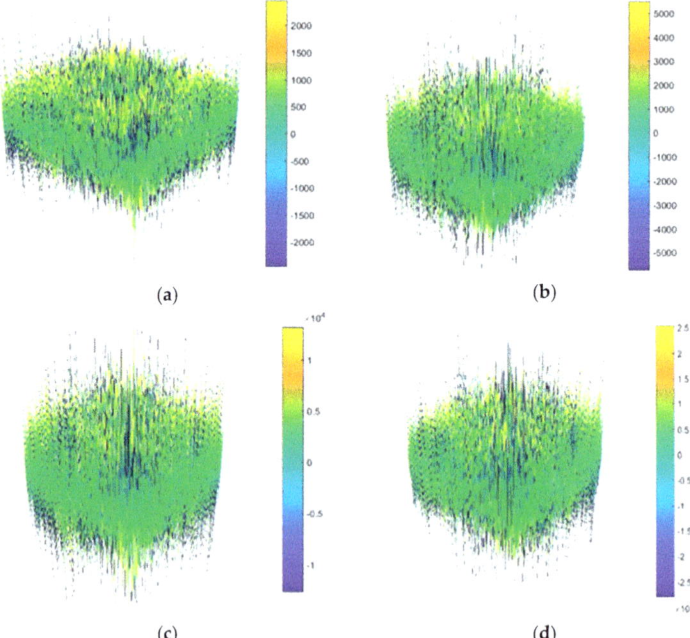

Figure 11. *Cont.*

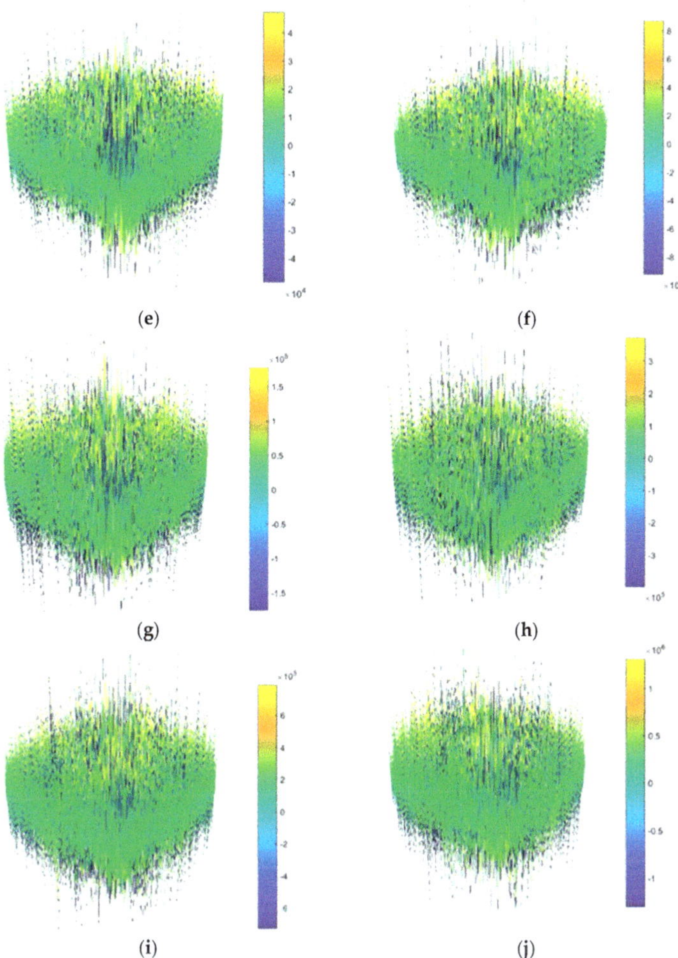

Figure 11. Three-dimensional images of encrypted DEM B. The numbering sequence is consistent with the encryption round.

From the two aforementioned encryption examples, we find that configurational entropy can help users choose the best-encrypted one according to specific requirements, e.g., the size of encrypted data should be as small as possible, and the encrypted image should be as complicated as possible. For instance, in consideration of transmission bandwidth, users can choose the encryption with the minimum S_A value. To enhance the complexity of the encrypted image, users can set larger and larger m and n values if possible.

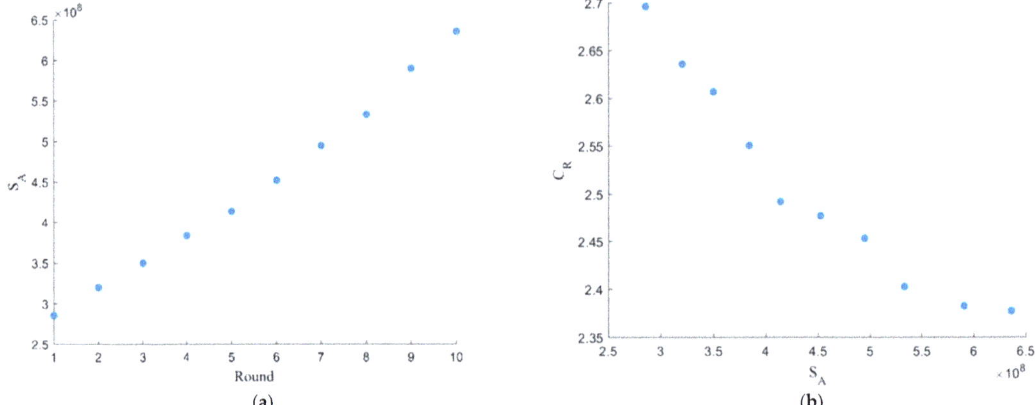

(a)   (b)

Figure 12. Scatter plots of m rounds compared to the S_A value of the encrypted DEM B and that of C_R compared to the S_A values.

4.2. Security Analysis

A good encryption algorithm should be capable of resisting all attacks. In this section, we perform a security analysis on the proposed encryption algorithm.

1. Key space and sensitivity analysis

A good encryption approach should be sensitive to the secret keys. In this study, the iteration times, (i.e., m and n) can be used as keys as well as the parameters r_0 and x_0 of a logistic map. Moreover, the precision of parameters of the logistic map can be used as keys as it can influence the performance of chaotic sequences. The key space is proportional to the parameter precision: m (≥ 1) and n (≥ 1). If the precision is 10^{-20}, the key space size can be at least $m \times 10^{40}$. Hence, the key space is big enough to resist brute-force attacks. Moreover, using keys (r_0, x_0) only to recover the original image is very difficult as the range of pixel values is changed after using the proposed encryption algorithm. Figure 13 shows two decrypted DEM A with wrong keys.

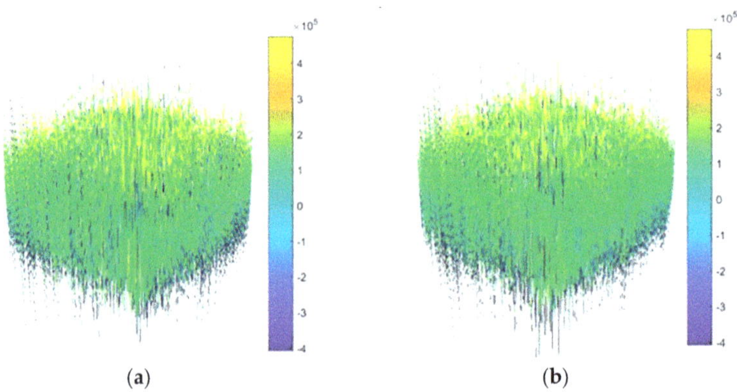

(a)   (b)

Figure 13. Three-dimensional images of decrypted DEM A: (**a**) with keys $r_0 = 3.7004182$, $x_0 = 0.28$, $r_0 = 3.8994119$, and $y_0 = 0.86$; (**b**) with keys $r_0 = 3.8777651$, $x_0 = 0.21$, $r_0 = 3.7276262$, and $y_0 = 0.27$.

2. Classical attacks

Attackers have many methods of attack. Four classical types of attacks [7] are listed as follows:

- Selected plaintext: The opponent chooses a plaintext string and constructs the cipher-text string when temporary access to the encryption machine is granted.
- Selected cipher text: The opponent obtains a ciphertext string and constructs the corresponding plaintext string when temporary access to the encryption machine is granted.
- Known plaintext: The opponent owes a plaintext string and its corresponding cipher-text.
- Ciphertext only: The opponent owes a ciphertext string

The selected plaintext attack is considered the most powerful one. The proposed encryption approach is highly sensitive to the initial parameters for a logistic map. Moreover, at the fusion phase, the encryption data are related to not only the one in the confusion phase but also the one predicted by the three-point prediction technique used at the diffusion stage. Moreover, different encrypted numerical raster data are derived from various former ones because m and n are variable. This means that the encrypted data are able to resist the chosen plaintext attack, indicating that it can resist the remaining attacks.

4.3. Decryption Results with True Keys

To decrypt the encrypted DEMs A and B, the true keys tabulated in Tables 3 and 5 are used. The decryption of an image is the inverse process of its encryption. With true keys, the reconstructed DEMs A and B are illustrated in Figure 14.

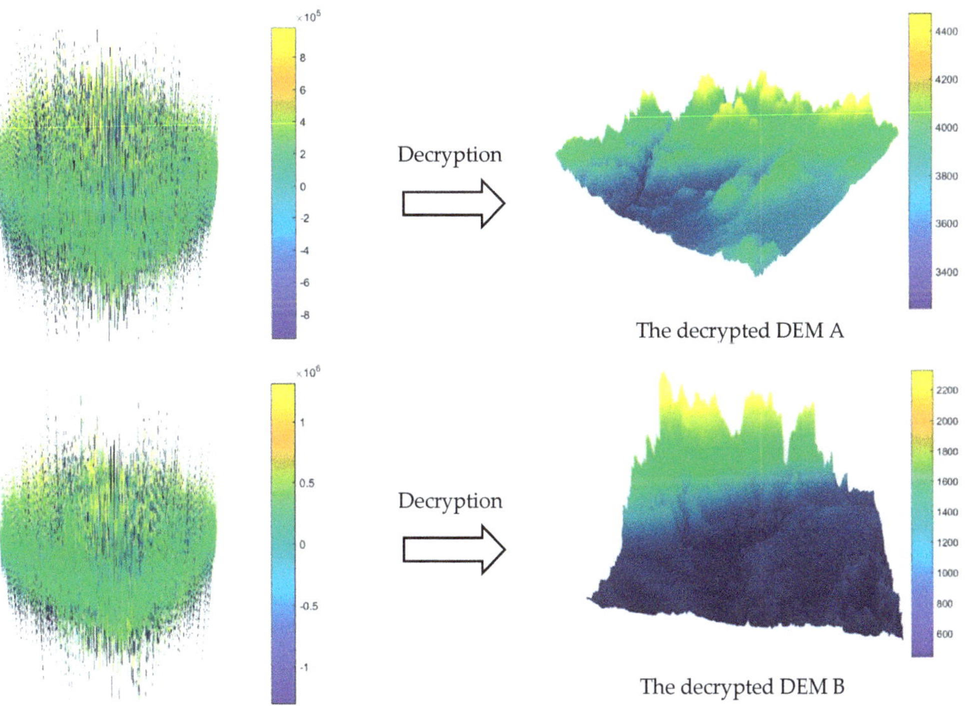

Figure 14. The decryption results of DEMs A and B with the use of true keys.

5. Conclusions

DEM is a digital representation of terrain information. Information security for DEMs is an important topic due to the openness of computer and network communication. By using encryption, the information from DEMs can be well protected. In this study, an algorithm based on chaos system and linear prediction is proposed. To optimize the proposed encryption algorithm, configurational entropy is employed. At the confusion stage, the one with the maximum relative configurational entropy different from the original is selected for the diffusion stage, where the one with the maximum absolute configurational entropy is chosen for the sake of obtain the best encryption performance and the one with the minimum absolute configurational entropy is chosen to reduce the burden on transmission and storage. Two DEMs are taken as experimental data and encrypted 10 times. From the experimental results and analysis, we draw the following major conclusions

- The proposed encryption algorithm is valid, and its security is high.
- Configurational entropy is helpful for optimizing the encryption process.

On the other hand, three areas are recommended for future research. The first is to investigate the effects of different predictors in the diffusion phase of an encryption performance. The second is to explore multiscale DEM encryption with the help of absolute configurational entropy. Finally, more advanced chaos systems and watermark signature techniques [36–39] are expected to be employed as one part of this study to provide excellent performance in only one encryption round.

Author Contributions: Conceptualization, X.C.; methodology, X.C.; software, X.C.; validation, X. Cheng and Z.L.; formal analysis, X.C.; investigation, X.C.; resources, X.C.; data curation, X.C.; writing—original draft preparation, X.C.; writing—review and editing, Z.L.; visualization, X.C.; supervision, Z.L.; project administration, Z.L.; funding acquisition, Z.L. All authors have read and agreed to the published version of the manuscript.

Funding: This study was funded by the Research Grant Council of Hong Kong SAR, China (grant number 15221918) and the Natural Science Funding Council of China (grant number 41930104).

Institutional Review Board Statement: Not applicable.

Informed Consent Statement: Not applicable.

Data Availability Statement: Two experimental DEMs can be downloaded from the website of NASA's Shuttle Radar Topography Mission (http://srtm.csi.cgiar.org/) according to the geographical extent tabulated in Table 1. All encrypted DEMs are available upon request.

Acknowledgments: Special thanks are given to Na Ren of Nanjing Normal University, China, for her constructive comments.

Conflicts of Interest: The authors declare no conflict of interest.

References

1. Lian, S. *Multimedia Content Encryption: Techniques and Applications*, 1st ed.; CRC Press: Boca Raton, FL, USA, 2008; pp. 7–10.
2. Uhl, A.; Pommer, A. *Image and Video encryption: From Digital RIGHTS management to Secured Personal Communication*, 1st ed.; Springer: New York, NY, USA, 2004; pp. 11–22.
3. Shannon, C.E. Communication theory of secrecy system. *Bell Labs Tech. J.* **1949**, *28*, 656–715. [CrossRef]
4. Lian, S. A block cipher based on chaotic neural networks. *Neurocomputing* **2009**, *72*, 1296–1301. [CrossRef]
5. Huang, C.K.; Nien, H.H. Multi chaotic systems based pixel shuffle for image encryption. *Opt. Commun.* **2008**, *282*, 347–350. [CrossRef]
6. Masood, F.; Ahmad, J.; Shah, S.A.; Jamal, S.S.; Hussain, I. A novel hybrid secure image encryption based on julia set of fractals and 3D Lorenz chaotic map. *Entropy* **2020**, *22*, 274. [CrossRef]
7. Zhang, L.; Liao, X.; Wang, X. An image encryption approach based on chaotic maps. *Chaos Soliton Fract.* **2005**, *24*, 759–765. [CrossRef]
8. Xiang, T.; Wong, K.; Liao, X. Selective image encryption using a spatiotemporal chaotic system. *Chaos* **2007**, *17*, 023115. [CrossRef]
9. Zhu, Z.; Zhang, W.; Wong, K.W.; Yu, H. A chaos-based symmetric image encryption scheme using a bit-level permutation. *Inf. Sci.* **2011**, *181*, 1171–1186. [CrossRef]
10. Lian, S.; Sun, J.; Wang, Z. Security analysis of a chaos-based image encryption algorithm. *Physica. A.* **2005**, *351*, 645–661. [CrossRef]
11. Wong, K.W.; Kwok, B.; Law, W.S. A fast image encryption scheme based on chaotic standard map. *Phys. Lett. A.* **2008**, *372*, 2645–2652. [CrossRef]

12. Li, H.; Wang, Y.; Zuo, Z. Chaos-based image encryption algorithm with orbit perturbation and dynamic state variable selection mechanisms. *Opt. Lasers Eng.* **2019**, *115*, 197–207. [CrossRef]
13. Farah, M.B.; Guesmi, R.; Kachouri, A.; Samet, M. A novel chaos based optical image encryption using fractional Fourier transform and DNA sequence operation. *Opt. Laser Technol.* **2020**, *121*, 105777. [CrossRef]
14. Chai, X.; Fu, X.; Gan, Z.; Zhang, Y.; Lu, Y.; Chen, Y. An efficient chaos-based image compression and encryption scheme using block compressive sensing and elementary cellular automata. *Neural. Comput. Appl.* **2020**, *32*, 4961–4988. [CrossRef]
15. Wang, Y.; Wong, K.W.; Liao, X.; Xiang, T.; Chen, G. A chaos-based image encryption algorithm with variable control parameters. *Chaos Soliton Fract.* **2009**, *41*, 1773–1783. [CrossRef]
16. Praveenkumar, P.; Amirtharajan, R.; Thenmozhi, K.; Rayappan, J. Triple chaotic image scrambling on RGB–a random image encryption approach. *Secur. Commun. Netw.* **2015**, *8*, 3335–3345. [CrossRef]
17. Wei, X.; Guo, L.; Zhang, Q.; Zhang, J.; Lian, S. A novel color image encryption algorithm based on DNA sequence operation and hyper-chaotic system. *J. Syst. Softw.* **2012**, *85*, 290–299. [CrossRef]
18. Guan, Z.; Huang, F.; Guan, W. Chaos-based image encryption algorithm. *Phys. Lett. A.* **2005**, *346*, 153–157. [CrossRef]
19. Shannon, C.E. A mathematical theory of communication. *Bell Labs Tech. J.* **1948**, *27*, 379–423. [CrossRef]
20. Wu, Y.; Zhou, Y.; Saveriades, G.; Again, S.; Noonan, J.P.; Natarajan, P. Local Shannon entropy measure with statistical tests for image randomness. *Inf. Sci.* **2013**, *222*, 323–342. [CrossRef]
21. Gao, P.; Li, Z.; Zhang, H. Thermodynamics-based evaluation of various improved Shannon entropies for configurational information of gray-level images. *Entropy* **2018**, *20*, 19. [CrossRef]
22. Kaufman, M. *Principles of Thermodynamics*, 1st ed.; CRC Press: Boca Raton, FL, USA, 2019; pp. 71–92.
23. Huettner, D.A. Net energy analysis: An economic assessment. *Science* **1976**, *192*, 101–104. [CrossRef] [PubMed]
24. Lebowitz, J. Macroscopic laws, microscopic dynamics, time's arrow and Boltzmann's entropy. *Physica A* **1993**, *194*, 1–27. [CrossRef]
25. Benson, H. *Entropy and the Second Law of Thermodynamics*, 1st ed.; University Physics, Wiley: New York, NY, USA, 1996; pp. 417–439.
26. Boltzmann, L. Weitere studien über das wärmegleichgewicht unter gasmolekülen [Further studies on the thermal equilibrium of gas molecules]. *Sitzungsber. Akad. Wiss.* **1872**, *66*, 275–370.
27. Cushman, S. Calculating the configurational entropy of a landscape mosaic. *Landsc. Ecol.* **2016**, *31*, 481–489. [CrossRef]
28. Gao, P.; Zhang, H.; Li, Z. A hierarchy-based solution to calculate the configurational entropy of landscape gradients. *Landsc. Ecol.* **2017**, *32*, 1133–1146. [CrossRef]
29. Cheng, X.; Li, Z. Using Boltzmann entropy to Measure Scrambling Degree of Grayscale Images. In Proceedings of the IEEE 5th International Conference on Cryptography, Security and Privacy (CSP), IEEE, Zhuhai, China, 8–10 January 2021.
30. May, R.M. Simple mathematical models with very complicated dynamics. *Nature* **1976**, *261*, 459–467. [CrossRef]
31. Huffman, D.A. Method for the construction of minimum-redundancy codes. *Proc. IEEE* **1952**, *40*, 1098–1101. [CrossRef]
32. Sneyers, J.; Wuille, P. FLIF: Free lossless image format based on MANIAC compression. In Proceedings of the IEEE International Conference on Image Processing (ICIP), Phoenix, AZ, USA, 25–28 September 2016.
33. Ratakonda, K.; Ahuja, N. Lossless image compression with multiscale segmentation. *IEEE Trans. Image Process* **2002**, *11*, 1228–1237. [CrossRef] [PubMed]
34. Ziv, J.; Lempel, A. A universal algorithm for sequential data compression. *IEEE Trans. Inf. Theory* **1977**, *23*, 337–343. [CrossRef]
35. Martín, G. Range encoding: An algorithm for removing redundancy from a digitised message. In Proceedings of the Video and Data Recording Conference, Southampton, UK, 24–27 July 1979; Institution of Electronic and Radio Engineers: London, UK.
36. Lan, R.; He, J.; Wang, S.; Gu, T.; Luo, X. Integrated chaotic systems for image encryption. *Signal Process.* **2018**, *147*, 133–145. [CrossRef]
37. Chai, X.; Gan, Z.; Yuan, K.; Chen, Y.; Liu, X. A novel image encryption scheme based on DNA sequence operations and chaotic systems. *Neural. Comput. Appl.* **2019**, *31*, 219–237. [CrossRef]
38. Qi, G. Modelings and mechanism analysis underlying both the 4D Euler equations and Hamiltonian conservative chaotic systems. *Nonlinear Dyn.* **2019**, *95*, 2063–2077. [CrossRef]
39. Liu, X.; Wang, J.; Luo, Y. Lossless DEM watermark signature based on directional wavelet. In Proceedings of the 2nd International Congress on Image and Signal Processing, Tianjin, China, 17–19 October 2009.

Article

Facies Heterogeneity and Lobe Facies Multiscale Analysis of Deep-Marine Sand-Shale Complexity in the West Crocker Formation of Sabah Basin, NW Borneo

Muhammad Jamil [1,2,*], Numair Ahmed Siddiqui [1], Abdul Hadi Bin Abd Rahman [1], Noor Azahar Ibrahim [1], Mohd Suhaili Bin Ismail [1], Nisar Ahmed [1], Muhammad Usman [1,3], Zain Gul [1] and Qazi Sohail Imran [1]

[1] Department of Geosciences, Universiti Teknologi PETRONAS, Seri Iskandar 32610, Malaysia; numair.siddiqui@utp.edu.my (N.A.S.); ar.abdul.hadi@gmail.com (A.H.B.A.R.); azahar.ibrahim@utp.edu.my (N.A.I.); suhaili.ismail@utp.edu.my (M.S.B.I.); nisarpu12@gmail.com (N.A.); usman.pu@outlook.com (M.U.); zain_17008726@utp.edu.my (Z.G.); qazi_17007588@utp.edu.my (Q.S.I.)
[2] Department of Earth Sciences, COMSATS University Islamabad, Abbottabad 22060, Pakistan
[3] Department of Earth and Environmental Sciences, University of Milan-Bicocca, Piazza Della Scienza 4, 20126 Milan, Italy
* Correspondence: jamil287@gmail.com

Featured Application: A generalized conceptual model for the characterization of a deep-marine siliciclastic complex deposition with respect to the integrated submarine fan and lobe architecture, which are essential for understanding the subseismic lithological heterogeneities in potential petroleum reservoirs of a deep marine environment.

Citation: Jamil, M.; Siddiqui, N.A.; Rahman, A.H.B.A.; Ibrahim, N.A.; Ismail, M.S.B.; Ahmed, N.; Usman, M.; Gul, Z.; Imran, Q.S. Facies Heterogeneity and Lobe Facies Multiscale Analysis of Deep-Marine Sand-Shale Complexity in the West Crocker Formation of Sabah Basin, NW Borneo. *Appl. Sci.* **2021**, *11*, 5513. https://doi.org/10.3390/app11125513

Academic Editor: Jianbo Gao

Received: 20 March 2021
Accepted: 14 April 2021
Published: 15 June 2021

Publisher's Note: MDPI stays neutral with regard to jurisdictional claims in published maps and institutional affiliations.

Abstract: Deepwater lobes constitute a significant volume of submarine fans and are primarily believed to exhibit a simple sheet geometry. However, recent studies interpret the geometries of these deep-marine lobes as distinct with respect to the complexity of the facies and their distribution. Hence, a conceptual model of deep-marine sediments is essential to discuss the deep-marine sediments associated with the fan and lobe architecture. The present study highlights the facies heterogeneity and distribution of various lobe elements at a multiscale level by considering a case study of the West Crocker Formation of Sabah in northwest Borneo. The formation was logged on a bed-to-bed scale from recently well-exposed sections, with a total vertical thickness of more than 300 m. The lithological characteristics, bed geometry, sedimentary textures and structures of individual beds were used to categorize the rock units into nine sedimentary lithofacies: five sandstone lithofacies (S1–S5), one hybrid bed facies (H), two siltstone facies (Si1 and Si2) and one shale or mudstone facies (M). These facies were grouped into four facies associations (FA1–FA4), which were interpreted as lobe axis (FA1), lobe off-axis (FA2), lobe fringe (FA3) and distal fringe to interlobe (FA4) facies associations. This study is applicable for the distribution of lobes and their subseismic, multiscale complexities to characterize the potential of hydrocarbon intervals in deep-marine sand-shale system around the globe.

Keywords: deep-marine lobe–fan multiscale analysis; sedimentary facies and facies association; subseismic lithological complexities; northwest Borneo; sand–shale depositional system; West Crocker Fan

1. Introduction

Deep-marine siliciclastic deposition is primarily influenced by several factors, including the rate, type and source of sediments, sea level changes and tectonic settings [1–3]. These deposits are mainly present at the basin floor, constituting various submarine fans, which are considered one of the major hydrocarbon producing systems around the globe [4,5]. However, these fan deposits are highly complex due to variations in geometry, internal architecture and vertical and lateral distribution [6–8]. A range of processes related to sediment transportation and accumulation control the overall depositional characteristics of deep-marine sediments. Gravity-driven flows are one of the major sediment transport processes in a submarine depositional environment [9,10]. These flows principally encompass two endmembers: turbidity currents (frictional flow) and debris flow (cohesive flow) [9,11,12]. Such sedimentary processes and gravity flows result in the development of submarine fan and lobe systems. The classification of deep-marine fan and lobe deposits, with respect to sedimentary processes, requires the spatial distribution, thickness of individual units, sedimentary structures and variation in grain size [13,14].

Deep-marine lobes are considered a vital component of submarine fans [15–17], and they are radial features with thin apexes but distribute laterally like a fan toward the distal end [18,19]. However, the lobe deposition is more complex in terms of internal heterogeneity and the distribution of facies [20–22]. Various subenvironments (the lobe axis, off-axis, lobe fringe and distal lobe fringe) have been assigned to these lobe deposits with respect to their thicknesses and facies associations [19,23]. The excessive input of siliciclastic sediments from a shallow marine environment result in the significant internal heterogeneity and complex distribution of sediments in deep-marine lobes [20–22,24]. Later, because of substantial uplift and denudation, these gigantic sand deposits are exposed on the surface [25]. The study area selected for this project is present in the Malaysian part of Borneo, named Sabah, which contains extensive exposures of deep-marine fan deposits stratigraphically termed as the West Crocker Formation. In the case of our study area, with recent infrastructure development (Pan Borneo Highway construction in East Malaysia), numerous new outcrop sections were exposed as fresh roadcuts, which paved the way for the detailed analysis of facies heterogeneity and the distribution of various lobe elements in the West Crocker submarine fan.

The recent literature suggests that the sedimentary facies of deep-marine deposition are significantly diverse and complex when compared with the previous classic Bouma model [26]. Late Paleogene deep-marine sediments of the West Crocker Formation along the Pan Borneo Highway in West Sabah are mainly comprised of thin to thick and massive bedded sandstones, with some siltstone and shale units [27]. Although the previous studies describe Late Paleogene sediments in terms of several components of a submarine fan based on individual outcrops [27–30], this study highlights the multiscale heterogeneity to interpret the distribution of deep-marine lobe complex systems in the West Sabah Basin.

The study area included five studied locations of the Crocker fan, representing the sand–shale complex in West Sabah and having a total vertical thickness of more than 300 m, principally on the roadsides from Kota Kinabalu to Telipok/Tuaran (Figure 1). The key objectives of the study included (1) analyzing the facies and facies distribution of several outcrop sections of the West Crocker Formation; (2) interpreting the differences in stacking patterns and architectural elements of the studied sections; and (3) evaluating the characteristics of various components of the submarine lobe complex. These sedimentological details were supportive to determine the depositional characteristics and distribution of lobe elements in the deep marine environment. This research work is intended to address the following research questions: What are the main facies heterogeneities and facies associations in the Oligocene West Crocker Formation? How we can relate these facies and facies associations into the multiscale lobe elements of the lobe axis, off-axis, lobe fringe and distal lobe? How could these lobe elements be effective for interpreting the individual lobe and lobe complexes for each outcrop section? The purpose of this study is to analyze the lobe architecture and the development of thickening and coarsening upward patterns, which are interpreted as a part of the individual lobe or lobe elements.

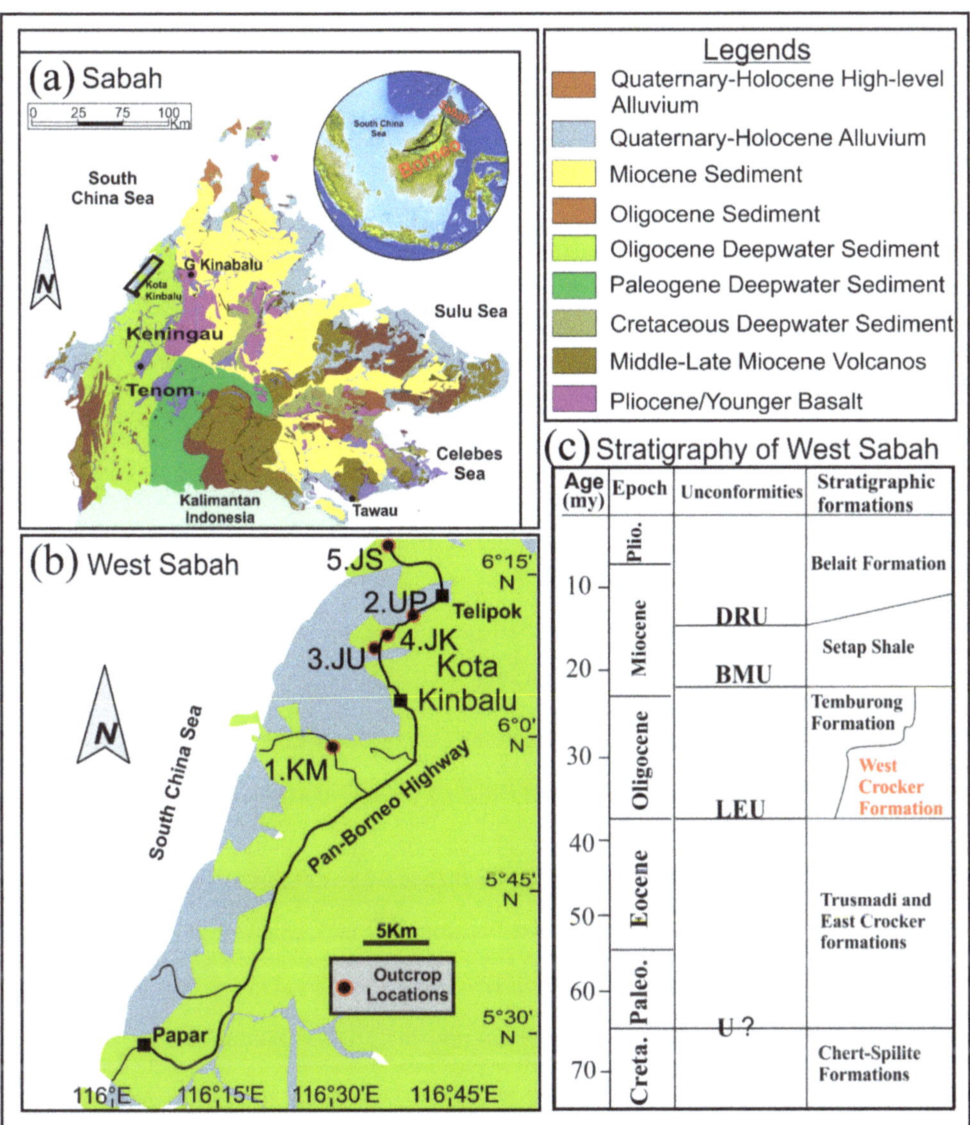

Figure 1. Location of the study area. (**a**) Regional map of Borneo bounded by the South China Sea in the west and the Celebes Sea and Sulu Sea to the east, with Sabah being in the northwest part of Borneo [31,32]. The study area is marked with a black rectangle in Sabah. (**b**) Map of the outcrop locations (1. The Kampung Madpai section (KM), 2. Prima University section (UP), 3. Jalan UMS section (JU), 4. Jalan UMS behind KFC section (JK), and 5. The Jalan Sulaman section (JS)), mainly located along roadsides in the area from Kota Kinabalu to Telipok. (**c**) Generalized stratigraphy of West Sabah with the Oligocene age of the West Crocker Formation, where late Eocene unconformity (LEU) is present at the base while the top of the West Crocker Formation is marked by base Miocene unconformity (BMU) [28].

2. Geological Background

The northwest Sabah Basin is considered as one of the major Tertiary depositional systems of northwest Borneo, having two distinct phases of sedimentation. The older deposition is termed as the Rajang Group, mainly comprised of the Paleocene to Eocene Trusmadi and East Crocker formations. These deposits were later uplifted and eroded to form late Eocene unconformity (LEU). After this LEU unconformity, the second phase of deposition started from the late Eocene to early Miocene epochs (West Crocker and Temburong formations) [28,29,33]. The present study focuses only on the second phase of deep-marine Tertiary deposition (late Paleogene) in the northwest Sabah Basin.

2.1. Deep Marine Environment, Processes and Lobe Complex

Deep-marine siliciclastic deposits are vital for the petroleum industry, with respect to hydrocarbon exploration, with a gradual increase in exploration for huge petroleum discoveries in a large volume of deep-water sediments [34]. The development of these deep-marine deposits is the result of various sedimentary processes, which resulted in numerous architectural elements and sedimentary facies [35–38]. In the deep marine environment, the components of a submarine fan are dependent on the distribution and variation of density flows and flow processes. Low-density turbidity flows are common in all subenvironments of deep-marine systems but are mostly abundant in the distal part of a submarine fan [39]. High-density flow processes are commonly associated with the feeder channels and distributary channels of a lobe in a submarine fan system. The variation in the thickness of sand units is also responsible for a variety of facies associations and depositional environments, such as the massive or thick-bedded sandstone with rare shale unit being most likely associated with the proximal lobe deposition. The lateral variation in the sedimentary succession of the individual lobe can be depicted from a decrease in the thickness of sand units with respect to neighboring shales, which represents the distal part of a submarine fan. The thickening up stacking patterns of lobes could be the result of the progradation of individual lobes [15,39–41].

In order to understand the paleoenvironments and facies analysis, the classical Bouma sequence has limited applications for deep-marine lobe systems [26,42]. Certain terminologies emerged in the past decade to refine the classification of deep-marine sediments. For example, the term "hybrid bed" is a product of the deceleration or transition of the turbidity current to mixed turbidite events [11,43,44]. Flow transformation is characterized by the erosion of underlying rock units within the feeder channels and the axial or proximal part of the lobe. The transition of flow from turbulent to laminar results in the development of heterogeneity in the form of hybrid beds, owing to the deceleration and expansion of flow [11,44,45].

Recent investigations explained that the deep-marine sand sheet or fan deposits included a feeder channel with several individual lobes [19,46]. The relative age of each lobe may vary as the older lobe may be overlapped by the younger one (Figure 2), and this overlapping of lobes may continue in a lobe complex [19,47]. The lobe complex can be classified based on the relative position of the feeder channel. Those present close to the feeder channel are termed as proximal lobes (Figure 2), the middle part of lobe complex is called the medial lobes, while those farther away from the medial lobes are labeled as distal lobes [24].

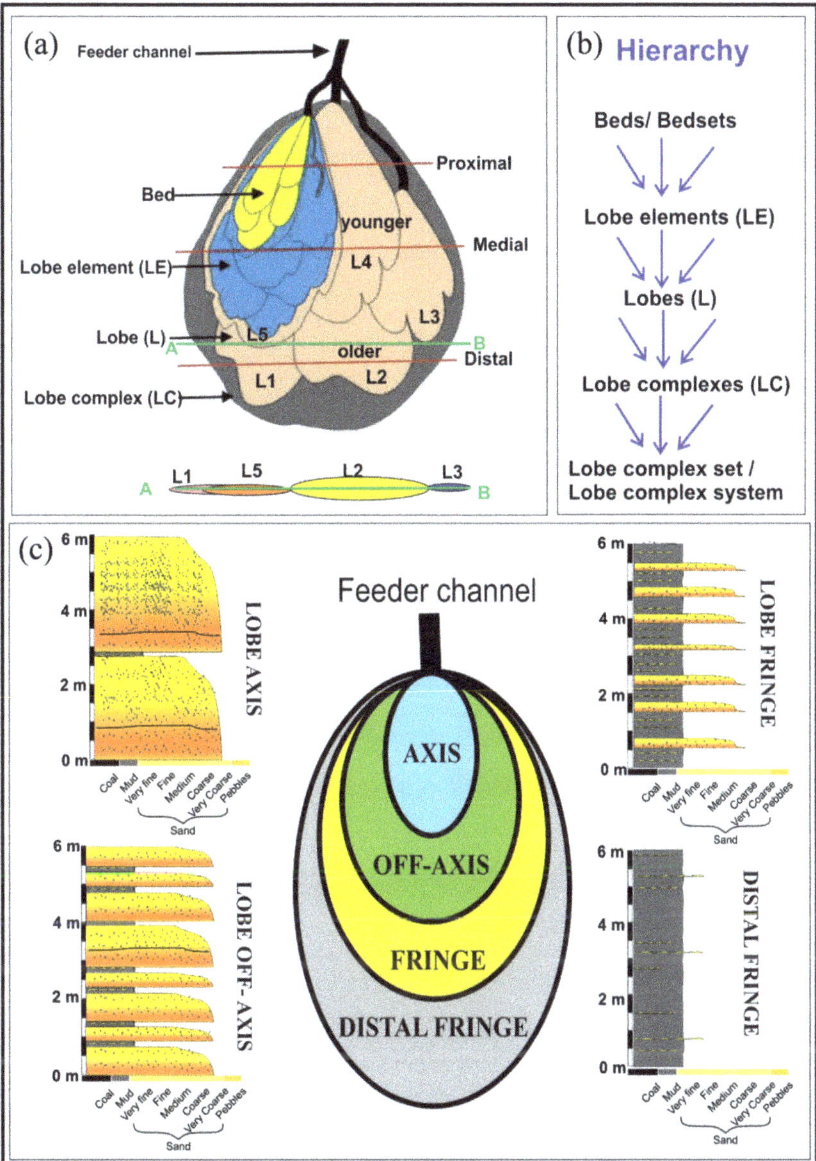

Figure 2. Lobe terminology used for the discussion of facies heterogeneity and multiscale analysis of the lobe complex system. (**a**) Classification of the lobe complex into proximal, medial, and distal parts of the individual lobes [24,46]. (**b**) Hierarchy of the lobe system, where the smallest unit is the bed or bed set while the largest unit is known as the lobe complex set [47] or lobe complex system. (**c**) Characterization of the lobe, with a feeder channel into the lobe axis, off-axis, fringe and distal fringe, each with representative logs [14,19,48].

Lobes are divided into four subenvironments—the lobe axis, lobe off-axis, lobe fringe (both frontal and lateral) and distal lobe fringe—on the basis of the amount of sand, amalgamated surfaces and sedimentary facies [17,45,49,50]. The lobe axis is predominantly composed of structureless, thick-to-massive bedded sandstone with amalgamation, indicating greater depositional rates with high energy turbidity currents [17,48]. The growth of the lobe off-axis mainly results from the deposition of medium-to-thick bedded sandstone, representing relatively low deposition rates and low energy turbidity deposits [17]. The deposition of the lobe fringe is mainly characterized by thin-bedded sandstone (fine grained and rippled) with the hybrid event beds, which are created due to the transformation of flow [17,45,48]. However, the distal lobe fringe or interlobe only contain thin-bedded siltstone and thick-to-massive shale units [14,48].

2.2. Paleogene and Neogene Geology of West Sabah

Borneo has a complex geological history of sedimentation and deformation, especially in the Tertiary period, when a large volume of sediments was transported from southern Borneo, namely the Schwaner Mountains and the Tin Belt, resulting in the huge thickness of deep-marine deposits [29,30,51]. The development of Borneo is associated with tectonic subduction, along with the obduction of ophiolite rocks and the collision of tectonic fragments with the continental part of the Sunda Plate, resulting in the closure of paleobasins [31,52–54]. The Borneo Accretionary Orogen is present in the center of South East Asia, which is bounded by the subduction of the Pacific and Indian plates with a passive continental margin of the South China Sea. The Borneo Accretionary Orogen is currently active, as the subduction of the Dangerous Grounds under the Borneo Block is still continuous [55].

Northern Borneo comprises the Sabah Basin at the geological complex junction between Sunda, Celeb, Sulu and the South China Sea, where Tertiary sediments are exposed due to the Sabah orogenic belt, which resulted in the closure of the South China Sea [51,56–60]. The post-orogenic foreland Sabah Basin is mainly comprised of marine sediments, where the depositional processes were disrupted by several tectonic events in the form of unconformities. These unconformable surfaces are well-preserved in the Paleogene and Neogene stratigraphic record of West Sabah [61]. The Top Crocker Unconformity (TCU) or Base Miocene Unconformity (BMU) is the major unconformity separating the Late Paleogene West Crocker Formation from the Neogene Setap Shale [62]. The northwest Sabah Basin is mainly comprised of the Crocker fold and thrust belt (CFTB), which is also termed as the Crocker Range [29,63,64]. The Crocker fold and thrust belt was developed due to the collision of continental plates and largely consists of siliciclastic sediments of the deep marine environment [52].

2.3. West Crocker Formation

The West Crocker Formation crops out in the form of several vertical to subvertical rock sections around Kota Kinabalu and generally in the West Sabah [61,65]. The late Paleogene Crocker sediments were deposited by erosion of the early Paleogene rocks. The thickness of the late Paleogene sediments varies from at least 1000 m to more than 2000 m, and the lithologies include sandstone, shales and siltstones [30,33,66]. Late Paleogene sediments mainly consist of sand-dominated debris flow deposits and heterolith siltstone mudstones, having all components of the inner, middle and outer fan environments [27], while at a few studied sections, the formation is interpreted to be only a middle-to-outer fan system [67]. These sediments are mainly sand-rich facies deposited by high density turbidity currents; however, they also contain low-density turbidites and mass-transport deposits like slumps and contorted layers [68,69].

High-density turbidity flows result in texturally immature, poorly sorted and angular fragments in siliciclastic rocks [30,70–72]. Sedimentary structures like flute marks, cross bedding, convolution, parallel lamination, amalgamation and dish structures have been reported [27]. Water escape dish structures and convolution are the result of rapid deposition [73] and are termed as soft sediment deformation structures (SSDSs). These deformation structures are inferred to be seismites, which are representative of active tectonic settings [74]. The early Paleogene Rajang Group was eroded and resedimented to form the late Paleogene rocks of the West Crocker Formation [72].

3. Materials and Methods

3.1. Geological Fieldwork

The dataset included sedimentological logs and lithological details from outcrops. Standard field geological operations were followed to delineate the detailed sedimentary evaluation of the West Crocker Formation in West Sabah. The best-exposed sections were selected for the detailed sedimentological description of rock units with the help of available geological maps, google satellite imagery as well as several reconnaissance field visits. Measurements of the vertical thicknesses of the beds, identification of numerous sedimentary structures and grain size variations and descriptions of the geometries of rock units were noted for understanding complexities. These details were quite helpful to discuss the facies heterogeneity and lobe systems of deep-marine multiscale sedimentary successions. The methodology was used for classifying the deep-marine sediments into sedimentary lithofacies based on variations in bed thickness, grain size and types of sedimentary features. These sedimentary facies were grouped into facies associations, which were interpreted to be part of the submarine lobe environment.

3.2. Field Sedimentary Logging and Facies Analysis

The dataset comprised of detailed sedimentological characterization of the outcrops, including (1) the Kampung Madpai (KM) section, (2) the Jalan Universiti Prima (UP) section, (3) the Jalan UMS (JU) roadside, (4) the Jalan UMS behind KFC (JK) section and (5) the Jalan Sulaman (JS) section around Kota Kinabalu (Figure 1). These logs contained the particulars of individual rock units, including the bed thickness, lithological character, sedimentary structures and types of bed contacts. These details were investigated to analyze, evaluate and interpret the complexity of the deep-marine exposed sections. The field sedimentary details were applied to interpret the facies analysis and facies association, which were correlated with the submarine lobe architecture (e.g., lobe element, lobe complex and composite lobe system).

3.3. Sandstone Thickness Analysis and Trends

The pattern of thickness of rock units varied considerably, like how thin-bedded sand units were related to distinct elements of lobe fringes while thick-bedded sands were linked to the proximal part of the lobe. These thin or thick beds were indicative of the flow conditions, such as interbedded, thin-bedded sandstones and siltstone representing low-density turbidity flows while thick-bedded or massive sandstones indicating high-density flow conditions. These thick-bedded sand units were quite established by feeder channels in the axial or off-axial parts of the lobes [75,76]. However, lobe progradation could generally be linked with thickening up cycles, or it could be the onset of a new individual lobe. The medial and distal frontal fringe lobe were associated with hybrid units and were the result of a downward dip of high-density turbidites or low-density flows, representing the lower part of the prograding lobe succession [77].

Deepwater lobes are explained as simple radial deposits which are fine and thin in morphology at the initiation point or feeder channel, but they are more complex with respect to their geometry and facies characterization. These lobes are further classified by the relative thickness of the sand and shale beds. The lobe complex is the larger entity of lobe deposition that is primarily comprised of a feeder channel and individual lobes, having a variety of morphologies and geometric distributions [19,23,46].

4. Results and Interpretations

4.1. Stratigraphic Distribution of Outcrop Sections

Geological field logs explained the stratigraphic distribution of each outcrop in the study area. Recently exposed road cuts and fresh exposures were selected to study the West Crocker Formation. In general, the stratigraphy was sand-dominated sections with multimeter sand beds present throughout the outcrops. It is pertinent to mention here the Kampung Madpai outcrop (KM) contained mainly massive sand units with thin shale laminae, and massive shale beds were completely absent, representing the inner part of the submarine fan. The basal part of the Prima University section (UP) contained both shale and sand beds, and the middle part of the section contained massive sandstone intervals while the upper part of the outcrop contained thick-to-massive sandstone (Figure 3) with little influx of shale, representing the middle part of the submarine fan.

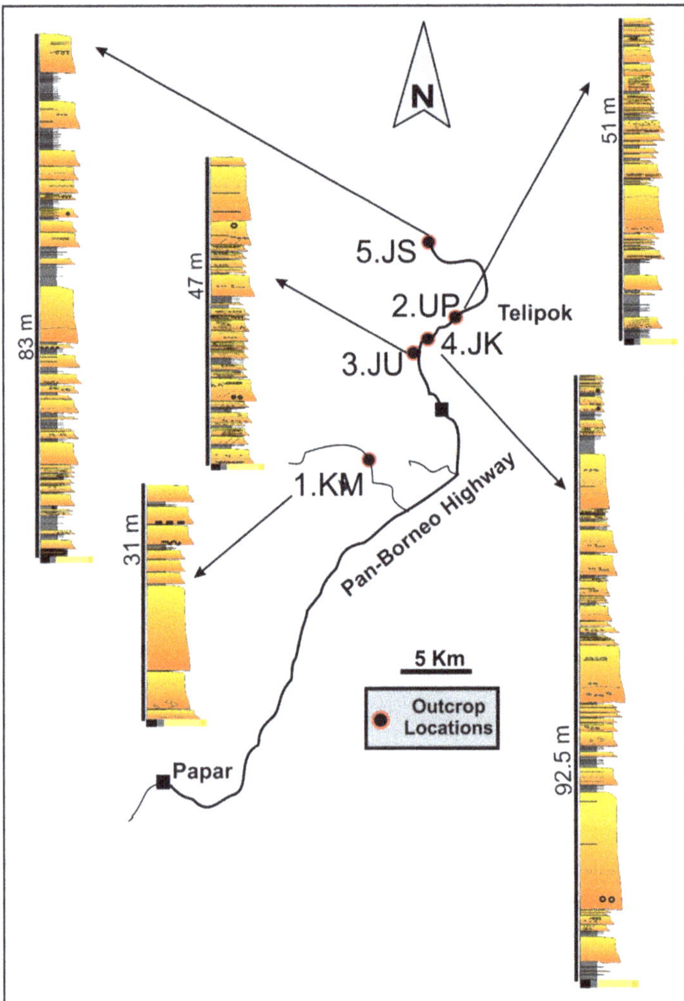

Figure 3. Stratigraphic distribution of the selected exposed outcrops that represent the multiscale sand–shale complex system. These lithological heterogeneities are interpreted as various components of the submarine fan–lobe architecture.

The lower part of the Jalan UMS road section (JU) comprised thick-bedded to massive sandstone beds, while the upper part of the section predominantly consisted of only massive multimeter sandstone beds, representing a high influx of sand, which is characteristic of a proximal fan environment. These massive beds were also common in the lower part of the Jalan UMS behind KFC section (JK), and the upper part was characterized by massive sandstone and shale intervals. The Jalan Sulaman section (JS) is a classic example of massive sandstone beds with alternate massive shale representing the cyclicity in the lobe–fan deposition.

4.2. Facies Analysis and Depositional Environment

A facies is defined as a rock unit comprising one or more beds which has specific characteristics, such as composition, bed thickness or texture. These facies are distinctive rock units which have been developed by a geological process and are indicative of certain conditions of sedimentation [42]. The clastic sedimentary rock units present in the study area were classified based on sedimentary structures and lithology, in which the sandstones are denoted with sandstone lithofacies (S), hybrid event beds (H), siltstone lithofacies (Si) and mudstone or shale lithofacies (M). These lithofacies were numbered according to each type of sedimentary facies.

4.2.1. S1 Facies: Graded Coarse-to-Fine-Grained Sandstone

The physical characteristics of this facies included thick-to-massive bedded sand units, mainly poorly sorted, some beds have normal grading and fine-to-coarse-grained sandstone units. The thicknesses of individual sand units ranged from 30 cm to more than 100 cm. Many sand units in this facies had thicknesses more than 2 m, which were often amalgamated. Based on amalgamation structures, the facies were interpreted as a result of multiple depositional events and a high-energy environment [11]. Moreover, a high vertical thickness and multimeter individual sand units were the result of a high sediment influx in a basin, where the lower part was characterized by high-density flow deposition, Ta division [78] or F5 and F8 facies [79].

4.2.2. S2 Facies: Ungraded Coarse Sandstone (Structureless)

The sandstone units were moderately sorted, having coarse to very coarse grain sizes (Figure 3). The thicknesses of the sand beds ranged from thick-bedded to massive (more than 30 cm up to 5 m). Most of the units had no grading and limited variation in grain size, which were termed as structureless and moderately sorted. Sand beds are often amalgamated showing tabular geometry and mainly lack any sedimentary structures. The facies was deposited by high-density turbidity currents, containing a traction carpet and classified as the S2 type [18,80], the lower part of Ta division by [78], and the flow is termed as dense sandy and gravely flow [79].

4.2.3. S3 Facies: Parallel Laminated Fine-to-Medium-Grained Sandstone

The thicknesses of sand units fluctuated from thin- to medium-bedded, while the grain size ranged from fine- to medium-grained sand. These beds exhibited parallel laminations and, in a few cases, laminated muddy sandstone were present. The parallel laminations (Tb) were often present above the massive structureless (Ta) units, indicating high energy conditions. The parallel stratification (Figure 4) indicated the near-bed suspension generated by progressive turbulent flow, where the rate of deceleration was relatively sluggish [81]. The facies was classified as the S3 type of sediment [80], with the Tb after Bouma [78] and F7 and F9 facies [79] representing high density turbidity currents [82,83].

4.2.4. S4 Facies: Ripple-Laminated Sandstone

The facies included fine- to very fine-grained sandstone showing ripple cross-lamination. These units were thin- to thick-bedded sands. The height of the ripple lamination may have varied from 4 to 10 cm (Figure 4C), and length ranged from 10 cm to 32 cm. The deposited rock unit indicated the lower flow regime and was marked as Tc [78] and F9 facies by [79]. These cross laminations were interpreted to be the result of a change in flow regime from higher energy to transitional or a low energy environment and loss of flow confinement. These facies are more frequently found in lobe off-axis settings [84].

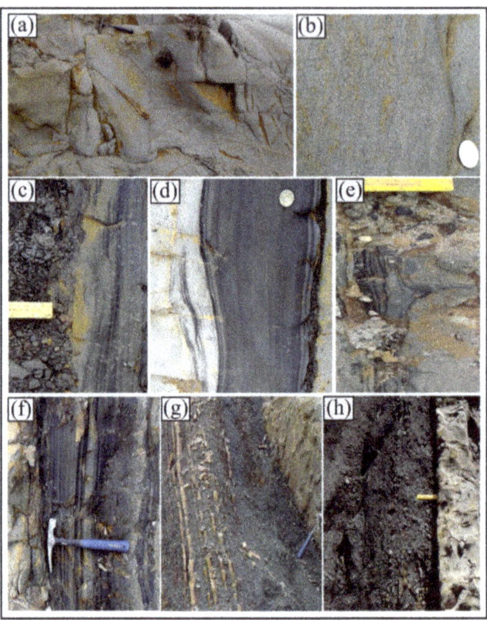

Figure 4. Sedimentary facies. (**a**) Amalgamated massive sand with floating mud clasts in the Jalan UMS road section, interpreted to be S1 lithofacies. (**b**) Massive, coarse-grained sandstone at the Jalan UMS road section, which belongs to the S2 facies. (**c**) Parallel lamination S3 facies and cross-lamination S4 facies in the Jalan UMS road section. (**d**) Parallel laminated S4 facies at the Jalan UMS road section. (**e**) Flame structure S5 facies. (**f**) Laminated siltstone facies Si1 at the Jalan UMS KFC section. (**g**) Laminated muddy siltstone Si2 facies at the Jalan UMS KFC section. (**h**) Massive dark shale or mudstone M facies at the University of Prima Condo road section.

4.2.5. S5 Facies: Medium- to Fine-Grained Soft Sediment Deformation Units

Convoluted lamination due to the deformation of unlithified sand units [74] is a typical characteristic of this sedimentary succession (Figure 5). The deformation of sandstone units varied from gentle to moderately strong, which indicates variation in the degree of deformation. Flame structures are also present in a few units, representing the facies at the hydraulic jump interprets to be a part of the proximal lobe [39]. Dewatering of unlithified clastic units due to the upward movement of fluids and some particles which had deformed the overlying strata can also be present [85]. The phenomenon of the generation of a deformational structure (Figure 5) is related with the fluidization process that develops the instabilities in the gravity flows, or it may also be related to seismic activity [74].

4.2.6. H Facies: Hybrid Event Beds

Hybrid beds are characterized by intermediate flow behavior comprised of two arrangements of lower mud-deficient sand overlain by a mud-rich sand interval, and various terms were assigned like slurry flows [18], transitional flow deposit [9], linked debrite [86], hybrid event [87] and matrix-rich sand [88]. The hybrid beds are often termed as bi- or tripartite beds, depending on the characteristics of the underlying and overlying units, and vary significantly from the downslope of the channelized to the unconfined area. Lateral lobe fringes are predominantly low-density turbidites and have hybrid events [23,45].

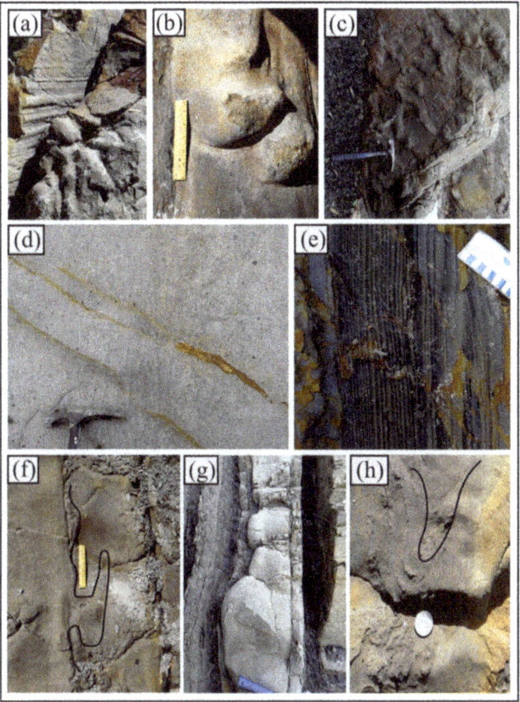

Figure 5. Sedimentary structures. (**a**) Tool marks in the Kampung Madpai section. (**b**) Load casts in the Kampung Madpai section. (**c**) Flute casts at the base of the sandstone in the Jalan UMS KFC outcrop. (**d**) Massive coarse-grained sandstone with dewatering in the Jalan UMS KFC section. (**e**) Ripple and parallel laminations in the Jalan UMS KFC section. (**f**) Flame structure in the Jalan Sulaman outcrop. (**g**) Load structure in the Jalan Sulaman section. (**h**) Convolute lamination in the Jalan Sulaman outcrop.

A great variety of lithofacies can be prevalent within the hybrid units and are also variable within the individual beds over a scale of centimeters to meters. These hybrid beds contain both the characteristics of turbidite and debrite within the same depositional event. The scales of thickness of hybrid beds vary considerably from tens of centimeters to more than a meter, which is associated with the influx of sediments deposited within the single event of hybrid flow. The hybrid beds in the study area (Universiti Prima road section) consisted of only three divisions (H1, H3 and H5) of hybrid event deposition [43]. The basal structureless graded sandstone (H1), overlain by a banded sand unit (H2), was composed of both sand and shale (irregular) bands. The third division was more chaotic (H3), having patches of sand with more mud, with the fourth subdivision having a laminated sand mud unit (H4) capped by a clayey shale unit (H5).

4.2.7. Si1 Facies: Laminated Siltstone

The facies represent the siltstone units, which are laminated siltstone and range in thickness from 6 to 17 cm. The major lithology was siltstone in the form of thin to laminated units with interbeds of shale or mudstone. Fine sand units and silt laminations are common in this facies, alternating with mudstone or shale lamination. The traction fallout and low energy depositional environment, or the diluted turbidity currents in the hemipelagic settings [11], are the possible explanations of these heterolithic facies, which were deposited from a suspension during a lower flow regime. This facies is equivalent to the Bouma Td division, representing a low-density flow deposit (Table 1).

Table 1. Summary of sedimentary facies, with their descriptions, outcrop locations and interpretations.

No.	Facies	Description	Sedimentary Log	Location	Interpretation
1	S1 facies (sandstone)	Thick to massive sandstone Normal grading Amalgamated		Basal and upper part of the Jalan UMS road section	Rapid accumulation High-density currents Ta Bouma
2	S2 facies (sandstone)	Structureless sand-stoneUngraded Amalgamated Coarse- to very coarse-grained		Basal part of the Kampung Madpai section Middle part of the Jalan UMS road section	Lower part of the Ta Bouma facies Sandy and gravely flow [79]
3	S3 facies (sandstone)	Thin- to medium-bedded Parallel-laminated Fine- to medium-grained		Lower part of the Jalan UMS road section Middle part of the Prima University section	Tb Bouma facies F7 and F9 facies [79]
4	S4 facies (sandstone)	Ripple lamination Fine- to very fine-grained Thin- to thick-bedded		Basal part of the Jalan UMS road section Basal part of the Prima University section	Tc Bouma facies F9 Mutti facies Lower flow regime
5	S5 Facies (sandstone)	Soft sediment deformation Thickly to massively bedded Medium- to coarse-grained		Lower part of the Kampung Madpai section Middle part of the Jalan UMS section	Proximal part of the lobe Tc Bouma facies
6	H facies (hybrid event)	Bipartite or tripartite beds Rich in mud and broken clasts		Lower part of the Prima University section	Transitional flow Intermediate flow behavior
7	Si1 facies (siltstone)	Siltstone units Very thin to thin units Rare interbeds of shale or mudstone		Lower and upper parts of the Jalan UMS KFC section	Suspension fallout Bouma Td facies Low density
8	Si2 facies (siltstone)	Higher mud content in siltstone Laminations are discontinuous		Upper part of the Jalan UMS KFC section Lower part of the Prima University section	Dilute sediment gravity flow Td–Te Bouma facies
9	M facies (mudstone or shale)	Mainly shale Thickly to massively bedded Lacking internal structures		Upper part of the Jalan UMS KFC section Basal part of the Prima University section	Te Bouma facies Mud turbidites Final deposition of sediment gravity flow

4.2.8. Si2 Facies: Laminated Muddy Siltstone

Laminated to medium-bedded siltstone beds with shale or mudstone layers were included in this sedimentary facies. The amount of mud or argillaceous material was relatively higher than the siltstone units. The siltstone lamination could be discontinuous due to more shale material, where these lithological characteristics are associated with lobe fringes or distal lobe settings [19]. Numerous individual lobes are usually separated by muddy siltstone intervals. Suspension fallout occurs due to a low energy of flow from a relatively dilute sediment gravity flow. It is also interpreted as change in swiftness of flow and a lower sediment influx. The shale input increases in this type of sedimentary facies, which indicates the energy conditions equivalent to Bouma Td–Te facies. Argillaceous sediments present in the turbidity currents were finally settling down in the lower flow regime.

4.2.9. Mudstone (M) or Massive Shale Facies

The massive shale or mudstone facies predominantly contained thick-bedded to massive shale units although, it may have held a little influx of silt laminae. However, the thickness of shale or mud was considerably larger than the silt laminae, which indicates the strong influx of shale or mud in the sedimentary basin. These mudstone facies represent the lateral lobe settings [19] that separate the individual lobe or lobe complex. A massive mudstone interval could also be evident from the most distal part of the lobe environment. The term "interlobe" is also used for massive shale intervals to differentiate between deposition of the multiple lobes in a submarine environment [22,24,41]. The mudstone or shale primarily lacked any internal structures. The facies was equivalent to Bouma Te facies or T6 or T7 Stow's classification [89,90] and was termed as mud turbidites. These units represent the final deposition from the phase of sediment gravity flow [91].

4.3. Facies Associations and Lobe Complexity

In this section, the outcrop sections are discussed with respect to various thickening or thinning cycles based on the range of thickness of the individual sand units. These cycles or trends are quite useful to relate the outcrop sections with lobe elements and lobe progradation, aggradation or cessation. The dynamics of the lobe in a deep marine system were quite evident from the thickening or thinning trends. Additionally, the lithological units were categorized into lobe elements, which were grouped into lobes and further into lobe complexes. Several individual sedimentary facies were identified for any rock formations, which were later grouped and categorized into facies associations [42]. Several facies associations were identified based on the facies analysis of lithological beds, including sandstone, siltstones and shales or mudstones. These facies associations are essentially connected with various components of submarine lobe deposits [17] equivalent to the proximal lobe or axial lobe (FA1), lobe off-axis (FA2), lobe fringe (FA3) and distal lobe fringe or interlobe (FA4).

4.3.1. Facies Association 1 (FA1): Lobe Axis

Lobe axis facies association is characterized by massive sandstone units, generally having thicknesses of more than 100 cm. The thickness of an individual sand unit may go up to more than 800 cm. These units are often structureless as there is no grading and only a minor change in grain size within the sand beds. Multimeter massive sandstone with an amalgamation structure is the characteristic of this facies. S2 and S5 facies and occasionally S1 facies are included in this facies association. The association of this facies is interpreted as unconfined lobe settings with lobe axis and lobe off-axis alternate beds that are stacked together [92,93], having amalgamated bodies (Figure 6) representing the proximal part of the lobe system. The facies is associated with high-density turbidity currents, where the huge amount of sand influx with rare or no argillaceous content is indicative of a lobe axis depositional environment.

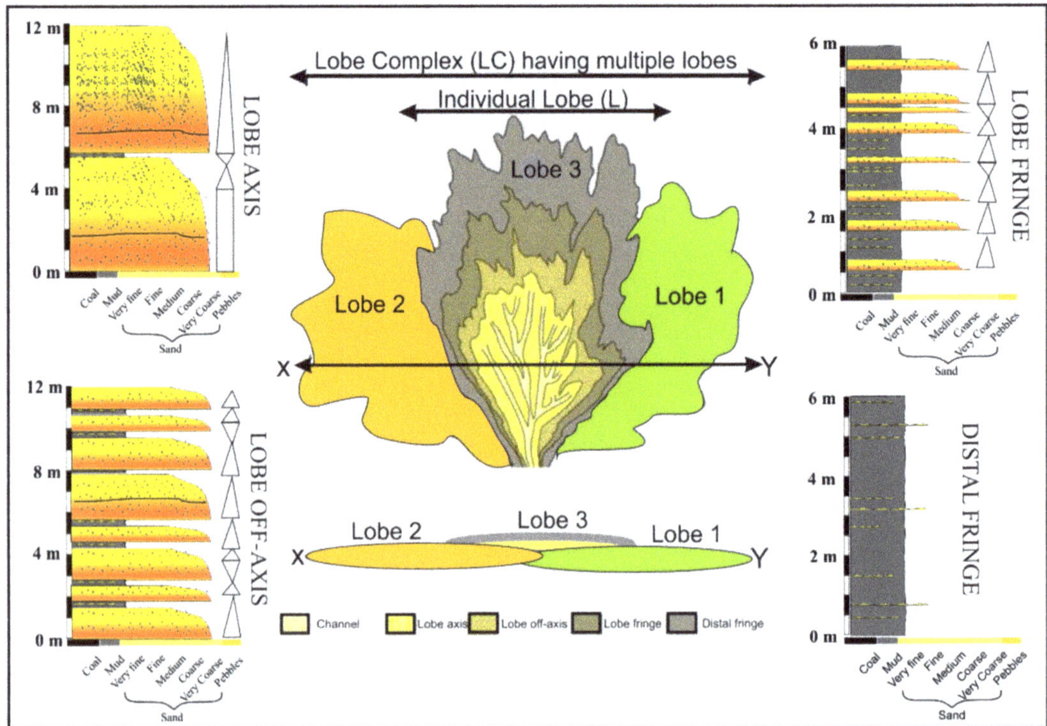

Figure 6. Lobe architecture and facies association of lobe settings. Thickening and thinning cycles are also marked on the bed scale for better understating of deep-marine lobe complex systems. L1 is the old event of lobe deposition, while L3 is the younger event partially overlapping the older lobe L1 and L2. Each lobe is further classified into axis (yellow color), off-axis (brown), fringe (brownish gray) and distal fringe (gray) from sand to shale or mud alterations [48].

4.3.2. Facies Association 2 (FA2): Lobe Off-Axis

The facies denote mainly sandstone units, which are medium- to thick-bedded and massive and where the average thickness was about 42.5 cm, while most thickness values ranged from 10 cm to 250 cm. There was a significant decrease in the sand-to-mud ratio and a lesser degree of amalgamation in the sand units. Hybrid event beds having greater sand bed thicknesses were also contained in this lobe off-axis [11] facies association. Sedimentary features like load casts and tool marks are common in this facies association. Some thick sand units are characterized by being massive or structureless, which is mostly associated with S1, S2 and S3 facies as well as with very rare S4 facies. Soft sediment deformation S5 is relatively common in this facies association. Amalgamated sand units also exist within this facies association. The abundant S3, S4 and Si1 facies represent the lobe off-axis deposition [50].

4.3.3. Facies Association 3 (FA3): Lobe Fringe

The lobe fringe is primarily characterized by muddy units, which comprise most of the percentage of sedimentary rocks. The main feature is rhythmic sandstone and mudstone units, which ranges in thickness from a few centimeters to tens of centimeters. The average thickness value was about 10.8 cm, where the thicknesses of the bed units ranged from 1 to 18 cm. The facies association included the S3, S4, Si1, Si2 and M facies. S1 facies are quite rare in this facies association. Furthermore, hybrid event beds (H) with a lower sand thickness are associated with lobe fringe deposits [49]. The rock units have sharp contact and are relatively continuous laterally. A complete Bouma sequence (from Ta to Te) could be present, but typically, the basal massive sequence (Ta) is usually absent in the lobe fringe settings [26,94]. Thin-bedded sand and shale interbeds with a high fraction of mud and good lateral continuity are interpreted to be a part of lobe fringe deposits.

4.3.4. Facies Association 4 (FA4): Distal Lobe or Interlobe

All types of clastic units, like sandstones (thin-bedded fine to very fine-grained), siltstones and medium- to thick-bedded mudstones or shales were present in this facies association (Figure 7). However, mudstone or shale units mainly comprised this association. The thicknesses of most of sandstone units were less than 10 cm, and the average value of the sand bed thickness was only 4.1 cm. The thickness of the shale units (M facies) was significantly higher (more than 230 cm) compared with other sandstone and siltstone facies. Owing to thin-bedded fine to very fine-grained deep-marine units having quite good lateral thicknesses and high fractions of thick mudstone or shale units, they were interpreted to be interlobe and lobe distal fringe facies associations. The slow hemipelagic to pelagic deposition was the result of low-density turbidity currents.

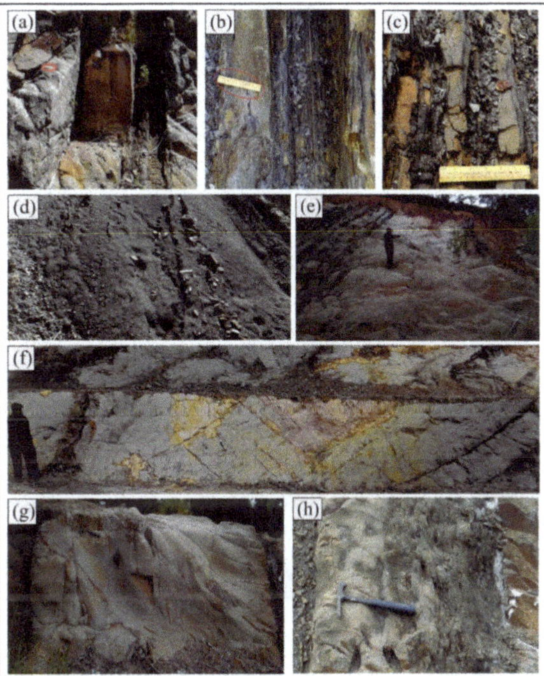

Figure 7. Facies associations. (**a**) Medium- to thick-bedded sandstone lobe fringe FA3 facies association in the Kampung Madpai section. (**b**) Medium- to thin-bedded sandstone of distal lobe fringe associated with the FA4 facies association in the Kampung Madpai outcrop. (**c**) Thin-bedded sandstone in the Jalan UMS road section, interpreted to be the distal lobe fringe to interlobe facies association FA4. (**d**) Mudstone facies interbedded with a thin sand unit, representative of distal lobe to interlobe FA4 settings in the Prima University Condo road outcrop. (**e**) Medium- to thick-bedded sandstone overlain by a massive sand unit of lobe axis FA2 in the Jalan UMS KFC outcrop. (**f**) Massive unit with amalgamation in the Jalan UMS road section, interpreted to be an FA1 facies association. (**g**) Massive sand with mudclasts and amalgamation in the Jalan UMS road section, belonging to the FA1 facies association. (**h**) Hybrid sand body in the Prima University road section, interpreted to be a lobe off-axis FA2 association.

5. Discussion

5.1. Thickening and Thinning Multiscale Trends

Deep-marine sedimentary successions are characterized by multiscale thinning or thickening upward successions in exposed sections. These patterns were sometimes quite evident as we moved stratigraphically in the younging direction. Generally, lobe deposition is characterized by thickening or coarsening upward cycles, whereas channel setting is mostly linked with a thinning and fining upward sequence [17,95]. However, lobe progradation is considered a thickening (Figure 8) or coarsening sequence that has variations in the rate of sediment influx, resulting in a variety of sedimentary facies and their associations [14], while an individual thinning sequence may also be developed due to starvation of the deep-marine lobe system toward the lobe fringe or lateral lobes [39,50]. These thickening and thinning sand units represent unconfined lobe settings, and these thickening sandstone cycles are related to lobe axis and lobe off-axis facies associations [11,15,88,96].

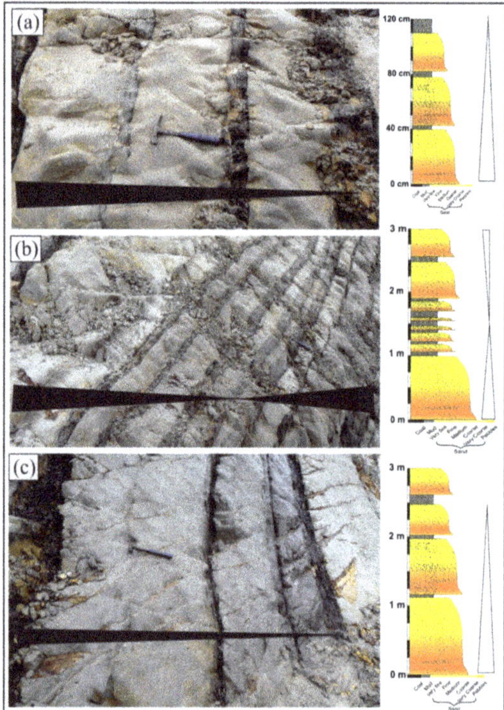

Figure 8. Thinning and thickening cycles in the Jalan UMS road section, where the vertical thicknesses of the cyclic patterns ranged only from 1.2 to 2.7 m, which is indicative of a bed set or a lobe element. (**a**) Thinning pattern interpreted to be a lobe element. (**b**) Thinning and then thickening sequence, where each pattern represents the geometry of an individual lobe element. (**c**) Thinning trend with a vertical thickness of about 2.6 m in the set of beds.

A thinning upward sequence was observed in the Jalan UMS road section (Figure 8) because of lobe abandonment, while at one location, a thinning and then thickening cycle was observed in the outcrop, representing the cessation of a relatively older lobe and subsequently followed by the development of a newer, younger lobe [15,97]. Lobe thickness variation was significant in the Jalan UMS behind KFC section, where three cycles were observed. First was the thinning cycle, where the thicknesses of the individual sand units gradually decreased while the shale content increased as the stratigraphic order became younger (Figure 9). Another event of thinning and then thickening followed by two cycles of thickening was present, which were interpreted to be progradational lobe geometry. A thinning upward cycle was also marked in the Jalan Sulaman outcrop that was distinctive of lobe desertion.

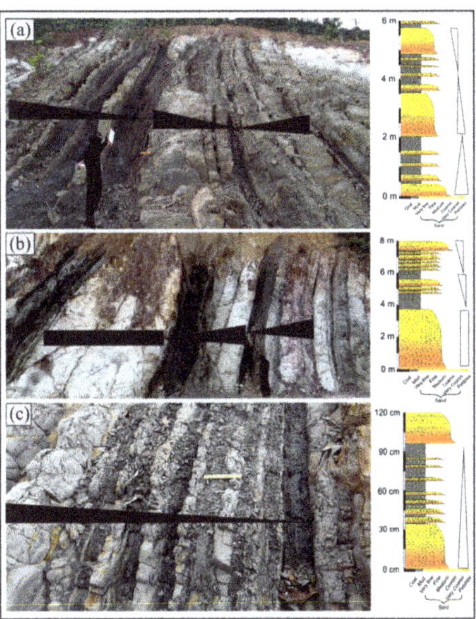

Figure 9. Thinning and thickening patterns (**a**) marked by three cycles—two thinning and one thickening—each of about a 2 m vertical thickness in the Jalan UMS behind KFC outcrop. (**b**) Massive sand unit overlain by two thickening trends in the Jalan UMS behind KFC outcrop. (**c**) One cycle of thinning with a vertical thickness of only 1 m in the Jalan Sulaman outcrop.

5.2. Distribution of the Lobe Complex

A submarine lobe system is a vital constituent of deep-marine fans. These lobes are characterized by geometries which are quite useful for interpreting the geological processes related to fan deposition [15]. Tectonically active regions are generally characterized by coarse sand units, representing the development of deep-marine fans having less than 10 km radial exposure on relatively higher slope angles, where a fan lobe system is frequently surrounded by shale cover [39].

These submarine fans are composed of numerous lobe complexes. The lobe architecture consists of a composite hierarchy from a smaller unit of a lobe element to a larger unit, which is termed as a composite lobe system or lobe complex set [22,24]. The lobe element is essentially comprised of one or more beds, with the thickness extending from a few decimeters to more than a meter [39]. The individual lithologies or beds collectively form an element of a lobe, while the group of lobe elements constitutes an individual lobe [15,24]. These lobes are characterized by interbedded sandstone and shale, with a collective range in size of several meters in thickness, combined to form a lobe complex (Figure 10) or also termed as stacked composite lobes [15,24]. Lobe components or sand-rich lobe complexes are separated from each other by a thick to massive hemipelagic to pelagic shale unit [22]. It is pertinent to mention here that each lobe component consists of one or more sandstone or shale beds, which enabled us to characterize the deep-marine deposits at a meter scale level. These high-resolution lithological observations could not be achieved by using seismic data. Hence, the study of lobe architecture at a lobe element scale caters to the idea of subseismic reservoir heterogeneity in submarine lobe–fan systems.

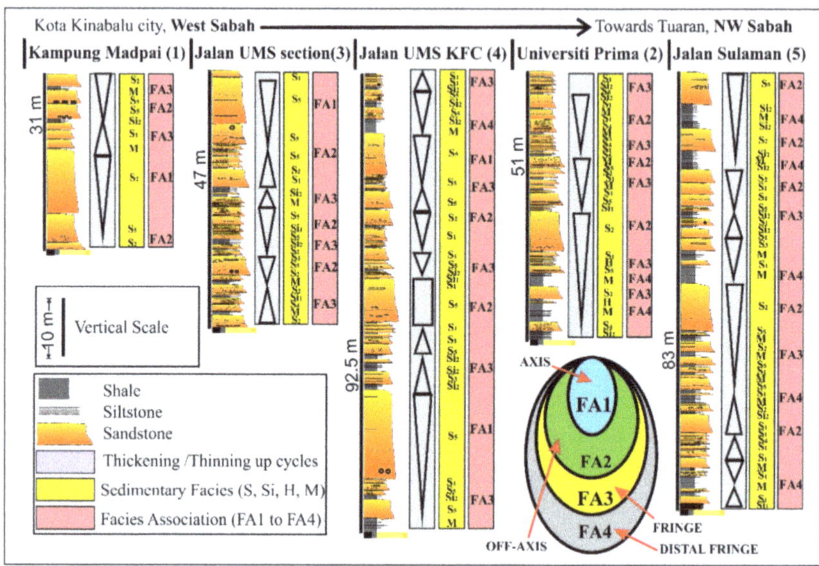

Figure 10. The distribution of facies, facies associations and thickening and thinning cycles in the exposed sections from Kota Kinabalu to Tuaran in northwest Sabah.

The results presented in this study reveal that there are multiple feeder channels in West Sabah's deposition, where the number of feeder channels and lobe complexes increased toward northwest Sabah. Multiple feeder channels resulted in three to four lobe complexes (LC), each of which was classified into individual lobes (L) and further into lobe elements (LE) (Figure 11). The thick to massive shale separated the individual lobe complexes.

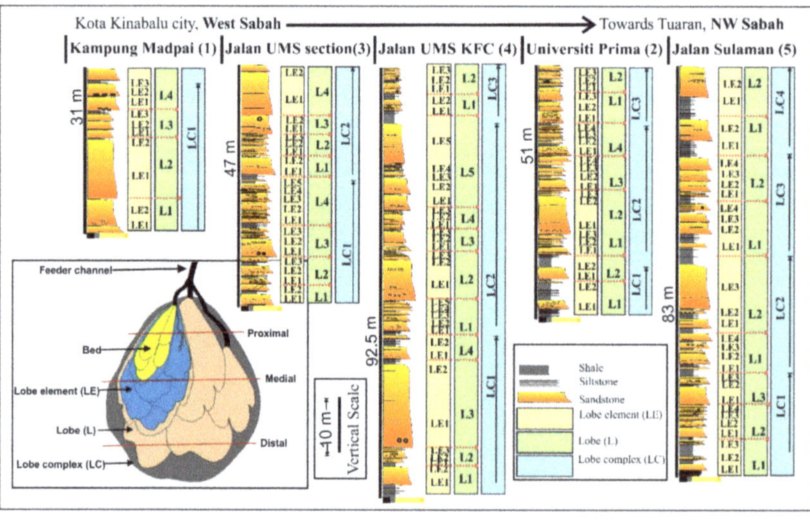

Figure 11. Distribution of lobe elements, individual lobes and lobe complexes in the studied outcrop sections [17,24].

5.3. Submarine Fan–Lobe System

The present study highlights that all the exposed sections were interpreted to be proximal to medial fan depositions, which were further classified into a lobe hierarchy. It incorporates the concept of a submarine fan and lobe system with respect to the multiscale analysis of sand–shale complexes in a deep marine environment. The individual lithological bed or bed set at a centimeter-to-meter scale is termed as a lobe element, which is the basic building block for the whole lobe–fan architecture. These lobe elements combine to form multimeter lobes, which are the cyclic or repetitive structures in a lobe complex or in a composite lobe system, which are tens of meters in thickness, while the composite lobes eventually constitute a smaller portion of the submarine fan at a scale of hundreds of meters. A complete fan system is present at a km scale over a large depositional area in a sedimentary basin.

6. Conclusions

The results highlight the facies analysis and facies association linked with the architectural elements of lobes in the submarine fan deposits of West Sabah. Based on these results, the following conclusions are drawn:

1. Although the West Crocker Formation is mainly considered to have sand-rich deposits (Crocker sands), the formation also contains massive shale and siltstone units. All types of sedimentary facies related to sandstone, siltstone and mudstone and could be termed as a sand–shale system. This variety of sediments shows more heterogeneity in lithological characteristics than previously thought;
2. The sedimentary facies were grouped into four facies associations, which were linked to the lobe architecture of deep-marine systems. These facies associations are discussed as components of individual lobes, namely the lobe axis, lobe off-axis, lobe fringe and distal fringe;
3. The thicknesses of individual sandstone units are helpful for interpreting several thickening and thinning multiscale sequences, which are characteristics of lobe progradation and lobe abandonment. These cycles of thickness variations represent the multiple tabular sand bodies of a lobe complex.
4. The deep-marine lobe deposits can be classified into beds or bed sets, which constitute the lobe elements. These lobe elements are grouped into individual lobes, which are broadly categorized into lobe complexes. These lobe element to lobe complex nomenclature can be identified on all exposed sections of West Sabah, where the individual thicknesses of lobe elements highly vary from as small as 1–3 m up to a large thickness of 8–10 m in the studied sections having multiscale sand-shale complex.
5. The West Crocker Formation is interpreted as a lobe complex set in which multiple lobe complexes are present, with their individual lobes and lobe elements based on bed-to-bed sedimentary analysis and supporting the multiscale modeling of deep ocean sediments.
6. The lobe complex sets are more developed in northwest Sabah, while West Sabah has a lower number of lobe complexes. This distribution of lobe complexes also verifies that the paleocurrent direction is mainly from the south to the north, where the feeder channels form multiple lobe complexes in northwest Sabah.
7. The detailed facies and lobe architecture depict reservoir heterogeneities in deep-marine siliciclastic rocks, which are usually interpreted as single homogeneous sand units by seismic data. Hence, the present study highlights the subseismic lithological complexities in deep-marine depositional settings.
8. The alternate lobe off-axis and lobe axis distributions, interpreted as unconfined lobe settings, could be applicable for several unconfined deep-marine sedimentary successions around the globe which are potential sites of exploration of natural resources.

Author Contributions: Conceptualization, M.J., N.A.S. and A.H.B.A.R.; methodology, M.J., N.A., M.S.B.I. and N.A.I; software, M.J., M.U., A.H.B.A.R., and Z.G.; validation, M.J., N.A.S., A.H.B.A.R., and N.A.I.; formal analysis, M.J., Q.S.I. and M.U.; investigation, M.J., N.A.S. and Z.G.; resources, N.A.S., A.H.B.A.R. and N.A.I.; data curation, M.J. and N.A.; writing—original draft preparation, M.J.; writing—review and editing, N.A.S., A.H.B.A.R. and M.U.; visualization, M.J., M.U. and N.A.; supervision, N.A.S., M.S.B.I. and A.H.B.A.R.; project administration, N.A.I. and N.A.; funding acquisition, N.A.S., A.H.B.A.R. and N.A.I. All authors have read and agreed to the published version of the manuscript.

Funding: This research was funded by the Petroleum Research Fund (PRF) cost number 0153AB-A33, awarded to Eswaran Padmanabhan and the Fundamental Research Grant of the Ministry of Higher Education (MoHE) Malaysia (project ID 16880, reference code FRGS/1/2019/STG09/UTP/03/1) for data analysis and detailed fieldwork. The first phase of geological field was supported by Yayasan UTP Malaysia (YUTP) grant number 015LC-017, awarded to Noor Azahar Ibrahim.

Institutional Review Board Statement: Not applicable.

Informed Consent Statement: Not applicable.

Data Availability Statement: The dataset would be available from the first author, if required for reference purposes.

Acknowledgments: The authors are thankful to the faculty and staff of the Department of Geoscience at UTP Malaysia for their support in arranging the geological fieldwork. The comments and feedback from anonymous reviewers improved the manuscript.

Conflicts of Interest: The authors declare no conflict of interest.

References

1. Lorentzen, S.; Augustsson, C.; Jahren, J.; Nystuen, J.P.; Schovsbo, N.H. Tectonic, sedimentary and diagenetic controls on sediment maturity of lower Cambrian quartz arenite from southwestern Baltica. *Basin Res.* **2019**, *31*, 1098–1120. [CrossRef]
2. Covault, J.A.; Graham, S.A. Submarine fans at all sea-level stands: Tectono-morphologic and climatic controls on terrigenous sediment delivery to the deep sea. *Geology* **2010**, *38*, 939–942. [CrossRef]
3. Pickering, K.T.; Bayliss, N.J. Deconvolving tectono-climatic signals in deep-marine siliciclastics, Eocene Ainsa basin, Spanish Pyrenees: Seesaw tectonics versus eustasy. *Geology* **2009**, *37*, 203–206. [CrossRef]
4. Kontakiotis, G.; Karakitsios, V.; Cornée, J.-J.; Moissette, P.; Zarkogiannis, S.D.; Pasadakis, N.; Koskeridou, E.; Manoutsoglou, E.; Drinia, H.; Antonarakou, A. Preliminary results based on geochemical sedimentary constraints on the hydrocarbon potential and depositional environment of a Messinian sub-salt mixed siliciclastic-carbonate succession onshore Crete (Plouti section, eastern Mediterranean). *Mediterr. Geosci. Rev.* **2020**, *2*, 247–265. [CrossRef]
5. Leila, M.; Moscariello, A.; Kora, M.; Mohamed, A.; Samankassou, E. Sedimentology and reservoir quality of a Messinian mixed siliciclastic-carbonate succession, onshore Nile Delta, Egypt. *Mar. Pet. Geol.* **2020**, *112*, 104076. [CrossRef]
6. Richards, M.; Bowman, M.; Reading, H. Submarine-fan systems i: Characterization and stratigraphic prediction. *Mar. Pet. Geol.* **1998**, *15*, 689–717. [CrossRef]
7. Zhang, J.; Wu, S.; Hu, G.; Fan, T.-e.; Yu, B.; Lin, P.; Jiang, S. Sea-level control on the submarine fan architecture in a deepwater sequence of the Niger Delta Basin. *Mar. Pet. Geol.* **2018**, *94*, 179–197. [CrossRef]
8. Zhang, J.-J.; Wu, S.-H.; Fan, T.-E.; Fan, H.-J.; Jiang, L.; Chen, C.; Wu, Q.-Y.; Lin, P. Research on the architecture of submarine-fan lobes in the Niger Delta Basin, offshore West Africa. *J. Palaeogeogr.* **2016**, *5*, 185–204. [CrossRef]
9. Kane, I.A.; Pontén, A.S. Submarine transitional flow deposits in the Paleogene Gulf of Mexico. *Geology* **2012**, *40*, 1119–1122. [CrossRef]
10. Liu, F.; Zhu, X.; Li, Y.; Xue, M.; Sun, J. Sedimentary facies analysis and depositional model of gravity-flow deposits of the Yanchang Formation, southwestern Ordos Basin, NW China. *Aust. J. Earth Sci.* **2016**, *63*, 885–902. [CrossRef]
11. Mueller, P.; Patacci, M.; Di Giulio, A. Hybrid event beds in the proximal to distal extensive lobe domain of the coarse-grained and sand-rich Bordighera turbidite system (NW Italy). *Mar. Pet. Geol.* **2017**, *86*, 908–931. [CrossRef]
12. Mulder, T.; Alexander, J. The physical character of subaqueous sedimentary density flow and their deposits. *Sedimentology* **2001**, *48*, 269–299. [CrossRef]
13. Palozzi, J.; Pantopoulos, G.; Maravelis, A.G.; Nordsvan, A.; Zelilidis, A. Sedimentological analysis and bed thickness statistics from a Carboniferous deep-water channel-levee complex: Myall Trough, SE Australia. *Sediment. Geol.* **2018**, *364*, 160–179. [CrossRef]
14. Starek, D.; Fuksi, T. Distal turbidite fan/lobe succession of the Late Oligocene Zuberec Fm.–architecture and hierarchy (Central Western Carpathians, Orava–Podhale basin). *Open Geosci.* **2017**, *9*, 385–406. [CrossRef]
15. Macdonald, H.A.; Peakall, J.; Wignall, P.B.; Best, J. Sedimentation in deep-sea lobe-elements: Implications for the origin of thickening-upward sequences. *J. Geol. Soc.* **2011**, *168*, 319–332. [CrossRef]

16. Mulder, T.; Etienne, S. Lobes in deep-sea turbidite systems: State of the art. *Sediment. Geol.* **2010**, *229*, 75–80. [CrossRef]
17. Prélat, A.; Hodgson, D.; Flint, S. Evolution, architecture and hierarchy of distributary deep-water deposits: A high-resolution outcrop investigation from the Permian Karoo Basin, South Africa. *Sedimentology* **2009**, *56*, 2132–2154. [CrossRef]
18. Lowe, D.R.; Guy, M. Slurry-flow deposits in the Britannia Formation (Lower Cretaceous), North Sea: A new perspective on the turbidity current and debris flow problem. *Sedimentology* **2000**, *47*, 31–70. [CrossRef]
19. Spychala, Y.T.; Hodgson, D.M.; Prélat, A.; Kane, I.A.; Flint, S.S.; Mountney, N.P. Frontal and lateral submarine lobe fringes: Comparing sedimentary facies, architecture and flow processes. *J. Sediment. Res.* **2017**, *87*, 75–96. [CrossRef]
20. Bell, D.; Kane, I.A.; Pontén, A.S.M.; Flint, S.S.; Hodgson, D.M.; Barrett, B.J. Spatial variability in depositional reservoir quality of deep-water channel-fill and lobe deposits. *Mar. Pet. Geol.* **2018**, *98*, 97–115. [CrossRef]
21. Clare, M.A.; Talling, P.J.; Challenor, P.; Malgesini, G.; Hunt, J. Distal turbidites reveal a common distribution for large (>0.1 km^3) submarine landslide recurrence. *Geology* **2014**, *42*, 263–266. [CrossRef]
22. Zhang, L.-F.; Dong, D.-Z. Thickening-upward cycles in deep-marine and deep-lacustrine turbidite lobes: Examples from the Clare Basin and the Ordos Basin. *J. Palaeogeogr.* **2020**, *9*, 1–16. [CrossRef]
23. Spychala, Y.T.; Hodgson, D.M.; Lee, D.R. Autogenic controls on hybrid bed distribution in submarine lobe complexes. *Mar. Pet. Geol.* **2017**, *88*, 1078–1093. [CrossRef]
24. Zhang, L.-F.; Pan, M.; Li, Z.-L. 3D modeling of deepwater turbidite lobes: A review of the research status and progress. *Pet. Sci.* **2020**, *17*, 317–333. [CrossRef]
25. Mutti, E.; Bernoulli, D.; Lucchi, F.R.; Tinterri, R. Turbidites and turbidity currents from Alpine 'flysch' to the exploration of continental margins. *Sedimentology* **2009**, *56*, 267–318. [CrossRef]
26. Peakall, J.; Best, J.; Baas, J.H.; Hodgson, D.M.; Clare, M.A.; Talling, P.J.; Dorrell, R.M.; Lee, D.R. An integrated process-based model of flutes and tool marks in deep-water environments: Implications for palaeohydraulics, the Bouma sequence and hybrid event beds. *Sedimentology* **2020**, *67*, 1601–1666. [CrossRef]
27. Zakaria, A.A.; Johnson, H.D.; Jackson, C.A.L.; Tongkul, F. Sedimentary facies analysis and depositional model of the Palaeogene West Crocker submarine fan system, NW Borneo. *J. Asian Earth Sci.* **2013**, *76*, 283–300. [CrossRef]
28. Jackson, C.A.L.; Zakaria, A.A.; Johnson, H.D.; Tongkul, F.; Crevello, P.D. Sedimentology, stratigraphic occurrence and origin of linked debrites in the West Crocker Formation (Oligo-Miocene), Sabah, NW Borneo. *Mar. Pet. Geol.* **2009**, *26*, 1957–1973. [CrossRef]
29. Jamil, M.; Abd Rahman, A.H.; Siddiqui, N.A.; Ibrahim, N.A.; Ahmed, N. A contemporary review of sedimentological and stratigraphic framework of the Late Paleogene deep marine sedimentary successions of West Sabah, North-West Borneo. *Bull. Geol. Soc. Malays.* **2020**, *69*, 53–65. [CrossRef]
30. Lambiase, J.J.; Tzong, T.Y.; William, A.G.; Bidgood, M.D.; Brenac, P.; Cullen, A.B. The West Crocker formation of northwest Borneo: A Paleogene accretionary prism. *Spec. Pap. Geol. Soc. Am.* **2008**, *436*, 171–184.
31. Usman, M.; Siddiqui, N.A.; Mathew, M.; Zhang, S.; El-Ghali, M.A.K.; Ramkumar, M.; Jamil, M.; Zhang, Y. Linking the influence of diagenetic properties and clay texture on reservoir quality in sandstones from NW Borneo. *Mar. Pet. Geol.* **2020**, *120*, 104509. [CrossRef]
32. Usman, M.; Siddiqui, N.A.; Zhang, S.; Ramkumar, M.; Mathew, M.; Sautter, B.; Beg, M.A. Ichnofacies and sedimentary structures: A passive relationship with permeability of a sandstone reservoir from NW Borneo. *J. Asian Earth Sci.* **2020**, *192*, 103992. [CrossRef]
33. Hall, R.; Nichols, G. Cenozoic sedimentation and tectonics in Borneo: Climatic influences on orogenesis. *Geol. Soc. Lond. Spec. Publ.* **2002**, *191*, 5–22. [CrossRef]
34. Banerjee, A.; Ahmed Salim, A.M. Stratigraphic evolution of deep-water Dangerous Grounds in the South China Sea, NW Sabah Platform Region, Malaysia. *J. Pet. Sci. Eng.* **2021**, *201*, 108434. [CrossRef]
35. Cullis, S.; Patacci, M.; Colombera, L.; Bührig, L.; McCaffrey, W.D. A database solution for the quantitative characterisation and comparison of deep-marine siliciclastic depositional systems. *Mar. Pet. Geol.* **2019**, *102*, 321–339. [CrossRef]
36. Mayall, M.; Jones, E.; Casey, M. Turbidite channel reservoirs—Key elements in facies prediction and effective development. *Mar. Pet. Geol.* **2006**, *23*, 821–841. [CrossRef]
37. Prather, B.E. Controls on reservoir distribution, architecture and stratigraphic trapping in slope settings. *Mar. Pet. Geol.* **2003**, *20*, 529–545. [CrossRef]
38. Zhang, L.; Pan, M.; Wang, H. Deepwater Turbidite Lobe Deposits: A Review of the Research Frontiers. *Acta Geol. Sin. Engl. Ed.* **2017**, *91*, 283–300. [CrossRef]
39. Postma, G.; Kleverlaan, K. Supercritical flows and their control on the architecture and facies of small-radius sand-rich fan lobes. *Sediment. Geol.* **2018**, *364*, 53–70. [CrossRef]
40. Hodgson, D.M.; Kane, I.A.; Flint, S.S.; Brunt, R.L.; Ortiz-Karpf, A. Time-transgressive confinement on the slope and the progradation of basin-floor fans: Implications for the sequence stratigraphy of deep-water deposits. *J. Sediment. Res.* **2016**, *86*, 73–86. [CrossRef]
41. Rodríguez-Cañero, R.; Jabaloy-Sánchez, A.; Navas-Parejo, P.; Martín-Algarra, A. Linking Palaeozoic palaeogeography of the Betic Cordillera to the Variscan Iberian Massif: New insight through the first conodonts of the Nevado-Filábride Complex. *Int. J. Earth Sci.* **2018**, *107*, 1791–1806. [CrossRef]
42. Reading, H.G. Clastic facies models, a personal perspective. *Bull. Geol. Soc. Den.* **2001**, *48*, 101–115.

43. Fonnesu, M.; Haughton, P.; Felletti, F.; McCaffrey, W. Short length-scale variability of hybrid event beds and its applied significance. *Mar. Pet. Geol.* **2015**, *67*, 583–603. [CrossRef]
44. Yang, T.; Cao, Y.; Friis, H.; Liu, K.; Wang, Y. Origin and evolution processes of hybrid event beds in the Lower Cretaceous of the Lingshan Island, Eastern China. *Aust. J. Earth Sci.* **2018**, *65*, 517–534. [CrossRef]
45. Kane, I.A.; Pontén, A.S.M.; Vangdal, B.; Eggenhuisen, J.T.; Hodgson, D.M.; Spychala, Y.T. The stratigraphic record and processes of turbidity current transformation across deep-marine lobes. *Sedimentology* **2017**, *64*, 1236–1273. [CrossRef]
46. Doughty-Jones, G.; Mayall, M.; Lonergan, L. Stratigraphy, facies, and evolution of deep-water lobe complexes within a salt-controlled intraslope minibasin. *Aapg Bull.* **2017**, *101*, 1879–1904. [CrossRef]
47. Groenenberg, R.M.; Hodgson, D.M.; Prelat, A.; Luthi, S.M.; Flint, S.S. Flow–deposit interaction in submarine lobes: Insights from outcrop observations and realizations of a process-based numerical model. *J. Sediment. Res.* **2010**, *80*, 252–267. [CrossRef]
48. Hansen, L.A.S.; Hodgson, D.M.; Pontén, A.; Bell, D.; Flint, S. Quantification of Basin-Floor Fan Pinchouts: Examples From the Karoo Basin, South Africa. *Front. Earth Sci.* **2019**, *7*, 12. [CrossRef]
49. Kuswandaru, G.Y.; Amir Hassan, M.H.; Matenco, L.C.; Taib, N.I.; Mustapha, K.A. Turbidite, debrite, and hybrid event beds in submarine lobe deposits of the Palaeocene to middle Eocene Kapit and Pelagus members, Belaga Formation, Sarawak, Malaysia. *Geol. J.* **2019**, *54*, 3421–3437. [CrossRef]
50. Prelat, A.; Hodgson, D. The full range of turbidite bed thickness patterns in submarine lobes: Controls and implications. *J. Geol. Soc.* **2013**, *170*, 209–214. [CrossRef]
51. Hall, R. Contraction and extension in northern Borneo driven by subduction rollback. *J. Asian Earth Sci.* **2013**, *76*, 399–411. [CrossRef]
52. Hazebroek, H.P.; Tan, D.N. Tertiary tectonic evolution of the NW Sabah continental margin. *Bull. Geol. Soc. Malays.* **1993**, *33*, 195–210. [CrossRef]
53. Siddiqui, N.A.; Ramkumar, M.; Rahman, A.H.A.; Mathew, M.J.; Santosh, M.; Sum, C.W.; Menier, D. High resolution facies architecture and digital outcrop modeling of the Sandakan formation sandstone reservoir, Borneo: Implications for reservoir characterization and flow simulation. *Geosci. Front.* **2019**, *10*, 957–971. [CrossRef]
54. Siddiqui, N.A.; Mathew, M.J.; Ramkumar, M.; Sautter, B.; Usman, M.; Abdul Rahman, A.H.; El-Ghali, M.A.K.; Menier, D.; Shiqi, Z.; Sum, C.W. Sedimentological characterization, petrophysical properties and reservoir quality assessment of the onshore Sandakan Formation, Borneo. *J. Pet. Sci. Eng.* **2020**, *186*, 106771. [CrossRef]
55. Wang, P.C.; Li, S.Z.; Guo, L.L.; Jiang, S.H.; Somerville, I.D.; Zhao, S.J.; Zhu, B.D.; Chen, J.; Dai, L.M.; Suo, Y.H.; et al. Mesozoic and Cenozoic accretionary orogenic processes in Borneo and their mechanisms. *Geol. J.* **2016**, *51*, 464–489. [CrossRef]
56. Chang, S.-P.; Jamaludin, S.N.F.; Pubellier, M.; Zainuddin, N.M.; Choong, C.-M. Collision, mélange and circular basins in north Borneo: A genetic link? *J. Asian Earth Sci.* **2019**, *181*, 103895. [CrossRef]
57. Clennell, B. Far-field and gravity tectonics in Miocene basins of Sabah, Malaysia. *Geol. Soc. Lond. Spec. Publ.* **1996**, *106*, 307–320. [CrossRef]
58. Hutchison, C.S. *Geology of North-West Borneo: Sarawak, Brunei and Sabah*; Elsevier: Amsterdam, The Netherlands, 2005; pp. 61–241.
59. Morley, C. Major unconformities/termination of extension events and associated surfaces in the South China Seas: Review and implications for tectonic development. *J. Asian Earth Sci.* **2016**, *120*, 62–86. [CrossRef]
60. Rangin, C.; Bellon, H.; Benard, F.; Letouzey, J.; Muller, C.; Sanudin, T. Neogene arc-continent collision in Sabah, northern Borneo (Malaysia). *Tectonophysics* **1990**, *183*, 305–319. [CrossRef]
61. Madon, M.; Kessler, F.L.; Jong, J.; Amin, M.K.A. "Fractured basement" play in the Sabah Basin?–the Crocker and Kudat formations as hydrocarbon reservoirs and their risk factors. *Bull. Geol. Soc. Malays.* **2020**, *69*, 157–171. [CrossRef]
62. Lunt, P. A new view of integrating stratigraphic and tectonic analysis in South China Sea and north Borneo basins. *J. Asian Earth Sci.* **2019**, *177*, 220–239. [CrossRef]
63. Hesse, S.; Back, S.; Franke, D. The deep-water fold-and-thrust belt offshore NW Borneo: Gravity-driven versus basement-driven shortening. *Geol. Soc. Am. Bull.* **2009**, *121*, 939–953. [CrossRef]
64. Mustafar, M.A.; Simons, W.J.; Tongkul, F.; Satirapod, C.; Omar, K.M.; Visser, P.N. Quantifying deformation in North Borneo with GPS. *J. Geod.* **2017**, *91*, 1241–1259. [CrossRef]
65. Madon, M. Sand injectites in the West Crocker Formation, Kota Kinabalu, Sabah. *Bull. Geol. Soc. Malays.* **2020**, *69*, 11–26. [CrossRef]
66. Leong, K. Geological setting of Sabah. In *The Petroleum Geology and Resources of Malaysia*; Petroliam Nasional Berhad (PETRONAS): Kuala Lumpur, Malaysia, 1999; pp. 473–497.
67. Leong, T.B.G.; Tahir, S.H.; Asis, J. Stratigraphy of Paleogene Sequences in Weston–Sipitang, Sabah. *Geol. Behav.* **2018**, *2*, 1–4.
68. Jackson, C.A.; Johnson, H.D. Sustained turbidity currents and their interaction with debrite-related topography; Labuan Island, offshore NW Borneo, Malaysia. *Sediment. Geol.* **2009**, *219*, 77–96. [CrossRef]
69. Jamil, M.; Rahman, A.H.A.; Siddiqui, N.A.; Ahmed, N. Deep marine Paleogene sedimentary sequence of West Sabah: Contemporary opinions and ambiguities. *War. Geol.* **2019**, *45*, 198–200.
70. Van Hattum, M.W.; Hall, R.; Pickard, A.L.; Nichols, G.J. Southeast Asian sediments not from Asia: Provenance and geochronology of north Borneo sandstones. *Geology* **2006**, *34*, 589–592. [CrossRef]
71. Van Hattum, M.W.A.; Hall, R.; Pickard, A.L.; Nichols, G.J. Provenance and geochronology of Cenozoic sandstones of northern Borneo. *J. Asian Earth Sci.* **2013**, *76*, 266–282. [CrossRef]

72. William, A.G.; Lambiase, J.J.; Back, S.; Jamiran, M.K. Sedimentology of the Jalan Salaiman and Bukit Melinsung outcrops, western Sabah: Is the West Crocker Formation an analogue for Neogene turbidites offshore? *Bull. Geol. Soc. Malays.* **2003**, *47*, 63–75. [CrossRef]
73. Stow, D.; Smillie, Z. Distinguishing between Deep-Water Sediment Facies: Turbidites, Contourites and Hemipelagites. *Geosciences* **2020**, *10*, 68. [CrossRef]
74. Koç-Taşgın, C.; Altun, F. Soft-sediment deformation: Deep-water slope deposits of a back-arc basin (middle Eocene-Oligocene Kırkgeçit Formation, Elazığ Basin), Eastern Turkey. *Arab. J. Geosci.* **2019**, *12*, 773. [CrossRef]
75. Hodgson, D.M.; Flint, S.S.; Hodgetts, D.; Drinkwater, N.J.; Johannessen, E.P.; Luthi, S.M. Stratigraphic Evolution of Fine-Grained Submarine Fan Systems, Tanqua Depocenter, Karoo Basin, South Africa. *J. Sediment. Res.* **2006**, *76*, 20–40. [CrossRef]
76. Johnson, S.D.; Flint, S.; Hinds, D.; De Ville Wickens, H. Anatomy, geometry and sequence stratigraphy of basin floor to slope turbidite systems, Tanqua Karoo, South Africa. *Sedimentology* **2001**, *48*, 987–1023. [CrossRef]
77. Hodgson, D.M. Distribution and origin of hybrid beds in sand-rich submarine fans of the Tanqua depocentre, Karoo Basin, South Africa. *Mar. Pet. Geol.* **2009**, *26*, 1940–1956. [CrossRef]
78. Bouma, A.H. Sedimentology of some flysch deposits. In *Agraphic Approach Facies Interpret*; Elsevier: Amsterdam, The Netherlands, 1962; Volume 168.
79. Mutti, E. *Turbidite Sandstones*; AGIP, Istituto di Geologia, Università di Parma: San Donato Milanese, Italy, 1992.
80. Lowe, D.R. Sediment gravity flows: II. Depositional models with special reference to the deposits of high-density turbidity currents. *J. Sediment. Petrol.* **1982**, *52*, 279–297.
81. Mutti, E.; Tinterri, R.; Benevelli, G.; Biase, D.d.; Cavanna, G. Deltaic, mixed and turbidite sedimentation of ancient foreland basins. *Mar. Pet. Geol.* **2003**, *20*, 733–755. [CrossRef]
82. Leclair, S.F.; Arnott, R.W.C. Parallel Lamination Formed by High-Density Turbidity Currents. *J. Sediment. Res.* **2005**, *75*, 1–5. [CrossRef]
83. Postma, G.; Kleverlaan, K.; Cartigny, M.J.B. Recognition of cyclic steps in sandy and gravelly turbidite sequences, and consequences for the Bouma facies model. *Sedimentology* **2014**, *61*, 2268–2290. [CrossRef]
84. Jobe, Z.R.; Lowe, D.R.; Morris, W.R. Climbing-ripple successions in turbidite systems: Depositional environments, sedimentation rates and accumulation times. *Sedimentology* **2012**, *59*, 867–898. [CrossRef]
85. Owen, G. Deformation processes in unconsolidated sands. *Geol. Soc. Lond. Spec. Publ.* **1987**, *29*, 11–24. [CrossRef]
86. Haughton, P.D.W.; Barker, S.P.; McCaffrey, W.D. 'Linked' debrites in sand-rich turbidite systems—origin and significance. *Sedimentology* **2003**, *50*, 459–482. [CrossRef]
87. Haughton, P.; Davis, C.; McCaffrey, W.; Barker, S. Hybrid sediment gravity flow deposits—Classification, origin and significance. *Mar. Pet. Geol.* **2009**, *26*, 1900–1918. [CrossRef]
88. Terlaky, V.; Arnott, R.W.C. Matrix-rich and associated matrix-poor sandstones: Avulsion splays in slope and basin-floor strata. *Sedimentology* **2014**, *61*, 1175–1197. [CrossRef]
89. Stow, D. Deep-sea clastics: Where are we and where are we going? *Geol. Soc. Lond. Spec. Publ.* **1985**, *18*, 67–93. [CrossRef]
90. Stow, D.A.V.; Shanmugam, G. Sequence of structures in fine-grained turbidites: Comparison of recent deep-sea and ancient flysch sediments. *Sediment. Geol.* **1980**, *25*, 23–42. [CrossRef]
91. Van Daele, M.; Meyer, I.; Moernaut, J.; De Decker, S.; Verschuren, D.; De Batist, M. A revised classification and terminology for stacked and amalgamated turbidites in environments dominated by (hemi)pelagic sedimentation. *Sediment. Geol.* **2017**, *357*, 72–82. [CrossRef]
92. Bábek, O.; Mikuláš, R.; Zapletal, J.; Lehotský, T. Combined tectonic-sediment supply-driven cycles in a Lower Carboniferous deep-marine foreland basin, Moravice Formation, Czech Republic. *Int. J. Earth Sci.* **2004**, *93*, 241–261. [CrossRef]
93. Shi, G.; Huang, C.; Jiang, S.; Wang, H.; Liang, C.; Yue, J.; Song, G. Late Paleozoic gravity flow depositional systems in the Mandula Basin of the Solonker Belt, Inner Mongolia, China: Towards a volcanic-associated submarine environment. *Int. J. Earth Sci.* **2020**, *109*, 1613–1637. [CrossRef]
94. Pickering, K.T. Two types of outer fan lobe sequence, from the late Precambrian Kongsfjord Formation submarine fan, Finnmark, North Norway. *J. Sediment. Res.* **1981**, *51*, 1277–1286. [CrossRef]
95. Chakraborty, P.P.; Mukhopadhyay, B.; Pal, T.; Dutta Gupta, T. Statistical appraisal of bed thickness patterns in turbidite successions, Andaman Flysch Group, Andaman Islands, India. *J. Asian Earth Sci.* **2002**, *21*, 189–196. [CrossRef]
96. Terlaky, V.; Arnott, R.W.C. The Control Of Terminal-Splay Sedimentation On Depositional Patterns and Stratigraphic Evolution In Avulsion-Dominated, Unconfined, Deep-Marine Basin-Floor Systems. *J. Sediment. Res.* **2016**, *86*, 786–799. [CrossRef]
97. Bayet-Goll, A.; de Carvalho, C.N. Architectural evolution of a mixed-influenced deltaic succession: Lower-to-Middle Ordovician Armorican Quartzite in the southwest Central Iberian Zone, Penha Garcia Formation (Portugal). *Int. J. Earth Sci.* **2020**, *109*, 2495–2526. [CrossRef]

Article

Modeling Snow Depth and Snow Water Equivalent Distribution and Variation Characteristics in the Irtysh River Basin, China

Liming Gao [1,2,3,*], Lele Zhang [1,2,3,*], Yongping Shen [4], Yaonan Zhang [4], Minghao Ai [4] and Wei Zhang [4]

[1] College of Geography Science, Qinghai Normal University, Xining 810008, China
[2] Qinghai Provincial Key Laboratory of Physical Geography and Environmental Processes, Xining 810008, China
[3] MOE Key Laboratory of Tibetan Plateau Land Surface Processes and Ecological Conservation, Xining 810008, China
[4] Northwest Institute of Eco-Environment and Resources, Chinese Academy of Sciences, Lanzhou 730000, China; shenyp@lzb.ac.cn (Y.S.); yaonan@lzb.ac.cn (Y.Z.); aimh@lzb.ac.cn (M.A.); zhangw06@lzb.ac.cn (W.Z.)
* Correspondence: gaogaotahj@163.com (L.G.); zhang1986lele@163.com (L.Z.)

Abstract: Accurate simulation of snow cover process is of great significance to the study of climate change and the water cycle. In our study, the China Meteorological Forcing Dataset (CMFD) and ERA-Interim were used as driving data to simulate the dynamic changes in snow depth and snow water equivalent (SWE) in the Irtysh River Basin from 2000 to 2018 using the Noah-MP land surface model, and the simulation results were compared with the gridded dataset of snow depth at Chinese meteorological stations (GDSD), the long-term series of daily snow depth dataset in China (LSD), and China's daily snow depth and snow water equivalent products (CSS). Before the simulation, we compared the combinations of four parameterizations schemes of Noah-MP model at the Kuwei site. The results show that the rainfall and snowfall (SNF) scheme mainly affects the snow accumulation process, while the surface layer drag coefficient (SFC), snow/soil temperature time (STC), and snow surface albedo (ALB) schemes mainly affect the melting process. The effect of STC on the simulation results was much higher than the other three schemes; when STC uses a fully implicit scheme, the error of simulated snow depth and snow water equivalent is much greater than that of a semi-implicit scheme. At the basin scale, the accuracy of snow depth modeled by using CMFD and ERA-Interim is higher than LSD and CSS snow depth based on microwave remote sensing. In years with high snow cover, LSD and CSS snow depth data are seriously underestimated. According to the results of model simulation, it is concluded that the snow depth and snow water equivalent in the north of the basin are higher than those in the south. The average snow depth, snow water equivalent, snow days, and the start time of snow accumulation (STSA) in the basin did not change significantly during the study period, but the end time of snow melting was significantly advanced.

Keywords: snow depth; snow water equivalent; ERA-Interim; CMFD; Noah-MP model; microwave remote sensing; Irtysh River Basin

Citation: Gao, L.; Zhang, L.; Shen, Y.; Zhang, Y.; Ai, M.; Zhang, W. Modeling Snow Depth and Snow Water Equivalent Distribution and Variation Characteristics in the Irtysh River Basin, China. *Appl. Sci.* **2021**, *11*, 8365. https://doi.org/10.3390/app11188365

Academic Editor: Jianbo Gao

Received: 7 July 2021
Accepted: 5 September 2021
Published: 9 September 2021

Publisher's Note: MDPI stays neutral with regard to jurisdictional claims in published maps and institutional affiliations.

1. Introduction

Snow plays an important role in the climatic system due to its high reflectivity, low thermal conductivity, and high melting latent heat, which directly affect the surface energy balance, and has obvious feedback, regulation, and indication effects on regional and global climate change [1–4]. It is also an important part of the global water cycle and an important source of fresh water [5]. In addition, the losses caused by floods, avalanches, and other disasters caused by snowmelt to industrial and agricultural production as well as the loss of people's lives and property cannot be ignored. Therefore, accurate snow cover simulation has important significance for water resources development, climate change, and geological disaster prediction.

Modeling is an important means to study snow cover change [6]. Snow models can be generally divided into two categories: one is an empirical model based on simple statistical methods; the other is physical models based on the energy and mass balance processes [7–9]. The advantage of the empirical model is that it requires fewer input parameters. Therefore, it has been widely used to simulate snow and glacier melting in Northern Europe, the Alps, the Greenland ice sheet, the Tibetan Plateau, and other regions [10,11]. Some hydrological models, such as Snowmelt Runoff Model (SRM) [12,13] and the HBV model [14,15], also use an empirical model to describe the melting process of ice, snow and glacier. These snowmelt runoff simulations also achieved good results [16–19]. However, the empirical model simulation accuracy decreases with the improvement in time resolution, and it is impossible to describe the spatial variation of snow surface ablation [20]. Compared with the empirical model, the snow model based on energy balance can better reflect the physical process, the exchange of energy and water between snow cover and atmosphere, the snow melt infiltration, the dynamic change in snow surface albedo, the compaction of snow cover, and other processes [21–23]. Therefore, physical models have a wide range of applications. There are many snowmelt models based on energy balance, such as the Utah Energy Balance model (UEB) [24] and the SNOWPACK model [25,26]. Some hydrological models, such as VIC [27] and WEB-DHM [28], also use physical models to describe snowmelt runoff. Land surface models, such as CLM [29], Noah-MP [30,31] and SURFEX [32], have continuously evolved according to the requirements of atmospheric and hydrological disciplines and can also effectively simulate snow processes. Wrzesien et al. [33] combined the Weather Research and Forecasting (WRF) regional climate model with the Noah-MP model to simulate the snow cover fraction (SCF) and snow water equivalent (SWE) over the central Sierra Nevada Mountains and demonstrated that the models can be an efficient approach to simulate snow processes.

Irtysh River is the second largest river in Xinjiang, and it is also an international river. It flows through China, the Republic of Mongolia, Kazakhstan, and Russia, which plays an important role in the social and economic development of these countries [23]. Snow has an important contribution to the hydrological process in the basin. However, due to the lack of systematic observation, there is little research on snow cover in the basin. Wu et al. [34] used a UEB model to simulate the snowmelt process at a site in the upper reaches of Irtysh River Basin. Wu et al. [35] coupled the WRF model with the temperature-index model to simulate snow melt in the Kayiertesi River Basin, which is in the upper reaches of the Irtysh River Basin. Zhang et al. [36] used a stable isotope technique to analyze the influence of snow melt water on regional hydrological processes in the upper reaches of the Irtysh River Basin. Wu et al. [37] relied on the Geomorphology-Based Ecohydrological Model (GBEHM) to simulate snowmelt processes of a river basin in the Altai Mountains of northwestern China. However, these studies mainly focused on small parts of the Irtysh River Basin, and there is a lack of research on the snow cover process in the whole basin.

Based on the above background, this study used two sets of high-resolution meteorological forcing data sets as drivers to simulate the spatial–temporal change in snow cover in the Irtysh River Basin from 2000 to 2018 by using the Noah-MP model. The main objective of this study is to obtain the dynamic change process of snow cover in the Irtysh River Basin in recent decades. The rest of the paper is arranged as follows: The overview of the study area and the models, data and statistical methods used in this study are introduced in Section 2. In Section 3, the simulation results of the Noah-MP model were verified at a single site, and the parameterization scheme suitable for the study area was selected and the long time series snow cover process in the whole study area was simulated. In Section 4, we discuss the possible reasons for the simulation errors and the shortcomings of this study. The conclusions are presented in Section 5.

2. Materials and Methods

2.1. Study Area

Irtysh River is the largest tributary of Ob River. It originates from the Altai Mountains, crosses the Chinese border, and flows west through Zaysan Lake and northwest across eastern Kazakhstan. The total length of Irtysh River is 4248 km, and total area of the basin is 1.64 million km^2 [38]. The upper reaches are above the border between China and Kazakhstan, the middle reaches are above the border between Kazakhstan and Russia, and the lower reaches are from the border between Kazakhstan and Russia to the confluence of the Ob River. Our study area is located in the Irtysh River Basin of China (Figure 1), with a river length of 633 km and a basin area of 4.53×10^4 km^2. The annual average precipitation of the basin is 200–500 mm, and the annual average runoff at the estuary is 95 billion m^3. The basin is higher in the northeast and lower in the southwest, with an average elevation of 1790 m. It has a temperate continental climate in the middle temperate zone, with long and cold winters and short and cool summers. The average annual temperature is about 4 °C. The water vapor in the basin mainly comes from the Atlantic Ocean, the precipitation is more in winter and summer than in spring and autumn, and there is more snowfall than rainfall. The runoff is mainly supplied by snow melting, precipitation, and ice melting. The proportion of snow melting water is the largest, accounting for 45%, while rainfall and glacier melting water account for 26% and 7.7%, respectively. The snow cover period lasts from November to April of the next year, and the snow cover period is longer in the areas with higher elevations [39]. The snow cover is thick, and the maximum snow cover thickness can even reach more than 1 m in some years.

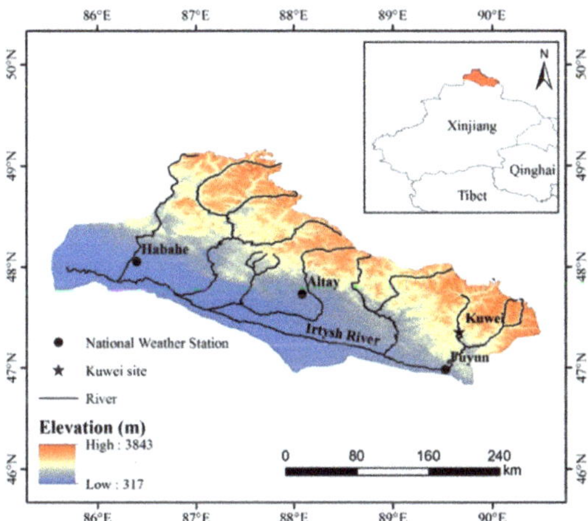

Figure 1. Geographical location of the Irtysh River Basin.

In the Irtysh River Basin, the National Meteorological Administration of China has set up three meteorological observation stations in Altay, Habahe, and Fuyun. The observations include temperature, relative humidity, wind speed, and precipitation. The observations of Altay station also include downward shortwave radiation. In the upper reaches of the basin, the Kuwei comprehensive meteorological observation station (47°21′9.1″ N, 89°39′43.22″ E; altitude of 1379 m) was set up in 2011. At the Kuwei site, meteorological observations include temperature, wind speed, wind direction, relative humidity, precipitation, downward and upward shortwave radiation, and longwave radiation; snow observations include snow depth, snow water equivalent, and snow temperature; and soil observations include soil temperature, soil moisture, and soil heat flux. The specifications of these observation instruments are presented in Table 1.

Table 1. Specifications of the observations and the instruments at Kuwei site.

Observations	Instruments	Accuracy
Air temperature	1000 Ω PRT, IEC 751 1/3 Class B	±0.4 °C
Wind speed	R.M. YOUNG 05103	±0.3 m/s
Wind direction	R.M. YOUNG 05103	±3°
Relative humidity	HUMICAP 180R	±2%
Precipitation	Geonor T-200B	±0.1 mm
Radiation	Kipp and Zonen CNR4	±1%
Snow depth	Campbell SR50A	±1 cm
Snow water equivalent	Snow pillow	±1 mm
Snow temperature	Campbell SI-111 (USA)	±0.5 °C
Soil temperature	Hydra	±0.1 °C
Soil moisture	Campbell CS616/CS625 (USA)	±0.1%
Soil heat flux	Thermopile	±5%
Data logger	Campbell CR1000 (USA)	-

2.2. Model Description

Noah-MP is a new land surface model developed on the basis of the Noah model [30,31]. Compared to Noah, Noah-MP adds 12 physical processes and provides multiple alternative parameterization schemes for each physical process (Table 2). The physical processes directly related to snow cover include snow surface albedo (ALB) and rainfall and snowfall (SNF). Snow/soil temperature time scheme (STC) is a solver option used to solve heat conduction equations and also has a great impact on snow cover [40]. You et al. [40] also proposed that surface layer drag coefficient (SFC) is also closely related to snow cover process.

Table 2. Alternative parameterization schemes for 12 physical processes in Noah-MP model.

Physical Process	Short Name	Parameterization Schemes
Vegetation model	DEVG	1. prescribed (table LAI, shdfac = FVEG); 2. dynamic; 3. table LAI, calculate FVEG 4. table LAI, shdfac = maximum
Canopy stomatal resistance	CRS	1. Ball-Berry; 2. Jarvis
Soil moisture factor for stomatal resistance	BTR	1. Noah; 2. CLM; 3. SSiB
Runoff and groundwater	RUN	1. SIMGM; 2. SIMTOP; 3. Schaake96; 4. BATS
Surface layer drag coefficient	SFC	1. M-O; 2. Chen97
Supercooled liquid water	FRZ	1. NY06; 2. Koren99
Frozen soil permeability	INF	1. NY06; 2. Koren99
Radiation transfer	RAD	1. gap = F (3D, cosz); 2. gap = 0; 3. gap = 1-veg
Snow surface albedo	ALB	1. BATS; 2. CLASS
Rainfall and snowfall	SNF	1. Jordan91; 2. BATS; 3. Noah
Lower boundary of soil temperature	TBOT	1. zero-flux; 2. Noah
Snow/soil temperature time scheme	STC	1. semi-implicit; 2. fully implicit

2.3. Dataset

ERA-Interim [41] and CMFD [42–44] were used as driving data for the Noah-MP land surface model, respectively. ERA-Interim data were downloaded from the European Centre for Medium-Range Forecasts (https://apps.ecmwf.int/ (accessed on 4 May 2020)). Air temperature, dew point temperature, and wind speed are real-time data with a time resolution of 6 h. Radiation and precipitation are forecast data, and 3 h time resolution can be obtained through processing. There are 11 kinds of spatial resolution available; the highest resolution is 0.125 × 0.125°, the lowest resolution is 3 × 3°, and the data resolution selected in this study is 0.125 × 0.125°. The CMFD data has a temporal resolution of 3 h and a spatial resolution of 0.1 × 0.1°. The data can be downloaded from the National Tibetan Plateau Third Pole Environment Data Center (http://data.tpdc.ac.cn/ (accessed on 26 April 2020)), and the detailed description of the data can also be obtained from the website.

In addition to meteorological data, land use data is also needed for Noah-MP model operation. In this study, we selected the global land use data developed by Tsinghua University (http://data.ess.tsinghua.edu.cn/ (accessed on 10 May 2020)), and the spatial resolution of the data was 30 m [45]. By resampling, we obtained land use data with the same resolution as ERA-Interim and CMFD data.

A gridded dataset of snow depth at Chinese meteorological stations (GDSD) was used to evaluate the simulation accuracy of the Noah-MP model at watershed scale. GDSD data was obtained by interpolation based on the snow depth data observed by more than 700 meteorological observation stations in China [46]. This interpolation method divides the 200 km range into one unit, calculates the orientational relationship (O), distance (D), and correlation coefficient (C) of all observation stations in each unit, and finally determines the interpolation weight of each grid point based on the relationship between O, D, and C. This interpolation method fully considers the spatial representation of snow depth at each station and its functional relationship with the snow depth at surrounding stations. The gridded snow depth obtained by this interpolation method was also compared with the snow depth data obtained by arithmetic average method and inverse distance weight method. The results show that the difference of snow depth data obtained by the three methods is very small. The GDSD data has a temporal resolution of about 5 days and a spatial resolution of $0.5 \times 0.5°$. This data can be downloaded from the National Cryosphere Desert Data Center (http://www.ncdc.ac.cn/ (accessed on 31 May 2021)) and a detailed description of the data can also be found on this website.

The error of snow depth simulated by the Noah-MP model was also compared with two sets of snow depth data retrieved based on microwave remote sensing. The first was the long-term series of daily snow depth dataset in China (LSD) released by Che and Dai [47], and the second was China's daily snow depth and snow water equivalent products (CSS) released by Jiang et al. [48]. These two sets of data were both produced using SMMR, SSM/I, and SSMIS satellite remote sensing brightness temperature data with a spatial resolution of 25 km × 25 km and a temporal resolution of 1 day. The LSD data can be downloaded from the National Tibetan Plateau Data Center (http://data.tpdc.ac.cn/ (accessed on 23 May 2021)) and the CSS data can be downloaded from the National Cryosphere Desert Data Center (http://www.ncdc.ac.cn/ (accessed on 31 May 2021)).

2.4. Statistical Method

Several statistical indicators were used to represent the characteristics of snow cover in the study area and the accuracy of simulation results or meteorological data. These indicators are listed as follows:

(1) Snow year

The snow year is considered to be the time from the beginning of snow accumulation in a year to the next year before the snow starts to accumulate. According to the characteristics of snow cover in the Irtysh River Basin, we regard 1 September to 30 August of the following year as a snow year.

(2) Mean deviation (*MD*) and root mean squared error (*RMSE*)

Mean deviation (*MD*) and root mean squared error (*RMSE*) are used to evaluate the accuracy of model simulation results or weather-driven data. The calculation formulas of MD and RMSE are as follows:

$$MD = \frac{1}{n} \sum_{i=1}^{n} (RD_i - O_i) \tag{1}$$

$$RMSE = \sqrt{\frac{1}{n} \sum_{i=1}^{n} (RD_i - O_i)^2} \tag{2}$$

In the above formula, RD_i is the meteorological value recorded by meteorological forcing data or the snow parameter value simulated by the model at *i*th time, O_i is the observed meteorological element or snow parameter value at ith time, and n is the number of samples. The closer the MD and $RMSE$ values are to 0, the higher the accuracy of meteorological forcing data or model simulation results are. If the MD value is greater than 0, it means meteorological forcing data or simulation results are overestimated, and if the MD value is less than 0, it means underestimated.

(3)   Linear slope and Mann–Kendall test

The linear slope is used to indicate trends of snow depth, snow water equivalent, snow days, and other snow parameters. The Mann–Kendall (M–K) test is used to determine the significance of the trends. When the statistic $p > 0.1$, the change trend of the time series is not significant; otherwise, the change trend is significant. The calculation process of the M–K method can be found in the published literature [49].

3. Results

3.1. Testing Noah-MP Model at Kuwei Site

In order to evaluate the simulation effect of the Noah-MP model on a single point, we first drive the model based on the meteorological observation data of Kuwei site from September 2013 to April 2014, and verify it based on the observed snow depth and snow water equivalent. As can be seen from Table 2, Noah-MP can combine more than 20,000 optional parameterization schemes. You et al. [40] tested these parameterization schemes at a site in the Altai Mountains, which is also located in the Irtysh River Basin, and the results show SFC and STC have the greatest influence on the simulation results of snow depth and snow water equivalent. At the Kuwei site, we also tested the SFC and STC parameterization schemes, and obtained four simulation results (Figure 2a,b). It can be seen from Figure 2a,b that different SFC and STC schemes have little influence on the simulation results during the snow accumulation period, but have a great influence on the simulation results during the ablation period. During the ablation period, the effect of SFC scheme on the simulation results was smaller than that of STC; when STC uses a fully implicit scheme, the error of simulated snow depth and snow water equivalent is greater than that of a semi-implicit scheme. The ALB and SNF parameterizations schemes were also tested at the Kuwei site. From the results obtained (Figure 2c–f), ALB mainly affects the melting process of snow, while SNF mainly affects the accumulation process. However, compared with STC, ALB and SNF have much less influence on the snow cover process. Furthermore, the simulated snow depth of all combined schemes is less than the observed value. In this study, SR50 sensor with a resolution of 1 cm is used for snow depth observation at the Kuwei site. However, in actual monitoring, especially in mountainous areas with complex terrain, the error of the SR50 sensor may be much higher than 1 cm. This may also be an important reason for the difference between the Noah-MP-simulated and the observed snow depth during the snow accumulation and melting period.

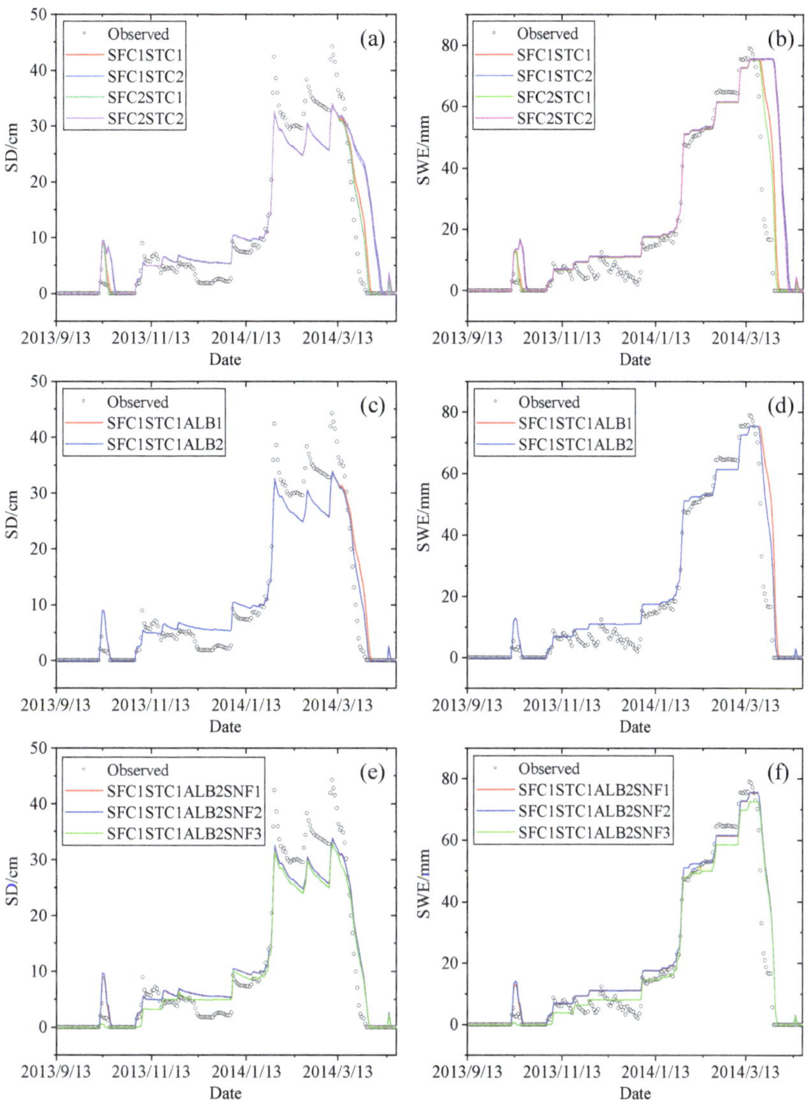

Figure 2. Modeling snow depth (**a,c,e**) and snow water equivalent (**b,d,f**) using Noah-MP model at Kuwei site. SFC1STC1 means the surface layer drag coefficient uses the M-O scheme, and snow/soil temperature time uses the semi-implicit scheme. Similarly, the parameterization schemes selected by SFC1STC2, SFC2STC1, SFC2STC2, SFC1STC1AB1, SFC1STC1AB2, SFC1STC1AB2SNF1, SFC1STC1AB2SNF2, and SFC1STC1AB2SNF3 can be obtained.

3.2. Testing Noah-MP Model in the Irtysh River Basin

From the simulation results at the Kuwei site, the simulation accuracy of SFC1STC1 and SFC2STC1 is better than that of SFC1STC2 and SFC2STC2. At another site in the Irtysh River Basin, You et al. [40] also proposed that the SFC1STC1 scheme has the best simulation accuracy for snow depth and snow water equivalent. Combined with the test results of this study at the Kuwei site, we chose the SFC1STC1AB2SNF1 scheme to simulate the snow depth and snow water equivalent at the whole basin, and the other eight schemes adopted the default selection of the model. Considering that the Noah-MP model requires a long time to reach equilibrium state [50–52], this study refers to the method proposed by You et al. [53], and uses the forcing data from 1 January 2000 to 30 August 2001 to spin-up the model. Through the simulation, we get the simulation results of snow depth and snow water equivalent at a 3 h time scale in the Irtysh River Basin from September 2001 to December 2018. In order to evaluate the accuracy of the model simulation results, we process the snow depth data from all sources to the same time resolution as the GDSD data, and give the time series of the average snow depth in the Irtysh River Basin (Figure 3). As can be seen from Figure 3c,d, the accuracy of snow depth simulated by the Noah-MP model is distinctly higher than that obtained based on microwave remote sensing inversion in the Irtysh River Basin. In years with small snow depth, the snow depth recorded by LSD and CSS data is highly consistent with GDSD data. However, in years with high snow depth, such as 2002, 2006, 2008, 2009, 2010, and 2012, LSD and CSS snow depth are seriously underestimated. The snow depth series simulated based on the CMFD and ERA-Interim data were in good agreement with the GDSD data. Through the calculation results of MD and RMSE (Table 3), it is found that the MD and RMSE values between Noah_CMFD and GDSD are smaller than those between Noah_ERA and GDSD. Therefore, the results obtained by using CMFD as the driving simulation in Irtysh River Basin are the most accurate.

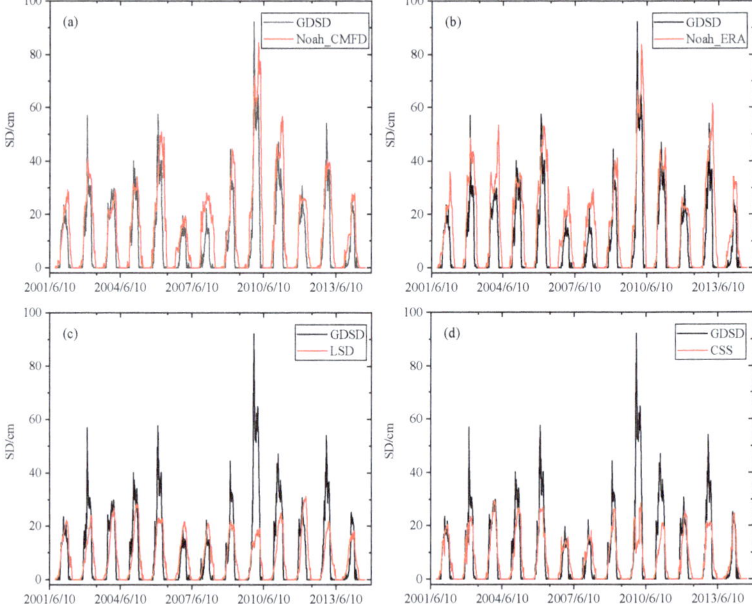

Figure 3. Average snow depth (SD) of each five-day period in the Irtysh River Basin from September 2001 to August 2014 based on GDSD data and CMFD simulation (**a**), GDSD data and ERA-Interim simulation (**b**), GDSD and LSD data (**c**), GDSD and CSS microwave remote sensing data (**d**). Noah_CMFD represents the SD series by Noah-MP simulation with CMFD as the driving data and Noah_ERA represents the SD series by Noah-MP simulation with ERA-Interim as the driving data.

Table 3. *MD* and *RMSE* values between Noah_CMFD, Noah_ERA, and GDSD snow depth.

	Noah_CMFD vs. GDSD	**Noah_ERA vs. GDSD**
MD	5.07	10.27
RMSE	6.47	11.32

3.3. SD and SWE Distribution and Variation Characteristics in the Irtysh River Basin

Based on the simulation results of the Noah-MP model, the annual maximum snow depth and snow water equivalent are calculated, and the spatial distribution of annual average maximum snow depth and snow water equivalent in the Irtysh River Basin is given (Figure 4). It can be seen from Figure 4 that the annual average maximum snow depth and snow water equivalent simulated based on CMFD and ERA-Interim data have good consistency in spatial distribution. Both snow depth and snow water equivalent are high in the north and low in the south of the basin. This spatial distribution feature is consistent with the topography of the basin. In the north of the basin, the altitude is high and the temperature is relatively low, which is conducive to the accumulation of snow. In the south of the basin, the altitude is relatively low, the temperature is relatively high, and the snow is easier to melt. Based on CMFD and era interim data, we also give the spatial distribution characteristics of the annual average precipitation in the basin (Figure 5). It can be seen from Figure 5 that the precipitation in the north of the basin is much higher than that in the south. High altitude and higher precipitation are the main reasons for the higher snow depth and snow water equivalent in the north of the basin than that in the south.

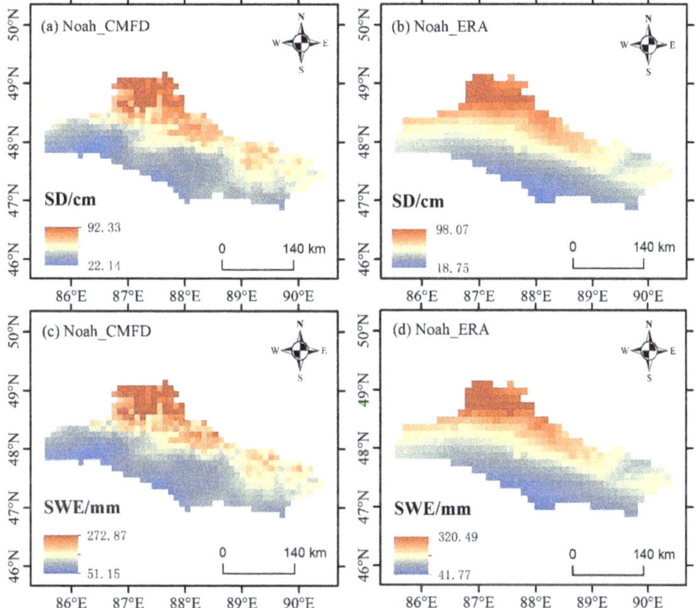

Figure 4. Annual average maximum snow depth and snow water equivalent from 2001 to 2018 based on Noah-MP simulations. (**a,c**) represent SD and SWE with CMFD as the drive, (**b,d**) represent the results obtained with ERA-Interim as the drive.

When analyzing the temporal variation characteristics of snow cover, because the accuracy of snow depth based on ERA-Interim simulation is slightly lower than that based on CMFD simulation, only the simulation results based on CMFD data are selected. In addition to the average maximum snow depth (SD_{max}) and snow water equivalent (SWE_{max}), we also selected the average snow days, the average start time of snow accumulation (STSA), and the end time of snow melting (ETSM) to analyze the variation characteristics of snow from 2001 to 2017. The linear slope and M–K test were used to determine the trend of these time series (Figure 6). As can be seen from Figure 6, the maximum snow depth, snow water equivalent, and snow days in the Irtysh River Basin showed an insignificant decreasing trend from 2001 to 2017. The start time of snow accumulation was delayed, but the change trend was not significant, while the end time of snow melting was significantly advanced.

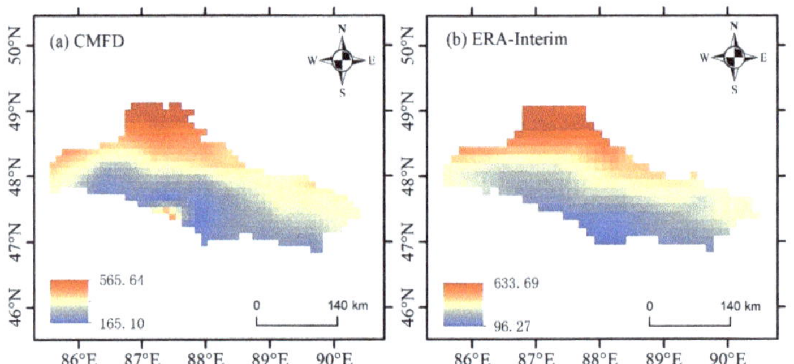

Figure 5. Spatial distribution of annual average precipitation based on CMFD (**a**) and ERA-Interim (**b**) data in the Irtysh River Basin.

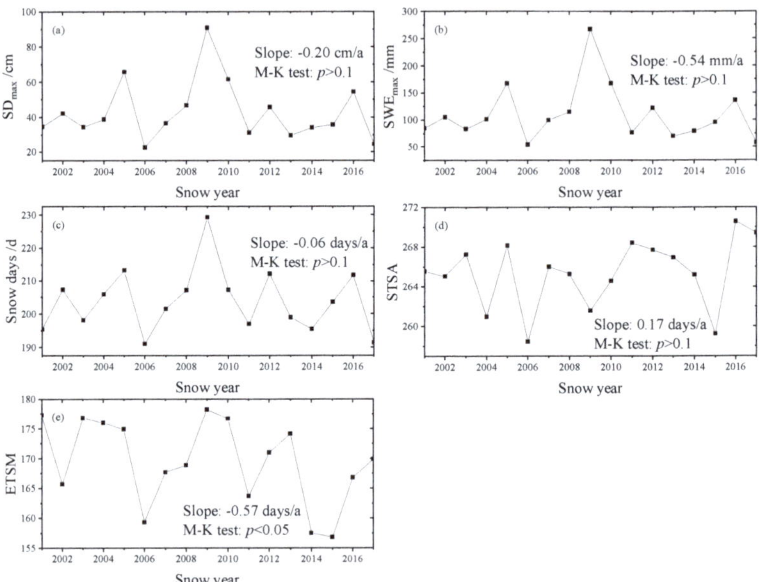

Figure 6. Average SD_{max} (**a**), SWE_{max} (**b**), snow days (**c**), STSA (**d**), and ETSM (**e**) in the Irtysh River Basin from 2001 to 2017.

4. Discussion

4.1. The Influence of Data Quality Uncertainty on Simulation Results

In previous studies, Guenther et al. [9] and Zhang et al. [54] analyzed the factors that affect the accuracy of snow cover simulation by land surface process model, and found that the uncertainty of forcing data has a greater impact on the simulation results than the structure and parameterization scheme of the model itself. In this study, meteorological station observation data were used to evaluate the accuracy of CMFD and ERA-Interim data. Since the data of Habahe, Altay, and Fuyun stations are used in the production of CMFD data, only the observation data of Kuwei station were selected. Meteorological data from CMFD and ERA-Interim were extracted based on the longitude and latitude of the Kuwei site. Scatter plots were drawn based on the CMFD, ERA-Interim, and the observed data, and the accuracy of the two meteorological forcing data was evaluated using MD and RMSE statistical parameters (Figure 7). It can be seen from Figure 7 that there are some deviations between CMFD, ERA-Interim and the observed temperature, wind speed, relative humidity, precipitation, and downward shortwave and longwave radiation. On the one hand, the reason for this phenomenon lies in the difference in spatial range between grid points and stations; on the other hand, the error of meteorological forcing data itself is also an important reason. From the calculated MD and RMSE values, the accuracy of CMFD temperature, wind speed, and downward shortwave and longwave radiation data is higher than ERA-Interim data. Although the accuracy of ERA-Interim relative humidity and precipitation is slightly better than that of CMFD at the Kuwei site, considering that the CMFD precipitation and relative humidity data were generated through fusion of remote sensing products, reanalysis datasets, and in situ station data [42], it is considered that CMFD also has high accuracy at the watershed scale. This is also the reason why the modeled snow depths by using the CMFD data are more consistent with GDSD data.

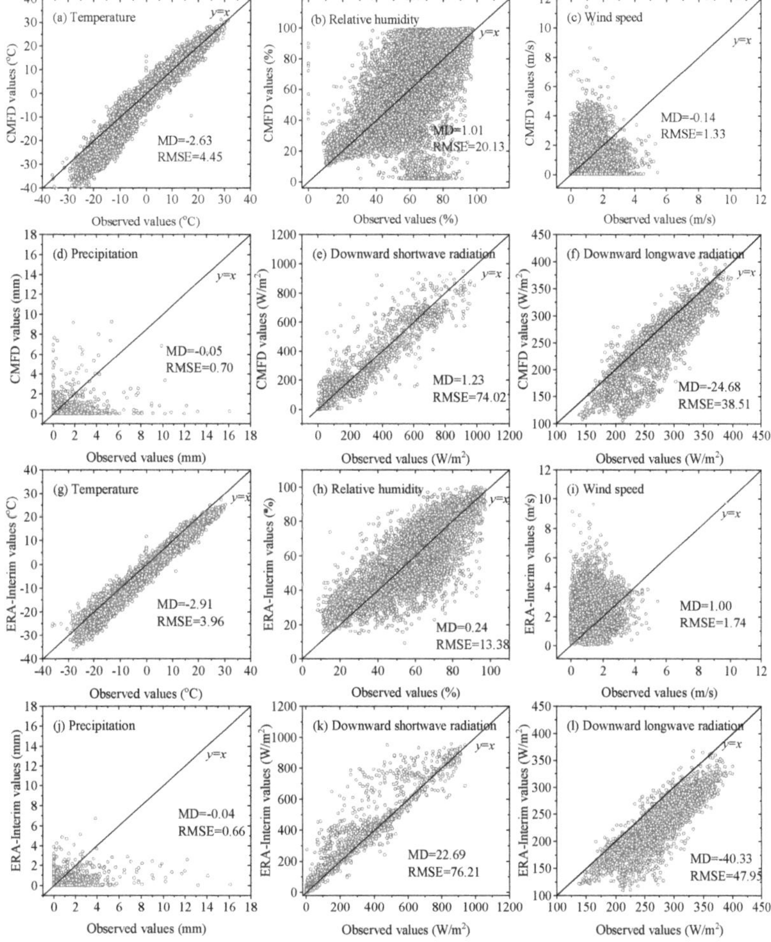

Figure 7. Scatter plot based on the hourly CMFD, ERA-Interim, and the observed temperature (**a,g**), relative humidity (**b,h**), wind speed (**c,i**), precipitation (**d,j**), downward shortwave radiation (**e,k**), downward longwave radiation (**f,l**) data during the study period.

4.2. Limitations of This Study

In high-latitude mountainous areas, wind blowing snow is also a factor that cannot be ignored. Wind blowing snow includes material migration and sublimation, which have great influence on the secondary distribution of snow in space [55,56]. In previous studies, the minimum wind speed threshold for wind blowing snow was generally set at 7 m/s [57], and when the wind speed is higher than the threshold, blowing snow will occur. Through the analysis of the daily maximum wind speed at the Altay, Habahe, and Fuyun meteorological stations in the study area from 2001 to 2018 (Figure 8), it was found that there are many days when the daily maximum wind speed of the three stations exceeds the wind blowing snow threshold. However, the Noah-MP model lacks the consideration of the wind blowing snow process, which may also be an important reason for the deviation between the snow depth simulated in this study and GDSD data.

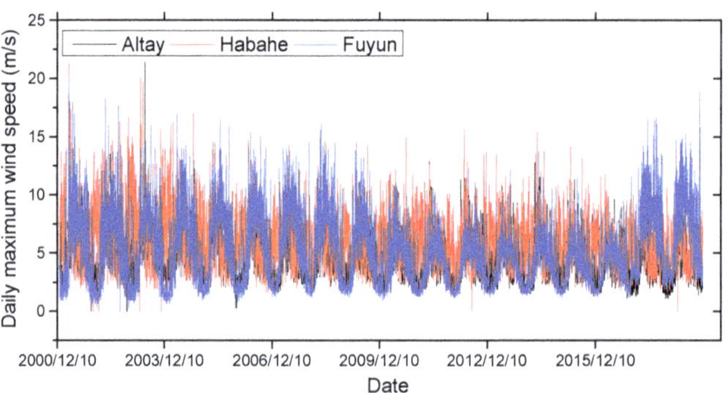

Figure 8. Daily maximum wind speed at the Altay, Habahe, and Fuyun sites from 2001 to 2018.

5. Conclusions

In this study, we tested the Noah-MP model for snow accumulation and melting process modeling at the Kuwei site in the Irtysh River Basin, and simulated the snow cover process by using CMFD and ERA-Interim as forcing data at the whole basin from 2000 to 2018. The simulation results were also compared with the gridded dataset of snow depth at Chinese meteorological stations (GDSD) and snow depth obtained from microwave remote sensing (LSD and CSS data). The main findings are as follows:

(1). STC, SFC, and ALB schemes mainly affect the snow melting process, while SNF mainly affects the accumulation process. Among the four schemes, STC has the greatest impact on the accuracy of snow cover simulation. When STC use the semi-implicit scheme, the overall simulation accuracy is better than that of the fully implicit scheme.

(2). CMFD and ERA-Interim as the forcing data can accurately simulate the snow accumulation and melting process of the whole basin, and the results of CMFD simulation are more accurate than those of ERA-Interim simulation. The main reason is that the data accuracy of CMFD is higher than that of ERA-Interim.

(3). In the years with low snow depth, the snow depth retrieved based on microwave remote sensing is in good agreement with the observed snow depth. However, in the years with high snow depth, such as 2002, 2004, 2008, 2009, 2010, and 2012, the snow depth retrieved by remote sensing is seriously underestimated.

(4). Spatially, the snow depth and snow depth equivalent in the north of Irtysh River Basin are higher than those in the south, mainly because the altitude and precipitation in the north are higher than those in the south. The snow depth, snow water equivalent, snow days, and the start time of snow accumulation (STSA) in the basin did not change significantly from 2001 to 2017. However, the end time of snow melting was obviously advanced.

Author Contributions: Conceptualization, L.Z.; methodology, L.G.; software, L.G. and M.A.; validation, L.G.; formal analysis, L.G. and L.Z.; investigation, L.G., L.Z. and W.Z.; resources, Y.S.; data curation, L.G. and W.Z.; writing—original draft preparation, L.G. and L.Z.; writing—review and editing, Y.Z.; supervision, Y.S.; project administration, L.Z.; funding acquisition, L.Z. All authors have read and agreed to the published version of the manuscript.

Funding: This study is supported by the National Natural Science Foundation of China (grant number 42001060, 42171467 and 41705139), Key Special Subject of the National Key Research and Development Program of China (grant number 2019YFC1510502) and the Natural Science Foundation of Qinghai Province (grant number 2021-ZJ-947Q).

Institutional Review Board Statement: Not applicable.

Informed Consent Statement: Not applicable.

Data Availability Statement: The data supporting this study are available from the authors.

Conflicts of Interest: The authors declare no conflict of interest.

References

1. Kang, S.; Guo, W.; Zhong, X.; Xu, M. Changes in the mountain cryosphere and their impacts and adaptation measures. *Clim. Chang. Res.* **2020**, *16*, 143–152.
2. Kevin, J.P.W.; Kotlarski, S.; Scherrer, S.C.; Schär, C. The Alpine snow-albedo feedback in regional climate models. *Clim. Dyn.* **2017**, *48*, 1109–1124.
3. Henderson, G.R.; Peings, Y.; Kushner, P.J. Snow–atmosphere coupling in the Northern Hemisphere. *Nat. Clim. Chang.* **2018**, *8*, 954–963. [CrossRef]
4. Kokhanovsky, A.; Lamare, M.; Danne, O.; Brockmann, C.; Dumont, M.; Picard, G.; Arnaud, L.; Favier, V.; Jourdain, B.; Meur, E.L.; et al. Retrieval of snow properties from the Sentinel-3 Ocean and Land Colour Instrument. *Remote Sens.* **2019**, *11*, 2280. [CrossRef]
5. Barnett, T.P.; Adam, J.C.; Lettenmaier, D.P. Potential impacts of a warming climate on water availability in snow-dominated regions. *Nature* **2005**, *438*, 303–309. [CrossRef]
6. Richard, E.; Pierre, E. Parameter sensitivity in simulations of snowmelt. *J. Geophys. Res.* **2004**, *109*, 2107–2117.
7. Winkler, M.; Schellander, H.; Gruber, S. Snow water equivalents exclusively from snow depths and their temporal changes: The ΔSNOW model. *Hydrol. Earth Syst. Sci.* **2021**, *25*, 1165–1187. [CrossRef]
8. Saloranta, T.M. Operational snow mapping with simplified data assimilation using the seNorge snow model—ScienceDirect. *J. Hydrol.* **2016**, *538*, 314–325. [CrossRef]
9. Guenther, D.; Markel, T.; Essery, R.; Strasser, U. Uncertainties in snowpack simulations—Assessing the impact of model structure, parameter choice and forcing data error on point-scale energy-balance snow model performance. *Water Resour. Res.* **2019**, *55*, 2779–2800. [CrossRef]
10. Liu, S.; Xie, Z.; Song, G.; MA, L.; Ageta, Y. Mass balance of Kangwure (flat-top) Glacier on the north side of Mt. Xixiabangma, China. *Bull. Glacier. Res.* **1996**, *14*, 37–43.
11. Braithwaite, R.J.; Zhang, Y. Sensitivity of mass balance of five Swiss glaciers to temperature changes assessed by tuning a degree-day model. *J. Glaciol.* **1994**, *46*, 7–14. [CrossRef]
12. Abudu, S.; Saydi, M.; King, J.P. Application of snowmelt runoff model (SRM) in mountainous watersheds: A review. *Water Sci. Eng.* **2012**, *2*, 123–136.
13. Martinec, J.; Rango, A. Parameter values for snowmelt runoff modelling. *J. Hydrol.* **1986**, *84*, 197–219. [CrossRef]
14. Firouzi, S.; Sadeghian, M.S. Application of Snow Melt Runoff Model in a Mountainous Basin of Iran. *J. Geosci. Environ. Prot.* **2016**, *4*, 74. [CrossRef]
15. Huang, S.; Eisner, S.; Magnusson, J.O.; Lussana, C.; Yang, X.; Beldring, S. Improvements of the spatially distributed hydrological modelling using the HBV model at 1 km resolution for Norway. *J. Hydrol.* **2019**, *577*, 123585. [CrossRef]
16. Tiwari, S.; Kar, S.C.; Bhatla, R.; Bansal, R. Temperature index based snowmelt runoff modelling for the Satluj River basin in the western Himalayas. *Met. Apps.* **2018**, *25*, 302–313. [CrossRef]
17. Ma, H.; Cheng, G. Snowmelt runoff simulation in Gongnaisi River Basin using of SRM. *Chin. Sci. Bull.* **2003**, *48*, 2088–2093.
18. Latif, Y.; Ma, Y.; Ma, W.; Muhammad, S.; Adnan, M.; Yaseen, M.; Fealy, R. Differentiating Snow and Glacier Melt Contribution to Runoff in the Gilgit River Basin via Degree-Day Modelling Approach. *Atmosphere* **2020**, *11*, 1023. [CrossRef]
19. Osuch, M.; Wawrzyniak, T.; Nawrot, A. Diagnosis of the hydrology of a small Arctic permafrost catchment using HBV conceptual rainfall-runoff model. *Hydrol. Res.* **2019**, *50*, 459–478. [CrossRef]
20. Zhang, Y.; Liu, S.; Ding, Y. Spatial variation of degree-day factors on the observed glaciers in western China. *Acta. Geogr. Sin.* **2006**, *61*, 89. [CrossRef]
21. Touzeau, A.; Landais, A.; Morin, S.; Arnaud, L.; Picard, G. Numerical experiments on vapor diffusion in polar snow and firn and its impact on isotopes using the multi-layer energy balance model Crocus in SURFEX v8. 0. *Geosci. Model. Dev.* **2018**, *11*, 2393–2418. [CrossRef]

22. Sauter, T.; Arndt, A.; Schneider, C. COSIPY v1.2—An open-source coupled snowpack and ice surface energy and mass balance model. *Geosci. Model. Dev. Discuss.* **2020**, *2020*, 1–25.

23. Gao, L.M.; Zhang, Y.N.; Shen, Y.P.; Zhang, L.L. Analysis of water and heat transfer in snow layer during snowmelt period in Irtysh River Basinbased on energy balance theory. *J. Glaciol. Geocryol.* **2016**, *38*, 323–331.

24. Tarboton, D.G.; Luce, C.H. *Utah Energy Balance Snow Accumulation and Melt Model (UEB): Computer Model Technical Description and User's Guide*; Utah Water Research Laboratory and USDA Forest Service Intermountain Research Station: Logan, UT, USA, 1996.

25. Bartelt, P.; Lehning, M. A physical SNOWPACK model for the Swiss avalanche warning: Part I: Numerical model. *Cold. Reg. Sci. Technol.* **2002**, *35*, 123–145. [CrossRef]

26. Lehning, M.; Bartelt, P.; Brown, B. A physical SNOWPACK model for the Swiss avalanche warning: Part III: Meteorological forcing, thin layer formation and evaluation. *Cold. Reg. Sci. Technol.* **2002**, *35*, 169–184. [CrossRef]

27. Liang, X.; Wood, E.F.; Lettenmaier, D.P. Modeling ground heat flux in land surface parameterization schemes. *J. Geophys. Res.* **1999**, *104*, 9581–9600. [CrossRef]

28. Shrestha, M.; Wang, L.; Koike, T.; Xue, Y.; Hirabayashi, Y. Improving the snow physics of WEB-DHM and its point evaluation at the SnowMIP sites. *Hydrol. Earth Syst. Sci.* **2010**, *14*, 2577–2594. [CrossRef]

29. Toure, A.M.; Rodell, M.; Yang, Z.L.; Beaudoing, H.; Kim, E.; Zhang, Y.; Kwon, Y. Evaluation of the snow simulations from the Community Land Model, version 4 (CLM4). *J. Hydrometeorol.* **2016**, *17*, 153–170. [CrossRef]

30. Niu, G.Y.; Yang, Z.L.; Mitchell, K.E.; Chen, F.; Ek, M.B.; Barlage, M.; Kumar, A.; Manning, K.; Niyogi, D.; Rosero, E.; et al. The community Noah land surface model with multi-parameterization options (Noah-MP): 1. Model description and evaluation with local-scale measurements. *J. Geophys. Res.* **2011**, *116*, D12109. [CrossRef]

31. Yang, Z.L.; Niu, G.Y.; Mitchell, K.E.; Chen, F.; Ek, M.B.; Barlage, M.; Longuevergne, L.; Manning, K.; Niyogi, D.; Tewari, M.; et al. The community Noah land surface model with multiparameterization options (Noah-MP): 2. Evaluation over global river basins. *J. Geophys. Res.* **2011**, *116*, D12110. [CrossRef]

32. Vionnet, V.; Brun, E.; Morin, S.; Boone, A.; Faroux, S.; Moigne, P.L.; Martin, E.; Willemet, J.M. The detailed snowpack scheme Crocus and its implementation in SURFEX v7.2. *Geosci. Model. Dev.* **2012**, *5*, 73–791. [CrossRef]

33. Wrzesien, M.L.; Durand, M.T.; Pavelsky, T.M.; Howat, L.M.; Margulis, S.A.; Huning, L.S. Comparison of methods to estimate snow water equivalent at the mountain range scale: A case study of the California Sierra Nevada. *J. Hydrometeorol.* **2017**, *18*, 1101–1119. [CrossRef]

34. Wu, X.; Wang, N.; Shen, Y.; He, J.; Zhang, W. In-situ observations and modeling of spring snowmelt processes in an Altay Mountains river basin. *J. Appl. Remote Sens.* **2014**, *8*, 214–233. [CrossRef]

35. Wu, X.; Shen, Y.; Wang, N.; Pan, X.; Zhang, W.; He, J.; Wang, G. Coupling the WRF model with a temperature index model based on remote sensing for snowmelt simulations in a river basin in the Altay Mountains, north-west China. *Hydrol. Process.* **2016**, *30*, 3967–3977. [CrossRef]

36. Zhang, W.; Kang, S.C.; Shen, Y.P.; He, J.Q.; Chen, A.A. Response of snow hydrological processes to a changing climate during 1961 to 2016 in the headwater of Irtysh River Basin, Chinese Altai Mountains. *J. Mt. Sci.* **2017**, *11*, 2295–2310. [CrossRef]

37. Wu, X.; Zhang, W.; Li, H.; Long, Y.; Pan, X.; Shen, Y. Analysis of seasonal snowmelt contribution using a distributed energy balance model for a river basin in the Altai Mountains of northwestern China. *Hydrol. Process.* **2021**, *35*, e14046. [CrossRef]

38. Huang, F.; Xia, Z.; Li, F.; Guo, L.; Yang, F. Hydrological changes of the Irtysh River and the possible causes. *Water Resour. Manag.* **2012**, *26*, 3195–3208. [CrossRef]

39. Liu, M.; Xiong, C.; Pan, J.; Wang, T.; Shi, J.; Wang, N. High-resolution reconstruction of the maximum snow water equivalent based on remote sensing data in a mountainous area. *Remote Sens.* **2020**, *12*, 460. [CrossRef]

40. You, Y.H.; Huang, C.L.; Zhang, Y.; Hou, J.L. Sensitivity evaluation of snow simulation to multi-parameterization schemes in the Noah-MP Model. *Adv. Earth Sci.* **2019**, *34*, 356–365.

41. Dee, D.P.; Uppala, S.M.; Simmons, A.J.; Berrisford, P.; Poli, P.; Kobayashi, S.; Andrae, U.; Balmaseda, M.A.; Balsamo, G.; Bauer, P.; et al. The ERA-Interim reanalysis: Configuration and performance of the data assimilation system. *Q. J. R. Meteorol. Soc.* **2011**, *137*, 553–597. [CrossRef]

42. He, J.; Yang, K.; Tang, W.; Lu, H.; Qin, J.; Chen, Y.Y.; Li, X. The first high-resolution meteorological forcing dataset for land process studies over China. *Sci. Data* **2020**, *7*, 25. [CrossRef] [PubMed]

43. Yang, K.; He, J.; Tang, W.J.; Qin, J.; Cheng, C. On downward shortwave and longwave radiations over high altitude regions: Observation and modeling in the Tibetan Plateau. *Agric. For. Meteorol.* **2010**, *150*, 38–46. [CrossRef]

44. Yang, K.; He, J. China Meteorological Forcing Dataset (1979–2018). National Tibetan Plateau Data Center. 2019. Available online: http://data.tpdc.ac.cn/en/data/8028b944-daaa-4511-8769-965612652c49/ (accessed on 26 April 2020).

45. Gong, P.; Liu, H.; Zhang, M.; Li, C.; Wang, J.; Huang, H.; Clinton, N.; Li, L.; Ji, L.; Li, W.; et al. Stable classification with limited sample: Transferring a 30-m resolution sample set collected in 2015 to mapping 10-m resolution global land cover in 2017. *Sci. Bull.* **2019**, *64*, 370–373. [CrossRef]

46. Ma, L. Gridded Data Set of Snow Depth at Chinese Meteorological Stations in 1951–2015. National Cryosphere Desert Data Center. 2020. Available online: http://www.ncdc.ac.cn/portal/metadata/9d6375ee-ef40-4f2e-8340-87748ecd4d95 (accessed on 31 May 2021).

47. Che, T.; Dai, L. Long-Term Series of Daily Snow Depth Dataset in China (1979–2020). National Tibetan Plateau Data Center. 2015. Available online: https://data.tpdc.ac.cn/zh-hans/data/df40346a-0202-4ed2-bb07-b65dfcda9368/ (accessed on 23 May 2021).

48. Jiang, L.; Yang, J.; Dai, L.; Li, X.; Qiu, Y.; Wu, S.; Li, Z. China's Daily Snow Water Equivalent 25 km Spatial Resolution Products from 1980 to 2020. National Cryosphere Desert Data Center. 2020. Available online: http://www.ncdc.ac.cn/portal/metadata/63c5cebb-587d-42cf-bd81-6f1325f1e165 (accessed on 31 May 2021).
49. Feng, W.; Lu, H.; Yao, T.; Yu, Q. Drought characteristics and its elevation dependence in the Qinghai–Tibet plateau during the last half-century. *Sci. Rep.* **2020**, *10*, 14323. [CrossRef]
50. Cai, X.; Yang, Z.L.; David, C.H.; Niu, G.Y.; Rodell, M. Hydrological evaluation of the Noah-MP land surface model for the Mississippi River Basin. *J. Geophys. Res.* **2014**, *119*, 23–38. [CrossRef]
51. Chen, F.; Barlage, M.; Tewari, M.; Rasmussen, R.; Jin, J.; Lettenmaier, D.; Livneh, B.; Lin, C.; Miguez-Macho, G.; Niu, G.; et al. Modeling seasonal snowpack evolution in the complex terrain and forested Colorado Headwaters region: A model inter-comparison study. *J. Geophys. Res.* **2014**, *119*, 13795–13819. [CrossRef]
52. Gao, Y.; Li, K.; Chen, F.; Jiang, Y.; Lu, C. Assessing and improving Noah-MP land model simulations for the central Tibetan Plateau. *J. Geophys. Res.* **2015**, *120*, 9258–9278. [CrossRef]
53. You, Y.; Huang, C.; Yang, Z.; Zhang, Y.; Bai, Y.; Gu, J. Assessing Noah-MP Parameterization Sensitivity and Uncertainty Interval Across Snow Climates. *J. Geophys. Res.* **2020**, *125*, e2019JD030417. [CrossRef]
54. Zhang, Y. Multivariate Land Snow Data Assimilation in the Northern Hemisphere: Development, Evaluation and Uncertainty Quantification of the Extensible Data Assimilation System. Ph.D. Thesis, The University of Texas at Austin, Austin, TX, USA, May 2015.
55. Yang, J.; Yau, M.K.; Fang, X.; Pomeroy, J.W. A triple-moment blowing snow-atmospheric model and its application in computing the seasonal wintertime snow mass budget. *Hydrol. Earth Syst. Sci.* **2010**, *14*, 1063–1079. [CrossRef]
56. Palm, S.P.; Kayetha, V.; Yang, K.; Pauly, R. Blowing snow sublimation and transport over Antarctica from 11 years of CALIPSO observations. *Cryosphere* **2017**, *11*, 2555–2569. [CrossRef]
57. Gordon, M.; Simon, K.; Taylor, P.A. On snow depth predictions with the Canadian land surface scheme including a parametrization of blowing snow sublimation. *Atmos.-Ocean* **2006**, *44*, 239–255. [CrossRef]

 **applied sciences**

 **MDPI**

Article

Ice Elevation Change Based on GNSS Measurements along the Korth-Traverse in Southern Greenland

Thomas Hitziger [1], **Luisa Näke** [2] and **Karel Pavelka** [3,*]

[1] Department of Mechanics and Numerical Methods, Institute of Civil and Structural Engineering, Brandenburg University of Technology Cottbus-Senftenberg, 03046 Cottbus, Germany
[2] Department of Civil and Mechanical Engineering, Technical University of Denmark, 2800 Kongens Lyngby, Denmark
[3] Department of Geomatics, Faculty of Civil Engineering, Czech Technical University in Prague, 16000 Prague, Czech Republic
* Correspondence: pavelka@fsv.cvut.cz

Abstract: In 1912, a Swiss expedition led by meteorologist Alfred de Quervain crossed the Greenland ice sheet on a route from Disko Bay to Tasiilaq. Based on that, in 2002, a series of geodetic expeditions carried out by W. Korth and later by T. Hitziger began along the same traverse as in 1912, with the last measurements taken in May 2021. The statically collected GPS/GNSS data provide very accurate elevation changes at 36 points along the almost 700 km long crossing over a period of 19 years. According to this, there is a maximum increase of 2.1 m in the central area and a decrease of up to 38.7 m towards the coasts (influence Ilulissat Isbræ). By using kinematic GNSS measurements, there is a very dense profile with a spacing of a few meters. The comparison of those measurements is performed using crossing points or minimum distances and gives equivalent results for both methods. It is shown that local ice topography is preserved, and thus gaps in data sets can be caught. Areas of accumulation and ablation on the ice sheet can be identified, showing the widespread influence of outlet glaciers up to 200 km. The data can be used for direct verification of altimetry data, such as IceSat. Both IceSat elevations and their changes can be compared.

Keywords: Greenland ice sheet; monitoring; GNSS; expedition; Jakobshavn Isbræ; Helheim Glacier; IceSat; climate change; glacier profile

Citation: Hitziger, T.; Näke, L.; Pavelka, K. Ice Elevation Change Based on GNSS Measurements along the Korth-Traverse in Southern Greenland. *Appl. Sci.* **2022**, *12*, 12066. https://doi.org/10.3390/app122312066

Academic Editor: Jianbo Gao

Received: 19 October 2022
Accepted: 22 November 2022
Published: 25 November 2022

Publisher's Note: MDPI stays neutral with regard to jurisdictional claims in published maps and institutional affiliations.

1. Introduction

This article is dedicated to the geodesist, polar explorer, and friend Wilfried Korth. He was the project initiator and scientific leader for a long time. The article is also based on his results, so he is mentioned here in memory as an additional author (Figure 1).

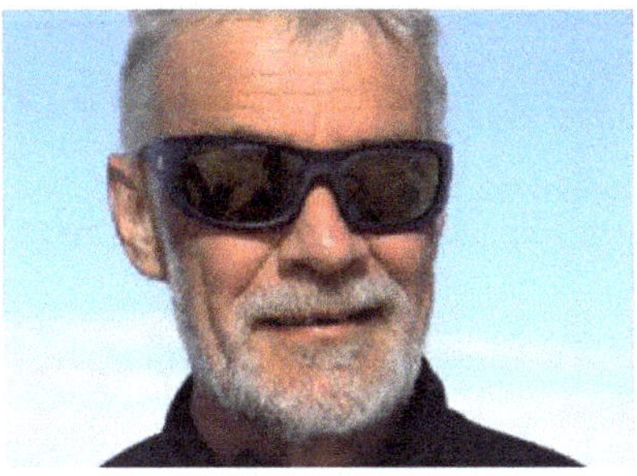

Figure 1. Project initiator Wilfried Korth (1959–2019) on his last Greenland Expedition in 2017.

The processes and consequences of climate change have been discussed for many years. Leaving aside catastrophic changes such as volcanic eruptions or earthquakes, it is indisputable that never in the recent history of the Earth, i.e., in the last million years, have there been fundamentally very rapid changes in the living conditions on our planet [1]. However, the climate today is changing rapidly. Geodetic measurements can make important and precise contributions to the monitoring of changes. In addition to remote sensing technique, which uses a wide variety of technologies, there are also ground-based measurements. These serve as "ground truth" for remote sensing, but because of their accuracy, they can also be used independently.

In the 1970s, only ground-based measurements, often obtained during scientific expeditions or from measuring stations, could be used to monitor the Greenland ice sheet. Aerial methods were also used, but only for coastal parts of Greenland. Here it is worth recalling the pioneering expeditions that began exploring inland Greenland more than a century ago.

One of the first was certainly the expedition of Fridtjof Nansen (1888–1889), who was the first to cross the southern part of Greenland on skis [2]. He brought back a wealth of scientific information and meteorological measurements and proved that the entire Greenland interior was covered by an ice sheet. Another important expedition was undertaken by the Swiss Alfred de Quervain in 1912. De Quervain, a meteorologist, crossed Greenland with three other expedition members considerably further north than Nansen [3]. He was shortly followed by Alfred Wegener and Lauge Koch [4,5].

After the First World War, expeditions were more frequent and much better prepared technically. Airplanes also began to be used for research after the First World War. The systematic mapping of the coastline by the Danish Geodetic Survey in 1931–1934 was significant. The mapping work was carried out using photogrammetry from an aircraft. Today there are thousands of unique photographs in the Danish Airbase project database, which serve as a source of information on the historical state of glaciation [6,7]. Several US military airfields were built in Greenland during World War II, some of which were converted to civilian airfields after the war and are still in use today [8]. Germany also built a small meteorological base in Greenland, but it was destroyed by an American air raid [9]. After the war, economic development began in Greenland, but other US military bases were also built in Greenland during the Cold War. More intensive research on the Greenland ice sheet took place after the fall of the Iron Curtain in the 1990s.

Modern instruments, expedition equipment, and technical support were available, as well as the possibility of using satellite data. The significant progression of global warming and the rapid melting of the western and southern parts of the Greenland ice sheet, in particular, increased interest in research activities [10–12]. Combined data sources and non-traditional technologies like drones, for example, were used in research. Today, drones are the most popular, which allow very detailed measurements in smaller areas, e.g., tracking the movement of a glacier face or capturing the surface with cm resolution [13–15]. Special remote sensing satellites have been used for a long time, since the 1970s, but it's only relatively recently that some data has been free of charge and freely downloadable.

Geodetic satellites monitor gravity changes, radar satellites can use InSAR technology to determine displacements or create digital surface models, and optical satellite systems can help monitor the extent of glaciation [13,16]. Fast and accurate GNSS instruments can monitor the height or movement of glaciers [17–19]. In the context of Arctic polar research, it is worth remembering Nansen's unique polar expedition on the Fram ship (1893–1896); this was followed in 2019 by an international expedition aboard the modern research ship Polarstern. The aim was the comprehensive mapping of the Arctic and, in particular, research on global warming [20]. Glacier changes related to global warming have been investigated in many other scientific papers [21,22].

Twenty years ago, in the summer of 2002, geodesist Wilfried Korth (Figure 1) started a climate research project in Greenland. The main objective was to determine elevations and their changes along a profile across the Greenland ice sheet. A 700 km traverse was surveyed between Tasiilaq on the east coast and Ilulissat on the west coast of Greenland (see Figure 2). After his tragic death, however, some members of his expeditions continued his work. This provided another valuable amount of information on the changing Greenland ice sheet. The results from all the expeditions are summarized in the following text.

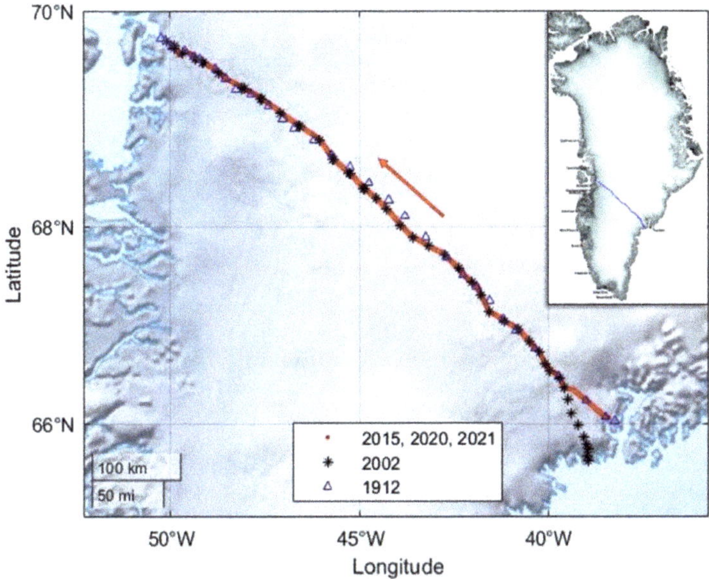

Figure 2. Map of Greenland [17] and route of Greenland Korth Expedition (GKE) with walking direction, camps from 2002 and historical camps from 1912. The blue line approximately marks the catchment area of the Helheim and Ilulissat (Jakobshavn) glaciers and the top of the Greenland ice sheet.

This route was first successfully crossed by the Swiss Alfred de Quervain in 1912. Even if the accuracy of his measurements was only relatively low compared to today's possibilities, the large time difference of more than a hundred years naturally tempts a comparison [18], which is especially interesting in the strongly changing marginal area of the ice sheet.

Meanwhile, during the eight expeditions since 2002, the profile was surveyed five times completely and three times partially with high accuracy (see Table 1). The process results in surface elevations with a measurement accuracy of 3–5 cm, from which annual surface changes are derived with similar accuracy. The measurements were carried out between the end of July and the beginning of September. During this period, the summer thaw was ending, while the winter snowfall had not yet begun. It is, therefore, the time of the year when the seasonal variations in ice elevations reach their minimum.

Table 1. Overview of geoscientific Expeditions on the historic route.

Year	Scientific Director	Method of Measurement	Remark
1912	A. de Quervain	barometric	39 camps; accuracy in the coastal area +/−3–5 m
2002	W. Korth	GPS static	34 positions; +/−3 cm
2006	W. Korth	GPS static	34 positions; +/−3 cm
2010	W. Korth	GNSS static	34 positions; +/−2 cm
2012	W. Korth	GNSS static	only east coast; 17 positions +/−2 cm
2015	W. Korth	GNSS kinematic and static	continuous profile; spot spacing 2–6 m; 700 km; +/−3 cm
2017	W. Korth	GNSS kinematic	continuous pr.; spot spacing 2–6 m; approx. 180 km; +/−3 cm
2020	T. Hitziger	GNSS kinematic	continuous pr.; spot spacing 2–6 m; approx. 500 km; +/−3 cm
2021	T. Hitziger/J. Heim	GNSS kinematic	continuous pr.; spot spacing 2–6 m; approx. 680 km; +/−3 cm

Based on GNSS technology progress since the 2015 expedition, the measurement program was changed. Unlike in previous years, not only were the profile points determined but the measurements were carried out continuously along the entire route at 1-second intervals. Thus, for the first time, a 700 km long profile with a point spacing of less than 2 m is available. The possibilities for comparison with satellite data have thus improved enormously.

Measurements on the ice sheet can be carried out in very different ways with today's technical possibilities. However, extreme problems occur, especially in the marginal areas, which lead to limitations: the use of (heavy) snowmobiles is hardly possible because of the numerous crevasses, which are often blown, and impossible in the large areas with melt-water rivers, gullies, and ice humps. But this concerns the most interesting area, about 20–30% of the planned route.

As a logistical alternative, skis and pulkas (freight sledges) were used on all expeditions, and the routes were covered on foot. What appears at first glance to be an increased risk is, on closer inspection, a gain in safety. On some expeditions, kites were used as towing devices on the glacier plateau. In good winds, it made the journey faster. The comparatively low travel speed, on the other hand, is not a measuring problem because the aim is to keep the distances between the measuring points as short as possible. Of course, this type of expedition requires the willingness of the participants to face the physical demands. But this has never been different throughout the history of polar research, from the expeditions of the pioneers to the present day.

2. Materials and Methods

The basic measurement in this project was the use of GNSS. The theory of GNSS is described in many technical articles, as well as the development of accuracy [23–25]. In high geographical latitudes, the integration of the GLONASS navigation system proves to be advantageous [26,27].

In our case, for the static GNSS measurements from 2002 to 2015, different generations of Trimble antennas and receivers were used. Since 2015, additional kinematic GNSS measurements have been performed using the NavXperience 3G + C antenna with the Trimble R7 receiver in 2015. Subsequent expeditions used the combined Trimble R10 and R12 systems. Portable GNSS units were used for orientation on all expeditions. Signals in the L- and G-band range of GPS and GLONASS (later also BeiDou and Galileo) were received. The accuracies of the campaigns are shown in Table 1. All measured coordinates are used with ellipsoidal heights.

During the nearly 40-day expedition, field logs were made of antenna heights as well as sled lowering depths and how they changed throughout the day. In addition to these geodetic records, weather data and density measurements were also noted (2017, 2020, 2021).

2.1. Static GNSS Measurements

Static GNSS measurements were taken approximately every 20 km at the respective overnight camps in 2002, 2006, 2010, 2012, and 2015, with each expedition member reaching the camp established in 2002 to ensure comparability. Upon arrival, the antenna was set up, aligned, and connected to the receiver (see Figure 3). The system is powered by solar cells. The measurement time is between 8 and 12 h, and the equipment is stored at a sufficient distance to avoid interference with the signal. The earlier measurements were made using ground stations on both coasts (Kangerlussuaq and Kulusuk and Tasiilaq, Kangerlussuaq and Ilulissat, respectively), and later, precision was achieved using Precise Point Positioning (PPP) by correcting the orbits afterward. WGS84 was used as a reference frame.

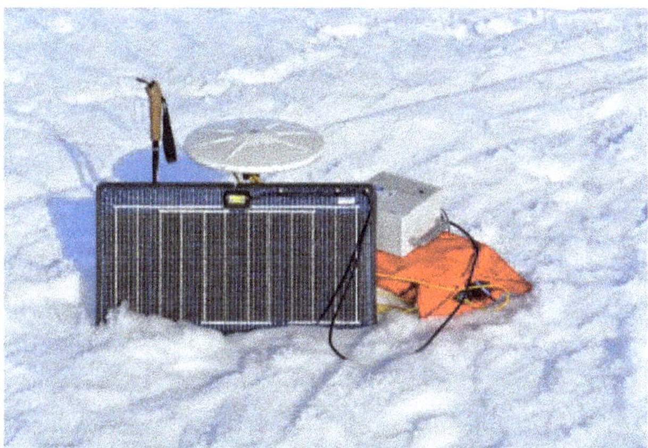

Figure 3. Static GNSS measurement during GKE 2015.

2.2. Kinematic GNSS Measurements

Kinematic GNSS measurements took place in 2015, 2017 (east coast only), 2020 (about 500 km), and 2021. The antenna was mounted on the pulka of an expedition member, and there were second-by-second recordings of the individual GNSS points. The 2 systems, GPS and GLONASS, were used (see Figure 4).

(a)   (b)

Figure 4. Kinematic GNSS measurement during GKE 2020 (**a**) on the pulka. (**b**) Trimble R12.

During post-processing, the TEQC software quality check was performed to verify the quality of the obtained data and to adjust the approximate position in the header of the observation files *.yyO of the RINEX data [26]. Because Greenland is a remote location and the technology has evolved, no extension systems or ground stations were used as reference stations for the kinematic measurements. Precise Point Positioning (PPP) in the International Terrestrial Reference Frame (ITRF2014) is used to achieve the precision of the data. Natural Resources Canada (NRCan) is used as the provider. Uploading is done through a web interface, and corrected positions are sent by mail. Using accurate ephemerides and clock corrections, the position can be determined to the nearest centimeter [28]. It takes 13 days to calculate the final corrections, so this period should be weighted between data collection and precision. Care should also be taken to use the correct evaluation method (static/kinematic) with the associated data, to always have enough satellites available, and to minimize the individual error sigma.

In the next step, the plate kinematics are considered, using 01.06.2015 (00:00:00) as the reference date, and the corresponding displacements and rotations of the North American plate are included in the plate motion model according to ITRF2014. This allows us to compensate for the effect of the Glacial Isostatic Adjustment (GIA) in Greenland [29].

After the data have been specified and reduced, they are further processed with Matlab. An overview of the program flow can be seen in Appendix A. First, the entries for the sled sinking depth and the antenna height are taken from the field book records. This is followed by a temporal sort and subsequent low-pass filtering of the data using Gaussian filtering to minimize the influence of noise. After a parameter study, a filter order of m = 50 is used as a target for the local topography of the Greenland ice sheet [30–32]. Due to the second-by-second measurement points and the largely tall jump-free relief, more distant points can also be considered for smoothing.

Two different principles are used to compare the GNSS kinematic data. First, crossing points are investigated, which requires a linear interpolation of two data points at the same position coordinates. Second, the principle of minimum distances between each data point of the 2015 expedition and subsequent expeditions is considered [32–36].

2.2.1. Crossing Point Comparison

For this purpose, the data are converted into the appropriate format so that they can be read and processed by the Linux-based program Generic Mapping Tool (GMT). The crossing points are determined as a linear interpolation between two different years. It should be noted that these are calculated values and not measured values. However, the advantage is that the position coordinates match exactly, and local unevenness has less influence. Since there are no large jumps on the Greenland ice sheet, the method is well suited. The x2sys package included in GMT is used for the calculation. The obtained crossing points are transformed to UTM coordinates in order not to neglect the curvature of the Earth. Then, the crossing points are assigned to the continuous track of the 2015 expedition by using the closest data point in each case. This method is sufficiently accurate over the entire track of nearly 700 km. In each case, the distances within a UTM zone are searched. The altitude differences previously calculated with GMT can now be visualized and analyzed.

2.2.2. Comparison of the Minimum Distances

For the comparison over the minimum distances, UTM coordinates are also used, and the holding times, which are caused, e.g., by pauses, are eliminated. As tolerance for the elimination of values, the distance of 0.005 m is used. Thus, the total matrix can be slimmed down considerably, and the computation time is shortened enormously. With this method, the distances of the position coordinates of an expedition to those of a following expedition are determined, and afterward, the respective data point with the smallest distance to the reference distance (here: 2015) is assigned. The calculation is very time-consuming and can be significantly shortened by using multiple processors via parallel computing in Matlab. After each point is assigned a minimum distance to a point in the follow-up measurement, data that are above tolerance are truncated. For the Greenland ice sheet, this was chosen for 5 m after the completion of the parameter study.

In addition, an adjustment to the data was made for the crossing point comparison. In the seasonally comparable expeditions in 2015 and 2020, almost 200 km were missing on the west coast because the expedition had to be aborted prematurely. However, in the following May 2021, the route could be walked completely, so the elevation component of the coordinates on the west coast is shifted to the connection point. The further one moves away from the endpoint of the 2020 expedition, the greater the uncertainty in the result becomes. Finally, the differences obtained are plotted and illustrated using Matlab's mapping toolbox.

3. Results

3.1. Static GNSS Measurements

Static measurements were made at 36 points spaced about 20 km apart. Figure 5 shows the mean annual elevation change at these points.

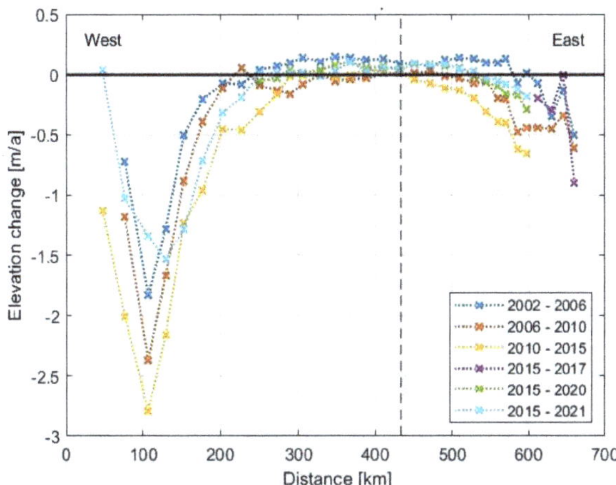

Figure 5. Annual elevation changes in the camps from 2002.

The catchment areas of the two glaciers Helheim (East coast in Figure 5) and Ilulissat Isbræ (West coast in Figure 5) are clearly visible. The watershed is located at km 420 (in Figure 5) and represents the highest point along the route. The 2002, 2006, 2010, and 2015 measurements seem to indicate an acceleration of mass loss, but the 2020/21 measurement does not confirm this. It appears that longer time series are needed to identify more reliable trends.

Figure 6 shows the absolute elevation changes at the camps between 2002 and 2021. While there is hardly any increase in the accumulation area (max. 2.1 m), there is an elevation loss of max. 38.7 m in the reservoirs in the Ilulissat Isbræ catchment. An elevation decreases of max. 6.4 m in the comparable area in the Helheim Glacier catchment and 10.1 m in the marginal area of the ice sheet on the east coast can be seen in Figure 6.

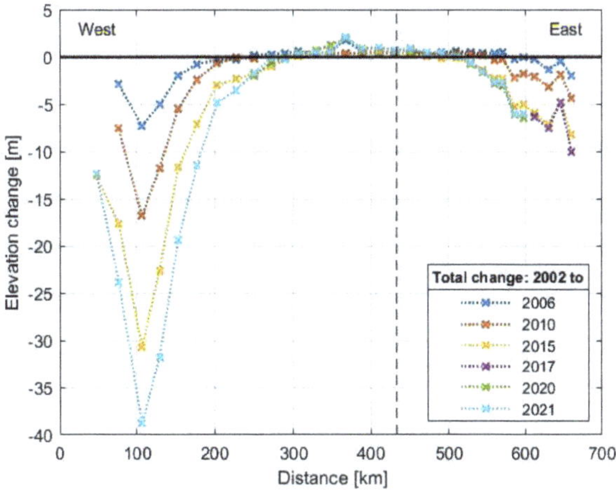

Figure 6. Total elevation changes in the camps from 2002.

3.2. Kinematic GNSS Measurements

3.2.1. Profile Comparison

First, the elevation profiles are compared. These retain their rough but finer details, which can be seen when magnified (Figure 7). As expected, the ice elevation decrease is more pronounced along the coasts, which is enhanced by the two outlet glaciers Helheim (fastest flowing outlet glacier on the east coast of Greenland at approx. 30 m/d and Jakobshavn Isbræ (the most productive glacier on the west coast) since the expedition route lies within the influence of these [28]. However, when local ice elevation topography is considered, slight terrain elevations show a larger elevation change than the adjacent depressions (Figure 7). Overall, individual values fluctuate up to +/−30 cm per year around a sectionally stable mean or median.

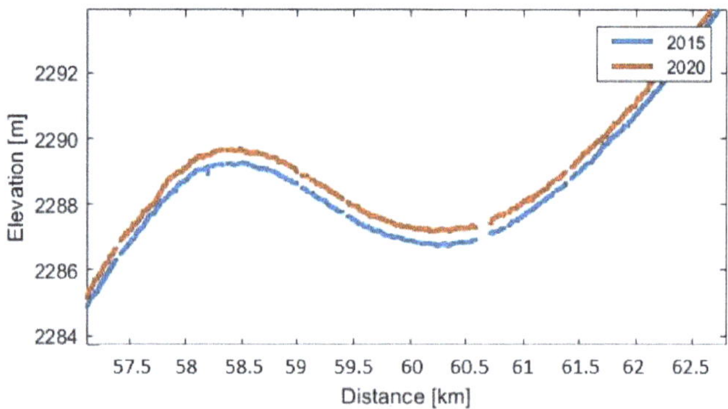

Figure 7. Comparison of a selected part of the profile from 2015 and 2020.

3.2.2. Elevation Change Comparison

With the help of the crossing point comparison, between 300 and 3000 intersections could be found, depending on the comparative section of the respective expedition, providing a dense network of data over the entire route. On the sections covered by skis and pulka, the density is significantly higher than in the sections covered by the kite. In practice, it is easier to generate crossing points in these areas because the speed traveled is lower. Particularly in the marginal area of the ice sheet, the variance of measured values is larger when comparing the expeditions' data, which is mainly due to the surface topography, as it is characterized by meltwater channels (see Figure 8) at the time of most expeditions (except 2021). In addition, the change from the minimum distance calculation is added here. After applying the previously described cutoff rule with a tolerance of 5 m, significantly more comparison points remain than for the crossing points. Occasionally, measurement gaps occurred during the expeditions, so no comparison is possible at these points.

Figure 8. Meltwater channels during the expedition 2020 on the east coast.

The elevation change produces very similar results with both methods (Figure 9). Statistical values such as the mean and median deviate only by 2–3 cm in selected sections. However, the method is not generally valid in this form. Only because of the known topography of the ice sheet with few slopes does it remain very reliable. However, it can be assumed that the error in the method of minimum distances is larger than in the method of the crossing point comparison.

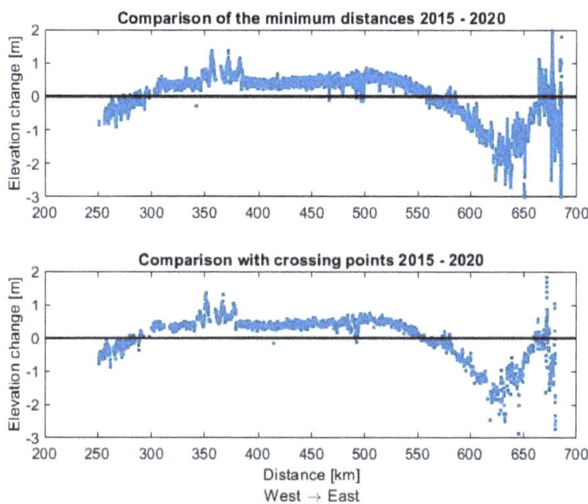

Figure 9. Comparison of the minimum distances method with 5 m tolerance (**top**) and the crossing point method (**bottom**).

3.2.3. Seasonal Changes during the Winter

The 2020 expedition took place in August/September and the next in the following May 2021 so that the seasonal changes could be observed over the winter. In Figure 10, these changes are shown along the profile, with the elevation component almost constant in the ice center. Towards the coasts, an increase due to precipitation of up to 2 m is observed. The west coast could not be investigated in more detail because the expedition had to be terminated prematurely.

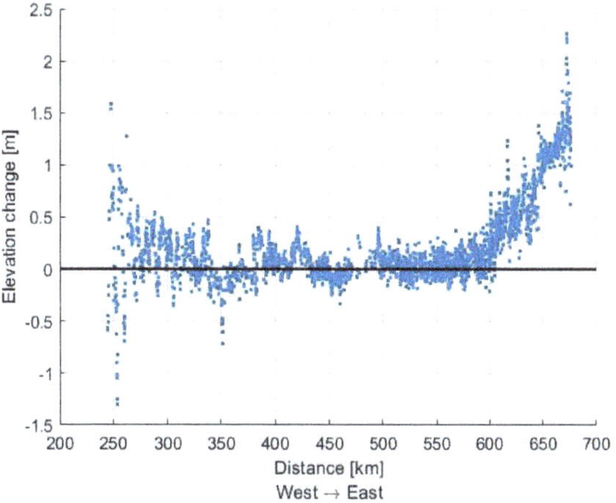

Figure 10. Elevation changes between before winter (August 2020–September 2020) and after winter (May 2021). Difference (blue points) = 2020–2021.

3.2.4. Modification of Missing Parts

For the kinematic data, complete profiles were only measured in 2015 and 2021, with the 2021 expedition taking place as early as May rather than between late July and early September as all previous expeditions had. An expedition took place in the previous season, covering about 500 km, so a link to the data from 2021 is made in this step to obtain a complete and seasonally comparable data set. Figure 11 shows the crossing point comparison for both 2015 and 2020 as well as 2015 and 2021, where the data density is not quite as high due to flooded sections.

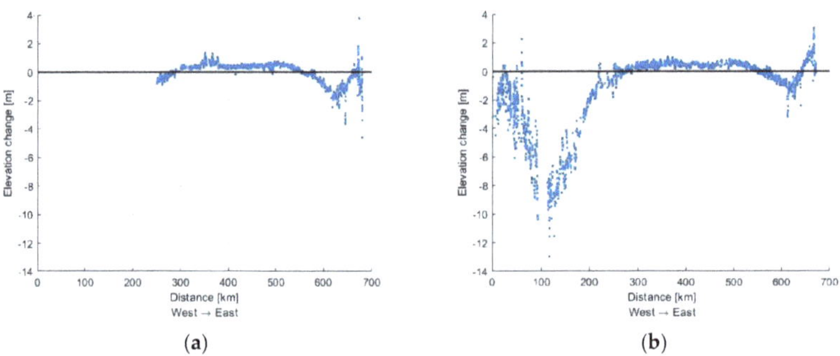

Figure 11. Comparison of crossing points (**a**) 2015–2020 and (**b**) 2015–2021.

The change in elevation over the winter from 2020 to 2021 is accounted for by appending to the endpoint of the 2020 data and the 2021 data set and shifted by the difference of -0.7448 m to get more realistic results for the coastal area (Figure 11). The farther the data is from the connection point, the larger the inaccuracy becomes.

In Figure 12, as in the static measurements, it can be seen that the influence of the two heads of glaciers is clearly visible in the data set. A larger ice elevation decrease is expected near the coast, which is amplified by calving the glaciers. Near the east coast, the decrease is somewhat delayed, which is probably related to the damming effect of the Schweizerland Alps (mountains on the east coast of Greenland, as de Quervain called them).

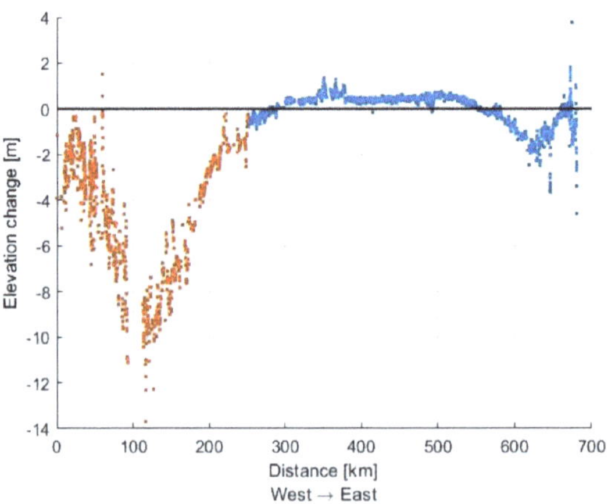

Figure 12. Comparison of crossing points along the expedition route with modified elevation change at the west coast. Blue color = 2020; orange color = 2021.

3.2.5. Accumulation and Ablation

The data show accumulation and ablation along the profile (see Figure 13). As expected, there is accumulation in the central part of the ice sheet and ablation toward the coasts. It should be noted that the seasonal change shifts the equilibrium line. It also illustrates the influence of the Helheim Glacier and the Jakobshavn Isbræ, with a catchment area of up to 200 km inland. For comparison, the ice velocity in Greenland was determined in [17] by Sentinel-1. The extensive catchment areas of the glaciers in relation to the expedition route are shown in Figure 14.

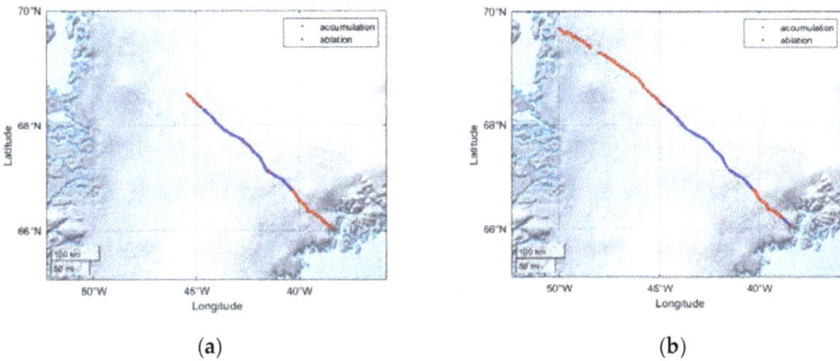

(a)   (b)

Figure 13. Accumulation (blue) and ablation (red) along the expedition route based on data from (a) 2015–2020. (b) 2015–2021.

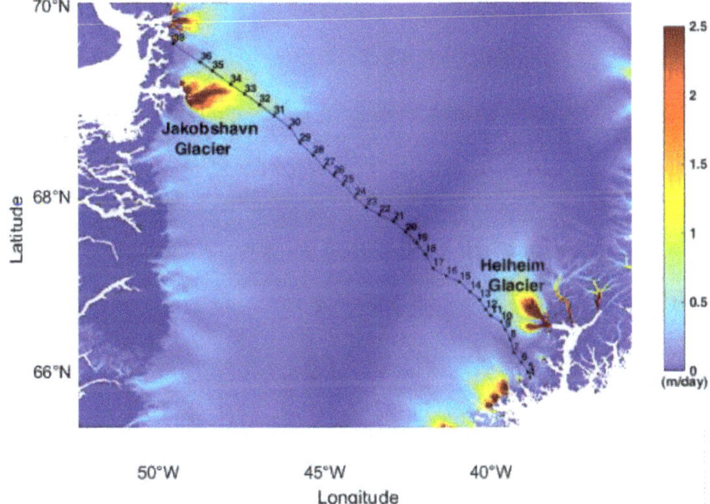

Figure 14. Ice velocity from synthetic aperture radar of Sentinel-1 acquired over October 2015–September 2016 with camps of the Greenland Korth Expedition route. The direction of glacier flow is from the central axis of the Greenland ice sheet towards the coast [17].

3.2.6. Comparison with Other Data

Satellite altimetry provides a real information on glacier elevation. However, these are only partially usable. There are gaps of several kilometers between the ground satellite tracks (see Figure 15). A direct comparison of our traverse data with those of the satellite altimetry is only possible at the crossing points. The Geoscience Laser Altimeter System

(GLAS) is developed for the IceSat mission and has a precision of about 3 cm for a footprint with less than 80 m diameter [37].

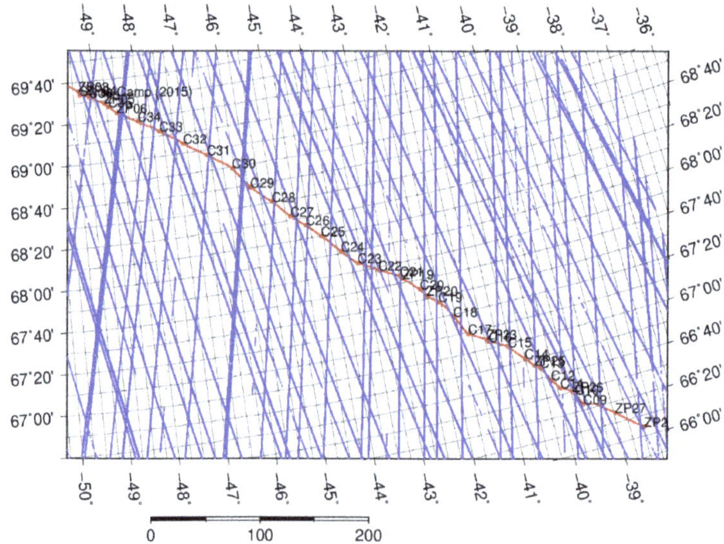

Figure 15. Coverage of the operation area by the IceSat mission (2003–2010). The diagonal line represents the profile measured by GKE on the ground.

The NASA IceSat satellite (Ice, Cloud, and land Elevation Satellite, 2003–2010) data were used. Comparing the profiles measured in this project with an elevation model derived from IceSat data [29], the qualitative difference between the two data sets becomes clearly visible (Figure 16). The elevation model is from 2010, based on the end of the IceSat mission and shows considerable deviations from the profile measurements due to the different spatial resolution. The IceSat provides a set of laser pulses, which have approximately 70 m spots on Earth's surface with a spacing of 170 m. The model from IceSat is interpolated and smoothed (blue line) compared to the profile measurement, which is more detailed (red line).

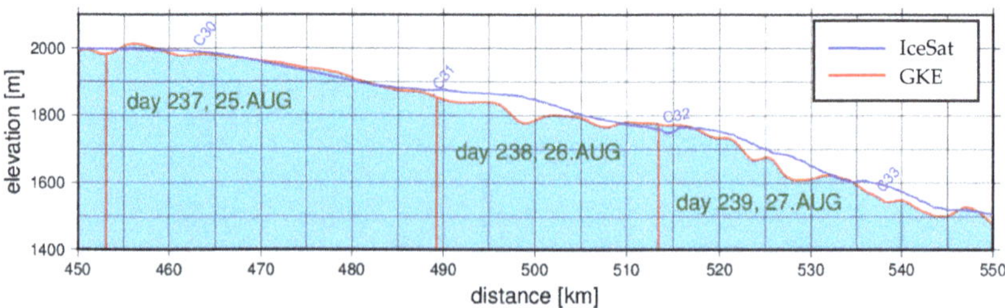

Figure 16. Part of the elevation profile from the west side (2015).

C30, C31, C32 and C33 are some of the points measured multiple times since 2002. The waves in the surface profile with amplitudes of up to 20 m are clearly visible. The blue line shows the heights derived from IceSat data; the red line shows the measured profile from this project.

4. Discussion–Error Influences

The aim was to demonstrate the possible link between terrestrial static or kinematic GNSS measurement and data derived from the GLAS device onboard the IceSat-1 satellite and, next, to verify the reliability and accuracy of the digital surface model based on IceSat-1 satellite measurements. The measured values were examined for random jumps using the difference quotient. There are jumps of a maximum of 1.2 cm, so that jump measured values cannot be identified as a significant source of error. Moreover, these are eliminated as much as possible by Gaussian filtering.

After PPP evaluation, the position accuracy is $+/-3$ cm, and the height accuracy is $+/-5$ cm for a single data set. For the investigations performed here, the elevation component is most relevant. Of course, there are some defined outliers; during measurement at these locations, a lower number of GNSS satellites occurred (PDOP). This influence is also visible in the dispersion of the elevation component (Figure 11, comparison km 680) around kilometer 680. This is classified as not trustworthy.

Regarding the measured antenna heights, a deviation of $+/-1$ cm is to be expected when it is attached to the pulkas. Likewise, variations of nearly 0 cm (in good conditions) to $+/-5$ cm (in uneven terrain) due to the ground conditions can be seen.

5. Conclusions

Geodetic-glaciological field work for monitoring glaciers and ice sheets is necessary even in the age of satellite technology. On the one hand, it is for the verification of the satellite data, but on the other hand, they also provide important results of their own.

The elevation changes determined during the Greenland Korth Expeditions (GKE) show a continuous melting process since 2002, mainly on the west coast. On the east side, the amount has increased from 20–40 cm/yr to 40–80 cm/yr. On the west side, the maximum annual ice loss has increased from 1.7 to 2.7 m/yr. Overall, the ice elevation at profile kilometer 100 has decreased by more than 35 m since 2002.

It turns out that long-term observations are always needed to make claims about climate change. Our observations are based on a historically short period of time, about 100 years. Nevertheless, it can be argued that we are now observing an enormous melting of Arctic ice, especially on the west coast of Greenland. It is not the purpose here to discuss causes or consequences, although this may have far-reaching implications for climate change, a possible change in the Gulf Stream, sea level rise and thus a significant impact on humanity. The aim was to demonstrate the possible link between terrestrial static or kinematic GNSS measurement and data derived from the GLAS (Geoscience Laser Altimeter System) device onboard the ICESat-1 satellite and, next, to verify the reliability and accuracy of the digital surface model based on ICESat-1 satellite measurements. A relatively good agreement was achieved, the differences being due to the different resolutions and the different terms of observation of the Greenland ice sheet.

Author Contributions: Conceptualization, T.H. and L.N.; methodology, T.H. and L.N.; software, L.N.; validation, T.H. and L.N.; investigation, T.H; data processing, L.N. and T.H.; writing—original draft preparation, T.H., L.N. and K.P.; writing—review and editing, T.H., L.N. and K.P.; visualization, T.H. and L.N.; supervision, K.P.; project administration, T.H. and K.P. All authors have read and agreed to the published version of the manuscript.

Funding: This research was funded by the Brandenburg University of Technology Cottbus-Senftenberg, Faculty of Architecture, Civil Engineering and Urban Planning. From the Czech part, the project was supported by a grant SGS22/049/OHK1/1T/11.

Institutional Review Board Statement: Not applicable.

Informed Consent Statement: Not applicable.

Data Availability Statement: The used data were collected during the expeditions and from open sources.

Acknowledgments: Our special thanks go to Wilfried Korth, who initiated the scientific project and actively accompanied it for many years. Unfortunately, he died in an accident in 2019. He was always involved in the expeditions carried out until then and had the research and data in his hands. Furthermore, we thank Carsten Grienitz from AllTerra Germany GmbH for the support with Trimble technology. Finally, we thank Frank Polte and Marco Schütze, our expedition members in 2020. From the Czech part, the project was supported by a grant SGS22/049/OHK1/1T/11.

Conflicts of Interest: The authors declare no conflict of interest. The funders had no role in the design of the study; in the collection, analyses, or interpretation of data; in the writing of the manuscript, or in the decision to publish the results.

Abbreviations

The following abbreviations are used in this manuscript:

GIA	Glacial Isostatic Adjustment
GKE	Greenland Korth Expedition
GMT	Generic Mapping Tool
ITRF	International Terrestrial Reference Frame
NRCan	Natural Resources Canada
PPP	Precise Point Positioning
RINEX	Receiver Independent Exchange Format
UTM	Universal Transverse Mercator

Appendix A

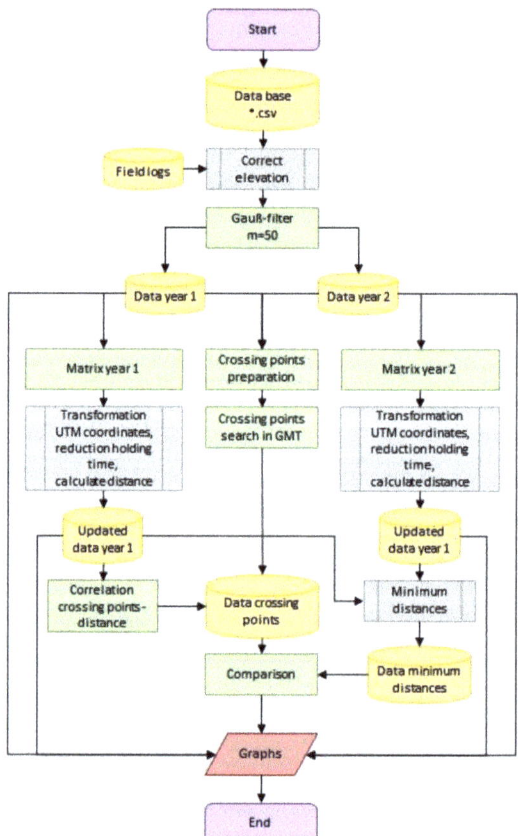

Figure A1. Program flow for kinematic analysis.

References

1. Wolff, E.; Fung, I.; Hoskins, B.; Mitchell, J.F.B.; Palmer, T.; Santer, B.; Shepherd, J.; Shine, K.; Solomon, S.; Trenberth, K.; et al. *Climate Change, Evidence, and Causes. An Overview from the Royal Society and the US National Academy of Sciences, Update 2020*; Royal Society and US National Academy of Sciences: Washington, DC, USA, 2020.
2. Hestmark, G. Fridtjof Nansen and the Geology of the Arctic. *Earth Sci. Hist.* **1991**, *10*, 168–212. [CrossRef]
3. Abplanalp, A. Alfred de Quervain. Available online: https://blog.nationalmuseum.ch/en/2020/02/de-quervain-greenland-1912 (accessed on 10 September 2022).
4. Hobbs, W.H. Lauge Koch. In *Encyclopaedia Arctica*; Dartmouth College Library: Hanover, NH, USA, 1864.
5. Wegener, E.; Loewe, F. *Greenland Journey, The Story of Wegener's German Expedition to Greenland in 1930–31 as told by Members of the Expedition and the Leader's Diary*; Translated from the Seventh German Edition by Winifred M. Deans; Blackie Son Ltd.: London, UK, 1939.
6. Bjørk, A.A.; Kjær, K.A.A. *The Greenland Ice Sheet—80 Years of Climate Change Seen from the Air*; Natural History Museum of Denmark, Faculty of Science, University of Copenhagen: Copenhagen, Denmark, 2014; 180p, ISBN 978-87-87519-46-5.
7. Bjørk, A.A.; Kjær, K.H.; Korsgaard, N.J.; Khan, S.A.; Kjeldsen, K.K.; Andresen, C.S.; Box, J.E.; Larsen, N.K.; Funder, S. An aerial view of 80 years of climate-related glacier fluctuations in southeast Greenland. *Nat. Geosci.* **2012**, *5*, 427–432. [CrossRef]
8. Technical Report. Greenland Bases. Air Force Historical Research Agency. Available online: https://www.afhra.af.mil/ (accessed on 12 May 2022).
9. Jensen, J.F.; Krause, T. Wehrmacht occupations in the new world: Archaeological and historical investigations in Northeast Greenland. *Polar Record* **2011**, *48*, 269–279. [CrossRef]
10. Slater, T.; Hogg, A.E.; Mottram, R. Ice-sheet losses track high-end sea-level rise projections. *Nat. Clim. Chang.* **2020**, *10*, 879–881. [CrossRef]
11. Mouginot, J.; Rignot, E.; Bjørk, A.A.; van den Broeke, M.; Millan, R.; Morlighem, M.; Noël, B.; Scheuchl, B.; Wood, M. Forty-six years of Greenland Ice Sheet mass balance from 1972 to 2018. *Proc. Natl. Acad. Sci. USA* **2019**, *116*, 9239–9244. [CrossRef]
12. Zwally, H.J.; Abdalati, W.; Herring, T.; Larson, K.; Saba, J.; Steffen, K. Surface Melt-Induced Acceleration of Greenland Ice-Sheet Flow. *Science* **2002**, *297*, 218–222. [CrossRef]
13. Pavelka, K.; Šedina, J.; Pavelka, K. Knud Rasmussen Glacier Status Analysis Based on Historical Data and Moving Detection Using RPAS. *Appl. Sci.* **2021**, *11*, 754. [CrossRef]
14. Pavelka, K.; Šedina, J.; Matoušková, E.; Hlaváčová, I.; Korth, W. Examples of different techniques for glaciers motion monitoring using InSAR and RPAS. *Eur. J. Remote Sens.* **2018**, *52*, 219–232. [CrossRef]
15. Bash, E.; Moorman, B.; Gunther, A. Detecting Short-Term Surface Melt on an Arctic Glacier Using UAV Surveys. *Remote Sens.* **2018**, *10*, 1547. [CrossRef]
16. Sasgen, I.; Wouters, B.; Gardner, A.S.; King, M.D.; Tedesco, M.; Landerer, F.W.; Dahle, C.; Save, H.; Fettweis, X. Return to rapid ice loss in Greenland and record loss in 2019 detected by the GRACE-FO satellites. *Commun. Earth Amp Environ.* **2020**, *1*, 1–8. [CrossRef]
17. Bezděk, A.; Kostelecký, J.; Sebera, J.; Hitziger, T. GNSS Profile from the Greenland Korth Expeditions in the Context of Satellite Data. *Appl. Sci.* **2021**, *11*, 1115. [CrossRef]
18. Riffeler, M. Eishöhenänderung in Grönland zwischen 1912 und 2010. Master's Thesis, Beuth Hochschule für Technik Berlin, Berlin, Germany, 2012.
19. Korth, W.; Hitziger, T.; Hofmann, U.; Pavelka, K. Monitoring of surface ice height changes in Greenland. Berichte zur Polar-und Meeresforschung 716, Polar Systems under Pressure. In Proceedings of the 27th International Polar Conference, Rostock, Germany, 25–29 March 2018; Alfred Wegener Institute for Polar and Marine Research: Bremerhaven, Germany, 2018. [CrossRef]
20. Shupe, M.D.; Rex, M.; Blomquist, B.; Persson, P.O.G.; Schmale, J.; Uttal, T.; Althausen, D.; Angot, H.; Archer, S.; Bariteau, L.; et al. Overview of the MOSAiC expedition: Atmosphere. *Elem. Sci. Anthr.* **2022**, *10*, 00060. [CrossRef]
21. Cooper, M.; Lewińska, P.; Smith, W.A.P.; Hancock, E.R.; Dowdeswell, J.A.; Rippin, D.M. Unravelling the long-term, locally heterogenous response of Greenland glaciers observed in archival photography. *Cryosphere* **2022**, *16*, 2449–2470. [CrossRef]
22. Lewinska, P.; Głowacki, O.; Moskalik, M.; Smith, W.A.P. Evaluation of structure-from-motion for analysis of small-scale glacier dynamics. *Measurement* **2021**, *168*, 108327. [CrossRef]
23. Yu, K.; Han, S.; Bu, J.; An, Y.; Zhou, Z.; Wang, C.; Tabibi, S.; Cheong, J.W. Spaceborne GNSS Reflectometry. *Remote Sens.* **2022**, *14*, 1605. [CrossRef]
24. Guerova, G.; Douša, J.; Dimitrova, T.; Stoycheva, A.; Václavovic, P.; Penov, N. GNSS Storm Nowcasting Demonstrator for Bulgaria. *Remote Sens.* **2022**, *14*, 3746. [CrossRef]
25. Chwedczuk, K.; Cienkosz, D.; Apollo, M.; Borowski, L.; Lewinska, P.; Guimarães Santos, C.A.; Eborka, K.; Kulshreshtha, S.; Romero-Andrade, R.; Sedeek, A.; et al. Challenges related to the determination of altitudes of mountain peaks presented on cartographic sources. *Geodetski Vestnik* **2022**, *66*, 49–59. [CrossRef]
26. Zheng, Y.; Zheng, F.; Yang, C.; Nie, G.; Li, S. Analyses of GLONASS and GPS+GLONASS Precise Positioning Performance in Different Latitude Regions. *Remote Sens.* **2022**, *14*, 4640. [CrossRef]
27. Godah, W.; Szelachowska, M.; Ray, J.D.; Krynski, J. Comparison of Vertical Deformation of The Earth's Surface Obtained Using GRACE-Based GGMS And GNNS Data—A Case Study Of South-Eastern Poland. *Acta Geodyn. Geomater.* **2020**, *17*, 198. [CrossRef]

28. Hesselbarth, A. *Statische und Kinematische GNSS-Auswertung Mittels Precise Point Positioning (PPP)*; Verlag der Bayerischen Akademie der Wissenschaften: München, Germany, 2011.
29. Altamimi, Z.; Métivier, L.; Rebischung, P.; Rouby, H.; Collilieux, X. ITRF2014 plate motion model. *Geophys. J. Int.* **2017**, *209*, 1906–1912. [CrossRef]
30. Näke, L. Vergleich kinematischer GNSS-Daten aus Ostgrönland. Bachelor-Thesis, Brandenburg University of Technology Cottbus, Senftenberg, Germany, 2020.
31. Hitziger, T.; Korth, W. Einfluss der lokalen Eistopographie auf die Qualität von Oberflächenhöhen aus Satellitendaten. In Proceedings of the Internationale Geodätische Woche Obergurgl 2019, Obergurgl, Austria, 10–16 February 2019; Hanke, K., Weinold, T., Eds.; Wichmann Verlag: Berlin/Offenbach, Germany, 2019; pp. 113–120.
32. Hitziger, T.; Näke, L. Vergleich kinematischer GNSS-Daten des grönländischen Eisschildes. In Proceedings of the Internationale Geodätische Woche Obergurgl, Obergurgl, Austria, 7–13 February 2021; Weinold, T., Ed.; Wichmann Verlag: Berlin/Offenbach, Germany, 2021; pp. 201–206.
33. Stempfhuber, W.; Korth, W.; Hitziger, T. Glacier surface and mass balance variation in Alaska and southern Greenland. Berichte zur Polar- und Meeresforschung 716, Polar Systems under Pressure. In Proceedings of the 27th International Polar Conference, Rostock, Germany, 25–29 March 2018; p. 176.
34. Korth, W.; Hitziger, T. Geodätisches Monitoring des Klimawandels in Grönland. In *Forum: Zeitschrift des BdVI, 45. Jahrgang, Heft 1/2019*; Universitätsbibliothek der LMU München: München, Germany, 2019; pp. 26–37. ISSN 0342-6165.
35. Joughin, I. Greenland rumbles louder as glaciers accelerate. *Science* **2006**, *311*, 1719–1720. [CrossRef] [PubMed]
36. Ewert, H. Auswertung von IceSat-Laseraltimeterdaten zur Untersuchung Glaziologischer Fragestellungen in Polaren Gebieten. Ph.D. Thesis, Technische Universität Dresden, Dresden, Germany, 2013.
37. Abshire, J.B.; Sun, X.; Riris, H.; Sirota, J.M.; McGarry, J.F.; Palm, S.; Yi, D.; Liiva, P. Geoscience Laser Altimeter System (GLAS) on the ICESat Mission: On-orbit measurement performance. *Geophys. Res. Lett.* **2005**, *32*, 1–4. [CrossRef]

Article

In Search of the 1654 Seismic Source (Central Italy): An Obscure, Strong, Damaging Earthquake Occurred Less than 100 km from Rome and Naples

Luigi Cucci and Francesca R. Cinti *

Istituto Nazionale di Geofisica e Vulcanologia, 00143 Roma, Italy; luigi.cucci@ingv.it
* Correspondence: francesca.cinti@ingv.it

Abstract: The M6.3 earthquake that occurred in southern Lazio (Central Italy) in 1654 is the strongest seismic event to have occurred in the area. However, our knowledge about this earthquake is scarce and no study has been devoted to the individuation of its causative source. The main purpose of this study is putting together all of the information available for this shock to provide reliable landmarks to identify its seismic source. To this end, we present and discuss historical, hydrological, geological, and seismological data, both reviewed and newly acquired. An important, novel part of this study relies on an analysis of the coseismic hydrological changes associated with the 1654 earthquake and on the comparison of their distribution with models of the coseismic strain field induced by a number of potential seismogenic sources. We find more satisfactory results when imposing a lateral component of slip to the faults investigated. In particular, oblique left-lateral sources display a better fit between strain and hydrological signatures. Finally, the cross-analysis between the results from modeling and the other pieces of evidence collected point to the Sora fault, with its trend variability, as the probable causative source of the 1654 earthquake.

Keywords: historical seismicity; earthquake environmental effects; coseismic hydrological changes; earthquake source modeling; central Italy

Citation: Cucci, L.; Cinti, F.R. In Search of the 1654 Seismic Source (Central Italy): An Obscure, Strong, Damaging Earthquake Occurred Less than 100 km from Rome and Naples. *Appl. Sci.* **2022**, *12*, 1150. https://doi.org/10.3390/app12031150

Academic Editor: Jianbo Gao

Received: 29 December 2021
Accepted: 18 January 2022
Published: 22 January 2022

Publisher's Note: MDPI stays neutral with regard to jurisdictional claims in published maps and institutional affiliations.

1. Introduction

Motivated by the limited knowledge concerning the M > 6 earthquake of 1654, this study is an attempt to understand the event that damaged the region of Lazio-Abruzzo in Central Italy, less than 100 km from Rome and Naples (Figure 1). Records on this earthquake are available but they are too old for seismogram data and are beyond the age limit for applying seismological analysis to robust historical documentation, including recognition of the causative fault that ruptured during the event. For these reasons, the approach used could not be solely based on direct data so validation through modeling was used as well.

The study area is located in the Central Apennines, an East verging, fold-and-thrust belt that developed during the Late Cretaceous to present Africa–Europe plate convergence [1,2]). The present-day landscape and tectonic setting of the region is the result of a long deformation history, characterized by cyclical extensional and contractional phases [3]. The regional seismicity and fault setting reflects the present-day NE–SW-oriented extensional regime characterizing the Central Apennines [4], with a broad and complex system of normal faults that dissect the belt and crosscut the pre-existing compressional structures.

Our analysis is based on the different typologies of direct historical coseismic data and geological/seismological data. Among the historical data, we utilize records extracted from the available seismic catalogues [5] and the hydrological and geological earthquake signatures newly acquired and derived from the archival research conducted in this study. Among the geological data, we take into account the active faults commonly considered as potential sources of the 1654 event. With regard to the seismological data, the distribution and the parameters of the present-day instrumental seismicity are considered.

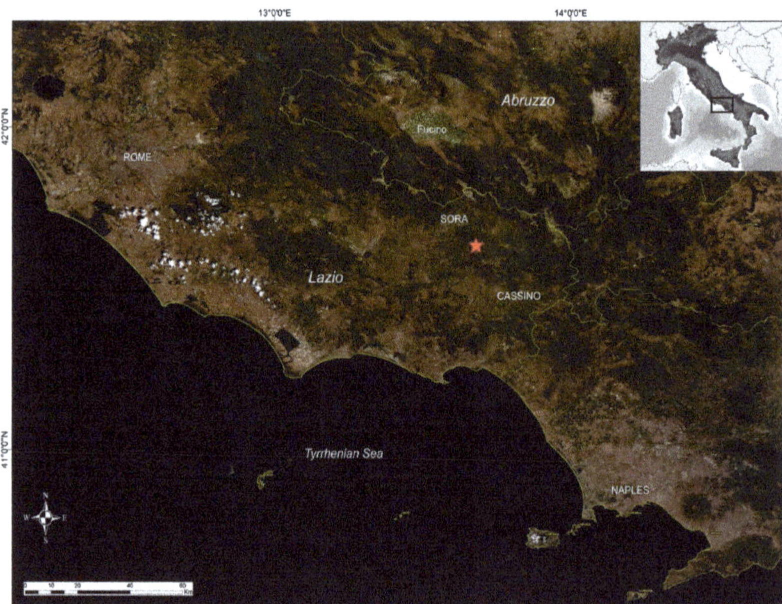

Figure 1. Epicentral location of the 1654 earthquake (red star). The cities of Rome and Naples are less than 100 km of distance from the source.

The 1654 earthquake had an MCS intensity of 9–10 and an Me of 6.33+/−0.14 [5], with the uncertainty in its epicentral location being +/−2.5 km. Although which fault caused this earthquake is still unknown, the 1654 earthquake occurred in a highly seismogenic region in Central Italy, which has been struck by medium to large earthquakes in present and historical times [5], with a ~200-year M6+ earthquakes average regional recurrence time [6]. The main aim of this study is to identify the source of the 1654 earthquake and to provide suggestions regarding the kinematics of the 1654 rupture mechanism. To this aim, we (1) modeled the intensity data (Boxer 4.0 code by Gasperini et al. [7], deriving different types of macroseismic sources; (2) calculated the strain fields imposed by all the potential faults; (3) analyzed which sources best matched the coseismic hydrological/geological signals that we collected; and (4) discriminated among the resulting potential faults considering the seismological imprint of the region as defined by the instrumental seismicity and by the geometry of the active faults in the area.

2. The 1654 Earthquake: Geological, Seismological, and Macroseismic Context

The earthquake hit during the early hours of the 24th of July, producing destructive effects (I = 9–10, M = 6.3) [5] over a vast area of the Southern Latium region between Sora and Cassino and widespread damage on the southern side of the Fucino area (Figures 1 and 2).

The map of the macroseismic intensities extends approximately 25 km from the epicenter, and the effects are differentiated, mainly because of different geological and topographical conditions of the villages involved. Six localities were almost completely destroyed and another twelve were heavily affected by the shock that was felt as far as Rome and Naples (IV–V MCS intensity). The relatively limited spatial distribution of the intensities associated with the 1654 earthquake partly reflects the paucity of official documents and historical sources available for this event. The economic marginality of the affected area (scarce productive activities, far from the main roads, no relevant center damaged) did not encourage the authorities to send investigators (scientists and technicians) to produce detailed descriptions of the event, neither did the subsequent local historiography show interest in this seismic disaster. Additionally, the fact that this is a 'border' earthquake that occurred between the two former states of the Vatican and the Kingdom of Naples (Figure 2) may have played a role in the general knowledge of the event and further investigation could help to extend the map of damage. Despite this and except for the ancient 1349 M6.8 earthquake of which the location and magnitude are still debated [8,9], the 1654 Sora event is the most powerful earthquake to have occurred south of latitude 42° S and within 150 km southeast of Rome, representing the local seismic maximum for the study area (Figure 3).

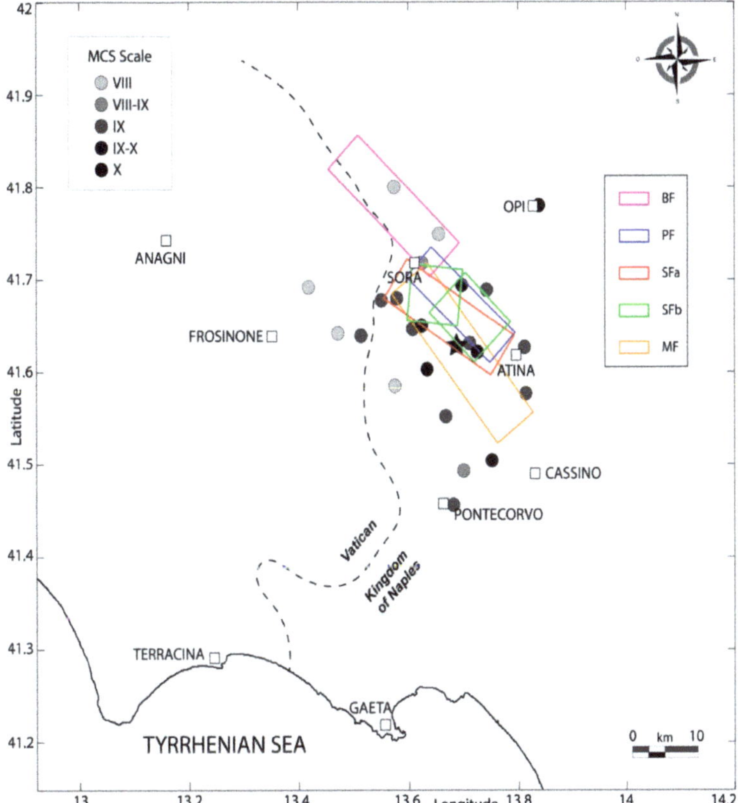

Figure 2. Map of the Mercalli–Cancani–Sieberg (MCS) intensities of the 1654 earthquake. A black star indicates the macroseismic epicenter of the event. The dashed line indicates the border between the two former states of the Vatican (West) and the Kingdom of Naples (East) in the XVII Century, at the time of the earthquake. Colored boxes are the modeled sources: BF, Balsorano fault; PF, Posta-Fibreno fault; SFa, Sora fault from ITHACA Working Group [12]; SFb, Sora fault from Boncio et al. [13]; MF, Macroseismic fault.

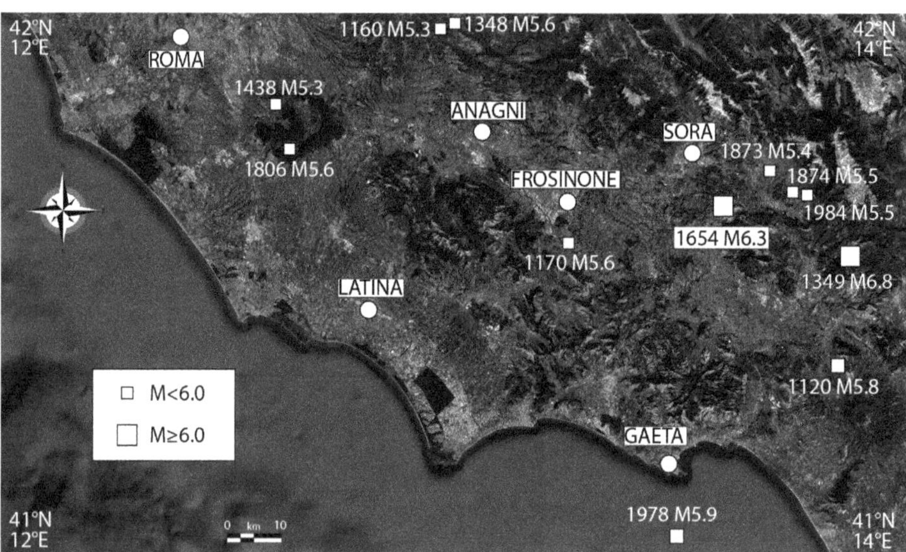

Figure 3. Historical seismicity [5] of the southern Lazio region. The 1654 earthquake is the strongest event to have occurred in this area, apart from the 1349 event.

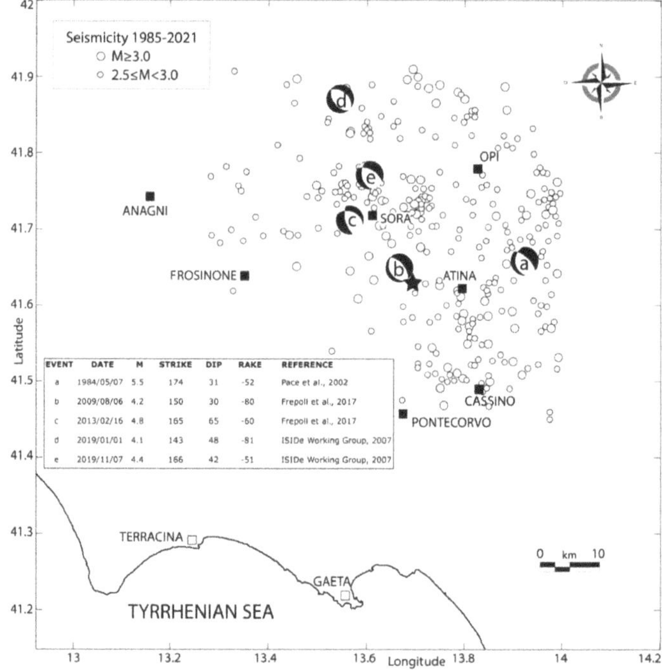

Figure 4. M ≥ 2.5 instrumental seismicity from 1985 [10] in the area of the 1654 earthquake (Latitude 41.45–41.90°, Longitude 13.30–14.00°). In the inset, we show the parameters of the five M > 4 instrumental earthquakes that occurred in the area since 1984. A black star indicates the macroseismic epicenter of the 1654 event.

The instrumental seismicity recorded in the area starting from 1985 [10] shows both low-to-moderate magnitude seismic sequences and diffuse swarm-like events, with the magnitude ranging from 0.4 to 4.8. Figure 4 reports the instrumental seismicity within 30 km from the 1654 epicenter, along with the focal mechanisms of the five most powerful ($M \geq 4.1$) seismic events in the study area that testify to a predominant normal faulting with an oblique left-lateral component [10,11].

The 1654 earthquake area contains at least three main active tectonic lineaments [12], hereinafter referred to as the Balsorano fault (BF), Posta-Fibreno fault (PF), and Sora fault (SFa and SFb) (Figure 2 and Table 1), that are generally indicated as potential sources for this event because of their location and geometry. The three faults are closely spaced and they belong to the western system of active faults of the Central Apennines [13]. They have a NNW–SSE average strike, dip towards WSW, and dip–slip to normal–oblique kinematics. They remain poorly investigated and are reported with differences in the mapping due to uncertainties on their longitudinal continuity, and their characterization is debated among the authors. Recently, evidence of Upper Pleistocene–Holocene activity has been collected along the BF and PF [14,15] while direct evidence of recent activity of the SF is unavailable.

Table 1. List of the five potential sources of the 1654 earthquake that represent the input to the modeling of the coseismic static strain.

Source	Length (km)	Width (km)	Min Depth (km)	Max Depth (km)	Strike°	Dip°	Rake°	Seismic Moment (Dyne cm)	Ref.
BF Balsorano	16.0	12.0	1	11.4	134	60	$-50/-90/-130$	2.9×10^{25}	[12]
PF Posta-Fibreno	13.0	10.4	1	10.0	133	60	$-50/-90/-130$	2.2×10^{25}	[12]
SFa Sora	16.6	11.5	1	11.4	125	60	$-50/-90/-130$	2.9×10^{25}	[12]
SFb Sora	17.0	14.4	1	13.5	115	60	$-50/-90/-130$	3.7×10^{25}	[13]
MF Macroseismic	19.6	10.3	5	13.9	142	60	$-50/-90/-130$	3.2×10^{25}	[7]

3. Effects of the Earthquake on the Natural Environment

The CFTI Catalogue of Strong Earthquakes in Italy [16] reports that this earthquake had two effects on the natural environment: a wide surface fracturing along Monte Corvo (M. Corvo) and a large landslide in Roccasecca (Figure 5 and Table 2). We re-positioned the fracture (number 4) that was misplaced in the Catalogue. In fact, Guidoboni et al. [16] located the fracture in Pontecorvo whilst the original source reports it at Monte del Corvo, 10 km NW of Sora. The landslide (number 7) reasonably occurred along the steep slope north of Roccasecca, where a scarf is still visible. We performed an in-depth round of investigation in local and national libraries and archives to seek new data of this type; the search in coeval chronicles, letters, and diaries and in later reports allowed us to add five new observations (Figure 5 and Table 2). Most of these new data concern hydrological changes that occurred immediately after the event. An increase in the discharge of the Gari River close to its springs in Cassino (number 1), a decreased and turbid flow of the Liri River in Isola Liri (numbers 2 and 3), and a decrease in discharge in the Fucino area (number 5) were reported. A fifth significant and previously unknown datum (number 6) derives from direct observation of fractures affecting the ancient structure of the Cathedral of Sora and it is indirectly inferred from the analysis of the church reconstruction history [17,18]. Some of the fractures of walls and the basement are preserved and appear aligned and adjacent to the trace of the SF. Moreover, an intriguing piece of evidence regarding the Cathedral is that its northwest end (presbyterium) is presently accessed through three steps as it is higher than the rest of the building [17,18]; however, according to the description of Bishop Giovannelli in 1618, it does not result in a higher position. This change possibly reflects a local ground deformation corresponding to the fractured zone.

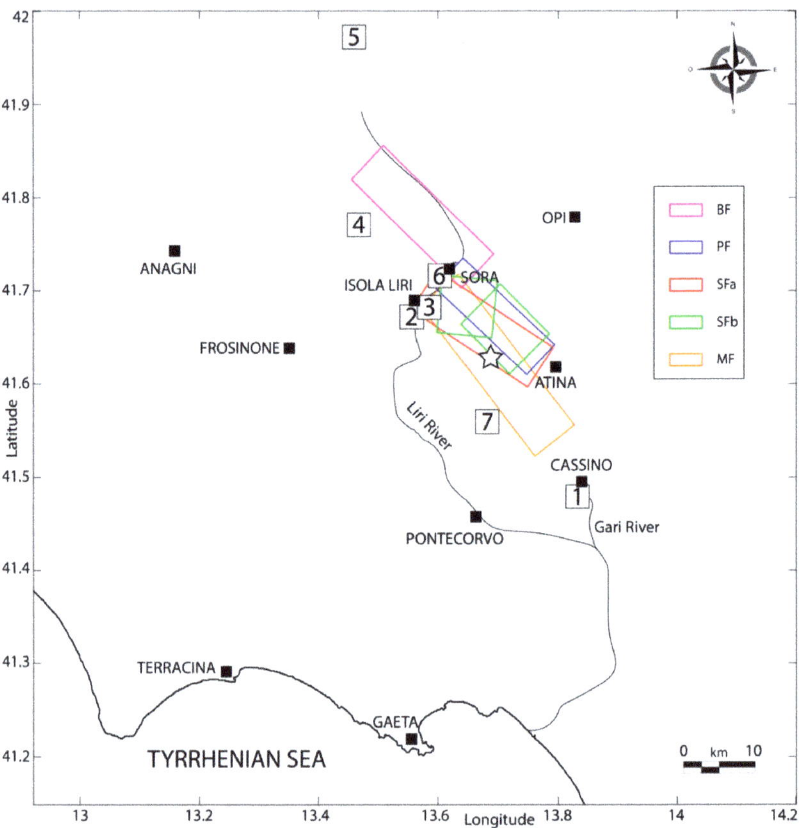

Figure 5. Distribution of the effects of the 1654 earthquake on the natural environment (see Table 1 and text for the description of the effects). Colored boxes are the modeled sources: BF, Balsorano fault; PF, Posta-Fibreno fault; SFa, Sora fault from ITHACA Working Group [12]; SFb, Sora fault from Boncio et al. [13]; MF, Macroseismic fault.

Table 2. List of the effects observed in the natural environment following the 1654 earthquake (progressive number corresponds to Figure 5). The epicentral distance is calculated from the location of the Italian seismic Catalogue (red star in Figure 1) [5].

No	Locality	Lat°	Lon°	Epic. dist. (km)	Effects	References
1	Cassino	41.480	13.832	21.2	Increase in discharge of Gari springs	[19,20]
2	Isola del Liri	41.680	13.574	10.5	Decrease in flow from Liri River	[21]
3	Isola del Liri	41.678	13.571	10.6	Turbid water from Liri River	[21]
4	M. Corvo	41.772	13.468	23.5	Wide surface fracturing	[16,19,20]
5	Luco dei Marsi	41.973	14.461	41.8	Chasms and lowering of waters in the Fucino Lake area	[22]
6	Sora	41.723	13.615	11.3	Inferred coseismic fracturing in the flooring of the Cathedral	[17,18]
7	Roccasecca	41.554	13.669	9.1	Landslide	[16,23]

4. Source Modeling

The scarcity of records belonging to the 1654 event is reasonably due to the ca. 350-year age of the earthquake itself, more than due to the real absence of effects on the landscape and villages. However, we are still able to use the collected records to infer the 1654 source parameters. Indeed, an analysis of the geographic distribution and the type of earthquake effect (i.e., building damage, ground failures, and hydrological change) is a way to provide constraints on both the fault location and the deformation style.

In particular, the coseismic hydrological changes (increase or decrease in the discharge of springs and streamflows, the water level in wells, turbid flow from springs and rivers, and liquefaction) can be a valid alternative method to provide further constraints to estimate (or to confirm related hypotheses) the faulting style of major historical earthquakes of which the seismogenic source is unknown or in dispute, as is the case of the 1654 earthquake. The basic rationale is that such hydrological variations are explained by the coseismic static strain and pore pressure changes predicted by the poroelastic theory, as first proposed by Wakita [24]. Following this theory, an earthquake imposes a coseismic strain field that causes rocks to dilate or contract; the opening or closing of saturated cracks in rocks result in decreases or increases in the ground water discharge from springs and streams. The amplitude of the hydrological changes is proportional to the volumetric strain field, so that the groundwater discharge increases in areas that contract and decreases in areas that extend. Following this rationale, in recent decades, the character of the coseismic hydrological changes has often been found to be related to the style of faulting (Cucci [25] and references therein). The most important caveat regarding the use of hydrological changes in this kind of study is that local precipitation can influence the effect that is observed and it must be carefully investigated. In the case of the 1654 earthquake, the available reports confirm the absence of rainfall in the days preceding the event, which is reasonable as the earthquake occurred at the end of July—the driest period in peninsular Italy. For a complete review of the application and of the limits of this theory in seismogenic studies, see Cucci [25]. It is possible now to perform the calculations of the coseismic strain for the 1654 earthquake produced by the potential sources listed in Table 1 to verify the best fault solution fitting with the observed hydrological and geological effects. The static strain change induced by an earthquake can be calculated using a fault dislocation model. The calculations of the strain were made in an elastic half-space with uniform isotropic elastic properties following Okada [26], and using Coulomb 3.4 [27,28]. In particular, we investigated the deformation imposed by the BF, SFa, and PF [12] and by the Sora fault as proposed by Boncio et al. [13], referred to as SFb. The fifth modeled source (macroseismic fault, referred as MF) is derived from Boxer 4.0 [7], a code that computes the quantitative parameters of earthquakes from the inversion of macroseismic intensity data, which is routinely used for the parametrization of the historical events of the Catalogue. The considered faults are generally reported by the authors with dip–slip to normal–oblique kinematics; this is also confirmed by the focal mechanisms displayed in Figure 4. Thus, we first performed our modeling on pure normal sources (rake $-90°$, Figure 6a–e); then, a second round of calculation was carried out considering left-lateral oblique slip (rake $-50°$, Figure 6f–i,l). Finally, we tested the strain calculations on a set of normal sources with oblique right-lateral slip component (rake $-130°$, Figure 6m–q). The outputs of our calculations are plots of the volumetric strain at the free surface on the five selected individual sources; the plots are shown in Figure 6. We expect to find an increase in discharge in areas of compressional strain and a decrease in discharge in areas of dilatational strain.

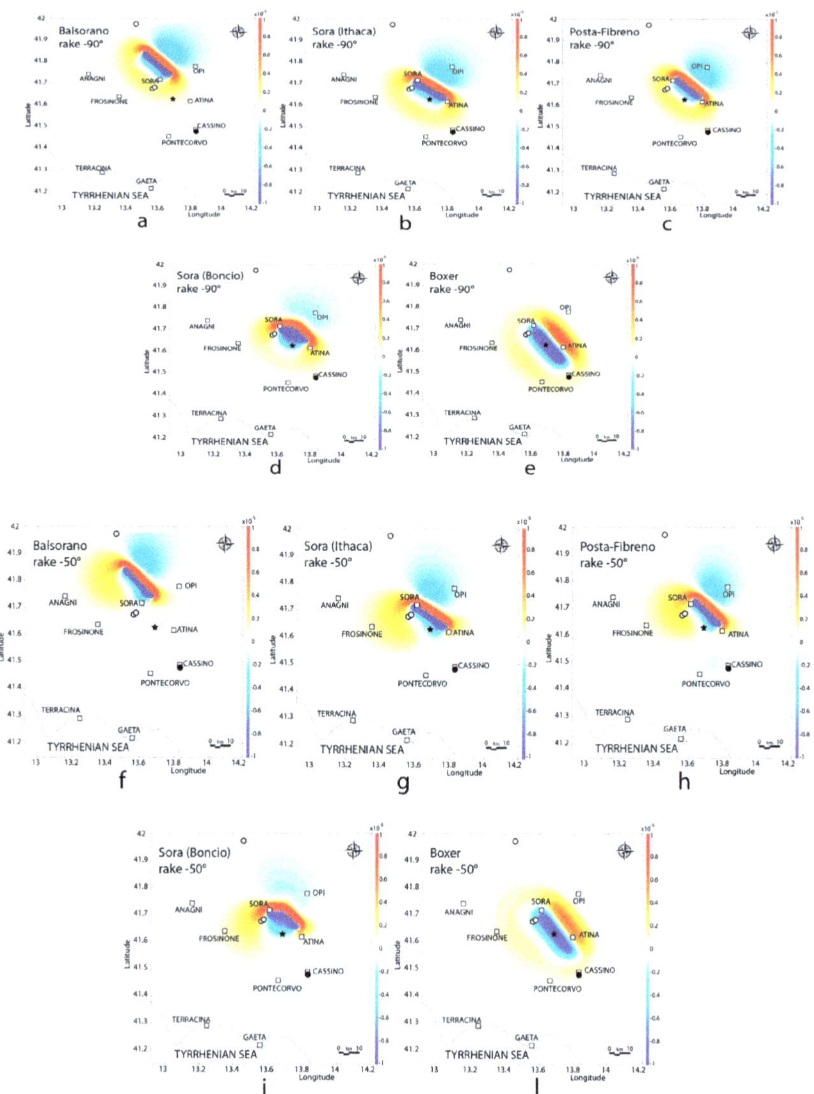

Figure 6. *Cont.*

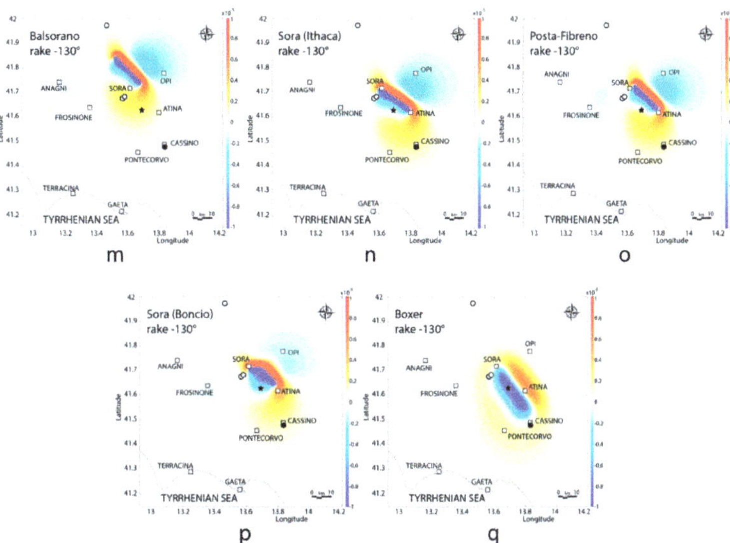

Figure 6. Comparison between the calculated coseismic strain fields along five potential sources and the observed hydrological effects produced by the 1654 earthquake. The calculation of strain was made using Coulomb 3.4 [27,28]. Plots '(**a–e**)' show the calculations for normal sources (rake −90°), plots '(**f–i,l**)' show the calculations for oblique left-lateral sources (rake −50°), and plots '(**m–q**)' show the calculations for oblique right-lateral sources (rake −130°). In the plots of strain, blue shading indicates areas in compression and red shading indicates areas in dilatation. Units: 10^{-5}. A red rectangle indicates the surface projection of the fault plane; a green line is the intersection of the updip projection of the fault with the surface. Streamflow changes are indicated by circles (black/discharge increase; white/discharge decrease).

5. Results and Discussion

A total of 15 plots of the volumetric strain at the free surface have been computed for the modeled faults, inferring a different sense of slip (Figure 6a–q). Being inferred by inversions of intensity data, the source MF (Figure 6e,l,q) obviously shows a good fit with the map of intensities and the location of the earthquake. However, the performance of the strain modeling of this source is limited, with no agreement between observed hydrological changes and the expected pattern of strain, independent of the style of faulting adopted. Additionally, there is no close association between the location of this source and the distribution of the other effects observed following the earthquake, all located northwest of the fault. When we impose pure normal kinematics to all of the seismogenic sources investigated (Figure 6a–e), we obtain a limited agreement between the predicted pattern of strain and the location of the hydrological changes. In particular, the noticeable increase in discharge of the Gari springs observed in Cassino constantly falls in an area of expected dilatation. Conversely, if we impose a lateral component of the slip on the five sources, we find more satisfactory solutions for data merging. In general, oblique left-lateral sources display a better fit between strain and hydrological signatures; in particular, PF (Figure 6h) and SF (Figure 6g,i) show the best fit for the observations in Cassino (increase in discharge/expected compression) and Isola Liri (decrease in discharge/expected dilatation).

The potential source of the 1654 earthquake is inferred through the cross-analysis between the results from modeling and the other pieces of evidence described above (see chapter 3). Though there is a fairly good fit from strain modeling shown by the Balsorano fault with rake −130° (only for the two observations in Isola Liri, Figure 6m), we exclude this source from the group of most likely causative faults of the 1654 earthquake because (1) there is no agreement between its location and the map of the intensities (see Figure 2);

(2) oblique-right lateral source seems to be an infrequent style of faulting in this sector of the Central Apennines based on seismological and geodetic data [11,29]; and (3) the epicenter of the 1654 quake is located 15 km from the southern tip of the BF. With regard to the Posta-Fibreno fault, its location fits better than the Balsorano fault when compared with the 1654 epicenter and distribution of macroseismic intensities, and an oblique left-lateral sense of slip along this fault (Figure 6h) displays a good fit with the distribution of the hydrological changes (effects 1, 2, and 3 in Figure 5 and Table 1). However, given the present mapping, the magnitude of a seismic event along this fault (M6.1–6.2, see also Table 2) would be underestimated when compared with the M6.3 magnitude presently reported for the 1654 event in the Catalogue (see Table 2). The results from the strain calculations suggest that the Sora fault, when modeled with a left-lateral component of slip, is the most probable candidate fault of the 1654 earthquake. In particular, the SFb source as traced in Boncio et al. [13] fits the most outstanding hydrological observations at Cassino and Isola del Liri as well as the evidence of fractures (effects 1, 2, 3, 4, and 6 respectively, in Figure 5 and Table 1). It is worth noting that the fractures we newly found at the Sora Cathedral would be located in the very near fault. Moreover, the magnitude of a seismic event along this source would coincide with the magnitude 6.3 quoted in the Catalogue for the 1654 earthquake.

6. Conclusions

As stated at the beginning of this manuscript, this study represents a first effort to search for new data, to merge the available information for an earthquake never studied despite its magnitude and heavy effects on the territory, and to provide some reliable landmarks for the individuation of its seismic source.

We collected five novel 1654 coseismic effects on the natural environment concerning hydrological changes (increase in discharge from a spring in Cassino and decrease in flow from Liri river in Isola Liri) and coseismic fracturing in Sora. We also shifted almost 40 km farther NNW and re-positioned one of the effects already documented (coseismic fracturing in Monte Corvo), thus enriching the picture of the natural and anthropic coseismic impact. Though it is a difficult task, due to the age of the event, the retrieval of further data on damage and natural effect as well as detailing of the fracturing would possibly consolidate the earthquake scenario and then support the source modeling.

In summary, the scenarios modeled on the basis of the collected evidence point to the Sora fault, with its trend variability, as the most probable candidate as the causative source of the 1654 event. However, the results of our model do not rule out the possibility of a complex fault rupture during the 1654 event, such as fault linkage between the Sora fault and the adjacent Posta-Fibreno fault, and possible slip along this latter source. The 1654 earthquake would be the most recent M $\geq$ 5.5 event along the Sora fault since no earlier damaging earthquake is reported in the historical catalogue for this area of study. Based on these reasons and on our results, the Balsorano fault would be silent for similar energetic events, raising its potential to generate a damaging earthquake in the area. Finally, this study confirms that the hydrological signatures of earthquake strains and field observations are valid supplemental data for estimating geometry and fault style even for early earthquakes limited by historical memory and historical reports, and they help to isolate the fault source within an active dense and complex system such as that of the Central Apennines.

Author Contributions: L.C.: Conceptualization, Methodology, Validation, Investigation, Formal analysis, Data Curation, Writing—Original Draft Preparation, Writing—Review and Editing, F.R.C.: Conceptualization, Methodology, Validation, Investigation, Writing—Original Draft Preparation, Writing—Review and Editing. All authors have read and agreed to the published version of the manuscript.

Funding: This research received no external funding.

Institutional Review Board Statement: Not applicable.

Data Availability Statement: The data presented in this study are openly available in [CPTI 15] at https://doi.org/10.13127/CPTI/CPTI15.3 (accessed on 22 December 2021), reference number [5]; [ISIDe] at https://doi.org/10.13127/ISIDE (accessed on 22 December 2021), reference number [10]; [ITHACA] at http://sgi2.isprambiente.it/ithacaweb/Mappatura.aspx (accessed on 22 December 2021), reference number [12]; [CFTI5Med] at https://doi.org/10.1038/s41597-019-0091-9 (accessed on 22 December 2021), reference number [16].

Acknowledgments: We thank the three anonymous reviewers for their helpful comments and suggestions. Thanks to Romina Rea of the Biblioteca Diocesi Sora-Cassino-Aquino-Pontecorvo for providing the historical documents. Thanks to Daniela Pantosti for revising a first version of this manuscript.

Conflicts of Interest: The authors declare no conflict of interest.

References

1. Malinverno, A.; Ryan, W.B. Extension in the Tyrrhenian Sea and shortening in the Apennines as result of arc migration driven by sinking of the litosphere. *Tectonics* **1986**, *5*, 227–245. [CrossRef]
2. Calamita, F.; Satolli, S.; Scisciani, V.; Esestime, P.; Pace, P. Contrasting styles of fault reactivation in curved orogenic belts: Examples from the Central Apennines (Italy). *Geol. Soc. Am. Bull.* **2011**, *123*, 1097–1111. [CrossRef]
3. Vai, G.B. Structure and stratigraphy: An overview. In *Anatomy of an Orogen: The Apennines and Adjacent Mediterranean Basins*; Vai, G.B., Martini, I.P., Eds.; Kluwer Academic Publ.: Dordrecht, The Netherlands, 2011; pp. 14–32. [CrossRef]
4. Montone, P.; Mariucci, M.T. The new release of the Italian contemporary stress map. *Geophys. J. Int.* **2016**, *205*, 1525–1531. [CrossRef]
5. Rovida, A.; Locati, M.; Camassi, R.; Lolli, B.; Gasperini, P.; Antonucci, A. *Catalogo Parametrico dei Terremoti Italiani (CPTI15)*, versione 3.0; Istituto Nazionale di Geofisica e Vulcanologia (INGV): Milano, Italy, 2021. [CrossRef]
6. Cinti, F.R.; Pantosti, D.; Lombardi, A.M.; Civico, R. Modeling of earthquake chronology from paleoseismic data: Insights for regional earthquake recurrence and earthquake storms in the Central Apennines. *Tectonophysics* **2021**, *816*, 229016. [CrossRef]
7. Gasperini, P.; Vannucci, G.; Tripone, D.; Boschi, E. The location and sizing of historical earthquakes using the attenuation of macroseismic intensity with distance. *Bull. Seism. Soc. Am.* **2010**, *100*, 2035–2066. [CrossRef]
8. Galli, P.A.C.; Naso, J.A. Unmasking the 1349 earthquake source (southern Italy): Paleoseismological and archaeoseismological indications from the Aquae Iuliae fault. *J. Struct. Geol.* **2009**, *31*, 128–149. [CrossRef]
9. Tertulliani, A.; Rossi, A.; Castelli, V.; Meletti, C.; D'Amico, V. Spunti e contrappunti di sismologia storica: 1349 annus horribilis, Abstract, 30° Convegno Nazionale Gruppo Nazionale di Geofisica della Terra Solida. *Sessione* **2011**, *1.1*, 14–17.
10. ISIDe Working Group. *Italian Seismological Instrumental and Parametric Database (ISIDe)*; Istituto Nazionale di Geofisica e Vulcanologia (INGV): Milano, Italy, 2007. [CrossRef]
11. Frepoli, A.; Cimini, G.B.; De Gori, P.; De Luca, G.; Marchetti, A.; Monna, S.; Montuori, C.; Pagliuca, N.M. Seismic sequences and swarms in the Latium-Abruzzo-Molise Apennines (central Italy): New observations and analysis from a dense monitoring of the recent activity. *Tectonophysics* **2017**, *712*, 312–329. [CrossRef]
12. ITHACA Working Group. *ITHACA (Italy Hazard from Capable Faulting), a Database of Active Capable Faults of the Italian Territory*; Via Vitaliano Brancati: Roma, Italy, 2019.
13. Boncio, P.; Lavecchia, G.; Pace, B. Defining a model of 3D seismogenic sources for Seismic Hazard Assessment applications: The case of central Apennines (Italy). *J. Seismol.* **2004**, *8*, 407–425. [CrossRef]
14. Dixit Dominus, G.; Maceroni, D.; Falcucci, E.; Galadini, F.; Gori, S.; Moro, M.; Saroli, M. Evidenze di Tettonica Attiva Lungo la Faglia Della val Roveto, Media Valle del Fiume Liri (Appennino Centrale), 39° GNGTS 2021. 2021. Available online: http://gngts.inogs.it/sites/default/files/Programma_GNGTS_2021.pdf (accessed on 22 December 2021).
15. Saroli, M.; Moro, M.; Cinti, F.R.; Montone, P. La faglia Val Roveto-Atina (Appennino Centrale): Evidenze di Attività Tettonica Quaternaria. Riassunti Estesi Delle Comunicazioni, 25° GNGTS 2006. 2006. Available online: https://www.google.com.hk/search?source=univ&tbm=isch&q=Evidenze+di+attivit%C3%A0+tettonica+Quaternaria.+Riassunti+e stesi+delle+comunicazioni.+2006,+25%C2%B0+GNGTS,+2006.&fir=v1gEIP8Vvfb4mM%252CzLugEHnwfetpsM%252C_%253 B53Hn7DwpbZJQJM%252Cfq6hDjtfwTEkzM%252C_%253B3JwXBawCu77dJM%252CQ3KOSPwqfndOZM%252C_%253BC3i YzBeFV2HSUM%252CQ3KOSPwqfndOZM%252C_&usg=AI4_-kSvspL0uqQzFeGaZVt8BweIGXt_bg&sa=X&ved=2ahUKEw jZqIjd6sT1AhXQZt4KHXCaBM4QjJkEegQIAhAC (accessed on 22 December 2021).
16. Guidoboni, E.; Ferrari, G.; Tarabusi, G.; Sgattoni, G.; Comastri, A.; Mariotti, D.; Ciuccarelli, C.; Bianchi, M.G.; Valensise, G. CFTI5Med, the new release of the catalogue of strong earthquakes in Italy and in the Mediterranean area. *Sci. Data* **2019**, *6*, 80. [CrossRef] [PubMed]
17. Marta, R. *La Cattedrale di Sora*; Inizio di un Restauro: Sora, Italy, 1982.
18. Squilla, G. La cattedrale di Sora dal 1100 al 1961 / d., prefazione di d. Tommaso Leccisotti—Casamari: Tipografia dell'Abbazia. 1961; 150p, Ill.; 22 cm.
19. Cicconio, E. Vera Relatione del Terremoto successo in Terra di Lavoro, con la desolazione di molte Terre, a 24 de Luglio 1654, Napoli. 1654; 8p.

20. Vivenzio, G. Istoria de' Tremuoti Avvenuti nella Provincia della Calabria Ulteriore, e Nella Città di Messina Nell'anno 1783 e di Quanto Nella Calabria fu Fatto per lo suo Risorgimento Fino al 1787 Preceduta da una Teoria, ed Istoria Generale de' Tremuoti, 2 voll. 1788. Available online: https://www.maremagnum.com/libri-antichi/istoria-de-tremuoti-avvenuti-nella-provincia-della-calabria/163406203 (accessed on 22 December 2021).

21. Biblioteca Apostolica Vaticana, Manoscritti, Barberiniani Latini, 4336, Relazione delli danni fatti dal terremoto seguito la notte delli 23 del cadente mese di giugno 1654. XVII sec.

22. Febonio, M. *Historiae Marsorum*; Apud Michaelem Monachum: Neapoli, Italy, 1678; Volume 145.

23. Archivio di Stato di Napoli, Regia Camera della Sommaria, Ruote, vol.52, cc.64-68, Consulta della Regia Camera della Sommaria Relativa alle Notizie Inviate dal Duca di Sora sui Danni Causati Nelle sue Terre dal Terremoto del 23 Luglio 1654, Napoli 28 Settembre 1654. Available online: http://www.cftilab.it/file_repository/pdf_T/003080-643001_T.pdf (accessed on 22 December 2021).

24. Wakita, H. Water wells as possible indicators of tectonic strain. *Science* **1975**, *189*, 553–555. [CrossRef] [PubMed]

25. Cucci, L. Insights into the geometry and faulting style of the causative faults of the M6.7 1805 and M6.7 1930 earthquakes in the Southern Apennines (Italy) from coseismic hydrological changes. *Tectonophysics* **2019**, *751*, 192–211. [CrossRef]

26. Okada, Y. Surface deformation due to shear and tensile faults in an half-space. *Bull. Seismol. Soc. Am.* **1985**, *75*, 1135–1154. [CrossRef]

27. Lin, J.; Stein, R.S. Stress triggering in thrust and subduction earthquakes, and stress interactions between the southern San Andreas and nearby thrust and strike-slip faults. *J. Geophys. Res.* **2004**, *109*, B02303. [CrossRef]

28. Toda, S.; Stein, R.S.; Richards-Dinger, K.; Bozkurt, S. Forecasting the evolution of seismicity in Southern California: Animation in building stress transfer. *J. Geophys. Res.* **2005**, *110*, B05S16. [CrossRef]

29. Carafa, M.M.C.; Galvani, A.; Di Naccio, D.; Kastelic, V.; Di Lorenzo, C.; Miccolis, S.; Sepe, V.; Pietrantonio, G.; Gizzi, C.; Massucci, A.; et al. Partitioning the ongoing extension of the central Apennines (Italy): Fault slip rates and bulk deformation rates from geodetic and stress data. *J. Geophys. Res. Solid Earth* **2020**, *125*, e2019JB018956. [CrossRef]

Article

Seismic Imaging of Complex Velocity Structures by 2D Pseudo-Viscoelastic Time-Domain Full-Waveform Inversion

Niloofar Alaei [1], Mehrdad Soleimani Monfared [1,2,*], Amin Roshandel Kahoo [1] and Thomas Bohlen [2]

[1] Department of Mining Petroleum and Geophysics Engineering, Shahrood University of Technology, Shahrood 3619995161, Iran; niloofar.alaei@shahroodut.ac.ir (N.A.); roshandel@shahroodut.ac.ir (A.R.K.)

[2] Geophysical Institute (GPI), Karlsruhe Institute of Technology (KIT), 76131 Karlsruhe, Germany; thomas.bohlen@kit.edu

* Correspondence: mehrdad.soleimani@partner.kit.edu or msoleimani@shahroodut.ac

Abstract: In the presented study, multi-parameter inversion in the presence of attenuation is used for the reconstruction of the P- and the S- wave velocities and the density models of a synthetic shallow subsurface structure that contains a dipping high-velocity layer near the surface with varying thicknesses. The problem of high-velocity layers also complicates selection of an appropriate initial velocity model. The forward problem is solved with the finite difference, and the inverse problem is solved with the preconditioned conjugate gradient. We used also the adjoint wavefield approach for computing the gradient of the misfit function without explicitly build the sensitivity matrix. The proposed method is capable of either minimizing the least-squares norm of the data misfit or use the Born approximation for estimating partial derivative wavefields. It depends on which characteristics of the recorded data—such as amplitude, phase, logarithm of the complex-valued data, envelope in the misfit, or the linearization procedure of the inverse problem—are used. It showed that by a pseudo-viscoelastic time-domain full-waveform inversion, structures below the high-velocity layer can be imaged. However, by inverting attenuation of P- and S- waves simultaneously with the velocities and mass density, better results would be obtained.

Keywords: complex velocity model; full waveform inversion; wave attenuation; preconditioned conjugate gradient; vibroseis sources

Citation: Alaei, N.; Soleimani Monfared, M.; Roshandel Kahoo, A.; Bohlen, T. Seismic Imaging of Complex Velocity Structures by 2D Pseudo-Viscoelastic Time-Domain Full-Waveform Inversion. *Appl. Sci.* **2022**, *12*, 7741. https://doi.org/10.3390/app12157741

Academic Editor: Jianbo Gao

Received: 27 June 2022
Accepted: 30 July 2022
Published: 1 August 2022

Publisher's Note: MDPI stays neutral with regard to jurisdictional claims in published maps and institutional affiliations.

1. Introduction

Imaging of geological complex structures in the subsurface can be used for geotechnical site characterization by geophysical methods. The term 'complex' is used for those subsurface earth models which cannot be easily imaged by conventional seismic imaging methods due to their complex velocity structures or geometry. Examples of complex structures can be steep dipping beds, intensive faulted or folded media, and earth models with strong velocity changes. In addition, the near-surface velocity anomalies can increase complexity of imaging problems, mostly due to the complexity in the simulation of the seismic wave propagation, or in other words, the complication caused by the propagation of the body waves through the complex near-surface layers [1]. The most realistic of such situations is the near-surface salt layers which can or cannot play a role as the caprock for petroleum reservoirs. In those cases, the fluids trapped in the layers beneath the salt have considerable effects on the elastic properties of the subsurface media. The better these properties are modeled, the more accurate an image of the subsurface will be obtained.

Conventional seismic imaging methods are no more reliable in solving imaging problems raised from complex geological media. High quality seismic imaging is needed in most exploration studies such as gas storage projects, geological hazard, CO_2 storage projects in target finding and monitoring, and also in geothermal resources. To obtain a high-quality seismic image, further investigation of obstacles to obtaining reasonable seismic images and developing reliable imaging methods are required. Considering the problems of seismic imaging in complex media, it was stated that poor seismic images from different regions mostly resulted from the application of inappropriate imaging algorithms [2]. The minor concerns were related to the data acquisition problems due to harsh topography, but the major issues are rooted in extreme complexity in subsurface media and poor quality in signal to noise ratio (SNR) [3]. The former could be resolved by adequate acquisition; however, the later requires deep investigation on developing proper imaging tools. In one study, it was proposed resolving obstacle partially by the common reflection surface (CRS) and the normal incidence point (NIP) tomography method [4]. However, the CRS still suffers in handling strong lateral velocity changes or geologically complicated media [5]. The reverse time migration (RTM) and the full waveform inversion (FWI) methods, as the latest introduced methods, deal with a vast majority of problems in seismic imaging [6,7]. However, these methods are still present issues in application to large field datasets, poor quality data with shortage in frequency content, and low SNR in the low frequency part of the data [8]. Challenges for FWI land applications consist of addressing the wavefield propagation from rough topography, low SNR of the low-frequency data, and determination of an appropriate source wavelet throughout the iterations by improving the velocities and model parameters [9].

The FWI employs an iterative procedure that is based on a forward modeling and inversion procedure to find the optimal parameters [10,11]. Some studies have been carried out to show the efficiency of FWI in the imaging of complex media [7], presented the application of the FWI method in the frequency domain on the wide-aperture onshore seismic data with a complex geological setting (thrust belt) [12], and applied the elastic frequency-domain FWI to the synthetic onshore Marmousi2 model [13]. They implemented a velocity-gradient starting model and a very low starting frequency to image the complex structure model. Reference [14] also tested the application of this strategy to the offshore versions of the synthetic Marmousi2 model. They successfully imaged the complex model using their strategy. Reference [8] presented a parallel 2D elastic frequency-domain FWI algorithm based on a discontinuous forward problem [15] that was applied to a realistic synthetic onshore case study. They obtained a high-resolution P- and S- wave velocity of the complex onshore structure using a joint inversion of the surface and body waves recorded by a wide aperture acquisition geometry. Reference [16] studied the application of the FWI method in the time-domain on the problem of subsalt imaging with the modified Flooding Technique and showed the difference between the results of elastic and acoustic FWI methods. These differences reveal that the result of the acoustic FWI algorithm on elastic data for the subsalt imaging problem is not reliable. The application of the multi-parameter viscoelastic FWI using a frequency-domain on synthetic data example was proved by [17]. The low-order finite element discontinuous Galerkin method was used to solve the forward problem which can be a good option when studying the complex topographies and high-velocity contrasts, and the quasi-Newton L-BFGS optimization was implemented to estimate the inverse of the Hessian matrix in order to decrease the computational cost and improve the reconstruction of the velocities, density, and attenuation parameters. Reference [9] implemented the FWI-SIMAT algorithm to investigate the capability of the acoustic FWI in the reconstruction of the Marmousi velocity model both in the time and frequency domain. Reference [18] used a developed FWI method in which a two-stage sequential approach (SFWI) was tested on the field datasets recorded in the Black Sea and in the shallow-water area of a river delta in the Atlantic Ocean to obtain detailed subsurface images containing rock formations that might be potential gas deposits. Most applications of the FWI methods on complex structures have been performed in the frequency domain or ignored seismic wave attenuation. Ignoring the viscous effect of the propagation media provides an unrealistic reconstructed S- wave velocity model, especially in the study of the complex geologic media [19]. Reference [20] showed that taking key elements properly into account, FWI produces a reliable high-resolution near-surface model that could not be otherwise recovered through traditional methods. Although few attempts have been reported that incorporate FWI for land studies [18]; however, they were convincing in providing acceptable seismic image. Therefore, it is supposed that deriving a processing workflow modified for accurate imaging of seismic data from complex regions would be promising in resolving the problem of low SNR and strong lateral velocity changes due to complexity in wave propagation media.

2. Problem Statement

It was shown that seismic imaging in seismic data with above mentioned properties is technically a challenging task due to several reasons. The first is complexity of the media. These complexities will introduce lateral velocity changes, make reduction in quality of data and reduce SNR of data. These problems prevent application of conventional imaging methods and require advanced methods, such as RTM and FWI, to be modified accordingly. The FWI method estimates subsurface properties affecting the seismic wave-field via minimizing the field data and synthetic seismogram generated from forward modeling. An ultimate FWI method should take attenuation and dispersions into account, which means considering the wave propagation medium as a viscoelastic medium. An appropriate choice of model parameterization is also very important in viscoelastic FWI. Various approaches are presented for FWI in viscous media in the frequency and time domain [21]. Shot parallelization, variable grids in the near future and better free surface implementation are also other compatibilities of an appropriate FWI method. Obviously, to make the inversion process converge to the correct and accurate response, the initial velocity model needs to be close enough to the real field velocity model. The focal issue here is to resolve the problem of imaging on data which contains a high-velocity layer and causes less energy of transmitted wavefield reach to the structures under this layer. Presence of steep dips, low SNR, and energy absorption by thick layers of evaporites—which dramatically reduce the quality of images in deeper parts—are obstacles in obtaining high quality images. Since the data suffer from reduction in quality due also to faults and variations in the thickness of the high-velocity layer, it is required that the FWI method modified accordingly in considering attenuation and wavelet estimations [22]. The lateral velocity changes due to the evaporites will reduce the sensitivity of the FWI method in reconstruction of the velocity models. Therefore, it is important to define appropriate initial velocity models. Furthermore, since the FWI package of the Karlsruhe institute of technology (KIT) could model the viscoelastic properties of the media in wave propagation simulation, it is assumed that the data quality will increase in regions with above mentioned problem [23]. The model parameterization and discretization of the media is also challenging in application of FWI method in such regions. Discretization should be flexible and appropriate for boundaries of abrupt changes in elastic properties of the media, which is the result of complex mud intrusions [24]. This complexity will also introduce problems in model parameterization, which needs to be optimized via parameter analysis. In this study, the performance of the 2D pseudo-viscoelastic FWI proposed by [20] to image a synthetic model with velocity complexity is investigated. A time-domain multi-parameter FWI is applied to reconstruct the P- wave velocity, S- wave velocity, and density models. The forward problem is solved using the finite difference method (FD) and the viscoelastic wave equation is discretized considering the convolutional perfectly matched layers (PMLs) absorbing boundary condition to prevent the edge effects. To solve the inverse problem, the preconditioned conjugate gradient (PCG) is used. The gradients are computed with the adjoint-state method. A simple model generated by the 1D linear gradient is considered as the initial model.

3. Theory

3.1. Forward problem

In this study, the stress–velocity equation of the wave equation in the time domain in an anisotropic viscoelastic medium with rheology described by a GSLS [25,26] is taken to solve the forward problem [27,28]:

$$\rho \frac{\partial v_i}{\partial t} = \frac{\partial \sigma_{ij}}{\partial x_j} + f_i \tag{1}$$

$$\dot{\sigma}_{ij} = \frac{\partial v_k}{\partial x_k}\left\{ M(1 + \tau^p) - 2\mu(1 + \tau^s) \right\} + 2\frac{\partial v_i}{\partial x_j}\mu(1 + \tau^s) + \sum_{l=1}^{L} r_{ijl} \quad if \quad i = j, \tag{2}$$

$$\dot{\sigma}_{ij} = \left(\frac{\partial v_i}{\partial x_j} + \frac{\partial v_j}{\partial x_i} \right)\mu(1 + \tau^s) + \sum_{l=1}^{L} r_{ijl} \quad if \quad i \neq j \tag{3}$$

$$\dot{r}_{ijl} = -\frac{1}{\tau_{\sigma l}}\left\{ (M\tau^p - 2\mu\tau^s)\frac{\partial v_k}{\partial x_k} + 2\frac{\partial v_i}{\partial x_j}\mu\tau^s + r_{ijl} \right\} \quad if \quad i = j, \tag{4}$$

$$\dot{r}_{ijl} = -\frac{1}{\tau_{\sigma l}}\left\{ (M\tau^s\left[\frac{\partial v_i}{\partial x_j} + \frac{\partial v_j}{\partial x_i}\right] + r_{ijl} \right\} \quad if \quad i \neq j. \tag{5}$$

where σ_{ij} denotes the i jth component of the stress tensor, v_i denotes the components of particle velocity, f_i is the components of external body force, ρ is density, M is the P- wave modulus, and μ is the S- wave modulus. r_{ijl} denotes the L memory variables $(l = 1, \ldots, L)$ which correspond to the stress tensor σ_{ij}, $\tau_{\sigma l}$, are the L stress relaxation times for P- and S- waves and τ^p, τ^s are the level of attenuation for P- and S- waves respectively. It is necessary to mention that the dot over symbols indicates partial differentiation with respect to time. The attenuation of rocks is defined by the seismic quality factor (Q):

$$Q(\omega, \tau_{\sigma l}, \tau) = \frac{1 + \sum_{l=1}^{L} \frac{\omega^2 \tau_{\sigma l}^2}{1 + \omega^2 \tau_{\sigma l}^2}\tau}{\sum_{l=1}^{L} \frac{\omega \tau_{\sigma l}}{1 + \omega^2 \tau_{\sigma l}^2}\tau} \tag{6}$$

where ω is the angular frequency, and the variable τ denotes

$$\tau = \frac{\tau_{\varepsilon l}}{\tau_{\sigma l}} \tag{7}$$

where $\tau_{\sigma l}$ is the stress relaxation time, and $\tau_{\varepsilon l}$ is the strain retardation time for the lth Maxwell body of the GSLS. With Equation (6), $L + 1$ parameters $\tau_{\sigma l}, \tau$ are obtained that describe a constant Q-spectrum within a limited frequency range by a limited number of Maxwell bodies [27]. The forward problem is solved by using a time-domain two-dimensional second order FD operator in time and space on a staggered grid [27]. To reduce the edge effects and reflections at the boundaries the CPMLs are implemented [29,30].

3.2. Inverse Problem

FWI is a non-linear optimization problem that needs an appropriate objective function to be minimized. The L2-norm of the data residuals as the objective function E is used in the presented study [28,31].

$$E = \sum_{s=1}^{n_s} \sum_{r=1}^{n_r} \sum_{j=1}^{n_c} \int_0^T \left(d_j\left(\vec{x}_s, \vec{x}_r, t \right), s_j\left(\vec{x}_s, \vec{x}_r, t, m \right) \right)^2 dt \tag{8}$$

where d_j denotes the observed data, and s_j is the synthetic data at receiver r at point $\vec{x}_r$. n_s and n_r are the number of sources and receivers respectively. n_c is the number of components and T is the recording time. The PCG method [32] is implemented to minimize the objective function by iteratively updating the model parameters m along the conjugate direction δc_n

$$\delta c_n = \delta m_n + \beta_n \delta c_{n-1} \tag{9}$$

At the first iteration step ($n = 1$), the model is updated along the steepest descent direction

$$m_2 = m_1 + \mu_1 \delta m_1 \tag{10}$$

The model is updated along the conjugate direction in all subsequent steps ($n > 1$)

$$m_{n+1} = m_n + \mu_n \delta c_n \tag{11}$$

where $\delta c_1 = \delta m_1$. μ_n denotes the step length that is estimated by a parabolic line search method [33–36]. The weighting factor beta is calculated using the Polak–Ribiere formulation:

$$\beta_n^{PR} = \frac{\delta m_n^T (\delta m_n - \delta m_{n-1})}{\delta m_{n-1}^T \delta m_{n-1}} \tag{12}$$

$\delta m_n = \frac{\partial E}{\partial m}$ denotes the gradients of material parameters that can be calculated using the adjoint state method [28,32,37,38]. The model parameters can be density ρ and unrelaxed P- and S-wave moduli π_u, μ_u for a viscoelastic medium assuming a constant a priori known quality factor Q. The gradients of the misfit function for the unrelaxed moduli of a grid cell at a point $\vec{x}''$ can be calculated by a zero-lag cross-correlation of the forward propagated s and the adjoint wavefield $s^\dagger$ are approximated [28].

$$\frac{\partial E}{\partial \pi_k} = -\int_0^T \left(\frac{\partial s_1^\dagger\left(\vec{x}'', T-t''\right)}{\partial x_1''} + \frac{\partial s_2^\dagger\left(\vec{x}'', T-t''\right)}{\partial x_2''} \right) \cdot \left(\frac{\partial s0_1\left(\vec{x}'', t\right)}{\partial x_1''} + \frac{\partial s0_2\left(\vec{x}'', t\right)}{\partial x_2''} \right) dt'' \Delta x''^3 \tag{13}$$

$$\frac{\partial E}{\partial \pi_{\mu_k}} = -\int_0^T \begin{aligned} &\left[-\left(\frac{\partial s_1^\dagger\left(\vec{x}'', T-t''\right)}{\partial x_2''} + \frac{\partial s_2^\dagger\left(\vec{x}'', T-t''\right)}{\partial x_1''} \right) \cdot \left(\frac{\partial s0_1\left(\vec{x}'', t''\right)}{\partial x_2''} + \frac{\partial s0_2\left(\vec{x}'', t''\right)}{\partial x_1''} \right) \right. \\ &\left. +2\left(\frac{\partial s_1^\dagger\left(\vec{x}'', T-t''\right)}{\partial x_1''} \frac{\partial s0_2\left(\vec{x}'', t''\right)}{\partial x_2''} + \frac{\partial s_2^\dagger\left(\vec{x}'', T-t''\right)}{\partial x_2''} \frac{\partial s0_1\left(\vec{x}'', t''\right)}{\partial x_1''} \right) \right] dt'' \Delta x''^3 \end{aligned} \tag{14}$$

$$\frac{\partial E}{\partial \rho_k} = \int_0^T \left(\frac{\partial s_1^\dagger\left(\vec{x}'', T-t''\right)}{\partial t''} \frac{\partial s0_1\left(\vec{x}'', t\right)}{\partial t''} + \frac{\partial s_2^\dagger\left(\vec{x}'', T-t''\right)}{\partial t''} \frac{\partial s0_2\left(\vec{x}'', t\right)}{\partial t''} \right) dt'' \Delta x''^3 \tag{15}$$

The parametrization considered in this study is (ρ, V_p, V_s). The gradients are calculated for these parameters using the chain rule. To change the parametrization from the parameters (ρ, π_u, μ_u) to (ρ, V_p, V_s) one can apply the chain rule according to the relations of unrelaxed moduli with the unrelaxed Lamé parameters ($\rho', \lambda_u', \mu_u'$) and seismic velocity parameters respectively (Equations (16) and (20)).

$$\rho = \rho', \pi_u = \lambda_u' + 2\mu_u', \text{ and } \mu_u = \mu_u' \tag{16}$$

The gradients for density and Lamé parameters can be expressed by

$$\frac{\partial E}{\partial \rho'} = \frac{\partial E}{\partial \rho} \frac{\partial \rho}{\partial \rho'} + \frac{\partial E}{\partial \pi_u} \frac{\partial \pi_u}{\partial \rho'} + \frac{\partial E}{\partial \mu_u} \frac{\partial \mu_u}{\partial \rho'} = \frac{\partial E}{\partial \rho} \tag{17}$$

$$\frac{\partial E}{\partial \lambda_u'} = \frac{\partial E}{\partial \pi_u} \frac{\partial \pi_u}{\partial \lambda_u'} + \frac{\partial E}{\partial \mu_u} \frac{\partial \mu_u}{\partial \lambda_u'} + \frac{\partial E}{\partial \rho} \frac{\partial \rho}{\partial \lambda_u'} = \frac{\partial E}{\partial \pi_u} \tag{18}$$

$$\frac{\partial E}{\partial \mu'_u} = \frac{\partial E}{\partial \pi_u}\frac{\partial \pi_u}{\partial \mu'_u} + \frac{\partial E}{\partial \mu_u}\frac{\partial \mu_u}{\partial \mu'_u} + \frac{\partial E}{\partial \rho}\frac{\partial \rho}{\partial \mu'_u} = 2\frac{\partial E}{\partial \pi_u} + \frac{\partial E}{\partial \mu_u}$$

$$= -\int_0^T [(\frac{\partial s_1^\dagger}{\partial x_2''} + \frac{\partial s_2^\dagger}{\partial x_1''}).(\frac{\partial s0_1}{\partial x_2''} + \frac{\partial s0_2}{\partial x_1''}) + 2(\frac{\partial s_1^\dagger}{\partial x_1''}\frac{\partial s0_1}{\partial x_1''}$$

$$+ \frac{\partial s_2^\dagger}{\partial x_2''}\frac{\partial s0_2}{\partial x_2''})]dt'' \Delta x''^3 \tag{19}$$

with the relations

$$\rho' = \rho \ \text{and} \ v'_p = \sqrt{\frac{\lambda + 2\mu}{\rho}} \rightarrow \lambda = \rho'(v'^2_p - 2v'^2_s) \tag{20}$$

$$v'_s = \sqrt{\frac{\mu}{\rho}} \rightarrow \mu = \rho'v'^2_s \tag{21}$$

one obtains

$$\frac{\partial E}{\partial v'_p} = 2\rho'v'_p\frac{\partial E}{\partial \lambda} \tag{22}$$

$$\frac{\partial E}{\partial v'_s} = -4\rho'v'_s\frac{\partial E}{\partial \lambda} + 2\rho'v'_s\frac{\partial E}{\partial \mu} \tag{23}$$

$$\frac{\partial E}{\partial \rho'} = \left(v'^2_p - 2v'^2_s\right)\frac{\partial E}{\partial \lambda} + v'^2_s\frac{\partial E}{\partial \mu} + \frac{\partial E}{\partial \rho} \tag{24}$$

It is worth noting that an approximated Hessian (after [39]) is applied as an appropriate preconditioning operator P to the gradient δm before updating the model parameters. The Hessian is calculated for each shot individually and will be applied to the gradient from each shot directly. A multi-scale inversion strategy is implemented to reduce the high nonlinearity at the beginning of the inversion and pass the cycle skipping problem [40].

4. Synthetic Data Example

In this section, a synthetic example is performed to investigate the capability of 2D pseudo-viscoelastic FWI in the time domain to image shallow complex structures using IFOS2D. The true model used to simulate the observed data for three parameters (P- wave and S- wave velocity model and density model) is generated inspired by a real model located in Iran, which contains large synclinal shape of evaporite layers with very high velocity, faulted in the left side and the thickness of the high-velocity layers varies through the section. The seismic velocity of this evaporite layers is between 3840 and 5420 m/s, according to the percent of the containing salt compare to anhydrite, depth and thickness of the layer, which is in the range investigated in different studies [41,42]. The surrounding carbonate and shale layers show velocities around 2800–3420 m/s. The main problem in the seismic data with the abovementioned problem is to image target layers below the high-velocity layer, which is supposed to be resolved by FWI method. Therefore, in our study, in the first step, we tried to build a synthetic model with same geometry and shape of the high-velocity layer. In the next step, we tried to select velocities for each layer according to the real velocity of the media. In this step, since the provided forward and reverse codes for FWI in this study are mainly used for near-surface data, rather than deep seismic; so to prevent instability in analysis, we scaled down all the velocities of the layers in the model with a constant value. Therefore, we modeled the high-velocity layer near the surface with velocity close to 600 m/s. It should not be considered as the real high-velocity layers in deep earth, but a downscaled version of that.

Due to high velocity, propagation of the surface and body waves through the complex near-surface layers would be more complicated. This example can test the capability of the FWI to image a complex velocity structure. The model space has a size of 400 grid points in the horizontal direction and 160 grid points in the vertical direction. Therefore, the actual dimension would be 50 × 20 m considering a grid spacing of 0.125 m. A total of 19 shots and a total of 73 receivers located at the constant depth of 0.2 m that record both horizontal and vertical components are used. A cubed sine wavelet with a center frequency of 31.25 Hz generated by a hammer source is used as the source signal. The CPML frame is marked by the black dashed line. A viscoelastic medium is considered in this example and approximated a constant quality factor of $Q_s = Q_p = 20$ in the analyzed frequency band up to 60 Hz (a high-cut frequency filter of 10, 20, 30, 40, 50, and 60 Hz is used in stages) with three relaxation mechanisms of a generalized standard linear solid. A minimum of five iterations are taken into account at each stage. A 1D linear gradient is used to build the background of the true model and the background is considered as the initial model for each parameter in inversion. All models are updated simultaneously during inversion. It is worth mentioning that the true and initial velocity models are built with a $\frac{v_p}{v_s}$ ratio of 1.5 and a total propagation time of 0.6 s is considered. Initially, we tried the V_p/V_s ratio of 1.5 because it is the minimum ratio which can be used as a reasonable value for sediments or soft rocks near the surface. In the following, we have selected the V_p/V_s ratio of 2.5 which is more realistic for our example. The PCG is carried out to solve the inverse problem.

5. Results

In this study, 314 iteration steps are calculated, and the inversion takes about 10 h when using a system with four cores with 3.1 GHz speed and 16 Gb of ram. The true, initial, and inverted P- wave velocity models are shown in Figure 1. The same order is given, for the S- wave velocity and density models in Figures 2 and 3, respectively. Figure 4 shows the vertical profiles through the P- wave and S- wave velocity and density models that are considered to compare the results with the true model in more detail. Vertical profiles through the models are obtained at $x = 25$ m. The reconstructed models are in good accordance with the true models, especially for the S- wave velocity model. In the inverted S- wave velocity model, the upper edge of the high-velocity layer can be seen more sharply compared to the two other models. The bottom edge of the high-velocity layer is reconstructed in each model but not in the accurate location. Some artifacts are seen in the low-velocity zone of the density and P- wave velocity models. Regarding the low sensitivity of surface waves with respect to the P- wave velocity and density model [41], inaccurate results of inversion for the P- wave velocity and density models can be expected, also because the amplitude of surface waves is much higher than the amplitude of P- waves. The sensitivity of surface waves with respect to the P- waves is low and it leads to an inaccurate P- wave velocity model at each iteration step.

Because the density model is in relation to the P- wave velocity model using an empirical relation. Therefore, it affects the density model and the result of these two models is not as accurate as of the inverted S- wave velocity model [42].

To assess the results precisely, the final synthetic shot gathers are compared with the observed data. The vertical velocity seismogram of the shot at $x = 9$ m is obtained and the seismograms for the initial models are calculated for trace 36 of the shot and compared with the seismogram of the observed and inverted model. The comparison of the synthetic and observed seismograms of the shot at $x = 9$ m is shown in Figure 5a and the comparison of the initial, observed, and inverted data for trace 36 at this shot is shown in Figure 5b. Each seismogram is normalized to its maximum amplitude. The comparison of the initial and inverted data indicates the good performance of the inversion method and application of the software IFOS2D (Inversion of Full Observed Seismograms (2D)) in reconstructing model parameters. The calculated data agreeably fit the observed data. Therefore, the inversion result is a model which better explains the observed data. In the following, the true and initial models are built considering the $\frac{v_p}{v_s}$ ratio of 2.5 that is more realistic in the case of studying the soft rocks near the surface. In this case, due to the increase in the velocity values, the wavelengths propagated through the medium are increased and the resolution is influenced by the wavelength. The high-velocity layer is not resolved with the P- wave velocity model. Therefore, to reconstruct the model, a broad bandwidth of the source signal is needed. A broad bandwidth signal cannot be generated by a hammer source, thus a vibroseis source can be used to generate a signal which has a higher center frequency and covers a broader frequency range than a cubed sine wavelet [43,44]. Since a

Ricker wavelet is similar to a Klauder wavelet generated by vibroseis source and is used in the synthetic seismic modeling, in the following a Ricker wavelet with a center frequency of 50 Hz is considered as the source signal.

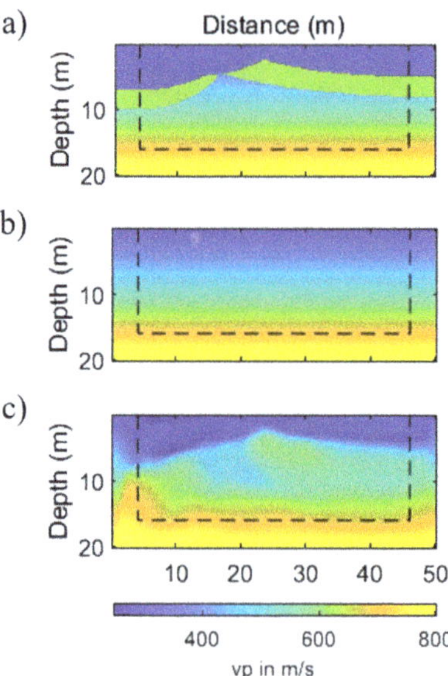

Figure 1. Multi-parameter synthetic example when using a low-frequency source signal: (**a**) the true P- wave velocity model for the calculation of the observed data, (**b**) the initial P- wave velocity model, and (**c**) the inverted P- wave velocity model. The CPML frame is marked by a thin black line.

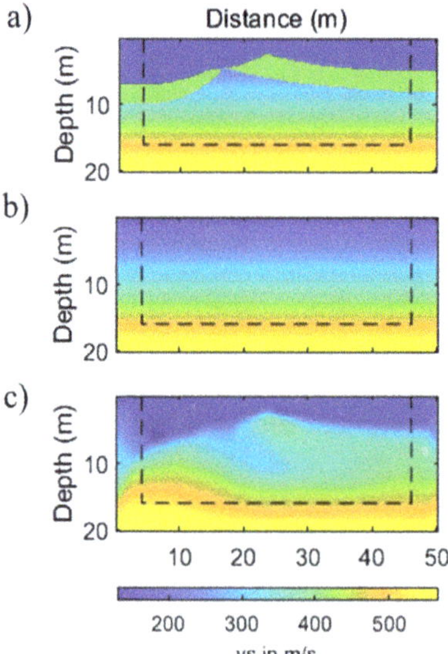

Figure 2. Multi-parameter synthetic example when using a low-frequency source signal: (**a**) the true S- wave velocity model for the calculation of the observed data, (**b**) the initial S- wave velocity model, and (**c**) the inverted S- wave velocity model. The CPML frame is marked by a thin black line.

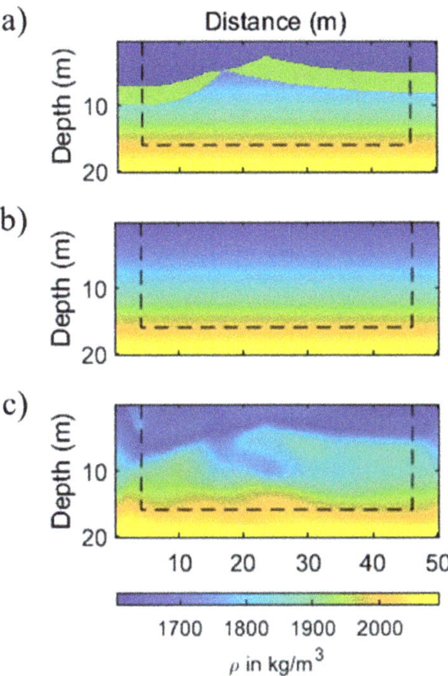

Figure 3. Multi-parameter synthetic example when using a low-frequency source signal: (**a**) the true density model for the calculation of the observed data, (**b**) the initial density model, and (**c**) the inverted density model. The CPML frame is marked by a thin black line.

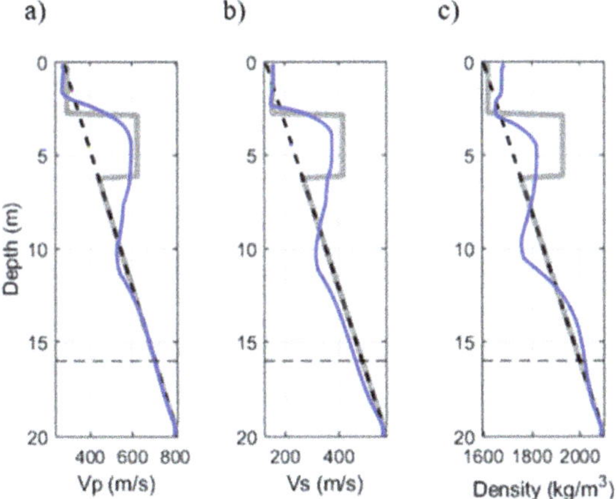

Figure 4. Model fitting when using a low-frequency source signal: (**a**) vertical profiles of the P- wave velocity model, (**b**) vertical profiles of the S- wave velocity model, and (**c**) vertical profiles of the density model. The true model is plotted with the grey line, the initial model is represented by the dashed black line and vertical profile at $x = 25$ m of the inverted models is the plotted blue line.

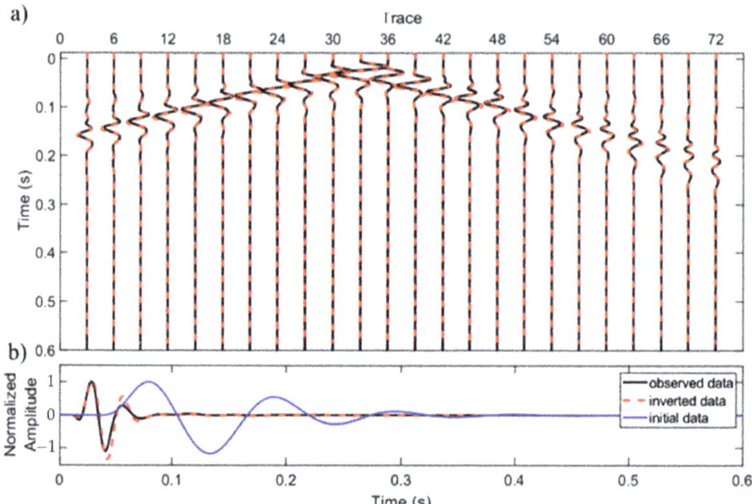

Figure 5. Fitting of the data in the shot at $x = 9$ m when using a low-frequency source signal. (**a**) Comparison of the vertical velocity observed and inverted seismograms. (**b**) Comparison of the normalized seismograms calculated for the initial, inverted, and observed data for trace 36.

Therefore, it can be said that in this study by considering a broad bandwidth signal as the source wavelet, the capability of the multi-parameter pseudo-viscoelastic FWI of the shallow-seismic wavefield is tested in the case of using a vibroseis source, too. A reflector is then added to the bottom of the true models (Figures 6a, 7a and 8a) at the depth of 15 m. A 1D model is also used for the initial and background of the true models (Figures 6b, 7b and 8b). Multi-parameter inversion is conducted for the parameters discretized at a 2D cartesian grid with the same grid spacing and the total propagation time as were used in the previous example. The high cut frequency filter, up to 100 Hz is applied progressively in the multi-scale strategy. In order to reduce the computational time, the number of receivers was reduced to 66 and the total of shots used in this test is 17. This test takes about 11 hours, and 317 iteration steps are calculated by using the same system as used in the previous test. In this example, the inverted S- wave velocity model (Figure 7c), is still better reconstructed than the P- wave velocity (Figure 6c) and the density (8c) models. The high-velocity layer is reconstructed sharper and more accurate compared to the inverted S- wave velocity model in the previous test. As can be seen in the vertical profile obtained for this model in Figure 9b, the velocity value of the high-velocity layer matches the value of the true high-velocity layer robustly. The velocity value of the low-velocity zone is obtained precisely too. There is an improvement in the results of inversion of the P- wave velocity and density models. As the artefacts in the low-velocity zone are decreased. In the presence of the reflector at the bottom of the model, the structure beneath the high-velocity layer is resolved with higher quality and resolution. In other words, the artefacts at the dipper parts of the models are significantly decreased too. Similar to the previous example, the final synthetic data nicely fits the observed data (Figure 10a). According to the zoomed comparison of the initial and inverted data for trace 36 (Figure 10b), misfit of the inverted and observed data is low.

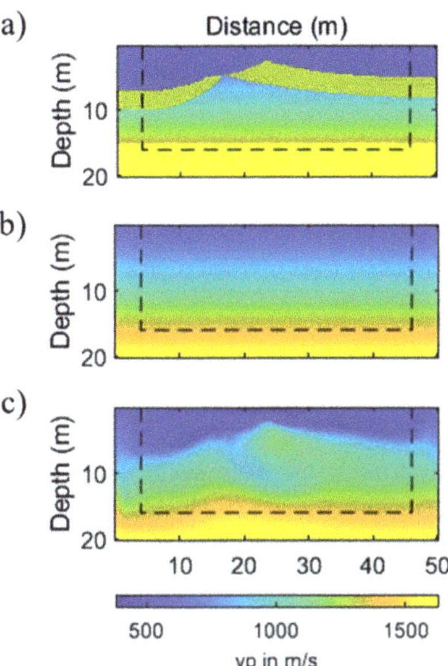

Figure 6. Multi-parameter synthetic example when using a high-frequency source signal: (**a**) the true P- wave velocity model for the calculation of the observed data, (**b**) the initial P- wave velocity model, and (**c**) the inverted P- wave velocity model. The CPML frame is marked by a thin black line.

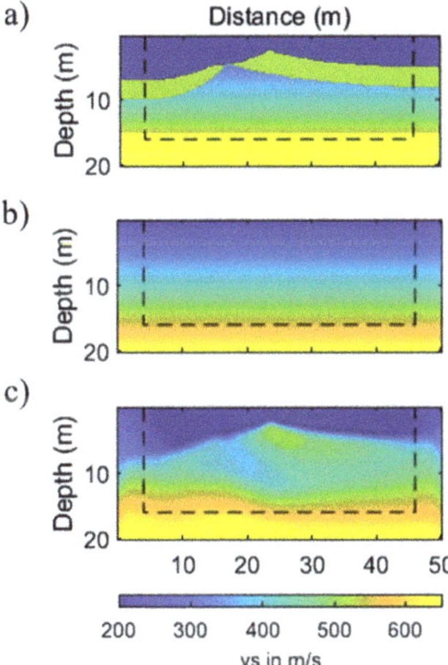

Figure 7. Multi-parameter synthetic example when using a high-frequency source signal: (**a**) the true S- wave velocity model for the calculation of the observed data, (**b**) the initial S- wave velocity model, and (**c**) the inverted S- wave velocity model. The CPML frame is marked by a thin black line.

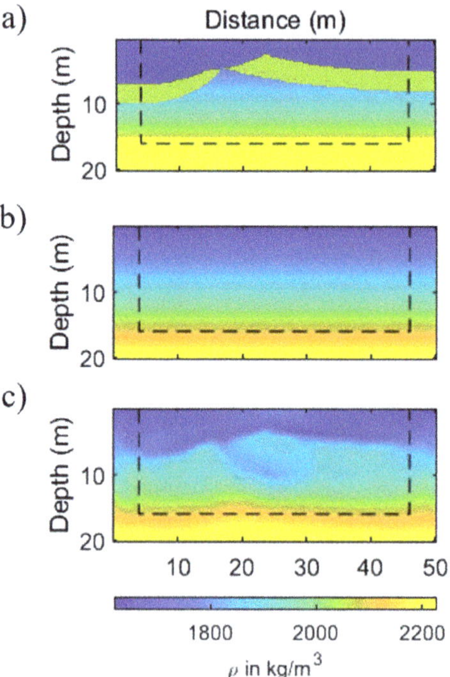

Figure 8. Multi-parameter synthetic example when using a high-frequency source signal: (**a**) the true density model for the calculation of the observed data, (**b**) the initial density model, and (**c**) the inverted density model. The CPML frame is marked by a thin black line.

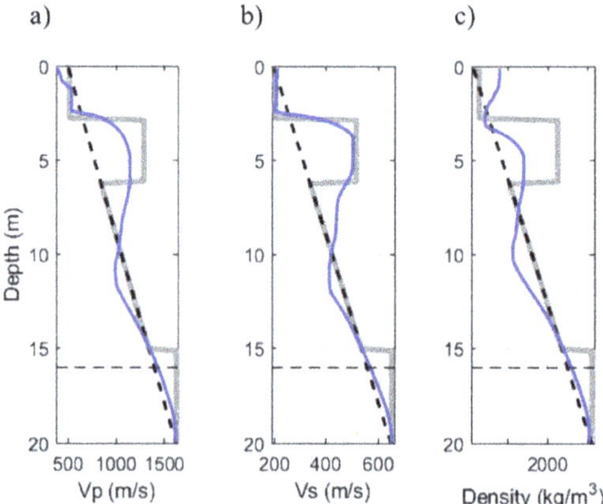

Figure 9. Model fitting when using a high-frequency source signal: (**a**) vertical profiles of the P- wave velocity model, (**b**) vertical profiles of the S- wave velocity model, (**c**) vertical profiles of the density model. The true model is plotted with the grey line, the initial model is represented by the dashed black line and vertical profile at $x = 25$ m of the inverted models is the plotted blue line.

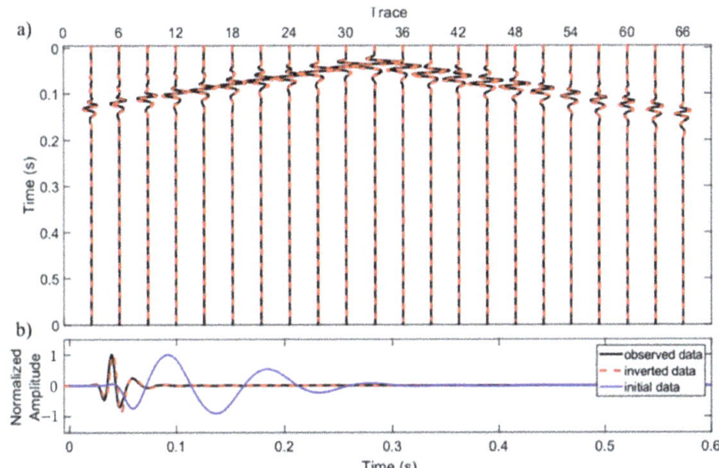

Figure 10. Fitting of the data in the shot at $x = 9$ m when using a high-frequency source signal. (**a**) Comparison of the vertical velocity observed and inverted seismograms. (**b**) Comparison of the normalized seismograms calculated for the initial, inverted, and observed data for trace 36.

6. Conclusions

In this study, a 2D multi-parameter pseudo-viscoelastic time domain is applied to a synthetic shallow complex velocity model where a dipping high-velocity layer near the surface with varying thicknesses is used as the case study and both surface and body waves are present. Investigation of these problems requires consideration of various aspects in the presented FWI methods. Some of these aspects that need to be considered in this workflow could be noise contamination, initial velocity model building, elastic and viscoelastic effects, Q factor estimation, and handling long offsets. The other concern about the presented FWI is the convergence speed and computational time of both the forward and inverse steps. Obviously, the size of the velocity model and observed data for near-surface application is not comparable with deep reflection data. The forward modeling step for generating synthetically predicted data from the initial model, back propagation, and computing the gradient, are time consuming steps in the proposed strategy. Thus, to speed up the processing time and increase the converge speed, the nonlinear conjugate gradient method was used. Defining the order of the finite difference operator, discretization, built-in wavelet, Q factor approximation, optimization method, and boundary condition definitions also need to be considered. The first synthetic example shows that when the velocity values in the model are not high, P- wave and S- wave velocity, and density models can be reconstructed well using a low frequency source signal. When the velocity values in the model are higher, the high-velocity layer cannot be resolved with the P- wave velocity model because of the large p-wavelength propagated through the medium. Therefore, the use of a wavelet with a broader bandwidth and higher center frequency can be the solution. In the second experiment, a Ricker wavelet is used to fulfill this issue. Both experiments provide satisfactory and reasonable results as the high-velocity layer near the surface is fairly reconstructed and the structures below this layer are also partially imaged. Reconstruction of the S- wave velocity model is more reliable and accurate compared to p-wave velocity and density models, due to less sensitivity of the surface waves with respect to the P- wave velocity and density parameters. This issue needs to be studied and improved in the future. However, it should be noted that a better image of subsurface structures would be obtained if attenuation of P- and S- waves are inverted simultaneously with the model parameters.

Author Contributions: Conceptualization, N.A., M.S.M. and T.B.; Methodology, T.B.; Software, N.A., M.S.M. and T.B.; Validation, N.A.; Formal analysis, N.A., M.S.M. and T.B.; Data curation, T.B.; Writing—original draft preparation, N.A.; Writing—review and editing, M.S.M.; Supervision, M.S.M., A.R.K. and T.B. All authors have read and agreed to the published version of the manuscript.

Funding: This research received no external funding.

Institutional Review Board Statement: Not applicable.

Informed Consent Statement: Not applicable.

Data Availability Statement: Not applicable.

Conflicts of Interest: The authors declare no conflict of interest.

References

1. Al-Ali, M.N.; Verschuur, D.J. An integrated method for resolving the seismic complex near-surface problem. *Geophys. Prospect.* **2006**, *54*, 739–750. [CrossRef]
2. Soleimani, M. Challenges of seismic imaging in complex media around Iran, from Zagros overthrust in the southwest to Gorgan Plain in the northeast. *Lead. Edge* **2017**, *36*, 499–506. [CrossRef]
3. Wittkamp, F.; Athanasopoulos, N.; Bohlen, T. Individual and joint 2-D elastic full-waveform inversion of Rayleigh and Love waves. *Geophys. J. Int.* **2019**, *216*, 350–364. [CrossRef]
4. Soleimani, M. Seismic image enhancement of mud volcano bearing complex structure by the CDS method, a case study in SE of the Caspian Sea shoreline. *Russ. Geol. Geophys.* **2016**, *57*, 1757–1768. [CrossRef]
5. Soleimani, M. Naturally fractured hydrocarbon reservoir simulation by elastic fractures modeling. *Petrol. Sci.* **2017**, *14*, 286–301. [CrossRef]
6. Pan, Y.; Gao, L.; Bohlen, T. Time-domain full-waveform inversion of Rayleigh and Love waves in presence of free-surface topography. *J. Appl. Geophys.* **2018**, *152*, 77–85. [CrossRef]
7. Ravaut, C.; Operto, S.; Improta, L.; Virieux, J.; Herrero, A.; Dell'Aversana, P. Multiscale imaging of complex structures from multifold wide-aperture seismic data by frequency-domain full-waveform tomography: Application to a thrust belt. *Geophys. J. Int.* **2004**, *159*, 1032–1056. [CrossRef]
8. Brossier, R. Seismic imaging of complex structures by 2D elastic frequency-domain full-waveform inversion. *Geophysics* **2009**, *74*, WCC63–WCC76. [CrossRef]
9. Singh, S.; Kanli, A.I.; Mukhopadhyay, S. Full waveform inversion in time and frequency domain of velocity modeling in seismic imaging: FWISIMAT a Matlab code. *Earth Sci. Res. J.* **2018**, *22*, 291–300. [CrossRef]
10. Tarantola, A. A strategy for nonlinear inversion of seismic reflection data. *Geophysics* **1986**, *51*, 1893–1903. [CrossRef]
11. Soleimani, M. Seismic imaging by 3D partial CDS method in complex media. *J. Pet. Sci. Eng.* **2016**, *143*, 54–64. [CrossRef]
12. Choi, Y.; Shin, C. Frequency-domain elastic full waveform inversion using the new pseudo-Hessian matrix: Experience of elastic Marmousi 2 synthetic data. *Bull. Seismol. Soc. Am.* **2008**, *98*, 2402–2415. [CrossRef]
13. Martin, G.S.; Wiley, R.; Marfurt, K.J. Marmousi2: An elastic upgrade for Marmousi. *Lead. Edge* **2006**, *25*, 156–166. [CrossRef]
14. Choi, Y.; Min, D.; Shin, C. Two-dimensional waveform inversion of multi-component data in acoustic elastic coupled media. *Geophys. Prospect.* **2008**, *56*, 863–881. [CrossRef]
15. Brossier, R.; Virieux, J.; Operto, S. Parsimonious finite-volume frequency-domain method for 2-D P-SV-wave modelling. *Geophys. J. Int.* **2008**, *175*, 541–559. [CrossRef]
16. Thiel, N.; Bohlen, T. *2d Acoustic Full Waveform Inversion of Submarine Salt Layer Using Dual Sensor Streamer Data*; Annual WIT Report; Annual WIT: Hamburg, Germany, 2016.
17. Brossier, R. Two-dimensional frequency-domain visco-elastic full waveform inversion: Parallel algorithms, optimization and performance. *Comput. Geosci.* **2011**, *37*, 444–455. [CrossRef]
18. Kurzmann, A.; Gaßner, L.; Shigapov, R.; Thiel, N.; Athanasopoulos, N.; Bohlen, T.; Steinweg, T. Real data applications of seismic full waveform inversion. In *High Performance Computing in Science and Engineering'17*; Springer: Cham, Switzerland, 2018; pp. 467–484.
19. Gao, L.; Pan, Y.; Bohlen, T. 2-D multiparameter viscoelastic shallow-seismic full-waveform inversion: Reconstruction tests and first field-data application. *Geophys. J. Int.* **2020**, *222*, 560–571. [CrossRef]
20. Groos, L.; Schäfer, M.; Forbriger, T.; Bohlen, T. The role of attenuation in 2D full-waveform inversion of shallow-seismic body and Rayleigh waves. *Geophysics* **2014**, *79*, R247–R261. [CrossRef]
21. Groos, L.; Schäfer, M.; Forbriger, T.; Bohlen, T. Application of a complete workflow for 2D elastic full-waveform inversion to recorded shallow-seismic Rayleigh waves. *Geophysics* **2017**, *82*, R109–R117. [CrossRef]
22. Bohlen, T.; Wittkamp, F. Three-dimensional viscoelastic time-domain finite-difference seismic modelling using the staggered Adams–Bashforth time integrator. *Geophys. J. Int.* **2016**, *204*, 1781–1788. [CrossRef]
23. Butzer, S.; Kurzmann, A.; Bohlen, T. 3D elastic full-waveform inversion of small scale heterogeneities in transmission geometry. *Geophys. Prospect.* **2013**, *61*, 1238–1251. [CrossRef]

24. Jetschny, S.; Bohlen, T.; Kurzmann, A. Seismic prediction of geological structures ahead of the tunnel using tunnel surface waves. *Geophys. Prospect.* **2011**, *59*, 934–946. [CrossRef]
25. Robertsson, J.O.; Levander, A.; Symes, W.W.; Holliger, K. A Comparative Study of Free-Surface Boundary Conditions for Finite-Difference Simulation of Elastic/Viscoelastic Wave Propagation. In *SEG Technical Program Expanded Abstracts*; SEG library: Houston, TX, USA, 1995; pp. 1277–1280.
26. Blanch, J.O.; Robertsson, J.O.A.; Symes, W.W. Modeling of a constant Q: Methodology and algorithm for an efficient and optimally inexpensive viscoelastic technique. *Geophysics* **1995**, *60*, 176–184. [CrossRef]
27. Bohlen, T. Parallel 3-D viscoelastic finite difference seismic modeling. *Comput. Geosci.* **2002**, *28*, 887–899. [CrossRef]
28. Groos, L. 2D Full Waveform Inversion of Shallow Seismic Rayleigh Waves. Ph.D Thesis, Karlsruher Institut für Technologie (KIT), Karlsruhe, Germany, 2013.
29. Komatitsch, D.; Martin, R. An unsplit convolutional perfectly matched layer improved at grazing incidence for the seismic wave equation. *Geophysics* **2007**, *72*, SM155–SM167. [CrossRef]
30. Martin, R.; Komatitsch, D. An unsplit convolutional perfectly matched layer technique improved at grazing incidence for the viscoelastic wave equation. *Geophys. J. Int.* **2009**, *179*, 333–344. [CrossRef]
31. Choi, Y.; Alkhalifah, T. Application of multi-source waveform inversion to marine streamer data using the global correlation norm. *Geophys. Prospect.* **2012**, *60*, 748–758. [CrossRef]
32. Köhn, D. Time Domain 2D Elastic Full Waveform Tomography. Ph.D. Thesis, Christian-Albrechts-Universität zu Kiel, Kiel, Germany, 2011.
33. Sourbier, F.; Operto, S.; Virieux, J.; Amestoy, P.; L'Excellent, J.Y. FWT2D: A massively parallel program for frequency-domain full-waveform tomography of wide-aperture seismic data—Part 1: Algorithm. *Comput. Geosci.* **2009**, *35*, 487–495. [CrossRef]
34. Sourbier, F.; Operto, S.; Virieux, J.; Amestoy, P.; L'Excellent, J.Y. FWT2D: A massively parallel program for frequency-domain full-waveform tomography of wide-aperture seismic data—Part 2: Numerical examples and scalability analysis. *Comput. Geosci.* **2009**, *35*, 496–514. [CrossRef]
35. Brossier, R. Imagerie Sismique à deux Dimensions des Milieux Visco-Élastiques par Inversion des Formes D'ondes: Développements Méthodologiques et Applications. Ph.D. Thesis, Université Nice Sophia Antipolis, Nice, France, 2009.
36. Soleimani, M.; Rafie, M. Imaging of seismic data in complex structures by introducing the partial diffraction surface stack method. *Studia Geophys. Geod.* **2016**, *60*, 644–661. [CrossRef]
37. Mora, P. Nonlinear two-dimensional elastic inversion of multioffset seismic data. *Geophysics* **1987**, *52*, 1211–1228. [CrossRef]
38. Tarantola, A. Theoretical background for the inversion of seismic waveforms, including elasticity and attenuation. *Pure Appl. Geophys.* **1988**, *128*, 365–399. [CrossRef]
39. Plessix, R.E.; Mulder, W.A. Frequency-domain finite-difference amplitude-preserving migration. *Geophys. J. Int.* **2004**, *157*, 975–987. [CrossRef]
40. Bunks, C.; Saleck, F.M.; Zaleski, S.; Chavent, G. Multiscale seismic wave-form inversion. *Geophysics* **1995**, *60*, 1457–1473. [CrossRef]
41. Kamberis, E.; Kokinou, E.; Koci, F.; Lioni, K.; Alves, T.M.; Velaj, T. Triassic evaporites and the structural architecture of the External Hellenides and Albanides (SE Europe): Controls on the petroleum and geoenergy systems of Greece and Albania. *Int. J. Earth Sci.* **2022**, *111*, 789–821. [CrossRef]
42. Roure, F.; Andriessen, P.; Callot, J.P.; Faure, J.L.; Ferket, H.; Gonzales, E.; Guilhaumou, N.; Lacombe, O.; Malandain, J.; Sassi, W.; et al. *The Use of Palaeo-Thermo-Barometers and Coupled Thermal, Fluid Flow and Pore-Fluid Pressure Modelling for Hydrocarbon and Reservoir Prediction in Fold and Thrust Belts*; Book chapter; Geological Society Publications: London, UK, 2010; Volume 348.
43. Socco, L.V.; Foti, S.; Boiero, D. Surface-wave analysis for building near-surface velocity models -Established approaches and new perspectives. *Geophysics* **2010**, *75*, 75A83–75A102. [CrossRef]
44. Athanasopoulos, N.; Manukyan, E.; Bohlen, T.; Maurer, H. Accurate reconstruction of shallow P-wave velocity model with time-windowed elastic full-waveform inversion. In Proceedings of the 80th EAGE Conference and Exhibition 2018, Copenhagen, Denmark, 11–14 June 2018; pp. 1–5.

Article

Stochastic Modeling of the Al Hoceima (Morocco) Aftershock Sequences of 1994, 2004 and 2016

Mohamed Hamdache [1], José A. Peláez [2,*], Dragomir Gospodinov [3], Jesús Henares [4], Jesús Galindo-Zaldívar [5,6], Carlos Sanz de Galdeano [5] and Boyko Ranguelov [7]

[1] Department of Seismological Survey, CRAAG (Center of Research in Astronomy, Astrophysics, and Geophysics), Bouzareah 16340, Algeria
[2] Department of Physics, University of Jaén, 23071 Jaén, Spain
[3] Faculty of Physics and Technology, Plovdiv University, 4002 Plovdiv, Bulgaria
[4] International University of La Rioja, 26006 Logroño, Spain
[5] Instituto Andaluz de Ciencias de la Tierra (CSIC-University of Granada), 18011 Granada, Spain
[6] Department of Geodynamics, University of Granada, 18011 Granada, Spain
[7] University of Mining and Geology "St. Ivan Rilski", 1700 Sofia, Bulgaria
* Correspondence: japelaez@ujaen.es

Abstract: The three aftershock sequences that occurred in Al Hoceima, Morocco, in May 1994 (Mw 6.0), February 2004 (Mw 6.4) and January 2016 (Mw 6.3) were stochastically modeled to investigate their temporal and energetic behavior. A form of the restricted trigger model known as the restricted epidemic type aftershock sequence (RETAS) was used for the temporal analysis of the selected series. The best-determined fit models for each sequence differ based on the Akaike information criteria. The revealed discrepancies suggest that, although the activated fault systems are close (within 10 to 20 km), their stress regimes change and shift across each series. In addition, a stochastic model was presented to study the strain release following a specific strong earthquake. This model was constructed using a compound Poisson process and depicted the progression of the strain release during the aftershock sequence. The proposed model was then applied to the data. After the RETAS model was used to evaluate the behavior of the aftershock decay rate, the best-fit model was obtained and integrated into the strain-release stochastic analysis. By detecting the potential disparities between the observed data and model, the applied stochastic model of strain release allows for a more comprehensive examination. Furthermore, comparing the observed and expected cumulative energy release numbers revealed some variations at the start of all three sequences. This demonstrates that significant aftershock clusters occur more frequently shortly after the mainshock at the start of the sequence rather than if they are assumed to occur randomly.

Keywords: point process modeling; RETAS model; aftershock energy release; Al Hoceima; Morocco

Citation: Hamdache, M.; Peláez, J.A.; Gospodinov, D.; Henares, J.; Galindo-Zaldívar, J.; Sanz de Galdeano, C.; Ranguelov, B. Stochastic Modeling of the Al Hoceima (Morocco) Aftershock Sequences of 1994, 2004 and 2016. *Appl. Sci.* **2022**, *12*, 8744. https://doi.org/10.3390/app12178744

Academic Editor: Jianbo Gao

Received: 13 July 2022
Accepted: 29 August 2022
Published: 31 August 2022

Publisher's Note: MDPI stays neutral with regard to jurisdictional claims in published maps and institutional affiliations.

1. Introduction

Seismic events can be classified into three main types based on their distribution over time [1]: (1) mainshock followed by a number of aftershocks decreasing in frequency, (2) slow build-up of seismicity leading to a type (1) sequence and (3) gradual increase and decay of seismicity without a distinct mainshock (seismic swarm), which occurs in areas with complex tectonic structures.

The decrease in aftershock occurrences caused by a strong earthquake can be studied using a wide range of methods, according to [2]. The Omori law model is the most typically adopted model [3], which [4] adapted into the modified Omori formula (MOF) by assuming that the fluctuation of the stress field of the mainshock initiates all the events in the sequence.

The trigger events are conditionally independent and follow a non-stationary stochastic Poisson process. Considering the complex behavior of some earthquake series, particularly in the presence of secondary events, [5] introduced the epidemic type aftershock sequence (ETAS) model, by increasing the capacity to generate secondary events for each event in the sequence. There are several triggering models between these two limit situations: the MOF and ETAS models being two of them [6,7]. The RETAS model [8] was developed by applying the principle of Bath's law [9,10] to the subsequences caused by principal events, such as the mainshock. It is worth mentioning that the magnitude difference between the mainshock and the strongest aftershock is commonly considered to be constant, ranging between 1.2 and 1.4 on average, depending on the criterion [11], although with a lot of variability between individual aftershock sequences [12].

In this study, we focus on three sequences of type (1) designated as the aftershock sequences of Al Hoceima 1994, 2004 and 2016, occurring near the city of Al Hoceima, in Morocco (Figures 1 and 2). The multifractal properties of these sequences have already been investigated in the framework of the spatial modeling of many seismic series in the Ibero–Maghrebian region [13]; however, even more temporal, energy and stress evaluations are required. Therefore, this study aims to examine three aftershock series using stochastic modeling.

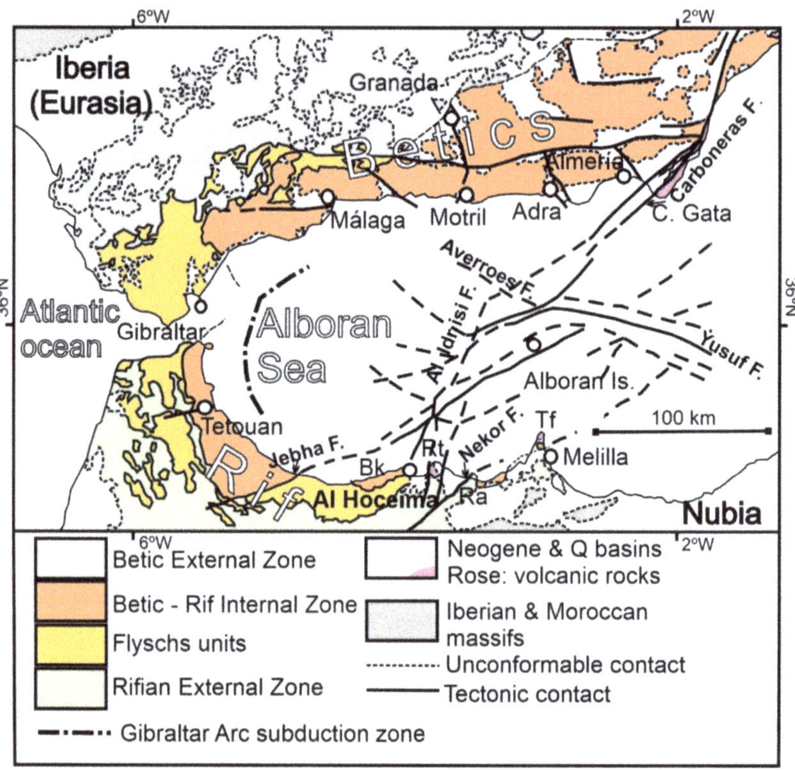

Figure 1. Geologic sketch map of the Betic-Rif region and location of the Al Hoceima study area. Bk: Bokoya Massif; Ra: Ras Afrou; Rt: Tas Tarf; Tf: Tres Forcas Cape. Internal Zone includes Sebtides and Ghomarides in the Rif and Nevado-Filabrides, Alpujarrides and Malaguides in the Betics.

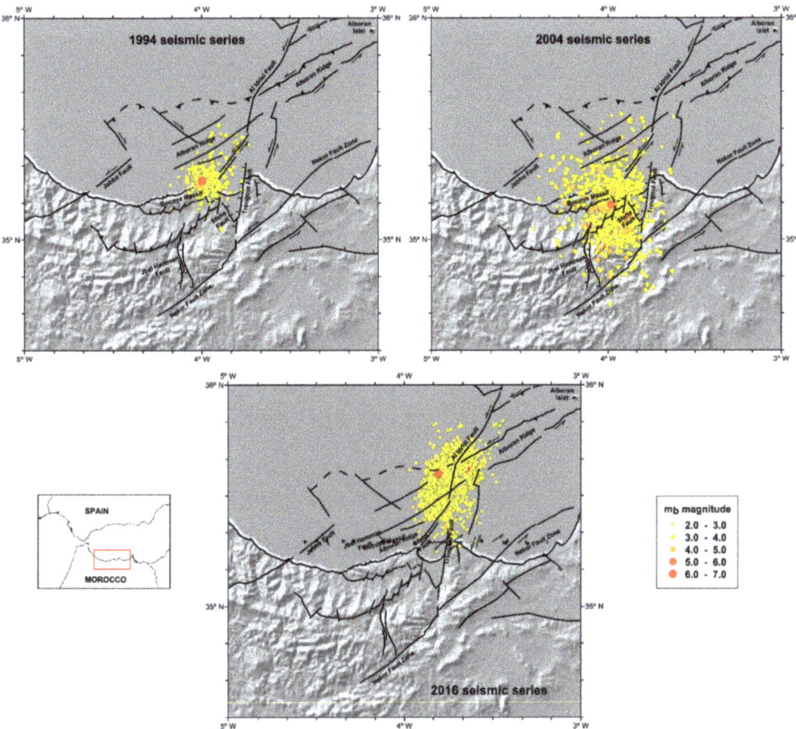

Figure 2. Seismicity recorded by the Spanish IGN included in the 1994, 2004 and 2016 seismic series from magnitude 2.0. Main tectonic features are displayed.

This contribution includes the first section, which describes the regional geological context. In the second section, as suggested by [14], the three aftershock sequences are analyzed, and the stress regime in each series is comprehensively described. The Gutenberg–Richter relationship analysis, performed in a later section, attempts to derive reliable threshold magnitude values and *b*-value estimates for each sequence. A stochastic point process modeling analysis was performed in the previous two sections. As described previously, the aftershock decay rate was comprehensively studied using the RETAS model. The identified best-fit model was then integrated into a stochastic analysis of strain release. A comparison between the real values of the cumulative energy release and the expected modeled values is also examined and addressed.

2. Geological Setting Overview

The Rif, along with the Betics, forms the westernmost alpine ranges of the Mediterranean Sea and are linked by the Gibraltar Arc. The Alboran Sea is in the center (Figures 1 and 2). The central-south region of the Alboran Sea and the eastern Rif Cordillera belong to the seismically active area of Al Hoceima. The Internal and External Zones separated by the Flysch Units tectonically constitute the Rif Cordillera. In addition, several late intramontane Neogene–Quaternary sedimentary basins emerged, some of which were linked to the Alboran Sea, forming the largest basin of the orogen [15].

The Internal Zone comprises Sebtide and Ghomaride superposed tectonic complexes formed by Paleozoic, Mesozoic and Cenozoic rocks, which are strongly affected by the Alpine Orogeny and have their equivalents in the Betics, called Alpujarride and Malaguide. Some of these complexes have undergone metamorphism and have been thrust over the Flysch Units and the External Zone in the Rif.

The Flysch Units, which are mostly Tertiary sedimentary rocks with some locally ultrabasic rocks, constitute the sedimentary basin that separates the Internal and External Zones and is underlain by oceanic crust. They thrust southward across the External Zone, which is formed by Mesozoic and Cenozoic sedimentary rocks that are mostly unmetamorphosed or, in some cases, have a low degree of metamorphism.

The Alboran Sea is primarily formed by Neogene and Quaternary sediments deposited on a basement that corresponds to the Internal Zone complexes [16]. Furthermore, Neogene to Quaternary volcanic rocks can be found in the central-eastern Alboran Sea and eastern Rif and Betics.

The main alpine deformations in this area occurred throughout the Oligocene and Miocene and continue into the present. The earliest stages of deformation were partially simultaneous with the process of western migration of the Betic and Rif Internal Zones, coinciding with the opening of the Alboran Sea and forming the Gibraltar Arc during a period of severe weakening of the continental crust. The new Alboran marine area corresponds to the western end of the Algero–Provençal Basin, which began to open at the end of the Oligocene [17], forming a new oceanic floor. In the Alboran Sea, the continental crust was markedly weaker and situated on the new oceanic floor on its eastern border.

Subduction processes, combined with NNW–SSE convergence and regional compression of the Iberian and Nubian plates [18–20], resulted in significant deformation in the northern and southern borders of the earliest Alboran Sea. These processes developed the Gibraltar Arc, with the uplift of the Rif and Betic Cordillera, which were radially deformed [21] around the Alboran Sea and undergoing a regional E–W compression.

Later, from the late Miocene, when the opening was nearly at its end ceasing the E–W compression, the general NNW–SSE compression was completely re-established [22]. The region then began to undergo folding (e.g., the Alboran Ridge anticline) and faulting (e.g., the Al Idrisi, Yusuf, Carboneras, Averroes, Jebha and Nekor faults, in addition to other minor faults).

Since the Miocene, important NNE–SSW sinistral strike-slip fault systems crossing the Alboran Sea (Trans Alboran shear zone; [23]) were formed, such as the Carboneras Fault, coming from Almería, Spain and the Nekor and Jebha faults, the last being renowned inland [24]. Later, the Al Idrisi and other conjugated NW–SE faults developed (e.g., the Averroes Fault), as well as some E–W faults and thrusts with a general ENE–SWS strike. The upper Neogene and Quaternary sediments of the central and eastern Alboran Sea are affected by these deformations, whereas the western Alboran Sea undergoes mud volcanic tectonics [25]. This regional geodynamic setting continues into the present.

Almost all the domains mentioned are present in the study area. The Internal Zone forms the inland Bokoya Massif, between Al Hoceima and Melilla to the west, as well as a smaller outcrop, Ras Afrou, on the coast between Al Hoceima and Melilla. The Flysch Units and External Zones, as well as several Neogene–Quaternary basins, appear south of the Internal Zones. Furthermore, Miocene volcanic rocks comprise most of the Raf Tarf and Tres Forcas capes. There are also significant faults (Figures 1 and 2), the most notable of which are located around the Nekor Basin limits, east of Al Hoceima, whose directions range from nearly N–S to NE–SW (e.g., the Trougout Fault, separating Ras Tarf volcanic rocks from Al Hoceima Bay). Some of these faults remain active offshore.

Evidently, all existing seismogenic faults, both onshore and offshore, are unknown, particularly because some are in the early stages of development [26].

3. Aftershock Sequences Description

The Alboran region, specifically the Al Hoceima region, has been the subject of numerous tectonic studies, including those by [27–31]. Owing to its location in the complex border zone between the Eurasian and Nubian plates, near the border between the eastern Rif Cordillera tip and the Alboran Sea, the Al Hoceima region is known to be the most seismically active sector in northern Morocco. Furthermore, because of its strong seismicity, it is one of the most seismically active sites in the western Mediterranean region.

The seismic database of the Spanish Instituto Geográfico Nacional (IGN) was used to assemble the data for this study, with no further processing or parameters from other local or regional agencies. This was performed to keep the database as homogeneous as possible, working in all cases with m_b magnitude.

The 1994 earthquake sequence (Figure 2) began on 26 May 1994, with a strong earthquake of magnitude Mw 6.0, which struck the coastal region near Al Hoceima. This event had a strong impact on the studied region of Al Hoceima [32]. The maximum felt intensity was VIII–IX (EMS-98), indicating an extended NNE–SSW corridor that accounted for over 80% of the damage reported. The magnitude of the earthquake was revised to Md 5.7 (Moroccan Scientific Institute), and the epicenter was relocated north of Al Hoceima at a focal depth of 13 km. According to [32], the distribution of aftershock epicenters in Figure 2 is largely scattered along a NNE–SSW trending cluster over an almost vertical plane.

Another seismic series struck the region on 24 February 2004, with a damaging mainshock of Mw 6.4 (Figure 2). This event occurred on land and caused severe damage. Ref. [33] estimated the maximum perceptible intensity around XI (EMS-98). In Al Hoceima and the surrounding area, nearly 630 people died, 926 were injured, and nearly 15,000 were left homeless. Ref. [34] relocated the 2004 sequence. According to [35] and other authors, the series epicenter occurred on a NE–SW trending strike-slip fault, while the presence of a NW–SE fault with conjugate NE–SW branches cannot be ruled out [34].

The third examined sequence (Figure 2) is linked to the major event on 25 January 2016 (Mw 6.3), whereas it is possible that the series began on January 21 with an event of Mw 5.1. Following these events, a major earthquake series with decreasing activity occurred in 2016 and 2017 [33,35,36]. The major event, with a maximum intensity of VI–VII, was felt over the Alboran area, particularly in Melilla, Spain, on the northern African coast, where extensive damage was reported, as well as in Al Hoceima [33,37].

The Spanish IGN found two distinct epicenter clusters in 2016, each with distinct tendencies. The first one is aligned NNE–SSW changing to N–S, with dominant strike-slip focal mechanisms, while the second one is to the northeast of the first, with a rounded shape and a dominant reverse focal mechanism solution. A NNE–SSW subvertical fault, roughly parallel to the elongation of the alignment and displaced west of the Al Idrisi Fault, is linked to the main NNE–SSW alignment [26].

Figure 3 shows the number of events per day for the selected series in the 150 days after the mainshock, while Table 1 provides the number of events in each series, the minimum recorded magnitude and other computed parameters, which will be discussed later. The 1994 Al Hoceima sequence had 263 recorded events with a magnitude above 2.0, which occurred until December 1994 (some early events have been included in the series); the 2004 sequence had 969 recorded events with a magnitude above 1.5, which occurred until February 2005; and the 2016 Al Hoceima sequence had 2577 recorded events with a magnitude above 0.8, which occurred until August 2016.

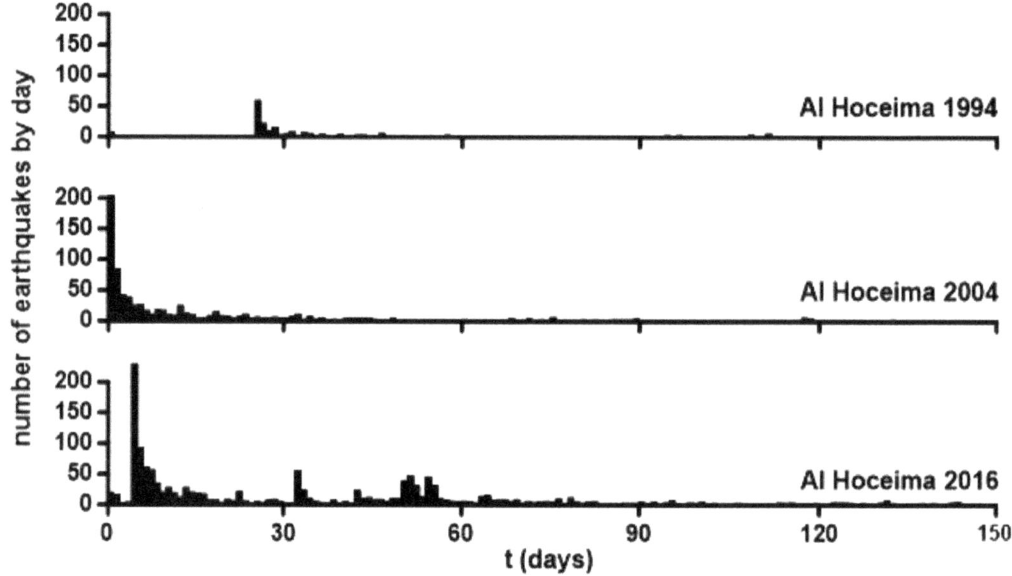

Figure 3. Temporal evolution of the studied series (magnitudes above 2.0). Starting of the x-axis does not always correspond to the occurrence of the main event.

Table 1. Number of events (n), minimum recorded magnitude (M_{min}), computed threshold magnitude (m_c) using the Maxc method and a and b-parameters.

Sequence	n	M_{min}	m_c	a ± σ	b ± σ	Sequence
1994	263	2.0	2.8	5.05 ± 0.02	1.01 ± 0.07	1994
2004	969	1.5	3.4	6.56 ± 0.02	1.14 ± 0.05	2004
2016	2577	0.8	2.0	4.86 ± 0.03	0.82 ± 0.02	2016

For the 1994 and 2004 sequences, Figure 3 reveals a direct decrease in the number of events per day over time, whereas the trend for the 2016 series is more complex, apparently because of the complexity of the rupture(s). There were several different phases in the late aftershock sequence. The recorded occurrences followed a deformation band with two unambiguous alignments with widths of less than 10 to 20 km for the first 30 days [26]. The main alignment, which is moved 5–10 km westward, appears to be spatially associated with the Al Idrisi Fault. A decrease in the seismic activity rate was observed over the next 30 days, resulting in an increase in the width of both alignments, which reached 10–20 km. A clear decrease in activity rate was observed at least 60 days after the mainshock, affecting a wider area than 15–25 km in width as [26] indicated.

4. Seismic Series Stress Regime

The IGN database, international agencies and numerous studies aimed at analyzing these seismic series were used to obtain earthquake focal mechanisms for the selected series. Figure 4 shows earthquakes with magnitudes greater than 3.0, from the 1994 Al Hoceima sequence, as well as the focal mechanism of aftershock events computed by [32] (Figure 4A); different solutions for the focal mechanism of the main shock (Figure 4B) from the Global Centroid Moment Tensor (GCMT), International Seismological Centre (ISC); and specific works by [38–41] (for both a pre-event and the main quake) and [29].

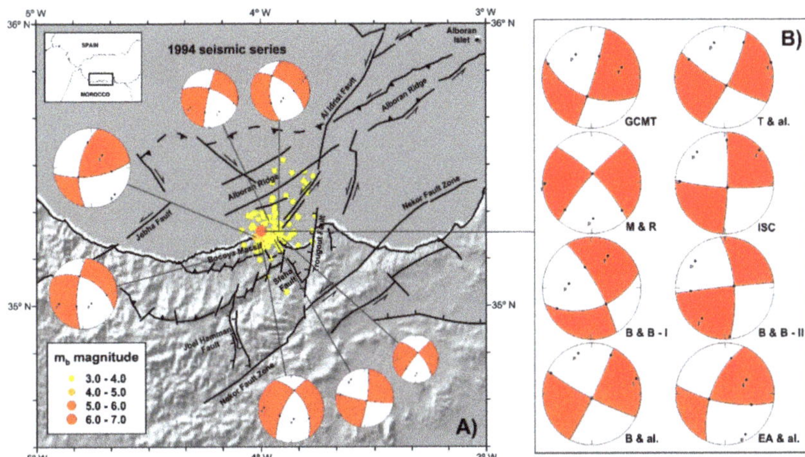

Figure 4. (**A**) 1994 seismic sequence showing earthquakes with magnitude above 3.0 and computed focal mechanisms for aftershocks [32]. (**B**) Different focal mechanism solutions for the mainquake. GCMT: Global Centroid Moment Tensor; T&al.: [40]; M&R: [38]; ISC: International Seismological Centre; B&B-I and B&B-II: [41], for a pre-event and the main quake, respectively; B&al.: [29]; EA&al.: [32,39].

Figure 5A depicts the distribution of events with magnitudes greater than 3.0, which were included in the 2004 series, as well as the estimated focal mechanisms of the largest events. In addition, Figure 5B shows the earthquakes and computed focal mechanisms of the so-called "eastern cluster" from [34].

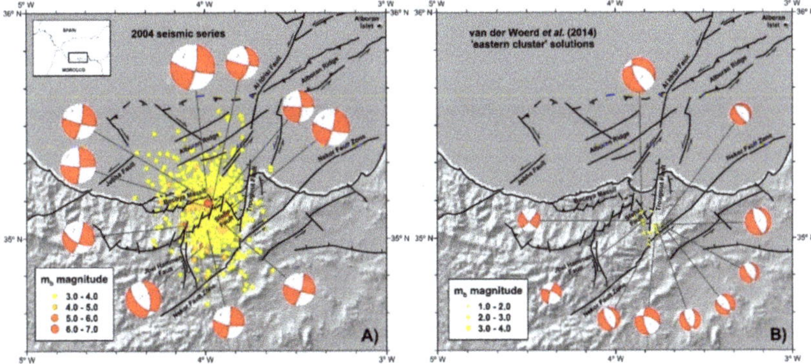

Figure 5. (**A**) 2004 seismic sequence showing earthquakes with magnitude above 3.0 and computed focal mechanisms for the biggest events. (**B**) Earthquakes with magnitude above 1.0 and focal mechanism solutions for the biggest events (m_b 1.9–3.0) of the "eastern cluster" studied by [34].

Figure 6 shows events with magnitudes more than 3.0 as well as estimated focal mechanism solutions for the 2016 Al Hoceima sequence.

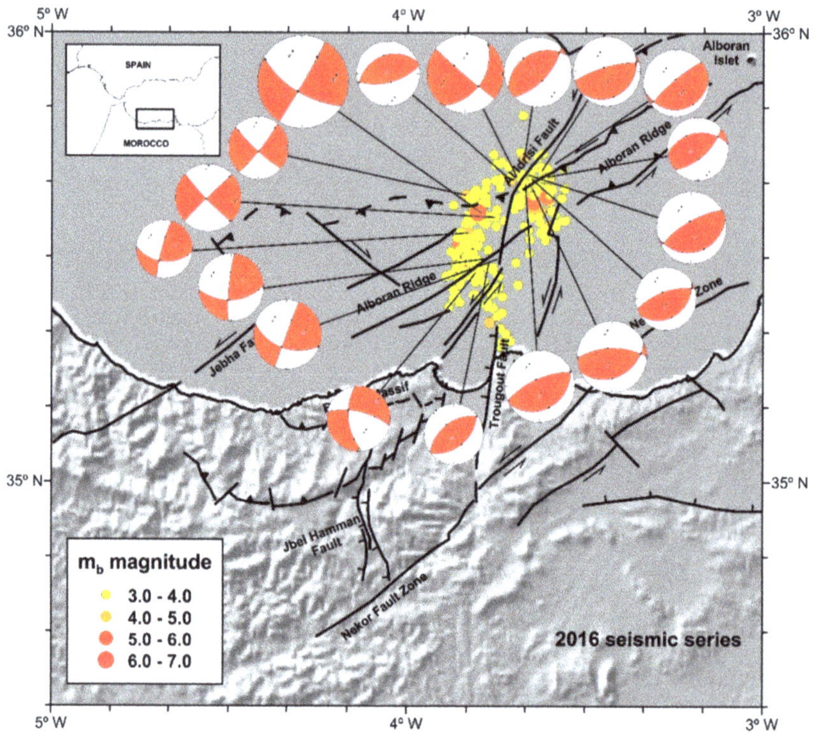

Figure 6. 2016 seismic sequence showing earthquakes with magnitude above 3.0 and computed focal mechanisms.

The stress pattern from the inversion of the available focal mechanism data was used to characterize the three seismic series. It is worth noting that for the inversion process, it was not necessary to select between two available nodal planes. To estimate the different parameters of the reduced stress tensor, we used the improved right dihedron method [42] combined with the iterative rotational dihedron method [43]. Our aim was to calculate the four parameters of the reduced stress tensor, σ_1, σ_2, σ_3 and the stress ratio, $R = (\sigma_2 - \sigma_3)/(\sigma_1 - \sigma_3)$.

According to [19], this method allows for estimating previous parameters and the extraction of filtered focal mechanism data by deleting nodal planes that are incompatible with the average stress regime. The compatible focal mechanisms and the calculated stress tensor produced at this point were then employed as the direct starting point for the rotational optimization technique.

The iterative grid-search rotational optimization process is based on a controlled grid search of the stress tensor using the Win-TensorTM code to reduce the so-called misfit (F5) [44]. According to [45], the nodal plane best explained by the stress tensor was chosen as the actual fault plane from the two planes of the focal mechanism. Consequently, the final inversion examines the focal planes that a uniform stress field best fits [46]. After the ultimate optimization, the omitted focal planes must be reconsidered without modifying the stress tensor. If this is the case, the data are re-entered into the database, the stress tensor is re-optimized, and the software runs another check for the rejected data.

This method was used to analyze the focal mechanism data for each seismic sequence. Table 2 and Figure 7 show the results of the stress inversion. Several authors (e.g., [19,44,47–49]) have used a similar approach to investigate the stress regime in other regions.

Table 2. Stress regime for the cases considered in Figure 7.

Sequence	σ_1	σ_2	σ_3	R	F5
1994 main event	327° N/13°	226° N/40°	072° N/47°	0.19	7.1
1994 aftershocks	132° N/44°	349° N/39°	242° N/20°	0.38	3.5
2004 all, without 'eastern' cluster	330° N/25°	149° N/65°	239° N/00°	0.59	2.6
2004 'eastern' cluster	020° N/79°	141° N/06°	232° N/09°	0.82	5.6
2016 main cluster	332° N/06°	224° N/72°	064° N/17°	0.74	1.6
2016 secondary cluster	330° N/23°	062° N/04°	161° N/67°	0.49	3.3

The variability in the stresses is highlighted in Figure 7 and Table 2. The different solutions of the mainshock, when considered together, and those determined for the aftershocks were not very distinct in the 1994 seismic series. A prolate stress ellipsoid with a near NW–SE horizontal (compressive stress regime) and roughly similar σ_2 and σ_3 values appeared in the mainshock. The slope then turns southeast, highlighting the NE–SW expansion with the horizontal σ_3 being more noticeable (extensive stress regime).

The seismic series from 2004 appears to be more uniform, with oblate stress ellipsoids and a noticeable NE–SW extension trend. While the main cluster shows NW–SE sub-horizontal σ_1 and NE–SW sub-horizontal σ_3 values, clearly indicating a strike-slip stress regime, in the "eastern cluster" [34], a cluster with a few low-energy earthquakes and σ_1 and σ_2 values becoming closer, and a NE–SW horizontal extension is dominant (extensional stress regime).

In the 2016 sequence, the computed stress regime for the main cluster agrees with that obtained in the 2004 main cluster (strike-slip stress regime); however, there is a noticeable increase in the axial ratio, where σ_1 and σ_2 magnitudes are closer, highlighting the well-defined NE–SW sub horizontal extensive stresses. In contrast, the secondary cluster is dominated by NW–SE compressive σ_1 stresses and a subvertical σ_3, suggesting thrusting (compressional stress regime). Furthermore, the secondary cluster stresses are like those obtained for the mainshock of 1994, which had a primary NNW–SSE σ_1 odd axis.

The maximum compressive horizontal stress (SHmax) in all solutions was between 142° N and 153° N, which is consistent with previous results for regional stresses (e.g., [18–20,50]). However, the extension directions are compatible with the Internal Zone movement of the Betic–Rif toward the southwest [26,51–53].

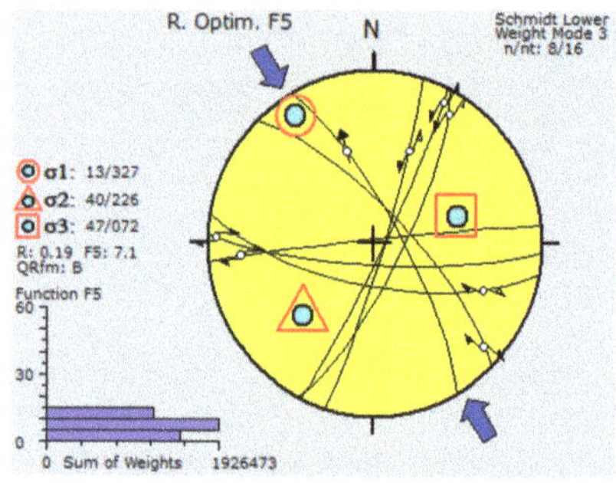

(**A**) 1994 series. Main shock.

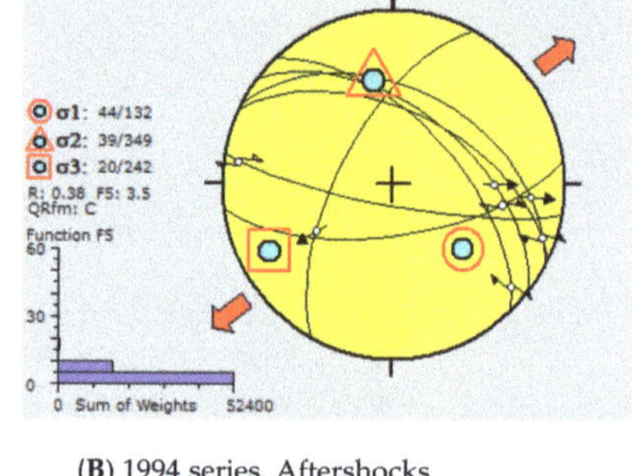

(**B**) 1994 series. Aftershocks.

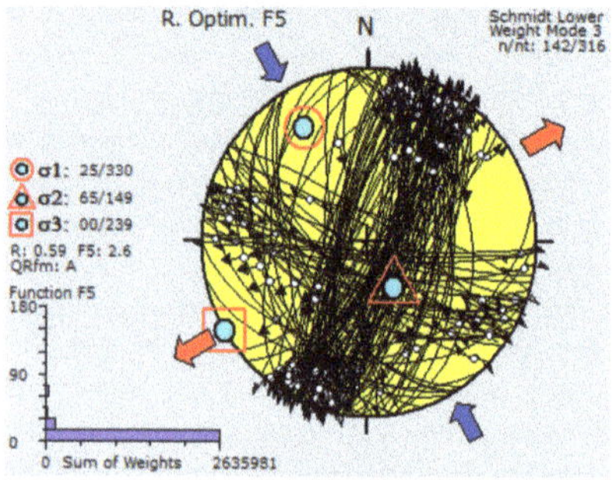

(**C**) 2004 series. Main cluster.

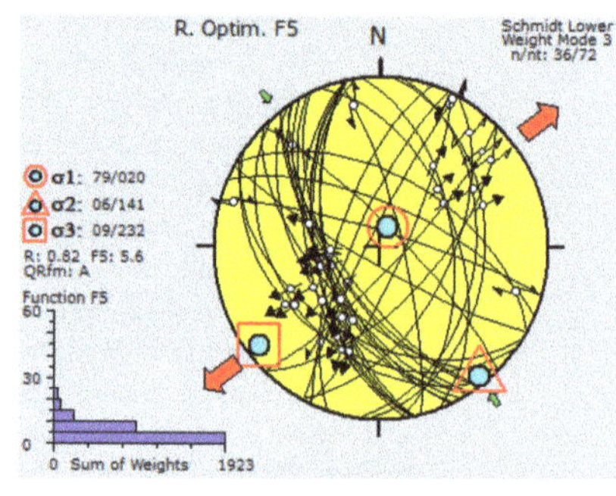

(**D**) 2004 series. Eastern cluster.

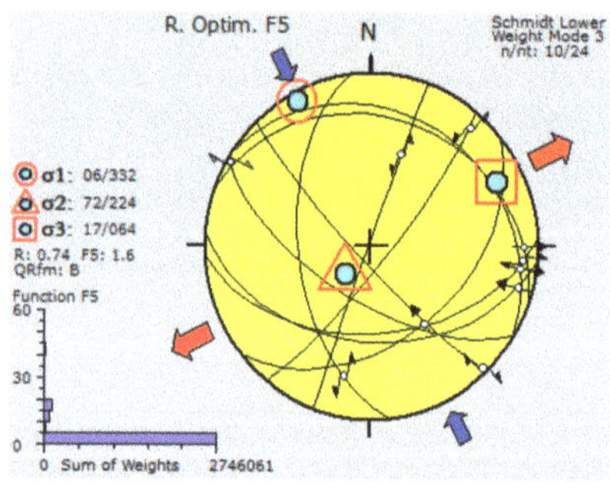

(**E**) 2016 series. Main cluster.

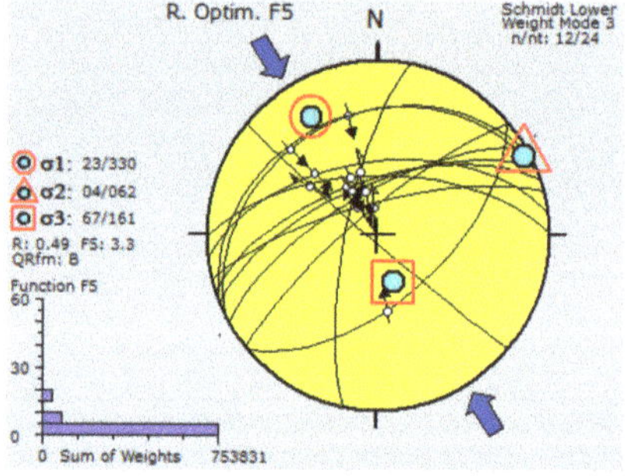

(**F**) 2016 series. Secondary cluster.

Figure 7. Stress regime computed from focal mechanisms. (**A**) For the 1994 mainshock, (**B**) for the 1994 aftershock sequence, (**C**) for the 2004 aftershock sequence, (**D**) for the 'eastern cluster' considered by [34], (**E**) for the 2016 'main cluster' and (**F**) for the 2016 'secondary cluster'.

5. Magnitude–Frequency Relationships

The Gutenberg–Richter recurrence relationship [54,55] is a frequently used approach for quantifying seismic activity in each region and has been shown to apply aftershock frequency–magnitude data. The equation is a reasonable approximation of the frequency–magnitude statistics that describe the correlation between earthquake occurrence frequency and magnitude

$$log_{10} N \left(\geq m \right) = a - bm \qquad m \geq m_c \tag{1}$$

where N ($\geq m$) is the number of events with magnitudes greater than or equal to m. For the estimation of both the a and b-values, it is widely recommended to use a complete dataset for all ranges of magnitude.

The threshold magnitude parameter m_c, is typically determined using one of two methods: a network-based [56,57] or catalog-based approach [56,57]. The first group uses the day-to-night ratio to calculate the earthquake frequency, if noise reduces the detection threshold at night [58,59]. The second set of approaches assumes that earthquake production is self-similar, allowing us to use a power law or the Gutenberg–Richter relationship to construct earthquake frequency–magnitude distributions. The most applied methods in this group reviewed by [60,61] are the maximum curvature (Maxc) method [62], entire magnitude range (EMR) method [60,63], median-based analysis of the segment slope (Mbase) [64], determination of b-value instability [65] and goodness-of-fit test (Gft) by [62], which was later modified by [66].

The threshold magnitude for each of the three sequences under consideration was thoroughly estimated in the current study. Although both the maximum curvature (Maxc), Gft, Gft at levels of 5% and 10% (Gft5% and Gft10%), and modified goodness of fit (mGft) methods were first investigated, the maximum curvature method produced better results, i.e., a better fit (Table 1). Threshold magnitude values of 2.8, 3.4 and 2.0 were obtained for the 1994, 2004 and 2016 aftershock sequences, respectively. The threshold value fluctuated over time, which was also studied. This temporal fluctuation is depicted in Figure 8 and was computed using a sliding-window method.

Prior to each new computation, a window of 20 events was shifted by five events [67]. The middle time of the associated window is supplied to each new threshold magnitude value. The window length was chosen as the balance between the need for temporal resolution and smoothness. Multiple tests were conducted first, changing the number of occurrences per window and the shift; however, neither aspect had a significant influence on the definition. The maximum likelihood approach was used and is considered one of the most reliable approaches for computing the b-value among the available methods. The estimator by [68], given below, was used to calculate the b-value:

$$b = \frac{log_{10}e}{\langle m \rangle - m_c} \tag{2}$$

$\langle m \rangle$ being the average value of the magnitude. An estimate of the standard deviation σ_b is obtained using the [69] relation, given as follows:

$$\sigma_b = 2.30 \, b^2 \sqrt{\frac{\sum_{i=1}^{N}(m_i - m)^2}{N(N-1)}} \tag{3}$$

According to [69,70], even when the b-value varies in time and/or space, this relationship provides a reliable approximation of b-uncertainty. When utilizing rounded magnitudes, the estimator by [68] is inaccurate but agrees with the modified distribution of [71]. Then, the improved estimator by [72], given below, is used to determine the maximum likelihood b-value:

$$b = \frac{log_{10}e}{\langle m \rangle - \left(m_c - \frac{\Delta m}{2} \right)} \tag{4}$$

The estimated b-value appraisal for the 1994, 2004 and 2016 aftershock sequences, using the maximum likelihood estimator by [72], with a bin width of 0.1 units, are 1.01 ± 0.07, 1.14 ± 0.05 and 0.82 ± 0.02, respectively (Table 1). Figure 9 shows the power law distributions for the three sequences as fitted by a straight line. The greatest magnitudes, corresponding to the mainshock, are not explained by the Gutenberg–Richter relationship and are thus regarded as outliers in all series. It is worth mentioning that the estimated maximum likelihood parameters are closely linked to and impacted by the previously determined threshold magnitude.

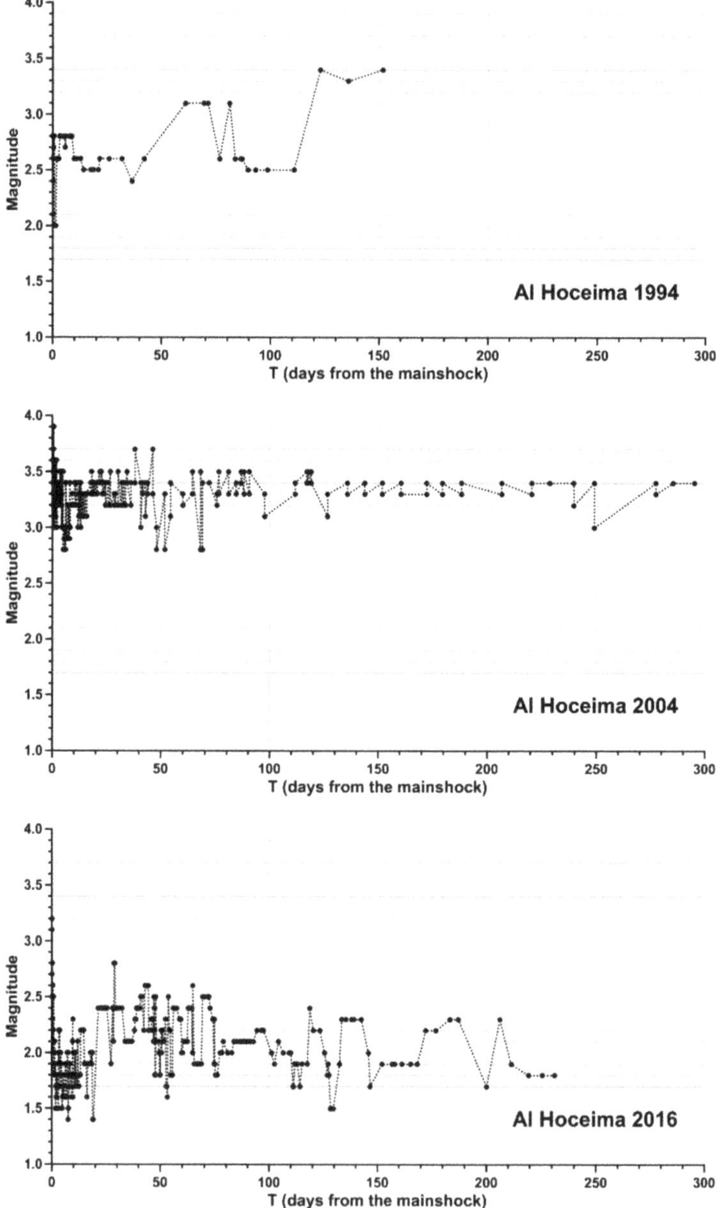

Figure 8. Temporal evolution of the threshold magnitude for the three sequences.

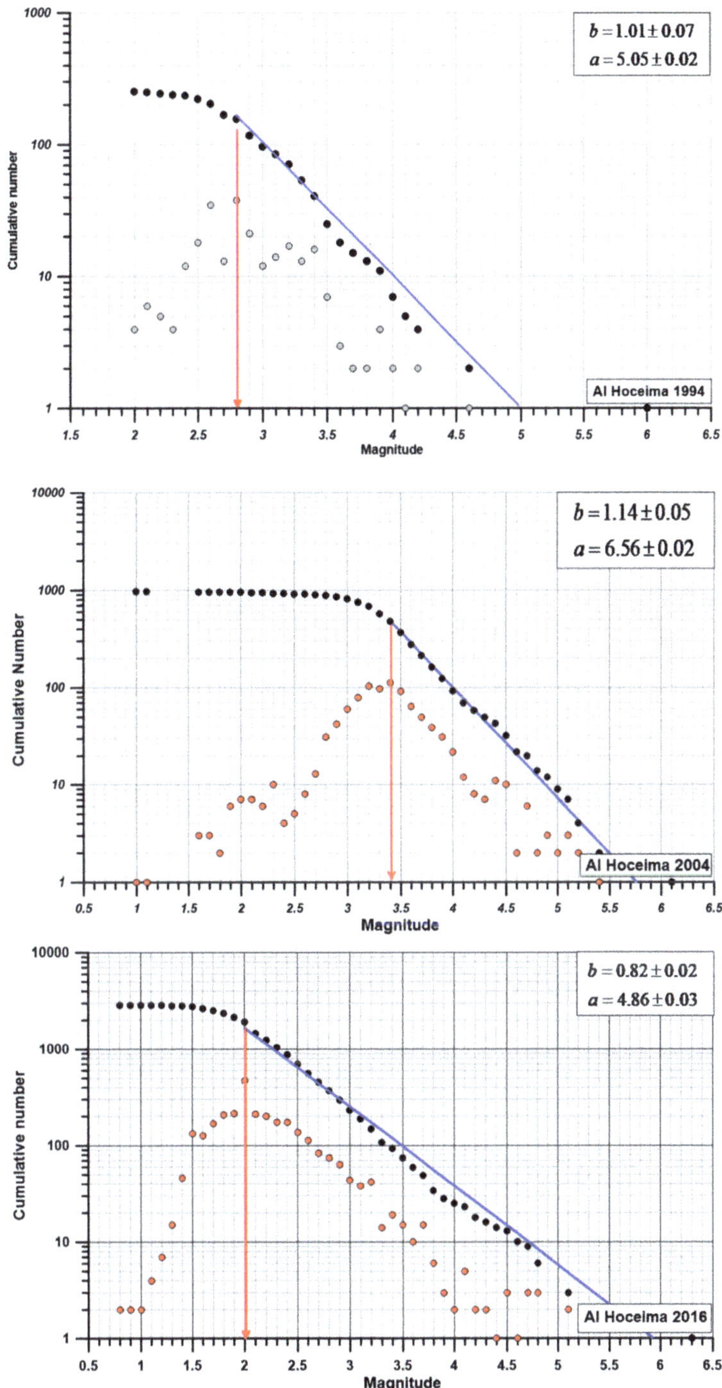

Figure 9. The cumulative number of earthquakes vs. magnitude (black points and linear fits) and the non-cumulative number of earthquakes (open points) for the three sequences. The threshold magnitude is highlighted.

6. Temporal Stochastic Modeling for the Al Hoceima Sequences

This section focuses on the point process modeling of the studied aftershock sequences. Many approaches for modeling this gradual decrease in aftershock frequency have been presented in the literature. The most commonly used model is the Omori law [3], which was improved by [4] into the modified Omori formula (MOF). According to [4], the decay rate of the aftershock per unit time is given as follows:

$$n(t) = \frac{K}{(t+c)^p} \tag{5}$$

where t is the time after the occurrence of the mainshock, and the parameter K is related to the mainshock event and threshold magnitude. The c parameter is a debatable number [73,74], with the early stages of imprecise detection of low events in the sequence having a significant effect [75]. Finally, the p parameter is a decay constant, and it is quite likely the most important parameter for understanding the behavior of the sequence. The p-value varies from sequence to sequence and typically ranges from 0.5 to 1.8 [73]. This variation may be related to tectonic activity in the area. However, the elements that influence the p-value remain unclear [76,77].

The quantity $n(t)$ allows us to connect with the point process model by considering it a conditional intensity.

$$n(t) \approx \lambda(t), \tag{6}$$

bearing in mind that [5]:

$$P \{an\ event\ occurs\ in\ (t, t+dt)|\Im_t\} = \lambda(t|\Im_t)\, dt + o(dt), \tag{7}$$

where $\Im_t$ denotes the internal history of the occurrence process at time t, $\lambda(t|\Im_t)$ is the conditional intensity function [5,78], and $o(dt)$, in the Landau notation, is a function of a lower order than the function λ, i.e., $o(dt)$ being negligible. The MOF model includes only the mainshock occurrence time because it is based on the concept that the entire relaxation process is driven by the stress changes induced by the mainshock alone. The aftershocks are conditionally independent and follow a non-stationary Poisson process.

The MOF model fits data for simple aftershock sequences well; however, secondary clustering is common when there are secondary aftershocks triggered by strong earthquakes in a sequence. The authors in [5] argued that aftershock clustering is a self-similar process in which all aftershocks might induce other aftershocks, with a triggering capacity dependent on their magnitudes, because of these complex situations involving one or more secondary events.

The model was named ETAS, and its conditional intensity function is given as follows [79]:

$$\lambda(t|\Im_t) = \mu + \sum_{i;\ t_i < t} \frac{k_0 e^{\alpha(m_i - m_c)}}{(t - t_i + c)^p} \tag{8}$$

where μ is the background seismicity rate. The internal history $\Im_t$ includes the time occurrence t_i (in days after the mainshock) and magnitude events m_i of all the events occurring before time t. The summation includes all events with occurrence times t_i and magnitudes equal to or greater than the lower cut-off m_c. The c and p model parameters were the same as those used in the MOF model. Furthermore, k_0 is a parameter that affects total aftershock productivity and is common to all aftershocks.

According to [80], every term of the summation in Equation (8) indicates the contribution of a prior event to the occurrence probability of subsequent events at time t. The exponential term is controlled by two factors: (a) the temporal rate decrease, as presented by the MOF model and (b) the exponential term chosen because the logarithm of the aftershock area and the magnitude of the mainshock present a linear relationship [81]. Parameter α measures the effect of the magnitude of the production of 'children' events, also called "descendants".

The MOF and ETAS models, as expressed by Equations (5) and (8), respectively, present two limited cases for modeling the temporal distribution of an aftershock sequence. Ref. [2] proposed a similar model, the RETAS model, based on the assumption that not all events in a series; however, only aftershocks with magnitudes greater than or equal to a threshold value, can directly cause the aftershocks of "descendants". This model allows for the inclusion of all potential models between the two limit cases of the MOF and ETAS models, resulting in the conditional intensity expressed as follows:

$$\check{}(t|\Im_t) = \mu + \sum_{\substack{i;\, t_i < t \\ m_i \geq M_{tr}}} \frac{k_0 e^{\alpha(m_i - m_c)}}{(t - t_i + c)^p} \tag{9}$$

It is worth noting that the RETAS model developed by [8] is based on Bath's law, which states that the difference between mainshock magnitude and the strongest aftershock magnitude, ranges between 1.2 to 1.4 units. According to this relationship, ref. [8] argued that the difference between the weaker primary event and the weakest event in the aftershock sequence must be at least 1.2 units, by applying this principle to the subsequences generated by the primary trigger model.

Furthermore, the RETAS model has the advantage of examining all potential models between the MOF and ETAS models because the triggering magnitude ranges from the threshold magnitude to the mainshock magnitude. The Akaike information criterion [82], abbreviated as AIC, was used to choose the best-fit model in our case, with the lower AIC value. This is expressed as follows:

$$\text{AIC} = -2 \max_\theta \log L(\theta; 0, T) + 2k'' \tag{10}$$

where k'' represents the number of parameters of the model, and $\log L$ is the logarithm of the likelihood function, given as follows:

$$\log L(\theta; 0, T) = \sum_{i=1}^{N} \log_{10} \check{}_\theta(t_i | \Im_{t_i}) - \int_0^T \check{}_\theta(s | \Im_s) ds \tag{11}$$

In the previous Equation (12), N is the number of earthquakes with magnitudes greater than or equal to m_c, occurring at times t_j ($j = 1, 2, \ldots, N$), during $[0, T]$. Due to the features of the described model, it was used in this study to investigate three aftershock sequences. Table 1 lists the magnitude thresholds used in this study.

As a result of the RETAS model analysis, Figure 10 displays the AIC value versus the triggering magnitude value. Analysis of this curve reveals the aftershock clustering type that is most common in the sequence. The AIC parameter was calculated for all the potential models by varying the triggering magnitude from the threshold magnitude to the magnitude of the mainshock. The best-fit model shows the AIC minimum values in Table 3 as well as the triggering magnitude and model parameters.

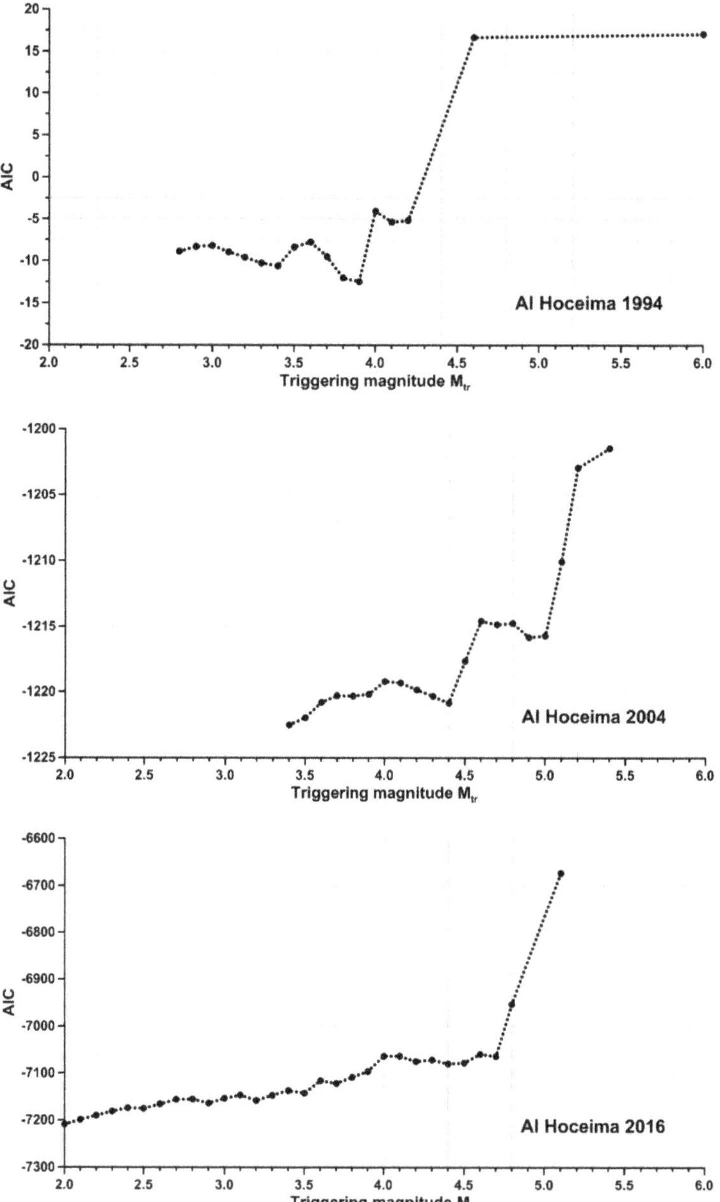

Figure 10. The AIC values of the RETAS model vs. the triggering magnitude for the three sequences.

Table 3. The computed parameters for the RETAS model.

Sequence	Mtr	Best Model	AIC	K_0	α	c	p
1994	3.9	RETAS	−12.4	1.727	0.0630	0.024	0.909
2004	3.4	ETAS	−1222.5	3.426	0.0035	0.082	1.070
2016	2.0	ETAS	−7209.0	1.477	0.0380	0.039	1.183

When examining the AIC values vs. M_{tr} curves (Figure 10), we can see that the best-fit model for the minimal AIC value was found to be for a M_{tr} value of 3.9 for the 1994 sequence. For the 2004 sequence, an M_{tr} value of 3.4 is the best-fit model obtained and, for the 2016 sequence, a magnitude of 2.0. The best-fit model became an ETAS model when the triggering magnitude M_{tr} coincided with the completeness threshold magnitude m_c in the last two M_{tr}. Analyzing the AIC vs. M_{tr} trigger magnitude curve for the 2016 sequence, a monotonic trend exists, resulting in a continuous increase in AIC values, the lowest of which is for $M_{tr} = m_c$.

According to the model presented in Figure 11, the estimated best model parameters shown in Table 3 were used to compute the expected cumulative number of earthquakes and error bounds. The computed values were compared with the observed cumulative number of earthquakes. Figure 11 shows that the estimated model matches the observed data well for all three sequences, with the observed values remaining inside the error boundaries throughout the whole sequence.

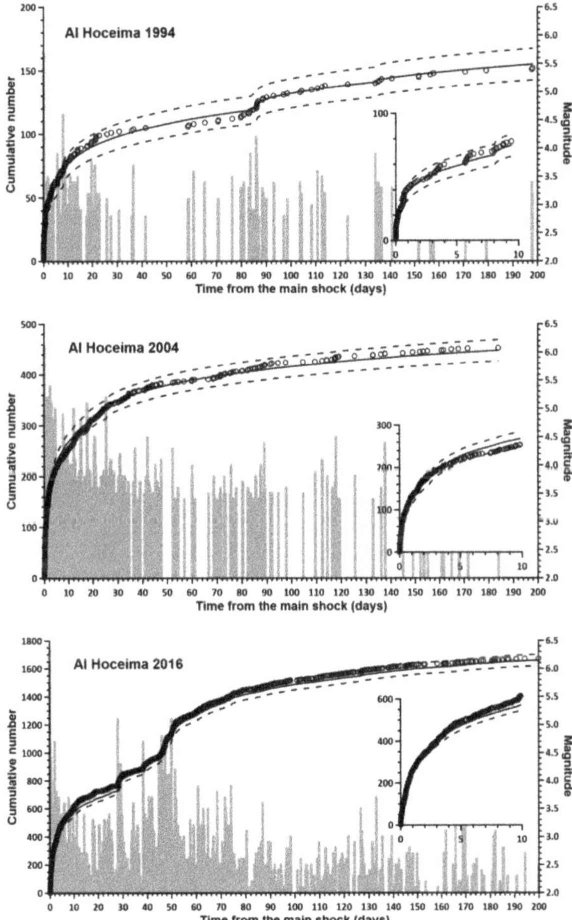

Figure 11. The cumulative number of events above magnitude m_c for the three sequences. Circles: observed values. Lines: mean $\pm\sigma$ fitted model. The magnitudes of the events included in the sequences are also depicted. Inset stand out the first ten days of the series.

7. Stochastic Model for the Energy Release

In the literature, several attempts have been made to improve earthquake time models by including data on other characteristics associated with event occurrences, such as space-time models or those that relate event occurrence time to a size value of the event (intensity, magnitude or energy). In this context, this section aims to investigate the relationship between the occurrence times $\{T_k, k = 1, \dots , n\}$ and released energy $\{E_k, k = 1, \dots , n\}$, providing information on the size of the events in the studied sequences. It is typically considered that the time of occurrence and scale of an event are unrelated.

As a result, our aim was to create an energetic stochastic model that incorporates these considerations and examines how well it represents the observed data. According to [83], marked point processes have similar counting processes [84,85]. The following is a description of this model. Given a Poisson process $\{N(t); t \geq 0\}$ with a rate $\lambda > 0$, it is assumed that the time T_k of each event is linked to a realization, which is a family $\{Y_k; k > 0\}$ of independent and identically distributed random variables called marks, with a probability distribution function equal to

$$G(y) = P\{Y_k \leq y\} \tag{12}$$

The second requirement is that the random variables are at the same time independent from $\{N(t); t \geq 0\}$. Then, according to [84,85], the stochastic model, typically called the compound Poisson process, is defined as follows:

$$Z(t) = \sum_{k=1}^{N(t)} Y_k; \quad t \geq 0 \tag{13}$$

Denoted by μ and γ^2 are the mean and variance of the marks Y_k; the moments of $Z(t)$ are then given as follows:

$$E[Z(t)] = \lambda\mu t \tag{14}$$

$$var[Z(t)] = \lambda\left(\gamma^2 + \mu^2\right)t \tag{15}$$

Consequently, in the case of a series of occurrence times and event sizes, the compound Poisson process can be used as a model process for random behavior. In the most general treatment of a compound point process, $\{N(t); t \geq 0\}$ is an inhomogeneous Poisson process with an intensity function $\{\lambda(t); t \geq 0\}$, and marks $\{Y_k; k > 0\}$ do not have to form an independent series of random variables. Equations (14) and (15) must be rewritten in the context of an inhomogeneous Poisson process with a rate that varies with time $\lambda = \lambda(t)$

$$E[Z(t)] = \mu \int_{t_0}^{t} \lambda(s)ds \tag{16}$$

$$var[Z(t)] = \left(\gamma^2 + \mu^2\right) \int_{t_0}^{t} \lambda(s)ds \tag{17}$$

The following relationship (18) between the log of the released energy and the magnitude is used [55,86] because the most common way of describing the size of an earthquake is by its magnitude.

$$log_{10}\sqrt{E} = 2.4 + 0.75\,M \tag{18}$$

The relationship described by this Equation (18), according to [86,87], is consistent with what is expected theoretically for classical crack models with a constant stress drop. This generalized energy–magnitude scaling equation works for various magnitude ranges.

As the cumulative released energy is a physical quantity, it is helpful to consider the series of released energies in the compound Poisson process scheme. It is worth noting that this method has several limitations. First, the energy is determined by the wide dynamic range of the released energy, as well as an instrumentation earthquake record. Thus, to reduce this variability, Equation (18) is transformed as follows:

$$E_k^{tr} = \frac{E_k}{E_0} = \frac{10^{4.8+1.5\,M_k}}{10^{4.8+1.5\,M_0}} = 10^{1.5\,(M_k-M_0)} \tag{19}$$

where E_k and M_k correspond to the energy and magnitude of the k-th event, respectively, and E_0 and M_0 are the energy and magnitude of the first event considered, respectively. E_k^{tr} is the Benioff's energy.

The approach described above was used to analyze the considered aftershock sequences. Considering the compound Poisson process, given as follows:

$$Z(t) = \sum_{k=1}^{N(t)} E_k^{tr} \tag{20}$$

Equations (16) and (17) were used to derive the estimation of the mean $E[Z(t)]$ and the variance $\mathrm{var}[Z(t)]$ for each series, considering the best-fit model derived and analyzed in the previous section. Figure 12 displays the expected cumulative energy and confidence bounds, mean plus/minus the standard deviation, according to the appropriate best-fit model for each aftershock sequence, and this is compared to the computed cumulative energy released in the sequences.

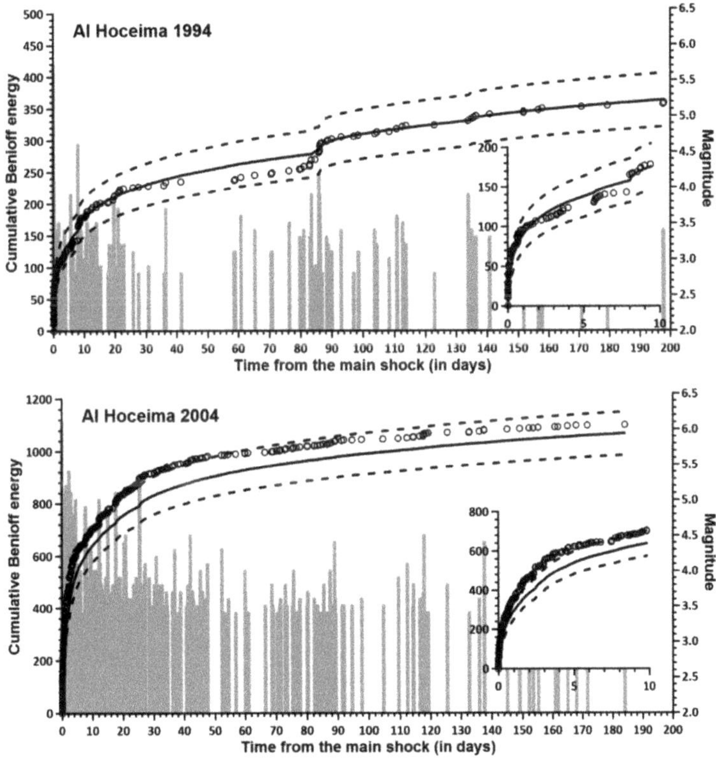

Figure 12. *Cont.*

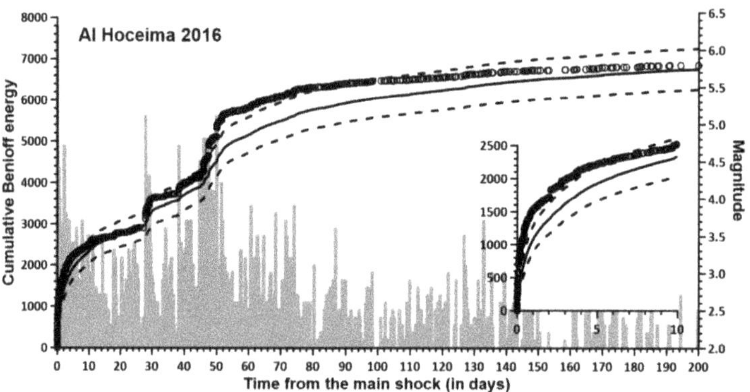

Figure 12. Cumulative released energy for the three sequences. Circles: observed values. Lines: mean ± σ fitted model. Magnitudes of the events included in the sequences are also depicted. The inset shows the first ten days of the series.

It should be noted that the stochastic model used follows the two previously quoted assumptions: both the independence of the magnitudes and the independence of the magnitudes with the occurrence time. Figure 12 shows that, for the 1994 sequence, no significant deviation of real data values from the model was observed, which supports the above assumptions. However, deviations of the real data values with the model were observed for the two other aftershock sequences.

For the 2004 sequence, the main deviation was observed at the beginning of the sequence, up to the 30th day, where the clustering of stronger aftershocks was recognized. This reveals that the assumption of the independence of magnitudes with occurrence times is not valid in this case. In addition, another deviation of the observed data out of the error bounds can be observed in the 2016 sequence, where an additional concentration of stronger aftershocks is recognized.

8. Discussion

In the present study, the stress pattern analysis revealed that all solutions had a maximum compressive horizontal stress (SHmax) between 142° N and 153° N, which is compatible with previous regional stress data as determined by previous studies. Furthermore, according to these results, the extension directions are clearly compatible with the Betic–Internal Rif's zone movement southwestward.

The results for all the seismic sequences share horizontal stresses, showing a NW–SE compression and a NE–SW extension as a result of the regional setting. Nonetheless, some significant differences can be observed in the axial ratios and local stress regimes because of the series location in the main shear zone crossing the Alboran Sea and the activated structures in the main and secondary clusters. The three series induced a different fault system, also hosting seismicity, from which the sequence began.

Concerning observed differences in the tectonic characteristics of the series, the 1994 sequence is mainly scattered along a NNE-SSW trending cluster over an almost vertical plane [32], the 2004 sequence is mainly scattered along a NE-SW strike slip fault, and the 2016 sequence is initially scattered along a NNE-SSW trending cluster over a subvertical strike-slip fault, then changing to N-S and finally distributed on a second rounded cluster with a dominant reverse focal mechanism solution [26]. The magnitude of completeness values considered in the current study were compared with those derived using Gft and mGft methods.

Figure 13 shows plots of residuals vs. the minimum magnitudes for the three series. The results obtained matched the estimated values derived using the maximum curvature (Maxc) method. All approaches yield a result of 2.8 for the 1994 series. For the 2004 series, the difference was of the order of 0.1 using mGft, Gft5% and Maxc and 0.2 between Gft10% and Maxc. However, in the case of the 2016 series, the difference varies between 0.2 and 0.3. The results are 1.9 using Gft10%, 2.2 for Gft5%, 2.3 for mGft and 2.0 for Maxc.

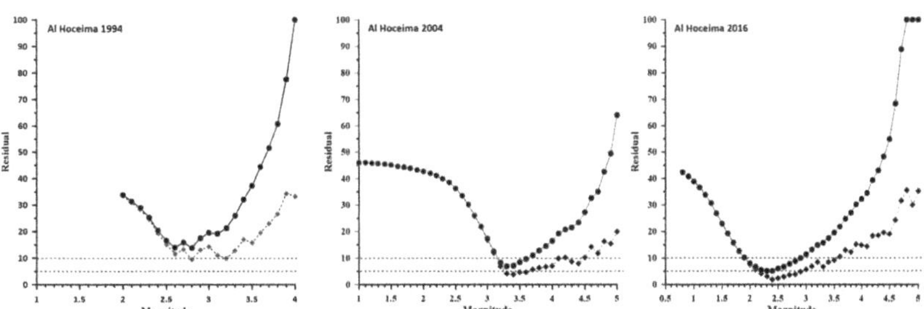

Figure 13. The obtained residuals vs. magnitude when fitting the observed data to the mGft power law (solid line) and theoretical distribution power law (dashed line). The 5% and 10% residual levels are shown as reference.

As discussed in this study, the RETAS model was used to study the gradual decay of the aftershock frequency based on the triggering magnitude M_{tr} assumption. The minimum value of the AIC parameter was used as a criterion for selecting the best-fit model. All possible models were estimated by varying the triggering magnitude M_{tr} from the threshold magnitude m_c (ETAS model) or m_{ms} (MOL model) to the mainshock magnitude. The results derived for the 1994 and 2004 series highlight and improve on the previous ones derived by [88]. For instance, for the 2004 aftershock sequence, as in [88], the minimum AIC value shows results for the best ETAS model. Considering that the 2016 sequence began on January 21 with an event of Mw 5.1, as suggested by some authors [26], the ETAS model was obtained as the best-fit model.

Stochastic modeling of the energy revealed that the energy released over the series was outlined well by the proposed model in the 1994 and 2004 series. However, there were minor discrepancies in the 2016 sequence and the computed cumulative released energy did not match the model well. We deduced from the analysis that large aftershock clusters occurred more frequently and quickly after the main shock than they would if they occurred at random. This observation holds true for the Zemmouri, Algeria aftershock series of 2003 [88].

9. Conclusions

Stochastic modeling was used to analyze three sequences of events that occurred in Al Hoceima, Morocco, May 1994, February 2004 and January 2016. The analysis of the behavior of the decay rate of the three series, together with the composite stress pattern and the obtained *b*-value, led us to conclude that the 2016 sequence is the most complex of the three series, likely because of two different fault systems being activated—the main one corresponding to a strike-slip stress and the second one to reverse faults.

The released energy analysis allowed us to characterize the occurrence of large aftershocks shortly after the mainshock better than it would if they occurred at random. However, further research is needed to estimate the recurrence period of such large occurrences, as well as the probability of exceeding a specific magnitude shortly after the mainshocks. This type of research could be conducted in other seismically active regions to investigate the behavior of the seismic series occurring in these areas. This could help to understand the characteristics of the earthquake generation process aimed at seismic forecasting studies.

It should be noted that earthquake forecasting is the ultimate challenge for seismologists because it accumulates scientific knowledge about the earthquake occurrence process and is an essential component of any efficient risk-mitigation strategy [89,90]. As stated by different authors (e.g., [80,91]), RETAS models can be applied to forecast the occurrence probability evolution of a certain sequence, providing the possibility to identify the type of clustering in future seismic series. This is an issue also related to, for instance, the fault distribution and possible fault interactions in the studied area.

Author Contributions: M.H. and J.A.P.: conceptualization, methodology, software, validation, formal analysis, data curation, writing, supervision, project administration and funding acquisition. D.G.: methodology and supervision. J.H.: software, validation, formal analysis, data curation and supervision. J.G.-Z.: validation, data curation, writing, supervision and funding acquisition. C.S.d.G.: validation, writing and supervision. B.R.: supervision. All authors have read and agreed to the published version of the manuscript.

Funding: This research was partially funded by the Consejería de Economía, Conocimiento, Empresa y Universidad, in the frame of the Programa Operativo FEDER Andalucía 2014–2020—Call made by the University of Jaén 2018 and projects FEDER DAMAGE, CGL2016-80687-R and B-RNM-301-UGR18, RNM148 and P18-RT-3275 (Junta de Andalucía).

Data Availability Statement: The seismicity data comes from the Spanish Instituto Geográfico Nacional seismicity catalog (available on the IGN website at http://www.ign.es/web/ign/portal/sis-catalogo-terremotos (accessed on 27 August 2022)) and quoted papers.

Acknowledgments: The authors are grateful to the editor and two anonymous reviewers for their comments, remarks and suggestions. This research was supported by the Algerian CRAAG and the Spanish Seismic Hazard and Active Tectonics research group. The authors also acknowledge the Spanish Instituto Geográfico Nacional for to share the earthquake data used in this study.

Conflicts of Interest: The authors declare that they have no known competing financial interest or personal relationships that could have appeared to influence the work reported in this paper.

References

1. Mogi, K. Some discussions on aftershocks, foreshocks and earthquake swarms. The fracture of a semi-infinite body caused by an inner stress origin and its relation to the earthquake phenomena. *Bull. Earthq. Res. Inst.* **1963**, *41*, 615–658.
2. Gospodinov, D.; Rotondi, R. Statistical analysis of triggered seismicity in the Kresna region of SW Bulgaria (1904) and the Umbria-Marche region of central Italy (1997). *Pure Appl. Geophys.* **2006**, *163*, 1597–1615. [CrossRef]
3. Omori, F. On the aftershocks of earthquake. *J. Coll. Sci. Imp. Univ. Tokyo* **1894**, *7*, 111–200.
4. Utsu, T. A statistical study on the occurrence of aftershocks. *Geophys. Mag.* **1961**, *30*, 521–605.
5. Ogata, Y. Statistical models for earthquake occurrence and residual analysis for point processes. *J. Am. Stat. Assoc.* **1988**, *83*, 9–27. [CrossRef]
6. Vere-Jones, D.; Davies, R.B. A statistical survey of earthquakes in the main seismic region of New Zealand. Part 2. Time Series Analyses. *N. Z. J. Geol. Geophys.* **1966**, *9*, 251–284. [CrossRef]
7. Vere-Jones, D. Stochastic models for earthquake occurrence (with discussion). *J. Roy. Statist. Soc. Ser.* **1970**, *32*, 1–62.
8. Gospodinov, D.; Rotondi, R. RETAS: A restricted ETAS model inspired by Bath's law. In Proceedings of the 4th International Workshop on Statistical Seismology, Kanagawa, Japan, 9–13 January 2006.
9. Bath, M. Lateral inhomogeneities in the upper mantle. *Tectonophysics* **1965**, *2*, 483–514. [CrossRef]
10. Bath, M. *Introduction to Seismology*; BirkhauserVerlag: Basel, Switzerland, 1973; 395p.
11. Ogata, Y. Exploratory analysis of earthquake clusters by likelihood-based trigger models. *J. Appl. Probab.* **2001**, *38A*, 202–212. [CrossRef]
12. Console, R.; Lombardi, A.M.; Murru, M.; Rhoades, D. Båth's law and the self-similarity of earthquakes. *J. Geophys. Res.* **2003**, *108*, 2128. [CrossRef]
13. Hamdache, M.; Henares, J.; Peláez, J.A.; Damerdji, J. Fractal analysis of earthquake sequences in the Ibero-Maghrebian region. *Pure Appl. Geophys.* **2019**, *176*, 1379–1416. [CrossRef]
14. Panzera, F.; Zechar, J.D.; Vogfjörd, K.S.; Eberhard, D.A.J. A revised earthquake catalogue for South Iceland. *Pure Appl. Geophys.* **2015**, *173*, 97–116. [CrossRef]
15. Chalouan, A.; Michard, A.; El Kadiri, K.; Negro, F.; de Lamotte, D.F.; Soto, J.I.; Saddiqi, O. The Rif Belt. In *Continental Evolution: The Geology of Morocco. Lecture Notes in Earth Sciences*; Michard, A., Saddiqi, O., Chalouan, A., Lamotte, D.F., Eds.; Springer: Berlin/Heidelberg, Germany, 2008; Volume 116, pp. 203–302.
16. Comas, M.C.; García-Dueñas, V.; Jurado, M.J. Neogene tectonic evolution of the Alboran Sea from MCS data. *Geo. Mar. Lett.* **1992**, *12*, 157–164. [CrossRef]
17. Boillot, G.; Montadert, L.; Lemoine, M.; Biju-Duval, B. *Les Margescontinentales Actuelles et Fossiles Autour de la France*; Elsevier Masson: Amsterdam, The Netherlands, 1984; 352p.

18. De Mets, C.; Gordon, R.G.; Argus, D.F. Geologically current plate motions. *Geophys. J. Int.* **2010**, *181*, 1–80. [CrossRef]
19. Peláez, J.A.; Henares, J.; Hamdache, M.; Sanz de Galdeano, C. A seismogenic zone model for seismic hazard studies in Northwestern Africa. In *Moment Tensor Solutions. A Useful Tool for Seismotectonics*; D'Amico, S., Ed.; Springer Natural Hazards: Berlin/Heidelberg, Germany, 2018; pp. 643–680.
20. Sparacino, F.; Palano, M.; Peláez, J.A.; Fernández, J. Geodetic deformation versus seismic crustal moment-rates: Insights from the Ibero-Maghrebian region. *Remote Sens.* **2020**, *12*, 952. [CrossRef]
21. Pedrera, A.; Ruiz Constán, A.; Galindo-Zaldívar, J.; Chalouan, A.; Sanz de Galdeano, C.; Marín Lechado, C.; Ruano, P.; Benmakhlouf, M.; Akil, M.; López Garrido, C.; et al. Is there an active subduction beneath the Gibraltar orogenic arc? Constraints from Pliocene to present-day stress field. *J. Geodyn.* **2011**, *52*, 83–96. [CrossRef]
22. Martínez-García, P.; Comas, M.; Lonergan, L.; Watts, A.B. From extension to shortening: Tectonic inversion distributed in time and space in the Alboran Sea, western Mediterranean. *Tectonics* **2017**, *36*, 2777–2805. [CrossRef]
23. Larouzière, F.; Bolze, J.; Bordet, P.; Hernández, J.; Montenat, C.; Ott d'Estevou, P. The Betic segment of the lithospheric Trans-Alboran shear zone during the Late Miocene. *Tectonophysics* **1988**, *152*, 41–52. [CrossRef]
24. Benkmakhlouf, M.; Galindo-Zaldívar, J.; Chalouan, A.; Sanz de Galdeano, C.; Ahmamou, M.; López-Garrido, A.C. Inversion of transfer faults: The Jebha-Chrafate fault (Rif, Morocco). *J. Afr. Earth Sci.* **2012**, *73–74*, 33–43. [CrossRef]
25. Sautkin, A.; Talukder, A.R.; Comas, M.C.; Soto, J.I.; Alekseev, A. Mud volcanoes in the Alboran Sea: Evidence from micro paleontological and geophysical data. *Mar. Geol.* **2003**, *195*, 237–261. [CrossRef]
26. Galindo Zaldívar, J.; Ercilla, G.; Estrada, F.; Catalán, M.; d'Acremont, E.; Azzouz, O.; Casas, D.; Chourak, M.; Vázquez, J.T.; Chalouan, A.; et al. Imaging the growth of recent faults: The case of 2016–2017 seismic sequence sea bottom deformation in the Alboran Sea (western Mediterranean). *Tectonics* **2018**, *37*, 2513–2530. [CrossRef]
27. Stich, D.; Mancilla, F.; Baumont, D.; Morales, J. Source analysis of the Mw 6.3 2004 Al Hoceima earthquake (Morocco) using regional apparent source time functions. *J. Geophys. Res.* **2005**, *110*, B06306.
28. Akoglu, M.; Cakir, Z.; Meghraoui, M.; Belabbes, S.; El Alami, S.O.; Ergintav, S.; Akyuz, H.S. The 1994–2004 Al Hoceima (Morocco) earthquake sequence: Conjugate fault ruptures deduced from InSAR. *Earth Planet. Sci. Lett.* **2006**, *252*, 467–480. [CrossRef]
29. Biggs, J.; Bergman, E.; Emmerson, B.; Funning, G.J.; Jackson, J.; Parson, B.; Wright, T.J. Fault identification for buried strike-slip earthquakes using INSAR: The 1994 and 2004 Al Hoceima, Morocco earthquakes. *Geophys. J. Int.* **2006**, *166*, 1347–1362. [CrossRef]
30. Galindo Zaldívar, J.; Chalouan, A.; Azzouz, O.; Sanz de Galdeano, C.; Anahnah, F.; Ameza, L.; Ruano, P.; Pedrera, A.; Ruiz-Constan, A.; Marín-Lechado, C.; et al. Are seismological and geological observations of the Al Hoceima (Morocco Rif) 2004 earthquake (M=6.3) contradictory? *Tectonophysics* **2009**, *475*, 59–67. [CrossRef]
31. Galindo Zaldívar, J.; Azzouz, O.; Chalouan, A.; Pedrera, A.; Ruano, P.; Ruiz Constán, A.; Sanz de Galdeano, C.; Marín Lechado, C.; López Garrido, A.C.; Anahnah, F.; et al. Extensional tectonics, graben development and fault terminations in the Eastern Rif (Bokoya-Ras Afrou área). *Tectonophysics* **2015**, *663*, 140–149. [CrossRef]
32. El Alami, S.O.; Tadili, B.A.; Cherkaoui, T.E.; Medina, F.; Ramdani, M.; Aït Brahim, L.; Harnafi, M. The Al Hoceima earthquake of May 26, 1994 and its aftershocks: A seismotectonic study. *Ann. Geophys.* **1998**, *41*, 519–537. [CrossRef]
33. Medina, F.; Cherkaoui, T.E. The south-western Alboran earthquake sequence of January-March 2016 and its associated coulomb stress changes. *Open J. Earthq. Res.* **2017**, *26*, 35–54. [CrossRef]
34. Van der Woerd, J.; Dorbath, C.; Ousadou, F.; Dorbath, L.; Delouis, B.; Jacques, E.; Tapponnier, P.; Hahou, Y.; Menzhi, M.; Frogneux, M.; et al. The Al Hoceima Mw 6.4 earthquake of 24 February 2004 and its aftershocks sequence. *J. Geodyn.* **2014**, *77*, 89–109. [CrossRef]
35. Kariche, J.; Meghraoui, M.; Timoulali, Y.; Cetin, E.; Toussaint, R. The Al Hoceima earthquake sequences of 1994, 2004 and 2016: Stress transfer and poroelasticity in the Rif and Alboran Sea region. *Geophys. J. Int.* **2018**, *212*, 42–53. [CrossRef]
36. Buforn, E.; Pro, C.; Sanz de Galdeano, C.; Cantavella, J.V.; Cesca, S.; Caldeira, B.; Udías, A.; Mattesini, M. The 2016 south Alboran earthquake (Mw = 6.4): A reactivation of the Ibero-Maghrebian region? *Tectonophysics* **2017**, *712*, 704–715. [CrossRef]
37. López Casado, C.; Garrido, J.; Delgado, J.; Peláez, J.A.; Henares, J. HVSR estimation of site effects in Melilla (Spain) and the damage pattern from the 01/25/2016 Mw 6.3 Alborán Sea earthquake. *Nat. Hazards* **2018**, *93*, S153–S167. [CrossRef]
38. Mezcua, J.; Rueda, J. Seismological evidence for a delamination process in the lithosphere under the Alboran Sea. *Geophys. J. Int.* **1997**, *129*, F1–F8. [CrossRef]
39. Medina, F.; El Alami, S.O. Focal Mechanisms and State of Stress in the Al Hoceima Area. *Bull. L'institut Sci. Sect. Sci. Terre* **2006**, *28*, 19–30.
40. Thio, H.K.; Song, X.; Saikia, C.; Helmberger, D.V.; Woods, B.B. Seismic source and structure estimation in the western Mediterranean using sparse broadband network. *J. Geophys. Res.* **1999**, *104*, 845–861. [CrossRef]
41. Bezzeghoud, M.; Buforn, E. Source parameters of the 1992 Melilla (Spain, MW = 4.8), 1994 Alhoceima (Morocco, MW = 5.8) and 1994 Mascara (Algeria, MW = 5.7) earthquakes and seismotectonic implications. *Bull. Seism. Soc. Am.* **1999**, *89*, 359–372.
42. Delvaux, D.; Sperner, B. New aspects of tectonic stress inversion with reference to the TENSOR program. *Geol. Soc. Lond. Spec. Publ.* **2003**, *212*, 75–100. [CrossRef]
43. Angelier, J.; Mechler, P. Sur une méthode graphique de recherche des contraintes principales également utilisable en tectonique et en sismologie: La méthode des diédres droits. *Bull. Soc. Géol. France XIX* **1977**, *7*, 1309–1318. [CrossRef]
44. Delvaux, D.; Barth, A. African stress pattern from formal inversion of focal mechanism data. *Tectonophysics* **2010**, *482*, 105–128. [CrossRef]

45. Hussein, H.M.; AbouElenean, K.M.; Marzou, I.A.; El-Nader, L.F.; Ghazala, H.; El Gabry, M.N. Present-day tectonic stress regime in Egypt and surrounding area based on inversion of earthquake focal mechanism. *J. Afr. Earth. Sci.* **2013**, *81*, 1–15. [CrossRef]

46. Gephart, J.W.; Forsyth, D.W. An improved method for determining the regional stress tensor using earthquake focal mechanism data: Application to the San Fernando earthquake sequence. *J. Geophys. Res.* **1984**, *89*, 9305–9320. [CrossRef]

47. Soumaya, A.; Ben Ayed, N.; Delvaux, D.; Ghanmi, M. Spatial variation of present-day stress field and tectonic regime in Tunisia and surroundings from formal inversion of focal mechanisms: Geodynamic implications for central Mediterranean. *Tectonics* **2015**, *34*, 1154–1180. [CrossRef]

48. Sawires, R.; Peláez, J.A.; Ibrahim, H.A.; Fat Helbary, R.E.; Henares, J.; Hamdache, M. Delineation and characterization of a new seismic source model for seismic hazard studies in Egypt. *Nat. Hazards* **2016**, *80*, 1823–1864. [CrossRef]

49. Hamdache, M.; Peláez, J.A.; Gospodinov, D.; Henares, J. Statistical features of the 2010 Beni-Ilmane, Algeria, aftershock sequence. *Pure Appl. Geophys.* **2018**, *175*, 773–792. [CrossRef]

50. Henares, J.; López Casado, C.; Sanz de Galdeano, C.; Delgado, J.; Peláez, J.A. Stress field in the Ibero-Maghrebianregion. *J. Seismol.* **2003**, *7*, 65–78. [CrossRef]

51. Sanz de Galdeano, C. Geologic evolution of the Betic Cordilleras in the Western Mediterranean, Miocene to the present. *Tectonophysics* **1990**, *172*, 107–119. [CrossRef]

52. Sanz de Galdeano, C. The E-W segments of the contact between the External and Internal Zones of the Betic and Rif Cordilleras and the E-W corridors of the Internal Zone (A combined explanation). *Estudios Geológicos* **1996**, *52*, 123–136. [CrossRef]

53. Galindo Zaldívar, J.; Jabaloy, A.; Serrano, I.; Morales, J.; González Lodeiro, F.; Torcal, F. Recent and present-day stresses in the Granada Basin (Betic Cordilleras): Example of a late Miocene-present-day extensional basin in a convergent plate boundary. *Tectonics* **1999**, *18*, 686–702. [CrossRef]

54. Gutenberg, B.; Richter, C.F. Frequency of Earthquake in California. *Bull. Seismol. Soc. Am.* **1944**, *34*, 185–188. [CrossRef]

55. Gutenberg, B.; Richter, C.F. *Seismicity of the Earth*; Princeton University Press: Princeton, NJ, USA, 1954; 310p.

56. Schorlemmer, D.; Woessner, J. Probability of detecting anearthquake. *Bull. Seismol. Soc. Am.* **2008**, *98*, 2103–2117. [CrossRef]

57. Mignan, A.; Werner, M.J.; Wiemer, S.; Chen, C.C.; Wu, Y.M. Bayesian estimation of the spatially varying completeness magnitude of earthquake catalogs. *Bull. Seismol. Soc. Am.* **2011**, *101*, 1371–1385. [CrossRef]

58. Rydelek, P.A.; Sacks, I.S. Testing the completeness of earthquake catalogs and the hypothesis of self-similarity. *Nature* **1989**, *337*, 251–253. [CrossRef]

59. Taylor, D.W.A.; Snoke, J.A.; Sacks, I.S.; Takanami, T. Non-linear frequency-magnitude relationship for the Hokkaido corner, Japan. *Bull. Seismol. Soc. Am.* **1990**, *80*, 605–609. [CrossRef]

60. Woessner, J.; Wiemer, S. Assessing the quality of earthquake catalogs: Estimating the magnitude of completeness and its uncertainty. *Bull. Seismol. Soc. Am.* **2005**, *95*, 684–698. [CrossRef]

61. Mignan, A.; Woessner, J. Estimating the Magnitude of Completeness for Earthquake Catalogs; Community Online Resource for Statistical Seismicity Analysis: 2012; 45p. Available online: http://www.corssa.org/en/articles/overview/ (accessed on 27 August 2022).

62. Wiemer, S.; Wyss, M. Minimum magnitude of completeness in earthquake catalogs: Examples from Alaska, the Western United States, and Japan. *Bull. Seismol. Soc. Am.* **2000**, *90*, 859–869. [CrossRef]

63. Ogata, Y.; Katsura, K. Analysis of the temporal and spatial heterogeneity of magnitude frequency distribution inferred from earthquake catalogs. *Geophys. J. Int.* **1993**, *113*, 727–738. [CrossRef]

64. Amorese, D. Applying a change-point detection method on frequency-magnitude distribution. *Bull. Seismol. Soc. Am.* **2007**, *97*, 1742–1749. [CrossRef]

65. Cao, A.M.; Gao, S.S. Temporal variation of seismic *b*-values beneath northeastern Japan island arc. *Geophys. Res. Lett.* **2002**, *29*, 48-1–48-3. [CrossRef]

66. Leptokaropoulos, K.M.; Karakostas, V.G.; Papadimitriou, E.E.; Adamaki, A.K.; Tan, O.; İnan, S. A Homogeneous Earthquake Catalog for Western Turkey and Magnitude of Completeness Determination. *Bull. Seismol. Soc. Am.* **2013**, *103*, 2739–2751. [CrossRef]

67. Wiemer, S. A software package to analyze seismicity: Zmap. *Seismol. Res. Lett.* **2001**, *72*, 373–382. [CrossRef]

68. Aki, K. Maximum likelihood estimate of b in the formula Log N = a − bM and its confidence limits. *Bull. Earthq. Res. Inst. Tokyo Univ.* **1965**, *43*, 237–239.

69. Shi, Y.; Bolt, B.A. The standard error of the magnitude frequency b value. *Bull. Seismol. Soc. Am.* **1982**, *72*, 1677–1687. [CrossRef]

70. Marzocchi, W.; Sandri, L. A review and new insights on the estimation of the *b*-value and its uncertainty. *Ann. Geophys.* **2003**, *46*, 1271–1282.

71. Márquez-Ramírez, V.H.; Nava, F.A.; Zúñiga, F.R. Correcting the Gutenberg-Richter b value for effects of rounding and noise. *Earthq. Sci.* **2015**, *28*, 129–134. [CrossRef]

72. Utsu, T. A method for determining the value of b in theformula Log n = a − bm showing the magnitude-frequency relation for earthquakes. *Geophys. Bull. Hokkaido Univ.* **1965**, *13*, 99–103. (In Japanese)

73. Utsu, T.; Ogata, Y.; Matsu'ura, R.S. The centenary of the Omori formula for a decay law of aftershock activity. *J. Phys. Earth* **1995**, *43*, 1–33. [CrossRef]

74. Enescu, B.; Mori, J.; Masatoshi, M.; Kano, Y. Omori-Utsu law c-values associated with recent moderate earthquakes in Japan. *Bull. Seismol. Soc. Am.* **2009**, *99*, 884–891. [CrossRef]

75. Kisslinger, C.; Jones, L.M. Properties of aftershocks in Southern California. *J. Geophys. Res.* **1991**, *96*, 11947–11958. [CrossRef]
76. Nyffengger, P.; Frolich, C. Recommendations for determining p values for aftershock sequence and catalogs. *Bull. Seismol. Soc. Am.* **1998**, *88*, 1144–1154.
77. Nyffengger, P.; Frolich, C. Aftershock occurrence rate decay properties for intermediate and deep earthquake sequences. Geophys. *Res. Lett.* **2000**, *27*, 1215–1218. [CrossRef]
78. Daley, D.J.; Vere Jones, D. *An Introduction to the Theory of Point Processes. Vol. I. Elementary Theory and Methods*; Springer: Berlin/Heidelberg, Germany, 2003; 471.
79. Zhuang, J.; Werner, M.J.; Hainzl, S.; Harte, D.; Zhou, S. Basic Models of Seismicity: Spatiotemporal Models; Community Online Resource for Statistical Seismicity Analysis. 2011. Available online: http://www.corssa.org/en/articles/overview/ (accessed on 27 August 2022).
80. Gospodinov, D.; Karakostas, V.; Papadimitriou, E. Seismicity rate modeling for prospective stochastic forecasting. The case of 2014 Kefalonia, Greece, seismic excitation. *Nat. Hazards* **2015**, *79*, 1039–1058. [CrossRef]
81. Utsu, T. Seismological evidence for anomalous structure of island arcs with special reference to the Japanese region. *Rev. Geophys.* **1971**, *9*, 839–890. [CrossRef]
82. Akaike, H. A new look at the statistical model identification. *IEEE Trans. Autom. Control* **1974**, *19*, 716–723. [CrossRef]
83. Gospodinov, D.; Rotondi, R. Exploratory analysis of marked Poisson processes applied to Balkan earthquake sequences. *J. Balkan Geophys. Soc.* **2001**, *4*, 61–68.
84. Ross, S.M. *Introduction to Probability Models*; Academic Press: Berkeley, CA, USA, 2010; 842p.
85. Taylor, H.M.; Karlin, S. *An Introduction to Stochastic Modeling*; Academic Press: Cambridge, MA, USA, 1984; 410p.
86. Tzanis, A.; Vallianatos, F. Distributed power-law seismicity changes and crustal deformation in the SW Hellenic arc. *Nat. Hazards Earth Syst. Sci.* **2003**, *3*, 179–195. [CrossRef]
87. Kanamori, H.; Anderson, D.L. Theoretical basis of some empirical relations in seismology. *Bull. Seismol. Soc. Am.* **1975**, *65*, 1073–1095.
88. Hamdache, M.; Peláez, J.A.; Talbi, A. Scaling properties of aftershock sequences in Algeria Morocco Region. In *Earthquake Research and Analysis—New Advances in Seismology*; D'Amico, S., Ed.; Intech Open: London, UK, 2013. [CrossRef]
89. Marzocchi, W.; Taroni, M.; Falcone, G. Earthquake forecasting during the complex Amatrice-Norcia seismic sequence. *Sci. Adv.* **2017**, *3*, e1701239. [CrossRef]
90. Jordan, T.H.; Chen, Y.-T.; Gasparini, P.; Madariaga, R.; Main, I.; Marzocchi, W.; Papadopoulos, G.; Sobolev, G.; Yamaoka, K.; Zschau, J. Operational earthquake forecasting. State of knowledge and guidelines for utilization. *Ann. Geophys.* **2011**, *54*, 4. [CrossRef]
91. Gospodinov, D.; Papadimitriou, E.E.; Karakostas, V.G.; Ranguelov, B. Analysis of relaxation temporal patterns in Greece through the RETAS model approach. *Phys. Earth Planet. Int.* **2007**, *165*, 158–175. [CrossRef]

applied sciences

Article

A Comprehensive Study of Local, Global, and Combined Optimization Methods on Synthetic Seismic Refraction and Direct Current Resistivity Data

Paul Edigbue [1], Ismail Demirci [2], Irfan Akca [2], Hamdan Ali Hamdan [3], Panagiotis Kirmizakis [1], Pantelis Soupios [1,*], Emin Candansayar [2], Sherif Hanafy [1] and Abdullatif Al-Shuhail [1]

1 Department of Geosciences, College of Petroleum Engineering and Geosciences, King Fahd University of Petroleum and Minerals, Dhahran 31261, Saudi Arabia
2 Department of Geophysical Engineering, Ankara University, Gölbaşı 06830, Turkey
3 Department of Applied Physics & Astronomy, University of Sharjah, Sharjah P.O. Box 27272, United Arab Emirates
* Correspondence: panteleimon.soupios@kfupm.edu.sa; Tel.: +966-(013)-860-2689

Abstract: Most geophysical inversions face the problem of non-uniqueness, which poses a challenge in the mapping and delineation of the subsurface anomalies. To tackle this challenge, a combined local and global optimization approach is considered for jointly inverting two-dimensional direct current resistivity (DCR) and seismic refraction (SR) data that aim to estimate the corresponding physical model parameters. In this combined approach, the output of the local optimization method is used to determine the search space and tuning parameters for the global optimization algorithm. The multi-objective genetic algorithm (non-dominated sorting genetic algorithm) was utilized to jointly optimize the objective functions of two different methods. Because the genetic algorithm is a population-based optimization method, it requires numerous forward calculations. To deal with the expected high computational cost associated with this approach, parallel computing was utilized for the forward function evaluations to reduce the run time of the entire process. The proposed approach was tested using synthetic two-dimensional resistivity and velocity models that had three different types of anomalies (dyke, positive, and combined positive and negative). The results showed an improvement in the anomaly delineation in the output of the combined local and global optimization method compared with the local optimization method. Additionally, similar synthetic models were tested using only the single objective global optimization algorithm (conventional global optimization), which showed promising anomaly delineation.

Keywords: individual inversion; joint inversion; seismic refraction; direct current resistivity; combined local and global optimization

Citation: Edigbue, P.; Demirci, I.; Akca, I.; Hamdan, H.A.; Kirmizakis, P.; Soupios, P.; Candansayar, E.; Hanafy, S.; Al-Shuhail, A. A Comprehensive Study of Local, Global, and Combined Optimization Methods on Synthetic Seismic Refraction and Direct Current Resistivity Data. *Appl. Sci.* **2022**, *12*, 11589. https://doi.org/10.3390/app122211589

Academic Editor: Jianbo Gao

Received: 5 October 2022
Accepted: 10 November 2022
Published: 15 November 2022

Publisher's Note: MDPI stays neutral with regard to jurisdictional claims in published maps and institutional affiliations.

1. Introduction

The primary goal of inverting geophysical data is to estimate the parameters of a model that will give a theoretical response similar to the field observations [1,2]. However, it is unusual to have a unique solution since most geophysical inverse problems are ill-posed. Previous studies included regularization constraints in the objective function to solve the problem of instability [3–6]. In some cases, with a regularization term in the objective function, the inversion still faces the challenge of non-uniqueness (i.e., ambiguity) associated with the inverse problem. To reduce uncertainties related to the inversion of a dataset belonging to a single geophysical method, many researchers have applied the concept of joint inversion [7–16] of more than one method, which provides better model resolution than individual inversion [17–19]. Joint inversion can even resolve ambiguity associated with the geophysical method(s) applied in an area with low physical properties contrast [20–22]. Geophysical joint inversion tries to optimize a single objective function formed by a weighted or arithmetic sum of the individual objective functions of corresponding methods [23]. However, the issue of parametric coupling (i.e., model integration) arises when we are dealing with the inversion of datasets acquired with more than one geophysical method [24].

In this study, we apply the structural model coupling approach, which involves the cross-gradient constraint method that is commonly used for geophysical inversion [8]. The basic idea using this approach is that the gradients of relevant model parameters are spatially correlated. The cross-gradient approach has been adapted for the joint inversion of different geophysical data [13–16,25–29]. Wang et al. [17] applied the cross-gradient algorithm in a joint inversion involving both the controlled-source audio magnetotelluric (CSAMT) and magnetic methods. Demirci et al. [15] formulated an objective function concerning weighted cross-gradient, which limits the dominance of one type of model parameter to another. Zhang et al. [30] utilized the cross-gradient constraint to impose a common structural framework in the joint inversion of EM and acoustic data, which reconstructs the structures satisfactorily.

Furthermore, Yin et al. [31] applied a cross-gradient technique to invert magnetotelluric (MT) and gravity data and tested their algorithm using both synthetic and real datasets. Finally, Jordi et al. [32] introduced a new approach to the cross-gradient constraint, which involved the use of an irregular grid in unstructured mesh in the finite element method. They used the method to invert the DCR and ground penetrating (GPR) data. All the above examples used the conventional or local optimization method that incorporated all acquired data within the same Jacobian using a unique objective function for all geophysical methods.

The local optimization techniques are applied iteratively to obtain an updated model that minimizes the objective function, which may not be the global solution to the inverse problem. The global optimization algorithm in geophysical inversion might be used to search for a solution space to avoid being stuck in the local minimum of the objective function. For instance, Liu et al. [33] applied the particle swarm optimization (PSO) algorithm in a parametric inversion involving magnetic data. Rani et al. [34] used the genetic price algorithm (GPA) to monitor the movement of contaminants in the subsurface. Additionally, [35] introduced a hybrid approach by using the results of the local optimization method as an input to the genetic algorithm for the modeling of the SR data. This novel algorithm by [35] overcame the problem of being stuck in the local minima. It optimized the computational cost of the genetic algorithm using the multicore parallel computing method.

Similarly, local and global (hybrid) optimization has been applied to complement each other to subdue their shortcomings [35–37]. Most previous work has considered an objective function from a different perspective to perform joint inversion using a global optimization method, such as when using a local optimization algorithm. For instance, Schwarzback et al. [38] and Ayani et al. [39] considered the objective function for the inversion of electromagnetic data in two terms. First, they tried to minimize the data misfit and the roughness of the model at the same time. Akca et al. [40] applied a non-dominated sorting genetic algorithm in a joint 1D parametric inversion involving magnetic resonance and vertical electrical sounding.

In previous studies, joint inversion/interpretation was applied by using either local or global optimization methods. Moreover, local and global (hybrid) optimization has been applied to complement each other to subdue their shortcomings. However, based on our knowledge, the joint modeling of different geophysical data by using the results from the inversion (local optimization) to constrain the search space of the global optimization approach has never been reported in the literature. Thus, a procedure that constrains the global part of the combined optimization algorithm by using the local optimization to define a close search space to the real model parameters was proposed and designed. In this way, the search space of the model parameters has been limited to a more reliable range, drastically reducing the computation time.

This study presents the combined local and global optimization approach to jointly model SR and DCR data. With this proposed approach, the multi-objective (i.e., integration of the DCR and SR misfit functions) global optimization algorithm is used for the first time for the two-dimensional joint inversion of two different geophysical datasets, i.e., SR and DCR data. Specifically, in this algorithm, the global part of the combined optimization algorithm is constrained by using the output of the local optimization to define a search space, significantly improving its run time and mitigating model instability. In addition to the improved computational cost that resulted from applying the combined local and global optimization methods, we made the multi-objective global optimization algorithm run on parallel computing. This process further optimized the computational cost and devised the optimum technique using the combined optimization algorithm. We tested and discussed the efficacy of our proposed algorithm using synthetic SR and DCR data.

2. Optimization Methods

2.1. Local Optimization Method

The individual and joint inversion of DCR and SR data are usually regularized with a smoothing function due to the non-uniqueness and instability associated with the inverse process. In the inversion of both DCR and SR methods, the data misfit can be formulated as follows:

$$E(m) = \|W_d(d_{obs} - f(m))\|^2, \tag{1}$$

$f(m)$ is a function that is used to describe forward modeling, d_{obs} is the observed field record, and W_d is the weight matrix used to adjust the data anomaly (e.g., high or low amplitudes). Usually, we use norms to quantify the misfit between the observed and calculated data, and the common one used for this type of inverse problem is the L_2 norm because we assume that the error in the data is Gaussian. Therefore, the objective functions for the DCR and SR data are given as follows:

$$\Phi(m_{dc}) = \|W_{dc}(d_{dc} - f_{dc}(m_{dc}))\|^2 + \alpha_{dc}\|\nabla^2 m_{dc}\|^2 \tag{2}$$

$$\Phi(m_{sr}) = \|W_{sr}(d_{sr} - f_{sr}(m_{sr}))\|^2 + \alpha_{sr}\|\nabla^2 m_{sr}\|^2 \tag{3}$$

where Φ is the misfit or objective function, d_{dc} and d_{sr} are the measured data, and f_{dc} and f_{sr} is the model response for the DCR and SR methods, respectively. Additionally, $S = \nabla^2 m_{sr}^2$ is the Laplacian of the model parameter that is transformed into the smoothness matrix by obtaining its Laplacian operator, while α_{dc} and α_{sr} are the regularization parameters that determine the level of the smoothness of resistivity and seismic models, respectively. The joint inversion offers conventional ways of integrating data from different geophysical methods in such a manner that the outcome models are consistent and similar. One of the methods used to accomplish this goal is to apply the cross-gradient constraint proposed by Gallardo et al. [8] in the objective function. The parallel spatial variation in the models (resistivity or seismic velocity) is required to satisfy the cross-gradient constraint [8,13,15]. This means that model anomalies or layer boundaries must essentially point in the same or opposite direction. Applying the cross-gradient constraint, Equations (2) and (3) become:

$$\Phi(m_{dc}, m_{sr}) = \left\|\begin{array}{c} W_{dc}(d_{dc} - f_{dc}(m_{dc})) \\ W_{sr}(d_{sr} - f_{sr}(m_{sr})) \end{array}\right\|^2 + \left\|\begin{array}{c} \alpha_{dc} \\ \alpha_{sr} \end{array}\right\|\left\|\begin{array}{c} \nabla^2 m_{dc} \\ \nabla^2 m_{sr} \end{array}\right\|^2 + \vec{c}\left\|\begin{array}{c} m_{dc} \\ m_{sr} \end{array}\right\|^2 \tag{4}$$

subject to $\vec{c}(m_{dc}, m_{sr}) = \vec{0}$. Where $\vec{c}(m_{dc}, m_{sr})$ is the cross-gradient constraint and can be defined as:

$$\vec{c}(m_{dc}, m_{sr}) = \nabla m_{dc}(x, z) \times \nabla m_{sr}(x, z) \tag{5}$$

Equation (4) can be minimized by applying the appropriate regularized local optimization algorithm similar to the approach used in [15]. Therefore, the model parameter correction vector can be expressed as:

$$\Delta m = G^{-1}n - G^{-1}B^T\left(BG^{-1}B^T\right)^{-1}\left[BG^{-1}n - B\Delta m_{i-1} + \vec{c}(m_{i-1})\right] \tag{6}$$

where G and n are defined as:

$$G = \left(J^T W^T W J + \alpha C^T C \right) \tag{7}$$

$$n = \left(J^T W^T W R - \alpha C^T C m^{i-1} \right) \tag{8}$$

In Equations (7) and (8), J is the Jacobean matrix, W is the weighting matrix, R is the data residual, C is the Laplacian operator, and B is the cross-gradient derivative. The terms given in Equations (7) and (8) may be rewritten for the joint inversion case as follows:

$$\Delta m = \begin{bmatrix} \Delta m_{res} \\ \Delta m_{seis} \end{bmatrix}, \quad G = \begin{bmatrix} G_{res} & 0 \\ 0 & G_{seis} \end{bmatrix}, \ and \ n = \begin{bmatrix} n_{res} \\ n_{seis} \end{bmatrix} \tag{9}$$

In this inversion approach, a resistivity model is created for the DCR method, while for the slowness model, the inverse of velocity is used for the SR method [15]. Conceptually, the joint inversion process requires the model discretization for all the geophysical methods to be structurally similar.

2.2. Global Optimization Method

The global optimization method applied in this study involves the application of the genetic algorithm (GA) for a single objective function case and non-dominated sorting genetic algorithm 2 (NSGA II) for the multi-objective joint optimization approach. The GA is a special case of evolutionary algorithm that simulates the process of biological evolution, and it is adapted to solve an optimization problem [41–43]. The process of the genetic algorithm starts from population initialization, which creates chromosomes (potential solutions with respect to the objective function) using the binary coding scheme. The binary coding scheme is conceptualized in a way that it constitutes a solution to the objective function of the inverse problem [43]. The binary coding system has a bit string or chromosome that describes each element or individual in the population. Each bit in the bit string represents a gene that can be assigned values of 0 or 1, also known as an allele [43]. Two individuals or parents are paired or selected from the initial population to produce two offspring. The higher the fitness value of an individual in the population, the higher the chance of being selected and the better its performance in the evolution loop. Among other selection methods, we apply the tournament selection because it practically depicts natural competition for mating rights among individuals in a population. Crossover involves exchanging information (genetic properties) between two paired models (parents) to create two new models (offspring). Three types of crossover options are available in the code used in this study: single point, two points, and scattered crossover. The final evolution operator used in the genetic algorithm is the mutation. This is the random alteration of genes in a chromosome to introduce diversity in the entire population of the genetic algorithm. This process is usually conducted using a probability index that is appropriately chosen based on the degree of randomness to be allowed and computation cost. The process described above is generally referred to as a single objective genetic algorithm; it is suitable for the inverse problem having one objective function (i.e., applied in the individual inversion of DCR or SR). All these GA evolution operators (i.e., selection, crossover, and mutation) are used to modify the solution parameters.

As mentioned above, the non-dominated sorting genetic algorithm 2 (NSGA II) is a variant of the GA commonly used to solve multi-objective optimization problems. The NSGA II procedures involve the creation of an offspring population (Q_t) that has an equal size to the initial population (P_t) using the selection, crossover, and mutation processes as shown in Figure 1. After that, (Q_t) and (P_t) are combined to produce (R_t), which is double the size of (P_t). Then, the non-domination (i.e., the optimum set of solutions in all the objective functions) sorting of the entire population (R_t) is performed, and the best non-dominated solutions are selected, which are indicated as F_1, F_2, $and\ F_3$ (levels of non-domination) in Figure 1. The topmost non-dominated solutions are accepted until the initial population size is reached; thereafter, the rest of the non-dominated solutions are rejected because the process cannot accept more than the initial population size. Sometimes only F_1 can satisfy the initial population size requirement, and that will be enough for the next generation (P_{t+1}); then, the rest are rejected (Figure 1). Consequently, the NSGA-II emphasizes both the non-dominated and less crowded points.

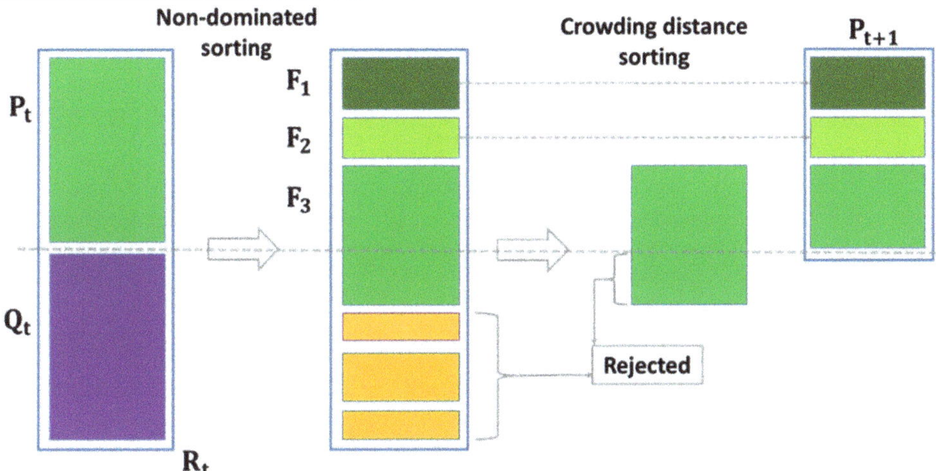

Figure 1. Illustration of the non-dominated sorting algorithm 2 (NSGA II). *Qt* is the offspring population; *Pt* is the initial population; *Rt* is the total population; and F_1, F_2, and F_3 are the levels of non-domination solutions.

One of the major challenges in applying the genetic algorithm to the inverse problems that require the simultaneous optimization of more than one objective function is preventing the domination of one objective function to another. Therefore, the multi-objective genetic algorithm was used to search for the optimum solutions that do not dominate (i.e., a set of optimum solutions in both DCR and SR objective functions) each other. The two objectives of the joint inversion of DCR and SR data using the multi-objective global optimization method can be represented similar to Equation (1) as:

$$\mathrm{GA}(\boldsymbol{m}_{dc}) = \left\| \boldsymbol{W}_{dc}(\boldsymbol{d}_{dc} - \boldsymbol{f}_{dc}(\boldsymbol{m}_{dc})) \right\|^2 \tag{10}$$

and

$$\mathrm{GA}(\boldsymbol{m}_{sr}) = \left\| \boldsymbol{W}_{sr}(\boldsymbol{d}_{sr} - \boldsymbol{f}_{sr}(\boldsymbol{m}_{sr})) \right\|^2 \tag{11}$$

The objective functions are used to map the decision variables (model parameters) into an objective space where we delineate solutions that are not dominated or perform optimally in both objectives. These sets of solutions (Pareto sets) align in a pattern known as Pareto optimal solutions or the Pareto front [44]. Basically, the term "Pareto optimality" is used to describe these sets of solutions, implying that no further optimal solution in both objectives can be obtained. Details about the concept of Pareto optimality and non-domination is discussed by [12]. The NSGA II algorithm sorts the set of solutions as they arrive at the Pareto optimal front. This technique was proposed by [45] to overcome some of the limitations observed in some previous evolutionary multi-objective algorithms. These limitations include computational intricacy, elitism problems, and parameter sharing specifications. The NSGA II only utilizes standard parameters of the genetic algorithm needs, and no extra parameters are needed for its multi-objective base optimization (see the appendix section for the concept of NSGA II).

2.3. Combined Local and Global Optimization Method

In the combined optimization algorithm, we used the output of the local optimization method to define the search space for the global optimization technique to speed up and reach the global solution faster. Since the local optimization has been constrained by smoothing terms (i.e., second terms of Equations (2) and (3)), the combined optimization algorithm is linearly constrained by applying the output of the local optimization algorithm to define the lower and upper bounds of the search space. Fundamentally, the search space is determined by modifying the range (minimum and maximum) of the model parameters obtained from the output of the local optimization algorithm. Depending on the quality of the output from the local optimization algorithm, scaling the model parameters up and down by 10 to 30% is recommended to obtain a good result. The variability (10–30%) of the model parameters was selected based on the expected variation in the modeled geophysical properties, such as velocity and resistivity, in Saudi Arabia. This process creates adequate diversity in the initial population for the global optimization algorithm. A flowchart illustrating the combined local and global optimization algorithm is shown in Figure 2. The summary of all terms we used to describe the combined optimization algorithms is presented in Table 1.

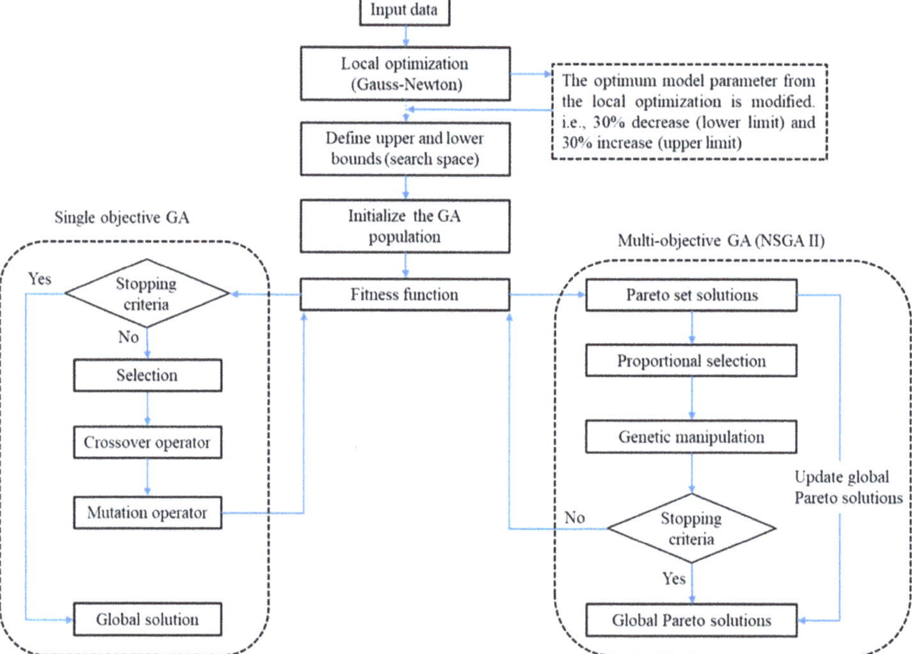

Figure 2. Flowcharts of the proposed combined local and global optimization algorithm.

Table 1. A summary of all the types of inversion applied in this study, their description, and abbreviations are presented below.

S/N	Inversion Type	Abbreviation	Description
1	Local optimization method	LOM	Optimization involves the derivatives of the objectives, e.g., Gauss–Newton.
2	Global optimization method	GOM	Non-derivation optimization, e.g., genetic algorithm.

Table 1. *Cont.*

S/N	Inversion Type	Abbreviation	Description
3	Individual inversion method	IIM	Inversion of a dataset from one geophysical method using the local optimization, e.g., SR data inversion.
4	Joint inversion method	JIM	Inversion of datasets from more than one geophysical method using the local optimization, e.g., SR and DCR data inversion.
5	Single-objective optimization	SOO	Processing of a dataset from one geophysical method using global optimization, e.g., SR data inversion.
6	Multi-objective optimization	MOO	Processing of dataset from more than one geophysical method using global optimization, e.g., SR and DCR data inversion.
7	Combined (local plus global) optimization method	CGO	Inversion of a dataset from either one or more geophysical methods using the combination of local (to define a search space) with global optimization, e.g., DCR, or SR and DCR data inversion.
8	(Global conventional) optimization method	GOM	Processing a dataset from either one or more geophysical methods using only the global optimization, e.g., DCR, or SR and DCR data inversion.

3. Synthetic Test

3.1. Synthetic Data

We examined the feasibility of the combined global optimization (CGO) algorithm by using synthetic earth models that simulated three near-surface scenarios. The first model comprised a dyke anomaly with a resistivity of 1250 ohm-m and a two-layered host environment with resistivities of 50 ohm-m and 250 ohm-m. Similarly, a dyke anomaly having a velocity of 2200 m/s and a two-layered host environment with layer velocities of 1000 m/s and 1500 m/s, respectively, was used for the corresponding velocity model (Figure 3a,d). The second model contained two blocks of positive anomalies, having a resistivity of 250 ohm-m and velocity of 2000 m/s, which was greater than the host environment as shown in Figure 3b,e. The third model was similar to the second one, with a higher parameter contrast (higher and lower than the host model anomalies) compared with the host rock, where the block parameters were set as 1250 ohm-m, 2500 m/s (Figure 3c,f). Generally, these models had the same profile length of 240 m in both methods, with a depth of 50 m in the DCR and 60 m in the SR methods. The DCR data were calculated for a setup with equally spaced 49 electrodes 5 m apart using the dipole–dipole (DD) array. The DD array was selected, since it had a fair penetration depth and a very good to excellent lateral resolution. The seismic survey layout was similar to the DCR profile, where we used 49 receivers with 5 m spacing and 13 sources with 20 m intervals in the SR forward calculations.

3.2. Synthetic Results

To obtain good local optimization (LOM) results, an appropriate regularization parameter was used in addition to adding the smoothing term in the objective function [46,47]. The regularization parameter was determined by obtaining the maximum value of the diagonal matrix in the singular value decomposition of the Jacobian matrix in both the individual and joint inversion of the DCR and SR data. The value of the regularization parameter was modified using a cooling approximation at each iteration. The inversion process started with the initial guess of the model parameters that improved with each iteration. This procedure was used to estimate the model parameter correction vector as applied by [15,16]. The LOM was terminated when there was no decrease in error and the RMS dropped below a certain threshold value (convergence criteria). In the combined global optimization (CGO) techniques, a forward modeling algorithm was used to estimate the theoretical data in both DCR and SR methods; thereafter, the calculated data were compared to the observed field data to compute their misfits. The data misfit (from Equations (10) and (11)) for the DCR method is presented in a simplified form as:

$$M_{dc} = \frac{100 \times \boldsymbol{W}_{dc} \times (\|d_{dc} - td_{dc}\|_2)}{\|d_{dc}\|_2} \tag{12}$$

where d_{dc} is the observed real data, $\boldsymbol{W}_{dc}$ is the weighting matrix of the real data, and td_{dc} is the theoretical data. The misfit function for the SR method is defined with a similar annotation as:

$$M_{sr} = \frac{100 \times \boldsymbol{W}_{sr} \times (\|d_{sr} - td_{sr}\|_2)}{\|d_{sr}\|_2} \tag{13}$$

The combined misfits form the objective function for the multi-objective optimization algorithm that is represented as:

$$M_{dc\ sr} = [M_{dc}\ M_{sr}] \tag{14}$$

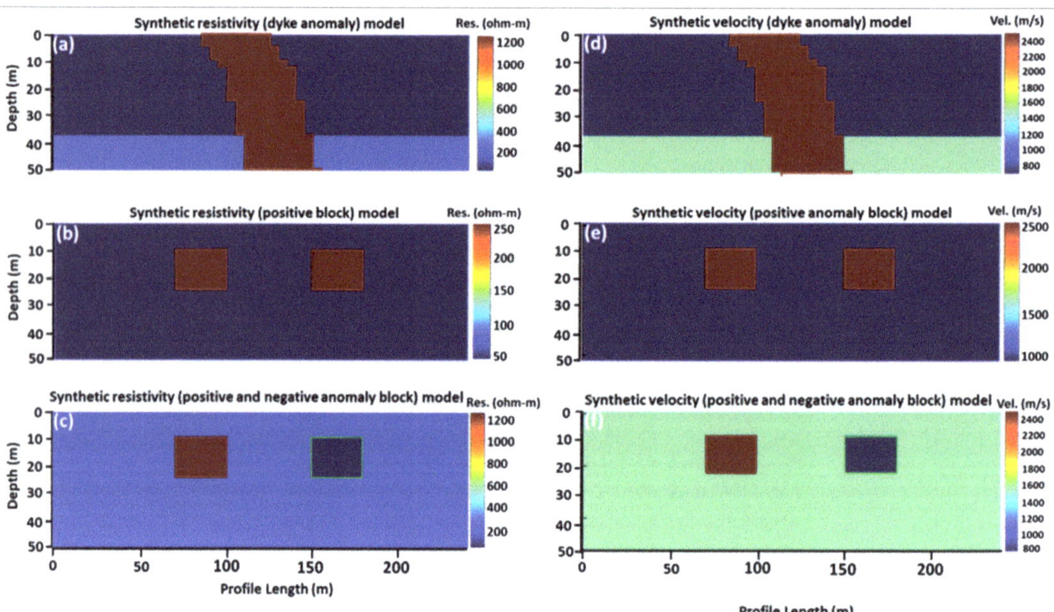

Figure 3. Synthetic models for (**a**) resistivity (dyke anomaly) model, (**b**) resistivity (positive anomaly) model, (**c**) resistivity (positive and negative anomalies) model, (**d**) velocity (dyke anomaly) model, (**e**) velocity (positive anomaly) model, and (**f**) velocity (positive and negative anomalies) model. The dark red and green lines are used to mark the anomaly boundaries.

Notice that for the multi-objective (MOO) algorithm in Equation (14), the individual misfits are not added together but are simply concatenated to make a two-column matrix of the misfits. In addition, selecting the number of populations and generations is important to improve the model resolution in the CGO algorithm for both methods. Conventionally, the population size of 50 is recommended for the GA, with the number of decision variables less than or equal to five. Otherwise, 200 populations should be used when the decision variable is greater than five. The CGO algorithm is terminated when the average change in the spread of Pareto solutions is less than the function tolerance (i.e., 1×10^{-4} for multi-objective optimization, MOO) and the specified number of generations (i.e., for single objective optimization, SOO).

The single objective genetic algorithm runs between 1000 and 1500 generations with an average population size of 250 individuals in both DC-resistivity and SR methods. The SR genetic algorithm runs were completed in about 748, 876, and 1194 min for the dyke, positive, and combined positive–negative anomalies synthetic models, respectively, while the processing times were measured as 176, 181, and 150 min for DC data. After applying the single objective (SOO) genetic algorithm in both geophysical methods (DCR and SR) separately, we performed the joint parameter estimation using the multi-objective (MOO) genetic algorithm. The non-dominated sorting genetic algorithm (NSGA II) used in the multi-objective global optimization showed that there were feasible solutions depicted by their Pareto optimal fronts (Figure 4). The compromised solution (the green star in Figure 4) was chosen and presented as the output of the MOO method. The NSGA II (for both geophysical methods) ran for about 3136, 2953, and 2891 min for the dyke, positive, and the combined positive and negative anomalies synthetic models, respectively. Figures 5–10 show the results of the local and combined optimization methods for the dyke anomaly model (Figures 5 and 6), positive anomaly (Figures 7 and 8), and the positive and negative anomaly models (Figures 9 and 10). The first column of each of the figures contains the inverted resistivity models, while the second column is the inverted velocity models. In addition, the first row in each figure contains the synthetic models, the second row is the individual (IIM)/single objective (SOO) inverted models, and the third row is the joint (JIM)/multi-objective (MOO) inverted models.

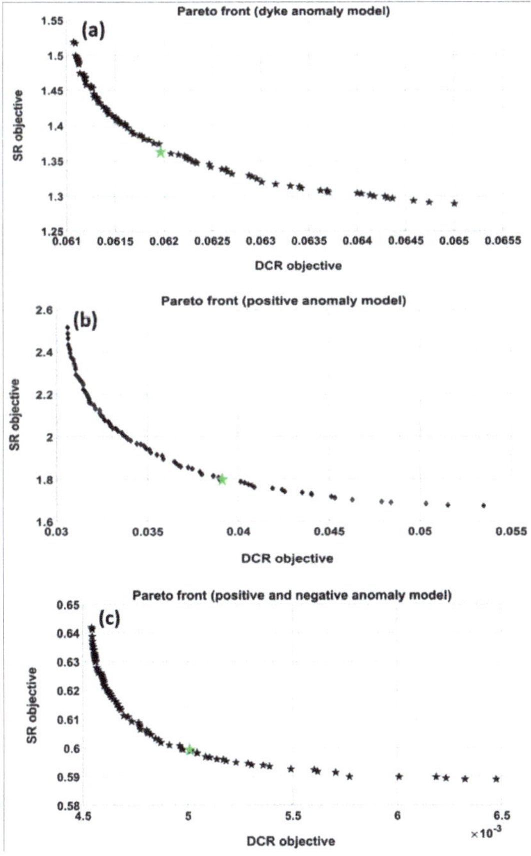

Figure 4. Plots of the Pareto fronts showing the compromise solution (green star shape) for (**a**) a dyke anomaly model, (**b**) a positive anomaly model, and (**c**) positive and negative anomalies model.

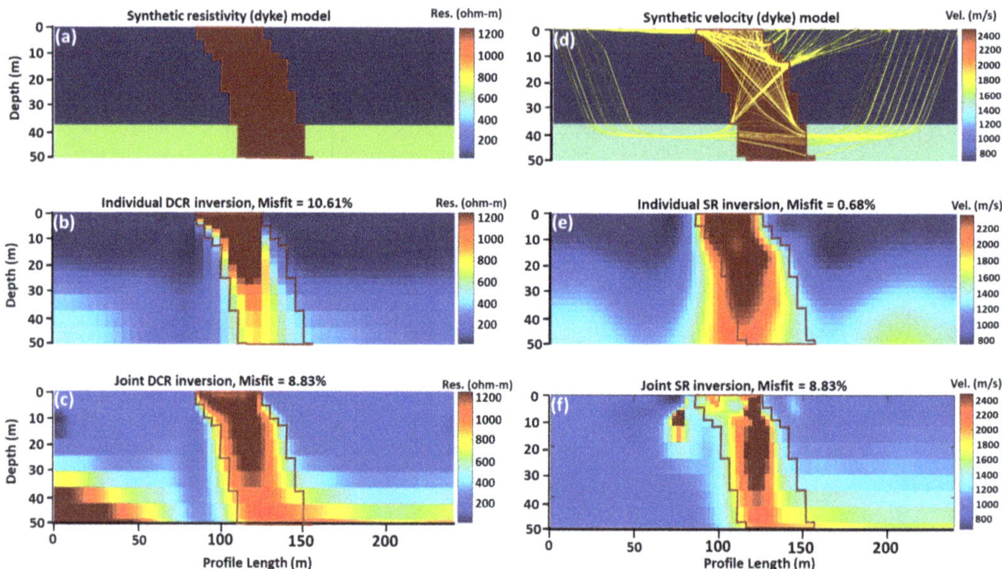

Figure 5. Dyke anomaly model inversion using local optimization method; (**a**) synthetic dc-resistivity model, (**b**) individual inverted resistivity model, (**c**) joint inverted resistivity model, (**d**) synthetic velocity model and its ray path coverage, (**e**) individual inverted velocity model, (**f**) joint inverted velocity model. The dark red line is used to mark the anomaly boundary.

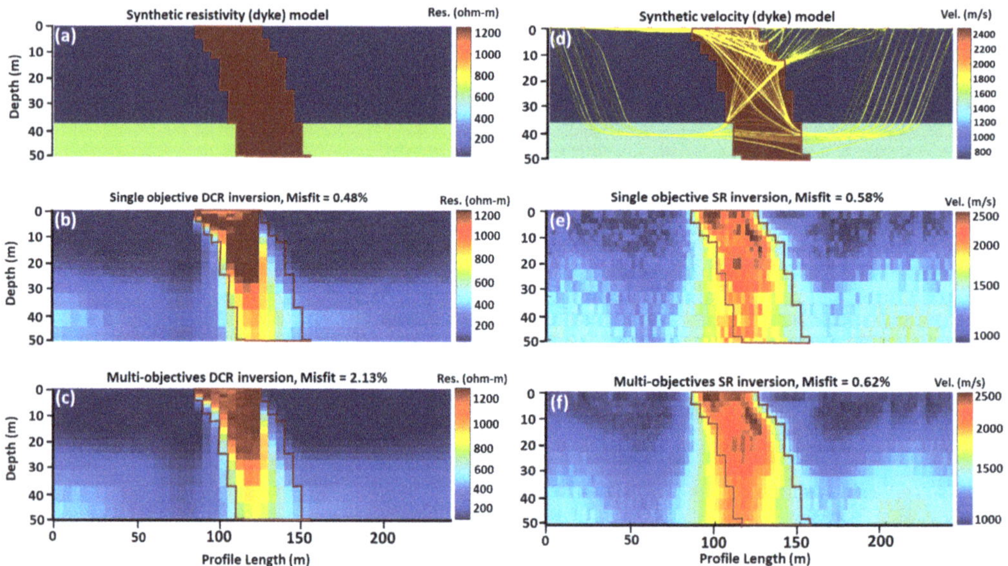

Figure 6. Dyke anomaly model inversion using combined optimization method; (**a**) synthetic dc-resistivity model, (**b**) single objective inverted resistivity model, (**c**) multi-objective inverted resistivity model, (**d**) synthetic velocity model and its ray path coverage, (**e**) single objective inverted velocity model, and (**f**) multi-objective inverted velocity model. The dark red line is used to mark the anomaly boundary.

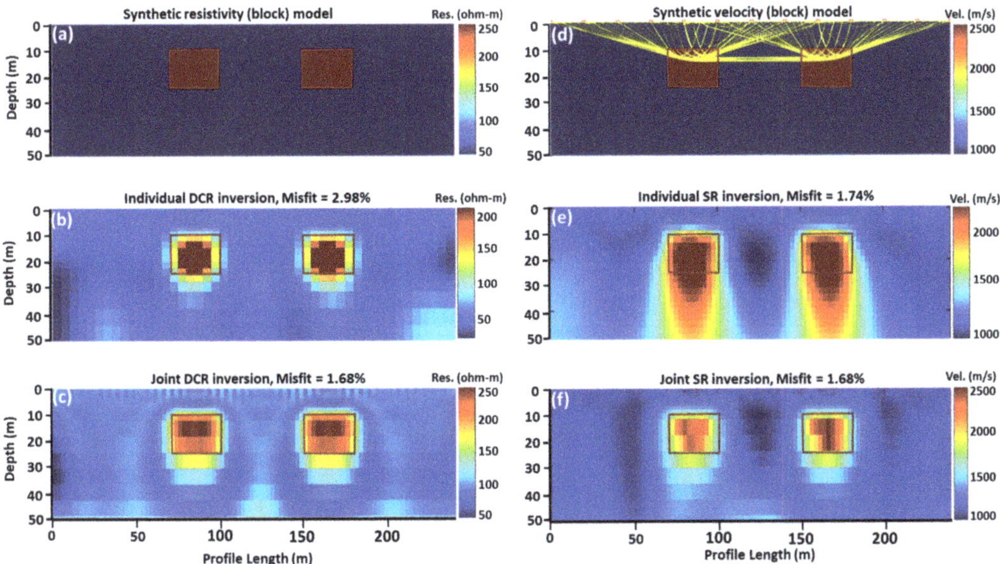

Figure 7. Positive anomaly model inversion using local optimization method; (**a**) synthetic dc-resistivity model, (**b**) individual inverted resistivity model, (**c**) joint inverted resistivity model, (**d**) synthetic velocity model and its ray path coverage, (**e**) individual inverted velocity model, and (**f**) joint inverted velocity model. The dark red line is used to mark the anomaly boundary.

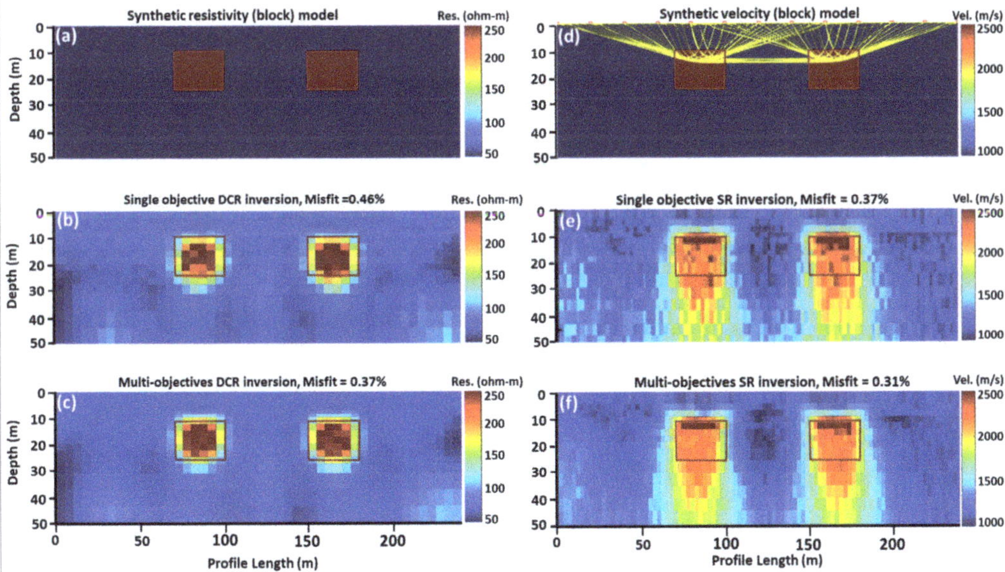

Figure 8. Positive anomaly model inversion using combined optimization method; (**a**) synthetic dc-resistivity model, (**b**) single objective inverted resistivity model, (**c**) multi-objective inverted resistivity model, (**d**) synthetic velocity model and its ray path coverage, (**e**) single objective inverted velocity model, and (**f**) multi-objective inverted velocity model. The dark red line is used to mark the anomaly boundary.

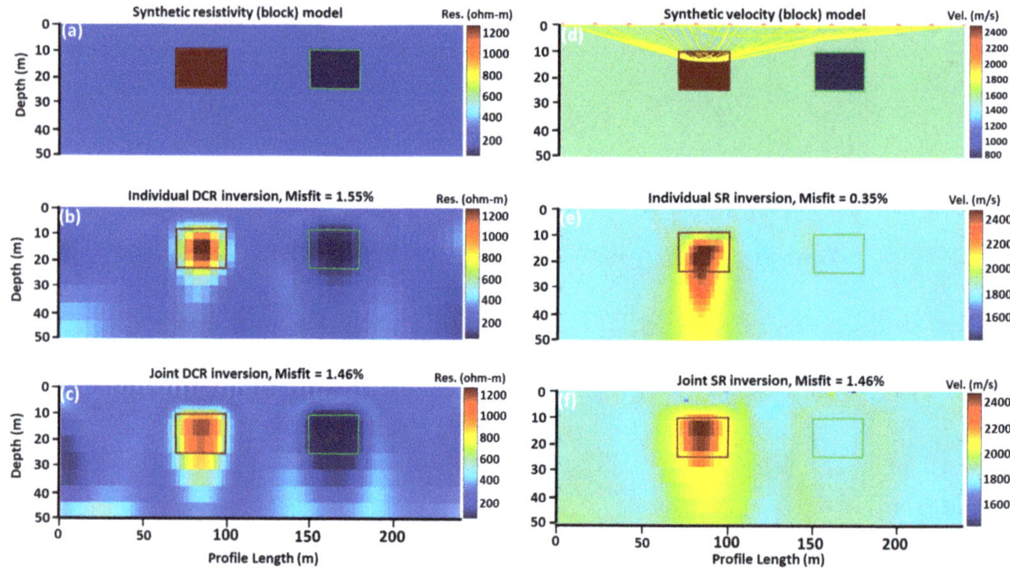

Figure 9. Positive and negative anomaly model inversion using local optimization method; (**a**) synthetic dc-resistivity model, (**b**) individual inverted resistivity model, (**c**) joint inverted resistivity model, (**d**) synthetic velocity model and its ray path coverage, (**e**) individual inverted velocity model, and (**f**) joint inverted velocity model. The dark red and green lines are used to mark the anomaly boundary.

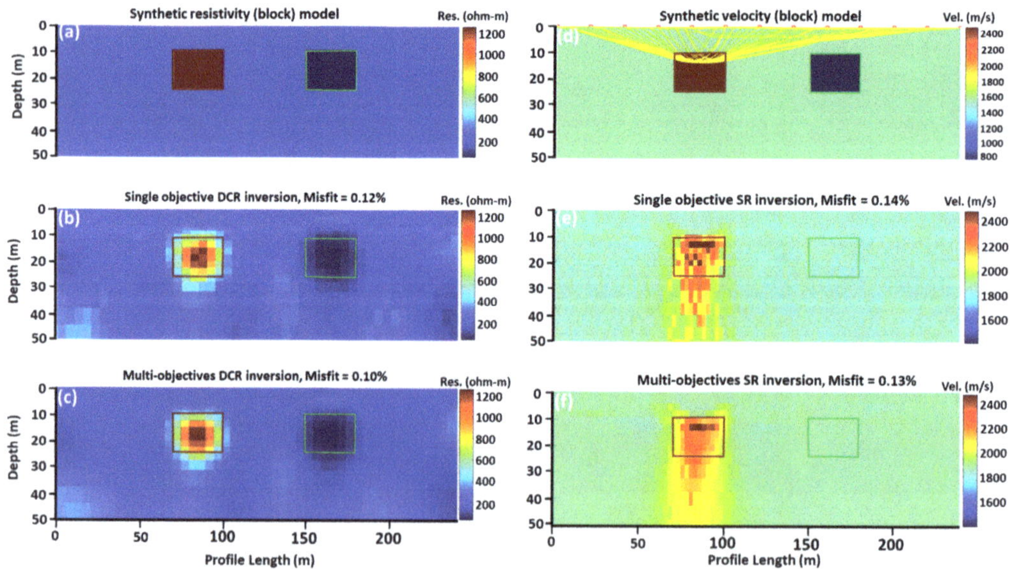

Figure 10. Positive and negative anomaly model inversion using the combined optimization method; (**a**) synthetic dc-resistivity model, (**b**) single objective inverted resistivity model, (**c**) multi-objective inverted resistivity model, (**d**) synthetic velocity model and its ray path coverage, (**e**) single objective inverted velocity model, and (**f**) multi-objective inverted velocity model. The dark red and green lines are used to mark the anomaly boundary.

To see the performance of the conventional global optimization (GOM) only using a similar misfit function as applied in the CGO method, we applied the genetic algorithm with a search space defined apart from the local optimization results using the same DCR and SR synthetic models. Applying the single objective genetic GOM algorithm to the resistivity model showed that the GOM technique provided unstable solutions in delineating both DCR and SR anomalies. For example, Figures 11–14 are the outputs of the genetic algorithm application on the dyke anomaly model (for both DCR and SR), the positive anomaly model, and the combined positive and negative anomaly model (for DCR only). Some of the artifacts observed in the results could probably be attributed to the absence of a constraint or model regularization [38]. The genetic algorithm was applied for 200 generations in the case of the SR method and 2500 generations for all DCR models inversion with a population size of ten times the amount of model parameters [35] in both geophysical methods. This inversion is feasible with the use of high performance and parallel computing (HPPC).

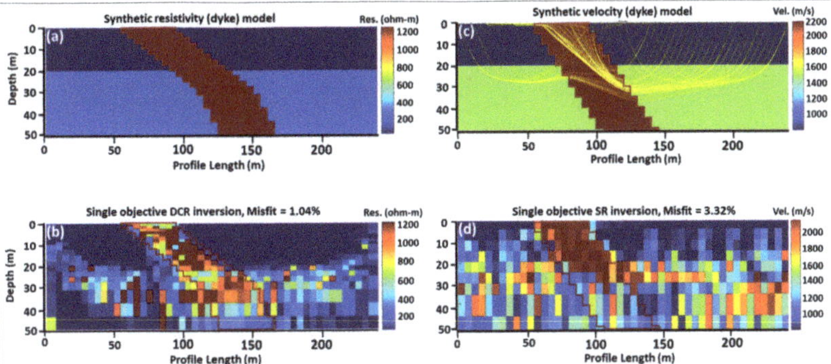

Figure 11. Dyke anomaly model inversion using the single objective global optimization method; (**a**) synthetic dc-resistivity model, (**b**) single object inverted resistivity model, (**c**) synthetic velocity model, and (**d**) single objective inverted velocity model. The dark red line is used to mark the anomaly boundary.

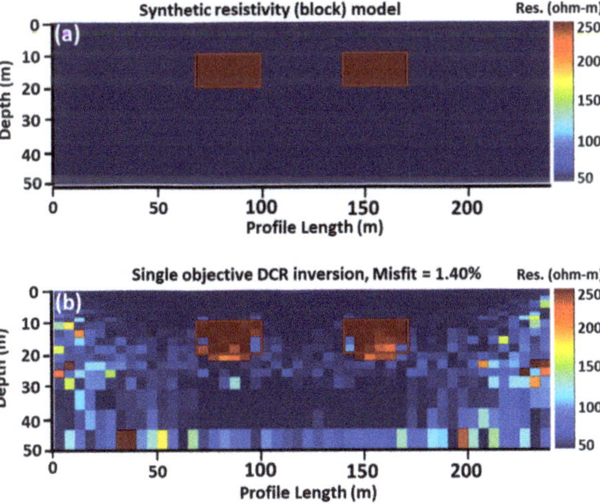

Figure 12. Positive anomaly model inversion using the single objective global optimization method; (**a**) synthetic dc-resistivity model and (**b**) single object inverted resistivity model. The dark red line is used to mark the anomaly boundary.

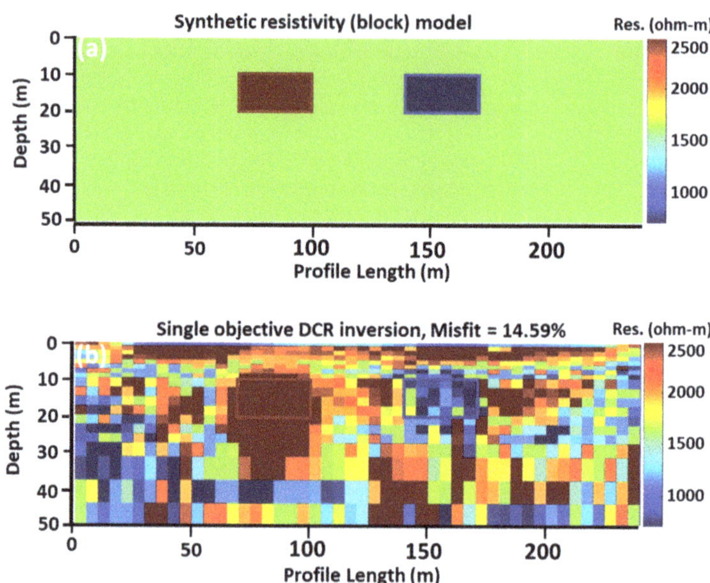

Figure 13. Positive and negative anomaly model inversion using the single objective global optimization method; (**a**) synthetic dc-resistivity model and (**b**) single object inverted resistivity model. The dark red/blue lines are used to mark the anomaly boundary.

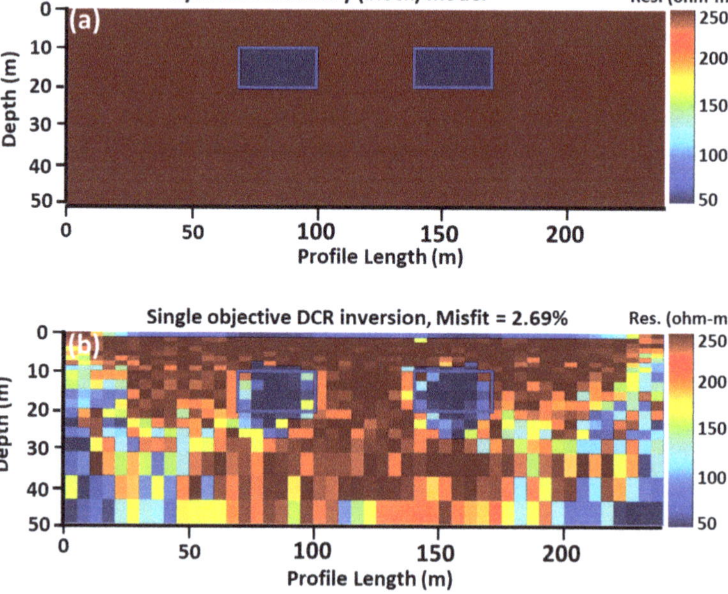

Figure 14. Negative anomaly model inversion using the single objective global optimization method; (**a**) synthetic dc-resistivity model and (**b**) single object inverted resistivity model. The dark blue line is used to mark the anomaly boundary.

4. Discussion

The local optimization algorithm performed optimally in delineating both resistivity and velocity anomalies regarding the tested models. This result is attributed to the application of the smoothing term and appropriate regularization parameters in the objective function to mitigate the effect of the non-uniqueness of the inverse problem. Tables 2–4 summarize the performance of both LOM and CGO algorithms in the inversion involving both DCR and SR methods. Generally, the results from Tables 2–4 show that the CGO improved the misfits compared with the LOM in both the DCR and SR methods but at a relatively high computation cost. Table 5 summarizes the results of the CGO for both (DCR and SR) synthetic models. Despite using a parallel computing approach, the GOM results showed a run time of 6642.15 to 13,962.00 min. This suggests that the most significant challenge with applying the GOM algorithm is the run time.

Table 2. Local and global optimization inversion results parameter for both DCR and SR (synthetic dyke anomaly model) methods.

	Inversion Type	Methods	No. of Iterations/ Generations	Time (min)	Misfit (%)	Target's Reconstruction (%)	
						Geometry	Amplitude
LOM	IIM	SR	10	1.69	0.68	80	95
		DC	8	0.81	10.61	70	97
	JIM	SR	10	10.07	8.83	60	95
		DC	10	10.07	8.83	80	97
CGO	SOO	SR	1000	747.85	0.58	85	100
		DC	1500	175.57	0.48	65	100
	MOO	SR	4000	3136.42	0.62	90	100
		DC	4000	3136.42	2.13	70	100

Table 3. Local and combined (local plus global) optimization inversion results parameters for both DCR and SR (synthetic positive anomaly model) methods.

	Inversion Type	Methods	No. of Iterations/Generations	Time (min)	Misfit (%)	Target's Reconstruction (%)	
						Geometry	Amplitude
LOM	IIM	SR	9	1.83	1.74	90	95
		DC	8	0.76	2.98	80	98
	JIM	SR	10	26.51	1.68	75	100
		DC	10	26.51	1.68	90	105
CGO	SOO	SR	1000	876.00	0.37	93	100
		DC	1500	181.09	0.46	85	100
	MOO	SR	4697	2953.45	0.31	95	100
		DC	4697	2953.45	0.37	95	100

Table 4. Local and global optimization inversion results parameters for both DCR and SR (synthetic positive and negative anomalies model) methods.

	Inversion Type	Methods	No. of Iterations/Generations	Time (min)	Misfit (%)	Target's Reconstruction (%)	
						Geometry	Amplitude
LOM	IIM	SR	11	2.034	0.31	80	90
		DC	8	0.93	1.55	70	90
	JIM	SR	8	20.08	1.46	95	95
		DC	8	20.08	1.46	98	95
CGO	SOO	SR	1500	1193.68	0.14	85	95
		DC	1500	150.22	0.12	80	95
	MOO	SR	3502	2891.43	0.13	97	100
		DC	3502	2891.43	0.10	80	100

Table 5. Summary of the convectional global optimization inversion results parameter for both DCR and SR (dyke, positive, positive and negative anomalies model) methods.

Method	Inversion Type	Cores	Population Size	Misfit (%)	Generations	Run Time (min)
DCR	Dyke	16	9500	1.04	2500	6642.15
SR	Dyke	16	4800	3.32	200	6803.45
DCR	Positive	6	9500	1.40	2500	13,962.00
DCR	Negative	6	9500	2.70	2500	11,426.68
DCR	Pos. & Neg.	4	9500	14.51	2500	13,929.15

The results showed that the CGO algorithm inherited some features of the model, such as the geometry and amplitude of the anomaly from the local optimization that reflects in its optimum performance. For example, the amplitude and geometry of the DCR in the Dyke model was reflected in the final output of global optimization (Figure 6). Notice that in Figures 9d and 10d, the ray path avoided the negative anomaly; thus, the ideal model structure cannot be reconstructed with high resolution. This scenario is peculiar with the application of the SR method regarding a low-velocity layer (e.g., cavity) surrounded by a high-velocity media [48].

In the CGO method, using the same size of the synthetic model from the LOM, we observed that obtaining a good inversion result for the SR model using a computer with 12 logical processors configuration (6 cores) takes a longer time than the DCR method. This is because the DCR method produces electrical perturbation and estimates the apparent resistivity of the model at once; however, the SR method first computes the travel time from one source to all receivers sequentially and thereafter repeats the same procedure for all other available sources. To optimize the computation time of both SOO and MOO, we made the misfit algorithm part of the code to run on parallel computing by using the built-in parallel computing (e.g., parfor loop) command in MATLAB. This process enhanced the computation cost of the hybrid global optimization algorithm. For instance, it took 189.75 min to obtain the same result as in Figure 8b (single-objective GA for the positive anomaly DCR model) without parallel computing, whereas it took 55.25 min (71% run time optimization) to obtain the same output with parallel computing. Similarly, running the single objective genetic algorithm for the positive anomaly velocity (SR) model for 100 generations took 664.56 min without parallel computing, whereas it took 75.66 min (89% run time optimization) to obtain the same result with parallel computing. The percentage of run time optimization depends on the population size, number of generations, and type of geophysical method involved. To make the synthetic test challenging for the CGO algorithm, we added 3% Gaussian noise to the data resulting from the positive anomaly model. Although the output did not match noise-free data perfectly, a larger portion of the anomaly was recovered (Figures 15 and 16).

Regarding the GA population size, we observed that the CGO algorithm involving the DCR and SR methods performed better with an increasing number of populations. For example, Figure 15 shows a graduate improvement in the model resolution as the number of populations increased in the DCR genetic algorithm result. Similarly, the CGO algorithm offered a better performance with increasing generations. Notwithstanding, increasing the number of populations and generations prolonged the computation run time (Figure 17).

Considering these criteria (number of generations, population size, and computation time), we applied an average population size of 500 and 500 generations for both DCR and SR combined global optimization inversions.

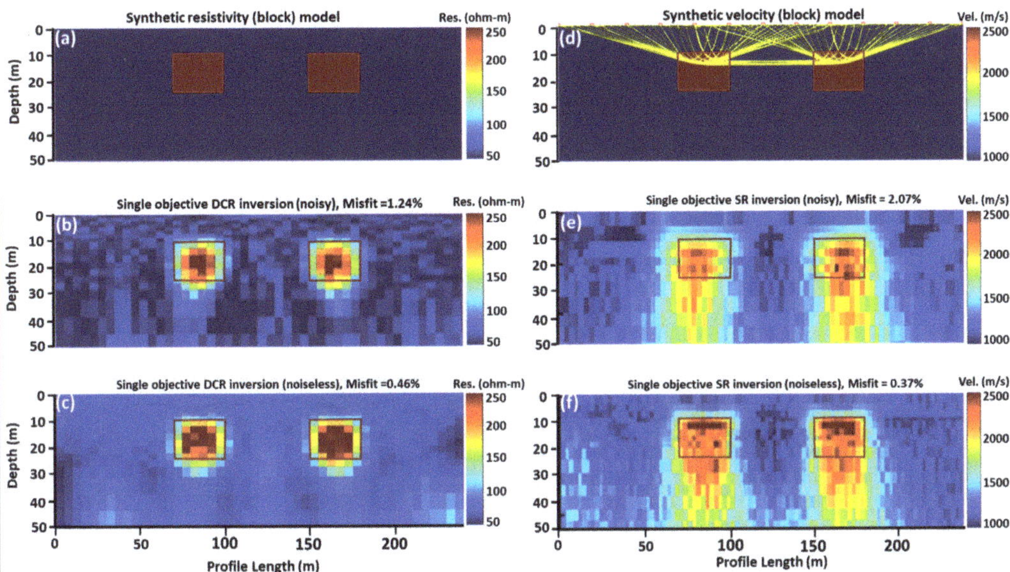

Figure 15. Test of the combined optimization algorithm using data (both DCR and SR) with added 3% gaussian noise; (**a**) synthetic dc-resistivity model, (**b**) single objective inverted resistivity model (noisy), (**c**) single objective inverted resistivity model (noiseless), (**d**) synthetic velocity model, (**e**) single objective inverted velocity model (noisy), and (**f**) single objective inverted velocity model (noiseless). The dark red lines are used to mark the anomaly boundary.

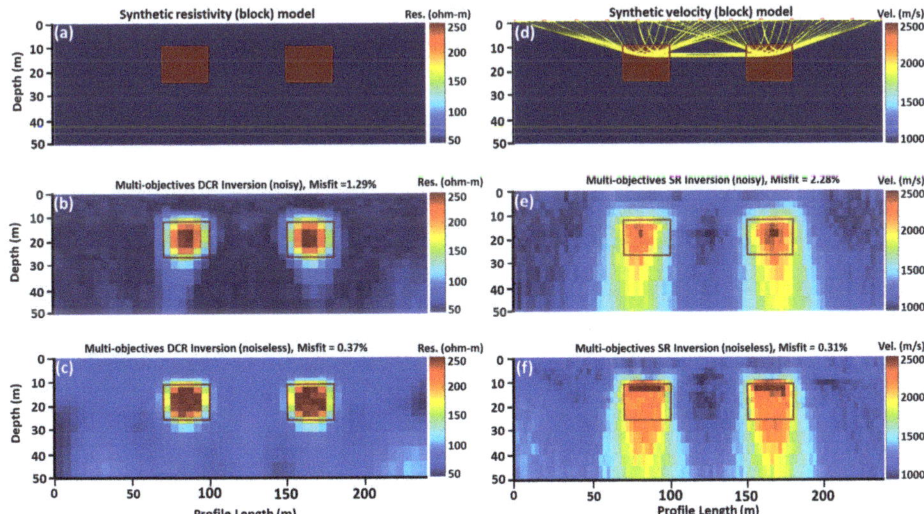

Figure 16. Test of the combined optimization algorithm using data (both DCR and SR) with added 3% gaussian noise; (**a**) synthetic dc-resistivity model, (**b**) multi-objectives inverted resistivity model (noisy), (**c**) multi-objectives inverted resistivity model (noiseless), (**d**) synthetic velocity model, (**e**) multi-objectives inverted velocity model (noisy), and (**f**) multi-objectives inverted velocity model (noiseless). The dark red lines are used to mark the anomaly boundary.

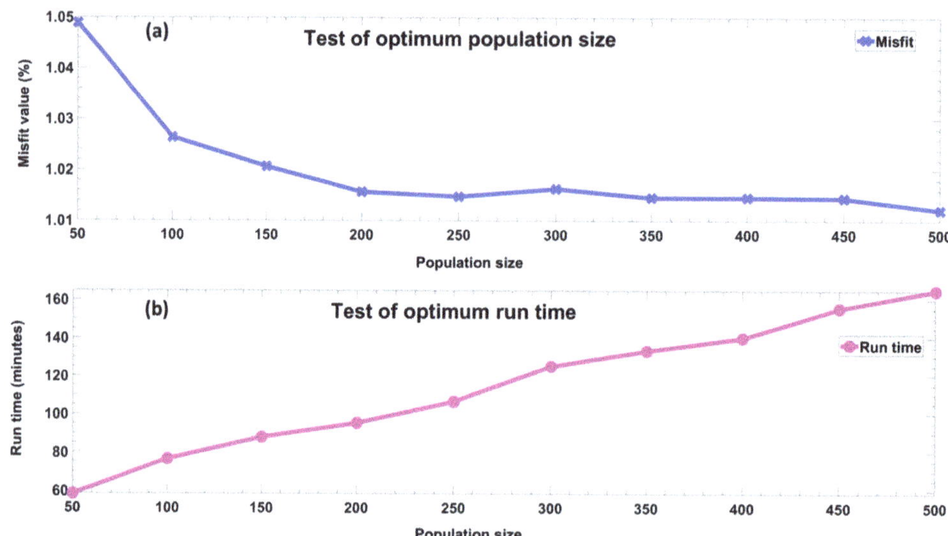

Figure 17. Effect of population size on DC-resistivity inversion using the genetic algorithm; (**a**) test of optimum population size that produces the most significant misfit and (**b**) test of optimum run time with respect to population size.

5. Conclusions

The proposed CGO used in this study begins with the application of a local optimization algorithm that requires the use of appropriate regularization parameters incorporated into the DCR and SR objective functions, and it is optimized (using the LOM) to obtain the best model, which will be used as an input model for the combined local–global optimization CGO method. This study applied this concept to obtain the best model parameters for both an individual and joint inversion of the DCR and SR geophysical methods. The CGO algorithm used to overcome the challenges associated with the separate application of LOM and GOM involved the application of the final output (optimum model parameter) of the LOM as the input for the CGO techniques. The global optimization part of the single objective CGO applied the GA to optimize the DCR and SR misfit functions (Equations (13) and (14), respectively) while the NSGA II was used to optimize the resultant misfit from the DCR and SR in the multi-objective optimization algorithm. Apart from the CGO method, which improved the computation run time, we made a part (the misfit function) of the CGO algorithm run on parallel computing. This approach not only contributed to the optimization of the CGO algorithm run time but also provided an opportunity to test the convectional GOM using a computer with 12 logical processor units (six cores). The CGO algorithm was tested with both resistivity and velocity models that had a dyke, two blocks (positive), and two blocks (positive and negative) anomalies. Generally, the CGO algorithm showed an improvement when compared with the local optimization output (Tables 2–4). However, the conventional GOM results showed instability in the delineation of the anomalies in all the tested SR and DCR models, and the model instability was probably due to the use of an unconstraint objective function. However, the CGO method overcame the challenge of model instability since it was linearly constrained by using the LOM's output to define the search space's lower and upper bounds. Additionally, the conventional GOM application remained computationally expensive (especially for the SR method) relative to the CGO techniques. Therefore, this study recommends applying a combined approach (local and global optimization algorithm) when characterizing the subsurface when both DCR and SR data are acquired.

Author Contributions: Conceptualization, P.S.; methodology, I.D., I.A., H.A.H., P.S., and E.C.; software, P.E., I.D., I.A., and H.A.H.; validation, I.D., I.A., and H.A.H.; formal analysis, P.E., I.D., I.A., H.A.H., P.S., and E.C.; resources, P.S.; data curation, P.E., I.D., I.A., H.A.H., P.K., and P.S.; writing—original draft preparation, P.E., I.D., I.A., H.A.H., P.K., and P.S.; writing—review and editing, P.E., I.D., I.A., H.A.H., P.K., P.S., E.C., S.H., and A.A.-S.; visualization, P.E., P.K., and P.S.; supervision, P.S.; project administration, P.S.; funding acquisition, A.A.-S. All authors have read and agreed to the published version of the manuscript.

Funding: This research was funded by the start-up grant SF18060 from the College of Petroleum Engineering and Geosciences (CPG) at King Fahd University of Petroleum and Minerals (KFUPM).

Institutional Review Board Statement: Not applicable.

Informed Consent Statement: Not applicable.

Data Availability Statement: Not applicable.

Conflicts of Interest: The authors declare no conflict of interest. The funders had no role in the design of the study; in the collection, analyses, or interpretation of data; in the writing of the manuscript; or in the decision to publish the results.

References

1. Oldenburg, D.W.; Ellis, R.G. Inversion of geophysical data using an approximate inverse mapping. *Geophys. J. Int.* **1991**, *105*, 325–353. [CrossRef]
2. Oldenburg, D.W.; Li, Y. Inversion for Applied Geophysics: A Tutorial. In *Near-Surface Geophysics*; Society of Exploration Geophysicists: Houston, TX, USA, 2005; pp. 89–150.
3. Farquharson, C.G.; Oldenburg, D.W. A comparison of automatic techniques for estimating the regularization parameter in non-linear inverse problems. *Geophys. J. Int.* **2004**, *156*, 411–425. [CrossRef]
4. Fomel, S. Shaping regularization in geophysical-estimation problems. *Geophysics* **2007**, *72*, R29. [CrossRef]
5. Gheymasi, H.M.; Gholami, A. A local-order regularization for geophysical inverse problems. *Geophys. J. Int.* **2013**, *195*, 1288–1299. [CrossRef]
6. Gündoğdu, N.Y.; Demirci, İ.; Demirel, C.; Candansayar, M.E. Characterization of the bridge pillar foundations using 3d focusing inversion of DC resistivity data. *J. Appl. Geophys.* **2020**, *172*, 103875. [CrossRef]
7. Haber, E.; Oldenburg, D. Joint inversion: A structural approach. *Inverse Probl.* **1997**, *13*, 63–77. [CrossRef]
8. Gallardo, L.A.; Meju, M.A. Characterization of heterogeneous near-surface materials by joint 2D inversion of dc resistivity and seismic data. *Geophys. Res. Lett.* **2003**, *30*, 1658. [CrossRef]
9. Gallardo, L.A.; Meju, M.A. Joint two-dimensional cross-gradient imaging of magnetotelluric and seismic traveltime data for structural and lithological classification. *Geophys. J. Int.* **2007**, *169*, 1261–1272. [CrossRef]
10. Linde, N.; Binley, A.; Tryggvason, A.; Pedersen, L.B.; Revil, A. Improved hydrogeophysical characterization using joint inversion of cross-hole electrical resistance and ground-penetrating radar traveltime data. *Water Resour. Res.* **2006**, *42*, W12404. [CrossRef]
11. Infante, V.; Gallardo, L.A.; Montalvo-Arrieta, J.C.; Navarro de León, I. Lithological classification assisted by the joint inversion of electrical and seismic data at a control site in northeast Mexico. *J. Appl. Geophys.* **2010**, *70*, 93–102. [CrossRef]
12. Moorkamp, M.; Heincke, B.; Jegen, M.; Roberts, A.W.; Hobbs, R.W. A framework for 3-D joint inversion of MT, gravity and seismic refraction data. *Geophys. J. Int.* **2011**, *184*, 477–493. [CrossRef]
13. Hamdan, H.A.; Vafidis, A. Joint inversion of 2D resistivity and seismic travel time data to image saltwater intrusion over karstic areas. *Environ. Earth Sci.* **2013**, *68*, 1877–1885. [CrossRef]
14. Bennington, N.L.; Zhang, H.; Thurber, C.H.; Bedrosian, P.A. Joint Inversion of Seismic and Magnetotelluric Data in the Parkfield Region of California Using the Normalized Cross-Gradient Constraint. *Pure Appl. Geophys.* **2015**, *172*, 1033–1052. [CrossRef]
15. Demirci, İ.; Candansayar, M.E.; Vafidis, A.; Soupios, P. Two dimensional joint inversion of direct current resistivity, radio-magnetotelluric and seismic refraction data: An application from Bafra Plain, Turkey. *J. Appl. Geophys.* **2017**, *139*, 316–330. [CrossRef]
16. Demirci, İ.; Dikmen, Ü.; Candansayar, M.E. Two-dimensional joint inversion of Magnetotelluric and local earthquake data: Discussion on the contribution to the solution of deep subsurface structures. *Phys. Earth Planet. Inter.* **2018**, *275*, 56–68. [CrossRef]
17. Wang, K.P.; Tan, H.D.; Wang, T. 2D joint inversion of CSAMT and magnetic data based on cross-gradient theory. *Appl. Geophys.* **2017**, *14*, 279–290. [CrossRef]
18. Vozoff, K.; Jupp, D.L.B. Joint Inversion of Geophysical Data. *Geophys. J. Int.* **1975**, *42*, 977–991. [CrossRef]
19. Autio, U.; Smirnov, M.Y.; Savvaidis, A.; Soupios, P.; Bastani, M. Combining electromagnetic measurements in the Mygdonian sedimentary basin, Greece. *J. Appl. Geophys.* **2016**, *135*, 261–269. [CrossRef]
20. Demirci, İ.; Gündoğdu, N.Y.; Candansayar, M.E.; Soupios, P.; Vafidis, A.; Arslan, H. Determination and Evaluation of Saltwater Intrusion on Bafra Plain: Joint Interpretation of Geophysical, Hydrogeological and Hydrochemical Data. *Pure Appl. Geophys.* **2020**, *177*, 5621–5640. [CrossRef]

21. Vafidis, A.; Soupios, P.; Economou, N.; Hamdan, H.; Andronikidis, N.; Kritikakis, G.; Panagopoulos, G.; Manoutsoglou, E.; Steiakakis, M.; Candansayar, E.; et al. Seawater intrusion imaging at Tybaki, Crete, using geophysical data and joint inversion of electrical and seismic data. *First Break* **2014**, *32*, 107–114. [CrossRef]

22. Shahrukh, M.; Soupios, P.; Papadopoulos, N.; Sarris, A. Geophysical investigations at the Istron archaeological site, eastern Crete, Greece using seismic refraction and electrical resistivity tomography. *J. Geophys. Eng.* **2012**, *9*, 749–760. [CrossRef]

23. Linde, N.; Doetsch, J. Joint Inversion in Hydrogeophysics and Near-Surface Geophysics. In *Integrated Imaging of the Earth: Theory and Applications*; John Wiley & Sons: Hoboken, NJ, USA, 2016; pp. 117–135.

24. Meju, M.A. Joint multi-geophysical inversion: Effective model integration, challenges and directions for future research. In Proceedings of the International Workshop on Gravity, Electrical & Magnetic Methods and Their Applications, Beijing, China, 10–13 October 2011; p. 37. [CrossRef]

25. Athanasiou, E.N.; Tsourlos, P.I.; Papazachos, C.B.; Tsokas, G.N. Combined weighted inversion of electrical resistivity data arising from different array types. *J. Appl. Geophys.* **2007**, *62*, 124–140. [CrossRef]

26. Hu, W.; Abubakar, A.; Habashy, T.M. Joint electromagnetic and seismic inversion using structural constraints. *Geophysics* **2009**, *74*, R99–R109. [CrossRef]

27. Bastani, M.; Hübert, J.; Kalscheuer, T.; Pedersen, L.B.; Godio, A.; Bernard, J. 2D joint inversion of RMT and ERT data versus individual 3D inversion of full tensor RMT data: An example from Trecate site in Italy. *Geophysics* **2012**, *77*, WB233–WB243. [CrossRef]

28. Gallardo, L.A.; Fontes, S.L.; Meju, M.A.; Buonora, M.P.; De Lugao, P.P. Robust geophysical integration through structure-coupled joint inversion and multispectral fusion of seismic reflection, magnetotelluric, magnetic, and gravity images: Example from Santos Basin, offshore Brazil. *Geophysics* **2012**, *77*, B237–B251. [CrossRef]

29. Lochbühler, T.; Doetsch, J.; Brauchler, R.; Linde, N. Structure-coupled joint inversion of geophysical and hydrological data. *Geophysics* **2013**, *78*, ID1–ID14. [CrossRef]

30. Zhang, Y.; Zhao, Z.; Nie, Z.; Liu, Q.H. Approach on Joint Inversion of Electromagnetic and Acoustic Data Based on Structural Constraints. *IEEE Trans. Geosci. Remote Sens.* **2020**, *58*, 7672–7681. [CrossRef]

31. Yin, C.; Sun, S.; Liu, Y.; Ren, X.; Wang, C. Joint inversion of geophysical data and applications. In Proceedings of the Fifth International Conference on Engineering Geophysics, Al Ain, United Arab Emirates, 21–24 October 2019; pp. 232–235. [CrossRef]

32. Jordi, C.; Doetsch, J.; Günther, T.; Schmelzbach, C.; Maurer, H.; Robertsson, J.O.A. Structural joint inversion on irregular meshes. *Geophys. J. Int.* **2020**, *220*, 1995–2008. [CrossRef]

33. Liu, S.; Liang, M.; Hu, X. Particle swarm optimization inversion of magnetic data: Field examples from iron ore deposits in China. *Geophysics* **2018**, *83*, J43–J59. [CrossRef]

34. Rani, P.; Piegari, E.; Di Maio, R.; Vitagliano, E.; Soupios, P.; Milano, L. Monitoring time evolution of self-potential anomaly sources by a new global optimization approach. Application to organic contaminant transport. *J. Hydrol.* **2019**, *575*, 955–964. [CrossRef]

35. Soupios, P.; Akca, I.; Mpogiatzis, P.; Basokur, A.T.; Papazachos, C. Applications of hybrid genetic algorithms in seismic tomography. *J. Appl. Geophys.* **2011**, *75*, 479–489. [CrossRef]

36. Akça, I.; Basokur, A.T. Extraction of structure-based geoelectric models by hybrid genetic algorithms. *Geophysics* **2010**, *75*, F15–F22. [CrossRef]

37. Chunduru, R.K.; Sen, M.K.; Stoffa, P.L. Hybrid optimization methods for geophysical inversion. *Geophysics* **2012**, *62*, 1196–1207. [CrossRef]

38. Schwarzbach, C.; Börner, R.-U.; Spitzer, K. Two-dimensional inversion of direct current resistivity data using a parallel, multi-objective genetic algorithm. *Geophys. J. Int.* **2005**, *162*, 685–695. [CrossRef]

39. Ayani, M.; MacGregor, L.; Mallick, S. Inversion of marine controlled source electromagnetic data using a parallel non-dominated sorting genetic algorithm. *Geophys. J. Int.* **2020**, *220*, 1066–1077. [CrossRef]

40. Akca, İ.; Günther, T.; Müller-Petke, M.; Başokur, A.T.; Yaramanci, U. Joint parameter estimation from magnetic resonance and vertical electric soundings using a multi-objective genetic algorithm. *Geophys. Prospect.* **2014**, *62*, 364–376. [CrossRef]

41. Holland, J.H. *Adaptation in Natural and Artificial Sytems; An Introductory Analysis with Applications to Biology, Control, and Artificial Intelligence*; Bradford Book: Denver, CO, USA, 1992; ISBN 978-0262581110.

42. Goldberg, D.E. *Genetic Algorithms in Search, Optimization and Machine Learning*; Addison-Wesley Longman Publishing Co., Inc.: Boston, MA, USA, 1989; Volume 7, ISBN 0201157675.

43. Sen, M.K.; Stoffa, P.L. *Global Optimization Methods in Geophysical Inversion*; Cambridge University Press: Cambridge, UK, 2013; ISBN 1139619519.

44. Zidan, A.; Li, Y.E.; Cheng, A. A Pareto Multi-Objective Optimization Approach for Anisotropic Shale Models. *J. Geophys. Res. Solid Earth* **2021**, *126*, e2020JB021476. [CrossRef]

45. Deb, K.; Pratap, A.; Agarwal, S.; Meyarivan, T. A fast and elitist multiobjective genetic algorithm: NSGA-II. *IEEE Trans. Evol. Comput.* **2002**, *6*, 182–197. [CrossRef]

46. Soupios, P.; Papazachos, C.B.; Vallianatos, F.; Papakostas, T. Numerical treatment and evaluation of inverse problems. *WSEAS Trans. Circuits Syst.* **2003**, *2*, 547–551.

47. Soupios, P.M.; Papazachos, C.B.; Juhlin, C.; Tsokas, G.N. Nonlinear 3-D traveltime inversion of crosshole data with an application in the area of the Middle Ural mountains. *Geophysics* **2001**, *66*, 627–636. [CrossRef]
48. Carollo, A.; Capizzi, P.; Martorana, R. Joint interpretation of seismic refraction tomography and electrical resistivity tomography by cluster analysis to detect buried cavities. *J. Appl. Geophys.* **2020**, *178*, 104069. [CrossRef]

Article

Investigating Limits in Exploiting Assembled Landslide Inventories for Calibrating Regional Susceptibility Models: A Test in Volcanic Areas of El Salvador

Chiara Martinello [1], Claudio Mercurio [1], Chiara Cappadonia [1,*], Miguel Ángel Hernández Martínez [2], Mario Ernesto Reyes Martínez [3], Jacqueline Yamileth Rivera Ayala [3], Christian Conoscenti [1] and Edoardo Rotigliano [1,*]

[1] Dipartimento di Scienze della Terra e del Mare (DiSTeM), University of Palermo, Via Archirafi 22, 90123 Palermo, Italy; chiara.martinello@unipa.it (C.M.); claudio.mercurio@unipa.it (C.M.); christian.conoscenti@unipa.it (C.C.)
[2] Escuela de Posgrado y Educación Continua, Facultad de Ciencias Agronómicas, University of El Salvador, Final de Av. Mártires y Héroes del 30 julio, San Salvador 1101, El Salvador; miguel.hernandez@ues.edu.sv
[3] Ministerio de Medio Ambiente y Recursos Naturales (MARN), Calle Las Mercedes, San Salvador 1101, El Salvador; mreyes@marn.gob.sv (M.E.R.M.); jacquelinerivera@marn.gob.sv (J.Y.R.A.)
* Correspondence: chiara.cappadonia@unipa.it (C.C.); edoardo.rotigliano@unipa.it (E.R.); Tel.: +39-091-238-64664 (C.C.); +39-091-238-64649 (E.R.)

Featured Application: This research deals with a very relevant topic in the framework of landslide susceptibility mapping, highlighting some very critical drawbacks in using a weak landslide inventory for regional-scale assessment. Tools and strategies for recognizing and approaching such limits are given.

Abstract: This research is focused on the evaluation of the reliability of regional landslide susceptibility models obtained by exploiting inhomogeneous (for quality, resolution and/or triggering related type and intensity) collected inventories for calibration. At a large-scale glance, merging more inventories can result in well-performing models hiding potential strong predictive deficiencies. An example of the limits that such kinds of models can display is given by a landslide susceptibility study, which was carried out for a large sector of the coastal area of El Salvador, where an apparently well-performing regional model (AUC = 0.87) was obtained by regressing a dataset through multivariate adaptive regression splines (MARS), including five landslide inventories from volcanic areas (Ilopango and Coatepeque caldera; San Salvador, San Miguel, and San Vicente Volcanoes). A multiscale validation strategy was applied to verify its actual predictive skill on a local base, bringing to light the loss in the predictive power of the regional model, with a lowering of AUC (20% on average) and strong effects in terms of sensitivity and specificity.

Keywords: incomplete landslide archives; MARS; Central America; validation procedures; regional-scale; debris flows

Citation: Martinello, C.; Mercurio, C.; Cappadonia, C.; Hernández Martínez, M.Á.; Reyes Martínez, M.E.; Rivera Ayala, J.Y.; Conoscenti, C.; Rotigliano, E. Investigating Limits in Exploiting Assembled Landslide Inventories for Calibrating Regional Susceptibility Models: A Test in Volcanic Areas of El Salvador. *Appl. Sci.* **2022**, *12*, 6151. https://doi.org/10.3390/app12126151

Academic Editors: Alessandro Simoni and Jianbo Gao

Received: 27 April 2022
Accepted: 15 June 2022
Published: 16 June 2022

Publisher's Note: MDPI stays neutral with regard to jurisdictional claims in published maps and institutional affiliations.

1. Introduction

Due to the subduction of the Cocos Plate under the Caribbean Plate along the Middle America Trench [1], El Salvador is characterized by intense tectonic activity and a number of active volcanoes, meaning that severe earthquakes and volcanic eruptions frequently affect the country. As a consequence, volcanic rocks (from Cenozoic hard rocks to pseudo-coherent recent ones) and their weathered products largely outcrop [2] along very highly steep slopes in this country. In particular, the tropical-humid climate setting of El Salvador [3], with a mean annual rainfall above 1846 mm and a temperature between 20 and 30 °C [4], is responsible for the intense weathering of the topsoil, resulting in poor geotechnical properties. At the same time, intense rainfall events associated with recurrent hurricanes frequently result in water saturation and neutral pression increasing in the regolithic mantle, causing the triggering of slope failures [5–7]. As a consequence, the volcanoes and the caldera's inner flanks are very frequently affected by landslides of a debris slide/flow type. These failures, in light of the high steepness of the initiation zones, very frequently take the form of very fast and long-runout debris flow phenomena, threatening those villages, which are set along the track channels or at the foot of the slopes. In recent years, the Nepaja (2020) [8] and San Vicente (2009) [9,10] disasters clearly illustrated this kind of phenomenon, resulting in widespread damage to houses and high numbers of injured and dead.

Differing from rockfall susceptibility studies [11–13], slide- and flow-type landslides are typically analyzed on a basin to regional scale, meaning that large inventories are required for robust modelling. In particular, a need arises to detect the potential initiation sites of landslides. To this end, landslide susceptibility modelling based on statistical analysis can offer a suitable approach for obtaining quantitative, objective, and validated prediction images of the potential triggering sites, which can then be processed with propagation tracking algorithms for a full hazard assessment.

Indeed, civil protection urgently requires regional-scale landslide susceptibility scenarios attempting to define statistically based national maps, eventually even exploiting limited but available landslide inventories for their calibration. To this end, grouping multiple clustered available datasets is frequently adopted as a solution to obtain landslide inventories large enough to train the statistical models. However, such landslide datasets can result in heterogeneity in terms of spatial distribution, the expertise of the operators, classification and mapping criteria, survey recognition methods and resolution (field, remote, reports), epoch and related triggering events, etc. It is worth noting that these limits could hamper the resolution and precision of the predictive models without giving clear effects down from standard validation procedures.

A number of landslide susceptibility studies have been conducted in the last fifteen years. In particular, post-Hurricane Mitch (1998) and post the 2001 earthquake, landslide inventories were processed through principal component analysis for assessing landslide susceptibility of an area in the extreme north-western sector of the country [14]. At the same time, regional susceptibility assessment studies in El Salvador have been carried out, exploiting the same 2001 seismically induced inventory (set on the epicentral area), both through binary logistic regression [15] and neural networks [16]. More recently, a regional landslide susceptibility scenario with a 30-arcsecond resolution was also proposed by applying a fuzzy-based heuristic approach [17]. Rotigliano et al. [5,6] and Mercurio et al. [7] extensively applied logistic regression and MARS to assess landslide susceptibility in two limited volcanic sectors (Ilopango caldera and San Vicente, respectively). Regarding the civil protection authorities, MARN (Ministerio de Medio Ambiente y Recursos Naturales) adopted a 1:50,000 scale landslide susceptibility map for the whole country [18] by applying the heuristic approach of [19,20]. However, all of the proposed regional models [14–17] were obtained through a calibration in very small sectors, with very weak, if any, validation procedures.

In this paper, an application to the El Salvador territory was carried out, aimed at suggesting approaches and strategies suitable for correctly investigating the actual quality of a susceptibility map obtained by calibrating a predictive model through a heterogeneous landslide inventory. In spite of its relevance, few other scientists have faced similar issues [21,22].

The susceptibility modelling was based on applying Multivariate Adaptive Regression Splines (MARS; [23]) and implemented by exploiting open source software (QGIS [24], SAGA GIS [25], RStudio [26]).

2. Materials and Methods

2.1. Study Areas

In this research, landslide susceptibility assessment was focused on the slopes of a set of volcano/caldera areas where debris flows are frequently activated (Figure 1): (i) the

Coatepeque area, which extends for about 82 km^2 east of the homonymous caldera lake; (ii) the San Salvador area, surrounding the homonymous volcano for about 144 km^2; (iii) the watershed inner basins of the Ilopango caldera, covering a total area of about 121 km^2; (iv) the San Vicente area, which includes the whole homonymous volcano, extending for about 287 km^2; and (v) the tip sector of the San Miguel volcano, covering a total area of about 11 km^2.

Figure 1. Setting of the study area.

Focusing on the five study areas, the lithologic units of the San Salvador formation are the most frequently outcropping rocks: Holocene pyroclastites named "Tobas color café", in the Coatepeque area (77%), "Tierra Blanca", in the Ilopango area (45%) and to a lesser extent in the San Salvador area. Accumulation cones dominate the San Miguel area (72%), while Pleistocene effusive rocks prevail in the San Salvador area (57%) largely outcropping also in the San Vicente area. In addition, acid pyroclastites of the Cuscatlán formation are widely diffused both in the Ilopango and the San Vicente areas. Finally, with very limited outcropping areas, the pyroclastic and effusive rocks of the Bálsamo formation are observed in the Coatepeque, Ilopango, and San Vicente areas.

2.2. Landslides Inventory and Related Triggering Rainfall Events

The main task of this research was to test the suitability of aggregated regional landslide archives in the evaluation of landslide susceptibility assessment. For this reason, a set of independent available debris flows/slides archives were exploited for training and validating a regional landslide susceptibility map. Archives from five different sectors of the El Salvador territory were considered, which, even in the same sector, were considered as un-uniform in terms of operators, methods (field/remote), and epoch (which means grouping debris flows/slides linked to multiple and/or different extreme rainfall). These landslide inventories were prepared in the framework of different studies (master's degree thesis, PhD thesis and so on, see Author Contributions), many of which have been part of the RIESCA project (Proyecto Regional de Formación Aplicada a los Escenarios de Riesgos con Vigilancia y Monitoreo de los Fenómenos Volcánicos, Sísmicos e Hidrogeológicos en Centro América). For this reason, the study areas were not a priori limited and as mentioned above, they were restricted to the sectors affected by the activation of the inventoried debris flows/slides: the Ilopango (ILO), Coatepeque (COA), San Miguel (SMG), San Vicente (SVC), and San Salvador (SSV) areas. The ILO and the COA inventories were mapped through systematic remote analysis and integrated by some random file checks, consisting of 38,525 and 1895 debris flows/slides, respectively. The SVC inventory included 4975 phenomena, which were remotely recognized according to an irregular spatial scheme. The debris flows/slides of the SMG (233 cases) were extracted by a historical simplified archive inventory, whilst the SSV inventory (382 cases) merged the results of some spot field surveys. At the same time, the expertise and perspective of the operators were different, with ILO and COA having been mapped in the framework of scientific research, all the other inventories coming from civil protection tasks and SMG collecting a number of historical reports. The main triggering events for these landslide scenarios were Hurricane Ida and the tropical depression 12 E (TD12 E). The tropical-humid climate setting of El Salvador produces, in the rainy season between May and October, very high rainfall amounts (above 1846 mm, on average) that, usually, occur in the form of intense storms. Therefore, rapid saturation of the regolithic mantle and powerful surface runoff trigger a huge number of landslides even in the case of a normal rainfall season [5].

Between November 7th and 8th, Hurricane Ida, and the low-pressure system 96 E, simultaneously struck the central area of El Salvador, with cumulated rainfall exceeding 300 mm/24 h in the Ilopango and San Vicente villages [5,6,27,28]. Floods and landslides lashed these areas, causing around 200 deaths and huge economic losses [9], with damages to cropland, rural houses, and roads. In particular, the most devastating debris flows were triggered from the north-western flank of the San Vicente Volcano, hitting the villages of Verapaz and Guadalupe [5,7].

Tropical depression 12 E affected El Salvador during the period from the 10th to 20th October. With a cumulative maximum of 1513 mm, equivalent to 42% of the mean annual rainfall of the period 1971–2000 [28], DT12 E was classified as the most severe meteorological event recorded in the region. Additionally, in this case, with 10% of the national territory affected, especially along the coastal plains and the volcanic mountains, El Salvador was heavily hit by the related floods and landslides, reporting 35 victims and an economic loss of more than USD nine hundred million [10,28].

The Coatepeque debris flows were triggered by the tropical depression (TD) 12 E in 2011. The same extreme rainfall event activated the debris flows/slides of the San Salvador dataset. Hurricane Ida was the trigger of the phenomena mapped in the San Vicente archive, while both TD12 E and Ida activated the debris flows/slides of the Ilopango dataset. Finally, the landslides of the San Miguel archive were triggered by several rainfall events from 2001 to 2018.

All of the mapped phenomena were individuated by exploiting Google Earth images, and the landslide identification point (LIP), which was generated for each of the mapped phenomena corresponding to the highest point along the landslide crown, was also taken as indicating the area that effectively represents the activation conditions for surface debris flows [5,6,29–33].

2.3. Model Building and Validation Strategy

2.3.1. Predictors and Mapping Units

The selection of a set of geo-environmental variables potentially expressing the landslide preparatory causes (Table 1) was based on widely adopted geomorphological criteria [5–7,34–37]. In particular, outcropping lithology (GEO) and soil use (USE) were derived from an available thematic map [38] and a remote survey, respectively. By processing a 10 m pixel digital terrain model (DTM), the following continuous variables were derived: elevation (ELE), steepness (STP), plan (PLN), and profile (PRF) curvatures, topographic wetness index (TWI), and aspect, the latter expressed in terms of easternness (EASTNS) and northernness (NORTHNS). In addition, the landform classification (LCL) categorical variable was obtained. In this way, a set of three categorial and seven continuous variables was prepared.

Table 1. Details of the selected geo-environmental variables.

Factor	Acronym	Description of Source Parameter	Units	References
Elevation	ELE	Raster of elevation distribution	m	
Landform classification	LCL	Outcome of an automated procedure that recognize landforms on a gridded elevation distribution (TPI)		Wilson and Galland [39]
Slope gradient	STP	Highest first derivative of elevation	degree	Burrough and McDonell [40]
Northerness	NORTHNS	Cosine of aspect (Direction of steepest downwards slope from each cell to its neighbors)		Wilson and Galland [39] (Aspect)
Easterness	EASTNS	Sine of aspect (Direction of steepest downwards slope from each cell to its neighbors)		Wilson and Galland [39] (Aspect)
Plan curvature	PLN	Second derivative of elevation, computed along the horizontal plane	rad/m	Zevenbergen and Thorne [41]
Profile curvature	PRF	Second derivative of elevation, computed along the direction of the highest slope gradient	rad/m	Zevenbergen and Thorne [41]
Topographic wetness index	TWI	Calculated as ln[A/tanβ], where A and β, computed on each cell, correspond to the area of upslope drained cells and the slope gradient, respectively	m	Beven and Kirkby [42]
Lithological map	GEO	Geolithological map of the study area, modified from original geological map		modified from Schmidt-Thomé [43]
Soil use	USE	Land use map derived from 2002 satellite images and filed survey		

With regard to lithology, based on the geomechanical expected response, the outcropping lithologies were grouped as soft, medium, and hard rocks and very soft, soft, medium, and hard soils. On the basis of the landslide distribution in the study areas, very soft and hard soils account for more than 80% of the mapped cases. The very low number of landslides recognized in soft soils has to be ascribed to the very limited extension of the outcropping areas.

All of the controlling factors were arranged in 10×10 m raster layers. The same grid cell structure was then adopted as the susceptibility mapping unit, assigning a stable/unstable status depending on the intersection of LIPs. In fact, according to a number of debris flow susceptibility assessment studies (e.g., [7,22,29,31,32,36,37,44–50]), we considered the instability conditions of each inventoried landslides to be effectively captured in the highest crown 10×10 m pixel. In order to optimize the final selected predictors that were included in the MARS modelling procedure, the variance inflation factor (VIF) [51] test was performed for multicollinearity analysis through the continuous variables.

2.3.2. Modelling and Validation Tools

Multivariate adaptive regression splines (MARS; [23]), which was successfully applied in a number of recent landslide and soil erosion susceptibility studies [5–7,35,37,52–59], was then applied to regress the outcome (stable/unstable status) onto the covariates set from the controlling factor layers. MARS is a non-parametric regression method that exploits the splitting of each independent variable into hinge functions to boost the maximum likelihood-based adaptation skill of the logistic regression method, according to:

$$y = f(x) = \alpha + \sum_{i=1}^{N} \beta_i h_i(x) \tag{1}$$

where y is the dependent variable (the outcome) predicted by the function $f(x)$, α is the model intercept, and β_i is the coefficient of the h_i basis functions, given the N number of basis functions. MARS analysis was performed by exploiting the "earth" R-package [60].

The MARS statistical modelling of landslide susceptibility conditions requires the random extraction of a sample made of a balanced number of stable and unstable cases to be split into calibration and validation subsets: the first is exploited for regressing the outcome status on the set of covariates that express the adopted controlling factors, while the latter furnishes the unknown-to-model target pattern whose status has to be blindly predicted. In a pixel-based method, where the number of stable cases is typically largely greater than the unstable, balanced samples are obtained by merging all the positives to an equal number of randomly extracted negatives. To account for any potential unrepresentativeness of the extracted negatives, by adopting recurrent random selection routines, multiple samples were produced. Similarly, to control the influence of the specific cases which feed the calibration subsets, multiple (75/25%) calibration/validation splitting was applied to each sample as well. In this way, one hundred samples were split one hundred times so that each pixel was classified ten thousand times, allowing us to estimate the model resolution and precision. Finally, to fully evaluate the prediction skill of the model, the regression coefficients gained in the calibration/validation subset were applied to the whole investigated area.

Receiver operating curve (ROC) [61–63] and confusion matrices analyses were the tools employed to investigate the model's accuracy. In particular, ROC plot analysis is based on evaluating true- versus false-positive rates for decreasing susceptibility scores, with a larger area under the curve (AUC) [64,65] attesting to more effective classifications. The score at the maximum gradient of the ROC is then used as an optimized cut-off [66] for building a binarized (positive/negative-observed/predicted) confusion matrix. In this way, the accuracy of the model can be evaluated both with score-independent (ROC_AUC) and -dependent (ACC) indices.

2.3.3. Research Design and Model Building Strategy

In the following, we will refer to a super area (ALL), considering that it is obtained by merging all the positive and negative cases of each of the five sectors (volcanic areas), the latter defining five local datasets (ILO, COA, SMG, SVC, and SSV).

It is worth noting that, in light of the number of causes that have been here claimed as responsible for the inventory incompleteness, a different approach from Steger et al. [21] was designed for evaluating the influence of the bias landslide inventory. In particular, to explore the topic of the research, the following model building procedure was designed by submitting the hypothesizing of completeness of the inventory to a strict validation procedure.

First, a grand model (ALL) was prepared by applying the typical approach aimed at obtaining a regional model from the available landslide inventories, including in the processed dataframe the whole set of positives and negatives from the five sectors. To maintain control over the variability of the negatives and the calibration/validation subset assignment of positives, a suite of one thousand multiple datasets were obtained by randomly extracting one hundred sets of negatives and submitting each dataset to ten random calibration/validation (75/25%) splitting processes.

Once the grand model was prepared, it was first validated with respect to the spatial distribution of the landslides in the whole super area (ALL_ALL), according to a self-validation scheme [5–7,32,35–37,46,67,68]. The validation performance of the grand model was then locally evaluated by restricting the validation dataset to a single sector in turn (e.g., ALL_ILO). For comparison, independent local models (e.g., LOC_ILO) were prepared for the five sectors by limiting the application of the modelling procedure to every single dataset and applying a local self-validation scheme. Finally, five one-leave-out models were prepared by applying the same above-described procedure but adopting a 4/1 sectors calibration/validation splitting in the modelling scheme; a local validation was then obtained, by assessing the predictive skill in recognizing the specific positives and negatives of the extracted (left-out) target sector. In the following, these models are referred to as OLO models (e.g., OLO_ILO).

Table 2 provides a summary of the prepared models, including the specification of the main characteristics.

Table 2. Adopted model building scheme for the tested models. Green and orange dots represent calibration and validation cases, respectively, on the schematized five sectors.

Type	Calibration	Validation	Graphic Example
ALL_ALL	75% randomly extracted balanced subset from the ALL * dataset	conjugate 25% randomly extracted balanced subset from the ALL dataset	
ALL_target	100% randomly extracted balanced subset from the ALL* dataset	100% randomly extracted balanced subset from a single target ** sector	e.g., ALL_4
OLO_target	100% randomly extracted balanced subset from a [ALL-target] *** dataset	100% randomly extracted balanced subset from the subtracted target ** sector	e.g., OLO_3
LOC_target	75% randomly extracted balanced subset from a target ** sector dataset	conjugate 25% randomly extracted balanced subset from a target ** sector dataset	e.g., LOC_5

* ALL: the sum of the positive and negative cases of the five sectors. ** target: the sum of positive and negative cases of a single sector. *** [ALL-target]: the difference between ALL and a target.

According to the main task of the research, the ALL_ALL is considered as the model that one can take as representative for a regional prediction image. At the same time, the imported models (ALL_local), in re-defining the validation set on a local basis, could furnish a useful warning in case the performance of the grand models is actually locally misleading. The local models give an estimation of the reference performance that the imported model (ALL or OLO) should achieve to be considered more informative. Finally, the one-leave-out modelling procedure simulates the results of applying the model to totally unknown sectors (such as a hypothetical sixth unknown volcanic area in our research).

3. Results

For each of the models described above, the results of the validation are reported both in Figures 2–4, where ROC curves and related AUCs are drawn, and in Table 3, where binarized positive/negative status comparisons between predicted/observed target cases are given.

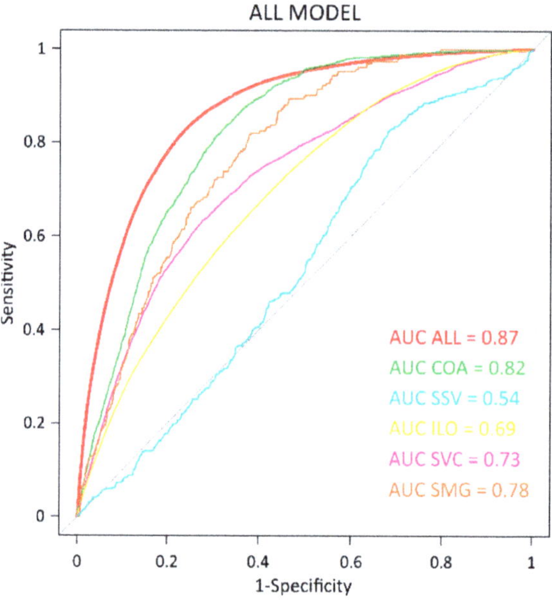

Figure 2. ROC plots and relative AUC values for the ALL models.

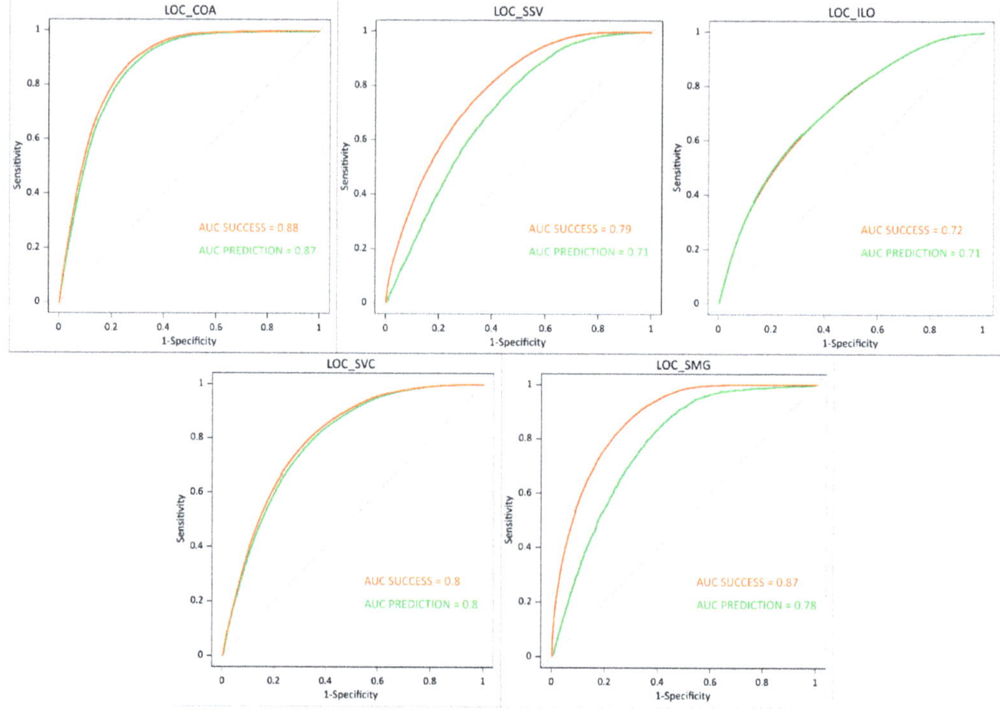

Figure 3. ROC plots and relative AUC values for the LOC models.

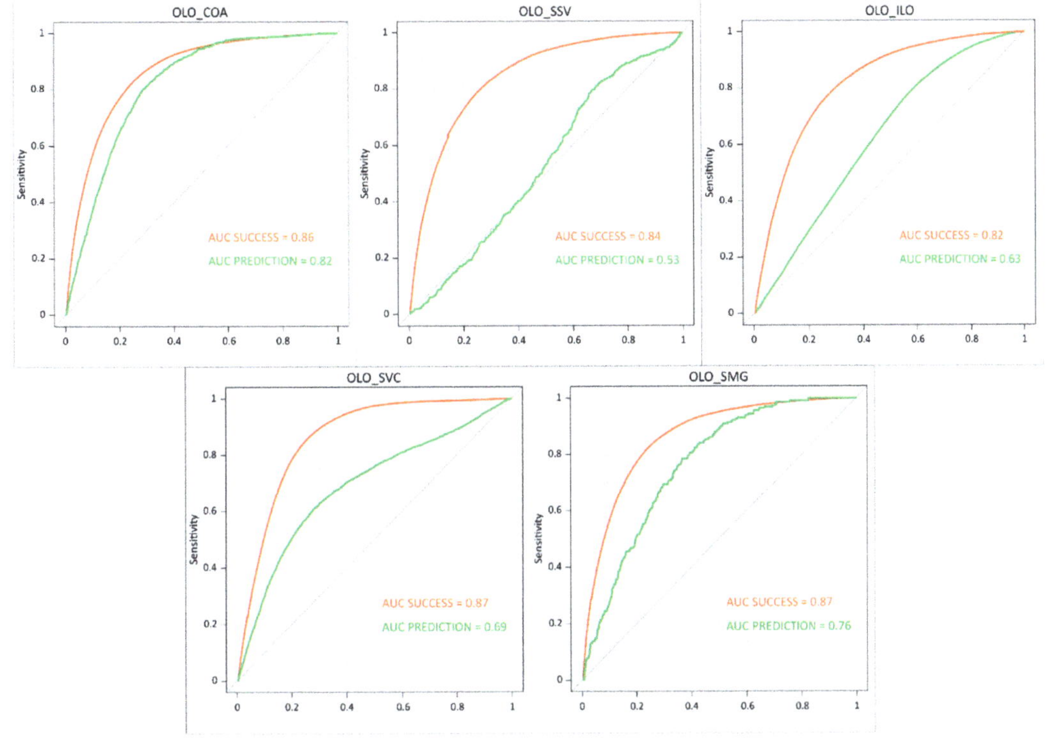

Figure 4. ROC plots and relative AUC values for the OLO models.

Table 3. Validation results (confusion matrices) for the sixteen models.

		Count	Positives	Negatives	TN	FN	FP	TP	ACC	Sensitivit	Specificit	AUC
ALL	ALL	6,311,320	46,010	6,265,310	4,786,221	8022	1,479,089	37,988	0.76	0.82	0.76	0.87
	COA	806,671	1895	804,576	698,607	967	105,969	928	0.87	0.44	0.87	0.82
	SSV	1,429,050	382	1,428,668	1,369,074	367	59,594	15	0.96	0.04	0.96	0.54
	ILO	1,161,436	38,525	1,122,911	378,750	4171	744,161	34,354	0.36	0.89	0.34	0.69
	SVC	2,794,399	4975	2,789,424	2,221,036	2295	568,388	2680	0.80	0.54	0.80	0.73
	SMG	119,964	233	119,731	118,754	222	977	11	0.99	0.05	0.99	0.78
LOC	COA	806,471	1895	804,576	590,261	219	214,315	1676	0.73	0.88	0.73	0.88
	SSV	1,429,050	382	1,428,668	839,269	66	589,399	316	0.59	0.83	0.59	0.78
	ILO	1,161,436	38,525	1,122,911	737,214	13392	385,697	25,133	0.66	0.65	0.66	0.72
	SVC	2,794,399	4975	2,789,424	1,880,683	1038	908,741	3937	0.67	0.79	0.67	0.80
	SMG	119,964	233	119,731	79,805	25	39,926	208	0.67	0.89	0.67	0.87
OLO	COA	806,471	1895	804,576	622,805	562	181,771	1333	0.77	0.70	0.77	0.82
	SSV	1,429,050	382	1,428,668	1,343,953	361	84,715	21	0.94	0.05	0.94	0.53
	ILO	1,161,436	38,525	1,122,911	455,548	7448	66,7363	31,077	0.42	0.81	0.41	0.63
	SVC	2,794,399	4975	2,789,424	2,044,869	2021	744,555	2954	0.73	0.59	0.73	0.69
	SMG	119,964	233	119,731	119,002	229	729	4	0.99	0.02	0.99	0.76

The performance of the ALL_ALL model is very high, with excellent AUC and accuracy (0.87 and 0.76, respectively) and highly satisfactory sensitivity (0.82) and specificity (0.76). Comparing these values to the ones obtained in importing the grand model into the specific sectors (ALL_local), satisfactory to excellent AUC and ACC values still hold, with the exception of ILO and SSV. However, lower sensitivity and higher specificity were recorded for all the models, with the exception of ILO. It is worth noting that only the SVC imported local model still performs with acceptable scores for all the main indices (sensitivity, specificity, ACC, AUC). At the same time, the local models are in general characterized by higher (0.8–0.9) AUC values, with a much more balanced sensitivity/specificity ratio, as a result of higher sensitivity and lower specificity. Again, the opposite behavior is observed for ILO.

Finally, the one-leave-out models confirm the general trend of performance indices' variation, which was observed for ALL_local validations.

With regard to the role of the predictors, the results obtained from the local modelling highlight two very different responses (Figure 5): SMG and SSV are fully controlled by elevation and steepness, whilst ILO, COA, and SVC also required the discriminating contribution of either landform classification (COA and SVC) or outcropping lithology (ILO and SVC) or soil use (for COA and ILO). Elevation, steepness, outcropping lithology, and soil use are all selected by the ALL grand model.

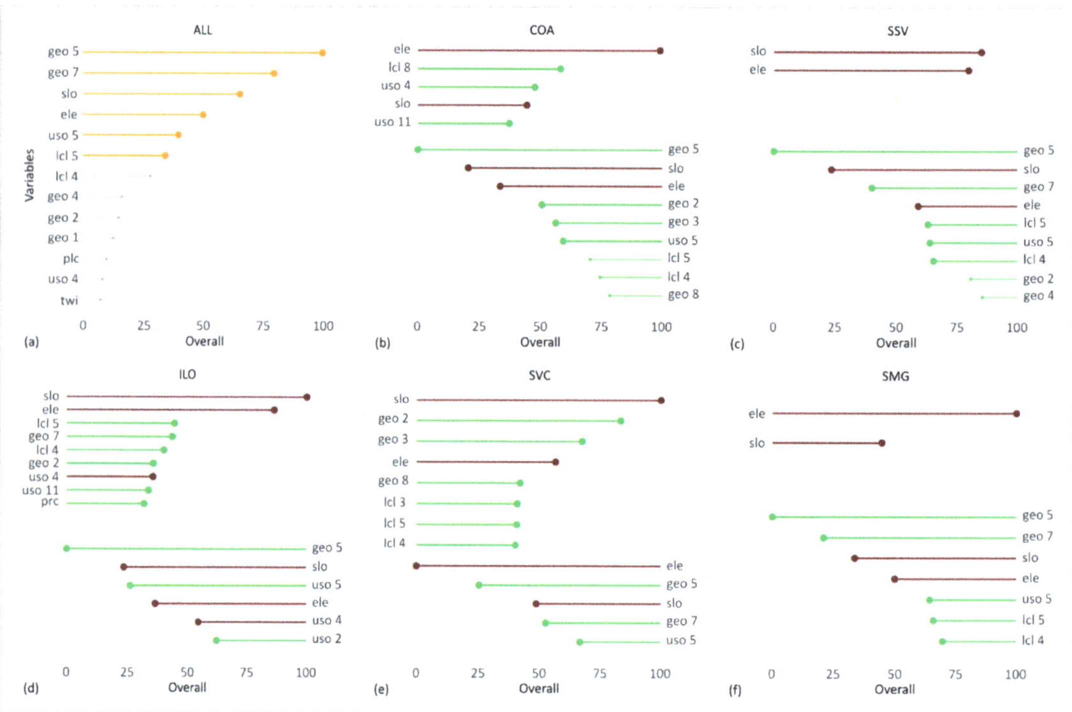

Figure 5. The most important variables for the ALL model (**a**) and the LOC (left side) and the OLO (right side) models (**b–f**). The common variables for the LOC and the OLO models are presented in amaranth, while the different variables are presented in green. Thin lines are used for variables with a lower overall (minor than 30 out of 100). Here are the acronyms used: geo 2 = soft rock; geo 3 = hard rock; geo 4 = medium rock; geo 5 = very soft soil; geo 6 = soft soil; geo 8 = medium soil; lcl 3 = valleys; lcl 4 = plains; lcl 5 = open slopes; lcl 8 = midslope ridges; uso 2 = forest; uso 4 = crop and pasture; uso 5 = permanent crop; uso 11 = shrub vegetation.

In Figure 6, a comparison between the ALL and LOC landslide susceptibility maps for two representative sectors (Ilopango and San Salvador) is given, highlighting either coherent or incoherent spatial patterns among the models for the two sectors.

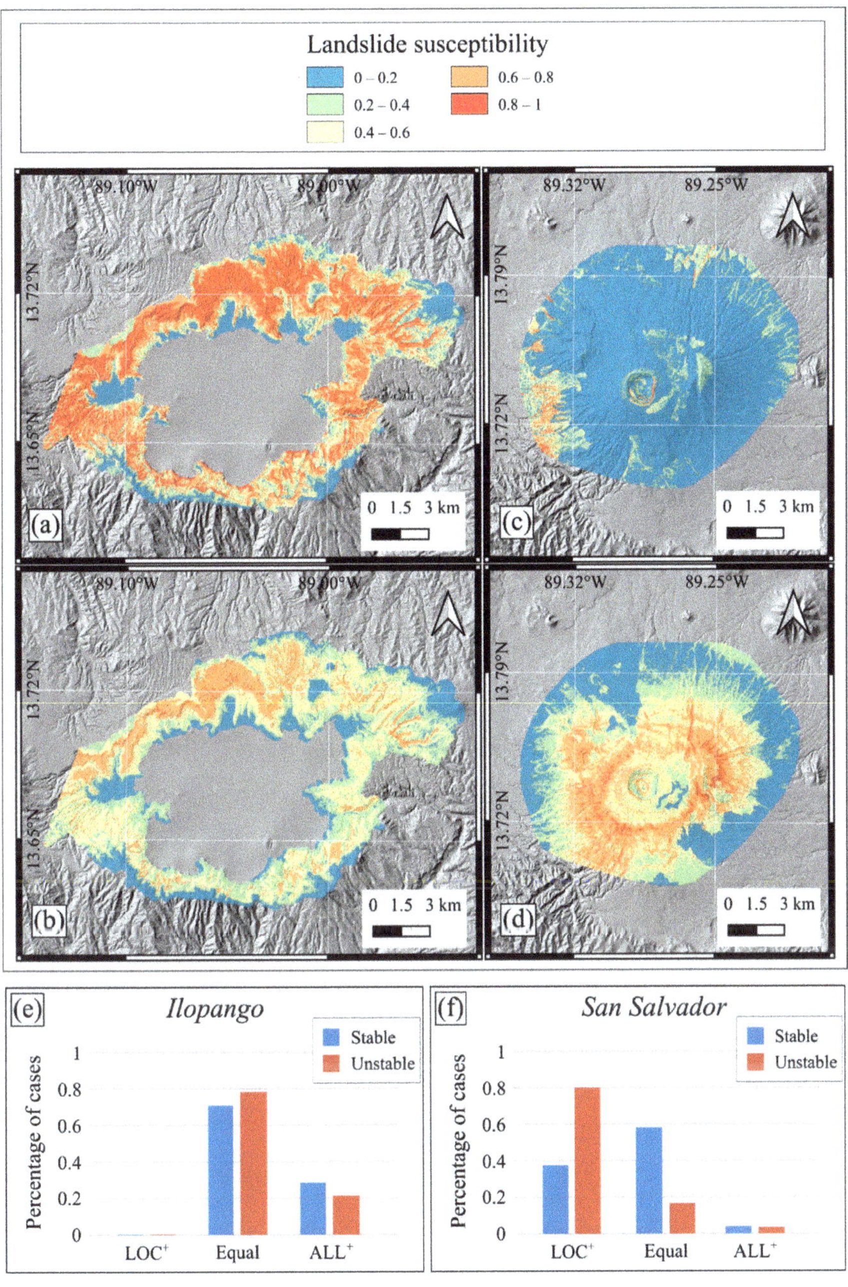

Figure 6. ALL (**a**,**c**) and LOC (**b**,**d**) landslide susceptibility maps for Ilopango (on the right side) and San Salvador (on the left side) sectors. The histograms (**e**,**f**) show the percentage of observed (stable/unstable) cases, when (i) LOC model assigns a higher susceptibility with respect to the ALL model (LOC$^+$), (ii) both the models assign the same predicted status (Equal) or (iii) the ALL model sets higher susceptibility with respect to the LOC model (ALL$^+$).

4. Discussion

The local landslide distribution in five different volcanic sectors was predicted both from imported (both ALL and OLO models) and locally calibrated models. The latter resulted in smoothly (with the exception of SSV) higher AUC values, with a proportional decrease in the cut-off-dependent accuracy, but driven by a marked sensitivity increase and a slight specificity decrease. In particular, the greater the LIP% incidence of a single sector, the higher the TPR decrease recorded for the imported models. A relevant exception that was highlighted by the results is the very odd behavior of ILO, whose local model produced a worse performance in recognizing its own positives. At the same time, in terms of scoring and status prediction, ALL and LOC models can result in different prediction images.

The ILO sector includes the great majority of landslides (83.7%) and, in light of its limited extension (18.4%), the maximum ratio between unstable and stable pixels. When trying to discriminate the status of the ILO pixels, on the basis of the ALL or OLO imported model, a better performance arises in positive detection when compared to the skill of the local model. This is due to the undifferentiated presence of positives and negatives in the same geomorphologic conditions, and this effect could have been enhanced by the severe triggering conditions (IDA tropical storm) that activated landslides even in less susceptible areas. In fact, the better performance of ALL and OLO relies on the circumstance that these models take their cases outside ILO, for positive and negative cases of OLO, or prevalently outside ILO, for the negatives of ALL. As a consequence, the local dataset confuses the binary discrimination whilst recurring for the outside pixels, which allowed us to better understand the unstable conditions. At the same time, for a more geomorphologically differentiated setting, the sub-catchment of ILO ("Arenal de Cujuapa"), Rotigliano et al. [5,6] obtained, with the same MARS modelling approach, higher AUC and accuracy values (0.83 and 0.73, respectively). Moreover, the same loss in the model performance was observed when trying to temporally predict the landslide inventory of 2003 (produced by a non-extreme rainfall triggering event) from the model calibrated with the same 2009 hurricane-induced inventory that was used in the present research.

Once the potentially hampering specific conditions of the ILO sector arose, a new grand model (ALL*) was tested excluding ILO from all sectors (which were reduced to four) and obtaining better locally imported results (Table 4). With the exception of SSV, these new imported models performed with similar, largely satisfactory AUCs to the local models and even higher sensitivity.

Table 4. Validation results (confusion matrices) for the ALL* models.

	Count	Positives	Negatives	TN	FN	FP	TP	ACC	Sensitivity	Specificity	AUC
ALL*_COA	806,471	1895	804,576	515,857	166	288,719	1729	0.64	0.91	0.64	0.85
ALL*_SSV	1,429,050	382	1,428,668	1,314,478	349	114,190	33	0.92	0.09	0.92	0.61
ALL*_SVC	2,794,399	4975	2,789,424	1,813,333	1026	976,091	3949	0.65	0.79	0.65	0.79
ALL*_SMG	119,964	233	119,731	37,646	8	82,085	225	0.32	0.97	0.31	0.75

5. Conclusions

On the basis of the obtained results, it is confirmed that grouping landslide inventories from different areas to increase the number of cases can lead to very unreliable results unless further validation tests are carried out. In particular, depending on both the number of landslides and frequency distribution of all the predictors in each of the grouped sectors, the grand model can be seen as having very high performance on average, but is very misleading and unstable on a local scale. In light of this effect, locally calibrated models can have better performance even if trained with a lower number of cases. This would typically lead to attaining a sense of security and considering the obtained prediction image as reliable for the study area, eventually suggesting that the obtained model also be exported to new neighboring unrecognized sectors (e.g., those between the five mapped ones). In this paper, a new approach was adopted, and related tools were proposed for verifying the inventory completeness hypothesis. This approach can be involved in any model building procedure so as to obtain warnings about the quality of the source data and its influence on the resolution of the derived susceptibility models.

Comparing grand to local models should be a standard procedure when assembling large landslide inventories, even in the case of secondary catchments in large basin-scale studies. The main factors controlling the performance of the grand model are the number of total pixels and the number of positives and the spatial distribution of the predictors. Two main factors hamper the accuracy and reliability of any grand model, based on a presence/absence method: depending on the relative spatial extension of the classes of each covariate, in light of the need to randomly extract the negatives to prepare balanced datasets, using the more diffused classes results in stable conditions; depending on the different levels of completeness of the merged landslide inventories, unstable conditions may come to light in the sectors or catchments with a higher number of mapped landslides. These two effects are much more severe for the categorical variables in the case of inhomogeneous geologic/geomorphologic settings, whilst DTM-derived variables are more unlikely to be so largely different as to mislead the modelling. It is worth noting that the limits produced by the qualitative and quantitative differences in the landslide inventories suggest that the adoption of presence-only methods is not suitable, also in light of the strong influence produced by any unrepresentativeness of the landslide inventories.

Optimizing susceptibility models for predicting new debris flow activation sites in volcanic areas is of crucial importance in El Salvador. In fact, under the triggering of the recurrent tropical storms which frequently strike the country, this kind of landslide rapidly evolves along the steep volcano flanks into very destructive debris flow phenomena hitting the hillside areas and causing damage and life losses. Investigating the reliability of prediction images for landslide activation constitutes a mandatory step in obtaining the starting base to be coupled with propagation algorithms for producing complete debris flow event scenarios.

Author Contributions: Conceptualization, C.M. (Chiara Martinello), C.M. (Claudio Mercurio), C.C. (Chiara Cappadonia), and E.R.; methodology, C.M. (Chiara Martinello), C.C. (Chiara Cappadonia), and E.R.; software, C.M. (Chiara Martinello) and C.M. (Claudio Mercurio); validation, C.M. (Chiara Martinello), C.M. (Claudio Mercurio), C.C. (Chiara Cappadonia), and C.C. (Christian Conoscenti); formal analysis, C.M. (Chiara Martinello), C.M. (Claudio Mercurio), C.C. (Chiara Cappadonia), and M.E.R.M.; investigation, M.E.R.M., M.Á.H.M. and C.M. (Claudio Mercurio); resources, C.M. (Claudio Mercurio) and M.Á.H.M.; M.E.R.M. and J.Y.R.A.; data curation, C.M. (Chiara Martinello), C.M. (Claudio Mercurio), M.E.R.M. and M.Á.H.M.; writing—original draft preparation, C.M. (Chiara Martinello), C.C. (Chiara Cappadonia), and C.M. (Claudio Mercurio); writing—review and editing, C.M. (Chiara Martinello), C.C. (Chiara Cappadonia), and E.R.; visualization, C.M. (Chiara Martinello), C.M. (Claudio Mercurio), and C.C. (Chiara Cappadonia); supervision, E.R., C.C. (Chiara Cappadonia) and C.C. (Christian Conoscenti); project administration, C.C. (Christian Conoscenti), J.Y.R.A. and E.R.; funding acquisition, J.Y.R.A., C.C. (Christian Conoscenti), and E.R. All authors have read and agreed to the published version of the manuscript.

Funding: This research was supported by the project RIESCA (Escenarios de Riesgos con Vigilancia y Monitoreo de los Fenómenos Volcánicos, Sísmicos e Hidrogeológicos en Centro América; coord. Prof. G. Giunta), funded by the Ministry of Foreign Affair of the Italian Government and realized by the University of Palermo, University of San Salvador and Ministerio de Medio Ambiente y Recursos Naturales—Gobierno de El Salvador.

Institutional Review Board Statement: Not applicable.

Informed Consent Statement: Not applicable.

Data Availability Statement: The landslide inventory, digital elevation model, soil use and geologic layers adopted for this study can be acquired from any of the authors. No remotely accessible databank is presently online.

Conflicts of Interest: The authors declare no conflict of interest.

References

1. DeMets, C. A New Estimate for Present-Day Cocos-Caribbean Plate Motion:Implications for Slip along the Central American Volcanic Arc. *Geophys. Res. Lett.* **2001**, *28*, 4043–4046. [CrossRef]
2. Agostini, S.; Corti, G.; Doglioni, C.; Carminati, E.; Innocenti, F.; Tonarini, S.; Manetti, P.; di Vincenzo, G.; Montanari, D. Tectonic and Magmatic Evolution of the Active Volcanic Front in El Salvador: Insight into the Berlín and Ahuachapán Geothermal Areas. *Geothermics* **2006**, *35*, 368–408. [CrossRef]
3. Von Köppen, W. *Handbuch Der Klimatologie in Fünf Bänden Das Geographische System der Klimate*; Gebrüder Borntraeger Verlagsbuchhandlung: Berlin, Germany, 1936.
4. The World Bank Group Climate Change Knowledge Portal. Available online: https://climateknowledgeportal.worldbank.org/ (accessed on 14 April 2022).
5. Rotigliano, E.; Martinello, C.; Hernandéz, M.A.; Agnesi, V.; Conoscenti, C. Predicting the Landslides Triggered by the 2009 96E/Ida Tropical Storms in the Ilopango Caldera Area (El Salvador, CA): Optimizing MARS-Based Model Building and Validation Strategies. *Environ. Earth Sci.* **2019**, *78*, 210. [CrossRef]
6. Rotigliano, E.; Martinello, C.; Agnesi, V.; Conoscenti, C. Evaluation of Debris Flow Susceptibility in El Salvador (CA): A Comparison between Multivariate Adaptive Regression Splines (MARS) and Binary Logistic Regression (BLR). *Hung. Geogr. Bull.* **2018**, *67*, 361–373. [CrossRef]
7. Mercurio, C.; Martinello, C.; Rotigliano, E.; Argueta-Platero, A.A.; Reyes-Martínez, M.E.; Rivera-Ayala, J.Y.; Conoscenti, C. Mapping Susceptibility to Debris Flows Triggered by Tropical Storms: A Case Study of the San Vicente Volcano Area (El Salvador, CA). *Earth* **2021**, *2*, 66–85. [CrossRef]
8. OCHA (Oficina para la Coordinación de Asuntos Humanitarios). *El Salvador: Deslizamiento Por Lluvias—Nejapa*; OCHA: New York, NY, USA, 2020.
9. MARN (Ministerio de Medio Ambiente y Recursos Naturales). *Síntesis de Los Informes de Evaluación Técnica de Las Lluvias Del 7 y 8 de Noviembre 2009 En El Salvador: Análisis Del Impacto Físico Natural y Vulnerabilidad Socio Ambiental*; MARN: San Salvador, El Salvador, 2010.
10. MARN. *Depresión Tropical 12E/Sistema Depresionario Sobre El Salvador y Otros Eventos Extremos Del Pacífico*; MARN: San Salvador, El Salvador, 2011.
11. Cappadonia, C.; Cafiso, F.; Ferraro, R.; Martinello, C.; Rotigliano, E. Rockfall Hazards of Mount Pellegrino Area (Sicily, Southern Italy). *J. Maps* **2021**, *17*, 29–39. [CrossRef]
12. Cafiso, F.; Cappadonia, C.; Ferraro, R.; Martinello, C. Rockfall hazard assessment of the Monte Gallo oriented nature reserve area (Southern Italy). In *IOP Conference Series: Earth and Environmental Science, Proceedings of the Mechanics and Rock Engineering, from Theory to Practice, Turin, Italy, 20–25 September 2021*; IOP Publishing Ltd.: Bristol, UK, 2021; Volume 833.
13. Cafiso, F.; Cappadonia, C. Landslide Inventory and Rockfall Risk Assessment of a Strategic Urban Area (Palermo, Sicily). *Rend. Online Soc. Geol. Ital.* **2019**, *48*, 96–105. [CrossRef]
14. Kopačková, V.; Šebesta, J. An approach for GIS-based statistical landslide susceptibility zonation: With a case study in the northern part of El Salvador. In Proceedings of the SPIE—The International Society for Optical Engineering, Florence, Italy, 17 September 2007; Volume 6749.
15. García-Rodríguez, M.J.; Malpica, J.A.; Benito, B.; Díaz, M. Susceptibility Assessment of Earthquake-Triggered Landslides in El Salvador Using Logistic Regression. *Geomorphology* **2008**, *95*, 172–191. [CrossRef]
16. García-Rodríguez, M.J.; Malpica, J.A. Assessment of Earthquake-Triggered Landslide Susceptibility in El Salvador Based on an Artificial Neural Network Model. *Nat. Hazards Earth Syst. Sci.* **2010**, *10*, 1307–1315. [CrossRef]
17. Kirschbaum, D.; Stanley, T.; Yatheendradas, S. Modeling Landslide Susceptibility over Large Regions with Fuzzy Overlay. *Landslides* **2016**, *13*, 485–496. [CrossRef]
18. MARN. *Memoria Técnica Para El Mapa de Susceptibilidad de Deslizamientos de Tierra En El Salvador*; MARN: San Salvador, El Salvador, 2004.
19. Mora, C.S.; Vahrson, W.G. Macrozonation Methodology for Landslide Hazard Determination. *Environ. Eng. Geosci.* **1994**, *31*, 49–58. [CrossRef]
20. Mora, S.; Vahrson, W.G. Determinación "A Priori" de la amenaza de deslizamientos en grandes áreas utilizando indicadores morfodinámicos. In *Memoria Del Primer Simposio Internacional Sobre Sensores Remotos Y Sistemas De Información Geografica (SIG) Para El Estudio De Riesgos Naturales*; IGAC: Bogotá, Colombia, 1992; pp. 259–273.
21. Steger, S.; Brenning, A.; Bell, R.; Glade, T. The Influence of Systematically Incomplete Shallow Landslide Inventories on Statistical Susceptibility Models and Suggestions for Improvements. *Landslides* **2017**, *14*, 1767–1781. [CrossRef]
22. Steger, S.; Mair, V.; Kofler, C.; Pittore, M.; Zebisch, M.; Schneiderbauer, S. Correlation Does Not Imply Geomorphic Causation in Data-Driven Landslide Susceptibility Modelling—Benefits of Exploring Landslide Data Collection Effects. *Sci. Total Environ.* **2021**, *776*, 145935. [CrossRef]
23. Friedman, J.H. Multivariate Adaptive Regression Splines. *Ann. Stat.* **1991**, *19*, 1–67. [CrossRef]
24. QGIS.org. *QGIS Geographic Information System*; QGIS Association, 2022; Available online: http://www.qgis.org (accessed on 26 April 2022).
25. Conrad, O.; Bechtel, B.; Bock, M.; Dietrich, H.; Fischer, E.; Gerlitz, L.; Wehberg, J.; Wichmann, V.; Böhner, J. System for Automated Geoscientific Analyses (SAGA) v. 2.1.4. *Geosci. Model Dev.* **2015**, *8*, 1991–2007. [CrossRef]
26. RStudio Team. *RStudio: Integrated Development for R*; RStudio Team: Boston, MA, USA, 2020.

27. CEPAL (Comisión Económica para América Latina y el Caribe). *El Salvador: Impacto Socioeconómico, Ambiental y de Riesgo Por La Baja Presión Asociada a La Tormenta Tropical Ida En Noviembre de 2009*; Cepal: Santiago, Chile, 2010.

28. CEPAL (Comisión Económica para América Latina y el Caribe). *Evaluación de Daños y Pérdidas En El Salvador Ocasionados Por La Depresión Tropical 12E*; Cepal: Santiago, Chile, 2011.

29. Rotigliano, E.; Agnesi, V.; Cappadonia, C.; Conoscenti, C. The Role of the Diagnostic Areas in the Assessment of Landslide Susceptibility Models: A Test in the Sicilian Chain. *Nat. Hazards* **2011**, *58*, 981–999. [CrossRef]

30. Lombardo, L.; Cama, M.; Maerker, M.; Rotigliano, E. A Test of Transferability for Landslides Susceptibility Models under Extreme Climatic Events: Application to the Messina 2009 Disaster. *Nat. Hazards* **2014**, *74*, 1951–1989. [CrossRef]

31. Costanzo, D.; Chacón, J.; Conoscenti, C.; Irigaray, C.; Rotigliano, E. Forward Logistic Regression for Earth-Flow Landslide Susceptibility Assessment in the Platani River Basin (Southern Sicily, Italy). *Landslides* **2014**, *11*, 639–653. [CrossRef]

32. Cama, M.; Lombardo, L.; Conoscenti, C.; Agnesi, V.; Rotigliano, E. Predicting Storm-Triggered Debris Flow Events: Application to the 2009 Ionian Peloritan Disaster (Sicily, Italy). *Nat. Hazards Earth Syst. Sci.* **2015**, *15*, 1785–1806. [CrossRef]

33. Lombardo, L.; Cama, M.; Conoscenti, C.; Märker, M.; Rotigliano, E. Binary Logistic Regression versus Stochastic Gradient Boosted Decision Trees in Assessing Landslide Susceptibility for Multiple-Occurring Landslide Events: Application to the 2009 Storm Event in Messina (Sicily, Southern Italy). *Nat. Hazards* **2015**, *79*, 1621–1648. [CrossRef]

34. Costanzo, D.; Rotigliano, E.; Irigaray, C.; Jiménez-Perálvarez, J.D.; Chacón, J. Factors Selection in Landslide Susceptibility Modelling on Large Scale Following the Gis Matrix Method: Application to the River Beiro Basin (Spain). *Nat. Hazards Earth Syst. Sci.* **2012**, *12*, 327–340. [CrossRef]

35. Vargas-Cuervo, G.; Rotigliano, E.; Conoscenti, C. Prediction of Debris-Avalanches and -Flows Triggered by a Tropical Storm by Using a Stochastic Approach: An Application to the Events Occurred in Mocoa (Colombia) on 1 April 2017. *Geomorphology* **2019**, *339*, 31–43. [CrossRef]

36. Martinello, C.; Cappadonia, C.; Conoscenti, C.; Agnesi, V.; Rotigliano, E. Optimal Slope Units Partitioning in Landslide Susceptibility Mapping. *J. Maps* **2020**, *17*, 152–162. [CrossRef]

37. Martinello, C.; Cappadonia, C.; Conoscenti, C.; Rotigliano, E. Landform Classification: A High-Performing Mapping Unit Partitioning Tool for Landslide Susceptibility Assessment—A Test in the Imera River Basin (Northern Sicily, Italy). *Landslides* **2022**, *19*, 539–553. [CrossRef]

38. Weber, H.S.; Wiesemann, G.; Lorenz, W. *Schmidt-Thome Mapa Geologico de La Republica de El Salvador/America Central, 1:100,000*; Bundesanstalt fur Geowissenschaften und Rhhstoffe: Hannover, Germany, 1978.

39. Wilson, J.P.; Gallant, J.C. Primary topographic attributes. In *Terrain Analysis: Principles and Applications*; Wilson, J.P., Gallant, J.C., Eds.; John Wiley & Sons: Hoboken, NJ, USA, 2000.

40. Burrough, P.A.; McDonnell, R.A. *Principle of Geographic Information Systems CODRA—Creating Opportunities to Develop Resilient Agriculture View Project Groundwater Governance in the Arab World: Taking Stock and Addressing the Challenges View Project*; Oxford University Press: Oxford, UK, 1998.

41. Zevenbergen, L.W.; Thorne, C.R. Quantitative Analysis of Land Surface Topography. *Earth Surf. Process. Landf.* **1987**, *12*, 47–56. [CrossRef]

42. Beven, K.J.; Kirkby, M.J. A Physically Based, Variable Contributing Area Model of Basin Hydrology. *Hydrol. Sci. Bull.* **1979**, *24*, 43–69. [CrossRef]

43. Schmidt-Thomé, M. *The Geology in the San Salvador Area (El Salvador, Central America), a Basis for City Development and Planning*; Geologisches Jahrbuch der BGR: Hanover, Germany, 1975.

44. Cama, M.; Conoscenti, C.; Lombardo, L.; Rotigliano, E. Exploring Relationships between Grid Cell Size and Accuracy for Debris-Flow Susceptibility Models: A Test in the Giampilieri Catchment (Sicily, Italy). *Environ. Earth Sci.* **2016**, *75*, 238. [CrossRef]

45. Mokhtari, M.; Abedian, S. Spatial Prediction of Landslide Susceptibility in Taleghan Basin, Iran. *Stoch. Environ. Res. Risk Assess.* **2019**, *33*, 1297–1325. [CrossRef]

46. Cama, M.; Lombardo, L.; Conoscenti, C.; Rotigliano, E. Improving Transferability Strategies for Debris Flow Susceptibility AssessmentApplication to the Saponara and Itala Catchments (Messina, Italy). *Geomorphology* **2017**, *288*, 52–65. [CrossRef]

47. Lombardo, L.; Bachofer, F.; Cama, M.; Märker, M.; Rotigliano, E. Exploiting Maximum Entropy Method and ASTER Data for Assessing Debris Flow and Debris Slide Susceptibility for the Giampilieri Catchment (North-Eastern Sicily, Italy). *Earth Surf. Process. Landf.* **2016**, *41*, 1776–1789. [CrossRef]

48. Nicu, I.C.; Asăndulesei, A. GIS-Based Evaluation of Diagnostic Areas in Landslide Susceptibility Analysis of Bahluieț River Basin (Moldavian Plateau, NE Romania). Are Neolithic Sites in Danger? *Geomorphology* **2018**, *314*, 27–41. [CrossRef]

49. Sameen, M.I.; Pradhan, B.; Bui, D.T.; Alamri, A.M. Systematic Sample Subdividing Strategy for Training Landslide Susceptibility Models. *Catena* **2020**, *187*, 104358. [CrossRef]

50. Erener, A.; Sivas, A.A.; Selcuk-Kestel, A.S.; Düzgün, H.S. Analysis of Training Sample Selection Strategies for Regression-Based Quantitative Landslide Susceptibility Mapping Methods. *Comput. Geosci.* **2017**, *104*, 62–74. [CrossRef]

51. Naimi, B. *Package "Usdm". Uncertainty Analysis for Species Distribution Models*. Available online: https://cran.r-project.org/web/packages/usdm/usdm.pdf (accessed on 26 April 2022).

52. Felicísimo, Á.M.; Cuartero, A.; Remondo, J.; Quirós, E. Mapping Landslide Susceptibility with Logistic Regression, Multiple Adaptive Regression Splines, Classification and Regression Trees, and Maximum Entropy Methods: A Comparative Study. *Landslides* **2013**, *10*, 175–189. [CrossRef]

53. Pourghasemi, H.R.; Rossi, M. Landslide Susceptibility Modeling in a Landslide Prone Area in Mazandarn Province, North of Iran: A Comparison between GLM, GAM, MARS, and M-AHP Methods. *Theor. Appl. Climatol.* **2017**, *130*, 609–633. [CrossRef]
54. Conoscenti, C.; Ciaccio, M.; Caraballo-Arias, N.A.; Gómez-Gutiérrez, Á.; Rotigliano, E.; Agnesi, V. Assessment of Susceptibility to Earth-Flow Landslide Using Logistic Regression and Multivariate Adaptive Regression Splines: A Case of the Belice River Basin (Western Sicily, Italy). *Geomorphology* **2015**, *242*, 49–64. [CrossRef]
55. Conoscenti, C.; Rotigliano, E.; Cama, M.; Caraballo-Arias, N.A.; Lombardo, L.; Agnesi, V. Exploring the Effect of Absence Selection on Landslide Susceptibility Models: A Case Study in Sicily, Italy. *Geomorphology* **2016**, *261*, 222–235. [CrossRef]
56. Liu, C.-C.; Luo, W.; Chung, H.-W.; Yin, H.-Y.; Yan, K.-W. Influences of the Shadow Inventory on a Landslide Susceptibility Model. *ISPRS Int. J. Geo-Inf.* **2018**, *7*, 374. [CrossRef]
57. Lay, U.S.; Pradhan, B.; Yusoff, Z.B.M.; Abdallah, A.F.B.; Aryal, J.; Park, H.-J. Data Mining and Statistical Approaches in Debris-Flow Susceptibility Modelling Using Airborne LiDAR Data. *Sensors* **2019**, *19*, 3451. [CrossRef]
58. Wang, L.J.; Guo, M.; Sawada, K.; Lin, J.; Zhang, J. Landslide Susceptibility Mapping in Mizunami City, Japan: A Comparison between Logistic Regression, Bivariate Statistical Analysis and Multivariate Adaptive Regression Spline Models. *Catena* **2015**, *135*, 271–282. [CrossRef]
59. Conoscenti, C.; Martinello, C.; Alfonso-Torreño, A.; Gómez-Gutiérrez, Á. Predicting Sediment Deposition Rate in Check-Dams Using Machine Learning Techniques and High-Resolution DEMs. *Environ. Earth Sci.* **2021**, *80*, 380. [CrossRef]
60. Milborrow, S. Notes on the Earth Package. 2014, pp. 1–60. Available online: http://mtweb.cs.ucl.ac.uk/mus/lib/R/earth/doc/earth-notes.pdf (accessed on 26 April 2022).
61. Fawcett, T. An Introduction to ROC Analysis. *Pattern Recognit. Lett.* **2006**, *27*, 861–874. [CrossRef]
62. Lasko, T.A.; Bhagwat, J.G.; Zou, K.H.; Ohno-Machado, L. The Use of Receiver Operating Characteristic Curves in Biomedical Informatics. *J. Biomed. Inform.* **2005**, *38*, 404–415. [CrossRef] [PubMed]
63. Goodenough, D.J.; Rossmann, K.; Lusted, L.B. Radiographic Applications of Receiver Operating Characteristic (ROC) Curves. *Diagn. Radiol.* **1974**, *110*, 89–95. [CrossRef]
64. Hosmer, D.W.; Lemeshow, S. *Applied Logistic Regression*; John Wiley & Sons, Inc.: Hoboken, NJ, USA, 2000; ISBN 0471722146.
65. Hanley, J.A.; McNeil, B.J. The Meaning and Use of the Area under a Receiver Operating Characteristic (ROC) Curve1. *Radiology* **1982**, *143*, 29–36. [CrossRef]
66. Youden, W.J. Index for Rating Diagnostic Tests. *Cancer* **1950**, *3*, 32–35. [CrossRef]
67. Chung, C.J.F.; Fabbri, A.G. Validation of Spatial Prediction Models for Landslide Hazard Mapping. *Nat. Hazards* **2003**, *30*, 451–472. [CrossRef]
68. Guzzetti, F.; Reichenbach, P.; Ardizzone, F.; Cardinali, M.; Galli, M. Estimating the Quality of Landslide Susceptibility Models. *Geomorphology* **2006**, *81*, 166–184. [CrossRef]

Article

Using Public Landslide Inventories for Landslide Susceptibility Assessment at the Basin Scale: Application to the Torto River Basin (Central-Northern Sicily, Italy)

Chiara Martinello [1], Claudio Mercurio [1], Chiara Cappadonia [1,*], Viviana Bellomo [1], Andrea Conte [1], Giampiero Mineo [1], Giulia Di Frisco [1], Grazia Azzara [1], Margherita Bufalini [2], Marco Materazzi [2] and Edoardo Rotigliano [1]

[1] Department of Earth and Marine Sciences, University of Palermo, 90123 Palermo, Italy; chiara.martinello@unipa.it (C.M.); claudio.mercurio@unipa.it (C.M.); viviana.bellomo01@unipa.it (V.B.); andrea.conte@unipa.it (A.C.); giampiero.mineo@unipa.it (G.M.); giulia.difrisco@unipa.it (G.D.F.); grazia.azzara@unipa.it (G.A.); edoardo.rotigliano@unipa.it (E.R.)

[2] School of Science and Technology, Geology Division, University of Camerino, 62032 Camerino, Italy; margherita.bufalini@unicam.it (M.B.); marco.materazzi@unicam.it (M.M.)

* Correspondence: chiara.cappadonia@unipa.it; Tel.: +39-091-238-64664

Abstract: In statistical landslide susceptibility evaluation, the quality of the model and its prediction image heavily depends on the quality of the landslide inventories used for calibration. However, regional-scale inventories made available by public territorial administrations are typically affected by an unknown grade of incompleteness and mapping inaccuracy. In this research, a procedure is proposed for verifying and solving such limits by applying a two-step susceptibility modeling procedure. In the Torto River basin (central-northern Sicily, Italy), using an available regional landslide inventory (267 slide and 78 flow cases), two SUFRA_1 models were first prepared and used to assign a landslide susceptibility level to each slope unit (SLU) in which the study area was partitioned. For each of the four susceptibility classes that were obtained, 30% of the mapping units were randomly selected and their stable/unstable status was checked by remote analysis. The new, increased inventories were finally used to recalibrate two SUFRA_2 models. The prediction skills of the SUFRA_1 and SUFRA_2 models were then compared by testing their accuracy in matching landslide distribution in a test sub-basin where a high-resolution systematic inventory had been prepared. According to the results, the strong limits of the SUFRA_1 models (sensitivity: 0.67 and 0.57 for slide and flow, respectively) were largely solved by the SUFRA_2 model (sensitivity: 1 for both slide and flow), suggesting the proposed procedure as a possibly suitable modeling strategy for regional susceptibility studies.

Keywords: landslide susceptibility; public landslide inventory; MARS; landslide incompleteness

Citation: Martinello, C.; Mercurio, C.; Cappadonia, C.; Bellomo, V.; Conte, A.; Mineo, G.; Di Frisco, G.; Azzara, G.; Bufalini, M.; Materazzi, M.; et al. Using Public Landslide Inventories for Landslide Susceptibility Assessment at the Basin Scale: Application to the Torto River Basin (Central-Northern Sicily, Italy). *Appl. Sci.* **2023**, *13*, 9449. https://doi.org/10.3390/app13169449

Academic Editor: Jianbo Gao

Received: 20 July 2023
Revised: 10 August 2023
Accepted: 18 August 2023
Published: 21 August 2023

1. Introduction

Landslide susceptibility assessment can be performed by applying statistical methods to model the dependence between a set of predictors and an outcome expressing the stable/unstable status of a mapping unit [1–4]. The reliability of a predictive model strongly relies on the completeness and representativeness of the landslide inventory that is used for calibration [5–9]. In particular, regional landslide susceptibility studies require the use of landslide inventories, which are typically available only from public administrations. In fact, such a big database is typically the result of long-term cumulative reported cases that are mapped following warnings from local municipality offices, transportation companies, or even citizens. As a matter of fact, the reported landslide cases are clustered around urban areas and the infrastructural axis. For this reason, this kind of inventory suffers from an unknown grade of incompleteness and inaccuracy. The number of cases is also too large for an accurate check to be performed by regional authorities. Both multiple typologies and landslide polygons are frequently corrected. These limits are obviously much more marked in agricultural and pastoral areas [10,11], where the potential interest for urban development is not infrequent. On the other hand, regional landslide databases allow the available landslide inventories to be immediately obtained, thereby saving time and resources from mapping [12].

Thus, defining a useful way to increase the quality of regional landslide inventories is a goal of research focused on landslide susceptibility evaluation but also of public administrations. In fact, the latter, generally determine landslide risk by crossing the inventoried phenomena (and their typological/geometrical characteristics) and the exposed vulnerable areas (e.g., urbanized sectors or communication routes). In addition, support for territory management, planning, and safety measures is mainly defined based on geo-hydrological hazards. In this sense, public administrations have made various efforts to obtain more correct and complete landslide inventories [13,14].

In light of the abovementioned issues, a need arises to find possible modeling procedures for regional landslide susceptibility assessment that are capable of both detecting and solving the potential limits induced by poor calibration inventories. However, studies aimed at evaluating the effects of incomplete inventories are nowadays focused on the models' performance [7] or the variables' importance [5,10]. In this research, a procedure for using regional landslide inventories to prepare reliable and accurate susceptibility models is proposed. By applying the approach suggested by Martinello et al. [7], the potential limits of a susceptibility model calibrated with the source inventory were first identified. By systematically checking a portion of the study area, an enrichment of the original calibration landslide inventory was then obtained. A new model was then recalibrated and its accuracy evaluated and compared with that of the source model.

The research was carried out in the context of the SUFRA project, a challenging project that involves the analysis and evaluation of all types of landslide susceptibilities (slide, flow, rapid flow, fall-topple, and lateral spread) for the whole regional territory of Sicily ($\sim$26,000 km^2). It is the first project focused on landslide susceptibility evaluation at the regional scale, and it will be used by the public administration for territorial planning and civil protection aims. Considering the short duration of the project (only two years), we were forced to base our analysis on the landslide inventories already available with the Sicilian public administration. At the same time, in the context of the PNRR project GeoSciences IR, the research was focused on defining strategies to increase the overall quality of public landslide inventories, thus optimizing costs, resources, and time.

2. Materials and Methods

The available slide and flow inventories of the Torto River basin (420 km^2, central-northern Sicily), which were prepared by the "Dipartimento Regionale dell'Autorità di Bacino del Distretto Idrografico Sicilia" (the so-named P.A.I. inventories), and a set of twelve geo-environmental predictors were used to produce two basin-scale susceptibility models (for slides and flows, respectively) by applying multivariate adaptive regression splines (MARS). The obtained first-level landslide susceptibility maps were used for checking 30% of mapping units in which no landslides of P.A.I. were present and defining their stable/unstable status with respect to flow and slide movements. The checked archives were used for integrating the main inventories (the P.A.I. inventories) in order to obtain second-level landslide susceptibility maps. Once all landslide susceptibility maps were produced (first level and second level), the accuracy of the obtained maps was verified by validating high-resolution flow/slide archives detected for a small sub-basin (Sciara) of the Torto catchment.

The research was implemented using open-source geographical information system software (GIS; Quantum GIS [15], GRASS GIS [16], and SAGA GIS [17]) and the Rstudio statistical platform [18].

2.1. Study Area

The Torto River extends for 423 km^2 in the northern section of Sicily (Italy, Figure 1a) between two mountain ranges, namely, the Madonie Mountains at the east and the Termini Mountains at the west, and the Tyrrhenian Sea. The geomorphological setting of the study area is the result of tectonic and selective erosion, karstification, and deep-seated gravitation slope deformation [19,20].

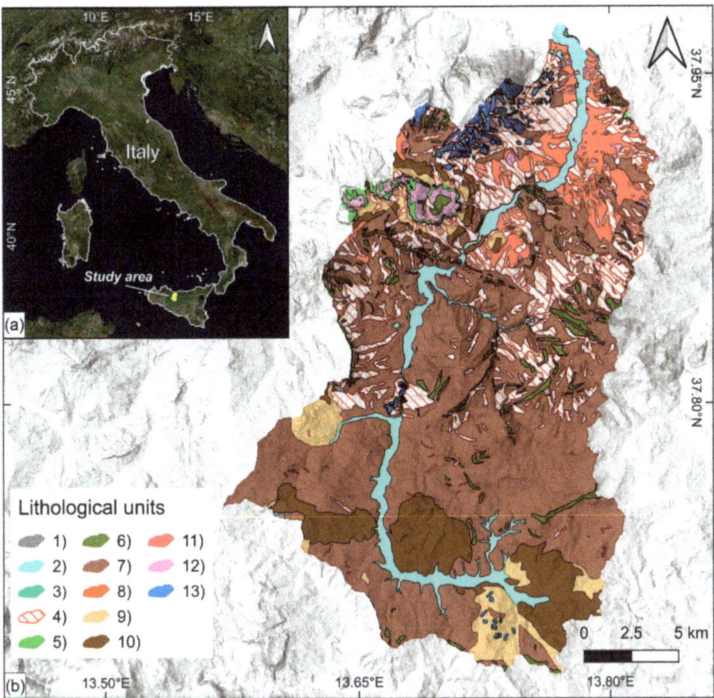

Figure 1. (**a**) Location of the Torto River basin. (**b**) Bedrock lithology map of the study area. (1) Anthropic deposits; (2) alluvial deposits; (3) alluvial fan and talus deposit; (4) colluvium and old landslide deposits; (5) evaporitic rocks; (6) sandstones; (7) Flysch Numidico pelites; (8) Flysch Numidico sandstones/conglomerates; (9) "Terravecchia" pelites; (10) "Terravecchia" sandstones/conglomerates; (11) "Varicolori" clays; (12) calcareous and clayey marls; (13) lithoid units.

In fact, the study area falls within the central-western section of the Sicilian fold and thrust belt, which is the result of the retreat of the subduction hinge of the Ionian oceanic lithosphere and the postcollisional convergence between Africa and Europe [21–24]. This complex structural setting results in a multiduplex system where the basin tectonic units overthrust platform tectonic units across subhorizontal surfaces with prevalent S–SW transport direction and components of northward back-thrusting. In the area, Sicilide units and the Numidian Flysch are widely outcropped, while Imerese basin units mainly represent the basal body. However, Plio-Quaternary high-angle faults create new contacts between the carbonatic Imerese successions and Cenozoic clayey rocks belonging to the Numidian Flysch, which are sometimes overthrust by the Sicilide units [19,20] (Figure 1b).

According to the geological setting, the study area is characterized by a hilly landscape modeled by gravitational movements and water erosion, whilst carbonate reliefs [20,25] are affected by gravitational (mainly falls) and karstic processes. Mount San Calogero is the highest relief of the area (1370 m s.l.m).

The climate of the Torto River basin is classified as the Mediterranean type, with rainfall concentrated mainly in the winter semester, while the summer period is characterized by almost drought conditions. The mean annual rainfall is around 600 mm, while the mean temperature value is about 15 °C.

2.2. Landslide Inventory

Starting from the available P.A.I. (Piano stralcio di bacino per l'Assetto Idrogeologico) landslide archives prepared by the "Dipartimento Regionale dell'Autorità di Bacino del Distretto Idrografico Sicilia", slide, flow, and complex inventories were distinct and submitted to remote checking. In fact, frequently, single phenomena are typically grouped into large polygons in these inventories, and, moreover, their boundaries are not so accurate (Figure 2a).

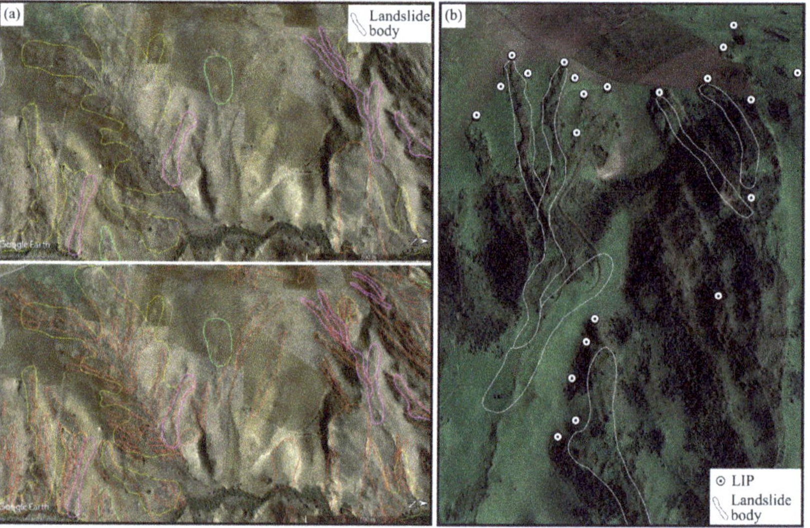

Figure 2. (**a**) Top image: landslides as mapped in the original P.A.I. inventory (yellow polygons are complex landslides, purple polygons are flows, red polygons are slides, and green polygons are diffused erosional areas); bottom image: mapping of single phenomena (red polygons). (**b**) Example of P.A.I.-driven mapping: original P.A.I. landslide inventory (polygons) and checked P.A.I. landslide inventory (LIPs).

In order to propose a landslide susceptibility evaluation technique with statistical methods, it is necessary to discriminate every individual landslide and, when needed, reinterpret the type of movement [26,27]. It is worth noting that the single phenomena were checked only inside the P.A.I. landslide polygons. This means that instead of a systematic (and complete) inventory, P.A.I.-driven mapping was produced (Figure 2b). The reason for this choice lies in the aim of the research, i.e., testing a good practice where available regional public landslide inventory can be used to obtain basin-scale susceptibility maps. In this way, according to Hungr et al. [26], for complex landslides, each component of the phenomenon was defined so that only two different inventories were obtained at the end of the mapping: the slide (78 cases) and the flow (267 cases) archives. In fact, it was assumed here that rotational and translational slides share their slope susceptibility conditions to a large extent. With regard to checking the P.A.I. inventory, the more frequently observed flaws (12 cases) concerned large earth-flows, which were misclassified as (rotational) slides.

Two examples of these very diffused landslide types are given in Figure 3. The landslide identification point (LIP), which corresponds to the highest point along the crown of the landslide area, was assumed as diagnostic in potentially marking unstable slope conditions [27–31].

Figure 3. (**a**) Rotational slide/flow landslide affecting the slope of the A19 motorway; (**b**) multiple rotational slide/flow landslides affecting the slope of the SS120 national road.

2.3. Mapping Units and Landslide Conditioning Factors

Considering the type of phenomena analyzed and the scale of the landslide susceptibility evaluation, we decided to employ slope units as mapping units (SLU). In fact, for the purpose of the project, we needed to detect the activation area but also include the potential area of propagation and arrest of the phenomena. According to the literature [6,9,32], SLUs have been demonstrated to be more geomorphologically adequate to represent all landslide phases (for the flow and slide phenomena) as it is assumed the complete landslide kinematic (initiation, propagation, and accumulation) occurs inside. For this research, SLUs were delimited by applying the r.watershed [33,34] GRASS GIS module using the 2000 contributing area threshold. By overlapping the SLUs with the landslide inventories, the stable/unstable status with respect to the slide and flow phenomena was defined for each slope unit depending on whether it hosts at least one LIP.

Geo-environmental predictors were selected on the basis of the expected direct or proxied role in landslides [7,27,35] (Table 1): outcropping lithology (LITO), land use (obtained by the Corine Land Cover 2018-USE), elevation (ELE—10 m), landform classification (LCL), steepness (SLO), aspect (expressed as northerness and easterness), plan (PLN), and profile (PRF) curvatures, topographic wetness index (TWI), and stream power index (SPI).

For the continuous variables, a multicollinearity analysis was carried out using the variance inflation factor (VIF) obtained by applying the "usdm" R-package [36]. No multicollinearity emerged between the selected predictors. However, considering that specific modeling procedures were implemented separately for flow and slide, the SPI predictor was excluded for the slide model, while the TWI variable was excluded from the flow model.

Table 1. Details of the employed geo-environmental variables (modified from [7,27]).

Acronym	Description of Predictor	References	Potential Proxy Significance
ELE	Distribution of elevation		Mean annual rainfall
LCL	Morphological classification of the territory based on the variation in elevation with respect to the neighbouring areas	[37]	Morphological setting
SLO	The first derivative of elevation	[38]	Speed of the water and potential underlying rupture surfaces [6,27]
N	Cosine of aspect (direction in which the slope degrades more rapidly)	[39]	Seasonal wet/dry cycles of soils [40]
E	Sine of aspect (direction in which the slope degrades more rapidly)	[39]	Seasonal wet/dry cycles of soils [40]
PLN	The second derivative of elevation, computed along the horizontal plane	[41]	Activation and propagation of landslides [42]
PRF	The second derivative of elevation, computed along the direction of the highest slope gradient	[41]	The direction of flow [42]
TWI	Calculated as $\ln[A/\tan\beta]$, where A and β, computed on each cell, corresponds to the area of upslope drained cells and the slope gradient, respectively	[43]	Potential infiltration or saturated soil thickness [6,27]
SPI	Natural logarithm of the catchment area multiplied by the tangent of the slope gradient	[44]	Proxy of the intensity of surface water erosion [6]
LITO	Original geological map		Physical–mechanical properties of rocks [27]
USE	CORINE land cover (2018)		Potential hydrological and surface hydric erosion induced disturbances [27]

Each variable was then characterized inside the SLUs by zonal statistics as deciles for the continuous variables and as relative frequencies for the categorical ones.

2.4. Statistical Model, Validation Tools, and Model-Building Strategies

The multivariate adaptive regression splines (MARS; [45]) method was used for all modeling procedures as it has been confirmed to be very effective in modeling nonlinear components of the relationship between landslides and their causative factors [6,46].

MARS is a nonparametric regression method that splits each independent variable into branches (optimizing their number based on the characteristics of the variable itself and the correlation with the distribution of other predictors). Each branch is defined by a hinge function (a function used for defining a nonlinear relationship between y and x) and the relative knot. The derived structures (hinge function and knots) identify a basis function that can take the shape of a simple linear regression (when the basis function corresponds to the model intercept, set to a constant value of 1) or more complex geometry (when the basis function is the product of one or more hinge functions associated with different covariates).

In this way, hinge functions boost the maximum-likelihood-based adaptation skill of the logistic regression method, according to

$$y = f(x) = \alpha + \sum_{i=1}^{N} \beta_i h_i(x) \tag{1}$$

where y is the dependent variable (the outcome) predicted by the function $f(x)$, α is the model intercept, and β_i is the coefficient of the hi basis functions given the N number of base functions. For other information about the method, please refer to [6,27,35,47–49]. For this research, MARS analysis was performed using the "earth" R-package [50].

Due to the fact that the MARS method is based on a presence–absence approach, a random extraction of negative cases in the same number as the positive cases was carried out. The random selection of negative cases and the subsequent modeling was replicated one-hundred times to evaluate the independence of the results (resolution and precision) from the specific choice of the negative cases [6,27]. On the other hand, to verify the prediction skill of the models, each balanced dataset was randomly split using 75% for calibration and the remaining 25% for validation [51].

AUC value (area under the curve) in the ROC (receiver operating characteristics) [52–54] was employed to evaluate the prediction skill of the model according to Hosmer and Lemeshow [55]. At the same time, the Youden index optimized score cut-off [56] was obtained from the ROC plots to set confusion matrices and calculate the related validation indices (sensitivity, specificity, and accuracy). Nested applications of the Youden index cut-off were employed to define the different cut-offs of four susceptibility levels in an objective way: S1 (low), S2 (moderate), S3 (high), and S4 (very high).

In Figure 4, the model-building strategy employed in this research is synthetically shown. Once the P.A.I. inventory was checked and the relative LIPs extracted, a first model named SUFRA_1 was obtained and validated, both for slide and flow landslides. Thus, each SLU was classified according to the resulting susceptibility score classes.

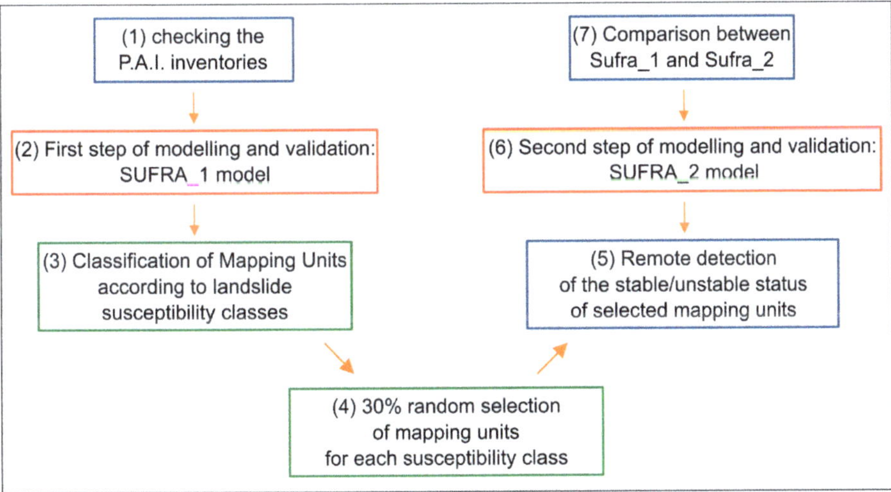

Figure 4. Synthetic scheme of the adopted model-building procedures.

To test the quality of the prediction images in predicting a high-resolution unknown landslide inventory, a second validation was performed in the small Sciara sub-basin (~21 km²), where a new systematic inventory for flow and slide was prepared using remote surveys. The Sciara sub-basin was selected because, in light of its geomorphological setting, it is largely representative of the landslide susceptibility in the whole Torto basin area. Then, 30% of unrecognized P.A.I. SLUs were randomly extracted for each susceptibility class and submitted to remote detection of stable/unstable status with respect to flow and slide movements. Thus, using both the checked P.A.I inventory and the 30% systematically mapped one, two new (slides/flows) SUFRA_2 models were prepared. Finally, the performance of the models was evaluated both with respect to the whole Torto basin (P.A.I. checked inventories) and the Sciara basin.

3. Results

In Figure 5, the ROC plots for the SUFRA_1 models, both for the validation in the whole Torto basin and the Sciara sub-basin, are shown. The AUC values for SUFRA_1 models were outstanding for validation in the Torto basin (Figure 5a,b). However, the values decreased when the validation was focused on the Sciara sub-basin with respect to the systematic inventories (Figure 5c,d). This lowering was more marked for the flow model whose performance went from outstanding to good (0.77).

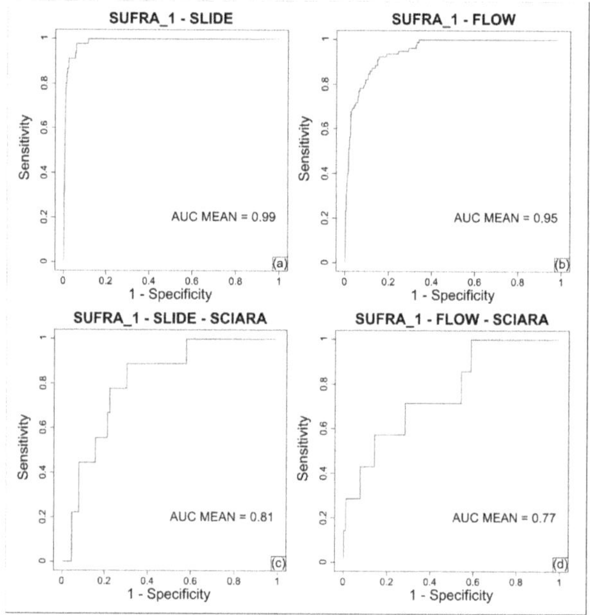

Figure 5. ROC plots of the two SUFRA_1 models validated in the whole Torto River basin (**a**,**b**) and in the Sciara sub-basin (**c**,**d**). AUC mean values were computed through one-hundred replicates given by extraction of different random negatives.

Confusion matrices (Table 2) confirmed these behaviors, with very high values of sensitivity (Sens. values of 1 and 0.98 for slide and flow model, respectively). However, a limited specificity (Spec. values of 0.69 and 0.67 for slide and flow model, respectively) resulted due to the high number of false positives (FPs) produced. These very low values of specificity also affected the accuracy (Acc.), which showed just sufficient values (~0.7).

Table 2. Confusion matrix of the SUFRA_1 models in the Torto basin and in the Sciara sub-basin.

		Positive Cases	Negative Cases	TN	FN	FP	TP	Acc.	Sens.	Spec.
Torto Area	SUFRA_1 Slide	45	968	666	0	302	45	0.70	1	0.69
	SUFRA_1 Flow	78	935	627	1	308	77	0.69	0.98	0.67
Sciara Area	SUFRA_1 Slide	9	90	70	3	20	6	0.77	0.67	0.78
	SUFRA_1 Flow	7	92	72	3	20	4	0.77	0.57	0.78

On the other hand, the validation in the Sciara sub-basin revealed that the sensitivity suffered in the prediction images produced for both the slide and flow models when a systematic high-resolution archive was detected. This limit was more evident for the flow model for which the sensitivity was markedly insufficient (<0.6).

The ROC plots relative to the validation of the SUFRA_2 models for slide and flow movements are shown in Figure 6. In this case, outstanding AUC values (>0.9) were achieved for both the whole Torto basin (Figure 6a,b) and the Sciara sub-basin (Figure 6c,d).

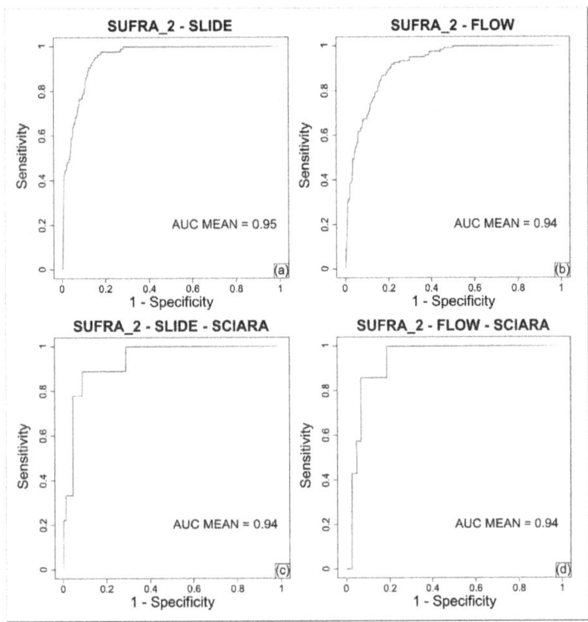

Figure 6. ROC plot of the two SUFRA_2 models validated in the whole Torto River basin (**a,b**) and in the Sciara sub-basin (**c,d**). AUC mean values were computed through one-hundred replicates given by extraction of different random negatives.

Confusion matrices (Table 3) confirmed the high performance in validation within a coeval/homogeneous inventory of calibration with sensitivity values of 1 for slide and 0.95 for flow. Again, the specificity was just over 0.7 due to the high number of FPs produced. However, the validation in the Sciara sub-basin confirmed the better performance of the prediction images produced: the sensitivity was 1 for both flows and slides and, at the same time, the specificity was 0.75 for slides and 0.8 for flows; better values of accuracy (0.77 and 0.82 for slides and flows, respectively) were consequently obtained.

Table 3. Confusion matrix of the SUFRA_2 models in the Torto basin and in the Sciara sub-basin.

		Positive Cases	Negative Cases	TN	FN	FP	TP	Acc.	Sens.	Spec.
Torto Area	SUFRA_2 Slide	85	928	682	0	246	85	0.76	1	0.73
	SUFRA_2 Flow	122	891	643	6	248	116	0.75	0.95	0.72
Sciara Area	SUFRA_2 Slide	9	90	67	0	25	9	0.77	1	0.74
	SUFRA_2 Flow	7	92	74	0	18	7	0.82	1	0.80

4. Discussion

The validation results of the SUFRA_1 models in the whole Torto River basin showed outstanding AUC values but with limited specificity compared to the very high values of sensitivity. Considering that the false positives are not only errors but also future positives, these results gave us a warning about the accuracy of the predicted landslide scenario. The validation in the Sciara sub-basin, where new systematic inventories for flow and slide were detected, showed that the quality of the prediction images produced was inaccurate. In fact, the sensitivity dramatically decreased here, especially for the flow model, clearly reflecting the limited skill of the models to detect new unknown phenomena. Considering the geomorphological setting of the Sciara sub-basin is representative of a very large part of the Torto River catchment, the limits of SUFRA_1 were considered relevant. On the other hand, the SUFRA_2 models maintained outstanding AUC values with very high sensitivity and good specificity and, differently from SUFRA_1, the new models still showed outstanding AUC values in the Sciara basin. More importantly, the sensitivity reached the maximum performance with good to excellent specificity. The false-positive rates still suggest the basin is characterized by relevant proneness to both flow- and slide-type slope failures. The same high-model performance was observed for both the landslide typologies, confirming that the goodness of this model procedure is independent of the landslide typology and number of cases (provided the inventory is representative).

According to our test, the proposed two-step approach is suitable for optimizing landslide susceptibility evaluation when the source inventory is affected by incompleteness or mapping inaccuracy. In fact, the second step of mapping (the susceptibility level-driven checking) permitted us to increase the quality of the calibration inventory and to cost-effectively correct the potential misleading results of the SUFRA_1 models. Obviously, the percentage of slope units checked (30% in this test) is not a standard but needs to be tuned case by case. At the same time, the selection of a single test sub-basin could be insufficient in the case of a more articulated geomorphological setting of the whole study area, and criteria for selecting the number and extension of such sectors need to be optimized (see [7] for a deeper inside of this issue). Indeed, different criteria for selecting the checking areas to improve the original inventory could be also explored. In our study, we precautionarily decided to maintain the same percentage of random extraction for each SUFRA_1 susceptibility class.

5. Conclusions

The research we conducted was focused on detecting a useful way to use public landslide regional inventory in statistical landslide susceptibility evaluation at a basin scale. In the Torto River basin, the original P.A.I. inventories of slide and flow movements were submitted to remote checking to produce more accurate archives that are suitable for statistical modeling. The proposed procedure seems to be robust in strengthening weak inventories, maximizing cost-effectiveness in regional landslide susceptibility studies. In fact, the proposed procedure simply requires, together with a first susceptibility model, a status slope unit check for a small percentage of the study area and systematic mapping in one or more smaller subareas. The study was focused on slide and flow landslide typologies, but the strategies of analysis can also be helpful for increasing landslide archives and related resolution of landslide susceptibility maps for any other type of landslide (such as falls, topples, and deep-seated typologies) with the aim of identifying areas to be analyzed at a larger scale through the application of empirical or analytical models for rockfalls (e.g., [57–59]) or to assess the magnitude and deformations rate for other slower and more complex landslides (e.g., [60,61]).

Author Contributions: Conceptualization, C.M. (Chiara Martinello), C.C. and E.R.; methodology, C.M. (Chiara Martinello), C.C. and E.R.; software, C.M. (Chiara Martinello), C.M. (Claudio Mercurio), V.B., A.C., G.M., G.A., G.D.F. and M.B.; validation, C.M. (Chiara Martinello), C.M. (Claudio Mercurio), C.C. and E.R.; formal analysis, C.M. (Chiara Martinello); software, C.M. (Chiara Martinello), C.M. (Claudio Mercurio), V.B., A.C., G.M., G.A. and G.D.F.; investigation, C.M. (Claudio Mercurio), V.B., A.C., G.M., G.A., G.D.F. and M.B.; resources, C.M. (Claudio Mercurio), V.B., A.C., G.M., G.A. and G.D.F.; data curation, C.M. (Chiara Martinello) and C.C.; writing—original draft preparation, C.M. (Chiara Martinello), C.C. and E.R.; writing—review and editing, C.M. (Chiara Martinello), C.C., M.M. and E.R.; visualization, C.M. (Chiara Martinello), C.C., M.M. and E.R.; supervision, C.M. (Chiara Martinello), C.C., M.M. and E.R; project administration, E.R.; funding acquisition, E.R. All authors have read and agreed to the published version of the manuscript.

Funding: This research received no external funding.

Institutional Review Board Statement: Not applicable.

Informed Consent Statement: Not applicable.

Data Availability Statement: The landslide inventory, digital terrain model, soil use, and geologic layers adopted for this study can be requested from any of the authors. In addition, the DTM can be visualized and downloaded using the WCS server of the SITR webgis with the following link (accessed on 15 June 2023): https://map.sitr.regione.sicilia.it/gis/services/modelli_digitali/mdt_20 13/ImageServer/WCSServer. The adopted soil use map is the one available from the Corine coverage; it can be downloaded from the following link (accessed on 15 June 2023): https://land.copernicus. eu/paneuropean/corine-land-cover/clc2018?tab=download.

Acknowledgments: The authors are very grateful to Geol. Ignazio Giuffré for giving strong support for the drone surveys. The research whose results are presented and discussed here was carried out in the framework of the SUFRA (SUscetibilità da FRAna) project, funded by the Basin Authority of the Hydrographic District of Sicily (E. Rotigliano) and the PNRR project GeoSciences IR, funded by the Ministry for University and Research with Next-Generation EU funds (M4C2—Investment 3.1 Fund for construction of an integrated system of research and innovation infrastructures). At the same time, this research is also the result of a collaboration between different universities (the University of Palermo and the University of Camerino) in light of the Work Group of AIGeO (Italian Association of Physical Geography and Geomorphology) "Environmental and Applied Geomorphology".

Conflicts of Interest: The authors declare no conflict of interest.

References

1. Carrara, A.; Cardinali, M.; Guzzetti, F.; Reichenbach, P. Gis Technology in Mapping Landslide Hazard. In *Geographical Information Systems in Assessing Natural Hazards*; Springer: Dordrecht, The Netherlands, 1995; pp. 135–175. [CrossRef]
2. Crozier, M.J.; Glade, T. *A Review of Scale Dependency in Landslide Hazard and Risk Analysis*; Wiley: Hoboken, NJ, USA, 2012; ISBN 9780471486633.
3. Fell, R.; Whitt, G.; Miner, T.; Flentje, P. Guidelines for Landslide Susceptibility, Hazard and Risk Zoning for Land Use Planning. *Eng. Geol.* **2008**, *102*, 83–84. [CrossRef]
4. Brabb, E.E. Innovative Approaches to Landslide Hazard and Risk Mapping. In Proceedings of the 4th International Symposium on Landslides, Toronto, ON, Canada, 16–21 September 1984; pp. 307–324.
5. Steger, S.; Mair, V.; Kofler, C.; Pittore, M.; Zebisch, M.; Schneiderbauer, S. Correlation Does Not Imply Geomorphic Causation in Data-Driven Landslide Susceptibility Modelling–Benefits of Exploring Landslide Data Collection Effects. *Sci. Total Environ.* **2021**, *776*, 145935. [CrossRef] [PubMed]
6. Martinello, C.; Cappadonia, C.; Conoscenti, C.; Rotigliano, E. Landform Classification: A High-Performing Mapping Unit Partitioning Tool for Landslide Susceptibility Assessment—A Test in the Imera River Basin (Northern Sicily, Italy). *Landslides* **2022**, *19*, 539–553. [CrossRef]

7. Martinello, C.; Mercurio, C.; Cappadonia, C.; Hernández Martínez, M.Á.; Reyes Martínez, M.E.; Rivera Ayala, J.Y.; Conoscenti, C.; Rotigliano, E. Investigating Limits in Exploiting Assembled Landslide Inventories for Calibrating Regional Susceptibility Models: A Test in Volcanic Areas of El Salvador. *Appl. Sci.* **2022**, *12*, 6151. [CrossRef]

8. Harp, E.L.; Keefer, D.K.; Sato, H.P.; Yagi, H. Landslide Inventories: The Essential Part of Seismic Landslide Hazard Analyses. *Eng. Geol.* **2011**, *122*, 9–21. [CrossRef]

9. Lima, P.; Steger, S.; Glade, T. Counteracting Flawed Landslide Data in Statistically Based Landslide Susceptibility Modelling for Very Large Areas: A National-Scale Assessment for Austria. *Landslides* **2021**, *18*, 3531–3546. [CrossRef]

10. Steger, S.; Brenning, A.; Bell, R.; Glade, T. The Influence of Systematically Incomplete Shallow Landslide Inventories on Statistical Susceptibility Models and Suggestions for Improvements. *Landslides* **2017**, *14*, 1767–1781. [CrossRef]

11. Petschko, H.; Bell, R.; Glade, T. Effectiveness of Visually Analyzing LiDAR DTM Derivatives for Earth and Debris Slide Inventory Mapping for Statistical Susceptibility Modeling. *Landslides* **2016**, *13*, 857–872. [CrossRef]

12. Bufalini, M.; Materazzi, M.; De Amicis, M.; Pambianchi, G. From Traditional to Modern 'Full Coverage' Geomorphological Mapping: A Study Case in the Chienti River Basin (Marche Region, Central Italy). *J. Maps* **2021**, *17*, 17–28. [CrossRef]

13. Restele, L.O.; Hidayat, A.; Saleh, F.; Iradat Salihin, L.M. Landslide Hazard Assessments and Their Application in Land Management in Kendari, Southeast Sulawesi Province, Indonesia. *J. Degrad. Min. Lands Manag.* **2023**, *10*, 4349–4356. [CrossRef]

14. Martinello, C.; Bufalini, M.; Cappadonia, C.; Rotigliano, E.; Materazzi, M. Combining Multi-Typologies Landslide Susceptibility Maps: A Case Study for the Visso Area (Central Italy). *J. Maps* **2023**, *19*, 1–10. [CrossRef]

15. QGIS Association. QGIS.org QGIS Geographic Information System 2022. Available online: http://www.qgis.org (accessed on 20 August 2023).

16. *GRASS Development Team Geographic Resources Analysis Support System (GRASS) Software*; Version 8.0; Open Source Geospatial Foundation: Chicago, IL, USA, 2022.

17. Conrad, O.; Bechtel, B.; Bock, M.; Dietrich, H.; Fischer, E.; Gerlitz, L.; Wehberg, J.; Wichmann, V.; Böhner, J. System for Automated Geoscientific Analyses (SAGA) v. 2.1.4. *Geosci. Model Dev.* **2015**, *8*, 1991–2007. [CrossRef]

18. RStudio Team RStudio: Integrated Development for R. 2020. RStudio: Integrated Development for R. RStudio, PBC, Boston, MA. Available online: http://www.rstudio.com/ (accessed on 20 August 2023).

19. Cappadonia, C.; Confuorto, P.; Sepe, C.; Di Martire, D. Preliminary Results of a Geomorphological and DInSAR Characterization of a Recently Identified Deep-Seated Gravitational Slope Deformation in Sicily (Southern Italy). *Rend. Online Soc. Geol. Ital.* **2019**, *49*, 149–156. [CrossRef]

20. Catalano, R.; Avellone, G.; Basilone, L.; Contino, A.; Agate, M. Note Illustrative Della Carta Geologica d'Italia Alla Scala 1: 50.000 Del Foglio 609 "Termini Imerese", Con Allegata Carta Geologica in Scala 1: 50.000. 2011. Available online: https://www.isprambiente.gov.it/Media/carg/note_illustrative/596_609_CapoPlaia_Termini.pdf (accessed on 20 August 2023).

21. Faccenna, C.; Piromallo, C.; Crespo-Blanc, A.; Jolivet, L.; Rossetti, F. Lateral Slab Deformation and the Origin of the Western Mediterranean Arcs. *Tectonics* **2004**, *23*, 1–21. [CrossRef]

22. Parrino, N.; Pepe, F.; Burrato, P.; Dardanelli, G.; Corradino, M.; Pipitone, C.; Morticelli, M.G.; Sulli, A.; Di Maggio, C. Elusive Active Faults in a Low Strain Rate Region (Sicily, Italy): Hints from a Multidisciplinary Land-to-Sea Approach. *Tectonophysics* **2022**, *839*, 229520. [CrossRef]

23. Sulli, A.; Gasparo Morticelli, M.; Agate, M.; Zizzo, E. Active North-Vergent Thrusting in the Northern Sicily Continental Margin in the Frame of the Quaternary Evolution of the Sicilian Collisional System. *Tectonophysics* **2021**, *802*, 228717. [CrossRef]

24. Parrino, N.; Burrato, P.; Sulli, A.; Gasparo Morticelli, M.; Agate, M.; Srivastava, E.; Malik, J.N.; Di Maggio, C. Plio-Quaternary Coastal Landscape Evolution of North-Western Sicily (Italy). *J. Maps* **2023**, *19*, 2158889. [CrossRef]

25. Agnesi, V.; De Cristofaro, D.; Di Maggio, C.; Macaluso, T.; Madonia, G.; Messana, V. Morphotectonic Setting of the Madonie Area (Central Northern Sicily). *Mem. Soc. Geol. Ital.* **2000**, *55*, 373–379.

26. Hungr, O.; Leroueil, S.; Picarelli, L. The Varnes Classification of Landslide Types, an Update. *Landslides* **2014**, *11*, 167–194. [CrossRef]

27. Martinello, C.; Cappadonia, C.; Rotigliano, E. Investigating the Effects of Cell Size in Statistical Landslide Susceptibility Modelling for Different Landslide Typologies: A Test in Central–Northern Sicily. *Appl. Sci.* **2023**, *13*, 1145. [CrossRef]

28. Mokhtari, M.; Abedian, S. Spatial Prediction of Landslide Susceptibility in Taleghan Basin, Iran. *Stoch. Environ. Res. Risk Assess.* **2019**, *33*, 1297–1325. [CrossRef]

29. Nicu, I.C.; Asăndulesei, A. GIS-Based Evaluation of Diagnostic Areas in Landslide Susceptibility Analysis of Bahluieţ River Basin (Moldavian Plateau, NE Romania). Are Neolithic Sites in Danger? *Geomorphology* **2018**, *314*, 27–41. [CrossRef]

30. Sameen, M.I.; Pradhan, B.; Bui, D.T.; Alamri, A.M. Systematic Sample Subdividing Strategy for Training Landslide Susceptibility Models. *Catena* **2020**, *187*, 104358. [CrossRef]

31. Erener, A.; Düzgün, H.S.B. Landslide Susceptibility Assessment: What Are the Effects of Mapping Unit and Mapping Method? *Environ. Earth Sci.* **2012**, *66*, 859–877. [CrossRef]

32. Chung, C.-C.; Li, Z.-Y. Rapid Landslide Risk Zoning toward Multi-Slope Units of the Neikuihui Tribe for Preliminary Disaster Management. *Nat. Hazards Earth Syst. Sci.* **2022**, *22*, 1777–1794. [CrossRef]

33. Ehlschlaeger, C. Using the AT Search Algorithm to Develop Hydrologic Models from Digital Elevation Data. In Proceedings of the International Geographic Information System (IGIS) Symposium, Baltimore, MD, USA; 1989; pp. 275–281.

34. Metz, M.; Mitasova, H.; Harmon, R.S. Efficient Extraction of Drainage Networks from Massive, Radar-Based Elevation Models with Least Cost Path Search. *Hydrol. Earth Syst. Sci.* **2011**, *15*, 667–678. [CrossRef]

35. Mercurio, C.; Martinello, C.; Rotigliano, E.; Argueta-Platero, A.A.; Reyes-Martínez, M.E.; Rivera-Ayala, J.Y.; Conoscenti, C. Mapping Susceptibility to Debris Flows Triggered by Tropical Storms: A Case Study of the San Vicente Volcano Area (El Salvador, CA). *Earth* **2021**, *2*, 66–85. [CrossRef]
36. Naimi, B. Package "Usdm". Uncertainty Analysis for Species Distribution Models. *R-Cran* **2017**, *18*, 1–19.
37. Guisan, A.; Weiss, S.B.; Weiss, A.D. GLM versus CCA Spatial Modeling of Plant Species Distribution. *Plant Ecol.* **1999**, *143*, 107–122. [CrossRef]
38. Burrough, P.A.; McDonnell, R.A. *Principle of Geographic Information Systems*; Oxford University Press Inc.: New York, NY, USA, 1998; ISBN 0-19-823366-3.
39. Wilson, J.P.; Gallant, J.C. Primary Topographic Attributes. In *Terrain Analysis: Principles and Applications*; Wilson, J.P., Gallant, J.C., Eds.; John Wiley & Sons: Hoboken, NJ, USA, 2000.
40. Auslander, M.; Nevo, E.; Inbar, M. The Effects of Slope Orientation on Plant Growth, Developmental Instability and Susceptibility to Herbivores. *J. Arid Environ.* **2003**, *55*, 405–416. [CrossRef]
41. Zevenbergen, L.W.; Thorne, C.R. Quantitative Analysis of Land Surface Topography. *Earth Surf. Process Landf.* **1987**, *12*, 47–56. [CrossRef]
42. Ohlmacher, G.C. Plan Curvature and Landslide Probability in Regions Dominated by Earth Flows and Earth Slides. *Eng. Geol.* **2007**, *91*, 117–134. [CrossRef]
43. Beven, K.J.; Kirkby, M.J. A Physically Based, Variable Contributing Area Model of Basin Hydrology. *Hydrol. Sci. Bull.* **1979**, *24*, 43–69. [CrossRef]
44. Florinsky, I.V. *Digital Terrain Analysis in Soil Science and Geology*; Academic Press: Cambridge, MA, USA, 2012.
45. Friedman, J.H. Multivariate Adaptive Regression Splines. *Ann. Stat.* **1991**, *19*, 1–67. [CrossRef]
46. Mercurio, C.; Calderón-Cucunuba, L.P.; Argueta-Platero, A.A.; Azzara, G.; Cappadonia, C.; Martinello, C.; Rotigliano, E.; Conoscenti, C. Predicting Earthquake-Induced Landslides by Using a Stochastic Modeling Approach: A Case Study of the 2001 El Salvador Coseismic Landslides. *ISPRS Int. J. Geoinf.* **2023**, *12*, 178. [CrossRef]
47. Felicísimo, Á.M.; Cuartero, A.; Remondo, J.; Quirós, E. Mapping Landslide Susceptibility with Logistic Regression, Multiple Adaptive Regression Splines, Classification and Regression Trees, and Maximum Entropy Methods: A Comparative Study. *Landslides* **2013**, *10*, 175–189. [CrossRef]
48. Mohammed, S.; Jouhra, A.; Enaruvbe, G.O.; Bashir, B.; Barakat, M.; Alsilibe, F.; Cimusa Kulimushi, L.; Alsalman, A.; Szabó, S. Performance Evaluation of Machine Learning Algorithms to Assess Soil Erosion in Mediterranean Farmland: A Case-Study in Syria. *Land Degrad Dev.* **2023**, *34*, 2896–2911. [CrossRef]
49. Tian, M.; Li, L.; Xiong, Z. A Data-Driven Method for Predicting Debris-Flow Runout Zones by Integrating Multivariate Adaptive Regression Splines and Akaike Information Criterion. *Bull. Eng. Geol. Environ.* **2022**, *81*, 222. [CrossRef]
50. Milborrow, S. Notes on the Earth Package. *Retrieved Oct.* **2014**, *31*, 2017.
51. Chung, C.J.F.; Fabbri, A.G. Validation of Spatial Prediction Models for Landslide Hazard Mapping. *Nat. Hazards* **2003**, *30*, 451–472. [CrossRef]
52. Fawcett, T. An Introduction to ROC Analysis. *Pattern Recognit. Lett.* **2006**, *27*, 861–874. [CrossRef]
53. Goodenough, D.J.; Rossmann, K.; Lusted, L.B. Radiographic Applications of Receiver Operating Characteristic (ROC) Curves. *Radiology* **1974**, *110*, 89–95. [CrossRef]
54. Lasko, T.A.; Bhagwat, J.G.; Zou, K.H.; Ohno-Machado, L. The Use of Receiver Operating Characteristic Curves in Biomedical Informatics. *J. Biomed. Inform.* **2005**, *38*, 404–415. [CrossRef] [PubMed]
55. Hosmer, D.W. Lemeshow, Stanley. In *Applied Logistic Regression*; John Wiley & Sons, Inc.: Hoboken, NJ, USA, 2000; ISBN 0471722146.
56. Youden, W.J. Index for Rating Diagnostic Tests. *Cancer* **1950**, *3*, 32–35. [CrossRef] [PubMed]
57. Cappadonia, C.; Cafiso, F.; Ferraro, R.; Martinello, C.; Rotigliano, E. Analysis of the Rockfall Phenomenon Contributing to the Evolution of a Pocket Beach Area Using Traditional and Remotely Acquired Data (Lo Zingaro Nature Reserve, Southern Italy). *Remote Sens.* **2023**, *15*, 1401. [CrossRef]
58. Jia, Y.; Song, G.; Wang, L.; Jiang, T.; Zhao, J.; Li, Z. Research on Stability Evaluation of Perilous Rock on Soil Slope Based on Natural Vibration Frequency. *Appl. Sci.* **2023**, *13*, 2406. [CrossRef]
59. Cappadonia, C.; Cafiso, F.; Ferraro, R.; Martinello, C.; Rotigliano, E. Rockfall Hazards of Mount Pellegrino Area (Sicily, Southern Italy). *J. Maps* **2021**, *17*, 29–39. [CrossRef]
60. Delchiaro, M.; Della Seta, M.; Martino, S.; Nozaem, R.; Moumeni, M. Tectonic Deformation and Landscape Evolution Inducing Mass Rock Creep Driven Landslides: The Loumar Case-Study (Zagros Fold and Thrust Belt, Iran). *Tectonophysics* **2023**, *846*, 229655. [CrossRef]
61. Rouhi, J.; Delchiaro, M.; Della Seta, M.; Martino, S. New Insights on the Emplacement Kinematics of the Seymareh Landslide (Zagros Mts., Iran) Through a Novel Spatial Statistical Approach. *Front. Earth Sci. (Lausanne)* **2022**, *10*, 869391. [CrossRef]

*applied
sciences*

Article

Would Forest Regrowth Compensate for Climate Change in the Amazon Basin?

Nafiseh Haghtalab [1,*], Nathan Moore [2,*] and Pouyan Nejadhashemi [3]

1   Department of Geography and Anthropology, Kennesaw State University, Kennesaw, GA 30144, USA
2   Department of Geography, Environment and Spatial Sciences, Michigan State University,
    East Lansing, MI 48824, USA
3   Department of Biosystems and Agricultural Engineering, Michigan State University,
    East Lansing, MI 48824, USA; pouyan@msu.edu
*   Correspondence: nhaghtal@kennesaw.edu (N.H.); moorena@msu.edu (N.M.)

Abstract: Following potential reforestation in the Amazon Basin, changes in the biophysical characteristics of the land surface may affect the fluxes of heat and moisture behavior. This research examines the impacts of potential tropical reforestation on surface energy and moisture budgets, including precipitation and temperature. The study is novel in that while most studies look at the opposite driver (deforestation), this one examines the impact of potential forest rehabilitation on atmospheric behavior using WRF.V3.9 (weather research and forecast model). We found that forest rehabilitation across the Amazon Basin can make the atmosphere cooler with more moisture and latent heat (LH), especially during May-November. For instance, the mean seasonal temperature decreased significantly by about 1.2 °C, indicating the cooling effects of reforestation. Also, the seasonal precipitation increased by 5 mm/day in reforested areas. By reforestation, the mean monthly LH also increased as much as 50 W m^{-2} in August in certain areas, while available moisture to the atmosphere increased by 27%, indicating possible causal mechanisms between increased LH and precipitation and emphasizing the mechanisms that were identified between the onset of the wet season and forest cover. Therefore, it is likely that forest regrowth across the basin leads to, if not reverses regional climate change, at least slowing down the rate of changes in the climate.

Keywords: reforestation; land-atmosphere interactions; Amazon basin; heat and moisture fluxes; WRF

Citation: Haghtalab, N.; Moore, N.;
Nejadhashemi, P. Would Forest
Regrowth Compensate for Climate
Change in the Amazon Basin? *Appl.
Sci.* **2022**, *12*, 7052. https://doi.org/
10.3390/app12147052

Academic Editor: Joao Carlos
Andrade dos Santos

Received: 13 June 2022
Accepted: 11 July 2022
Published: 13 July 2022

1. Introduction

The land surface plays an important role in global energy, the hydrologic cycle, and carbon balance. Land cover change (LCC) directly alters surface-absorbed solar radiation, longwave radiation, and atmospheric turbulence. These alterations lead to changes in fluxes of momentum, heat, and water vapor through the mediation of albedo, evapotranspiration (ET), roughness, and CO_2 [1,2]. Land cover changes through atmospheric feedback can have a striking impact on the local, regional, and even global mean climate as well as climatic extremes and variability [3].

While 25 to 35% of Amazon precipitation is related to regional moisture recycling [4], during the rainy season, moist air from the basin travels along the Andes and provides precipitation over the La Plata basin too [5,6] through tele-connection processes. Therefore, any changes to land surface biophysical characteristics, even at the local scale, may alter the climate over the entire basin.

LCC in the Amazon basin has been studied to be one of the driving forces for climate change [7,8]. It affects the energy, carbon and water balance, and land-atmosphere interactions. It alters evapotranspiration and the hydrologic cycle more broadly which further affects Amazon rainforest stability [9], primarily through a reduction in moisture recycling [10,11]. Such changes have been investigated across the Amazon basin using global and regional climate models: notably, via complete deforestation scenarios e.g., [12–16] or scenarios ranging from low to extreme conversion of forest e.g., [17,18].

The conversion of forest to cropland in the Amazon Basin has resulted in a decrease in precipitation (P) [15], a decrease in ET [19,20], an increase in temperature (T) [18], and also indirectly intensifies fire occurrence [21]. Due to deforestation, the onset of the rainy season has also delayed 11 days, on average, over the last thirty years across the highly deforested areas in the state of Rondonia, Brazil [19]. In addition, the length of the dry season has been increased by one month in some areas [22–26] and drought conditions have also been exacerbated as a result of deforestation [27–29].

The spatial scale of LCC from local to regional to global is very important in land-atmosphere interaction analysis [30,31]. The most recent deforestation in the Amazon basin occurred at small-scale patches (less than 1 ha) during 2008–2014 [27]. In addition, the temporal scale of analysis is also important in understanding the magnitude and amplitude of the effects. For instance, Ref. [28] found that the impact of land surface variability on climate is more apparent at monthly timescales than at other timescales. Ref. [29] analyzed the interactions between clouds, rains, and the underlying land surface through biosphere processes in southwestern Rondônia, Brazil. They found that land-atmosphere interactions are higher during the dry season (May–November) than the wet season (December-April). They also hypothesized more complex interactions between cloudiness, moisture transport, and fluxes during the wet season.

When considering the effects of LCC at the basin scale, the land-atmosphere interaction is more intense [22]. For instance, Ref. [30] used IPCC CMIP3 models and found an increase in the annual mean temperature between 0.1 and 3.8 °C and a decrease in the annual precipitation of about 10–30% which could lead to changes in seasonality. Also, Ref. [11] argued that upon reaching 40% reduction in Amazon forest cover, wet and dry season rainfall totals may reduce by 12% and 21%, respectively. However, the magnitude and the location of rainfall changes is uncertain [31,32].

Ref. [14] also used a GCM to capture the climate response to Amazon deforestation. They found that the sensitivity of climate to LCC depends on the initial tree cover and type of irrigation. Using satellite observations to assess crop responses to drought in the basin, Refs. [33,34] found that due to reduced cloud cover, droughts induce a "greening-up" although other researchers have rejected this hypothesis, e.g., [35–37]. According to Ref. [35], analysis and model simulations of the impacts of Amazon deforestation over the past 40 years showed that more than 90% of studies agree on the sign of change which is a reduction in rainfall. But the amplitude, magnitude, and predictability are inconsistent since they highly depend on the spatio-temporal scale of analysis [15,36–43].

Even if the regional impacts of deforestation on precipitation patterns have been studied intensively e.g., [8,21,28,44–47], the reverse effects are still unclear. Therefore, in this study, we aim to examine the extent to which potential Amazon Forest regrowth may influence fluxes, precipitation, and temperature patterns during both wet (December–April) and dry seasons (May–November). We should note that wet and dry seasons are not consistent across the domain, but these timespans are a practical compromise for analysis.

Thus, in this research, we examined the sensitivity and magnitude of changes to the surface energy budget, including precipitation, due to potential new growth forests across the Amazon Basin (Figure 1). Our prescribed reforestation scenario using the Weather Research and Forecasting model (WRF)V3.9 is designed to answer the following questions: (a) How might forest regrowth contribute to changes in fluxes, temperature, and precipitation amounts across the basin at monthly and seasonal timescales; (b) what are the spatio-temporal patterns of changes; and (c) Do any tele-connected processes develop due to forest rehabilitation?

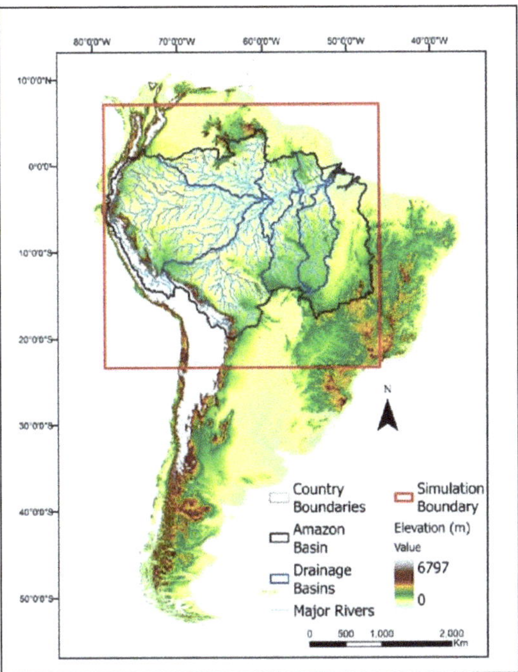

Figure 1. Geographic location of the Amazon Basin. The red box indicates our simulation boundary.

2. Materials and Methods

2.1. Study Area and Simulation Domain

Figure 1 shows the topography of the Amazon Basin along with our simulation boundary. The Amazon Basin extends through Brazil, Peru, Colombia, Ecuador, and Bolivia covering about 6 million km^2. The rainiest part of the basin is located on the eastern edge of the Andes Cordillera [48,49]. The Amazon Basin contains more than 20% of the world's fresh water and is a hot-spot for ecosystem diversity. The forest biomass holds an estimated 100 billion tons of carbon [50].

The basin's climate varies from continuously rainy in the northwest to long dry seasons in the east and south [51,52], where more conversion to agriculture has occurred. This is referred to as the "Arc of Deforestation". The basin's climate is controlled by atmosphere-ocean-land coupling as well as moisture recycling through evapotranspiration [53]. The El Nino Southern Oscillation (ENSO) decreases the Amazon River flow on the eastern side of the basin during El Nino years [54] while, during La Nina years, flooding increases [55]. The Southern American Monsoon System brings rainfall to the southern portion of the basin with the maximum rainfall during DJF (December-January-February) [56]. During JJA (June-July-August) the South American Convergence Zone (SACZ) contributes to the precipitation variability across the south of the Basin [57]. During MAM (March-April-May), rainfall is dominated by the Intertropical Convergence Zone (ITCZ), which is highly variable [58].

2.2. Data

We forced WRF with ESA 2009 land cover data which was reclassified based on US Geological Survey land cover classes to match the WRF settings and mosaicked to account for differences in resolution. The land cover was kept constant over the simulation years; this is a prescribed simulation, so we needed to control for annual land cover variations from our analysis. We choose 2009 to be consistent with our boundary layer data starting in 2009. For vertical boundary conditions, ERA_Interim with 80 km spatial resolution and 60 vertical levels, and 6-hourly temporal resolution for 2009, 2013, and 2014 were used to force the model. These years are among the most recent ENSO-neutral years and the data was more homogenous in terms of extreme events and outliers than other neutral years.

Due to the lack of adequate and robust observational information on precipitation and temperature that poses great difficulties in validating our climate model outputs, we used Tropical Rainfall Measuring Mission (TRMM) with a $0.25°$ spatial resolution and MODIS Land-Surface Temperature with a 1 km spatial resolution to validate the simulated temperature. All data were resampled based on the model output resolution.

2.3. WRF Model Setup

WRF.3.9 (ARW) is a three-dimensional, non-hydrostatic climate model that is widely used for atmospheric research. Simulations were initialized at 00:00 UTC and the first 15 days were considered spin-up and were removed from the analysis. Early trials using longer spin-up proved to be computationally expensive and unlikely to significantly affect the sensitivity tests. The horizontal grid spacing was 16 km, with 38 levels of vertical levels up to 1000 m. The thickness of the lowest atmospheric layer is about 50 m on smooth topography. At this resolution, cumulus parameterization is necessary to resolve convection, clouds, and precipitation properly [59]. Table 1 summarizes WRF parameterizations that were used in this study. SSTs (sea surface temperature) came from ERA data to be time-consistent with the vertical boundary conditions.

Table 1. WRF parameterizations.

Parameter	Scheme Option
Longwave radiation scheme	Rapid Radiative Transfer Model
Shortwave radiation	Dudhia scheme
Surface layer	Fifth-generation Pennsylvania State University–National Center for Atmospheric Research Mesoscale Model (MM5) scheme.
Cumulus scheme	Kain–Fritsch
Mp_physics	WSM6 Hong and Lim
LSM	NOAH
PBL	Yonsei University scheme

To quantify the model performance, we calculated the root-mean-square error (RMSE) and the systematic error (percent bias; PBias) on the areal basin mean of daily data. We also mapped the differences between the model outputs and observations at monthly and seasonal timescales to estimate model performance and examine the errors spatially. We resampled our observations based on the simulation outputs to eliminate spatial resolution discrepancies in our data and comparison.

2.4. Land Cover Change Scenario

The last 50 years have witnessed a rapid conversion of forest to pasture and soy agriculture, driven by new road building. For deforested areas, this has brought reduced soil moisture, higher SH, seasonally bare soils, higher albedos, and lowered zero-plane displacement heights. Figure 2 shows maps of current and reforested land cover that was used in this study to analyze the sensitivity of the atmosphere to deforestation across the Amazon Basin. In this study, only conversion from cropland to forest has been considered; cropped cerrado was not changed. Every grid cell which was primarily cropland has been replaced by mature evergreen rainforest (although this is complex in the southeastern domain). This conversion is dominant along the arc of deforestation and on the main stem of the Amazon River.

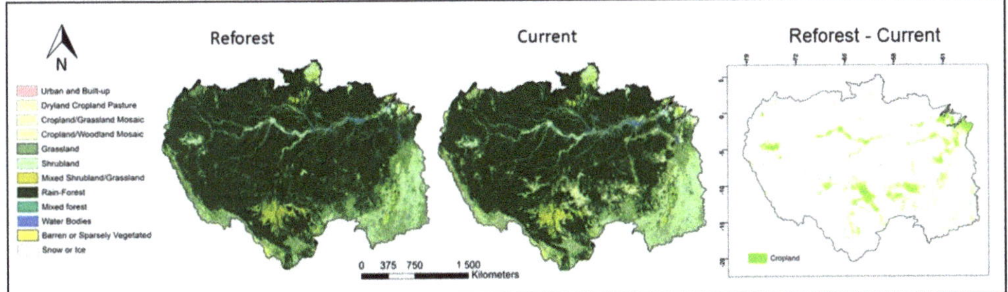

Figure 2. ESA land covers that were used in the simulation. In the Reforest map, all croplands are replaced by evergreen broadleaf forests. The highlighted areas on the difference map (right image) indicate reforested regions.

3. Results and Discussion

3.1. Model Validation

Figure 3 shows RMSE and Pbias errors for both precipitation and temperature. We validated the simulated precipitation against Brazilian Federal hydro-meteorological network (ANA) rain gauge measurements and TRMM reanalysis precipitation data and compared basin-wide averages. As stated before, due to high levels of missing values in ANA data, we removed them from our analysis. They are shown in this image only to highlight the shortcomings of some ANA data.

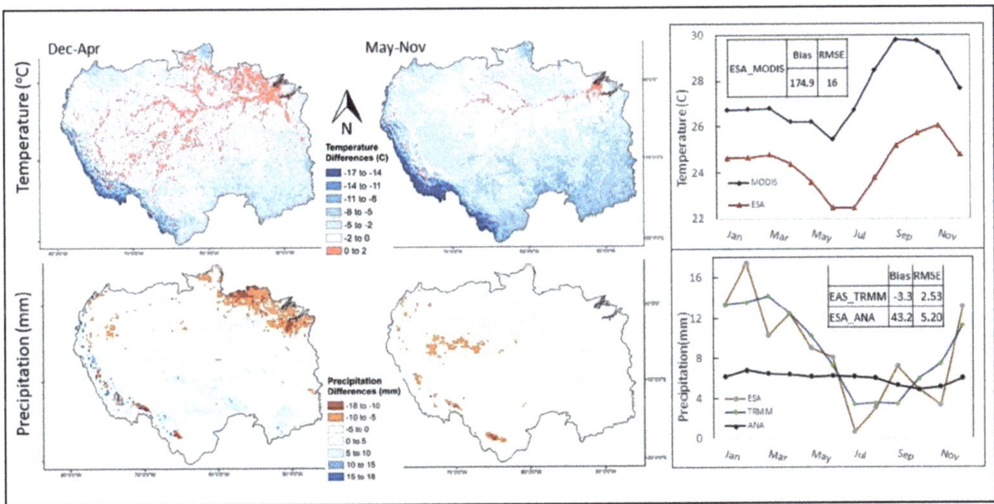

Figure 3. Difference maps between the simulated precipitation and simulated temperature, forced with reforested and current LCC on the left. On the right, the mean monthly temperature and precipitation (averaged over the basin) from observation and model output, along with the errors in the inset boxes.

Looking at temperature, the model performed very well with deviations at most 2 degrees centigrade cooler than the observations for most of the basin. Only at high altitudes over complex terrain on the edges did the model underestimate the temperature by up to $-17\,°C$. This error is consistent with WRF's well-known cold bias at high altitudes [60]. Also, along water bodies, the model simulated up to 2 degrees warmer than observations. Our model performed well in simulating the precipitation, too. Due to complex interactions between cloudiness, the land surface, and precipitation in the Amazon Basin [61] during the wet season (December–April), the model overestimates precipitation for the arc of deforestation by up to 5 mm/day compared to the observations. In terms of basin average, the temperature is simulated with the same spatial pattern as MODIS temperature but 1 °C cooler. Simulated precipitation shows broadly the same pattern as TRMM precipitation. The RMSE and Bias are reported in Figure 3 which are minimal and acceptable.

3.2. Sensitivity of Fluxes and Precipitation to Land Cover Change across the Basin

The results that are shown here are averaged across the three years of simulation. To assess the impacts of regrowth on fluxes and precipitation, we applied a Student *t*-test for each season spatial time series at each grid point (over space and time). In this test, the null statistical hypothesis is that the reforested and current population had the same mean [44]. Each grid point that could reject the null hypothesis at a 95% significance level is considered to have experienced a significant impact from the reforestation process. Although we used ENSO-neutral years, there exists interannual variability across the three years, and both positive and negative changes resulted from the model in response to reforestation.

3.2.1. Heat Flux

Figures 4 and 5 show the effects of LCC on LH and sensible heat (SH) (only significant changes are shown here). According to Figure 4, the LH has increased by 30 Wm^{-2} during May–November and by 15 Wm^{-2} during December–April despite some extreme increases in the north side of the region. We found no pronounced negative changes in the domain-averaged mean SH across the region with reforestation. As the land surface has a complex relationship with the atmosphere, SH did not show significant sensitivity to changes in the land surface biophysical characteristics at a seasonal scale. There is only the northeast area of the basin which shows a significant decreasing trend for SH with reforestation. This decrease is the highest in December–April which is geographically consistent with the highest increase in LH during the same time period.

Next, we looked at monthly changes. For regions with added tree cover, the LH has increased by 20, 50, and 30 Wm^{-2} in July, August, and September, respectively. SH shows a decrease of 10 Wm^{-2} in August and September at the same location. These months are in the dry season, therefore, an increase in the LH can provide more moisture to the environment if other criteria are met. By adding more vegetation cover through reforestation or forest rehabilitation, the transpiration rate and surface roughness increased leading to an increase in the LH and a decrease in SH. Since July has the highest LAI in the basin and it decreases toward the end of the year, we found the highest influence of LCC on exchanges of both SH and LH starting in July.

The effects of LCC on the temperature are spatially different in May–November and December–April. Reforestation decreased the surface temperature by about 1.2 °C in the northeast part of the basin and about 0.2 °C on the west side of the basin (Figure 6), far from the reforested areas. The increased ET drives a significant increase in the cloud cover that gets advected westward. The cooling effect of reforestation is clearer on a monthly scale, especially in Aug and Sept by about 2 °C. This finding is consistent with Ref. [38] who found 2 °C warmer air temperatures as a result of deforestation, as well as Ref. [18] who found 0.3 °C warmer surface temperatures due to deforestation of the Xingu region along the arc of deforestation.

3.2.2. Moisture Flux and Precipitation

Our results showed that reforestation significantly increased the domain-averaged available moisture to the atmosphere (QFX) (Figure 7), mostly during May–November, by 27%. The maximum increase in moisture flux occurred in August and September, about 0.03 g m^{-2} s^{-1}, especially in the arc of deforestation which has had significant widespread deforestation. However, other heavily deforested areas of the basin (along the rivers in the centroid of the basin, and near Iquitos) did not exhibit significant changes in moisture flux. These regions receive much more rainfall and have virtually no dry season. The QFX value of $0.01 \text{ g m}^{-2}\text{ s}^{-1}$ in the difference panel of Figure 7 converts to approximately 25 mm/month of precipitation, which is at the upper end of the RMSE that was measured by global ET products [62].

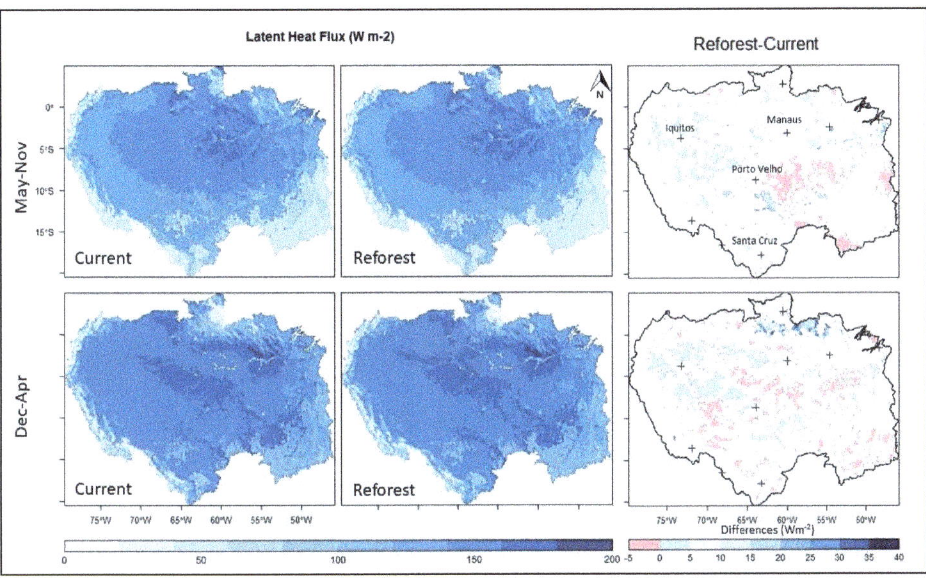

Figure 4. Simulated LH, forced with current and reforested land cover on the left. On the right, the difference between the two simulated LHs at a 95% significance level. Plus signs indicate major cities.

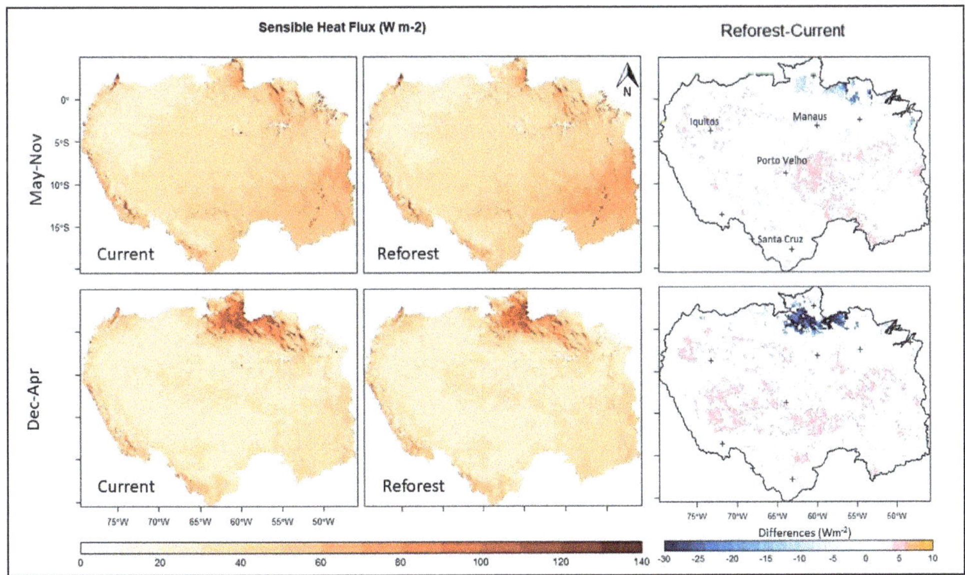

Figure 5. Simulated SH, forced with current and reforested land cover on the left. On the right, the difference between the two simulated SHs at a 95% significance level. Plus signs indicate major cities.

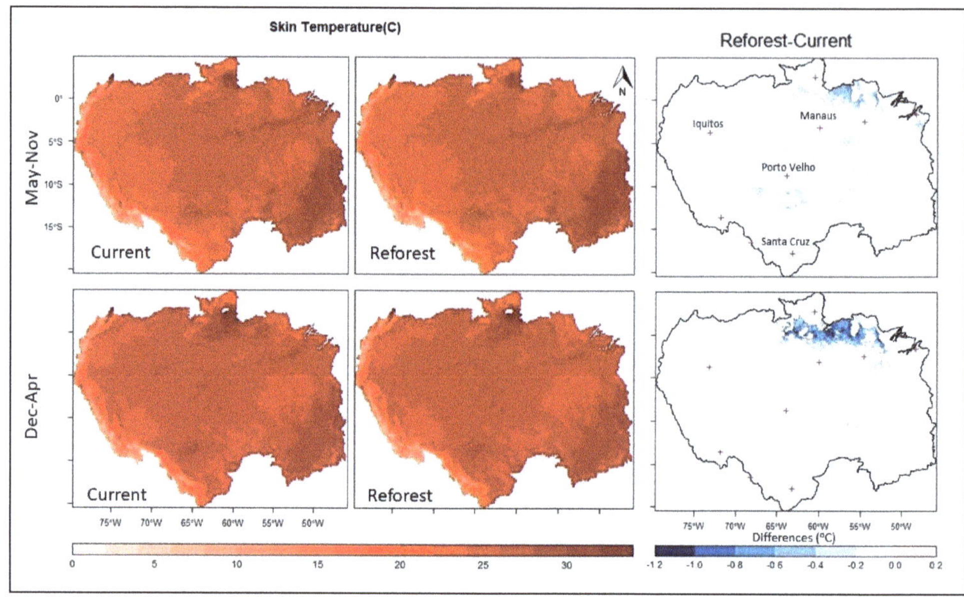

Figure 6. Simulated temperature, forced with current and reforested land cover on the left. On the right, the difference between the two simulated temperatures is at a 95% significance level. Plus signs indicate major cities.

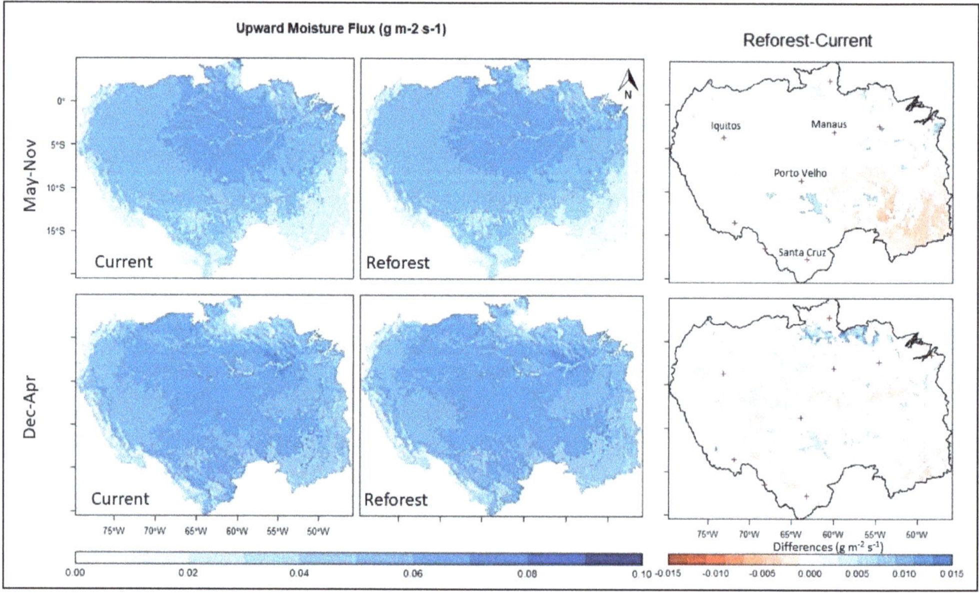

Figure 7. Simulated QFX, forced with current and reforested land cover on the left. On the right, the difference between the two simulated QFXs is at a 95% significance level. Plus signs indicate major cities.

Simulated precipitation data showed that the mean seasonal precipitation increased with forest regrowth by 5 mm/day (Figure 8). During May–November, these changes are spatially located on the west side of the region where the moisture gets transferred making more cloud fractions indicating the tele-connection impacts of reforestation on precipitation, as discussed in Ref. [63]. According to Ref. [19] precipitation is produced by both large and small-scale forcings, including thunderstorms and the development of deep convection at a larger scale and through shallow convection at a local scale. During December–April across the basin, Rossby waves can propagate northward and produce precipitation. Squall lines originating on the northeast coast of South America transport moisture and precipitation west toward the Andes. At larger scales, although the positioning and strength of the ITCZ control different precipitation regimes in the region, El Nino can affect the Walker-type circulations and can thus affect the spatial distribution of rainfall [64,65]. Therefore, the amount of rainfall is likely more dependent on synoptic-scale forcings such as the ITCZ and Walker-type cells and less on localized reforestations. Reforestation provides moisture, but larger processes typically initiate rainfall.

Thus, following potential reforestation in the Amazon Basin, changes in the biophysical characteristics of the land surface can affect the fluxes of heat and moisture behavior. As such, forest rehabilitation across the Amazon Basin can make the atmosphere cooler with more moisture and LH, especially during May–November. In addition, some laterally translated features suggest that land cover creates perturbations that get advected elsewhere, and large patterns also exist that suggest continent/synoptic-scale processes are being modified as a result of deforestation. This suggests complex interactions between climate and LCC that we will explore in future work.

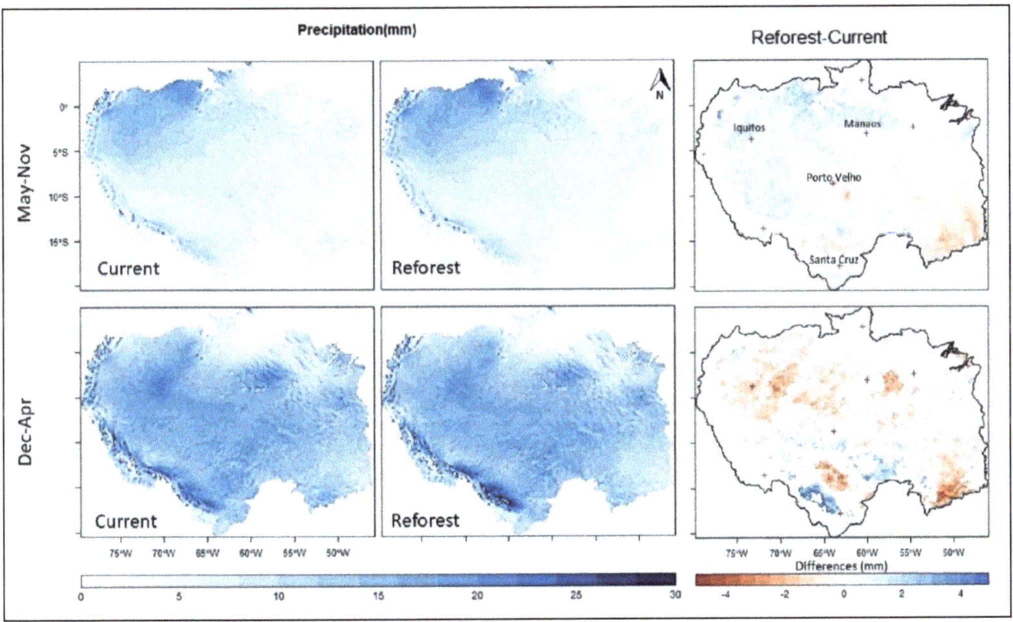

Figure 8. Simulated precipitation, forced with current and reforested land cover on the left. On the right, the difference between the two simulated precipitations is at a 95% significance level. Plus signs indicate major cities.

4. Conclusions

This paper examines the regional-scale impacts of potential reforestation on the energy and moisture budgets and precipitation across the Amazon Basin. Through the analysis of changes in regional moisture and heat fluxes, we presented results from regional simulations showing that the land surface and atmosphere are interacting tightly across the basin. We found several principal outcomes. First, the effects of reforestation on the atmosphere were more evident during May–November than December-April. Second, spatial patterns of the changes in fluxes due to reforestation were consistent with the pattern of LCC, with minimal tele-connected impacts. Third, the effects of forest regrowth on the atmosphere were more evident on a monthly time scale. For instance, although at the seasonal scale, the changes in SH were minimal, at the monthly scale, it simulated a decrease by 10 W m^{-2}. Forest regrowth enhances LH in the region due to an increase in the transpiration rate and surface roughness. In addition, the highest LAI in July highlights the highest influence of LCC on exchanges of both SH and LH starting in July.

Fourth, the mean seasonal temperature decreased by up to 1.2 °C, which is consistent with several studies, e.g., [18,46,66]. This decrease in temperature is more obvious in the northeastern side of the basin during December–April. Fifth, reforestation also increased the mean monthly LH by as much as 50 W m^{-2} in August in certain areas, while available moisture to the atmosphere increased by 27%. Other studies found equivalent scale results but due to deforestation e.g., [18,49,67]. Sixth, seasonal precipitation increased by 5 mm/day in reforested areas in both May-Nov and Dec-Apr, illustrating the causal mechanisms between increased LH and precipitation and emphasizing the mechanisms identified between wet season start and forest cover [68,69]. Precipitation also increased in the western side of the region, where is constantly wet, by forest regrowth. This indicates tele-connected influence of vegetation recovery on the atmosphere behavior.

Our results show that by altering the land surface biophysical characteristics—in this case, reforestation—temperature, LH and SH fluxes, moisture at the surface, and precipitation are strongly modified. With a higher proportion of LH, PBL cools down, increases its humidity, and becomes shallower. This further affects the transfer of moisture and energy from the surface to the boundary layer, even influencing transfer to the free atmosphere. Although unavailable, parameters for young moist forests would improve these simulations further. Due to tele-connection mechanisms, changing the exchange of energy and moisture balance between the PBL and the free atmosphere influences tropical convection, impacting the intensity of high-level tropical outflow and providing a mechanism that could affect the extratropics [70]. Consequently, changes in the surface fluxes of energy and moisture due to LCC causes impacts beyond the areas of disturbances. Thus, it would be reasonable if deforestation forces disturbances in the general circulation, including the Hadley and Walker-type circulations; the mechanisms for these disturbances are illustrated in Ref. [67].

Future work needs to focus on identifying the coupling strength of land cover changes to atmospheric processes to identify areas where rainfall is most sensitive to changes in the land surface and examining the extent to which changes in the regional scale can alter the circumstances at the larger scale. Also, different time scales from hourly to daily to monthly evaluations should be considered to distinguish the sensitivity of time-sensitive processes such as cloud formation and convection, which determine the amount and timing of precipitation to reforestation.

Author Contributions: Conceptualization, N.H.; methodology, N.H., N.M. and P.N.; software, N.H.; validation, N.H.; formal analysis, N.H. and N.M.; investigation, N.H., N.M. and P.N.; resources, N.H. and N.M.; data curation, N.H. and N.M.; writing—original draft preparation, N.H.; writing—review and editing, N.H., N.M. and P.N.; visualization, N.H., N.M. and P.N.; supervision, N.H., N.M. and P.N.; project administration, N.M.; funding acquisition, N.M. All authors have read and agreed to the published version of the manuscript.

Funding: This research was funded by NSF INFEWS/T3, grant number 1639115 and Partial support also came from the Department of Geography, Environment, and Spatial Sciences at Michigan State University.

Institutional Review Board Statement: Not applicable.

Informed Consent Statement: Not applicable.

Data Availability Statement: The data that support the funding of this study were acquired from different resources. TRMM reanalysis data are openly available at https://disc.gsfc.nasa.gov/information?keywords=precipitation&page=1&project=TRMM, accessed on 1 February 2020, MODIS temperature data are openly available at https://lpdaac.usgs.gov/products/mod11a1v006/, accessed on 1 February 2020, and ESA land cover data are openly available at http://www.esa-landcover-cci.org/?q=node/164, accessed on 1 February 2020, ERA-Interim boundary data were acquired from NCAR's Cheyenne repository which is available upon NCAR's permission. ANA observations data were received from collaborators in Brazil and can be available upon their permission. The simulated data can be available upon reasonable request by the co-corresponding author (NH).

Acknowledgments: We would like to acknowledge high-performance computing support from Cheyenne (DOI:10.5065/D6RX99HX) provided by NCAR's Computational and Information Systems Laboratory, sponsored by the National Science Foundation. Any opinions, findings, conclusions, or recommendations expressed in this material are those of the authors and do not necessarily reflect the views of the NSF.

Conflicts of Interest: The authors declare no conflict of interest.

References

1. Pielke, R.A.; Pitman, A.; Niyogi, D.; Mahmood, R.; McAlpine, C.; Hossain, F.; Goldewijk, K.K.; Nair, U.; Betts, R.; Fall, S.; et al. Land Use/Land Cover Changes and Climate: Modeling Analysis and Observational Evidence. *Wiley. Interdiscip. Rev. Clim. Chang.* **2011**, *2*, 828–850. [CrossRef]
2. Alkama, R.; Cescatti, A. Climate Change: Biophysical Climate Impacts of Recent Changes in Global Forest Cover. *Science* **2016**, *351*, 600–604. [CrossRef] [PubMed]
3. Malhi, Y.; Aragao, L.E.O.C.; Galbraith, D.; Huntingford, C.; Fisher, R.; Zelazowski, P.; Sitch, S.; McSweeney, C.; Meir, P. Hipoacusia Tubotimp'Anica. Concepto Fisiopatol'Ogico. *Proc. Natl. Acad. Sci. USA* **2008**, *106*, 20610–20615. [CrossRef] [PubMed]
4. Dirmeyer, P.A.; Schlosser, C.A.; Brubaker, K.L. Precipitation, Recycling, and Land Memory: An Integrated Analysis. *J. Hydrometeorol.* **2009**, *10*, 278–288. [CrossRef]
5. Alejandro Martinez, J.; Dominguez, F. Sources of Atmospheric Moisture for the La Plata River Basin. *J. Clim.* **2014**, *27*, 6737–6753. [CrossRef]
6. Arraut, J.M.; Satyamurty, P. Precipitation and Water Vapor Transport in the Southern Hemisphere with Emphasis on the South American Region. *J. Appl. Meteorol. Climatol.* **2009**, *48*, 1902–1912. [CrossRef]
7. Aragão, L.E.O.C.; Anderson, L.O.; Fonseca, M.G.; Rosan, T.M.; Vedovato, L.B.; Wagner, F.H.; Silva, C.V.J.; Silva Junior, C.H.L.; Arai, E.; Aguiar, A.P.; et al. 21st Century Drought-Related Fires Counteract the Decline of Amazon Deforestation Carbon Emissions. *Nat. Commun.* **2018**, *9*, 536. [CrossRef]
8. Sampaio, G.; Nobre, C.; Costa, M.H.; Satyamurty, P.; Soares-Filho, B.S.; Cardoso, M. Regional Climate Change over Eastern Amazonia Caused by Pasture and Soybean Cropland Expansion. *Geophys. Res. Lett.* **2007**, *34*, 17709. [CrossRef]
9. Marengo, J.A.; Nobre, C.A.; Tomasella, J.; Cardoso, M.F.; Oyama, M.D. Hydro-Climate and Ecological Behaviour of the Drought of Amazonia in 2005. *Philos. Trans. R. Soc. Lond. B Biol. Sci.* **2008**, *363*, 1773–1778. [CrossRef]
10. Betts, R.A.; Cox, P.M.; Collins, M.; Harris, P.P.; Huntingford, C.; Jones, C.D. The Role of Ecosystem-Atmosphere Interactions in Simulated Amazonian Precipitation Decrease and Forest Dieback under Global Climate Warming. *Theor. Appl. Climatol.* **2004**, *78*, 157–175. [CrossRef]
11. Spracklen, D.V.; Arnold, S.R.; Taylor, C.M. Observations of Increased Tropical Rainfall Preceded by Air Passage over Forests. *Nature* **2012**, *489*, 282–285. [CrossRef] [PubMed]
12. Zhang, K.; de Almeida Castanho, A.D.; Galbraith, D.R.; Moghim, S.; Levine, N.M.; Bras, R.L.; Coe, M.T.; Costa, M.H.; Malhi, Y.; Longo, M.; et al. The Fate of Amazonian Ecosystems over the Coming Century Arising from Changes in Climate, Atmospheric CO2, and Land Use. *Glob. Chang. Biol.* **2015**, *21*, 2569–2587. [CrossRef] [PubMed]
13. Lawrence, D.; Vandecar, K. Effects of Tropical Deforestation on Climate and Agriculture. *Nat. Clim. Chang.* **2014**, *5*, 27–36. [CrossRef]
14. Marengo, J.A.; Espinoza, J.C. Extreme Seasonal Droughts and Floods in Amazonia: Causes, Trends and Impacts. *Int. J. Climatol.* **2016**, *36*, 1033–1050. [CrossRef]
15. Chambers, J.Q.; Artaxo, P. Deforestation Size Influences Rainfall. *Nat. Clim. Chang.* **2017**, *7*, 175–176. [CrossRef]

16. Sampaio, G.; Borma, L.S.; Cardoso, M.; Alves, L.M.; von Randow, C.; Rodriguez, D.A.; Nobre, C.A.; Alexandre, F.F. Assessing the Possible Impacts of a 4 °C or Higher Warming in Amazonia. In *Climate Change Risks in Brazil*; Springer: Cham, Switzerland, 2019; pp. 201–218. [CrossRef]
17. Meehl, G.A.; Washington, W.M.; Collins, W.D.; Arblaster, J.M.; Hu, A.; Buja, L.E.; Strand, W.G.; Teng, H. How Much More Global Warming and Sea Level Rise? *Science* **2005**, *307*, 1769–1772. [CrossRef]
18. Badger, A.M.; Dirmeyer, P.A. Climate Response to Amazon Forest Replacement by Heterogeneous Crop Cover. *Hydrol. Earth Syst. Sci.* **2015**, *19*, 4547–4557. [CrossRef]
19. Oliveira, P.T.S.; Nearing, M.A.; Moran, M.S.; Goodrich, D.C.; Wendland, E.; Gupta, H. V Trends in Water Balance Components across the Brazilian Cerrado. *Water Resour. Res.* **2014**, *50*, 7100–7114. [CrossRef]
20. Spera, S.A.; Galford, G.L.; Coe, M.T.; Macedo, M.N.; Mustard, J.F. Land-Use Change Affects Water Recycling in Brazil's Last Agricultural Frontier. *Glob. Chang. Biol.* **2016**, *22*, 3405–3413. [CrossRef]
21. da Silva, R.R.; Werth, D.; Avissar, R. Regional Impacts of Future Land-Cover Changes on the Amazon Basin Wet-Season Climate. *J. Clim.* **2008**, *21*, 1153–1170. [CrossRef]
22. Silvério, D.V.; Brando, P.M.; Macedo, M.N.; Beck, P.S.A.; Bustamante, M.; Coe, M.T. Agricultural Expansion Dominates Climate Changes in Southeastern Amazonia: The Overlooked Non-GHG Forcing. *Environ. Res. Lett.* **2015**, *10*, 104015. [CrossRef]
23. Aragão, L.E.O.C.; Malhi, Y.; Barbier, N.; Lima, A.; Shimabukuro, Y.; Anderson, L.; Saatchi, S. Interactions between Rainfall, Deforestation and Fires during Recent Years in the Brazilian Amazonia. *Philos. Trans. R. Soc. B Biol. Sci.* **2008**, *363*, 1779–1785. [CrossRef] [PubMed]
24. Butt, N.; De Oliveira, P.A.; Costa, M.H. Evidence That Deforestation Affects the Onset of the Rainy Season in Rondonia, Brazil. *J. Geophys. Res. Atmos.* **2011**, *116*, 2–9. [CrossRef]
25. Costa, M.H.; Pires, G.F. Effects of Amazon and Central Brazil Deforestation Scenarios on the Duration of the Dry Season in the Arc of Deforestation. *Int. J. Climatol.* **2010**, *30*, 1970–1979. [CrossRef]
26. Marengo, J.A.; Souza, C.M.; Thonicke, K.; Burton, C.; Halladay, K.; Betts, R.A.; Alves, L.M.; Soares, W.R. Changes in Climate and Land Use Over the Amazon Region: Current and Future Variability and Trends. *Front. Earth Sci.* **2018**, *6*, 228. [CrossRef]
27. Knox, R.; Bisht, G.; Wang, J.; Bras, R.; Knox, R.; Bisht, G.; Wang, J.; Bras, R. Precipitation Variability over the Forest-to-Nonforest Transition in Southwestern Amazonia. *J. Clim.* **2011**, *24*, 2368–2377. [CrossRef]
28. Bagley, J.E.; Desai, A.R.; Harding, K.J.; Snyder, P.K.; Foley, J.A. Drought and Deforestation: Has Land Cover Change Influenced Recent Precipitation Extremes in the Amazon? *J. Clim.* **2014**, *27*, 345–361. [CrossRef]
29. Alves, L.M.; Marengo, J.A.; Fu, R.; Bombardi, R.J. Sensitivity of Amazon Regional Climate to Deforestation. *Am. J. Clim. Chang.* **2017**, *06*, 75–98. [CrossRef]
30. D'almeida, C.; Vörösmarty, C.J.; Hurtt, G.C.; Marengo, J.A.; Dingman, S.L.; Keim, B.D. The Effects of Deforestation on the Hydrological Cycle in Amazonia: A Review on Scale and Resolution. *Int. J. Climatol.* **2007**, *27*, 633–647. [CrossRef]
31. Pitman, A.J.; Lorenz, R. Scale Dependence of the Simulated Impact of Amazonian Deforestation on Regional Climate. *Environ. Res. Lett.* **2016**, *11*, 094025. [CrossRef]
32. Kalamandeen, M.; Gloor, E.; Mitchard, E.; Quincey, D.; Ziv, G.; Spracklen, D.; Spracklen, B.; Adami, M.; Aragaõ, L.E.O.C.; Galbraith, D. Pervasive Rise of Small-Scale Deforestation in Amazonia. *Sci. Rep.* **2018**, *8*, 1600. [CrossRef] [PubMed]
33. Huete, A.R.; Didan, K.; Shimabukuro, Y.E.; Ratana, P.; Saleska, S.R.; Hutyra, L.R.; Yang, W.; Nemani, R.R.; Myneni, R. Amazon Rainforests Green-up with Sunlight in Dry Season. *Geophys. Res. Lett.* **2006**, *33*, L06405. [CrossRef]
34. Saleska, S.R.; Didan, K.; Huete, A.R.; da Rocha, H.R. Amazon Forests Green-Up during 2005 Drought. *Science* **2007**, *318*, 612. [CrossRef] [PubMed]
35. Magrin, G.O.; Marengo, J.A.; Boulanger, J.-P.; Buckeridge, M.S.; Castellanos, E.; Alfaro, E.; Anthelme, F.; Barton, J.; Becker, N.; Bertrand, A.; et al. Central and South America Coordinating Lead Authors: Lead Authors: Contributing Authors: Review Editors: To the Fifth Assessment Report of the Intergovernmental Panel on Climate Change. In *Contribution of Working Group II to the Fifth Assessment Report of the Intergovernmental Panel on Climate Change*; Cambridge University Press: Cambridge, UK; New York, NY, USA, 2014; pp. 1499–1566.
36. Dirmeyer, P.A. An Evaluation of the Strength of Land–Atmosphere Coupling. *J. Hydrometeorol.* **2002**, *2*, 329–344. [CrossRef]
37. Silva Dias, M.A.F.F.; Rutledge, S.; Kabat, P.; Silva Dias, P.L.; Nobre, C.; Fisch, G.; Dolman, A.J.; Zipser, E.; Garstang, M.; Manzi, A.O.; et al. Cloud and Rain Processes in a Biosphere-Atmosphere Interaction Context in the Amazon Region. *J. Geophys. Res. D Atmos.* **2002**, *107*, 8072. [CrossRef]
38. Joetzjer, E.; Douville, H.; Delire, C.; Ciais, P. Present-Day and Future Amazonian Precipitation in Global Climate Models: CMIP5 versus CMIP3. *Clim. Dyn.* **2013**, *41*, 2921–2936. [CrossRef]
39. Brando, P.M.; Goetz, S.J.; Baccini, A.; Nepstad, D.C.; Beck, P.S.A.; Christman, M.C. Seasonal and Interannual Variability of Climate and Vegetation Indices across the Amazon. *Proc. Natl. Acad. Sci. USA* **2010**, *107*, 14685–14690. [CrossRef]
40. Xu, L.; Samanta, A.; Costa, M.H.; Ganguly, S.; Nemani, R.R.; Myneni, R.B. Widespread Decline in Greenness of Amazonian Vegetation Due to the 2010 Drought. *Geophys. Res. Lett.* **2011**, *38*, L07402. [CrossRef]
41. Spracklen, D.V.; Garcia-Carreras, L. The Impact of Amazonian Deforestation on Amazon Basin Rainfall. *Geophys. Res. Lett.* **2015**, *42*, 9546–9552. [CrossRef]
42. Lejeune, Q.; Davin, E.L.; Guillod, B.P.; Seneviratne, S.I. Influence of Amazonian Deforestation on the Future Evolution of Regional Surface Fluxes, Circulation, Surface Temperature and Precipitation. *Clim. Dyn.* **2015**, *44*, 2769–2786. [CrossRef]

43. Lima, L.S.; Coe, M.T.; Soares Filho, B.S.; Cuadra, S.V.; Dias, L.C.P.; Costa, M.H.; Lima, L.S.; Rodrigues, H.O. Feedbacks between Deforestation, Climate, and Hydrology in the Southwestern Amazon: Implications for the Provision of Ecosystem Services. *Landsc. Ecol.* **2014**, *29*, 261–274. [CrossRef]
44. Nobre, P.; Malagutti, M.; Urbano, D.F.; De Almeida, R.A.F.; Giarolla, E. Amazon Deforestation and Climate Change in a Coupled Model Simulation. *J. Clim.* **2009**, *22*, 5686–5697. [CrossRef]
45. Lean, J.; Warrilow, D.A. Simulation of the Regional Climatic Impact of Amazon Deforestation. *Nature* **1989**, *342*, 411–413. [CrossRef]
46. Nobre, C.A.; Sellers, P.J.; Shukla, J. Amazonian Deforestation and Regional Climate Change. *J. Clim.* **1991**, *4*, 957–988. [CrossRef]
47. Werth, D.; Avissar, R. The Local and Global Effects of Amazon Deforestation. *J. Geophys. Res.* **2002**, *107*, 8087. [CrossRef]
48. Moore, N.; Arima, E.; Walker, R.; Ramos da Silva, R. Uncertainty and the Changing Hydroclimatology of the Amazon. *Geophys. Res. Lett.* **2007**, *34*, L14707. [CrossRef]
49. Hasler, N.; Werth, D.; Avissar, R. Effects of Tropical Deforestation on Global Hydroclimate: A Multimodel Ensemble Analysis. *J. Clim.* **2009**, *22*, 1124–1141. [CrossRef]
50. Walker, R.; Moore, N.J.; Arima, E.; Perz, S.; Simmons, C.; Caldas, M.; Vergara, D.; Bohrer, C. Protecting the Amazon with Protected Areas. *Proc. Natl. Acad. Sci. USA* **2009**, *106*, 10582–10586. [CrossRef]
51. Medvigy, D.; Walko, R.L.; Avissar, R.; Medvigy, D.; Walko, R.L.; Avissar, R. Effects of Deforestation on Spatiotemporal Distributions of Precipitation in South America. *J. Clim.* **2011**, *24*, 2147–2163. [CrossRef]
52. Espinoza, J.C.; Chavez, S.; Ronchail, J.; Junquas, C.; Takahashi, K.; Lavado, W. Rainfall Hotspots over the Southern Tropical Andes: Spatial Distribution, Rainfall Intensity, and Relations with Large-Scale Atmospheric Circulation. *Water Resour. Res.* **2015**, *51*, 3459–3475. [CrossRef]
53. Paccini, L.; Espinoza, J.C.; Ronchail, J.; Segura, H. Intra-Seasonal Rainfall Variability in the Amazon Basin Related to Large-Scale Circulation Patterns: A Focus on Western Amazon–Andes Transition Region. *Int. J. Climatol.* **2018**, *38*, 2386–2399. [CrossRef]
54. Saatchi, S.S.; Houghton, R.A.; Dos Santos Alvalá, R.C.; Soares, J.V.; Yu, Y. Distribution of Aboveground Live Biomass in the Amazon Basin. *Glob. Chang. Biol.* **2007**, *13*, 816–837. [CrossRef]
55. Sombroek, W. Spatial and Temporal Patterns of Amazon Rainfall. *AMBIO A J. Hum. Environ.* **2001**, *30*, 388–396. [CrossRef]
56. Davidson, E.A.; de Araújo, A.C.; Artaxo, P.; Balch, J.K.; Brown, I.F.; Bustamante, M.M.C.; Coe, M.T.; DeFries, R.S.; Keller, M.; Longo, M.; et al. The Amazon Basin in Transition. *Nature* **2012**, *481*, 321–328. [CrossRef] [PubMed]
57. Angelini, I.M.; Garstang, M.; Davis, R.E.; Hayden, B.; Fitzjarrald, D.R.; Legates, D.R.; Greco, S.; Macko, S.; Connors, V. On the Coupling between Vegetation and the Atmosphere. *Theor. Appl. Climatol.* **2011**, *105*, 243–261. [CrossRef]
58. Marengo, J.A. Interdecadal Variability and Trends of Rainfall across the Amazon Basin. *Theor. Appl. Climatol.* **2004**, *78*, 79–96. [CrossRef]
59. Coe, M.T.; Costa, M.H.; Botta, A.; Birkett, C. Long-Term Simulations of Discharge and Floods in the Amazon Basin. *J. Geophys. Res.* **2002**, *107*, 8044. [CrossRef]
60. Vera, C.; Silvestri, G.; Liebmann, B.; González, P. Climate Change Scenarios for Seasonal Precipitation in South America from IPCC-AR4 Models. *Geophys. Res. Lett.* **2006**, *33*, 13707. [CrossRef]
61. Carvalho, L.M.V.; Jones, C.; Liebmann, B.; Carvalho, L.M.V.; Jones, C.; Liebmann, B. The South Atlantic Convergence Zone: Intensity, Form, Persistence, and Relationships with Intraseasonal to Interannual Activity and Extreme Rainfall. *J. Clim.* **2004**, *17*, 88–108. [CrossRef]
62. Fu, R.; Dickinson, R.E.; Chen, M.; Wang, H.; Fu, R.; Dickinson, R.E.; Chen, M.; Wang, H. How Do Tropical Sea Surface Temperatures Influence the Seasonal Distribution of Precipitation in the Equatorial Amazon? *J. Clim.* **2001**, *14*, 4003–4026. [CrossRef]
63. Réveillet, M.; MacDonell, S.; Gascoin, S.; Kinnard, C.; Lhermitte, S.; Schaffer, N. Impact of Forcing on Sublimation Simulations for a High Mountain Catchment in the Semiarid Andes. *Cryosph.* **2020**, *14*, 147–163. [CrossRef]
64. Paca, V.H.d.M.; Espinoza-Dávalos, G.E.; Hessels, T.M.; Moreira, D.M.; Comair, G.F.; Bastiaanssen, W.G.M. The Spatial Variability of Actual Evapotranspiration across the Amazon River Basin Based on Remote Sensing Products Validated with Flux Towers. *Ecol. Process.* **2019**, *8*, 6. [CrossRef]
65. Souza, E.P.; Rennó, N.O.; Dias, M.A.F.S.; Souza, E.P.; Rennó, N.O.; Dias, M.A.F.S. Convective Circulations Induced by Surface Heterogeneities. *J. Atmos. Sci.* **2000**, *57*, 2915–2922. [CrossRef]
66. Snyder, P.K.; Delire, C.; Foley, J.A. Evaluating the Influence of Different Vegetation Biomes on the Global Climate. *Clim. Dyn.* **2004**, *23*, 279–302. [CrossRef]
67. Zhang, H.; Henderson-Sellers, A.; McGuffie, K. Impacts of Tropical Deforestation. Part II: The Role of Large-Scale Dynamics. *J. Clim.* **1996**, *9*, 2498–2521. [CrossRef]
68. Haghtalab, N.; Moore, N.; Heerspink, B.P.; Hyndman, D.W. Evaluating Spatial Patterns in Precipitation Trends across the Amazon Basin Driven by Land Cover and Global Scale Forcings. *Theor. Appl. Climatol.* **2020**, *140*, 411–427. [CrossRef]
69. Myneni, R.B.; Yang, W.; Nemani, R.R.; Huete, A.R.; Dickinson, R.E.; Knyazikhin, Y.; Didan, K.; Fu, R.; Negrón Juárez, R.I.; Saatchi, S.S.; et al. Large Seasonal Swings in Leaf Area of Amazon Rainforests. *Proc. Natl. Acad. Sci. USA* **2007**, *104*, 4820–4823. [CrossRef]
70. Richey, J.E.; Nobre, C.; Deser, C. Amazon River Discharge and Climate Variability: 1903 to 1985. *Science* **1989**, *246*, 101–103. [CrossRef]

Article

Evaluation and Driving Determinants of the Coordination between Ecosystem Service Supply and Demand: A Case Study in Shanxi Province

Yushuo Zhang [1,*], **Boyu Liu** [2,*] **and Renjing Sui** [1]

[1] School of Culture Tourism and Journalism Arts, Shanxi University of Finance and Economics, Taiyuan 030006, China; 222120203006@sxufe.edu.cn

[2] College of Mining Engineer, Taiyuan University of Technology, Taiyuan 030024, China

* Correspondence: zhangys@sxufe.edu.cn (Y.Z.); liuby10@mails.jlu.edu.cn (B.L.)

Abstract: Understanding the coordination relationship between ecosystem service (ES) supply and demand and elucidating the impact of driving factors is critical for regional land use planning and ecological sustainability. We use a large watershed area as a case to map and analyze ES supply, demand and the coordination relationship, and identify the associated socio-ecological driving variables. This study assessed the supply and demand of five ESs (crop production, water retention, soil conservation, carbon sequestration, and outdoor recreation) in 2000 and 2020, and evaluated the coordination between them employing the coupling coordination degree model (CCDM). Additionally, we utilized the geo-detector model (GDM) to identify driving determinants and their interactive effects on the spatial pattern of the coupling coordination degree (CCD) between ES supply and demand. The results showed that mountainous regions with abundant forest coverage were high-value areas for ES supply, while the ESs were predominantly required in city center areas within each basin area. From 2000 to 2020, there was a slight decline in ES supply and a significant increase in ES demand. Counties were grouped into four coordination zones in the study area: extreme incoordination, moderate incoordination, reluctant coordination, and moderate coordination. The number of counties with extreme incoordination linked to regions with a mountain ecosystem is increasing, where the ES supply is much greater than the demand. The moderate incoordination counties dominated by a cropland ecosystem exhibited slightly higher levels of ES supply than demand. The moderate and reluctant coordination were linked to counties with distinct ecological characteristics. Construction land played a major role in the characteristics of the CCD, followed by grassland. The interaction between construction land and all other factors significantly increased the influence on the CCD. These findings offered valuable insights for land managers to identify areas characterized by incoordination between ES supply and demand and understand associated factors to develop optimal ES management strategies.

Keywords: ecosystem services (ESs); ES supply; ES demand; coupling coordination degree model; geo-detector model; Shanxi Province

Citation: Zhang, Y.; Liu, B.; Sui, R. Evaluation and Driving Determinants of the Coordination between Ecosystem Service Supply and Demand: A Case Study in Shanxi Province. *Appl. Sci.* **2023**, *13*, 9262. https://doi.org/10.3390/app13169262

Academic Editor: Jianbo Gao

Received: 22 July 2023
Revised: 10 August 2023
Accepted: 12 August 2023
Published: 15 August 2023

1. Introduction

Ecosystem services (ESs) are defined as the benefits that humans directly or indirectly receive from ecosystems [1]. The Millennium Ecosystem Assessment (MA) [2], first conducted in 2005, established a framework for the global assessment of ecosystems. This framework divides indicators of ESs into four categories: provisioning, regulating, cultural, and supporting services. These categories are based on the connection between ESs and human well-being. Due to regional socio-economic factors and the rapid, high-intensity expansion of land for human use, ecological systems are facing continuous destruction [3]. This has caused a significant reduction in the ecosystem service supply (ESS), while the demand for a better living environment continues to grow [4]. Consequently, this has intensified the incoordination relationship between ESS and ecosystem service demand (ESD), negatively affecting sustainable development and human well-being [5]. Thus, it will be difficult to effectively manage and optimize regional ecosystems and encourage sustainable development via only the supply of ESs, while ignoring the human demand for ESs. Investigating the relationship between ESS and ESD not only addresses the challenges of sustainability arising from increasing human demand, but also establishes a solid theoretical foundation for adept and efficient ES management and utilization practices [6,7]. Therefore, understanding the relationship between ESS and ESD is essential for effective land use planning and decision-making processes that ensure the long-term resilience of ecosystems and social well-being.

As a closer relationship is established between ESs and human well-being, ES demand (ESD) gradually integrates with ESS. ESS is generally defined as the beneficial effect that ecosystems have on society [8]. The structure and function of ecosystems often result in simultaneous positive and negative changes in ESS [9]. However, there is no universally accepted definition of ESD. Currently, it is mainly defined from two perspectives. From a consumption perspective, ESD refers to the services provided by ecosystems that are useful to consumers, reflecting the actual demand for ESs [10]. From a preference perspective, ESD refers to the ecosystem services requested by social groups [11], reflecting not only the actual demand for ESs, but also the potential demand that cannot be met due to certain conditions. ESD includes the requirements, desires, or aspirations of human societies in relation to the benefits and contributions provided by ecosystems [12]. These needs arise from the dependence of human well-being and quality of life on the services and resources provided by natural ecosystems. ESD is driven by various factors, including population growth, economic activities, urbanization, and societal preferences. When the demand for ESs exceeds the natural capacity of ecosystems to provide them, it can lead to the overexploitation, degradation, or loss of those services. Conversely, if the supply of ES surpasses the demand, it can result in the underutilization or inefficient allocation of resources. Therefore, the relationship between ESS and ESD is crucial for creating a dynamic balance through which ecosystem products and services are transferred from ecosystems to social systems [13,14].

In recent years, an increasing number of studies have focused on evaluating the relationship between ESS and ESD, considering the coordination or conflict between regional social systems and ecosystems [15,16]. Thus far, these studies have primarily concentrated on quantifying and comparing the supply and demand of distinct ES indicators, such as water supply [17], flood regulation [18], air purification [19], and erosion control [20], and identifying the degree of mismatch, imbalance or incoordination between ESS and ESD. The majority of these investigations are centered around specific types of ecosystems, including forests [21], croplands [22], and urban regions [23]. These studies were mainly conducted in European contexts; however, in recent years, an increasing number of studies have been conducted in China, driven by ecological regionalization policies at different scales. Most previous studies assessed ESS, mainly focusing on quantifying and identifying the patterns and functions of ESs, as well as the effects of land use and land cover changes on ESs [24]. Compared to ESS, there are two main distinct approaches to ESD assessment: One is to evaluate the demand for each individual ES, and the other involves conducting a comprehensive assessment of the overall demand for ecosystem services. The land development index (LDI) has been widely used to comprehensively assess the ESD [25]. The rationale behind using the LDI to assess demand for ESD lies in its capacity to gauge the intensity of land development and its corresponding impact on the demand for these services [26]. Thus, ESS and ESD can be expressed from the perspective of the interaction between ecosystems and social systems; hence, the coordination relationship between ESS and ESD is a significant reflection of whether regional socio-economic structure and natural ecological backgrounds can develop in a coordinated manner [27].

Previous studies mainly analyzed temporal dynamic variations and spatial imbalances or mismatches in ESS and ESD [28,29], using methods such as modeling, mapping, participatory methods, etc. [30]. The concept of "coupling coordination" provides a framework for quantitating the coordination between ESS and ES [31,32]. The coupling coordination degree (CCD) is a measure used to assess the level of coordination and interdependence between different components or subsystems within a larger system [33]. In the context of ES, the coupling coordination degree model (CCDM) can be used to evaluate the level of coordination between the ecological system that provides services and the social system that demands and utilizes such services. It can also be applied to assess how effectively the ESS and ESD function together and whether their interactions are balanced and mutually beneficial. In recent years, many studies have applied the CCDM to analyze the relationship between ESS and ESD. For example, Guan et al. [34] provided valuable insights into the evolving characteristics of the spatial coupling between ESS and ESD using the CCDM. Li et al. [35] analyzed the dynamic characteristics of the supply and demand coupling of ESs in Lanzhou, China. Yang et al. [36] identified the coupling coordination relationship between sustainable development and ESs in Shanxi Province, China. Therefore, CCD is regarded as an efficient tool for researchers and policymakers to gain insights into the functioning of the coupled ecological and social systems and identify areas that require attention or intervention. Consequently, this helps us to understand the complex interactions between systems as well as design strategies for the sustainable management and conservation of ESs.

The coordination between ESS and ESD is influenced by the rapid development of regional socio-economic factors and the ecological factors, especially when socio-economic factors interact with natural and ecological factors [37]. Natural factors, land use/land cover, and socio-economic factors have been identified as the primary determinants of the relationship between ESS and ESD. For instance, Sun et al. [38] conducted an empirical study examining the correlations among 12 natural and socio-economic variables related to both ESS and ESD. Their study sheds light on the disparity between these two aspects within the United States. Wu et al. [16] analyzed the relationships between ES supply and demand and identified the effect of forest area on ESS, as well as the effects of per capita GDP, energy consumption per unit of GDP, and permanent population on ES demand in China. Peng et al. [39] systematically analyzed the impact of urbanization on ESs in metropolitan areas. Previous studies identified multiple influencing factors and encompassed different ESs. However, the majority of studies primarily focus on identifying the individual influencing factors of ESS or ESD. Alternatively, some studies have solely examined the influencing factors of ESS and ESD as substitutes for the actual relationship between the two [40]. As a result, there is a lack of studies investigating factors that directly impact the relationship between ESS and ESD. Research on the relative importance of socio-ecological drivers of coordination between ESS and ESD remains limited, and little attention has been given to the relationships between various drivers and the coordination between ESS and ESD, as well as associated spatial influences [41,42].

Shanxi Province is a typical resource-based area heavily reliant on coal and other resources for rapid economic growth. However, this excessive consumption of resources has created an exceedingly fragile ecosystem across the province. This increase in ecosystem degradation is having a negative effect on the economy and society, posing significant challenges to the sustainable development of Shanxi Province [43]. For example, intensive agriculture and mining activities are contributing to soil degradation in Shanxi. Erosion, the loss of topsoil, and contamination from pollutants can have serious consequences for agricultural productivity and ecosystem health. Unsustainable mining practices, particularly in coal mining, are causing land subsidence issues in Shanxi [44,45]. This phenomenon can lead to infrastructure damage, waterlogging, and the disruption of ecosystems [46]. In addition, some parts of Shanxi are vulnerable to desertification due to factors such as soil erosion, overgrazing, and unsustainable land use practices. This threatens agricultural productivity and ecosystem stability [47]. Since 2000, large-scale ecological restoration projects, such as the Natural Forest Protection Projects (NFPP) and the Grain for Green Program (GFGP), have significantly bolstered vegetation restoration in Shanxi Province. Efforts to address these ecosystem problems likely involve a combination of policy interventions, stricter environmental regulations, technological innovations, public awareness campaigns, and sustainable development practices.

Based on this context, this study aims to analyze the coupling coordination relationship between ESS and ESD and identify the associated socio-ecological driving variables in Shanxi Province. The objectives of this study are to (i) quantify and map the spatial distribution of the ES supply and demand, respectively; (ii) analyze the spatial–temporal characteristics of CCD between ESS and ESD based on CCDM; and (iii) determine the decisive influencing factors and the effects of interactions between factors using GDM. The results are anticipated to provide valuable information for achieving a harmonious balance between economic development and ecological restoration on a national scale within the provincial administrative units of China.

2. Materials and Methods

2.1. Study Area

Shanxi Province, located in the northern part of China (110°14′–114°33′ E, 34°34′–40°44′ N), covers an area of 156,700 km^2, accounting for 1.6% of the country's territory (Figure 1a). It consists of 107 counties and is characterized by a typical mountain plateau terrain. Its topography is complex and diverse, including mountains, hills, plateaus, basins, and platforms. Mountains and hills make up 80% of its area, with altitudes ranging from 208 m to 2988 m above sea level (Figure 1b). The study area falls within the temperate continental monsoon climate region, the annual average temperature is between 4 °C and 14 °C, and the average annual rainfall is 468 mm. The dominant land use/land cover types include cropland, grassland, forest, and construction land (Figure 1c). Shanxi Province straddles the Yellow River basin and the Haihe River basin, and the river system is a self-generated outflow. The total population of the study area accounts for 2.48% of the national population and 1.71% of China's total GDP [48].

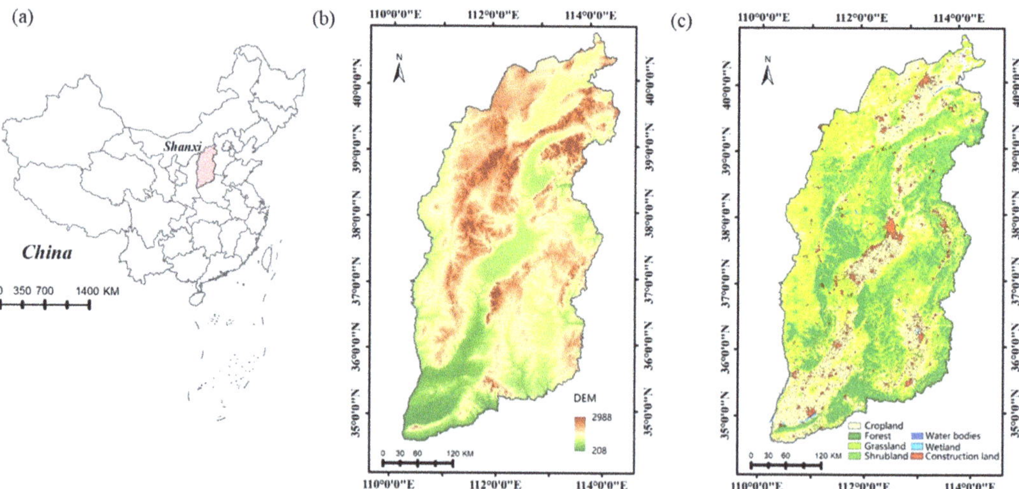

Figure 1. Study area location in China (**a**), elevation (**b**), and land use/land cover type of Shanxi Province (**c**).

Shanxi Province is one of the most important provinces with coal and mineral resources in China. As a region highly dependent on coal and mineral resources, its economic and social development is closely linked to its ecological environment [49]. However, excessive coal and mineral resource exploitation, the overexploitation of groundwater, rapid urbanization, and accelerated water and land resource exploitation have made the ecosystem in the study area extremely fragile. In 2019, the Chinese Government put forth the objectives of "ecological protection and high-quality development" for the Yellow River basin. As part of the Yellow River basin, in recent decades, Shanxi Province has faced daunting challenges in coordinating population, resources, ecosystems, and economic development.

2.2. Data Collection

In this study, we selected 2000 and 2020 as representative years to collect and analyze data. The data encompass both spatial and statistical information. The spatial data were processed at a grid cell resolution of 1 km × 1 km, while statistical data were aggregated at the county level. Table 1 provides a comprehensive list of the primary data required to calculate the ES indicators. To ensure consistency, all spatial data were transformed to a common spatial reference system, specifically the WGS84 coordinate system and Albers equal-area conic projection. The flowchart depicted in Figure 2 illustrates the methodology that we employed to achieve our study objectives.

Table 1. Datasets used in the study.

ES Variable	Data Type	Spatial Resolution	Data Source
Land use/land cover	Raster	30 m	National Geomatics Center of China (http://www.globallandcover.com/GLC30Download/index.aspx, accessed on 11 December 2020)
NDVI	Raster	250 m	National Aeronautics and Space Administration and United States Geological Survey (http://e4ftl01.cr.usgs.gov/MOLT/MOD13Q1.006/, accessed on 7 December 2020)
DEM	Raster	90 m	Geospatial Data Cloud (https://www.gscloud.cn/#page1, accessed on 20 December 2020)
Meteorological data	Numeric	Sites	China Meteorological Data Sharing Service System (http://www.escience.gov.cn/metdata/page/index.html, accessed on 5 November 2020)
Soil database	Raster	30 arc-second	Harmonized World Soil Database (http://www.fao.org/soils-portal/soil-survey/soil-maps-and-databases/harmonied-world-soil-datebse-v12/en/, accessed on 1 December 2020)
Administrative map	Vector	County	National Geomatics Center of China (http://ngcc.sbsm.gov.cn/ngcc/, accessed on 11 December 2020)
Crop yield, GDP and population	Numeric	County	Statistical Yearbook

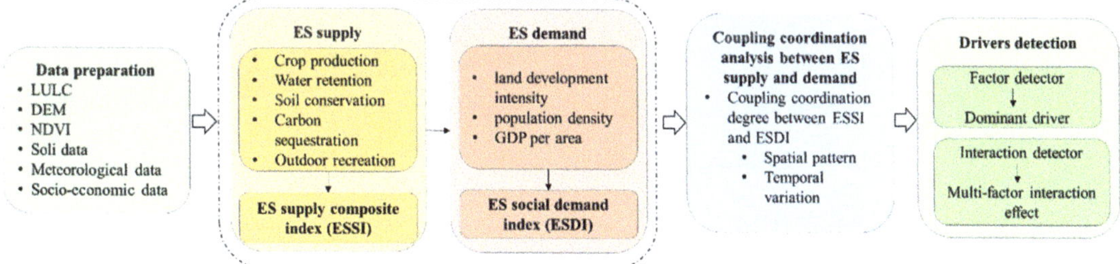

Figure 2. Flowchart of the proposed methodology.

2.3. Quantifying ES Supply

2.3.1. Selection of ES Indicators

The selection of appropriate ES indicators is a critical step in assessing ecosystem services. In this study, we followed certain criteria for selecting ES indicators: (i) aligning with the classification of ES according to the Millennium Ecosystem Assessment [2] to ensure comparability with other studies; (ii) considering existing case studies conducted in Shanxi Province and selecting ES indicators that are closely related to the natural, ecological, social, and economic conditions of the study area; and (iii) the availability of the primary data required for evaluating ES indicators. Based on these criteria, our study focused on five key ES indicators relevant to the study area. These included one provisioning service (crop production), three regulating services (water retention, soil conservation, and carbon sequestration), and one cultural service (outdoor recreation).

2.3.2. Calculation of the ES Indicators

To quantify the selected ES indicators, we employed existing and widely used assessment models originally developed for this purpose. Specifically, the assessment of crop production was based on annual crop yield data [50]. The water balance equation served as a proxy for measuring water retention. The Universal Soil Loss Equation (USLE) model was used to calculate soil conservation [51]. Net primary production (NPP) was used as a proxy for carbon sequestration [52,53] and was assessed using the Carnegie–Ames–Stanford Approach (CASA) model, a widely adopted approach for NPP estimation [54,55]. The spatial distribution of individual ES indicators was visualized via mapping in ArcGIS.

To analyze the relationships between the five ES indicators, we employed ArcGIS 10.2, which serves as a common spatial unit, to aggregate all ES indicators at a national level. According to Raudsepp-Hearne et al. [56], administrative boundaries are suitable for identifying socio-ecological systems in a landscape, as management decisions at this level influence the provision and consumption of ES. The specific models and processes used for assessing the ES indicators are summarized in Table 2.

2.3.3. Assessment of ES Supply Index

Since each ES has its own measurement unit, we individually standardized the ES values and then summarized them within each county to mitigate the influence of magnitude and variability. We employed min–max normalization to standardize the values of the five ES indicators [63,64]. This normalization method removes the units of the input data and scales them to a common range. After standardization, the standardized values were accumulated to obtain the ES supply index (ESSI), which represents the total ecosystem service supply. The calculation equation for ESSI is as follows:

$$ESS_{ij} = \frac{ES_{ij} - minES_j}{maxES_j - minES_j} \tag{1}$$

$$ESSI_i = \sum_{j=1}^{n} ESS_{ij} \tag{2}$$

where $ESSI_i$ is the ES supply index of county i; ESS_{ij} is the standardized value for ES j of county i; ES_{ij} is the initial value for ES j of county i; max ES_j is the maximum value of ES over 107 counties, and min ES_j denotes the minimum value of ES_j; n = 5.

Table 2. ES indicators from the MA categories and their quantitative methods, units, and ES variable requirements.

ES Indicator	Code	Model or Proxy	Unit	Assessment Process
Crop production	Cro	Crop yield per square kilometer	ton/km^2	Crop production was calculated by dividing the crop yield of each county by its territory to illustrate per-unit provision service.
Water retention	Wret	Water balance equation	m^3/km^2	$TQ = \sum_{i=1}^{j} (P_i - R_i - ET_i)$ TQ is water conservation, P_i is the annual average rainfall (mm), R_i is the annual average surface runoff (mm), and ET_i is the annual evapotranspiration (mm).
Soil conservation	Scon	USLE (Universal Soil Loss Equation)	t/(hm^2·a)	$\Delta A = R \times K \times L \times S(1 - C \times P)$ ΔA = soil conservation (t/ (hm^2·a)), R = rainfall erosivity index (MJ·mm/(hm^2·h·a)), K = soil erodibility factor (t·hm^2·h/ (MJ·mm·hm^2)), L S = slope length and steepness factor (unitless), C = cover and management factor (unitless), P = conservation practice factor (unitless). The parameters R were from Wischmeier and Smith [57], K from Williams [58], L S from McCool et al. [59] and Liu et al. [60], C from Cai et al. [61], and P from Kumar et al. [62].
Carbon sequestration	Cseq	CASA (Carnegie–Ames–Stanford approach)	kg C/km^2	$NPP = APAR \times \xi$ NPP = net primary productivity (g C/m^2), $APAR$ = absorbed photosynthetic active radiation (MJ/m), ξ = the utilization rate of light energy (g C/MJ).
Outdoor recreation	Rec	Tourists per square kilometer	persons/km^2	Outdoor recreation was calculated via the area of forest land in each county.

2.4. Quantifying ES Demand

The ES demand represents the human demand and preference for ecosystem products and services within a specific time period. In this study, we used a research method [26,65] to quantify ESD by considering land development intensity, population density, and gross domestic product (GDP) per area. This helped us to understand the coordination between the development and preservation of essential natural processes that sustain human well-being and environmental quality [29,66,67]. Specifically, land development intensity was measured as the percentage of construction land in the total land area. It reflects the intensity of the human consumption of ES. A higher percentage of construction land indicates a greater intensity of human land development in a given area, and consequently, a higher demand for ES. Population density serves as an indicator of the amount of ES demand. A higher population density corresponds to a greater ESD. GDP per area reflects the economic development of the region and indirectly indicates how much humans wish to consume or utilize ESs. Logarithmic methods were employed to remove fluctuations in the data. The ESD index is calculated using the following formula:

$$\mathrm{ESDI_i = D_i \times lg(P_i) \times lg(G_i)} \tag{3}$$

where $\mathrm{ESDI_i}$ is the ES demand index of each county i; Di, Pi, and Gi are the land development intensity (%), population density (person/$\mathrm{km^2}$), and GDP per area (yuan/$\mathrm{km^2}$) of county i, respectively.

2.5. Assessing Coordination between ES Supply and Demand

In this study, we employed the CCDM to investigate the interactive coordination relationship between ESS and ESD. The CCDM highlights the interdependence of ecological and social systems and aims to understand how they interact and mutually influence each other. By adopting this model, we can develop a holistic understanding of the relationship between ecosystems and human societies.

The CCDM recognizes the interconnectedness of ecological and social systems and emphasizes the importance of studying them together. This enables us to analyze the developmental pattern of these systems or indicators, progressing from disorder to order [68]. This representation provides insights into the overall effectiveness and synergistic impact between systems [69].

Mathematically, the CCDM is expressed as follows:

$$\mathrm{CCD_i = \sqrt{C_i \cdot T_i}} \tag{4}$$

$$\mathrm{C_i = \left\{ \left(S_{ESSCI_i} \cdot D_{ESSDI_i}\right) / \left[\left(S_{ESSCI_i} + D_{ESSDI_i}\right)/2\right]^2 \right\}^{1/2}} \tag{5}$$

$$\mathrm{T_i = \alpha \cdot S_{ESSCI_i} + \beta \cdot D_{ESSDI_i}} \tag{6}$$

where $\mathrm{CCD_i}$ represents the coupling coordination degree of county i ($0 \leq \mathrm{CCD_i} \leq 1$) between ESS and the ESSD; $\mathrm{C_i}$ refers to the coupling degree between ESS and the ESSD; $\mathrm{T_i}$ is the comprehensive development index of ESS and the ESSD; and $\mathrm{S_{ESSIi}}$ and $\mathrm{D_{ESDIi}}$ are the values of standardized ESSI and ESDI ($0 \leq \mathrm{SESSI_i} \leq 1, 0 \leq \mathrm{DESDI_i} \leq 1$). α and β are the weights to be determined; due to the equal importance of ESS and the ESSD in the coordination, α and β are given the same weight, that is, $\alpha = \beta = 0.5$. Referring to previous research [32], we divide the CCD into five levels: When $0 \leq \mathrm{CCD_i} \leq 0.20$, the ESS and ESSD are in extreme incoordination; when $0.20 < \mathrm{CCD_i} \leq 0.35$, they are in moderate incoordination; when $0.35 < \mathrm{CCD_i} \leq 0.55$, they are in reluctant coordination; when $0.55 < \mathrm{CCD_i} \leq 0.70$, they are in moderate coordination; and when $0.70 < \mathrm{CCD_i} \leq 1$, they are in superior coordination.

2.6. Driving Variables of ES Coordination

2.6.1. Critical Driving Variables

In this study, we selected socio-ecological variables to explain the spatiotemporal differences between ESS and ESD, based on relevant research [70,71]. Potential explanatory variables were chosen from three sources: (1) the variables used to quantify ESS or ESD in our study, (2) variables identified in the literature as directly or indirectly driving individual ESs and/or their associations [72], and (3) variables for which quantitative data were available. After considering these factors, we ultimately selected thirteen potential socio-ecological variables, including natural variables such as elevation (DEM), slope (SLOPE), average annual precipitation (PRE), and average annual temperature (TEM); ecological variables such as NDVI, percentage of crop land (CROP), percentage of forestland (FOREST), and percentage of grassland (GRASS); and socio-economic variables such as percentage of construction land (CON), total population (POP), GDP, proportion of urban population (URBAN) and distance to the nearest county center (COUNTY) (Table 3).

Table 3. Details of the driving variables for coupling coordination between ESSI and ESDI in this study.

Variable	Code	Description	Unit
Elevation	DEM	Derived from the SRTM3 global digital elevation model	Meter
Slope	SLOPE	Derived from the SRTM3 global digital elevation model	Degree
Precipitation	PRE	Annual trends of precipitation for the period 1956–2017	mm
Temperature	TEM	Annual trends of temperature for the period 1956–2017	°C
Normalized Difference Vegetation Index	NDVI	Vegetation cover	%
Cropland	CROP	County land area that is occupied by area that is classified as cropland	%
Forestland	FOREST	County land area that is occupied by area that is classified as forest	%
Grassland	GRASS	County land area that is occupied by area that is classified as grassland	%
Construction land	CON	County land area that is occupied by area that is classified as construction land	%
Population	POP	Annual total population	person
Economic level	GDP	Gross domestic product	yuan
Urbanization rate	URBAN	Urban population proportion	%
Distance to the nearest county	COUNTY	Distance to the nearest county center	km

2.6.2. Effects of Driving Variables on Coordination via Geo-Detector Model

The geo-detector model (GDM) can help identify the most influential factors or variables that contribute to specific spatial patterns, and reveal how different factors interact in a spatial context. It offers insights into the potential effects of human activities on ecosystems, water resources, or air quality. In this study, we employed the GDM to assess the spatial correlation between the explanatory variables and the dependent variables through spatial variance analysis (SVA) [73,74]. The GDM is a valuable analytical tool to identify and quantify the spatial associations between driving factors and specific outcomes. This insight can guide decision making by highlighting where interventions or ecological resource allocation should be focused for maximum impact. By employing statistical and spatial analytical techniques, the GDM enables researchers to identify dominant driving factors and their interactive effects, as well as explore spatial patterns and trends in complex geographical processes [75]. The fundamental assumption of the GDM is as follows: if an independent variable X significantly affects a dependent variable Y, then the spatial distributions of X and Y should exhibit similarity. SVA is used to compare the spatial consistency between the dependent variable and independent variables. Based on this comparison, the interpretation of independent variables in relation to the dependent variable can be quantified.

In this study, we utilized the "factor detector" module of GDM to identify the driving factor(s) that determine the distribution of CCD. This module identifies the extent to which the driving variables explain the spatial differentiation of CCD. The calculation results of the factor detector include the q-statistic and p-value. The q-statistic represents the influencing coefficient of the driving variable on CCD, with larger values indicating a stronger impact of the driving variable on CCD. The p-value indicates the significance level of the explanation, and a significance level of 0.1 (p-value < 0.1) is considered statistically significant. The formula for the factor detector is as follows:

$$q = 1 - \frac{\sum_{h=1}^{l} N_h \sigma_h^2}{N \sigma^2} \tag{7}$$

where q signifies the influencing coefficients of the driving variables for the ES (q -statistic), the values of which range from 0 to 1, where 0 corresponds to no correlation between the two and 1 to CCD's complete dependence on a driving variable. σ^2 is the variance of the CCD, and N is the size of CCD. The superposition of the driving variables and CCD forms L layers in CCD, which are indexed by h = 1, 2... l, and N_h and σ_h^2 represent the scale and variance of layer h, respectively.

The "interaction detector" module of GDM was used to examine whether two factors have a stronger or weaker effect on ESs than they do independently. The types of interactions between two variables are as follows:

Enhance: if $q (D_1 \cap D_2) > q (D_1)$ or $q (D_2)$

Enhance, bivariate: if $q (D_1 \cap D_2) > q (D_1)$ and $q (D_2)$

Enhance, nonlinear: if $q (D_1 \cap D_2) > q (D_1) + q (D_2)$

Weaken: if $q (D_1 \cap D_2) < q (D_1) + q (D_2)$

Weaken, univariate: $q (D_1 \cap D_2) < q (D_1)$ or $q (D_2)$

Weaken, nonlinear: if $q (D_1 \cap D_2) < q (D_1)$ and $q (D_2)$

Independent: if $q (D_1 \cap D_2) = q (D_1) + q (D_2)$

where the symbol "$\cap$" denotes the intersection between the layers D_1 and D_2. The attributes of layer $(D_1 \cap D_2)$ are determined by the combination of the attributes of layer D_1 and D_2 using a spatial overlay to form a new layer. $q (D_1)$, $q (D_2)$, and $q (D_1 \cap D_2)$ were calculated using Equation (1). By comparing the sum ($q (D_1) + q (D_2)$) of the factors' contribution to two individual attributes ($q (D_1)$, $q (D_2)$) with the contribution of the two attributes when combined ($q (D_1 \cap D_2)$), the interactive effects of the two factors can be defined using the above seven types.

3. Results

3.1. Spatial–Temporal Patterns of ES Supply and Demand

Using a quantitative method, we calculated ESSI and ESDI data for the 107 counties of Shanxi Province for the reference years 2000 and 2020. We normalized the two indices to a range of 0–1 and mapped them to facilitate comparisons (Figures 3 and 4).

The unique geographical location of Shanxi Province resulted in significant variations in natural conditions, leading to considerable heterogeneity in the counties' ability to provide ESs. The ESSI exhibited substantial variation across Shanxi Province in both 2000 and 2020, which was similar to the distribution characteristics of outdoor recreation (Figure 3). Areas with high ESS were dispersed across the mountainous regions of the study area, surrounded by counties with higher ESSI. Specifically, Mount Taiyue, Mount Zhongtiao, and Mount Wangwu, located in the southern parts of the province, have a particularly abundant supply of ESs. The areas with low supply were concentrated in the northwest edge of Shanxi and the western part of Mount Luliang. From 2000 to 2020, the spatial pattern of ESSI remained largely unchanged in most counties. However, there was a slight overall decrease in the level of ESS, and the ESSI of Yuanping and Xinzhou, located in the northwestern part of Shanxi Province, experienced significant increases. There was a clear upward trend in the number of counties with the lowest supply, increasing from

20.6% for the 107 counties in 2000 to 32.7% in 2020. However, the number of counties with the highest supply remained static.

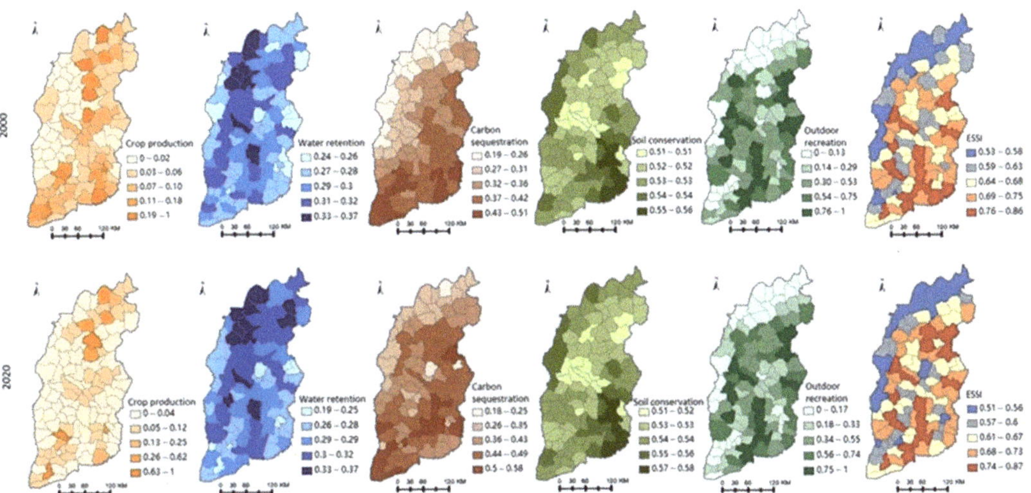

Figure 3. Spatial distributions of the key five ESs and ESSI across the 107 counties of Shanxi Province in 2000 and 2020.

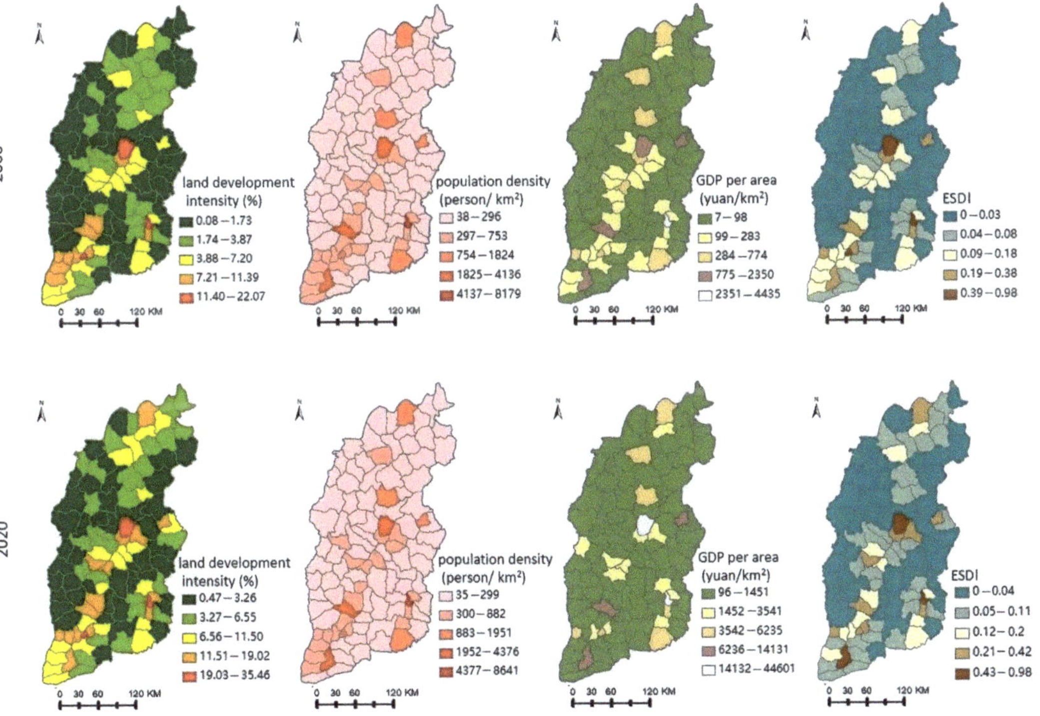

Figure 4. Spatial distributions of ES demand and the ESDI across the 107 counties of Shanxi Province in 2000 and 2020.

The spatial distribution hierarchy of ESDI was weaker compared to ESSI (Figure 4). In both 2000 and 2020, ESDI displayed spatial distribution characteristics with higher values in the central areas and lower values in the outer areas of the study area, which closely resemble the spatial distribution of construction land. The areas with the highest and higher ESS grades were primarily concentrated in central Shanxi, southeastern areas, and the northern plains, indicating a relatively concentrated distribution. The areas with the lowest demand were contiguous and distributed in mountainous regions such as Mount Luliang in the west and Mount Taihang in the east. From 2000 to 2020, due to population growth and economic development in the study area, there was a clear increase in ESDI. The number of counties with higher grades of ESS increased, while the counties with the lowest demand decreased. Some areas with medium demand in 2000 shifted to higher demand categories by 2020, indicating a transformation from relatively lower to higher grades. Overall, the spatial distribution of ESDI displayed noticeable differences between the outskirts and the middle regions of the study area.

3.2. Coupling Coordination Characteristics of ES Supply and Demand

Using the CCDM, we measured and mapped the CCD of ESSI and ESDI for the 107 counties of Shanxi Province in 2000 and 2020 (Figure 5). The CCD values ranged from 0 to 0.57 in 2000 and from 0 to 0.43 in 2020 (Table 4), indicating a relatively low level of coupling coordination.

There are four main types of coupling coordination relationships between ESSI and ESDI: extreme incoordination, moderate incoordination, reluctant coordination, and moderate coordination (Figure 5). Most counties belong to the extreme incoordination and moderate incoordination patterns, mainly located in the western and eastern parts of Shanxi. In 2000, the areas with reluctant coordination were primarily found in the Taiyuan Basin in central Shanxi, Linfen Basin, and the Yuncheng Basin in the southeastern parts. However, in 2020, reluctant coordination areas were sparsely distributed in only a few municipalities. Overall, there was a clear incoordination relationship between ES supply and demand in Shanxi.

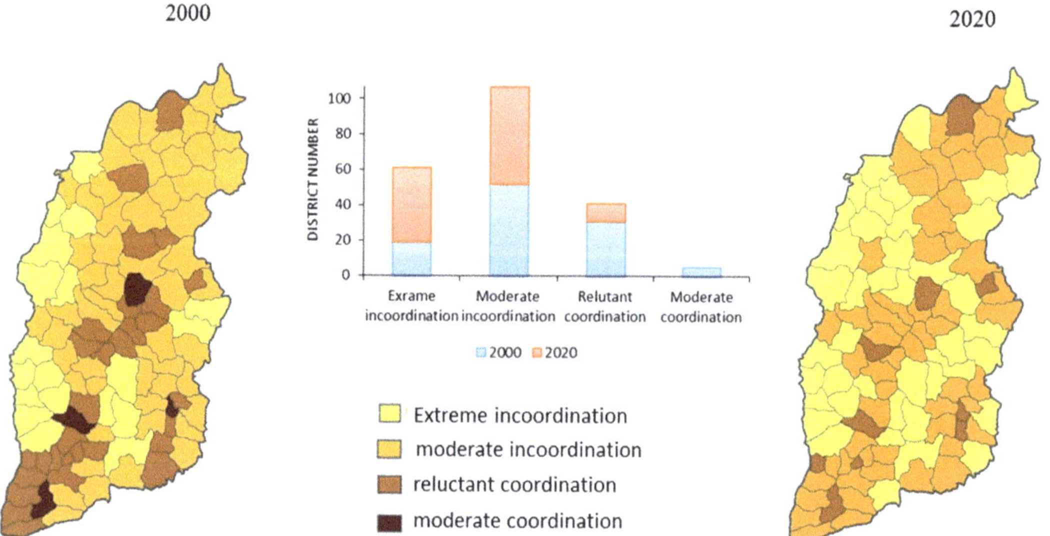

Figure 5. The coupling coordination degree between ES supply and demand of Shanxi Province in 2000 and 2020.

Table 4. The statistical values of standardized values of ES supply, ES demand, and coupling coordination degree in 2000 and 2020.

	ESSI		ESDI		CCD	
Year	2000	2020	2000	2020	2000	2020
Minimum value	0.528	0.512	0.649	0.638	0.002	0.059
Maximum value	0.859	0.870	0.979	0.980	0.566	0.427
Mean value	0.670	0.649	0.044	0.067	0.212	0.196

Between 2000 and 2020, the number of counties with extreme incoordination significantly increased from 29.0% to 36.5%. The number of reluctant coordination areas decreased from 34.6% to 24.3%, with fourteen counties transitioning from reluctant coordination to moderate incoordination. This indicated a substantial decline in the coupling coordination between ESSI and ESDI in Shanxi over the 20-year period. The main change characteristics were the negative transitions from relatively high coordination grades to incoordination grades. The most significant changes were observed in the shifts from moderate incoordination to extreme incoordination and from reluctant coordination to moderate incoordination, accounting for 20.31% and 16.28% of the counties, respectively.

In general, the analysis of coordination between ES supply and demand revealed coexisting states of coordination and incoordination, with incoordination being predominant in most counties. The difference in the spatial polarization of CCD in 2020 was more significant than in 2000. Overall, the coordination relationship between ES supply and demand in Shanxi Province deteriorated between 2000 and 2020.

3.3. Determining Drivers for the Coupling Coordination Degree between ES Supply and Demand

The GDM was utilized to identify the most influential socio-ecological drivers for the coupling coordination between ES supply and demand. The factor detection results for thirteen socio-ecological variables yielded the influencing coefficients (q-statistic values) and significance levels (p-values) (Figure 6).

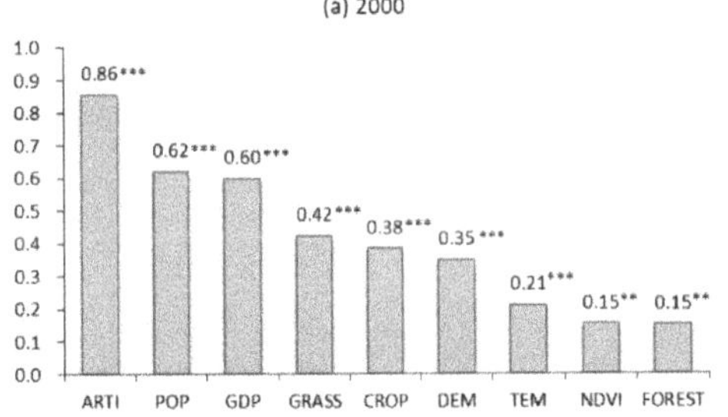

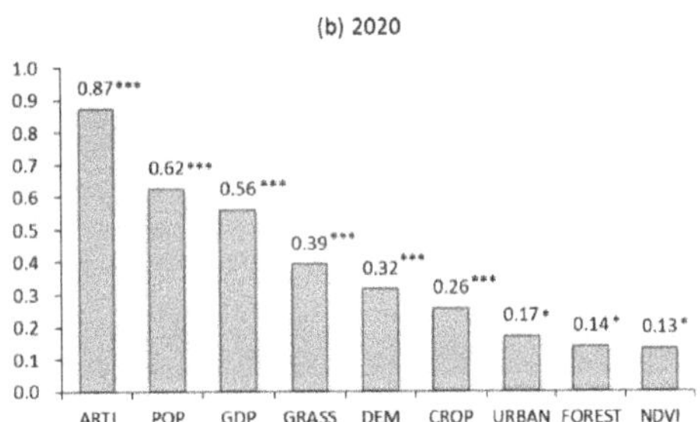

Figure 6. Factor-detected results of socio-ecological variables of CCD using GDM in 2000 (**a**) and 2020 (**b**). "***" $p < 0.001$, "**" $p < 0.01$, "*" $p < 0.05$. Panels (**a**) and (**b**) display the prominent influencing factors of CCD in 2000 and 2020, respectively. All factors are organized in descending order based on their influencing coefficients (q-statistic values).

In 2000 and 2020, the variables SLOPE, PRE, and COUNTY did not pass the significance tests. Meanwhile, the variable TEM was a statistically significant driving variable (p-value < 0.001) in 2000, but its significance level was greater than 0.1 in 2020. Conversely, the variable URBAN was not significant in 2000 (p-value > 0.1) but became a significant driving variable in 2020 (p-value < 0.05). Figure 6 presents the sorting results for the q-statistic values for significant variables (p-value ≤ 0.05), revealing their influencing coefficients on CCD in 2000 and 2020. The influencing coefficients of CON, POP, and GDP were greater than 0.5 and higher than the other factors in both 2000 and 2020. This demonstrates that these variables were the main drivers of the spatial pattern of CCD during both time periods. GRASS, CROP, and DEM were considered sub-high determinate variables based on their q-statistic values. Notably, the variables TEM, NDVI, and FOREST had weak effects on CCD in 2000, whereas in 2020, the variables URBAN, FOREST, and NDVI had a weak effect. This suggests that the impact of forest coverage on the coordination relationship was relatively small in the study area.

Furthermore, the interaction detector module of GDM assessed the influencing coefficients of any two socio-ecological variables (Figure 7) and compared them with their separate influencing coefficients (q-statistic values). The results reveal two interaction modes of socio-ecological variables on CCD: nonlinear enhancement and mutual enhancement. This indicates that the explanatory power of the interaction between any two variables for CCD is greater than that of any single variable. The interactive effects on CCD between CON and the other variables were the strongest, with q-statistic values exceeding 0.85. After interacting with POP and GDP in both 2000 and 2020, the q-statistic values of all variables were above 0.65, which was higher than the values for the separate effects of POP and GDP on CCD. It is important to note that SLOPE, PRE, and COUNTY did not exhibit significant effects on CCD in the single-factor detection results (Figure 6). However, after interacting with CON, POP, and GDP, the influencing coefficients of SLOPE, PRE, and COUNTY were 0.88, 0.67, and 0.68 in 2000, respectively, and similar results were observed in 2020. This suggests that even if individual socio-economic variables do not have a significant effect on the spatial distribution of CCD, they may play a key role via interaction with variables that have high influencing coefficients. Thus, SLOPE, PRE, and COUNTY were identified as important external driving factors for CCD between ESSI and ESDI. Additionally, this highlights the importance of considering the interactions between variables in understanding the spatial distribution of CCD.

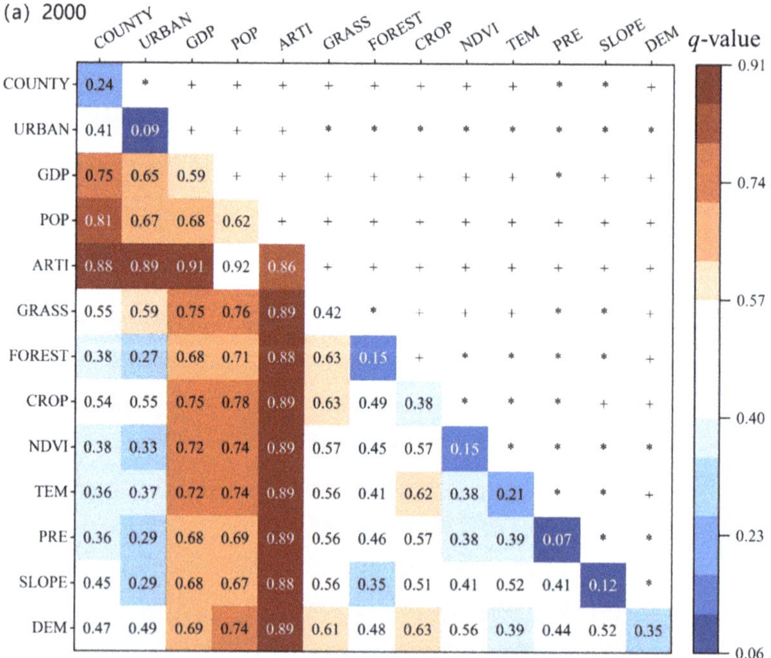

Figure 7. *Cont.*

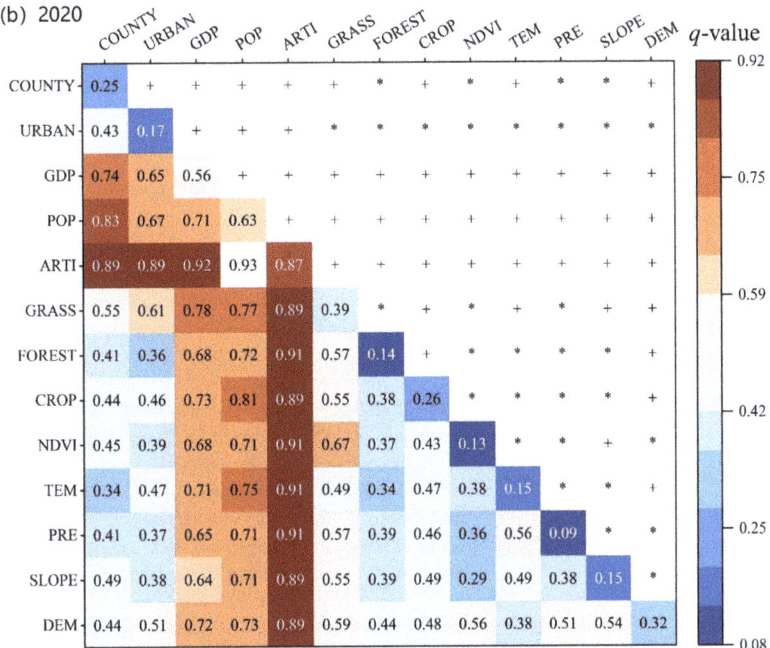

Figure 7. Interaction-detected results of socio-ecological variables of CCD using GDM for 2000 (**a**) and 2020 (**b**). "*" indicates nonlinear enhancement: $q (X1 \cap X2) > q (D1) + q (D2)$; "+" indicates bivariate enhancement: $q (D1 \cap D2) > q (D1)$ and $q (D2)$. A deeper shade of blue indicates a smaller interaction coefficient between X1 and X2 on CCD, and a darker shade of red signifies a larger interaction coefficient.

4. Discussion

4.1. Spatial–Temporal Characteristics of ESSI and ESDI

Our framework of analysis was used to make sense of relationships between ES supply (ESS) and ES demand (ESD) in a complex social–ecological system. Utilizing the ES supply index (ESSI), we have derived the spatiotemporal variation characteristics of ESS (Figure 3). These characteristics provide valuable insights into assessing the level of ES supply, as well as tracking its enhancement or decline in every county within Shanxi Province.

We found that high values of ESSI were predominantly observed in mountainous areas with dense forest coverage. In contrast, low values were mainly found in the Fen River basin, which is dominated by cropland and stretches from northeast to southwest. The Fen River basin is the most active area in terms of social and economic activities in Shanxi Province. The extensive human disturbance in this basin, characterized by high population density and economic growth, has led to the rapid expansion of construction land, causing a reduction in ecological land. As a result, the supply of ESs in the counties within this watershed is relatively low. Similarly, the northwestern parts of the study area, dominated by cropland and grassland, also exhibited relatively low ESSI values. This aligns with findings from previous studies [76]. This region represents a transition zone from scrub steppe to typical steppe, and factors such as increasing population, livestock, and desertification posed significant threats to the area for an extended period. Despite the implementation of the GFGP in these northwestern counties of Shanxi Province since 2000, the supply of local ESs continues to deteriorate [77]. Hence, the grass-planting-based GFGP in this region is required to improve efficiency and achieve ecological protection. Additionally, the western area of Mount Luliang exhibited relatively low ESSI values.

This can be attributed to Mount Luliang's location in the loess hilly and gully region, which is characterized by severe soil and water loss [78]. Moreover, these mountainous areas are contiguous to regions with concentrated mineral resources and frequent human activities. As of 2021, there are 91 coal mines in Luliang, accounting for 13.62% of the total number of coal mines in Shanxi. Extensive mining methods and inadequate management measures have imposed a burden on ecosystems, leading to a diminished supply level of regional ESs.

Concerning the ES demand index (ESDI), we carefully curated three indicators representing land use, population dynamics, and economic factors (Figure 4). These indicators collectively offer a comprehensive representation of ESD, encompassing preferences and requirements. We found that high values of ESDI were primarily distributed in the central Taiyuan Basin, northern Datong Basin, southern Linfen Basin and Yuncheng Basin, and southeastern Changzhi Basin, while low values were observed in the mountainous areas on the east and west sides. These basin regions, with Taiyuan, Datong, Linfen, Yuncheng, and Changzhi as their central cities, have experienced significant population and industrial concentration, resulting in high demand for ESs. Notably, the spatial pattern of ESDI closely resembles that of the degree of land use development (Figure 3). This finding is consistent with existing studies [79] that indicate a strong correlation between ESDI distribution and land use development degree, population density, and per capita GDP. It is a common phenomenon in many parts of China to expand the area of construction land in order to meet social and economic needs, particularly in areas with extensive human activity [16]. From 2000 to 2020, the average ESDI values significantly increased in the Datong Basin, decreased in the Taiyuan Basin and Yuncheng Basin, and the disparity in ESD values among the counties within these basins significantly decreased. Additionally, ESDI levels in Taiyuan city and Changzhi city district, which had the highest ESDI values in the whole study area, remained relatively stable between 2000 and 2020. This stability can be attributed to the slower economic and population growth in Taiyuan and Changzhi, reaching a stable socio-economic agglomeration state within Shanxi Province [16]. Overall, Shanxi Province exhibits clear spatial mismatch characteristics between ESS and ESD.

4.2. Spatial–Temporal Characteristics of CCD between ESSI and ESDI

The coordination mechanism between ESS and ESD primarily revolves around the harmonization of the ecosystem and social system [80]. It involves a feedback loop, where societal choices, such as land use and economic activities, impact the provision of ESs. In turn, the condition and health of the ecosystem shape the quality and quantity of services that can be supplied to meet social needs. Thus, it is rare to gain a clear spatial relationship between the biophysical supply of ESs and their demand; it is a rare achievement. This rarity primarily stems from the utilization of distinct measurement units for assessing supply and demand [78]. Furthermore, this situation underscores a current challenge within this field of study. In this study, we introduced the CCD model to address this challenge, and obtained the spatiotemporal variation characteristics of the coordination relationship between ESS and ESD in various counties of the Shanxi Province.

The majority of counties exhibited a state of incoordination in terms of CCD (Figure 5). Extreme incoordination was predominantly observed in Mount Taihang in the east, Mount Luliang in the west, and Mount Taiyue in the south of Shanxi Province. This can be attributed to the higher supply of the five key ESs compared to the demand in these counties (Figures 3 and 4). Most of these counties are situated in mountainous areas characterized by forest and grassland with high vegetation coverage [36]. Due to the implementation of ecological protection projects, there is minimal interference from human activities in these counties, resulting in little change in the type and quantity of land use. Furthermore, NFPP and GFGP contribute to increased vegetation coverage in these mountainous areas, enhancing the types and capacities of ESs. As a result, a high level of "lock-in effect" of regional ESS has been achieved [35,76], and ESS in these mountains is significantly weaker compared to the plain and basin areas. On the other hand, moderate incoordination was primarily observed in the Datong Basin, Xinding Basin, Taiyuan Basin, Linfen Basin, Yuncheng Basin, and Changzhi Basin, spanning from the north to south of the study area. This can be attributed to the extensive human activity in these basin regions and indicates the need for additional efforts to improve the coordination between ESS and ESD in these areas.

The coordination relationship between ESSI and ESDI exhibited a decrease in incoordination between 2000 and 2020, as evidenced by an increase in the number of extreme incoordination counties and a decrease in reluctant coordination and moderate coordination counties. For example, the relationship between ESSI and ESDI in Taiyuan, Linfen, Yuncheng, and Changzhi cities shifted from moderate coordination in 2000 to reluctant coordination in 2020. No other counties changed to a moderate coordination, resulting in the absence of moderate coordination counties/districts in the study area in 2020. This shift can be attributed to the slower growth rate of ESS compared to ESD in these cities. The increase in population, expansion of construction land, and economic growth have led to higher ESD in these districts [81]. Despite the increase in green spaces within the cities between 2000 and 2020 due to China's ecological civilization construction projects, it falls short of meeting the substantial demand for ES in urban production and daily life [82].

Overall, the distribution of CCD in 2020 exhibited noticeable spatial differences between the basin region and the mountain region compared to 2000. The coordination relationship between ESSI and ESDI demonstrates incoordination and spatially varies across Shanxi Province. The degree of incoordination intensified in nearly half of the counties from 2000 to 2020. In 2020, the CCD highlighted the spatial disparities between urban areas, agricultural areas, and forest–grassland areas from the perspective of land use, as well as the differences between valleys, basins, and mountains from the perspective of terrain.

4.3. Associations between CCD and Driving Covariates

In our study, we employed a geo-detector model (GDM) to capture the spatially response characteristics of the dependent variable in relation to the independent variable. This approach allowed us to analyze how different factors contribute to the observed spatial patterns of the coordination between ESS and ESD across the study area (Figures 6 and 7). We found that socio-economic factors had a greater impact on the coordination relationship between ESS and ESD than natural and ecological factors, emphasizing the significance of socio-economic factors in shaping the spatial pattern of this relationship. This finding aligns with previous research that highlighted the role of socio-economic variables as determinants of ES supply and demand distribution [43,83]. The dominant factors influencing the spatial pattern of CCD were identified as construction land, followed by population and GDP. The expansion of construction land, driven by extensive human activity, has significantly influenced land use patterns in terms of magnitude, type, and distribution [84]. The complex nature of human activity further complicates these relationships. However, these findings contrast with a study by Yang et al. [82] in China's Loess Plateau, which found that vegetation cover had the greatest positive effect on the relationship between ES supply and demand. This discrepancy could be attributed to the dominant influence of vegetation coverage on both ESS and ESD in the Loess Plateau.

Interestingly, the effect of grassland on CCD was second only to socio-economic factors, ranking below elevation and cropland. Grassland emerged as the vegetation cover factor with the strongest impact on CCD, indicating its significant role in shaping the spatial pattern of the coordination relationship between ESS and ESD in Shanxi Province. This finding is supported by previous studies that emphasized the influence of land use changes caused by socio-economic factors on ESs [85]. Notably, the effect of forestland and NDVI on CCD was much smaller compared to grassland, suggesting that the increase in grassland area through GFGP mainly influenced the relationship between ES supply and demand in Shanxi Province. According to data from the 2020 "Shanxi Province Third Land Survey Main Data Bulletin", grassland covers an area of 3.11 million hectares in

Shanxi Province, and is primarily distributed in Datong, Xinzhou, Luliang, Jinzhong, and Linfen cities, accounting for 73% of the province's grassland [86]. However, the grassland ecosystem in the study area remains fragile, with 70% of the grassland experiencing varying degrees of degradation and facing challenges such as insufficient protection and restoration, low utilization and management efficiency, and a lack of effective technological support. Therefore, the coordination degree between ESS and ESD could be improved by implementing quantitative and spatial adjustments to grassland planting policies.

In this study, the slope and precipitation factors were found to have no significant direct effect on CCD. However, after interacting with construction land, population, and GDP, slope and precipitation played an important role in shaping the spatial pattern of CCD. This indicates that socio-economic factors enhance the influence of natural factors on CCD. Generally, slope and precipitation influence the supply capacity and demand preferences for ESs by controlling the spatial distribution of human activity and landscapes [76], and previous studies demonstrated that precipitation facilitates coordination between ESS and ESD in arid and semi-arid areas [79]. The changes in terrain and precipitation affect both social and economic processes, leading to changes in coordination relationships. Previous research provided guidance on identifying factors that contribute to ES supply and demand mismatch [87], ESS, and ESD, including terrain ruggedness for supply and population density for demand [88].

The spatial distribution of the coordination relationship between ESS and ESD is primarily supported by the expansion of construction land, population growth, and economic benefits. However, this puts immense pressure on ES supply. Although forest and grassland coverage significantly increased between 2000 and 2020, the expansion of construction land has outpaced these gains. The growth rates of forestland and grassland in the study area were 4.2% and 4.3%, respectively, while the growth rate of construction land was as high as 105.4%. Since 2000, Shanxi Province has been the subject of economic system reforms, and thus rapid economic development, but the lack of awareness regarding ecological protection has resulted in the degradation of ESs. Consequently, the coordination relationship between ES supply and demand deteriorated somewhat between 2000 and 2020, indicating that the expansion of construction land and the concentration of population and industry have threatened the coordination between ESS and ESD in the counties of Shanxi. To address this issue, policymakers in Shanxi Province must make significant progress in promoting the coordination between ESS and ESD, implementing measures to ensure a dynamic balance between ES supply and demand.

4.4. Limitations

There are several limitations and uncertainties in this study. Firstly, the evaluation of five types of ES was limited in terms of reflecting the overall ESS level due to data availability and quality constraints [89]. In addition, the equal-weight superposition calculation method used to calculate total ESS may have overlooked significance of various ES types in Shanxi Province. Future studies should strive to include a more comprehensive range of ES indicators and consider their relative importance. Secondly, the selection of indicators to represent ESD focused on land development intensity, population density, and GDP per area, assuming that all types of ESs have the same demand. This oversimplification may not accurately reflect spatiotemporal changes in ESD. Further research is needed to refine the measurement of ESD and capture its dynamics. Thirdly, this study compared the changes in ESS and ESD between 2000 and 2020, overlooking the temporal volatility of ES. Incorporating temporal dynamics would enhance our understanding of the coordination relationship between ES supply and demand.

The use of county-level administrative boundaries was advantageous [90], providing official statistical data; however, this limits our understanding of causality and spatial heterogeneity in coordination relationships [91]. Shanxi Province has diverse topography, vegetation and natural geographical features, and there is a significant spatial mismatch between ESS and demand. Thus, it is difficult to reveal the variation of ESS at a local scale in the spatial units of county. In addition, spatial heterogeneity and scale effects impact the relationship between ESS and ESD [92]. Focusing only on a single scale tends to miss information about the correlation between scales, and the influence mechanism is inevitably one-sided. Future research should use a multi-scale analysis to capture the correlation between scales and comprehensively investigate influence mechanisms.

Additionally, our study assumed that ESD in a county is solely provided by the local ecosystem without considering the flow of ecosystem services across county boundaries. For example, water resources can originate from upstream regions, and food shortages in a city can be mitigated via food trade [93]. Future studies should account for the cross-boundary flow of ES, as services from neighboring ecosystems can also contribute to ESD. Furthermore, while GDM helped identify the strength of influence of socio-ecological factors on the spatial pattern of CCD, it was unable to capture whether this effect was positive or negative. Exploring the positive or negative nature of these effects would provide a more comprehensive understanding of the coordination relationship between ESS and ESD.

5. Conclusions

Shanxi Province has experienced a stage of rapid expansion of construction land; there has been a rapid transformation in the intensity, type, and pattern of land use, which has created an urgent need to optimize ESs, social progress, and economic development. Under this background, this study analyzed spatiotemporal changes in ES supply, demand, and their coordination relationship, and identified the relevant socio-ecological driving factors across 107 counties in Shanxi from 2000 to 2020.

The results reveal that the spatial pattern of ESS was closely linked to forest coverage, while ESD was closely related to the degree of land use development. The changes in ESS and ESD exhibited spatial heterogeneity. Over the study period, Shanxi Province experienced a slight decrease in ESS and an increase in ESD. The evaluation using the CCDM demonstrated significant incoordination between ESS and ESD in Shanxi, which worsened between 2000 and 2020. Based on the CCD of ESS and ESD, Shanxi Province was divided into four zones: extreme incoordination, moderate incoordination, reluctant coordination, and moderate coordination. The extreme incoordination zone was mainly located in mountainous regions, where ecosystems dominated by the eastern Taihang Mountain and western Luliang Mountain provided high levels of ESs but low ESD. The moderate incoordination zone was primarily found in basins with intensive human activity, where ecosystems dominated by cropland exhibited slightly higher levels of ESs compared to ESD. The reluctant coordination and moderate coordination zones were mainly located in central cities within basins, where the CCD between ESS and ESD significantly decreased. The spatial distribution of CCD was primarily influenced by construction land, population, and GDP, with grassland playing a secondary role, largely driven by the GFGP policy.

Accordingly, in order to promote the sustainable development of the counties, we propose the following recommendations: Firstly, the government should enhance ecological compensation policies for residents in mountainous regions, with a special focus on areas like Taihang Mountain and Luliang Mountain. This entails raising compensation for ecosystem service providers. Secondly, county and municipal district administrations should adopt advanced strategies for balanced ecological, social, and economic development. This will bolster land use efficiency and curb haphazard expansion of construction projects. Thirdly, policymakers and governments need to comprehensively assess the distinct impacts of various drivers on the interplay between ESS and ESD at both the local and county scales. This is particularly pertinent when formulating strategies for regional management approaches. These policy insights are applicable not just to Shanxi Province, but also to other regions endowed with coal and mineral resources. In future research, it is imperative to prioritize effective land management decisions that account for the interrelation between socio-ecological factors and the supply and demand of ecosystem services.

Author Contributions: Y.Z. and B.L. designed the study, collected data and conducted the research. B.L. analyzed data and participated in data processing and model calculation. Y.Z. wrote the first draft, and B.L. revised and edited the first draft. R.S. collected and processed the data. The remaining authors contributed to the discussion of results and revised the manuscript. All authors have read and agreed to the published version of the manuscript.

Funding: This research was funded by the Shanxi Province Basic Research Plan Project under grant No. 202203021212496, and the Science and Technology Innovation Project of University in Shanxi Province under grant No. 2020L0248. We thank the academic editors and anonymous reviewers for their kind suggestions and valuable comments.

Institutional Review Board Statement: Not applicable.

Informed Consent Statement: Not applicable.

Data Availability Statement: Not applicable.

Conflicts of Interest: The authors declare no conflict of interest.

References

1. Costanza, R.; D'Arge, R.; Groot, R.D.; Farber, S.; Grasso, M.; Hannon, B.; Limburg, K.; Naeem, S.; O'Neill, R.V.; Paruelo, J.; et al. The value of the world's ecosystem services and natural capital. *Ecol. Econ.* **1997**, *25*, 3–15. [CrossRef]
2. MA (Millennium Ecosystem Assessment). Overview of the Millennium Ecosystem Assessment. 2005. Available online: http://www.millenniumassessment.org/en/About.html# (accessed on 7 July 2017).
3. Chen, W.X.; Chi, G.Q.; Li, J.F. 2019. The spatial association of ecosystem services with land use and land cover change at the County level in China, 1995–2015. *Sci. Total Environ.* **2019**, *669*, 459–470. [CrossRef]
4. Sutton, P.C.; Costanza, R. Global estimates of market and non-market values derived from nighttime satellite imagery, land cover, and ecosystem service valuation. *Ecol. Econ.* **2002**, *41*, 509–527. [CrossRef]
5. Metzger, M.J.; Schröter, D. Towards a spatially explicit and quantitative vulnerability assessment of environmental change in Europe. *Reg. Environ. Change* **2006**, *6*, 201–216. [CrossRef]
6. Zagonari, F. Using ecosystem services in decision-making to support sustainable development: Critiques, model development, a case study, and perspectives. *Sci. Total Environ.* **2016**, *548–549*, 25–32. [CrossRef] [PubMed]
7. Cortinovis, C.; Geneletti, D. A performance-based planning approach integrating supply and demand of urban ecosystem services. *Landsc. Urban Plan.* **2020**, *201*, 103842. [CrossRef]
8. De Groot, R.S.; Alkemade, R.; Braat, L.; Hein, L.; Willemen, L. Challenges in integrating the concept of ecosystem services and values in landscape planning, management and decision making. *Ecol. Complex.* **2010**, *7*, 260–272. [CrossRef]
9. Bennett, E.M.; Peterson, G.D.; Gordon, L.J. Understanding relationships among multiple ecosystem services. *Ecol. Lett.* **2010**, *12*, 1394–1404. [CrossRef]
10. Burkhard, B.; Kroll, F.; Nedkov, S.; Muller, F. Mapping ecosystem service supply, demand and budgets. *Ecol. Indic.* **2012**, *21*, 17–29. [CrossRef]
11. Villamagna, A.M.; Angermeier, P.L.; Bennett, E.M. Capacity, pressure, demand, and flow: A conceptual framework for analyzing ecosystem service provision and delivery. *Ecol. Complex.* **2013**, *15*, 114–121. [CrossRef]
12. Wolff, S.; Schulp, C.J.E.; Verburg, P.H. Mapping ecosystem services demand: A review of current research and future perspectives. *Ecol. Indic.* **2015**, *55*, 159–171. [CrossRef]
13. Castillo-Eguskitza, N.; Martín-López, B.; Onaindia, M. A comprehensive assessment of ecosystem services: Integrating supply, demand and interest in the Urdaibai Biosphere Reserve. *Ecol. Indic.* **2018**, *93*, 1176–1189. [CrossRef]
14. Fu, B.J.; Tao, T.; Liu, Y.X.; Zhao, W.W. New developments and perspectives in physical geography in China. *Chin. Geogr. Sci.* **2019**, *29*, 363–371. [CrossRef]
15. Lorilla, R.S.; Kalogirou, S.; Poirazidis, K.; Kefalas, G. Identifying spatial mismatches between the supply and demand of ecosystem services to achieve a sustainable management regime in the Ionian Islands (Western Greece). *Land Use Policy* **2019**, *88*, 104171. [CrossRef]
16. Wu, X.; Liu, S.; Zhao, S.; Hou, X.; Xu, J.; Dong, S.; Liu, G. Quantification and driving force analysis of ecosystem services supply, demand and balance in China. *Sci. Total Environ.* **2019**, *652*, 1375–1386. [CrossRef]
17. Quintas-Soriano, C.; Castro, A.J.; García-Llorente, M.; Cabello, J.; Castro, H. From supply to social demand: A landscape-scale analysis of the water regulation service. *Landsc. Ecol.* **2014**, *29*, 1069–1082. [CrossRef]
18. Stürck, J.; Poortinga, A.; Verburg, P.H. Mapping ecosystem services: The supply and demand of flood regulation services in Europe. *Ecol. Indic.* **2014**, *38*, 198–211. [CrossRef]
19. Baró, F.; Haase, D.; Gómez-Baggethun, E.; Frantzeskaki, N. Mismatches between ecosystem services supply and demand in urban areas: A quantitative assessment in five European cities. *Ecol. Indic.* **2015**, *55*, 146–158. [CrossRef]
20. Castro, A.J.; Verburg, P.H.; Martín-López, B.; Garcia-Llorente, M.; Cabello, J.; Vaughn, C.C.; López, E. Ecosystem service trade-offs from supply to social demand: A landscape-scale spatial analysis. *Landsc. Urban Plan.* **2014**, *132*, 102–110. [CrossRef]

21. Blanco, V.; Holzhauer, S.; Brown, C.; Lagergren, F.; Vulturius, J.; Lindeskog, M.; Rounsevell, M.D.A. The effect of forest owner decision-making, climatic change and societal demands on land-use change and ecosystem service provision in Sweden. *Ecosyst. Serv.* **2017**, *23*, 174–208. [CrossRef]
22. Schulp, C.J.E.; Lautenbach, S.; Verburg, P.H. Quantifying and mapping ecosystem services: Demand and supply of pollination in the European Union. *Ecol. Indic.* **2014**, *36*, 131–141. [CrossRef]
23. Larondelle, N.; Lauf, S. Balancing demand and supply of multiple urban ecosystem services on different spatial scales. *Ecosyst. Serv.* **2016**, *22*, 18–31. [CrossRef]
24. Turpie, J.K.; Forsythe, K.J.; Knowles, A.; Blignaut, J.; Letley, G. Mapping and valuation of South Africa's ecosystem services: A local perspective. *Ecosyst. Serv.* **2017**, *27*, 179–192. [CrossRef]
25. Hitzhusen, F.J. *Economic Valuation of River Systems*; Edward Elgar: Cheltenham, UK, 2007.
26. Peng, J.; Yang, Y.; Xie, P.; Liu, Y.X. Zoning for the construction of green space ecological networks in Guangdong Province based on the supply and demand of ecosystem services. *Acta Ecol. Sin.* **2017**, *37*, 4562–4572.
27. Nedkov, S.; Burkhard, B. Flood regulating ecosystem services—Mapping supply and demand, in the Etropole municipality, Bulgaria. *Ecol. Indic.* **2012**, *21*, 67–79. [CrossRef]
28. Peña, L.; Casado-Arzuaga, I.; Onaindia, M. Mapping recreation supply and demand using an ecological and a social evaluation approach. *Ecosyst. Serv.* **2015**, *13*, 108–118. [CrossRef]
29. Wang, J.; Zhai, T.; Lin, Y.F.; Kong, X.; He, T. Spatial imbalance and changes in supply and demand of ecosystem services in China. *Sci. Total Environ.* **2019**, *657*, 781–791. [CrossRef]
30. Wei, H.; Fan, W.; Wang, X.; Lu, N.; Dong, X.; Zhao, Y.; Zhao, Y. Integrating supply and social demand in ecosystem services assessment: A review. *Ecosyst. Serv.* **2017**, *25*, 15–27. [CrossRef]
31. Xing, L.; Xue, M.; Hu, M. Dynamic simulation and assessment of the coupling coordination degree of the economy–resource– environment system: Case of Wuhan City in China. *Environ. Manag.* **2019**, *230*, 474–487. [CrossRef]
32. Li, W.; Wang, Y.; Xie, S.; Cheng, X. Coupling coordination analysis and spatiotemporal heterogeneity between urbanization and ecosystem health in Chongqing municipality, China. *Sci. Total Environ.* **2021**, *791*, 148311. [CrossRef]
33. Liu, W.; Zhan, J.; Zhao, F.; Wei, X.; Zhang, F. Exploring the coupling relationship between urbanization and energy eco-efficiency: A case study of 281 prefecture-level cities in China. *Sustain. Cities Soc.* **2020**, *64*, 102563. [CrossRef]
34. Guan, Q.; Hao, J.; Ren, G.; Li, M.u.; Chen, A.; Duan, W.; Chen, H. Ecological indexes for the analysis of the spatial–temporal characteristics of ecosystem service supply and demand: A case study of the major grain producing regions in Quzhou, China. *Ecol. Indic.* **2020**, *108*, 105748. [CrossRef]
35. Li, P.; Liu, C.; Liu, L.; Wang, W. Dynamic Analysis of Supply and Demand Coupling of Ecosystem Services in Loess Hilly Region: A Case Study of Lanzhou, China. *Chin. Geogr. Sci.* **2021**, *31*, 276–296. [CrossRef]
36. Yang, Z.; Zhan, J.; Wang, C.; Twumasi-Ankrah, M.J. Coupling coordination analysis and spatiotemporal heterogeneity between sustainable development and ecosystem services in Shanxi Province, China. *Sci. Total Environ.* **2022**, *836*, 155625. [CrossRef]
37. Huang, M.D.; Xiao, Y.; Xu, J.; Liu, J.Y.; Wang, Y.Y.; Gan, S.; Lv, S.X.; Xie, G.D. A Review on the Supply-Demand Relationship and Spatial Flows of Ecosystem Services. *J. Resour. Ecol.* **2022**, *13*, 925–935.
38. Sun, X.; Tang, H.; Yang, P.; Hu, G.; Liu, Z.; Wu, J. Spatiotemporal patterns and drivers of ecosystem service supply and demand across the conterminous United States: A multiscale analysis. *Sci. Total Environ.* **2020**, *703*, 135005. [CrossRef]
39. Peng, J.; Tian, L.; Liu, Y.X.; Zhao, M.Y.; Hu, Y.N.; Wu, J.S. Ecosystem services response to urbanization in metropolitan areas: Thresholds identification. *Sci. Total Environ.* **2017**, *607-608*, 706–714. [CrossRef]
40. Wang, L.; Gong, J.; Ma, S.; Wu, S.; Zhang, X.; Jiang, J. Ecosystem service supply–demand and socioecological drivers at different spatial scales in Zhejiang Province, China. *Ecol. Indic.* **2022**, *140*, 109058. [CrossRef]
41. Goldenberg, R.; Kalantari, Z.; Cvetkovic, V.; Mörtberg, U.; Deal, B.; Destouni, G. Distinction, quantification and mapping of potential and realized supplydemand of flow-dependent ecosystem services. *Sci. Total Environ.* **2017**, *593–594*, 599–609. [CrossRef] [PubMed]
42. Wilkerson, M.L.; Mitchell, M.G.E.; Shanahan, D.; Wilson, K.A.; Ives, C.D.; Lovelock, C.E.; Rhodes, J.R. The role of socio-economic factors in planning and managing urban ecosystem services. *Ecosyst. Serv.* **2018**, *31*, 102–110. [CrossRef]
43. Wang, S.F.; Zhuang, Y.N.; Cao, Y.G.; Yang, K. Ecosystem Service Assessment and Sensitivity Analysis of a Typical Mine–Agriculture–Urban Compound Area in North Shanxi, China. *Land* **2022**, *11*, 1378. [CrossRef]
44. Xu, M.J.; Feng, Q.; Zhang, S.R.; Lv, M.; Duan, B.L. Ecosystem Services Supply–Demand Matching and Its Driving Factors: A Case Study of the Shanxi Section of the Yellow River Basin, China. *Sustainability.* **2023**, *15*, 11016. [CrossRef]
45. Pan, H.H.; Wang, J.Q.; Du, Z.Q.; Wu, Z.T.; Zhang, H.; Ma, K.M. Spatiotemporal evolution of ecosystem services and its potential drivers in coalfields of Shanxi Province, China. *Ecol. Indic.* **2023**, *148*, 110109. [CrossRef]
46. Hu, B.A.; Kang, F.F.; Han, H.R.; Cheng, X.Q.; Li, Z.Z. Exploring drivers of ecosystem services variation from a geospatial perspective: Insights from China's Shanxi Province. *Ecol. Indic.* **2021**, *131*, 108188. [CrossRef]
47. Wang, J.F.; Li, Y.; Wang, S.; Li, Q.; Liu, X.L. Assessment of Multiple Ecosystem Services and Ecological Security Pattern in Shanxi Province, China. *Int. J. Environ. Res. Public Health* **2023**, *20*, 4819. [CrossRef]
48. National Bureau of Statistics of China. *China Statistical Yearbook (2022)*; China Statistics Press: Beijing, China, 2023.
49. Guo, S.; Ma, Y. Comprehensive evaluation for sustainable development capacity of resource-based region. *Chin. Popul. Resour. Environ.* **2017**, *27*, 72–79.

50. Yang, G.; Ge, Y.; Xue, H.; Yang, W.; Shi, Y.; Peng, C.; Du, Y.; Fan, X.; Ren, Y.; Chang, J. Using ecosystem service bundles to detect trade-offs and synergies across urban–rural complexes. *Landsc. Urban Plan.* **2015**, *136*, 110–121. [CrossRef]

51. Wischmeier, W.H.; Smith, D.D. *Predicting Rainfall Erosion Losses—A Guide to Conservation Planning*; Agricultural Handbook No. 537; US Department of Agriculture Science and Education Administation: Washington, DC, USA, 1978.

52. Peng, J.; Chen, X.; Liu, Y.X.; Lu, H.L.; Hu, X.X. Spatial identification of multifunctional landscapes and associated influencing factors in the Beijing-Tianjin-Hebei region, China. *Appl. Geogr.* **2016**, *74*, 170–181. [CrossRef]

53. Lyu, R.; Clarke, K.; Zhang, J.M.; Feng, J.L.; Jia, X.H.; Li, J.J. Spatial correlations among ecosystem services and their socio-ecological driving factors_ A case study in the city belt along the Yellow River in Ningxia, China. *Appl. Geogr.* **2019**, *108*, 64–73. [CrossRef]

54. Potter, C.S.; Randerson, J.T.; Field, C.B.; Matson, P.A.; Mooney, H.A.; Klooster, S.A. Terrestrial ecosystem production: A process model based on global satellite and surface data. *Glob. Biogeochem. Cycles* **1993**, *7*, 811–841. [CrossRef]

55. Crabtree, R.; Potter, C.; Mullen, R.; Sheldon, J.; Huang, S.; Harmsen, J.; Rodman, A.; Jean, C. A modeling and spatio-temporal analysis framework for monitoring environmental change using NPP as an ecosystem indicator. *Remote Sens. Environ.* **2009**, *113*, 1486–1496. [CrossRef]

56. Raudsepp-Hearne, C.; Peterson, G.D.; Bennett, E.M. Ecosystem service bundles for analyzing tradeoffs in diverse landscapes. *Proc. Natl. Acad. Sci. USA* **2010**, *107*, 5242–5247. [CrossRef] [PubMed]

57. Zhao, T.; Ouyang, Z.; Zheng, H.; Wang, X.; Hong, M. Forest ecosystem services and their valuation in China. *Nat. Resour.* **2004**, *19*, 480–491.

58. Williams, J.R. The erosion-productivity impact calculator (EPIC) model: A case history. *Philos. Trans. R. Soc. B* **1990**, *329*, 421–428.

59. Mccool, D.K.; Foster, G.R.; Mutchler, C.K.; Meyer, L.D. Revised Slope Length Factor for the Universal Soil Loss Equation. *Trans. ASAE* **1989**, *32*, 1571–1576. [CrossRef]

60. Liu, B.; Nearing, M.; Shi, P.; Jia, Z. Slope Length Effects on Soil Loss for Steep Slopes. *Soil. Sci. Soc. Am. J.* **2000**, *64*, 1759–1763. [CrossRef]

61. Cai, C.; Ding, S.; Shi, Z.; Huang, L.; Zhang, G. Study of applying USLE and geographical information system IDRISI to predict soil erosion in small watershed. *Soil Water Conserv.* **2000**, *14*, 19–24.

62. Kumar, A.; Devi, M.; Deshmukh, B. Integrated remote sensing and geographic information system based RUSLE modelling for estimation of soil loss in Western Himalaya, India. *Water Resour. Manag.* **2014**, *28*, 3307–3317. [CrossRef]

63. Mouchet, M.A.; Paracchini, M.L.; Schulp, C.J.E.; Stürck, J.; Verkerk, P.J.; Verburg, P.H.; Lavorel, S. Bundles of ecosystem(dis)services and multifunctionality across European landscapes. *Ecol. Indic.* **2017**, *73*, 23–28. [CrossRef]

64. Zhang, K.; Lü, Y.; Fu, B.; Yin, L.; Yu, D. The effects of vegetation coverage changes on ecosystem service and their threshold in the Loess Plateau. *Acta Geogr. Sinica* **2020**, *75*, 949–960.

65. Wang, M.; Bai, Z.; Dong, X. Land Consolidation Zoning in Shaanxi Province based on the Supply and Demand of Ecosystem Services. *China Land Sci.* **2018**, *32*, 73–80. [CrossRef]

66. Zhai, T.; Wang, J.; Jin, Z.; Qi, Y.; Fang, Y.; Liu, J. Did improvements of ecosystem services supply-demand imbalance change environmental spatial injustices? *Ecol. Indic.* **2020**, *111*, 106068. [CrossRef]

67. Ouyang, X.; Wang, Z.; Zhu, X. Construction of the ecological security pattern of urban agglomeration under the framework of supply and demand of ecosystem services using Bayesian network machine learning: Case study of the Changsha–Zhuzhou–Xiangtan urban agglomeration. China. *Sustainability* **2019**, *11*, 4616. [CrossRef]

68. Zhou, S.; Yang, L.; Gao, R.; Wang, X.; Gao, X.; Nie, W.; Xu, P.; Zhang, Q.; Wang, W. A comparison study of carbonaceous aerosols in a typical North China Plain urban atmosphere: Seasonal variability, sources and implications to haze formation. *Atmos. Envion.* **2017**, *149*, 95–103. [CrossRef]

69. Shi, T.; Yang, S.; Zhang, W.; Zhou, Q. Coupling coordination degree measurement and spatiotemporal heterogeneity between economic development and ecological environment—Empirical evidence from tropical and subtropical regions of China. *J. Clean. Prod.* **2020**, *244*, 118739. [CrossRef]

70. Fu, B.; Wang, S.; Liu, Y.u.; Liu, J.; Liang, W.; Miao, C. Hydrogeomorphic Ecosystem Responses to Natural and Anthropogenic Changes in the Loess Plateau of China. *Annu. Rev. Earth Planet. Sci.* **2017**, *45*, 223–243. [CrossRef]

71. Hu, M.; Li, Z.; Wang, Y.; Jiao, M.; Li, M.; Xia, B. Spatio-temporal changes in ecosystem service value in response to land-use/cover changes in the Pearl River Delta. *Resour. Conserv. Recycl.* **2019**, *149*, 106–114. [CrossRef]

72. Chen, W.; Chi, G.; Li, J. The spatial aspect of ecosystem services balance and its determinants. *Land Use Policy* **2020**, *90*, 104263. [CrossRef]

73. Wang, J.F.; Li, X.H.; Christakos, G.; Liao, Y.L.; Zhang, T.; Gu, X.; Zheng, X.Y. Geographical detectors-based health risk assessment and its application in the neural tube defects study of the Heshun region, China. *Int. J. Geogr. Inf. Sci.* **2010**, *24*, 107–127. [CrossRef]

74. Wang, J.; Zhang, T.; Fu, B. A measure of spatial stratified heterogeneity. *Ecol. Indic.* **2016**, *67*, 250–256. [CrossRef]

75. Zhang, Y.; Lu, X.; Liu, B.; Wu, D.; Fu, G.; Zhao, Y.; Sun, P. Spatial relationships between ecosystem services and socioecological drivers across a large-scale region: A case study in the Yellow River Basin. *Sci. Total Environ.* **2021**, *766*, 142480. [CrossRef]

76. Liu, H.; Tang, F.; Ding, Y.; Zhang, Y.; Guo, X.; Tan, J.; Cheng, Y. Temporal and spatial evolution characteristics of the coupling between county high-quality development and ecosystem services in Shanxi Province. *Arid. Zone Res.* **2022**, *39*, 1234–1245.

77. Ma, Y.; Qian, J.; Su, Z. Climate Change and Its Impact on Land Desertificatiin in Northwestern Shanxi Province. *Desert Res.* **2011**, *31*, 1585–1589.

78. Seppelt, R.; Dormann, C.F.; Eppink, F.V.; Lautenbach, S.; Schmid, S. A quantitative review of ecosystem service studies: Approaches, shortcomings and the road ahead. *Appl. Ecol.* **2011**, *48*, 630–636. [CrossRef]
79. Yang, M.; Zhao, X.; Wu, P.; Hu, P.; Gao, X. Quantification and spatially explicit driving forces of the incoordination between ecosystem service supply and social demand at a regional scale. *Ecol. Indic.* **2022**, *137*, 108764. [CrossRef]
80. Xin, R.; Skov-Petersen, H.; Zeng, J.; Zhou, J.; Li, K.; Hu, J.; Liu, X.; Kong, J.; Wang, Q. Identifying key areas of imbalanced supply and demand of ecosystem services at the urban agglomeration scale: A case study of the Fujian Delta in China. *Sci. Total Environ.* **2021**, *791*, 148173. [CrossRef]
81. Hou, W.; Zhou, W.; Li, J.; Li, C. Simulation of the potential impact of urban expansion on regional ecological corridors: A case study of Taiyuan, China. *Sustain. Cities Soc.* **2022**, *83*, 103933. [CrossRef]
82. Bryan, B.A.; Gao, L.; Ye, Y.; Sun, X.; Connor, J.D.; Crossman, N.D.; Stafford-Smith, M.; Wu, J.; He, C.; Yu, D.; et al. China's response to a national land-system sustainability emergency. *Nature* **2018**, *559*, 193–204. [CrossRef]
83. Ouyang, Z.; Zheng, H.; Xiao, Y.; Polasky, S.; Liu, J.; Xu, W.; Wang, Q.; Zhang, L.; Xiao, Y.; Rao, E.; et al. Improvements in ecosystem services from investments in natural capital. *Science* **2016**, *352*, 1455–1459. [CrossRef]
84. Schneider, A.; Mertes, C.M. Expansion and growth in Chinese cities, 1978–2010. *Environ. Res. Lett.* **2014**, *9*, 024008. [CrossRef]
85. Kim, J.H.; Jobbagy, E.G.; Jackson, R.B. Trade-offs in water and carbon ecosystem services with land-use changes in grasslands. *Ecol. Appl.* **2016**, *26*, 1633–1644. [CrossRef] [PubMed]
86. Shanxi Forestry and Grassland Bureau. Available online: http://lcj.shanxi.gov.cn/lcfm/lcgk/ (accessed on 16 June 2023).
87. Shen, J.; Li, S.; Wang, H.; Wu, S.; Liang, Z.; Zhang, Y.; Wei, F.; Li, S.; Ma, L.; Wang, Y.; et al. Understanding the spatial relationships and drivers of ecosystem service supply-demand mismatches towards spatially-targeted management of social-ecological system. *J. Clean. Prod.* **2023**, *406*, 136882. [CrossRef]
88. Schirpke, U.; Kohler, M.; Leitinger, G.; Fontana, V.; Tasser, E.; Tappeiner, U. Futureimpacts of changing land-use and climate on ecosystem services of mountain grassland and their resilience. *Ecosyst. Serv.* **2017**, *26*, 79–94. [CrossRef] [PubMed]
89. Liu, Y.; Bi, J.; Lv, J.S.; Ma, Z.W.; Wang, C. Spatial multi-scale relationships of ecosystem services: A case study using a geostatistical methodology. *Sci. Rep.* **2017**, *7*, 9486. [CrossRef]
90. Liu, Y.; Li, T.; Zhao, W.W.; Wang, S.; Fu, B. Landscape functional zoning at a county level based on ecosystem services bundle: Methods comparison and management indication. *J. Environ. Manag.* **2019**, *249*, 109315. [CrossRef]
91. Raudsepp-Hearne, C.; Peterson, G.D. Scale and ecosystem services: How do observation, management, and analysis shift with scale-lessons from Québec. *Ecol. Soc.* **2016**, *21*, 16. [CrossRef]
92. Cui, F.; Tang, H.; Zhang, Q.; Wang, B.; Dai, L. Integrating ecosystem services supply and demand into optimized management at different scales: A case study in Hulunbuir, China. *Ecosyst. Serv.* **2019**, *39*, 1200984. [CrossRef]
93. Ding, T.; Chen, J.; Fang, L.; Ji, J.; Fang, Z. Urban ecosystem services supply-demand assessment from the perspective of the water-energy-food nexus. *Sustain. Cities Soc.* **2023**, *90*, 104401. [CrossRef]

Article

A Real-Time BLE/PDR Integrated System by Using an Improved Robust Filter for Indoor Position

Shenglei Xu [1,2], Yunjia Wang [1,2,*], Meng Sun [2], Minghao Si [2] and Hongji Cao [2]

[1] Key Laboratory of Land Environment and Disaster Monitoring, MNR, China University of Mining and Technology, Xuzhou 221116, China; cumtxsl@cumt.edu.cn
[2] School of Environmental Science and Spatial Informatics, China University of Mining and Technology, Xuzhou 221116, China; msun@cumt.edu.cn (M.S.); hmsi@cumt.edu.cn (M.S.); hjcao@cumt.edu.cn (H.C.)
[*] Correspondence: wyj4139@cumt.edu.cn; Tel.: +86-132-252-31855

Abstract: Indoor position technologies have attracted the attention of many researchers. To provide a real-time indoor position system with high precision and stability is necessary under many circumstances. In a real-time position scenario, gross errors of the Bluetooth low energy (BLE) fingerprint method are more easily occurring and the heading angle of the pedestrian will drift without acceleration and magnetic field compensation. A real-time BLE/pedestrian dead-reckoning (PDR) integrated system by using an improved robust filter has been proposed. In the PDR method, the improved Mahony complementary filter based on the pedestrian motion states is adopted to estimate the heading angle reducing the drift error. Then, an improved robust filter is utilized to detect and restrain the gross error of the BLE fingerprint method. The robust filter detected the gross error at different granularity by constructing a robust vector changing the observation covariance matrix of the extended Kalman filter (EKF) adaptively when the application is running. Several experiments are conducted in the true position scenario. The mean position accuracy obtained by the proposed method in the experiment is 0.844 m and RMSE is 0.74 m. Compared with the classic EKF, these two values are increased by 38% and 18%, respectively. The results show that the improved filter can avoid the gross error in the BLE method and provide high precision and scalability in indoor position service.

Keywords: indoor position; robust filter; integrated system; BLE; PDR

Citation: Xu, S.; Wang, Y.; Sun, M.; Si, M.; Cao, H. A Real-Time BLE/PDR Integrated System by Using an Improved Robust Filter for Indoor Position. *Appl. Sci.* **2021**, *11*, 8170. https://doi.org/10.3390/app11178170

Academic Editor: Giovanni Petrone

Received: 18 June 2021
Accepted: 31 August 2021
Published: 3 September 2021

Publisher's Note: MDPI stays neutral with regard to jurisdictional claims in published maps and institutional affiliations.

1. Introduction

With the rapid development of technology and people's increasing demands for a better life, various applications based on location-based service (LBS) provide a great convenience for people's life. Position information is the key element for LBS to provide services. Global Navigation Satellite System (GNSS) can provide high accuracy position outdoors. However, the accuracy deteriorates significantly because GNSS signals are unreliable or blocked in indoor environments. To provide a reliable, stable position service in indoor environments, many types of indoor positioning technologies such as wireless fidelity (Wi-Fi) [1–3], Bluetooth low energy (BLE) beacons [4,5], radio frequency identification (RFID) [6,7], ultrasonic [8], infrared [9], ultra-wideband (UWB) [10,11], pseudolite [12,13], computer vision [14,15] had been proposed by experts and scholars.

Among them, the positioning techniques based on Wi-Fi are the most popular indoor positioning method due to the wide deployment of Wi-Fi routers in shopping malls, hospitals, airports, and stations. However, the process of Wi-Fi scanning is time-consuming and power-consuming. Many devices have optimized their systems to limit the scanning frequency. According to the latest google android development document [16], it is specified that the scanning frequency of Wi-Fi is limited no more than four times in two minutes after the Android Oreo system. It is difficult for the Wi-Fi-based positioning method to be applied to indoor position service applications with high real-time performance. Fortunately, Bluetooth-based techniques are another good choice to position due to the increased popularity of smartphones along with the development of the Internet of Things. Low cost, power-saving, and ease of deployment without affecting the infrastructures of buildings can avoid the shortcoming of Wi-Fi scanning [17]. Most importantly, the principle of the BLE positioning method is the same as that of Wi-Fi. Some of the current Wi-Fi positioning algorithms can be directly used for BLE positioning [18]. The fingerprint-based method of BLE is one of the popular approaches for the indoor position which can offer many advantages for the realizable position accuracy and is infrastructure-free. However, the radio signals of BLE can be affected by the multipath effect and device heterogeneity [19] which cause the signal to change during offline acquisition and online position, resulting in the error in the positioning process.

In short, all of these indoor position techniques mentioned above have both advantages and disadvantages. It is the focus of researchers to propose a fusion of indoor position methods to keep a balance among accuracy, coverage, cost, and complexity. Common fusion position methods include multimodal fingerprinting, triangulation-based fusion, and pedestrian dead reckoning (PDR)-based fusion [20]. The PDR-based fusion method which combines PDR with wireless localization methods is widely used in the literature. PDR is a self-positioning method that provides a relatively high accuracy position estimation based on the smartphone's built-in sensors, but it suffers from the drift problem, resulting in huge cumulative error for long-time positioning [21]. By contrast, wireless positioning method such as BLE fingerprint position can obtain absolute position without cumulative error but has poor accuracy position estimation. Fusing PDR with wireless localization methods which are often known as Bayes filter, Kalman Filter (KF), Extended Kalman filter (EKF), or particle filter (PF) can make up for both methods' shortcomings to provide high accuracy and stable position service. The PF has good performance in solving nonlinear problems but suffers from a high computational load. In our research, the fusion EKF algorithm was chosen to combine PDR with the BLE fingerprint position method in the real-time integrated system. To inhibit the outliers from the BLE fingerprint method, a robust filter based on EKF was proposed to compensate for the gross error in the real-time positioning procedure. Our contributions are as follows:

1. We found that the errors of the BLE fingerprint method are not only related to the signal fluctuation but are also affected by scanning numbers of BLE beacons after statistically analyzing the real-time signal data in a harsh environment. When the scanning BLE beacon numbers are few, coarse errors will more likely occur;
2. We found that the accuracy of the heading is also affected by the motion states of the pedestrian. An improved Mahony complementary filter is introduced to keep the heading angle stable by adaptively changing the control parameters in the filter after considering the different people's motion states;
3. To meet the demand of real-time position and considering the computational load of the smartphone, we adopt the EKF method to solve the nonlinear fusion problem to combine PDR with the BLE fingerprint position method to provide the real-time position service. To cope with the gross error caused by the BLE fingerprint method in a harsh environment, a robust filter based on the EKF was proposed. The robust filter detected the gross error at different granularity by constructing a robust vector changing the observation covariance matrix of the extended Kalman filter (EKF) adaptively when the application is running. The experimental results demonstrate that the proposed method has better performance at position accuracy and stability.

The remainder of this paper is organized as follows: Section 2 is about the related works of the BLE-based position, self-contained position, and fusion position algorithms in detail. Section 3 introduces the BLE fingerprint method and analyzes the error of the position method. PDR method and the heading estimation based on the motion state are also introduced in this section. Next, the fusion method EKF and a robust filter are presented as well. Finally, a diagram about the BEL/PDR integrated system localization framework is demonstrated. Section 4 describes the experiments and analyzes the results. Then the discussion of the paper and the conclusions are presented in Section 5 and Section 6.

2. Related Works

The BLE-based position systems have been widely used by utilizing the received signal strength (RSS) measurements. Zuo et al. pointed out that the radio signals of BLE had two features. One is that the signal can change dramatically in a small spatial change. The other is that the signal can be reported multiple times or not reported at all during a single scan. Both features will cause huge noises in both RSS and BLE beacon availability [22]. In addition to the features of the signal, multipath effect, device heterogeneity, and deployment are the sources of errors in the BLE fingerprint position [19,23]. To analyze the effects of dense deployment, Ng et al. proposed a high-resolution proximity detection using an adaptive scanning mechanism fusion with spontaneous differential evolution [24]. Tian et al. defined a coverage degree criterion by leveraging the Cramer Rao Lower Bound (CRLB) and the differential evolution algorithm to optimize the placement of Wi-Fi and BLE access points (APS) to hybrid two types of signals to fuse position based on the position performance analysis [23]. Andrew et al. applied three Bayesian filtering techniques to fit the BLE signal distance equation based on considering various errors and conduct comparison experiments to verify the significant modification in two environments [25]. Subhan et al. presented an in-depth experimental analysis of RSS and its effect on distance and position [26]. The position methods based on BLE and WiFi are the same, which can be divided into two categories: proximity detection, multilateration, and fingerprinting approaches. The research by Zhao et al. showed that under the same environment and conditions, the BLE-based position is more accurate than Wi-Fi because of its lower transmission power and unique channel hopping mechanism [27]. The key point of much research on proximity detection and multilateralism is a distance-based estimation [25,28]. Moreover, it is difficult to obtain an accurate distance model under complex environmental circumstances. The fingerprint method has become the popular approach in real-time positioning with many advantages of being infrastructure-free and easily realizable. RADAR was the first RSS-based fingerprint system that used a K-nearest neighbor (KNN) algorithm to estimate the location indoors [29]. Zuo et al. introduced an efficient and graph optimization-based way for estimating the beacon positions and the reference fingerprint map to combine range-based and fingerprint-based methods of BLE [21]. A self-adaptive weighted KNN algorithm PhaseFi proposed a deep network instead of the fingerprint database and estimated the position by a radial basis function probabilistic method [30].

The self-contained position method mainly contains two types: the data-driven inertial navigation method and the PDR method. The data-driven inertial navigation technology has been increasingly used in recent works which use the deep neural network with great potential in model-free generalization to regress pedestrian motion characteristics. It used inertial measurement units (IMU) in a short time and ground-truth motion trajectories to regress motion parameters (velocity and heading). Robust IMU double integration (RIDI) had made a breakthrough in coordinate frame normalization and used support vectors to regress a more accurate velocity vector [31]. IONet and RoNIN utilized trained neural networks to regress the magnitude of speed and the rate of heading angle change, showing the capability to obtain plain displacement [32,33]. ILIO demonstrated a network that regresses 3D displacement estimation and its uncertainty to tightly fuse the relative state measurement into a stochastic cloning EKF to solve for the pose, velocity, and sensor biases [34]. IDOL presented a two-stage, data-driven pipeline using a commodity smartphone that first estimates device orientations and then estimates device position to solve the problem of inaccurate orientation estimates [35]. Although the data-driven inertial navigation has a good performance in long-time tracking based on the sophisticated deep learning technology, it needs a large amount of data to train and extra equipment to get the ground-truth trajectory in advance, and for real-time positioning, the device will be under greater computational load. PDR position is a self-positioning method that consists of three key components of step detection, step length, and heading estimation that is a good choice for the real-time position. Lachapelle et al. proposed three-step length error models: Gaussian model, constant random model, and Gauss Markov model. He modeled the error of the gyroscope as a random constant deviation when establishing the PDR error model, so the heading error was considered to be linear with time [36]. Jahn et al. established the error models for four methods of measuring step length and discussed the systematic and random errors with the Taylor expansion [37]. You et al. proposed the multipoint positioning algorithm based on the received RSSI for calibration to eliminate the cumulative error existing in the traditional PDR system by analyzing the characteristics of walking postures [38].

For fusion methods, Li et al. [39] proposed an adaptive system noise EKF algorithm to develop an integrated Wi-Fi/PDR system. The proposed filter could determine the dynamic noise of the transition matrix according to the movement (straight or turning) of the pedestrian and reduce the computational complexity of the matching fingerprint database by using an affinity clustering algorithm. The positioning error could be reduced to 2.32 m by the experiment. Deng et al. [40] also presented a novel data fusion framework by using an EKF to integrate Wi-Fi localization with PDR. They developed a measurement model based on kernel density estimation to enable accurate Wi-Fi localization and adaptive measurement noise statistics estimation. The experiments show that the proposed method obtains comparable accuracy and greatly reduces computation cost compared with a particle filter. Atia et al. [41] utilized a grid-based nonlinear Bayesian filter algorithm to fuse the Wi-Fi, BLE, and inertial navigation system (INS) sensor information to develop a calibration-free hybrid indoor positioning methodology. The experiments demonstrated that the performance of the fusion method is much better than the BLE fingerprint position method. However, the harsh indoor environment will lead to coarse errors in the wireless localization method while different motion states of pedestrians affect the performance of PDR, which both affect the accuracy of the dynamic and observation models. Adaptive and robust filters can be employed to mitigate the effects of large errors in the dynamic and observation models, respectively. Yang et al. [42] proposed an adaptively robust filter based on a robust maximum-likelihood estimation to kinematic geodetic positioning and measurement. The method could not only balance the contribution between the updated parameters and measurement but also mitigated the influence of measurement outliers. Yang et al. [43] also presented an adaptively robust filter with multi adaptive factors based on the principles of the adaptive KF and bifactor robust estimations for correlated observations. The proposed filter is more flexible in controlling the disturbing effects of the state components compared to the classified adaptive factors. Chang et al. [44] proposed a robust KF using the Chi-squared test to detect measurement outliers. Li et al. [45] presented an adaptive and robust filter to combine the Wi-Fi and PDR information to develop an integrated system. The adaptive filter is based on scenario and motion state recognition and the robust filter is based on the Mahalanobis distance. The experiment results indicate that the proposed filter is better than the common EKF.

3. Materials and Methods

3.1. BLE Position Technology

The fingerprint position is mainly based on the similarity of the received signal strength (RSS) and the fingerprint database to obtain the position result. The RADAR system was the first RSS-based fingerprint system developed by Bahl et al. [29] for indoor localization by utilizing a KNN algorithm. Like any other fingerprint method, the BLE fingerprint position method consists of two stages: the offline data training stage and the online positioning stage. During the offline data training stage: people stand at the reference points (RPs) whose coordinates are known in advance to collect the RSS from the access points (APs) for some time. Then, the fingerprint database as shown in Equation (1) is constructed after computing the distribution of collected data. Suppose that there are n RPs and m APs in the position scenario.

$$\text{Fingerprint database} = \begin{bmatrix} (x_1, y_1) & \left\langle rssi_1^1, rssi_2^1, rssi_3^1, \ldots, rssi_m^1 \right\rangle \\ (x_2, y_2) & \left\langle rssi_1^1, rssi_2^1, rssi_3^1, \ldots, rssi_m^1 \right\rangle \\ \vdots & \vdots \\ (x_n, y_n) & \left\langle rssi_1^1, rssi_2^1, rssi_3^1, \ldots, rssi_m^1 \right\rangle \end{bmatrix} \tag{1}$$

During the online positioning stage, the position of the target is obtained by matching the real-time fingerprint to the database on certain algorithms. There are many matching algorithms. Considering the efficiency of real-time positioning and the simplicity of implementation, the classic matching method is K-nearest Neighbor (KNN). The basic idea of the KNN algorithm is to classify the target into the nearest sample class in the feature space. For the BLE fingerprint position method, the feature space is the fingerprint database. The process of BLE fingerprint KNN algorithms is shown as follows:

Step 1: When online positioning is carried out, a real-time fingerprint is collected by smartphone. The real-time fingerprint is expressed as Equation (2).

$$\text{Real time fingerprint} = [\langle rssi_1, rssi_2, \ldots, rssi_m \rangle] \tag{2}$$

Step 2: Then the Euclidean distance between the real-time fingerprint and fingerprint database can be calculated by the following Equation (3).

$$d_i = \sqrt{\left(rssi_1 - rssi_1^i\right)^2 + \left(rssi_2 - rssi_2^i\right)^2 + \cdots + \left(rssi_m - rssi_m^i\right)^2} \tag{3}$$

Step 3: Then we sort the n distance d_i in ascending order and choose the first K items to calculate the target position by averaging the K corresponding coordinates as shown in Equation (4).

$$(x, y) = \frac{1}{K} \sum_{i}^{K} (x_i, y_i) \tag{4}$$

The process of the fingerprint position method is shown in Figure 1.

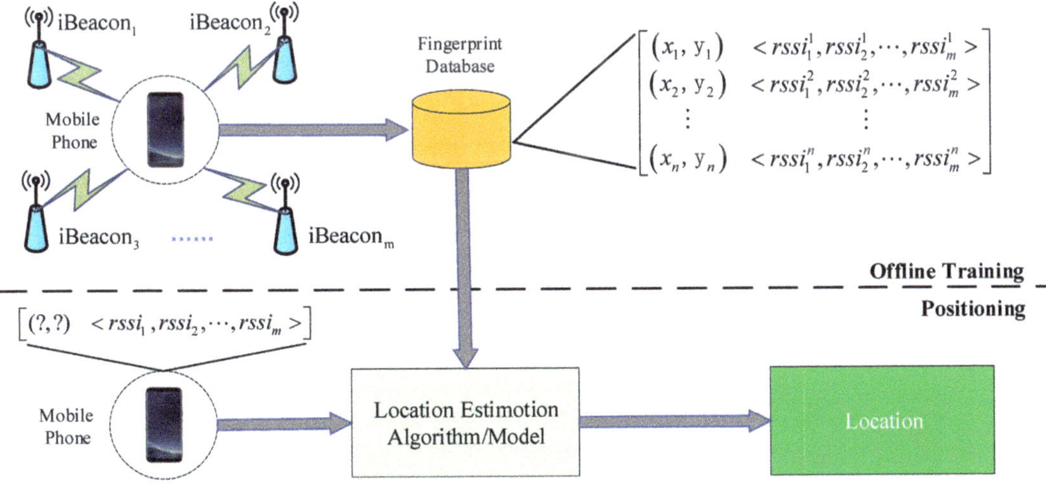

Figure 1. The process of the BLE fingerprint position method.

In addition to the general fingerprint position process described above, BLE real-time fingerprint also has its features. When we carry out a real-time BLE fingerprint position, the BLE module broadcasts data at a certain frequency. The broadcast data include the identity number, module name, mac address, received signal strength indicator (RSSI), and other information. The broadcast frequency is usually from 10 nanoseconds to 10 s, and the default broadcast frequency is 500 milliseconds. The common android devices such as smartphones provide a function to scanning the BLE signal for real-time fingerprint position. Generally, the continuous positioning with a time interval of 1 s is regarded as the real-time position. Due to the influence of the scanning mechanism, the number of the scanning BLE RSSI is often different in 1 s. In addition to the impact of signal fluctuations, the number of the scanning BLE RSSI will also cause errors in fingerprint position.

Here are the real-time RSSIs from 54 BLE APs in one position scenario at certain RP collected by smartphone HUAWEI P20 in 60 s. The collected data per second correspond to a real-time fingerprint data, totaling 60 fingerprint data. Then we count the scanning RSSI number of each fingerprint and Figure 2 shows the scanning RSSI number of each fingerprint in detail with a bar chart.

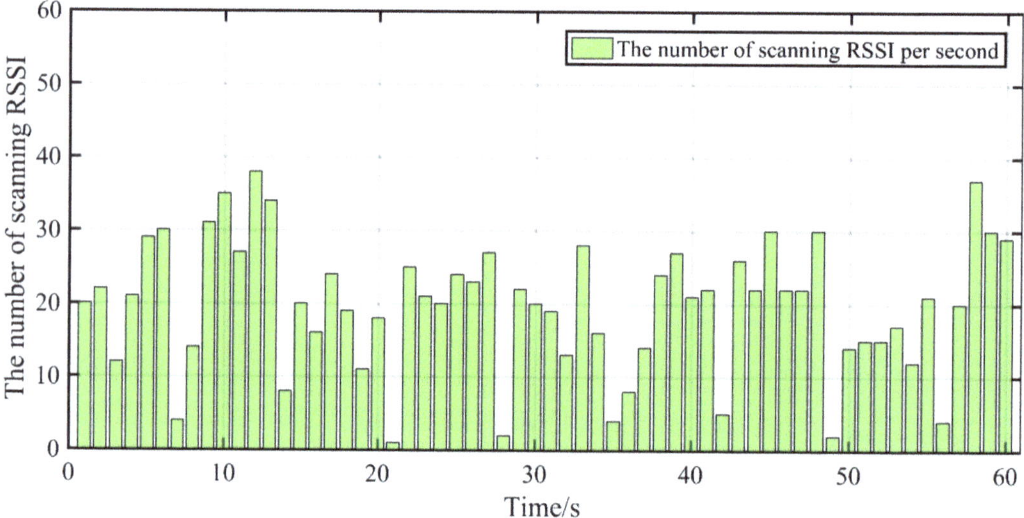

Figure 2. The scanning RSSI number of the per real-time fingerprint in 60 s.

As can be seen from Figure 2, there are differences in the scanning RSSI number per second. The scanning RSSI number per second is at least 1 and at most 37, with an average of 19.78 RSSI. The different number of the scanning RSSI directly affects the positioning accuracy of BLE fingerprint position. If the scanned number of APs is too small, it is easy to cause a coarse error in the BLE fingerprint position [46]. A robust filter is needed to restrain the coarse error.

3.2. PDR Technology

Pedestrian dead reckoning (PDR) is a self-positioning method for indoor navigation. The key technologies of PDR are step detection (or counting), step length estimation, and heading angle of pedestrian estimation. After a pedestrian step was detected, the position can be obtained using step length and heading angle based on the previous position [20]. The general process of PDR is illustrated as Equation (5):

$$\begin{cases} N_{k+1} = N_k + s_k \cdot cos\psi_k \\ E_{k+1} = E_k + s_k \cdot sin\psi_k \end{cases} \tag{5}$$

where N_k and E_k refer to the coordinate of the pedestrian at the north and east direction at time k, respectively. s_k is the step length and ψ_k is the heading angle at time k.

In the PDR method, step detection and step length are estimated according to the accelerometer readings. Many step detection algorithms have been proposed by researchers, including peak detection [47], threshold setting [29], zero velocity update [48], auto-correlation [49] and finite-state machine (FSM) [50]. Among them, the FSM method is easy to implement and more resistant to interference from errors. After detecting a step, step length is estimated by different models. Studies have shown that the step length is related to the acceleration, heights, and strides of different people. Linear models [51], constant models [52], and nonlinear models [53] are the most common methods which are used to estimate the step length. The step length estimated by different methods differs little. For simplicity, we choose the Weinberg model [54] to estimate the step length in our research and the expression for the model is as follows in Equation (6):

$$s_k = K \cdot \sqrt[4]{a_{max} - a_{min}} \tag{6}$$

where K is the scale factor of the step length, a_{max} and a_{min} are the maximum and minimum acceleration in one step cycle.

Apart from step detection and step length estimation, heading estimate is another important component of PDR. The compass [55] or the gyroscope [56] are usually used to estimate the heading angle. Because of the inherent sensor noise in the smartphone, the accuracy of the heading obtained by the compass is not high but it will not drift for a long time. In contrast, using the gyroscope to estimate the heading, the accuracy in a short time will be high, but it will suffer a drift problem. We also found that different motion states of pedestrians would affect the accuracy of heading estimation. To avoid the shortcomings of compass and gyroscope and take into account the different motion states of people, an improved Mahony complementary filter (AMMCF) based on the motion states had been proposed in our research. The parameter K_p in the filter can adaptively change based on the motion states which are judged according to the acceleration readings. The principle of PDR is illustrated in Figure 3.

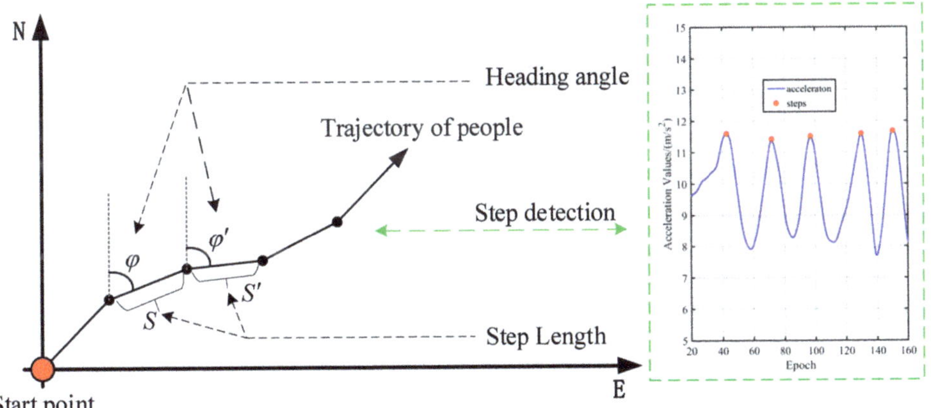

Figure 3. The principle of PDR.

3.3. An Improved Mahony Complementary Filter Based on the Motion States

The attitude of the device can be described by Euler angle, rotation matrix, and quaternion methods. The Euler angle is the angle between the three axes of the carrier coordinate system (CCS) and geographic coordinate system (GCS) which include pitch, roll, and yaw. The pitch, roll, and yaw represent rotation around x, y, and z axes as shown in Figure 4 and are denoted by the symbol θ, γ and ψ, respectively. The Euler angle method only needs three elements to store attitude information, which is simple, intuitive, and easy to understand, but it is easy to generate a universal joint deadlock phenomenon.

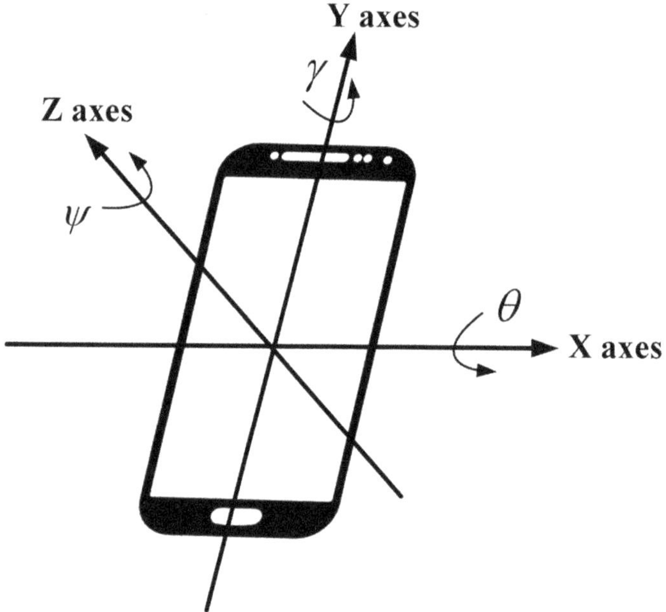

Figure 4. The attitude angle of the smartphone.

Different from the Euler angle, the Euler angle describes the attitude of the device at a certain moment, while the rotation matrix describes the motion process of the device rotation. In a GCS defined by the North-East-Up (NEU), the rotation matrix C_n^b between the two systems can be defined in Equation (7):

$$C_n^b = C_n^\gamma C_n^\theta C_n^\psi = \begin{bmatrix} cos\gamma & 0 & sin\gamma \\ 0 & 1 & 0 \\ -sin\gamma & 0 & cos\gamma \end{bmatrix} \begin{bmatrix} 1 & 0 & 0 \\ 0 & cos\theta & sin\theta \\ 0 & -sin\theta & cos\theta \end{bmatrix} \begin{bmatrix} cos\psi & sin\psi & 0 \\ -sin\psi & cos\psi & 0 \\ 0 & 0 & 1 \end{bmatrix} \tag{7}$$

where C_n^b represents the rotation matrix from GCS to CCS, C_n^ψ, C_n^θ, C_n^γ represent the corresponding matrices for yaw, pitch, and roll in order. As we know, the matrix C_n^b is the orthogonal matrix, so the rotation matrix C_b^n is the inverse matrix of C_n^b, which represents the rotation matrix from CCS to GCS. The matrix C_b^n can be defined as following Equation (8):

$$C_b^n = \left(C_n^b\right)^T = \left(C_n^b\right)^{-1} = \begin{bmatrix} cos\psi cos\gamma - sin\psi sin\theta sin\gamma & sin\psi cos\theta & cos\psi sin\gamma + sin\psi sin\theta cos\gamma \\ -sin\psi cos\gamma - cos\psi sin\theta sin\gamma & cos\psi cos\theta & -sin\psi sin\gamma + cos\psi sin\theta cos\gamma \\ -cos\theta sin\gamma & -sin\theta & cos\theta cos\gamma \end{bmatrix} \tag{8}$$

Because the rotation matrix will suffer universal joint deadlock phenomenon and need 9 parameters to store attitude information, the quaternion is the most commonly used method to calculate the attitude. Assuming that the quaternion vector is $Q = [q_0, q_1, q_2, q_3]^T$, the attitude angles expressed in quaternions are as follows in Equation (9):

$$\begin{cases} \psi = arctan\frac{2q_1q_2 - 2q_0q_3}{1 - 2q_1^2 - 2q_3^2}, \psi \in (0, 2\pi) \\ \theta = \arcsin(-2q_2q_3 - 2q_0q_1), \theta \in (-\frac{\pi}{2}, \frac{\pi}{2}) \\ \gamma = arctan\frac{-2q_1q_2 + 2q_0q_2}{1 - 2q_1^2 - 2q_2^2}, \gamma \in (-\pi, \pi) \end{cases} \tag{9}$$

The Mahony complementary filter (MCF) [39] utilizes the gyroscope to calculate the attitude angle of the device and the accelerometer and magnetometer are used to complement the accumulated error. When a gyroscope raw data $\omega = (\omega_x, \omega_y, \omega_z)^T$ was obtained, we can use the first-order Runger-Kutta method to obtain the quaternion update as shown in Equations (10) and (11):

$$Q[t + \Delta t] = Q[t] + \Delta t \cdot \Omega_w[t] \cdot Q[t] \tag{10}$$

$$\begin{pmatrix} q_0 \\ q_1 \\ q_2 \\ q_3 \end{pmatrix}_{t+\Delta t} = \begin{pmatrix} q_0 \\ q_1 \\ q_2 \\ q_3 \end{pmatrix}_t + \frac{\Delta t}{2} \cdot \begin{pmatrix} 0 & -\omega_x & -\omega_x & -\omega_x \\ \omega_x & 0 & \omega_z & -\omega_y \\ \omega_y & -\omega_z & 0 & \omega_x \\ \omega_z & \omega_y & -\omega_x & 0 \end{pmatrix} \cdot \begin{pmatrix} q_0 \\ q_1 \\ q_2 \\ q_3 \end{pmatrix}_t \tag{11}$$

where the Equation (11) is an expansion of the Equation (10), $Q[t]$ refers to the quaternion vector of time t, Δt represents tiny time and is often assigned of sampling cycles, $\Omega_w[t]$ is the corresponding matrix as shown in Equation (11), which is made up of gyroscope raw data at time t. As a result, we can obtain the latest quaternion vector by utilizing the gyroscope data based on the Runger-Kutta method. However, as time goes on, it will suffer a drift problem, resulting in huge cumulative error for long-time orientation. To reduce the drift problem, the acceleration and magnetometer field are used to compensate for the cumulative error. Supposed that the gyroscope error correction is $e = [e_x, e_y, e_z]^T$, it can be defined as:

$$e = e_a + e_m \tag{12}$$

where e_a, e_m are the error correction items calculated by accelerometer and magnetometer readings, respectively. They are expressed as $e_a = [e_{ax}, e_{ay}, e_{az}]^T$ and $e_m = [e_{mx}, e_{my}, e_{mz}]^T$, and can be obtained by Equation (13):

$$\begin{cases} e_a = a_n^b \times a \\ e_m = m_n^b \times m \end{cases} \tag{13}$$

where a, m are the normalized accelerometer and magnetometer readings. a_n^b, m_n^b are the normalized vectors of gravity acceleration and magnetic field after matrix C_n^b conversion. The symbol "$\times$" represents the vector cross product. After getting the error correction, the compensated gyroscope value $\omega' = (\omega'_x, \omega'_y, \omega'_z)$ can be calculated based on the proportional-integral (PI) method as follows in Equation (14):

$$\omega' = \omega + K_p \cdot e + K_i \cdot \int e \tag{14}$$

where K_p, K_i are the proportional and integral control parameters, respectively. The compensated gyroscope value ω' is plugged into the quaternion differential Equation (11) and the quaternion is updated. Then the attitude is obtained by Equation (9) based on the updated quaternion.

In general, K_p, K_i are fixed empirical values without considering the impact of pedestrian motion status. In our research, we proposed an improved MCF to change the control parameter K_p and K_i adaptively based on the pedestrian motion status. For simplicity, the pedestrian motion status can be divided into three types including static, walking, and running which can be judged by the standard deviation of triaxial acceleration modulus. The triaxial acceleration modulus can be calculated by Equation (15):

$$acc_{mod} = \|acc\| = \sqrt{acc_x^2 + acc_y^2 + acc_z^2} \tag{15}$$

where acc and $\|acc\|$ refer to the acceleration vector and the corresponding modulus, respectively. Figure 5 shows the performance of the acceleration modulus on different motion statuses. The dotted green line is an artificial boundary between different motion states. The acceleration modulus is collected for 1 min at the frequency of 50 Hz. People in the first 20 s are in a static state, people in the middle 20 s are in a walking state, and people in the last 20 s are in a running state.

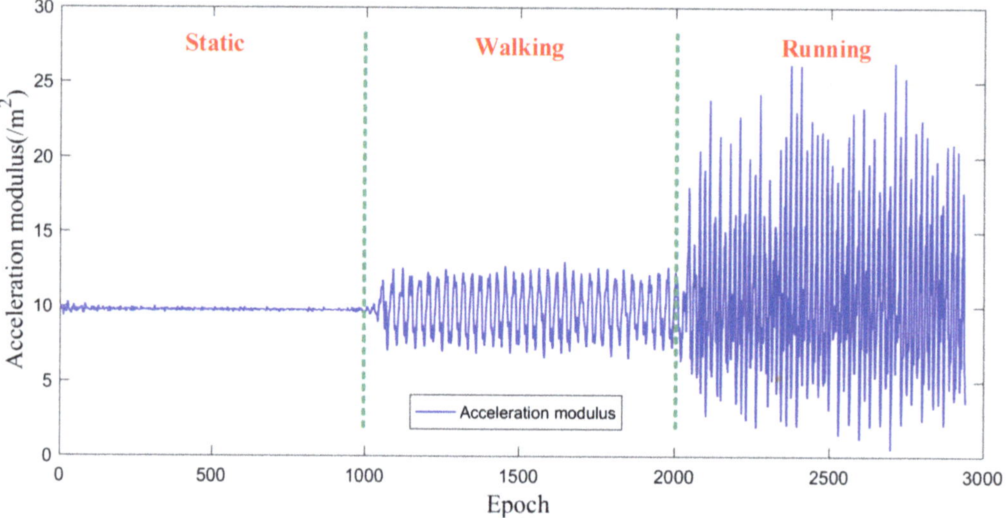

Figure 5. The acceleration modulus on different motion statuses.

We set up a sliding window to store the acceleration modulus and calculate the standard deviation in the window as shown in Equations (16) and (17).

$$\mu_{acc} = \frac{1}{n} \cdot \sum_{i=1}^{n} acc_{mod_i} \tag{16}$$

$$\sigma_{acc} = \sqrt{\frac{1}{n} \cdot \sum_{i=1}^{n} \left(acc_{mod_i} - \mu_{acc}\right)^2} \tag{17}$$

where μ_{acc} is the average acceleration modulus of the sliding window in size n, σ_{acc} refers to the standard deviation of the sliding window. Then, a moving average filter (MAF) is applied to process the σ_{acc} value to classify the motion state. Figure 6 demonstrates that the standard deviation of the acceleration modulus is smoother and easier to distinguish after MAF. Therefore, σ_{acc} can be used to judge different motion states of pedestrians.

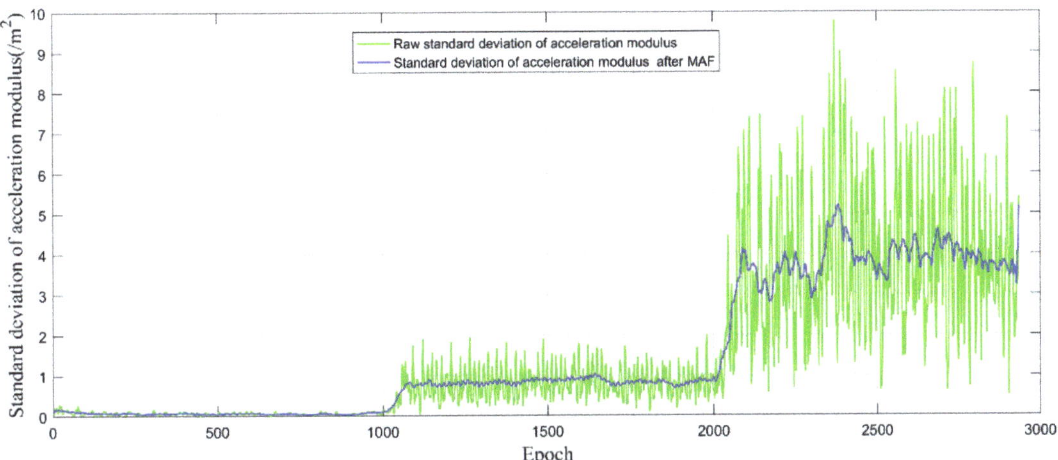

Figure 6. The curve of different motion states between the raw standard deviation of acceleration modulus and the standard deviation of acceleration modulus after MAF.

In our research, we proposed an improved MCF to change the control parameters K_p and K_i adaptively based on the pedestrian motion status. The parameters and K_i can be calculated by Equations (18) and (19):

$$K_p = \begin{cases} K_p \cdot (\sigma_{acc} + 1) & \sigma_{acc} < 0.05\ g \\ \dfrac{K_p}{(\sigma_{acc}+1)} & 0.05\ g < \sigma_{acc} < 0.2\ g \\ 0 & 0.2g < \sigma_{acc} \end{cases} \tag{18}$$

$$K_i = \begin{cases} K_i \cdot (\sigma_{acc} + 1) & \sigma_{acc} < 0.05\ g \\ \dfrac{K_i}{(\sigma_{acc}+1)} & 0.05g < \sigma_{acc} < 0.2\ g \\ 0 & 0.2\ g < \sigma_{acc} \end{cases} \tag{19}$$

The gyroscope is sensitive to changes in motion status which are often accompanied by changes in acceleration. The value of 0.2 g is approximately the average value of running which is almost the boundary from static, walking to fast motion. When the device is in a strenuous state, in which the acceleration mode is bigger than the value g, the attitude is mainly calculated by the gyroscope. The experiments show that the improved Mahony complementary filter based on the motion states has a better performance.

3.4. BLE/PDR Integrated System Based on EKF

PDR is a self-contained algorithm that can provide accurate position information at a short distance but it suffers accumulated errors. BLE fingerprint position accuracy is poor without cumulative error. The positioning result obtained by the BLE fingerprint method at present will not be affected by the previous result. To improve the positioning accuracy, continuity, and stability of the system, two positioning methods are often combined as an integrated system based on the Kalman filter (KF). Since the PDR position method is a non-linear algorithm, the Extended Kalman filter (EKF) is often utilized to replace the KF method to fuse two positioning methods. The hybrid position model mainly includes two models, one is the state transition model and the other is the observation model.

The state transition model is mainly to estimate the state vector which is composited of position coordinates, step length, and heading angle. The state vector is expressed by Equation (20):

$$X = [N, E, s, \psi] \tag{20}$$

The state transition model is expressed by Equation (21) at time k:

$$\begin{cases} N_k = N_{k-1} + s_{k-1} \cdot cos\psi_{k-1} + \omega_N \\ E_k = E_{k-1} + s_{k-1} \cdot sin\psi_{k-1} + \omega_E \\ s_k = s_{k-1} + \omega_s \\ \psi_k = \psi_{k-1} + \omega_\psi \end{cases} \tag{21}$$

where N_k and E_k are the position coordinates of the PDR position method in the north and east, respectively. s_k and ψ are the step length and the heading angle calculated by the PDR method at time k. Further, ω_N, ω_E, ω_s, ω_ψ are the corresponding process noise of the state vector. They conform to Gaussian distribution, and their variances are denoted by δ_N^2, δ_E^2, δ_s^2, δ_ψ^2, respectively. The state transition matrix A_k^- is expressed as Equation (22):

$$A_k^- = \begin{bmatrix} 1 & 0 & cos\psi_{k-1} & 0 \\ 0 & 1 & sin\psi_{k-1} & 0 \\ 0 & 0 & 1 & 0 \\ 0 & 0 & 0 & 1 \end{bmatrix} \tag{22}$$

The Jacobi matrix A_k of A_k^- is expressed as Equation (23):

$$A_k = \begin{bmatrix} 1 & 0 & cos\psi_{k-1} & -s_{k-1} \cdot sin\psi_{k-1} \\ 0 & 1 & sin\psi_{k-1} & s_{k-1} \cdot cos\psi_{k-1} \\ 0 & 0 & 1 & 0 \\ 0 & 0 & 0 & 1 \end{bmatrix} \tag{23}$$

The observation model is mainly to estimate the observation vector which is composited of position coordinates of BLE fingerprint position. The observation vector is expressed by Equation (24):

$$Z = \begin{bmatrix} N', E' \end{bmatrix} \tag{24}$$

The observation model is expressed as follow (25):

$$\begin{cases} N'_k = N'_{k-1} + \omega_{N'} \\ E'_k = E'_{k-1} + \omega_{E'} \end{cases} \tag{25}$$

where N'_k and E'_k are the position coordinates of the BLE fingerprint method in the north and east, respectively, $\omega_{N'}$ and $\omega_{E'}$ are the corresponding observation noise that conforms to Gaussian distribution and their variances are $\delta_{N'}^2$, $\delta_{E'}^2$. The observation matrix H_k is expressed as Equation (26):

$$H_k = \begin{bmatrix} 1 & 0 & 0 & 0 \\ 0 & 1 & 0 & 0 \end{bmatrix} \tag{26}$$

When the state vector and observation vector are obtained by BLE fingerprint and PDR methods, respectively, the EKF estimation is employed to update the state parameters through time as well as the observation parameters. In the EKF, the process of prior estimation is expressed as follows:

$$X_k = A_k^- \cdot \hat{X}_{k-1} \tag{27}$$

$$P_k^- = A_k \cdot P_{k-1} \cdot A_k^T + Q_k \tag{28}$$

The Gain matrix is expressed as (29):

$$G_k = P_k^- \cdot H_k^T \cdot \left(H_k \cdot P_k^- \cdot H_k^T + R_k \right) \tag{29}$$

Then, the state vector and the covariance matrix are updated according to the observations. The update process is written as follows:

$$\hat{X}_k = X_k + G_k \cdot (Z_k - H_k \cdot X_k) \tag{30}$$

$$P_k = P_k^- - G_k \cdot H_k \cdot P_k^- \tag{31}$$

where X_k and $\hat{X}_k$ are the prior and posterior state estimate vector, G_k is the gain matrix of EKF, P_k^- and P_k are the prior and posterior system covariance matrix, Q_k is the covariance matrix of the process noise, R_k is the covariance matrix of the observational noise vector. When the integrated system runs, the prior position, step length, and heading angle are obtained by the PDR method. The state vector composed of previous elements and the other observation vector obtained by the BLE position method is input into the fusion method. When the fusion method is cyclically executed, the corrected position results will be obtained.

3.5. A Robust Filter Model

The EKF model described above can effectively suppress the drift error of the PDR method and improve the overall positioning accuracy and stability, but the effect of suppressing the gross error of the BLE fingerprint position is poor. This paper proposes a robust filter model. In the EKF model, the innovation vector r_k is expressed as Equation (32):

$$r_k = Z_k - H_k \cdot X_k = [N_k' - N_k, E_k' - E_k] = [\Delta N, \Delta E] \tag{32}$$

where ΔN and ΔE represent the position coordinate difference in the north and east, respectively. From Equation (32), the r_k represents the position coordinate difference between the BLE and PDR methods.

When there are gross errors in the positioning method, there are the following situations:

1.   The difference ΔN exceeds the limit;
2.   The difference ΔE exceeds the limit;
3.   The innovation vector r_k exceeds the limit, ΔN and ΔE are within the acceptable range;
4.   The difference ΔN, ΔE, and the innovation vector r_k exceed the limit.

Therefore, to judge whether there is a gross error in the process of position, it is necessary to determine the distribution of the ΔN, ΔE and r_k.

In the EKF method, Z_k should conform to Gaussian distribution with mean $H_k X_k$ and covariance P_k^r as shown in Equation (33). The innovation vector r_k conforms to Gaussian distribution with mean 0, covariance P_k^r.

$$P_k^r = H_k \cdot P_k^- \cdot H_k^T + R_k \tag{33}$$

while the difference ΔN conforms to the Gaussian distribution with mean 0, covariance $\delta_{\Delta N}^2$ and the difference ΔE conforms to the Gaussian distribution with mean 0, covariance $\delta_{\Delta N}^2$.

$$\delta_{\Delta N}^2 = \begin{bmatrix} 1 & 0 \end{bmatrix} \cdot P_k^r \cdot \begin{bmatrix} 1 \\ 0 \end{bmatrix} \tag{34}$$

$$\delta_{\Delta E}^2 = \begin{bmatrix} 0 & 1 \end{bmatrix} \cdot P_k^r \cdot \begin{bmatrix} 0 \\ 1 \end{bmatrix} \tag{35}$$

$$\lambda_{\Delta N} = \Delta N \sim N\left(0, \delta_{\Delta N}^2\right) \tag{36}$$

$$\lambda_{\Delta E} = \Delta E \sim N\left(0, \delta_{\Delta E}^2\right) \tag{37}$$

The squared Mahalanobis distance M_k of r_k conforms to the chi-square distribution $\chi^2_{m,\alpha}$ with the freedom m which is the dimension of the observation Z_k.

$$\lambda_r = M_k^2 = r_k^T (P_k^r)^{-1} r_k \sim \chi^2_{m,\alpha} \tag{38}$$

The chi-square distribution $\chi^2_{m,\alpha}$ is constructed to determine whether the actual vector r_k calculated in the EKF exceeds the limit under the Gaussian assumption. Significance level α is the probability threshold and 5% is adopted in our research.

$$P_r \left[\lambda_r > \chi^2_{m,\alpha} \right] < \alpha \tag{39}$$

where P_r represents the probability of a random event that the probability of λ_r being larger than $\chi^2_{m,\alpha}$ is very small. Hence, if the actual r_k is larger than the α-quantile, the null hypotheses are rejected and it can be concluded that r_k exceeds the limit and Z_k has a gross error in the positioning. For ΔN and ΔE, when the actual measurements are larger than twice the corresponding standard deviation, the significance level is less than 5%, and it can be concluded that ΔN or ΔE or both exceed the limit and Z_k has a gross error.

When the vector Z_k has a gross error, the observational covariance matrix R_k should multiply a robust vector:

$$\overline{R}_k = \beta_k^T \cdot R_k \cdot \beta_k \tag{40}$$

where R_k represents vector. $\overline{R}_k$ is the modified observation covariance matrix. When the vector Z_k has a gross error in the first situation, the robust vector β_k is constructed as follows:

$$\beta_k = \left[\ \frac{\lambda_{\Delta N}}{\delta_{\Delta N}} \quad 1 \ \right]^T \tag{41}$$

When the vector Z_k has a gross error in the second situation, the robust vector β_k is constructed as follows:

$$\beta_k = \left[\ 1 \quad \frac{\lambda_{\Delta E}}{\delta_{\Delta E}} \ \right]^T \tag{42}$$

When the vector Z_k has a gross error in the third situation, the robust vector β_k is constructed as follows:

$$\beta_k = \left[\ \frac{\lambda_r}{\chi^2_{m,\alpha}} \quad \frac{\lambda_r}{\chi^2_{m,\alpha}} \ \right]^T \tag{43}$$

When the vector Z_k has a gross error in the fourth situation, the robust vector β_k is constructed as follows:

$$\beta_k = \left[\frac{\lambda_{\Delta N}}{\delta_{\Delta N}} + \frac{\lambda_r}{\chi^2_{m,\alpha}} \quad \frac{\lambda_{\Delta E}}{\delta_{\Delta E}} + \frac{\lambda_r}{\chi^2_{m,\alpha}} \ \right]^T ss \tag{44}$$

According to the above method, the robust filter can effectively restrain the gross error in the observation vector Z_k. In practice, it is possible for the matrix R_k to be modified even there is no gross error in the observation Z_k.

At last, we proposed the BLE/PDR integrated System localization framework in the following diagram of Figure 7. In the PDR position method, the gyroscope readings are used to compute the heading angle and the accelerations and magnetometer are utilized to compensate for accumulated error as well. To improve the accuracy of the heading angle, people's motion status is considered in correcting the control parameters of MCF. The accelerations are also used to detect the steps and estimate the step length. With the step length and heading angle, a position is computed by the PDR method. The other position is obtained by the BLE fingerprint method at the same time. Then, two positioning estimations are input into the Extend Kalman filter to resolve the fusion position. To improve the integrated system robustness and scalability, a robust filter is used to restrain the gross error of BLE observational position results.

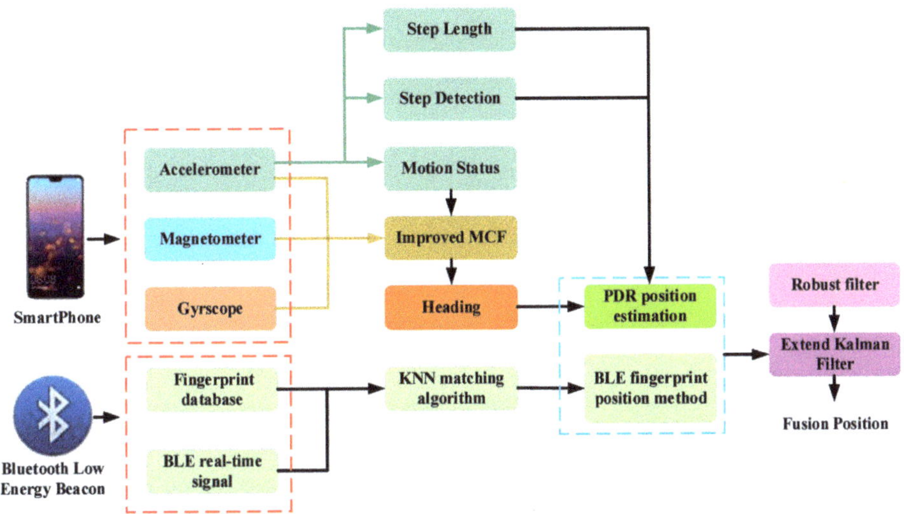

Figure 7. The BLE/PDR integrated system localization framework.

4. Experiments and Analysis

4.1. BLE Fingerprint Position Experiments and Error Analysis

The experiments were set up on the second floor of the C7 test site which has three floors in the 54th Research Institute of China Electronics Technology Group Corporation as shown in Figure 8. The second-floor test site is 24.92 m long and 27.49 m wide which consists of a rectangular corridor and several rooms. The center part of Figure 8 is a hollow space enclosed with glass. When the radio signal of BLE propagates, there is a severe multipath effect. There are 54 Bluetooth beacons installed on the whole C7 test site and 18 on each floor. The deployment of the Bluetooth beacon in Figure 8 is based on the principle of optimizing beacon placement by maximizing localization accuracy and satisfying a predefined coverage degree [23]. The HUAWEI P20 smartphone was chosen as the test device. In Figure 8, the blue triangle refers to the Bluetooth low energy beacon and there are 18 BLE beacons were installed on the second-floor test site. For the BLE fingerprint position method, all of the BLE beacons installed on the C7 test site were utilized to collect signals for the experiment. Another scenario called 331 test site had been chosen to conduct a comparison experiment. The test site is 21.72 m long and 7.75 m wide. The deployment of the Bluetooth beacon in the 331 test site is upon the same optimization solution as C7.

Then we carried out a BLE fingerprint positioning experiment in this position scenario. The BLE fingerprint database was constructed in advance. People stranded with the HUAWEI P20 at each known RP to collect data for 7 s, the time is set randomly. Then the data was processed per second into real-time fingerprint data. In the actual acquisition process, only 6 s of data were collected at some RPs. There were real-time 332 BLE fingerprint data collected at 48 RPs for fingerprint position. We calculated the positioning results through the KNN matching method and compared them with the true coordinates of RPs to obtain the position error. We analyzed the gross error of the BLE fingerprint method, whose error is larger than twice the standard deviation under the Gaussian distribution. Then we picked out the gross error caused by the few numbers of scanning RSSI which is less than 5 and we eliminate the corresponding fingerprint. Finally, we recalculate the position error, and both position errors were obtained and they were illustrated in Table 1 and Figure 9.

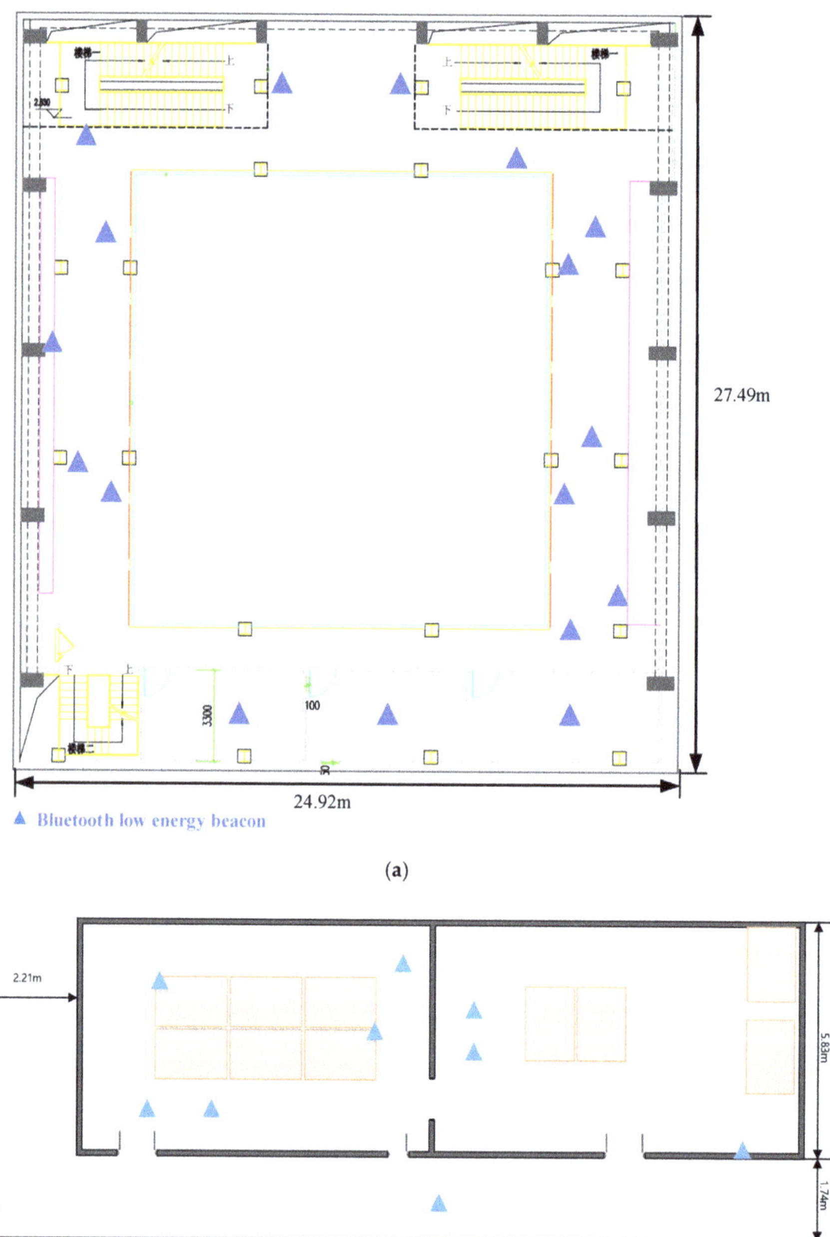

(a)

(b)

Figure 8. The two location scenarios. (**a**) The location scenario of the second floor of the C7 test site. (**b**) The location scenario of the 331 test site.

Table 1. Position error comparison/(m).

Method	Min	Max	Mean	RMSE
BLE fingerprint method	0.005	18.289	2.834	3.106
Eliminating the less scanned BLE number	0.005	18.289	2.312	2.042

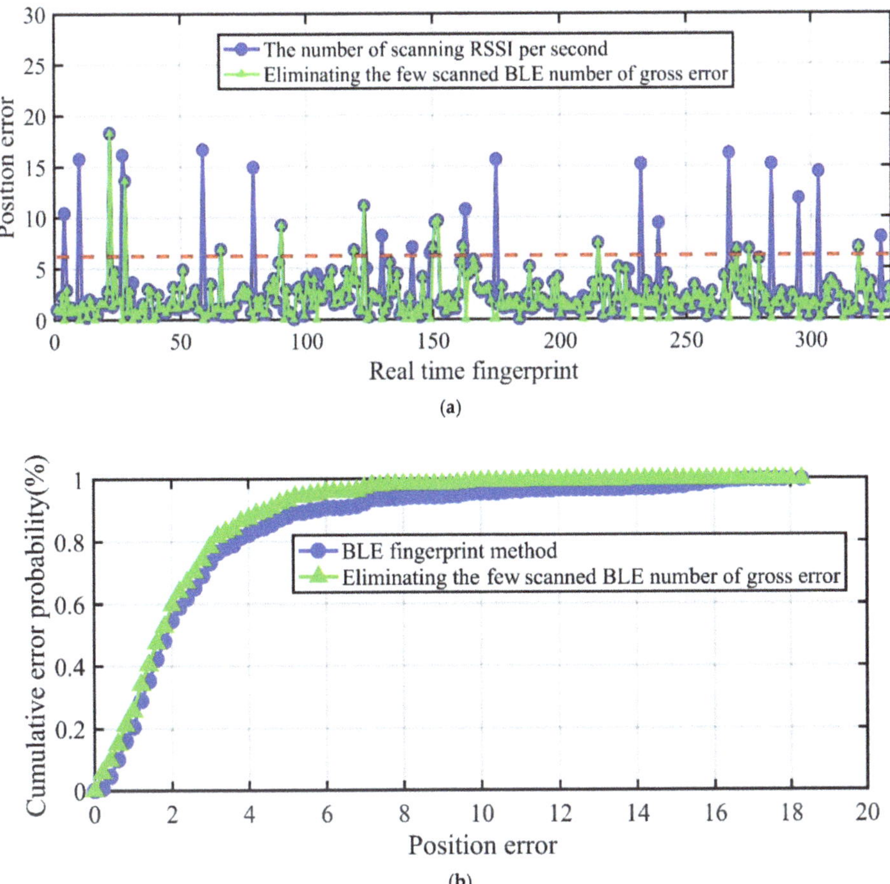

Figure 9. BLE fingerprint method and BLE fingerprint method after eliminating the few scanning numbers of RSSI. (**a**) Position error curves of two methods (**b**) Cumulative distribution of position error of two methods.

As shown in Table 1, after eliminating the few numbers of scanning RSSI real-time fingerprints, the mean position accuracy and the root-mean-square error (RMSE) were 2.312 m and 2.043 m, respectively. Compared with all data on the BLE fingerprint method, the position accuracy was reduced by 0.5220 m and RMSE decreased by 1.064 m. In Figure 8, the blue line refers to the result of all real-time fingerprint data and the green line refers to eliminating the few numbers of scanning RSSI real-time fingerprints. The red dashed line represents twice the standard deviation which means the position error that exceeds the red dashed line is a gross error. From Figure 9a, the gross error number shown in the green line is greatly reduced compared with the blue line. Figure 9b indicates that the method of eliminating the few numbers of scanning RSSI has a higher confidence level than the BLE fingerprint method. From the statistics of gross errors, the total number of gross errors is 31. Among them, the number of gross errors caused by the few numbers of scanning RSSI is 17, accounting for about 54.8%. Therefore, we can conclude that the causes of gross errors in the BLE fingerprint method are not only caused by signal fluctuations but also affected by the few numbers of scanning RSSI. It is necessary to find a robust filter to restrain the gross error of the BLE fingerprint method.

4.2. Heading Estimation Based on Motion States

The experiment about heading estimation was carried out in the same position scenario as shown in Figure 9.

In Figure 10, the red star refers to the start point and endpoint of the trajectory. The dark green line represents the trajectory and the arrow represents the direction. The current motion state is indicated on each trajectory which includes walking and running. The static state of the pedestrian is contained inside the red ellipse. The pedestrian walked along the dark green trajectory with different motion states.

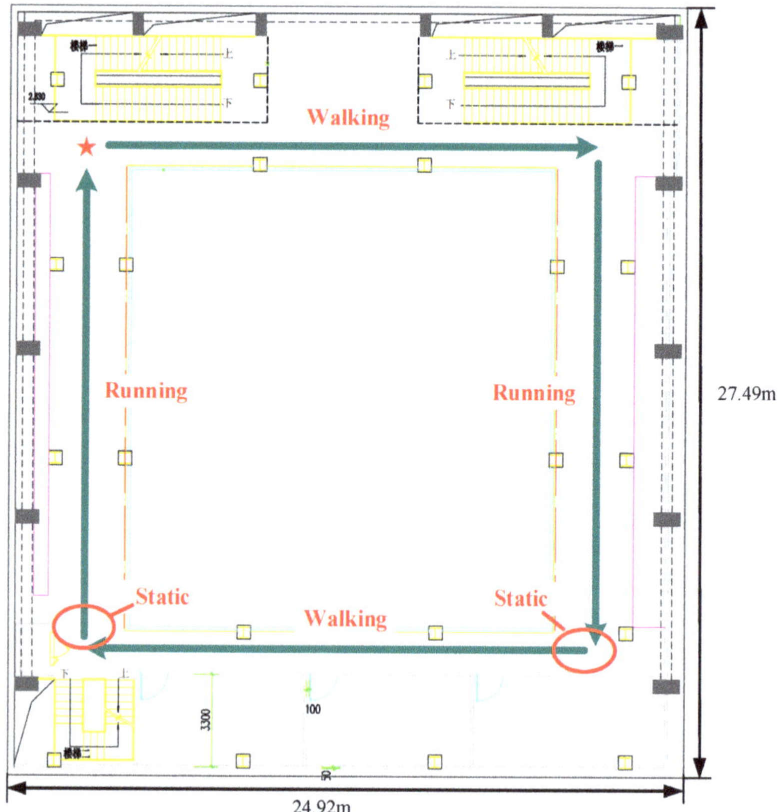

Figure 10. Trajectory and motion state of heading estimation.

Then the heading is estimated by different methods as shown in the following Figure 11.

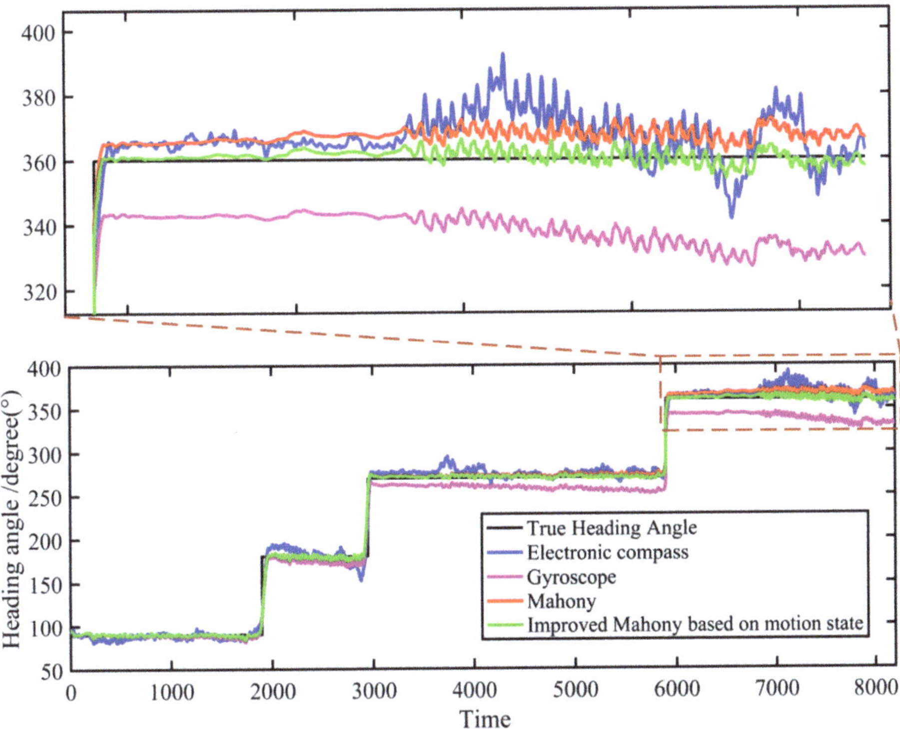

Figure 11. Heading angle estimated based on the four methods and partially enlarged view.

Four methods were utilized to estimate the heading angle. In Figure 11, the blue line refers to the heading angle estimated by the electronic compass while the magenta line, red line, and green line were represented as gyroscope-based, Mahony, and improved Mahony based on motion state methods, respectively. The black line denoted the true heading. From Figure 10, the accuracy of the heading angle estimated by the electronic compass was poor and suffered severe fluctuation, but it would be no drift for a long time. In contrast, the heading estimated by gyroscope did not have severe fluctuation but would suffer huge drift problems that would distort the heading.

For the MCF method, the control parameters K_p and K_i were given two values of 0.001 and 0.000001, respectively. The MCF performed well in the early state but would suffer little drift for a long time without the adaptive parameter adjustment based on the motion state. From the partially enlarged view of the last trajectory, the heading estimated by the improved Mahony based on the motion state was the most accurate, which fluctuated around the true angle and had no cumulative error. Finally, a diagram of different motion states of experiment trajectory is shown in Figure 12 which contained a semi-transparent layer of heading estimation for comparison. It could be concluded that the heading estimation is closely related to the motion state. When stationary, the heading estimation was relatively stable when walking or moving, and the heading would fluctuate.

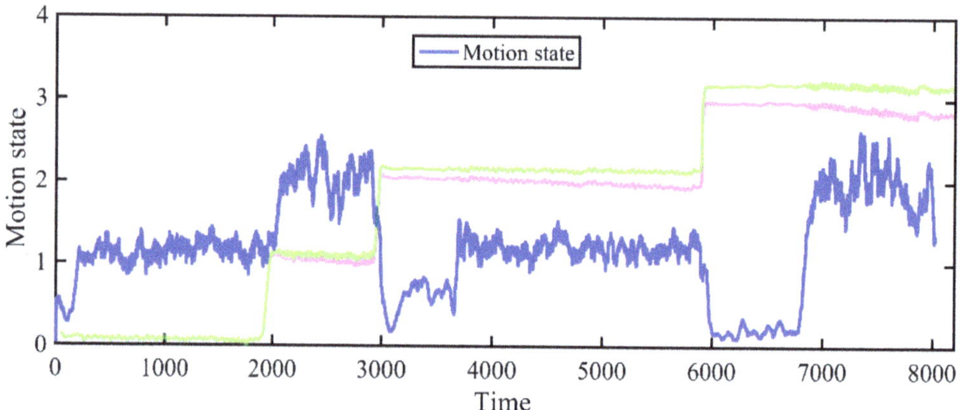

Figure 12. Comparison of Motion state and heading estimation.

4.3. BLE/PDR Integrated System Position Experiment

The BLE/PDR integrated system position experiment was carried out in the C7 test site. The BLE AP routers and the smartphone used in the experiment are the same as in the experiment of Section 3.1. The new trajectory was planned and was shown in Figure 13.

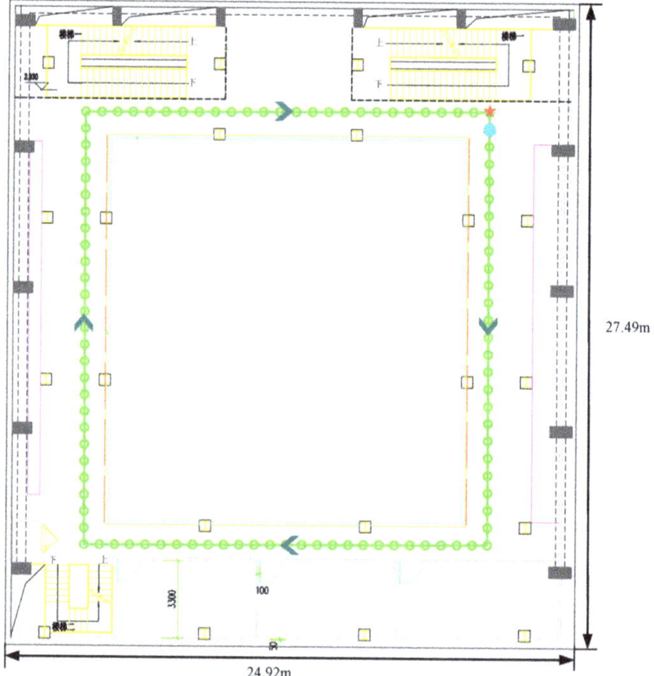

Figure 13. The trajectory of BLE/PDR integrated system position experiments in the C7 test site.

In Figure 13, the red star and blue cycle refer to the start point endpoint of the trajectory, respectively. The green dotted line represents the reference trajectory and the arrow represents the direction. During the experiment, the data sampling frequency of

PDR was set as 50 Hz and the smartphone scanned the BLE APs per second. The pedestrian started from the start point and reached the endpoint at a constant speed. The pedestrian held the smartphone level and walked 107 steps in total during the experiment. Another integrated system position experiment was conducted at the 331 test site. The comparison results of the different methods were shown in Figure 16.

To validate the efficiency of the robust filter method, another two methods, the BLE fingerprint method and the EKF method, were also utilized for the experiment. Position errors of the three methods were computed concerning the reference points for evaluation. Figure 14a showed the time series of the position errors and (b) showed the corresponding cumulative distribution errors.

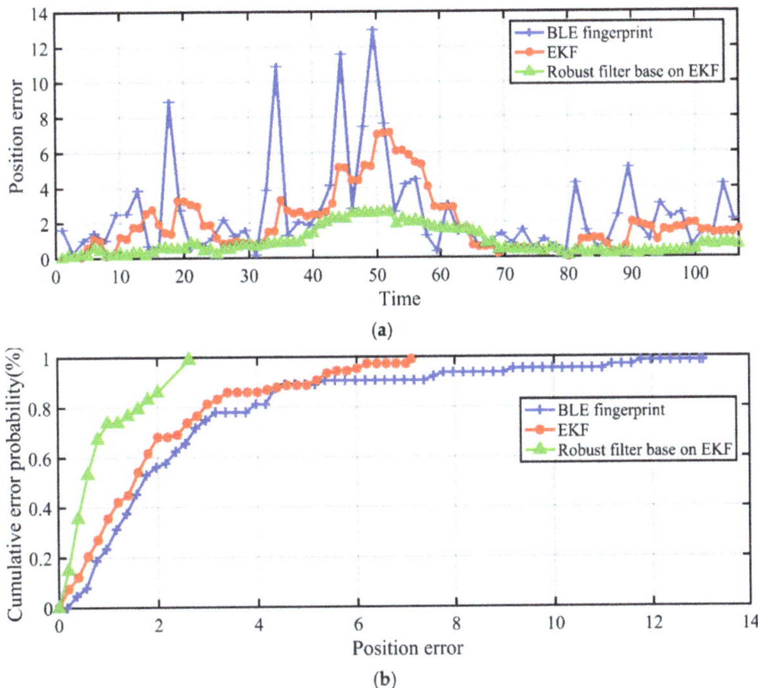

Figure 14. Comparison of Position errors and cumulative errors of three methods. (**a**) Time series of position errors, (**b**) cumulative errors of three methods.

In the above figure, the blue line refers to the BLE fingerprint method. The red line and green line represent the EKF and robust filter methods, respectively. From the value and distribution of the position error, the robust filter method denoted by the green line performs best. The cumulative distribution of the robust filter indicates that it has a higher confidence level than the other two methods. About 74% of the position error of points are lower than 1 m and about 86% of the position error of points are lower than 2 m. Compared with the common EKF, these two indicators are increased by 38% and 18%, respectively. Then, Table 2 shows the detail of the mean error and RMSE of the three methods.

Table 2. Position error comparison/(m).

Method	Min	Max	Mean	RMSE
BLE fingerprint method	0.165	13.033	2.647	2.727
Extend Kalman filter method	0	7.122	1.960	1.727
Robust filter base on EKF	0	2.641	0.844	0.745

As shown in Table 2. The mean position accuracy and the RMSE of the robust filter method are 0.844 m and 0.745 m respectively. Comparing with the EKF and BLE fingerprint methods, the position accuracies were reduced by 1.116 m and 1.803 m, and the RMSEs were decreased by 1.982 m and 0.982 m. From the perspective of max value, the max value of the robust filter is 2.641 m which is also significantly reduced. The figure and table mentioned above show that the proposed robust filter can not only reduce the position error but also improve the stability.

Figure 15 shows the trajectory of three methods in the true position scenario at the C7 test site. Another fusion method particle filter is also utilized to solve the result as a comparison experiment.

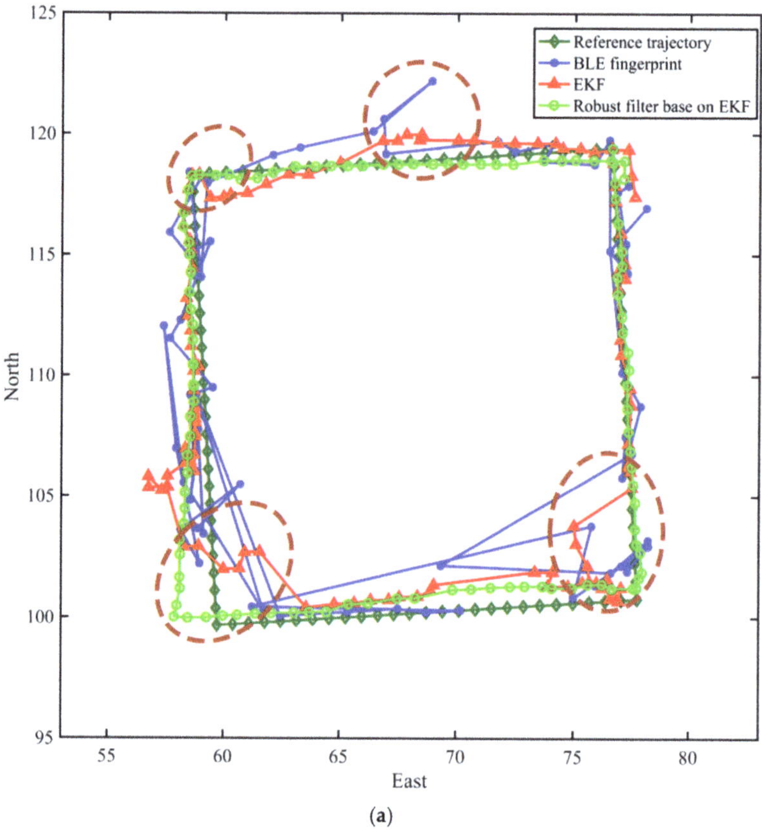

(**a**)

Figure 15. *Cont.*

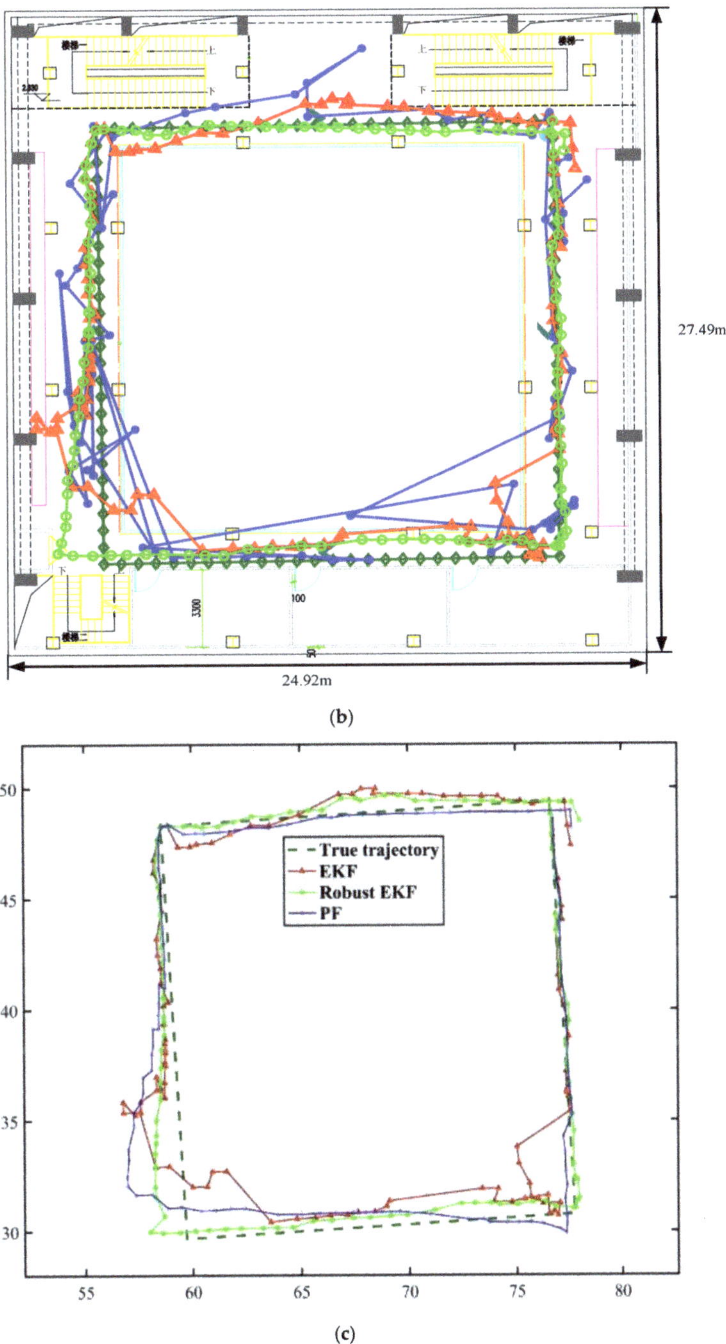

(b)

(c)

Figure 15. *Cont.*

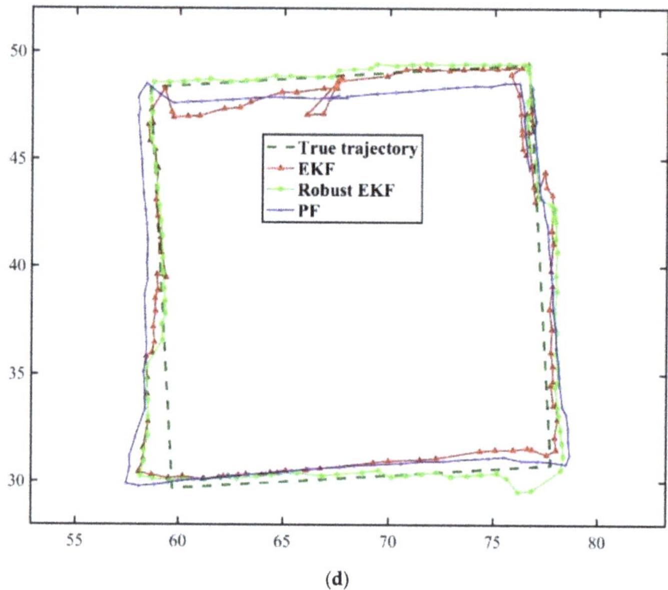

(d)

Figure 15. The trajectory of three methods in the true position scenario. (**a**) The trajectory and jump points, (**b**) the trajectory in the position scenario, (**c**) the comparison of different fusion methods with trajectory one, (**d**) the comparison of different fusion methods with trajectory two.

The blue line, red line, green line, and dark green line represent the trajectory of BLE fingerprint, EKF, robust filter methods, and reference trajectory, respectively. In Figure 15a, some jump points, which mean the gross error surrounded by the red circle, appear in the BLE fingerprint method or the EKF method. The robust filter can improve the observation matrix based on detecting the gross errors and get a smooth trajectory without jump points. We can see that the proposed robust filter performs better than the other two methods in (b). (c) and (d) are the results of different hybrid positioning methods under two trajectories. From the comparison of the curves with the real trajectory in (c) and (d), the proposed method is superior to the classical EKF. The particle filter also performs well in some positions compared to the proposed method. However, the particle filter requires the construction of a large number of particles requiring a heavy computational load. The proposed method is more suitable for real-time localization than the particle filter method. Another comparison of the experimental results of the integrated system was conducted in the 331 test site and the results were shown in Figure 16. From Figure 16, the green line that represents the proposed method is closer to the real track and smoother in some corners compared to other methods. Although the test paths are relatively short because of the extent of the experimental scenarios, the two methods of fusion position have different principles, in which PDR has high instantaneous accuracy but suffers cumulative errors, and the real-time Bluetooth fingerprint method can compensate for this shortcoming based on the robust filter. Therefore, even with the test paths becoming longer, there will be no cumulative error to make the path diverge. We can conclude that the proposed method makes it possible to have similar improvement under different circumstances.

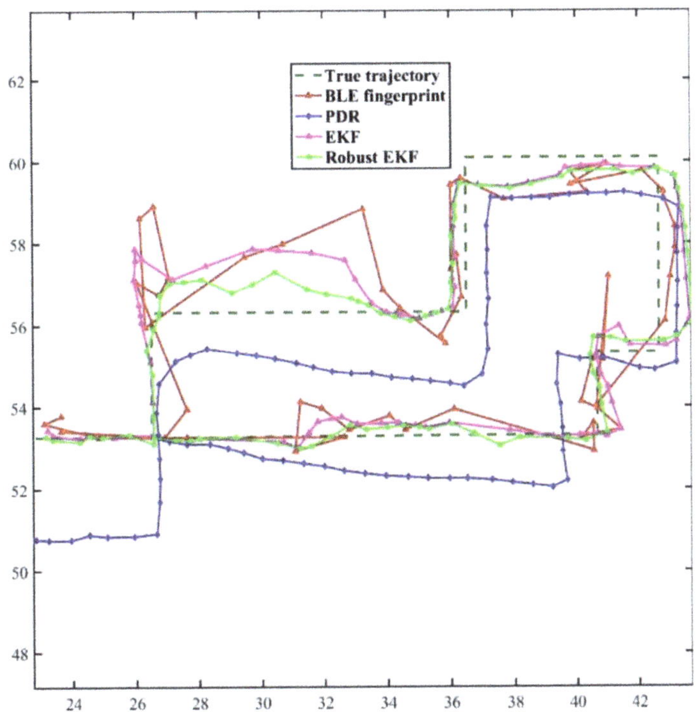

Figure 16. The trajectory of four methods in 331 test site.

5. Discussion

In the process of real-time positioning, poor position results may be obtained by the BLE fingerprint method. The heading of the device is affected by different motion states of people. Then we get rid of the bad results of the BLE fingerprint method when the scanning number of RSSI is smaller than 5. We correct the heading based on different motion states. Even if the BLE fingerprint method and PDR are integrated, the gross error is difficult to be suppressed. Compared with the adaptive and robust filter proposed in the research of Li et al. [27], the robust filter proposed in this paper considering the error distribution in more conditions and provides a robust vector instead of a numerical correction. The experimental results conducted on the true position scenario show that the proposed method can detect, suppress the gross errors and make the results smoother. The mean position accuracy of the proposed robust filter was 0.844 m and RMSE was 0.745. The experiments are in line with expectations. From Figure 14, we can find that the green points obtained by the robust filter are far away from the true position. The reason for this is the bad position results caused by the BLE fingerprint method at the previous few steps. The jump points in blue color have occurred continuously. The robust filter can only suppress gross errors but not eliminate them. If gross errors occur continuously, then the results will deviate from the true trajectory and it would take some time to converge. We will improve the method by considering the situation of continuous gross errors to get better results in future research work.

6. Conclusions

In this paper, we concentrated on the real-time BLE/PDR integrated System and fusion method. For BLE real-time fingerprint, we found that the position error of BLE is not only with the signal fluctuation but also with the scanned number of BLE APs. If the scanned number is too few, there will easily be gross errors. Next, we introduced the method of commonly used attitude methods and an improved Mahony complementary filter is proposed to estimate the heading angle under different motion states. Finally, a robust filter model was proposed to fusion the BLE/PDR methods because of the gross error in the BLE method. We conducted an experiment to validate the efficiency of the proposed method in the true position scenario. The mean position accuracy obtained by the robust filter was 0.844 m and RMSE was 0.745. The experiment showed that the proposed method has better performance in positioning accuracy and stability. The experimental scenario in this paper is surrounded by glass in the center, which has serious multi-path effects for the fingerprint positioning method and can easily cause coarse errors. The classic Kalman filter or Extended Kalman filter is not effective for coarse difference suppression. The proposed method can detect gross errors at different granularities and suppress them. The fusion methods based on the KNN and EKF are suitable for the high-real-time requirements of the positioning applications, keeping a balance between computational efficiency and position accuracy. The estimation of the heading angle is more stable based on the people's motion states. However, the effect of different pedestrian motion states on hybrid positioning was not analyzed in detail. In addition to the detection and suppression of coarse differences in the observation noise matrix, some modifications of the state transition matrix will be made based on the people's motion states. How the variance matrix of process noise adaptively changes according to the people's motion states will be investigated in-depth in future work. Combined with other positioning methods such as map matching, landmarks matching for multi-mode fusion positioning will be considered too.

Author Contributions: Conceptualization, S.X.; Formal analysis, S.X. and M.S. (Meng Sun); Funding acquisition, Y.W.; Methodology, S.X.; Project administration, Y.W.; Data curation, H.C. and M.S. (Minghao Si); Writing-original draft preparation, S.X. All authors have read and agreed to the published version of the manuscript.

Funding: This research was funded by the National Key Research and Development Program of China under grant number 2016YFB0502102.

Institutional Review Board Statement: Not application.

Informed Consent Statement: Not applicable.

Data Availability Statement: The experiment uses an internal data set and the data presented in this study are available on request from the corresponding author.

Conflicts of Interest: The authors declare no conflict of interest. The funding sponsors had no role in the design of the study; in the writing of the manuscript; and in the decision to publish the results.

Abbreviations

rssi	Received signal strength indicator
E_k	The east coordinate in the pedestrian dead reckoning at time k
N_k	The north coordinate in the pedestrian dead reckoning at time k
s_k	The step length at time k
ψ	The heading angle
θ	The pitch angle
γ	The roll angle
C_n^b	Refer to the rotation matrix from the geographic coordinate system to the carrier coordinate system
C_b^n	Refer to the rotation matrix from the carrier system to the carrier coordinate system
Q	The quaternion vector
q_i	The ith item of the quaternion vector
e	Refer to the error correction in the Runger-Kutta method
K_p	The proportional control parameters in the proportional-integral method
K_i	The integral control parameters in the proportional-integral method
ω	Refer to the gyroscope data in Section 3.3

m	Refer to the magnetometer data in Section 3.3
a	Refer to the accelerometer data in Section 3.3
g	Refer to the acceleration of gravity in Section 3.3
μ_{acc}	The average acceleration modulus
σ_{acc}	The standard deviation of acceleration modulus
X	Refer to the state vector in the EKF
ω_N	The process noise of the north coordinate in Section 3.4
ω_E	The process noise of the east coordinate in Section 3.4
ω_s	The process noise of the step length in Section 3.4
ω_ψ	The process noise of the heading angle in Section 3.4
δ_N^2	The variances of the north coordinate process noise
δ_E^2	The variances of the east coordinate process noise
δ_s^2	The variances of the step length process noise
δ_ψ^2	The variances of the heading angle process noise
A_k^-	The state transition matrix in the EKF
A_k	The Jacobi matrix of A_k^-
Z	Refer to the observation vector in the EKF
N_k'	The north coordinate of observation at time k
E_k'	The east coordinate of observation at time k
$\omega_{N'}$	The observation noise of the north coordinate
$\omega_{E'}$	The observation noise of the east coordinate
$\delta_{N'}^2$	The variances of the north coordinate observation noise
$\delta_{E'}^2$	The variances of the east coordinate observation noise
H_k	The observation matrix at time k
P_k^-	Refer to the prior system covariance matrix in the EKF
P_k	Refer to the posterior system covariance matrix in the EKF
G_k	Refer to the gain matrix of the EKF
Q_k	Refer to the covariance matrix of the process noise in the EKF
R_k	Refer to the covariance matrix of the observational noise vector in the EKF
ΔN	Refer to the position coordinate difference in the north
ΔE	Refer to the position coordinate difference in the east
r_k	The innovation vector consisting of ΔN and ΔE, which represents the position coordinate difference between two methods in the EKF
P_k^r	Refer to the covariance of the innovation vector in the EKF
$\lambda_{\Delta N}$	Refer to the distribution of the position coordinate in the north
$\lambda_{\Delta E}$	Refer to the distribution of the position coordinate in the east
λ_r	The distribution of the squared mahalanobis distance of the innovation vector
$\chi_{m,\alpha}^2$	The symbol of chi-square distribution with eh freedom m
α	The significance level of the distribution
β_k	The robust vector defined in the paper which is utilized to modify the observation noise covariance matrix
$\overline{R}_k$	Refer to the modified observation noise covariance matrix

References

1. Zhang, W.; Hua, X.; Yu, K.; Qiu, W.; Zhang, S.; He, X. A novel WiFi indoor positioning strategy based on weighted squared Euclidean distance and local principal gradient direction. *Sens. Rev.* **2019**, *39*, 99–106. [CrossRef]
2. Wei, S.; Min, X.; Hongshan, Y.; Hongwei, T.; Anping, L. Augmentation of Fingerprints for Indoor WiFi Localization Based on Gaussian Process Regression. *IEEE Trans. Veh. Technol.* **2018**, *67*, 10896–10905.
3. Song, X.; Fan, X.; Xiang, C.; Ye, Q.; Liu, L.; Wang, Z.; He, X.; Yang, N.; Fang, G. A Novel Convolutional Neural Network Based Indoor Localization Framework With WiFi Fingerprinting. *IEEE Access* **2019**, *7*, 110698–110709. [CrossRef]
4. Topak, F.; Pekeriçli, M.K.; Tanyer, A.M. Technological Viability Assessment of Bluetooth Low Energy Technology for Indoor Localization. *J. Comput. Civ. Eng.* **2018**, *32*, 04018034. [CrossRef]
5. Chen, L.; Pei, L.; Kuusniemi, H.; Chen, Y.; Kröger, T.; Chen, R. Bayesian Fusion for Indoor Positioning Using Bluetooth Fingerprints. *Wirel. Pers. Commun.* **2013**, *70*, 1735–1745. [CrossRef]
6. He, X.; Ye, D.; Peng, L.; Ruchuan, W.; Yizhu, L. An RFID Indoor Positioning Algorithm Based on Bayesian Probability and K-Nearest Neighbor. *Sensors* **2017**, *17*, 1806.
7. Zhang, D.; Yang, L.T.; Chen, M.; Zhao, S.; Guo, M.; Zhang, Y. Real-Time Locating Systems Using Active RFID for Internet of Things. *IEEE Syst. J.* **2016**, *10*, 1226–1235. [CrossRef]

8. Khyam, M.O.; Rahim, N.-A.; Li, X.; Ritz, C.; Guan, Y.L.; Ge, S.S. Design of Chirp Waveforms for Multiple-Access Ultrasonic Indoor Positioning. *IEEE Sens. J.* **2018**, *18*, 6375–6390. [CrossRef]
9. Lee, C.; Chang, Y.; Park, G.; Ryu, J.; Jeong, S.-G.; Park, S.; Park, J.W.; Lee, H.C.; Hong, K.-S.; Lee, M.H. Indoor positioning system based on incident angles of infrared emitters. In Proceedings of the 30th Annual Conference of IEEE Industrial Electronics Society, IECON 2004, Busan, Korea, 2–6 November 2004.
10. Musa, A.; Nugraha, G.D.; Han, H.; Choi, D.; Seo, S.; Kim, J. A decision tree-based NLOS detection method for the UWB indoor location tracking accuracy improvement. *Int. J. Commun. Syst.* **2019**, *32*, e3997. [CrossRef]
11. Yu, K.; Wen, K.; Li, Y.; Zhang, S.; Zhang, K. A Novel NLOS Mitigation Algorithm for UWB Localization in Harsh Indoor Environments. *IEEE Trans. Veh. Technol.* **2019**, *68*, 686–699. [CrossRef]
12. Kee, C.; Yun, D.; Jun, H. Precise calibration method of pseudolite positions in indoor navigation systems. *Comput. Math. Appl.* **2003**, *46*, 1711–1724. [CrossRef]
13. Li, X.; Zhang, P.; Huang, G.; Zhang, Q.; Guo, J.; Zhao, Y.; Zhao, Q. Performance analysis of indoor pseudolite positioning based on the unscented Kalman filter. *GPS Solut.* **2019**, *23*, 79. [CrossRef]
14. Aoran, X.; Ruizhi, C.; Deren, L.; Yujin, C.; Dewen, W. An Indoor Positioning System Based on Static Objects in Large Indoor Scenes by Using Smartphone Cameras. *Sensors* **2018**, *18*, 2229.
15. Mulloni, A.; Wagner, D.; Barakonyi, I.; Schmalstieg, D. Indoor Positioning and Navigation with Camera Phones. *IEEE Pervasive Comput.* **2009**, *8*, 22–31. [CrossRef]
16. Google. Wi-Fi Scanning Overview. Available online: https://developer.android.google.cn/guide/topics/connectivity/wifi-scan (accessed on 18 June 2021).
17. Zhou, C.; Yuan, J.-Z.; Liu, H.; Qiu, J. Bluetooth Indoor Positioning Based on RSSI and Kalman Filter. *Wirel. Pers. Commun.* **2017**, *96*, 4115–4130. [CrossRef]
18. Cao, H.; Wang, Y.; Bi, J.; Qi, H. An Adaptive Bluetooth/Wi-Fi Fingerprint Positioning Method based on Gaussian Process Regression and Relative Distance. *Sensors* **2019**, *19*, 2784. [CrossRef]
19. Wu, C.; Xu, J.; Yang, Z.; Lane, N.D.; Yin, Z. Gain without Pain: Accurate WiFi-based Localization using Fingerprint Spatial Gradient. *Proc. ACM Interact. Mob. Wearable Ubiquitous Technol.* **2017**, *1*, 1–19. [CrossRef]
20. Gu, F.; Hu, X.; Ramezani, M.; Acharya, D.; Khoshelham, K.; Valaee, S.; Shang, J. Indoor Localization Improved by Spatial Context—A Survey. *ACM Comput. Surv.* **2019**, *52*, 1–35. [CrossRef]
21. Kenn, H.; Behrens, N.; Kleiner, A. Optimizing indoor PDR performance with self-deployed position markers. In Proceedings of the 4th International Forum on Applied Wearable Computing 2007, Tel Aviv, Israel, 12–13 March 2007.
22. Zuo, Z.; Liu, L.; Zhang, L.; Fang, Y. Indoor Positioning Based on Bluetooth Low-Energy Beacons Adopting Graph Optimization. *Sensors* **2018**, *18*, 3736. [CrossRef] [PubMed]
23. Tian, Y.; Huang, B.; Jia, B.; Zhao, L. Optimizing AP and Beacon Placement in WiFi and BLE hybrid localization. *J. Netw. Comput. Appl.* **2020**, *164*, 102673. [CrossRef]
24. Ng, P.C.; She, J.; Park, S. High Resolution Beacon-Based Proximity Detection for Dense Deployment. *IEEE Trans. Mob. Comput.* **2017**, *17*, 1369–1382. [CrossRef]
25. Mackey, A.; Spachos, P.; Song, L.; Plataniotis, K.N. Improving BLE Beacon Proximity Estimation Accuracy Through Bayesian Filtering. *IEEE Internet Things J.* **2020**, *7*, 3160–3169. [CrossRef]
26. Subhan, F.; Khan, A.; Saleem, S.; Ahmed, S.; Imran, M.; Asghar, Z.; Bangash, J.I. Experimental analysis of received signals strength in Bluetooth Low Energy (BLE) and its effect on distance and position estimation. *Trans. Emerg. Telecommun. Technol.* **2019**. [CrossRef]
27. Zhao, X.; Xiao, Z.; Markham, A.; Trigoni, N.; Ren, Y. Does BTLE measure up against wifi? A comparison of indoor location performance. In Proceedings of the 20th European Wireless Conference, EW 2014, Barcelona, Spain, 14–16 May 2014; pp. 263–268.
28. Pakanon, N.; Chamchoy, M.; Supanakoon, P. Study on Accuracy of Trilateration Method for Indoor Positioning with BLE Beacons. In Proceedings of the 2020 6th International Conference on Engineering, Applied Sciences and Technology (ICEAST), Chiang Mai, Thailand, 1–4 July 2020; pp. 1–4.
29. Bahl, P.; Padmanabhan, V.N. RADAR: An in-Building RF-Based User Location and Tracking System. In Proceedings of the IEEE INFOCOM 2000 Conference on Computer Communications, Nineteenth Annual Joint Conference of the IEEE Computer and Communications Societies (Cat. No.00CH37064), Tel Aviv, Israel, 26–30 March 2000; Volume 2, pp. 775–784.
30. Wang, X.; Gao, L.; Mao, S.; Pandey, S. DeepFi: Deep learning for indoor fingerprinting using channel state information. In Proceedings of the 2015 IEEE Wireless Communications and Networking Conference (WCNC), New Orleans, LA, USA, 9–12 March 2015; Institute of Electrical and Electronics Engineers (IEEE): New York, NY, USA, 2015; pp. 1666–1671.
31. Yan, H.; Shan, Q.; Furukawa, Y. RIDI: Robust IMU Double Integration. In Proceedings of the European Conference on Computer Vision (ECCV), Munich, Germany, 8–14 September 2018.
32. Chen, C.; Lu, X.; Markham, A.; Trigoni, N. IONet: Learning to Cure the Curse of Drift in Inertial Odometry. In Proceedings of the AAAI Conference on Artificial Intelligence, New Orleans, LA, USA, 2–7 February 2018.
33. Herath, S.; Yan, H.; Furukawa, Y. RoNIN: Robust Neural Inertial Navigation in the Wild: Benchmark, Evaluations, & New Methods. In Proceedings of the 2020 IEEE International Conference on Robotics and Automation (ICRA), Paris, France, 31 May–1 August 2020; pp. 3146–3152.

34. Liu, W.; Caruso, D.; Ilg, E.; Dong, J.; Mourikis, A.I.; Daniilidis, K.; Kumar, V.; Engel, J. TLIO: Tight Learned Inertial Odometry. *IEEE Robot. Autom. Lett.* **2020**, *5*, 5653–5660. [CrossRef]
35. Sun, S.; Melamed, D.; Kitani, K. IDOL: Inertial Deep Orientation-Estimation and Localization. *arXiv* **2021**, arXiv:2102.04024.
36. Mezentsev, O.; Lachapelle, G.; Collin, J. Pedestrian dead reckoning—A solution to navigation in GPS signal degraded areas? *Geomatica* **2005**, *59*, 175–182.
37. Jahn, J.; Batzer, U.; Seitz, J.; Patino-Studencka, L.; Boronat, J.G. Comparison and evaluation of acceleration based step length estimators for handheld devices. In Proceedings of the 2010 International Conference on Indoor Positioning and Indoor Navigation, Zurich, Switzerland, 15–17 September 2010; pp. 1–6. [CrossRef]
38. You, Y.; Wu, C. Hybrid Indoor Positioning System for Pedestrians with Swinging Arms Based on Smartphone IMU and RSSI of BLE. *IEEE Trans. Instrum. Meas.* **2021**, *70*, 1–15. [CrossRef]
39. Li, X.; Wang, J.; Liu, C.; Zhang, L.; Li, Z. Integrated WiFi/PDR/Smartphone Using an Adaptive System Noise Extended Kalman Filter Algorithm for Indoor Localization. *ISPRS Int. J. Geo-Inf.* **2016**, *5*, 8. [CrossRef]
40. Deng, Z.A.; Hu, Y.; Yu, J.; Na, Z. Extended Kalman Filter for Real Time Indoor Localization by Fusing WiFi and Smartphone Inertial Sensors. *Micromachines* **2015**, *6*, 523–543. [CrossRef]
41. Atia, M.; Iqbal, U.; Givigi, S.; Noureldin, A.; Korenberg, M. Adaptive Integrated Indoor Pedestrian Tracking System Using MEMS sensors and Hybrid WiFi/Bluetooth-Beacons with Optimized Grid-based Bayesian Filtering Algorithm. In Proceedings of the 2015 International Technical Meeting of the Institute of Navigation, Dana Point, CA, USA, 26–28 January 2015.
42. Yang, Y.; He, H.; Xu, G. Adaptively robust filtering for kinematic geodetic positioning. *J. Geod.* **2001**, *75*, 109–116. [CrossRef]
43. Yang, Y.; Song, L.; Xu, T. Robust estimator for correlated observations based on bifactor equivalent weights. *J. Geod.* **2002**, *76*, 353–358. [CrossRef]
44. Chang, G. Robust Kalman filtering based on Mahalanobis distance as outlier judging criterion. *J. Geod.* **2014**, *88*, 391–401. [CrossRef]
45. Li, Z.; Liu, C.; Gao, J.; Li, X. An Improved WiFi/PDR Integrated System Using an Adaptive and Robust Filter for Indoor Localization. *Int. J. Geo-Inf.* **2016**, *5*, 224. [CrossRef]
46. Shaowei, L.; Xianghong, H.; Weining, Q.; Ying, S.; Kang, W.; Xuesheng, P. The effects of AP number on WiFi fingerprint positioning. *Eng. Surv. Mapp.* **2017**, *26*, 33–36.
47. Gu, F.; Khoshelham, K.; Shang, J.; Yu, F.; Wei, Z. Robust and Accurate Smartphone-Based Step Counting for Indoor Localization. *IEEE Sens. J.* **2017**, *17*, 3453–3460. [CrossRef]
48. Yuan, Z.; Haiyu, L.; You, L.; Naser, E.S. PDR/INS/WiFi Integration Based on Handheld Devices for Indoor Pedestrian Navigation. *Micromachines* **2015**, *6*, 793–812.
49. Rai, A.; Chintalapudi, K.; Padmanabhan, V.; Sen, R. Zee: Zero-effort crowdsourcing for indoor localization. In Proceedings of the 18th Annual International Conference on Mobile Computing and Networking, Istanbul, Turkey, 22–26 August 2012. [CrossRef]
50. Sun, M.; Wang, Y.; Xu, S.; Cao, H.; Si, M. Indoor Positioning Integrating PDR/Geomagnetic Positioning Based on the Genetic-Particle Filter. *Appl. Sci.* **2020**, *10*, 668. [CrossRef]
51. Ladetto, Q. On foot navigation: Continuous step calibration using both complementary recursive prediction and adaptive Kalman filtering. In Proceedings of the 13th International Technical Meeting of the Satellite Division of The Institute of Navigation (ION GPS 2000), Salt Lake City, UT, USA, 19–22 September 2000.
52. Vildjiounaite, E.; Malm, E.-J.; Kaartinen, J.; Alahuhta, P. Location Estimation Indoors by Means of Small Computing Power Devices, Accelerometers, Magnetic Sensors, and Map Knowledge. *Comput. Vis.* **2002**, *2414*, 211–224. [CrossRef]
53. Weinberg, H. Using the ADXL202 in pedometer and personal navigation applications. *Analog Devices AN-602 Appl. Note* **2002**, *2*, 1–6.
54. Liu, J.; Chen, R.; Pei, L.; Guinness, R.; Kuusniemi, H. A Hybrid Smartphone Indoor Positioning Solution for Mobile LBS. *Sensors* **2012**, *12*, 17208–17233. [CrossRef]
55. Wang, H.; Sen, S.; Elgohary, A.; Farid, M.; Youssef, M.; Choudhury, R.R. No need to war-drive: Unsupervised indoor localization. In Proceedings of the 10th International Conference on Mobile Systems, Applications, and Services, Low Wood Bay, UK, 26–28 June 2012. [CrossRef]
56. Mahony, R.; Hamel, T.; Pflimlin, J.-M. Nonlinear Complementary Filters on the Special Orthogonal Group. *IEEE Trans. Autom. Control* **2008**, *53*, 1203–1218. [CrossRef]

Article

Weighted Centrality and Retail Store Locations in Beijing, China: A Temporal Perspective from Dynamic Public Transport Flow Networks

Cong Liao [1,2], Teqi Dai [3,*], Pengfei Zhao [1,2] and Tiantian Ding [3,*]

[1] Institute of Remote Sensing and Geographical Information Systems, Peking University, No. 5 Yiheyuan Road, Haidian District, Beijing 100871, China; liaocong233@pku.edu.cn (C.L.); pfzhao@pku.edu.cn (P.Z.)
[2] Beijing Key Laboratory of Spatial Information Integration & Its Applications, Beijing 100871, China
[3] Beijing Key Laboratory for Remote Sensing of Environment and Digital City, Faculty of Geographical Science, School of Geography, Beijing Normal University, No. 19, Xinjiekouwai Street, Haidian District, Beijing 100875, China
* Correspondence: daiteqi@bnu.edu.cn (T.D.); 202021051041@mail.bnu.edu.cn (T.D.)

Citation: Liao, C.; Dai, T.; Zhao, P.; Ding, T. Weighted Centrality and Retail Store Locations in Beijing, China: A Temporal Perspective from Dynamic Public Transport Flow Networks. *Appl. Sci.* **2021**, *11*, 9069. https://doi.org/10.3390/app11199069

Academic Editor: Jianbo Gao

Received: 23 August 2021
Accepted: 21 September 2021
Published: 29 September 2021

Abstract: The spatial relationship between transport networks and retail store locations is an important topic in studies related to commercial activities. Much effort has been made to study physical street networks, but they are seldom empirically discussed with considerations of transport flow networks from a temporal perspective. By using Beijing's bus and subway smart card data (SCD) and point of interest (POI) data, this study examined the location patterns of various retail stores and their daily dynamic relationships with three weighted centrality indices in the networks of public transport flows: degree, betweenness, and closeness. The results indicate that most types of retail stores are highly correlated with weighted centrality indices. For the network constructed by total public transport flows in the week, supermarkets, convenience stores, electronics stores, and specialty stores had the highest weighted degree value. By contrast, building material stores and shopping malls had the weighted closeness and weighted betweenness values, respectively. From a temporal perspective, most retail types' largest correlations on weekdays occurred during the after-work period of 19:00 to 21:00. On weekends, shopping malls and electronics stores changed their favorite periods to the daytime, while specialty stores favored the daytime on both weekdays and weekends. In general, the higher store type level of the shopping malls correlates more to weighted closeness or betweenness, and the lower-level store type of convenience stores correlates more to weighted degree. This study provides a temporal analysis that surpasses previous studies on street centrality and can help with urban commercial planning.

Keywords: complex network; POI; smart card data; public transport flows; KDE; weighted centrality

1. Introduction

Location is a key factor for the commercial success of retail stores, as consumers tend to patronize stores that have higher access advantages [1,2]. The configuration of a city's transport network has been found to have significant impacts on the distribution of retail service activities [3–7]. Additionally, in urban planning and design, the locations of retail services are important for city growth and vitality [8]. Therefore, location analysis of retail stores is important for retail investment decisions and urban planning.

Location analysis has been increasingly applied to the retail sector with the growing computing power and the advent of big data [9]. While many location-allocation models have been developed and used for location decision making of retail stores [10,11], identifying the spatial pattern of retail locations is still an important and basic research task to date. Generally, many factors may affect the location retail stores, which made it a complex and multi-dimensional problem [12]. Among these factors, transportation is often regarded as a key element for retail locations. There is much empirical literature focusing on exploring the spatial relationships between physical street networks and retail stores. Various accessibility indices are optional to capture the convenience of retail stores in physical street networks [13,14]. Among them, the centrality features of a store are critical for the commercial competition of market areas according to the central place theory and spatial interaction theory [15]. Based on the approaches of space syntax or complex networks, the centrality features of a transport network can be measured by various centrality indices [16]. The multiple centrality assessment (MCA) model, which groups several indices together, has been applied to examine the relationship between street centrality and the spatial distributions of retail stores [17]. Different cities around the world have been examined, and the findings indicate that the centralities of the physical street network may well explain the retail distributions [18–21]. The study of Wang and Chen et al. [22] first examined the differences among location preferences for different types of retail stores. Later, new data sources as point of interest (POI) data were introduced [23]. As the relationship between various types of stores and multiple centrality indices of street networks across regions and cities were examined, it was revealed that different store types may correlate to different spatial networks centralities [23,24], which is helpful for retail location selection and planning.

While these previous studies have focused on examining the physical street network, few quantitative empirical studies have examined centrality in networks with transport flows. However, the location advantage of attracting transport flows is one important factor that influences the location selection of commercial services. The transport flows can reflect where people would like to go, and the correlation for retail stores is an important element for commercial development. According to the classic Hotelling model [25], the location strategy serves to obtain maximum flows, which are not necessarily geometric central points in space [26]. In addition, the flow network has a temporal attribute. Exploring temporal dynamics in flow networks may provide some possible insights into retail location patterns [27], as temporal factors such as store opening hours and individuals' travel time cannot be addressed by the static location analyses on the physical street network [28,29].

For retail location, it was recognized early that accessibility by public transport is a key issue for a store [30]. Several studies have verified that public transport has a substantial impact on retail patterns in city centers when compared to those of out-of-town malls [31–34]. In the big data era, public transport flow data become available from a smart card system and a number of studies have devised various weighted centrality indices to analyze the complex network of transport flows [35–37]. However, to the best of our knowledge, research on the relationships between retail store locations and their centrality in public transport flow network still lacking.

This paper aims to examine the relationship between weighted centrality indices and various retail stores from a temporal perspective. Beijing is chosen as the case city, in which public transport is well-developed. According to the 2020 Beijing transport development annual report released by the Beijing Transport Institute (http://www.bjtrc.org.cn/, accessed on 10 August 2021), the modal shares of public transport (bus and subway) in most urbanized areas of Beijing are more than 31%, which is greater than that of car and taxi (about 22% and 2.5%). In our study, the public transport flows are extracted from the bus and subway smart card data (SCD) of Beijing. The remainder of this study is organized as follows. Section 2 describes the study area and data preparation, and discusses the research methods. Section 3 presents the results. The last section discusses and summarizes the main findings.

2. Materials and Methods

2.1. Study Area and Data Preparation

Beijing is the capital of China and includes both urban and rural areas. As this study addresses public transport flows and retail activity, the analysis is conducted in the urban area of Beijing. Here, an area of approximately 38.64 km^2 within the sixth ring road of Beijing is selected as the case study area. The area covers most urbanized areas of Beijing.

The study area is divided into grid cells to conduct further analysis. The appropriate cell size of the study units may affect the results and computational complexity. In previous studies of retail stores and network centralities, a cell size of 1 km × 1 km has most commonly been used despite some variances [21]. Considering the road network density, this study selects a cell size of 1 km × 1 km (see Figure 1).

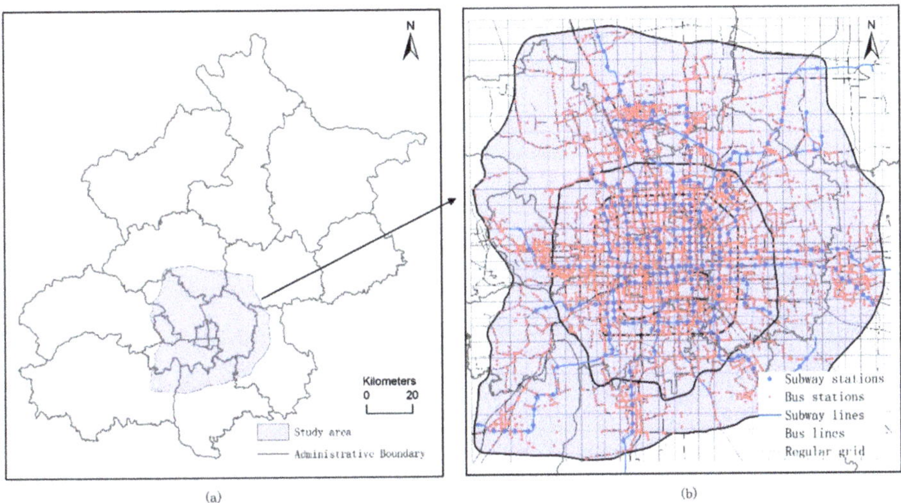

Figure 1. Case study area: (**a**) Beijing; (**b**) sixth ring road of Beijing.

Point of interest (POI) data are used to construct a dataset of retail stores. The POI data for 2018 are sourced from Autonavi (Gaode), which is a popular electronic navigation map in China that provides information on the names, location, and types of various retail stores. Based on previous studies and the classification of POI data [21,23,37], 72 subtypes of retail stores (as illustrated in Table 1) were extracted from the POI dataset. According to the Retail Type Categorization of China (RTCC), they were categorized within six major categories, including shopping malls, supermarkets, convenience stores, specialty stores, electronics stores, and building material stores. A total of 91,243 POI retail stores in Beijing were extracted. The distributions of the six types of POI are shown in Figure 2.

Table 1. Categories and total counts of POI.

Category	Sub-Category	Total Counts
Shopping malls	Shopping Plaza, Shopping Center, etc.	768
Supermarkets	Carrefour, Wal-Mart, Hualian, Watsons, etc.	12,756
Convenience stores	7-ELEVEN, Circle K, etc.	15,027
Specialty stores	Sports Store, Clothing Store, Franchise Store, Personal Care Items Shop, etc.	27,222
Electronics stores	Home Electronics Hypermarket, Digital Electronics, Mobile Handsets Sales, etc.	7894
Building material stores	Furniture Store, Kitchen Supply, Hardware Store, Lighting, Porcelain Market, etc.	27,576

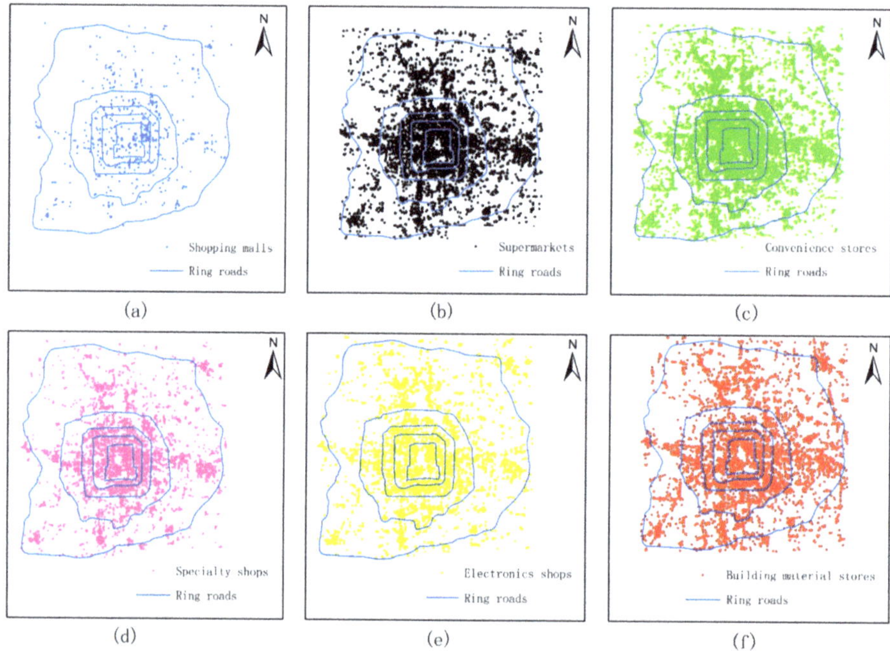

Figure 2. The POI distributions of six types of retail stores: (**a**) shopping malls; (**b**) supermarkets; (**c**) convenience stores; (**d**) specialty stores; (**e**) electronics stores; (**f**) building material stores.

According to the Beijing Statistical Yearbook in 2018, the public transport lines of Beijing sum to a total length of 19,881 km, including 637 km of metro lines. The annual passenger volume of public transport is 7038.18 million, which includes 3848.43 million metro passengers. Approximately 565 bus lines and 22 subway lines pass through the case study area, and there are more than 3000 bus stations and 259 subway stations within the sixth ring road of Beijing. Approximately 7.5 million bus and 2.5 million subway cards swipes are recorded each day. The modal shares of public transport in the urbanized areas of Beijing are more than 30%.

The public transit flow data used in this study were obtained from one week of bus and subway smart card data (SCD) from 19 April to 25 April 2015, which were obtained from the Beijing Public Transport Group. In recent years, two big events have serious impacts on public transit in Beijing. In 2014, Beijing started a price reform on its public transport system and adjusted public transportation fares to a higher level since then. Another event happened in 2019: the transport flows were much impacted by COVID-19. Therefore, the year 2015 may well reflect the stage of post-era of price reform and pre-era of COVID-19. We processed the data in two steps. First, the total flow for one week was accumulated by time periods of one day to capture temporal changes in transit flow. Various divisions of time periods have been used to aggregate the datasets in previous studies [38–40]. Considering the purpose of analysis and data features, the dataset was organized into seven periods based on two-hour intervals from 7:00 to 21:00 in the day. Then, we separated the weekly data according to weekdays and weekends to detect the differences in public traffic flow between working days and rest days. Sample records and selected fields of smart card data are shown in Table 2. All flow data were accumulated on the above-mentioned raster grid with a cell size of 1 km × 1 km and were based on 7 time periods. The aggregation process was completed in Python.

Table 2. Sample records of smart card data.

Time	Card Number	Type	Line Number	Vehicle Number	Boarding Station	Departure Station
20150813091012	46,343,397	1	751	95,740	17	11
20150813112013	80,245,649	1	609	83,601	5	8

2.2. Research Methods

In this paper, the SCDs of buses and subways are used to construct a network of public transport flows, and then, a weighted MCA model is used to calculate centralities for multiple time slices. The kernel density estimation (KDE) method is used to transform the centrality indices and the distribution of different types of retail stores to the same data framework.

Constructing a network is the basis for further complex network analysis. In this study, a weighted complex network is established according to public transport flows in the study area. Each raster grid is abstracted as a network node, and then, the transport flows between nodes are used as the weights of edges between nodes. The generated complex network has the topological characteristics of P-space, as all stops along a route can be connected if there is one line connecting two nodes [41].

2.2.1. Multiple Weighted Centrality Assessment Indices

Centrality indices provide a common and effective approach to analyze the spatial configurations of transport networks [42]. For a flow network, weighted complex indices have been developed and applied to public transport [43–45]. We select three critical indices in the MCA model to measure the characteristics of centrality: namely, weighted degree, weighted betweenness, and weighted closeness. These measures were computed by using the "networkx" package in Python [46].

Equation (1): weighted node degree centrality (WNDC). The unweighted degree is a basic indicator that is defined as the number of nodes that are connected to the focal node [47]. In a weighted network, WNDC is generally defined as the sum of weights and labeled as node strengths [48]. In this study, WNDC is defined as the traffic flow between network nodes on the constructed complex network that directly flows in or out of a node, which is formalized as follows:

$$\text{WNDC}_i^w = \sum_{j \in v(i)} w_{ij} \tag{1}$$

where w_{ij} represents the traffic flows between nodes i and j. Here, the WNDC value of node i is the total volume of the passenger O-D flows connected with node i.

Equation (2): weighted node betweenness degree (WNBC). The original indicator of betweenness refers to how often a node is traversed by the shortest paths connecting all pairs of nodes in the network [47]. In a weighted network, it has been suggested that the reciprocal link weights should be used to define the shortest path in a weighted graph, which reflects the ability to transmit through the chain or indicates whether a node is included in a path with a relatively large flow [49]. Here, the WNBC is adopted, which can be formalized as follows:

$$\text{WNBC}_i^w = \sum_{k \neq i \neq j \in N} \frac{\delta_{kj}(i)}{\delta_{kj}} \tag{2}$$

where δ_{kj} is the number of shortest paths between nodes j and k and $\delta_{kj}(i)$ is the number of these shortest paths through node i.

Equation (3): weighted node closeness centrality (WNCC). The original indicator of closeness is the average distance from a given starting node to all other nodes in the network [50]. It measures how close a node is to all other nodes along the shortest paths of the network. In a weighted network, WNCC considers both the number of intermediary nodes and the tie weights [51], which are defined as:

$$\text{WNCC}_i^w = \frac{n-1}{\sum_{j \in v(i)} d_{ij}}$$

(3)

where n is the total number of nodes in the network. d_{ij} is the shortest distance between nodes i and j. In a public transport flow network, d_{ij} is the minimum number of nodes to pass between nodes i and j. The weight in this case is defined in the same manner as that in the weighted betweenness.

2.2.2. Using KDE to Convert Density Values to a Grid Frame

The KDE method is used to convert the density values of retail stores and multiple centrality values to the same raster data frame to further perform correlation analysis. The advantage of KDE is that the density values at the middle locations of the raster grid are generated by considering the surrounding events [52,53]. For points that fall within the search range, different weights are assigned. The closer the point to the search center, the greater the weight, and vice versa. Equation (4) for estimating the kernel density at point x at the center of a grid is as follows:

$$\hat{f}(x) = \frac{1}{nh} \sum_{i=1}^{n} K\left(\frac{x - x_i}{h}\right)$$

(4)

where K is the kernel function, h is the bandwidth, and n is the total number of points within the bandwidth. In this study, the grid cell size is set at 1 km × 1 km, and a bandwidth is set at 5 km. The KDE tool in ArcGIS was used to obtain the density values.

3. Results

3.1. Distribution Characteristic of Retail Stores

Figure 3 shows the spatial distribution characteristics of the KDE values of six types of retail stores. The values are graded into five classes in the sub-figures, and the method of natural breaks is applied, which minimizes the sum of variance within the groups. A general pattern of higher values in the core area and lower values in the peripheral areas can be observed. Among the densities of the six types of retail stores, building material stores have the largest average density, which is followed by specialty stores, convenience stores, supermarkets, electronics stores, and shopping malls. For the high-density centers, building material stores, specialty stores, supermarkets, and electronics stores had multiple centers. Shopping malls and convenience stores showed a strong monocentric pattern.

3.2. Distribution Characteristics of Weighted Centrality

Figure 4 shows the spatial distributions of three weighted centrality indices: namely, weighted degree, weighted betweenness, and weighted closeness, based on the network constructed by the total public transport flows in the week. The lighter the color, the lower the centrality value. The degree values gradually decrease from the core to peripheral areas, and high values are mainly distributed within the fourth ring road. Betweenness presents a pattern with a high-value core and multiple secondary centers. The high-value core of the closeness is mainly distributed between the second east ring road and fourth east ring road. The closeness also exhibits a decreasing trend from the core to the peripheral areas, and the area with a high value covers a wider range.

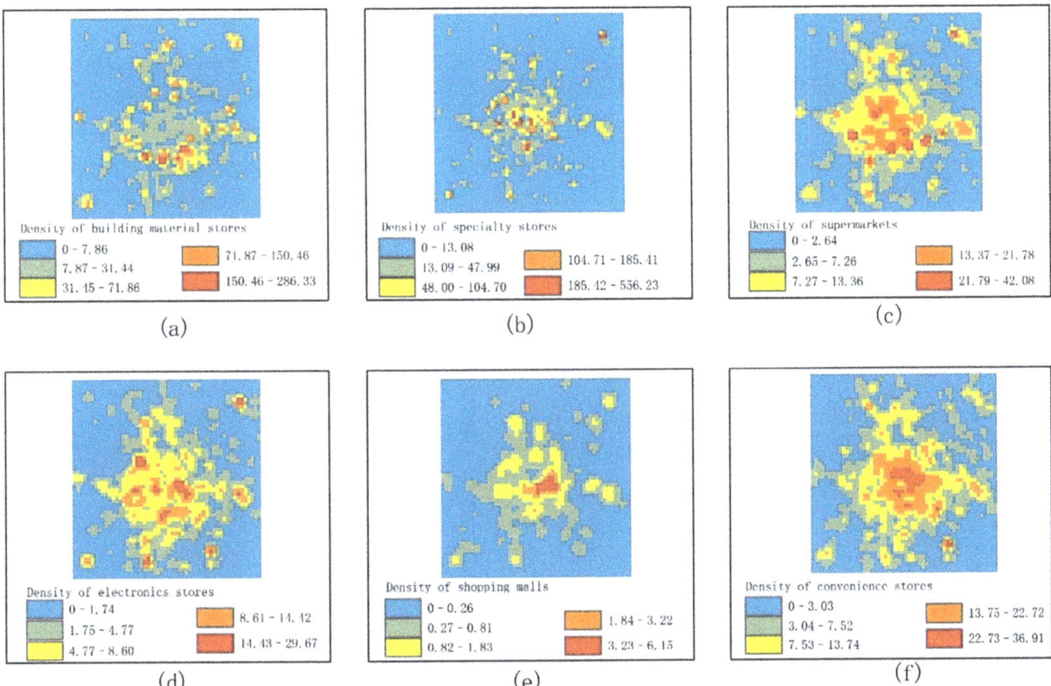

Figure 3. Density distributions of different types of retail stores determined by KDE: (**a**) building material stores; (**b**) specialty stores; (**c**) supermarkets; (**d**) electronics stores; (**e**) shopping malls; (**f**) convenience stores.

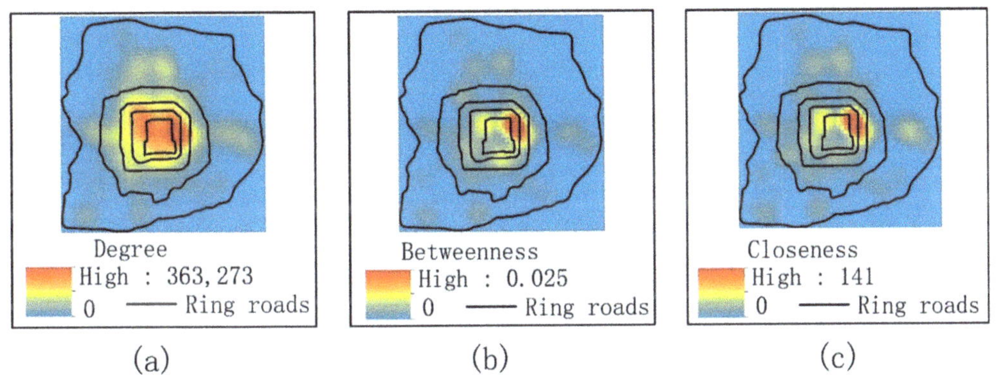

Figure 4. Spatial distributions of three weighted centrality indices of the total flow network: (**a**) degree; (**b**) betweenness; (**c**) closeness.

Figure 5 shows the temporal changes in the average values of the three weighted centrality indices on weekends and on weekdays. The horizontal axis represents time, and the points on the graph correspond to the median values of the different time periods.

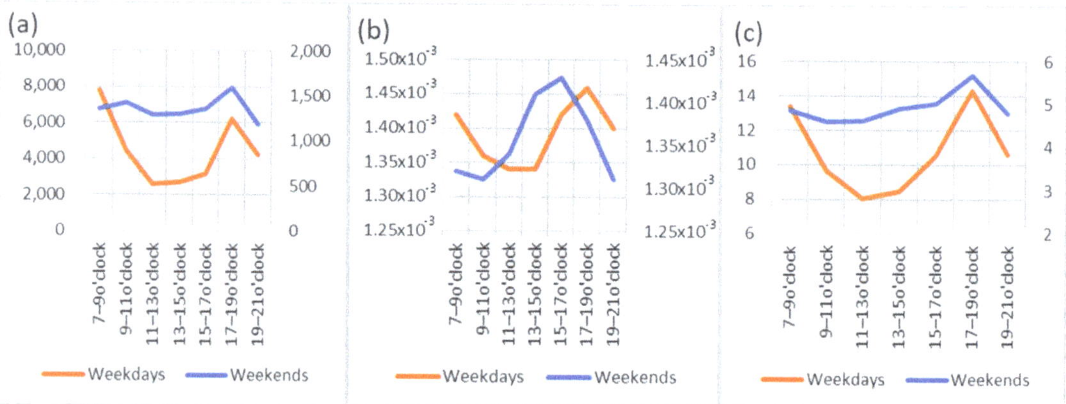

Figure 5. Temporal variations of the three weighted centrality indices on weekdays and weekends: (**a**) degree; (**b**) betweenness; (**c**) closeness.

The vertical axis represents the centrality values. Overall, three indices show quite different temporal patterns. For weekdays, the degree curve shows two peaks, which indicate a morning peak from 7:00 to 9:00 and an evening peak from 17:00 to 19:00, and the value of the early peak is greater than that of the late peak. The betweenness and closeness curve also show two peaks, but the late peak is greater than the early peak. The low point of three indices appeared at 11:00–13:00, and an extra low point appeared at 13:00–15:00 for betweenness.

Compared with weekdays, the weighted degree curve for weekends fluctuates mildly before 17:00. The peak appeared at 17:00–19:00 and then the low point appeared at 19:00–21:00. For betweenness, the curve of weighted betweenness for weekends shows a trend of high in the middle and low on both sides. The peak appeared at 15:00–17:00, which is earlier than the time of the evening peak for weekdays (17:00–19:00). Compared with weekdays, the range of fluctuation for the weighted closeness curve for weekends is smaller. The peak appeared at 17:00–19:00.

The spatial distributions of weighted centrality indices in seven periods of a day from 7:00 to 21:00 are calculated, and here, we present three of them, including the morning period from 7:00 to 9:00, noon period from 11:00 to 13:00, and after-work period from 19:00 to 21:00.

Figure 6 shows the spatial distributions on weekdays. In general, the core area of Beijing maintains an advantageous position in the networks with public transit flow. Although the distributions exhibit certain similarities for different time periods for the same index, there are some differences. The degree centrality values between the west second ring and west third ring road change with time, with a trend of increasing first and then decreasing. For betweenness, the two secondary centers between the west second ring road and west third ring road and the south second ring road and south third ring road change over time, with a trend of increasing first and then decreasing. The other sub-centers also exhibit minor changes with time. For closeness, peripheral areas change slightly with time, and the central area also exhibits minor changes with time. Figure 7 shows the spatial distributions on weekends. Compared with weekdays, the weighted centrality values change relatively smoothly over the weekends.

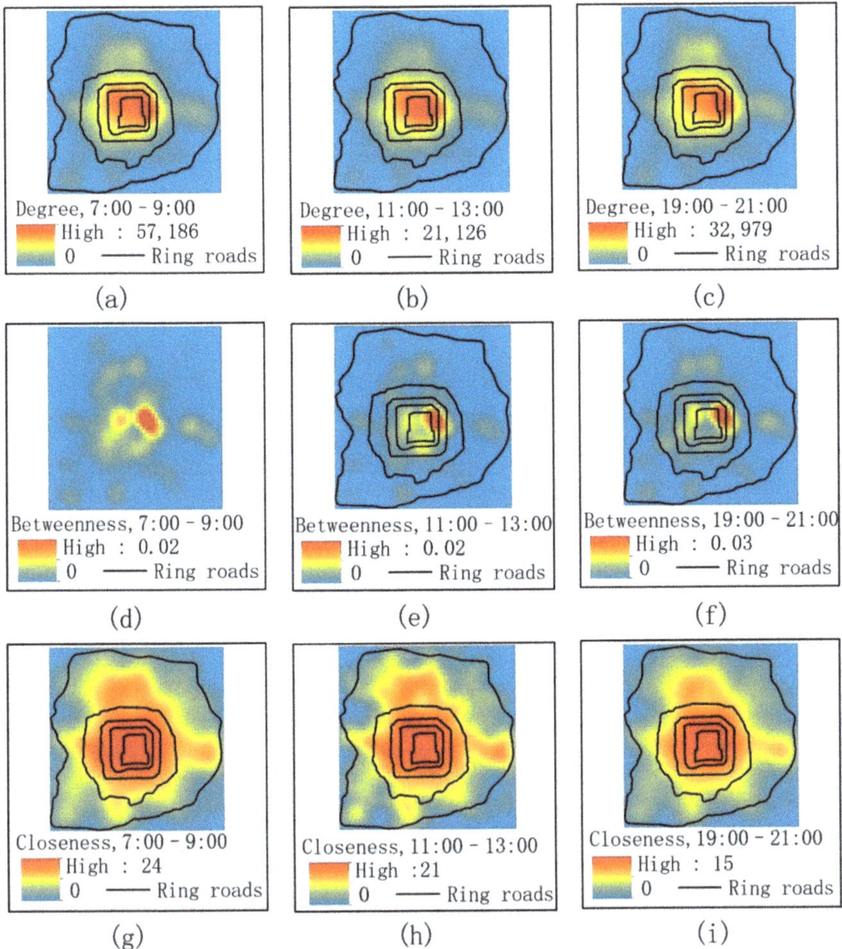

Figure 6. Spatial distributions of weighted degree centrality indices on weekdays: (**a**) degree, 7:00–9:00; (**b**) degree, 11:00–13:00; (**c**) degree, 19:00–21:00; (**d**) betweenness, 7:00–9:00; (**e**) betweenness, 11:00–13:00; (**f**) betweenness, 19:00–21:00; (**g**) closeness, 7:00–9:00; (**h**) closeness, 11:00–13:00; (**i**) closeness, 19:00–21:00.

3.3. Relationships between Retail Store Locations and Weighted Centrality from a Temporal Perspective

This section examines how the density distribution of retail stores may correlate with the weighted centrality indices. First, the flow network without temporal division is examined, which is constructed by the total public transit flows of the whole week. Table 3 shows the highest correlation coefficients between various retail stores and weighted centrality indices. Pearson's correlation analysis was conducted between the density of retail stores and weighted centrality indices.

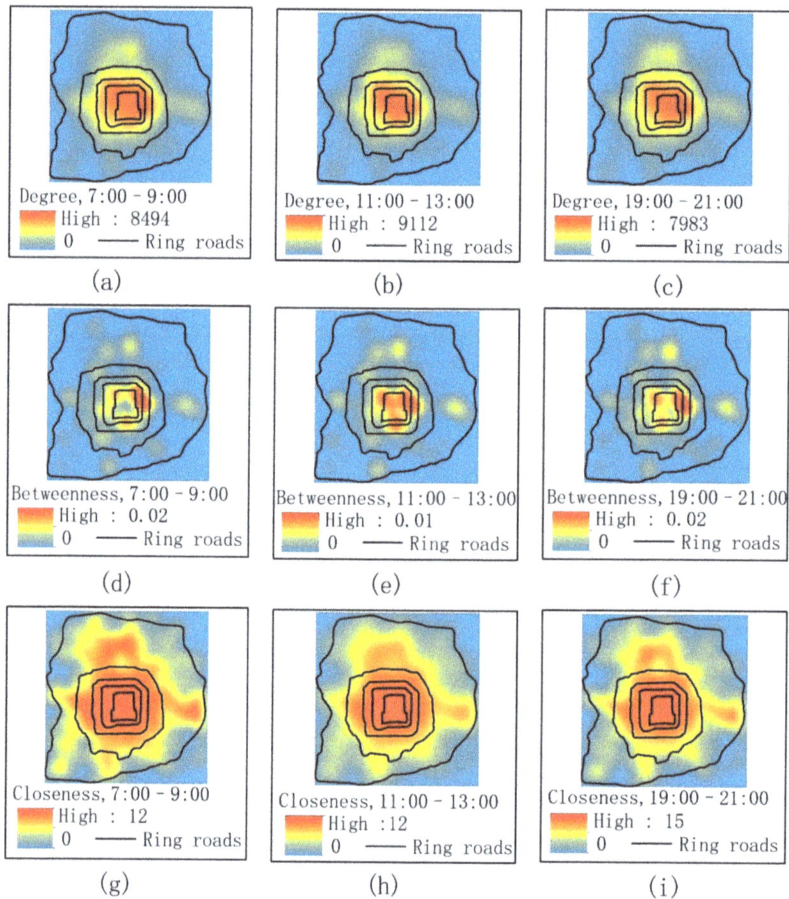

Figure 7. Spatial distributions of weighted degree centrality indices on weekends: (**a**) degree, 7:00–9:00; (**b**) degree, 11:00–13:00; (**c**) degree, 19:00–21:00; (**d**) betweenness, 7:00–9:00; (**e**) betweenness, 11:00–13:00; (**f**) betweenness, 19:00–21:00; (**g**) closeness, 7:00–9:00; (**h**) closeness, 11:00–13:00; (**i**) closeness, 19:00–21:00.

Table 3. Correlation coefficients of KDE values of stores and weighted centrality indices of total flow network.

Retail Types	Degree	Betweenness	Closeness
Shopping malls	0.770	0.785	0.580
Supermarkets	0.722	0.625	0.718
Convenience stores	0.812	0.747	0.740
Electronics stores	0.716	0.636	0.685
Specialty stores	0.553	0.485	0.413
Building material stores	0.261	0.211	0.371

First, most store types have rather high correlation coefficients with weighted centrality indices. Convenience stores, shopping malls, supermarkets, and electronics stores have strong correlations with all weighted centrality indices, with coefficients above 0.6. The highest correlation coefficients for each type of store are more than 0.7, and the highest coefficient is achieved by convenience stores (with values above 0.8). Specialty stores have the highest coefficient, exceeding 0.5. Only building material stores exhibit weak correlations with weighted centrality indices (the highest coefficient is less than 0.4), which is consistent with the previous findings by using street centrality indices [22–24], which implies that building material stores may be relatively less correlated to the public transport flow. These results indicate that most of the six types of retail stores are highly correlated to weighted centralities in the public transport flow network.

Second, four types, namely, supermarkets, convenience stores, electronics stores, and specialty stores, show the highest correlations with weighted degree. Only shopping malls show the highest correlation coefficients with weighted betweenness, with the highest correlation coefficient value reaching 0.785. This finding indicates that high-grade retail stores prefer nodes that are included in paths with relatively large flows. In comparison, it has been reported that betweenness performs well in previous physical street network studies [18,22–24]. This is also consistent with our findings, as betweenness in street networks reflects the frequency of the shortest paths passing through, while the weighted degree in this study directly reflects public traffic volume. The results indicate that transport volume has a significant impact on the location patterns of retail stores.

Then, the flow networks for different periods of a day with a distinction between weekdays and weekends are examined. Tables 4–6 show the temporal analysis results. Tables 4 and 5 show the correlation coefficients to the three weighted centrality indices across store types at different periods, and Table 6 shows the highest correlation coefficients for each store type across the periods of a day and the relative centrality indices. Tables 4 and 5 indicate that correlation coefficients vary across the day. For the relationship between weighted closeness and most retail stores on weekends, there is a continuous slight upward trend in the correlation coefficients with time. The relationships between the weighted degree and building material stores on weekdays are high in the morning and evening and low at noon. However, for the weighted degrees among specialty stores on weekdays, this pattern is reversed.

Table 4. Correlation coefficients of KDE values of stores and weighted centrality indices for each period on weekdays.

Centrality	Retail Types	7:00–9:00	9:00–11:00	11:00–13:00	13:00–15:00	15:00–17:00	17:00–19:00	19:00–21:00
degree	Shopping mall	0.763	0.786	0.775	0.772	0.768	0.768	0.784
	Supermarket	0.723	0.711	0.717	0.714	0.718	0.717	0.718
	Convenience store	0.807	0.810	0.812	0.809	0.809	0.808	0.812
	Specialty store	0.545	0.546	0.562	0.563	0.561	0.552	0.538
	Electronics store	0.715	0.711	0.710	0.708	0.709	0.710	0.721
	Building material store	0.266	0.253	0.252	0.248	0.253	0.257	0.264
betweenness	Shopping mall	0.771	0.786	0.814	0.811	0.815	0.754	0.775
	Supermarket	0.643	0.615	0.637	0.635	0.648	0.572	0.605
	Convenience store	0.751	0.738	0.762	0.763	0.774	0.710	0.724
	Specialty store	0.473	0.460	0.506	0.511	0.515	0.443	0.453
	Electronics store	0.638	0.620	0.650	0.643	0.656	0.581	0.618
	Building material store	0.222	0.215	0.215	0.208	0.220	0.177	0.204
closeness	Shopping mall	0.570	0.589	0.611	0.616	0.628	0.622	0.640
	Supermarket	0.715	0.727	0.741	0.744	0.748	0.742	0.750
	Convenience store	0.733	0.749	0.766	0.770	0.776	0.769	0.783
	Specialty store	0.406	0.420	0.439	0.443	0.454	0.444	0.455
	Electronics store	0.680	0.691	0.704	0.708	0.713	0.709	0.718
	Building material store	0.373	0.372	0.369	0.371	0.368	0.371	0.368

Table 5. Correlation coefficients of KDE values of stores and weighted centrality indices for each period on weekends.

Centrality	Retail Types	7:00–9:00	9:00–11:00	11:00–13:00	13:00–15:00	15:00–17:00	17:00–19:00	19:00–21:00
degree	Shopping mall	0.747	0.747	0.742	0.747	0.750	0.750	0.763
	Supermarket	0.749	0.733	0.724	0.719	0.720	0.727	0.734
	Convenience store	0.817	0.811	0.806	0.805	0.806	0.810	0.820
	Specialty store	0.540	0.554	0.559	0.563	0.564	0.557	0.550
	Electronics store	0.733	0.717	0.712	0.709	0.709	0.719	0.727
	Building material store	0.294	0.270	0.260	0.253	0.253	0.266	0.277

Table 5. *Cont.*

Centrality	Retail Types	7:00–9:00	9:00–11:00	11:00–13:00	13:00–15:00	15:00–17:00	17:00–19:00	19:00–21:00
betweenness	Shopping mall	0.752	0.758	0.769	0.753	0.729	0.766	0.769
	Supermarket	0.688	0.694	0.703	0.655	0.639	0.667	0.694
	Convenience store	0.765	0.776	0.790	0.754	0.734	0.770	0.777
	Specialty store	0.486	0.504	0.556	0.587	0.581	0.564	0.519
	Electronics store	0.672	0.677	0.685	0.644	0.630	0.657	0.683
	Building material store	0.266	0.267	0.266	0.220	0.213	0.226	0.256
closeness	Shopping mall	0.577	0.588	0.601	0.610	0.622	0.624	0.640
	Supermarket	0.726	0.728	0.735	0.740	0.744	0.744	0.752
	Convenience store	0.745	0.749	0.758	0.764	0.770	0.771	0.783
	Specialty store	0.413	0.423	0.433	0.441	0.451	0.451	0.460
	Electronics store	0.688	0.691	0.699	0.704	0.710	0.711	0.719
	Building material store	0.375	0.368	0.367	0.366	0.364	0.366	0.366

Table 6. Highest correlation coefficients of all periods.

Store Types	Weekdays			Weekends		
	Period	Centrality	Coefficient	Period	Centrality	Coefficient
Shopping malls	19:00–21:00	Betweenness	0.815	11:00–13:00	Betweenness	0.769
Supermarkets	19:00–21:00	Closeness	0.750	19:00–21:00	Closeness	0.752
Convenience stores	19:00–21:00	Degree	0.812	19:00–21:00	Degree	0.820
Specialty stores	11:00–13:00	Degree	0.563	13:00–15:00	Betweenness	0.587
Electronics stores	19:00–21:00	Degree	0.721	7:00–9:00	Degree	0.733
Building material stores	7:00–9:00	Closeness	0.373	7:00–9:00	Closeness	0.375

Table 6 shows that most of the highest correlation coefficients are rather large both on weekends and on weekdays. Compared with the results for the total flow network (Table 3), the values of the highest correlation coefficients here are larger, which means that analyses without time divisions may underestimate correlations. For the same store types, most types, except for shopping malls, have higher correlations on weekends. Most types show consistency in a preference for the highest centrality index from weekdays to weekends. Only the index type of specialty stores changes in degree on weekdays to betweenness on weekends, but its correlations are less than 0.7.

It is noteworthy that the three types of shopping malls, supermarkets, and convenience stores sell general commodities but differ in store size and diversity in their commodity types. For these three types, they all nearly achieve the highest correlations during the period of 19:00–21:00 for the whole week, while the only outlier is that the shopping mall type correlates more strongly to a different period of 11:00–13:00 on weekends. The same period implies that most consumers go shopping after work, but shopping behavior for malls on weekends may differ, as people may like to spend time in malls.

Another interesting result for the three types is that the highest centrality indices are different: convenience stores correlate best with degree, supermarkets correlate best with closeness, and shopping malls correlate best with betweenness. Recall that for the total flow network without periods in Table 3, the highest centrality index for the supermarket changed here from degree to closeness. In this case, the results of the total flow network may be misleading. Moreover, recall that the degree reflects the total traffic flow, the closeness reflects the closeness to all nodes in the flow network, and the betweenness reflects the traffic corridor. Thus, it can be inferred that the higher levels of store types are associated with higher correlations to the key structure of the flow network.

For the two types of specialty stores and electronics stores, both correlate best to degree centrality on weekdays, and neither correlate best with the period of 19:00–21:00 on weekends. These results imply that people may visit these types of stores after work on weekdays and may visit them at various periods in the daytime on weekends.

4. Discussion and Conclusions

This paper examines the relationships between the spatial distributions of six types of retail stores and their weighted centrality indices in the public transport flow network from the perspective of temporal dynamics. Three weighted node centrality indices were measured, e.g., degree, betweenness, and closeness. This study contributes to existing research on static physical street networks by analyzing the traffic flows of networks and their dynamic time processes.

The findings illustrate that generally, the distribution patterns of six types of retail stores are influenced by weighted street centrality significantly. Except for building material stores, all types of stores are highly correlated with weighted centrality indices. Among the three weighted centrality indicators, weighted degree is the best for four types of retail stores in terms of correlation coefficients and is followed by closeness and betweenness.

Temporal analysis can reveal more details and allow an inference of consumer behaviors. The correlation coefficients at different periods on weekdays and weekends vary over the time of day. For shopping malls, supermarkets, and convenience stores, the highest correlation coefficients on weekdays occur during the after-work period of 19:00 to 21:00. These may change on weekends for shopping malls, as shopping malls provide more than shopping services. Lower store levels correlate to degree centrality, that is, traffic volume itself, such as convenience stores and electronics stores. Higher store levels are correlated with the spatial characteristics of the flow network, such as closeness or betweenness. For specialty stores and electronics stores, people may visit these types of stores after work on weekdays and visit them at various times of the day on weekends.

This research provides a more comprehensive understanding of retail location analysis from a static physical street network to a dynamic flow network. Further research can be conducted to examine the following topics. As consumers may travel in a variety of traffic modes, the flow network that is based on various travel modes is needed to more comprehensively describe traffic flow information. In addition, there is a significant characteristic of disparity of centrality in different cities. Thus, it is necessary to identify the differences among different cities by conducting more case studies.

Author Contributions: C.L.: Conceptualization, Methodology, Writing—Original draft; T.D. (Teqi Dai): Writing—Review and editing, Supervision, Project administration, Funding acquisition; P.Z.: Software, Formal analysis, Data curation; T.D. (Tiantian Ding): Resources, Validation, Visualization. All authors have read and agreed to the published version of the manuscript.

Funding: This research received no external funding.

Institutional Review Board Statement: Not applicable.

Informed Consent Statement: Not applicable.

Data Availability Statement: Not applicable.

Conflicts of Interest: The funders had no role in the design of the study; in the collection, analyses, or interpretation of data; in the writing of the manuscript, or in the decision to publish the results. The authors declare no conflict of interest.

References

1. Taneja, S. Technology moves in Chain store age. *GeoJournal* **1999**, *75*, 136–138.
2. Roig-Tierno, N.; Baviera-Puig, A.; Buitrago-Vera, J.; Mas-Verdú, F. The retail site location decision process using GIS and the analytical hierarchy process. *Appl. Geogr.* **2013**, *40*, 191–198. [CrossRef]
3. Hillier, B.; Penn, A.; Hanson, J.; Grajewski, T.; Xu, J. Natural movement: Or, configuration and attraction in urban pedestrian movement. *Environ. Plan. B* **1993**, *20*, 29–66. [CrossRef]
4. Wang, F.; Guldmann, J.M. Simulating urban population density with a gravity-based model. *Socio-Econ. Plan. Sci.* **1996**, *30*, 245–256. [CrossRef]
5. Liu, Y.; Wang, H.; Jiao, L.; Liu, Y.; He, J.; Ai, T. Road centrality and landscape spatial patterns in Wuhan Metropolitan Area, China. *Chin. Geogr. Sci.* **2015**, *25*, 511–522. [CrossRef]

6. Nilsson, I.M.; Smirnov, O.A. Measuring the effect of transportation infrastructure on retail firm co-location patterns. *J. Transp. Geogr.* **2016**, *51*, 110–118. [CrossRef]

7. Rui, Y.; Yang, Z.; Qian, T.; Shoaib, K.; Nan, X.; Wang, J. Network-constrained and category-based point pattern analysis for Suguo retail stores in Nanjing, China. *Int. J. Geogr. Inf. Sci.* **2016**, *30*, 186–199. [CrossRef]

8. Glaeser, E.L.; Kolko, J.; Saiz, A. Consumer city. *J. Econ. Geogr.* **2001**, *1*, 27–50. [CrossRef]

9. Hernández, T.; Bennison, D. The art and science of retail location decisions. *Int. J. Retail Distrib. Manag.* **2000**, *28*, 357–367. [CrossRef]

10. Laporte, G.; Nickel, S.; Gama, F. *Introduction to Location Science*; Springer International Publishing: New York, NY, USA, 2015.

11. Jensen, P. Network-based predictions of retail store commercial categories and optimal locations. *Phys. Rev. E* **2006**, *74*, 035101. [CrossRef]

12. Sánchez-Saiz, R.M.; Ahedo, V.; Santos, J.I.; Gómez, S.; Galán, J.M. Identification of robust retailing location patterns with complex network approaches. *Complex Intell. Syst.* **2021**, *6*, 465.

13. Hansen, W. How accessibility shapes land use. *J. Am. Inst. Plan.* **1959**, *25*, 73–76. [CrossRef]

14. Luo, W.; Wang, F. Measures of spatial accessibility to healthcare in a GIS environment: Synthesis and a case study in Chicago region. *Environ. Plan. B Plan. Des.* **2003**, *30*, 865–884. [CrossRef]

15. Dawson, J.A. *Retail Geography*; Halsted Press: New York, NY, USA, 2013.

16. Casetti, E. Spatial analysis: Perspectives and prospects. *Urban Geogr.* **1993**, *14*, 526–537. [CrossRef]

17. Porta, S.; Crucitti, P.; Latora, V. The network analysis of urban streets: A dual approach. *Phys. A Stat. Mech. Appl.* **2006**, *369*, 853–866. [CrossRef]

18. Porta, S.; Strano, E.; Iacoviello, V.; Messora, R.; Latora, V.; Cardillo, A.; Wang, F.; Scellato, S. Street centrality and densities of retail and services in Bologna, Italy. *Environ. Plan. B Plan. Des.* **2009**, *36*, 450–465. [CrossRef]

19. Omer, I.; Goldblatt, R. Spatial patterns of retail activity and street network structure in new and traditional Israeli cities. *Urban Geogr.* **2016**, *37*, 629–649. [CrossRef]

20. Wang, S.; Xu, G.; Guo, Q. Street centralities and land use intensities based on points of interest (poi) in Shenzhen, China. *ISPRS Int. J. Geo-Inf.* **2018**, *7*, 425. [CrossRef]

21. Li, Q.; Zhou, S.; Wen, P. The relationship between centrality and land use patterns: Empirical evidence from five Chinese metropolises. *Comput. Environ. Urban Syst.* **2019**, *78*, 101356. [CrossRef]

22. Wang, F.; Chen, C.; Xiu, C.; Zhang, P. Location analysis of retail stores in Changchun, China: A street centrality perspective. *Cities* **2014**, *41*, 54–63. [CrossRef]

23. Lin, G.; Chen, X.; Liang, Y. The location of retail stores and street centrality in Guangzhou, China. *Appl. Geogr.* **2018**, *100*, 12–20. [CrossRef]

24. Han, Z.; Cui, C.; Miao, C.; Wang, H.; Chen, X. Identifying spatial patterns of retail stores in road network structure. *Sustainability* **2019**, *11*, 4539. [CrossRef]

25. Hotelling, H. Stability in competition. *Econ. J.* **1929**, *39*, 41–57. [CrossRef]

26. D'Aspremont, C.; Gabszewicz, J.J.; Thisse, J.-F. On hotelling's "Stability in Competition". *Econometrica* **1979**, *47*, 1145–1150. [CrossRef]

27. Malleson, N. Building temporal dynamism into applied GIS research. *Appl. Spat. Anal. Policy* **2019**, *12*, 1–3. [CrossRef]

28. Kim, H.M.; Kwan, M.P. Space-time accessibility measures: A geocomputational algorithm with a focus on the feasible opportunity set and possible activity duration. *J. Geogr. Syst.* **2003**, *5*, 71–91. [CrossRef]

29. Newing, A.; Clarke, G.P.; Clarke, M. Developing and applying a disaggregated retail location model with extended retail demand estimations. *Geogr. Anal.* **2015**, *47*, 219–239. [CrossRef]

30. Geuens, M.; Brengman, M.; S'Jegers, R. Food retailing, now and in the future. a consumer perspective. *J. Retail. Consum. Serv.* **2003**, *10*, 241–251. [CrossRef]

31. Castillo-Manzano, J.I.; López-Valpuesta, L. Urban retail fabric and the metro: A complex relationship. Lessons from middle-sized Spanish cities. *Cities* **2009**, *26*, 141–147. [CrossRef]

32. Cervero, R.; Kang, C.D. Bus rapid transit impacts on land uses and land values in Seoul, Korea. *Transp. Policy* **2011**, *18*, 102–116. [CrossRef]

33. Tsou, K.; Cheng, H. The effect of multiple urban network structures on retail patterns–A case study in Taipei, Taiwan. *Cities* **2013**, *32*, 13–23. [CrossRef]

34. Ganning, J.; Miller, M.M. Transit oriented development and retail: Is variation in success explained by a gap between theory and practice? *Transp. Res. Part D Transp. Environ.* **2020**, *85*, 102357. [CrossRef]

35. Gao, S.; Wang, Y.; Gao, Y.; Liu, Y. Understanding urban traffic-flow characteristics: A rethinking of between ness centrality. *Environ. Plan. B Plan. Des.* **2013**, *40*, 135–153. [CrossRef]

36. Ding, R.; Ujang, N.; Bin Hamid, H.; Manan, M.S.A.; Li, R.; Albadareen, S.S.M.; Nochian, A.; Wu, J. Application of complex networks theory in urban traffic network researches. *Netw. Spat. Econ.* **2019**, *19*, 1281–1317. [CrossRef]

37. Wang, L.-N.; Wang, K.; Shen, J.-L. Weighted complex networks in urban public transportation: Modeling and testing. *Phys. A Stat. Mech. Appl.* **2020**, *545*, 123498. [CrossRef]

38. Fransen, K.; Neutens, T.; Farber, S.; De Maeyer, P.; Deruyter, G.; Witlox, F. Identifying public transport gaps using time-dependent accessibility levels. *J. Transp. Geogr.* **2015**, *48*, 176–187. [CrossRef]

39. Boisjoly, G.; El-Geneidy, A. Daily fluctuations in transit and job availability: A comparative assessment of time-sensitive accessibility measures. *J. Transp. Geogr.* **2016**, *52*, 73–81. [CrossRef]
40. Gan, Z.; Feng, T.; Wu, Y.; Yang, M.; Timmermans, H. Station-based average travel distance and its relationship with urban form and land use: An analysis of smart card data in Nanjing City, China. *Transp. Policy* **2019**, *79*, 137–154. [CrossRef]
41. Lin, J.; Ban, Y. Complex network topology of transportation systems. *Transp. Rev.* **2013**, *33*, 658–685. [CrossRef]
42. Kuby, M.; Tierney, S.; Roberts, T.; Upchurch, C. *A Comparison of Geographic Information Systems, Complex Networks, and Other Models for Analyzing Transportation Network Topologies*; NASA/CR-2005-213522; Arizona State University: Tempe, AZ, USA, 2005.
43. Soh, H.; Lim, S.; Zhang, T.; Tse, C.-K. Weighted complex network analysis of travel routes on the Singapore public transportation system. *Phys. A Stat. Mech. Appl.* **2010**, *389*, 5852–5863. [CrossRef]
44. Huang, A.; Xiong, J.; Shen, J.; Guan, W. Evolution of weighted complex bus transit networks with flow. *Int. J. Modern Phys. C* **2015**, *27*, 1650064. [CrossRef]
45. Meng, Y.; Tian, X.; Li, Z.; Zhou, W.; Zhou, Z.; Zhong, M. Exploring node importance evolution of weighted complex networks in urban rail transit. *Phys. A Stat. Mech. Appl.* **2020**, *558*, 124925. [CrossRef]
46. Brandes, U. A faster algorithm for betweenness centrality. *J. Math. Sociol.* **2001**, *25*, 163–177. [CrossRef]
47. Freeman, L.C. Centrality in social networks: Conceptual clarification. *Soc. Netw.* **1978**, *1*, 215–239. [CrossRef]
48. Barrat, A.; Barthélemy, M.; Pastor-Satorras, R.; Vespignani, A. The architecture of complex weighted networks. *Proc. Natl. Acad. Sci. USA* **2004**, *101*, 3747–3752. [CrossRef] [PubMed]
49. Opsahl, T.; Agneessens, F.; Skvoretz, J. Node centrality in weighted networks: Generalizing degree and shortest paths. *Soc. Netw.* **2010**, *32*, 245–251. [CrossRef]
50. Sabidussi, G. The centrality index of a graph. *Psychometrika* **1966**, *31*, 581–603. [CrossRef]
51. Sun, X.; Wandelt, S.; Linke, F. Temporal evolution analysis of the European air transportation system: Air navigation route network and airport network. *Transportmetr. B Transp. Dyn.* **2015**, *3*, 153–168. [CrossRef]
52. Fotheringham, A.S.; Brunsdon, C.; Charlton, M. *Quantitative Geography: Perspectives on Spatial Data Analysis*; Sage Publications: London, UK, 2000.
53. Wang, F. *Quantitative Methods and Applications in GIS*; Taylor & Francis: Boca Raton, FL, USA, 2006.

Article

Towards Health Equality: Optimizing Hierarchical Healthcare Facilities towards Maximal Accessibility Equality in Shenzhen, China

Zhuolin Tao [1], **Qi Wang** [2] **and Wenchao Han** [3],*

[1] Faculty of Geographical Science, Beijing Normal University, No. 19, Xinjiekouwai Ave., Beijing 100875, China; taozhuolin@bnu.edu.cn

[2] Proficiency Skill Appraisal and Guidance Center of Natural Resources Ministry, Beijing 100830, China; 201221170050@mail.bnu.edu.cn

[3] Chinese Research Academy of Environmental Sciences, Beijing 100012, China

* Correspondence: 201631490013@mail.bnu.edu.cn

Abstract: Equal accessibility to healthcare services is essential to the achievement of health equality. Recent studies have made important progresses in leveraging GIS-based location–allocation models to optimize the equality of healthcare accessibility, but have overlooked the hierarchical nature of facilities. This study developed a hierarchical maximal accessibility equality model for optimizing hierarchical healthcare facilities. The model aims to maximize the equality of healthcare facilities, which is quantified as the variance of the accessibility to facilities at each level. It also accounts for different catchment area sizes of, and distance friction effects for hierarchical facilities. To make the optimization more realistic, it can also simultaneously consider both existing and new facilities that can be located anywhere. The model was operationalized in a case study of Shenzhen, China. Empirical results indicate that the optimal healthcare facility allocation based on the model provided more equal accessibility than the status quo. Compared to the current distribution, the accessibility equality of tertiary and secondary healthcare facilities in optimal solutions can be improved by 40% and 38%, respectively. Both newly added facilities and adjustments of existing facilities are needed to achieve equal healthcare accessibility. Furthermore, the optimization results are quite different for facilities at different levels, which highlights the feasibility and value of the proposed hierarchical maximal accessibility equality model. This study provides transferable methods for the equality-oriented optimization and planning of hierarchical facilities.

Keywords: health equality; spatial optimization; hierarchical healthcare facilities; maximal accessibility equality; 2SFCA

Citation: Tao, Z.; Wang, Q.; Han, W. Towards Health Equality: Optimizing Hierarchical Healthcare Facilities towards Maximal Accessibility Equality in Shenzhen, China. *Appl. Sci.* **2021**, *11*, 10282. https://doi.org/10.3390/app112110282

Academic Editor: Jianbo Gao

Received: 27 September 2021
Accepted: 27 October 2021
Published: 2 November 2021

Publisher's Note: MDPI stays neutral with regard to jurisdictional claims in published maps and institutional affiliations.

1. Introduction

Healthcare services are widely regarded as one of the essential public services that affect residents' health and well-being. Efficient and equal provision of healthcare services to the population is always at the center of the governance and planning of healthy cities [1]. From the spatial perspective, the distribution of healthcare facilities directly influences the accessibility of residents to healthcare services and the utilization of healthcare services, which in turn impact their respective health outcomes [2–4]. Accessibility is a multidimensional concept that is related to both spatial and non-spatial factors [5,6]. The concept of spatial accessibility is adopted in this study, which measures how easily and how many opportunities can be reached by residents from different locations [7]. Ensuring essential and equal accessibility to healthcare services is a key target of the Sustainable Development Goal proposed by the United Nations [8,9]. In China, both the central and municipal governments have set up strategies to promote the equalization of healthcare services [10,11]. The worldwide outbreak of the COVID-19 pandemic and its far-reaching impacts have significantly highlighted health and safety issues and the rational planning of healthcare resources [12,13].

However, the distribution of healthcare services decided in the traditional manner often do not provide equal accessibility to all [3,14–16]. There are significant disparities in healthcare accessibility across different locations or different socio-economic groups (e.g., natives vs. immigrants, high-income vs. local income, and the elderly vs. the young) [17–19], which has important spatial/social equity implications [20]. In Shenzhen, China, the study area, significant inequality in healthcare accessibility has also been revealed by existing studies [16,21]. The irrational distribution of healthcare facilities and inequality of accessibility make a strong call for the optimization of healthcare facilities.

Academia from fields such as public health and geography have paid increased attention to the optimization of healthcare facilities. A series of optimization models have been developed, which are usually known as the location–allocation models [22,23]. Typically, these models set up one or more objective functions and a set of constraints [24]. Efficiency and equality are the most important objectives for allocating public facilities to different sites [25]. However, most of existing studies focus on the efficiency objectives such as minimizing the numbers or cost of facilities, maximizing the coverage of facilities, and minimizing the travel cost between consumers and facilities [26,27]. By contrast, little attention has been paid to equality of facilities distribution in location–allocation studies, partially due to the difficulty in modelling and optimizing equality [4].

Recently, an innovative stream of studies has considered spatial/accessibility equality in the location–allocation analysis [28–31]. The maximal accessibility equality (MAE) model developed by Wang and Tang [28] is a novel and helpful method for researchers and practitioners who are interested in improving the equality in demanders' accessibility to public services (e.g., healthcare services). However, the development and implementation of the MAE model are still confined to single-level facilities. Comparatively little attention has been paid to the equality optimization of hierarchical (or multi-level) healthcare facilities. As existing studies [32–34] have demonstrated, spatial analysis of hierarchical facilities should account for more characteristics such as various service scopes, frictions of distance, and transport modes. Therefore, the existent MAE model is not suitable for analyzing accessibility to hierarchical facilities. Although hierarchical location–allocation problems have been studied for decades [35–37], few have addressed the equality issue or incorporated spatial accessibility into location–allocation analysis. There are still gaps in terms of simultaneously considering the hierarchical nature and accessibility equality optimization of healthcare facilities.

This study's contributions are threefold. First, it develops a hierarchical maximal accessibility equality (HMAE) model, which is hierarchy-sensitive and can act as a useful tool in the equality-oriented spatial optimization of hierarchical healthcare facilities or other hierarchical facilities. Second, this study provides a method that simultaneously accounts for both existing fixed facility locations and newly added locations that are flexible in the location–allocation analysis. This can make the optimized solution more feasible because the fixed resources/stocks of existing facilities are considered in the optimization. Third, online map application programming interface (API) is introduced to improve the accuracy of estimated travel time in location–allocation analysis. The proposed model maximizes the equality of spatial accessibility to healthcare facilities by minimizing the variation in accessibility across all locations. This is achieved by both adding new facilities and reallocating the resources at existing facilities. The method is valuable for the implementation of equality-oriented healthcare planning and policymaking. Shenzhen, which is one of the first cities to have highlighted the policy goal to achieve equality in healthcare services in China, was selected as the study area to demonstrate the feasibility and usefulness of the proposed model.

2. Literature Review

2.1. Classic Location–Allocation Models

Serving as a tool for people to analyze and optimize locations of facilities, location–allocation models have been present for more than five decades [22]. There are a set of such models that are termed as the classic location–allocation models [24,26]. In addition to the p-median model, the classic models also include the maximal covering location model, the location set covering model, and the p-center model. The covering models deal with the coverage of demanders (usually represented by discrete and aggregate locations) within a certain radius of each facility [38]. The maximal covering location model aims to maximize the coverage on the basis of a certain number of facilities [27], whereas the location set covering model is designated to achieve full coverage using the least number of facilities. The p-center model is different from the above models, aiming to minimize the maximal distance from each demand nodes to its nearest facility. It is also known as the minimax problem [24]. To some extent, the p-center model considers equity issues by improving the situation of the remotest demanders.

The classic models are tailored to approach various policy objectives and have engendered numerous applications [22,24]. The classic models have also been extended and improved in other instances, e.g., the gravity p-median model that incorporates a gravity rule into the p-median model [39], making them applicable in more complicated real-world contexts. The classic models, however, are faced with several drawbacks [4]. First, they fail to explicitly address the equity issue. Most of the existing location–allocation models only address efficiency-oriented objectives. Second, the assumptions of the spatial interaction between demanders and facilities in these models are relatively simple. Most notably, few have employed realistic accessibility measurement when considering how demanders reach facilities.

2.2. Hierarchical Location–Allocation Problems

Researchers have developed location–allocation models for hierarchical facilities. Hierarchical facilities are a type of facility that consist of multi-level facilities, facilities at each level that provide (totally or partially) different functions of service within different territories [36,40]. A healthcare facility is a typical type of hierarchical facility [32].

From a modelling perspective, hierarchical facilities can be classified according to their flow patterns, service varieties, spatial configurations, and optimization objectives [36,37]. Flow pattern is about the organization and delivery of services among different levels of facilities. The single-flow pattern assumes demanders at each node are serviced in facilities at the lowest-level, then transferred to facilities at higher levels, if necessary. The multi-flow pattern indicates that demanders can be allocated to facilities at any level. Service varieties determine whether the functions at a lower level can also be supplied at higher levels. According to spatial configurations, the service scopes of facilities at a lower level should be in accord with those at higher levels. The optimization objectives of hierarchical location–allocation models are mainly built on the basis of the classic models described above, e.g., the hierarchical p-median model, hierarchical maximal covering model, and hierarchical location set covering model [35,41,42]. Therefore, these models focus on the efficiency of facility configurations and more or less overlook equity/equality issues. In addition, most hierarchical location–allocation models also fail to address the complex interactions between the demand and supply. In other words, the two drawbacks of the classic models previously pointed out by Wang [4] persist in hierarchical models.

2.3. The Maximal Accessibility Equality Model

Aiming to address the equality issue, Wang and Tang [28] initiated a novel location–allocation model, termed the "maximal accessibility equality" (MAE) model. The MAE model quantifies the equality dimension of facilities' spatial configuration as the sum of squares of differences in the accessibility to different facilities. The optimal configuration would minimize the disparity in the accessibility to facilities [28]. The objective function of this problem represents how spatial equality is understood and quantified. Solving the problem thus addresses the policy concerns over equality in public services. The MAE model can be expressed and solved as a quadratic programming problem.

Tao et al. [29] applied the maximal equality model to analyze optimal configuration of residential facilities and introduced the particle swarm optimization heuristic algorithm to solve the model. Another study tried to extend it by selecting newly added facility locations rather than reallocating resources at existing or given locations [43]. Two studies further introduced a two-step procedure, with the first step to optimize locations of facilities, while the second step to optimize the respective sizes of the facilities [30,44]. Dai et al. [31] incorporated a random allocation mechanism into the MAE model to optimize educational opportunities. To date, however, few have paid attention to the equality optimization of hierarchical facilities. To achieve this goal, the MAE model needs to be extended to account for hierarchical nature of facilities.

Note that a few recent studies have made efforts to model the spatial accessibility to hierarchical healthcare facilities [32–34]. These studies adapted the spatial accessibility measurements, e.g., the two-step floating catchment area (2SFCA) method, in order to account for the hierarchical characteristics of facilities, including variable service scopes, different distance frictions, and different transport modes for facilities at various levels. Although these studies fail to improve and optimize the accessibility to hierarchical facilities, the above advancements in modelling accessibility to hierarchical facilities can help to develop a hierarchical version of MAE model. The current study makes further efforts to combine the measurement of accessibility to hierarchical facilities and the MAE model that optimizes the equality of accessibility.

3. Data and Methods

3.1. Study Area and Data

Shenzhen was chosen for the empirical analysis. It was selected because existing studies have revealed hierarchical features in healthcare facilities and in accessibility to these facilities in Shenzhen [32]. Shenzhen is one of the special economic zones and megacities in China. It is located in the Pearl River Delta region in China. Shenzhen has undergone rapid socio-economic development in the last four decades since China's reform and opening in 1978. By 2018, Shenzhen has 11 million permanent residents and 1997 square kilometers of land area. It is composed of 10 administrative districts, 55 sub-districts (or *jiedao* in Chinese), and 771 communities (or *Shequ* in Chinese) (see Figure 1). In Shenzhen, like in other Chinese cities, public healthcare facilities play a predominant role in the provision of healthcare services to the residents. According to the "Medical Regulations of Shenzhen Special Economic Zone" issued by Shenzhen's municipal government in 2016, healthcare facilities in the city are organized as a three-level system. They are the tertiary, secondary, and primary healthcare facilities from the top to the bottom. The primary facilities consist of the community health service centers (CHSCs) serving the city's 771 communities. The secondary and tertiary facilities serve the 10 districts and the city, respectively. The current hierarchical healthcare system in Shenzhen is not well established, wherein the referral system between various levels has not been formulated, and patients are free to choose healthcare facilities at various levels [16].

The data used in this study comprised three types:

(1) community-level population counts;
(2) point-level healthcare facilities with attribute information such as names, hierarchy, number of physicians, and addresses;
(3) the travel time between community centroids to healthcare facilities.

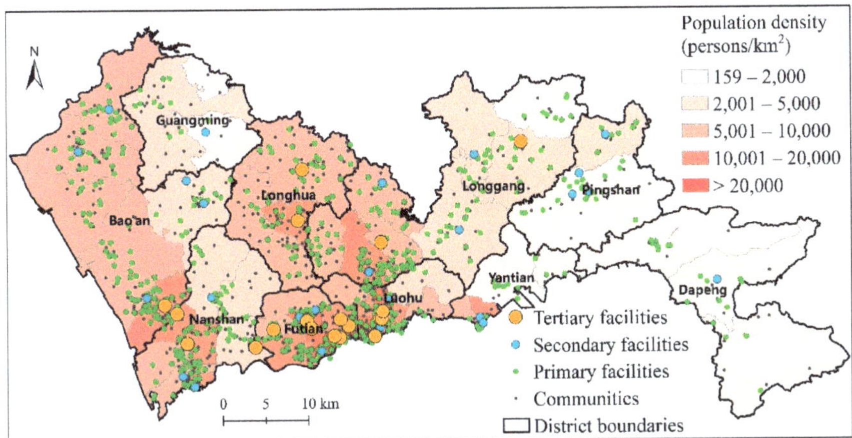

Figure 1. Location and distribution of population and healthcare facilities in Shenzhen.

The population counts are from the sixth population census of China, which was conducted in 2010 and is the most up to date of its kind available to the public. Centroids of communities and their geospatial information such as longitude and latitude coordinates are obtained via the geocoding API of Baidu Maps. The average population size of each community is 13,500. Each community is treated as a demand node in our ensuing analyses.

Detailed information concerning all the three levels' facilities is available at the official website of Shenzhen Municipal Health Commission [45]. As of December 2018, there were 19 tertiary healthcare facilities, 35 secondary healthcare facilities, and 612 primary healthcare facilities in Shenzhen. The average numbers of physicians in these facilities were 422, 182, and 6 physicians, respectively (Table 1).

Table 1. Basic statistics of the hierarchical healthcare facilities in Shenzhen.

Facility Levels	Number of Facilities	Total Physicians	Average Physicians
Primary facilities	612	3672	6
Secondary facilities	35	6377	182
Tertiary facilities	19	8012	474

The primary facilities are the largest in quantity and the smallest in the average number of physicians. They are widely distributed in all districts in Shenzhen (Figure 1). The primary facilities provide incredibly wide-ranging functions and service qualities that they usually become the first choice of patients residing in proximity. Therefore, the maximal equality model may be not applicable to these facilities. Furthermore, existing studies have found that the spatial accessibility to primary hospitals is relatively equal in Shenzhen [32]. Given this, only the tertiary and secondary facilities were considered in this study. In brief, tertiary and secondary healthcare facilities are different mainly in three aspects. First, tertiary facilities usually provide more complicated and higher-level services than secondary facilities. Second, tertiary facilities usually have larger coverage areas than secondary facilities, which can be reflected by the different catchment sizes in the model. Third, the average size (number of physicians) of tertiary facilities is significantly larger than that of secondary facilities.

The measurement of spatial accessibility relies on the travel times between demand nodes and different facility locations. Following existing studies [16,21], the travel times are estimated using the driving navigation API of Baidu Map [46], the most popular online map in China. The estimation is based on real-world transportation network, historical traffic congestion information, and the local driving rules. The departure time of these trips are assumed to be between 10 a.m. and 5 p.m. on weekdays. This is done to avoid extreme travel times during peak hours.

3.2. The Hierarchical Maximal Accessibility Equality (HMAE) Model

The maximal equality model was developed by Wang and Tang [28]. It aims to achieve equal accessibility by minimizing the variation in the spatial accessibility to healthcare facilities from different demand nodes. In this study, the original MAE model was further adapted into a hierarchical version. The objective function can be expressed as Equation (1).

$$minimize \sum_{i}^{m} P_i \left(A_i^l - \frac{\sum_i^m P_i A_i^l}{\sum_i^m P_i} \right)^2 \tag{1}$$

where A_i^l is the spatial accessibility at demand node i to facilities at level l, P_i is the population, and m is the number of demand nodes (i.e., communities in this study). The function means the population-weighted sum of the difference between the accessibility at each node and the population-weighted average accessibility. In the original study, spatial accessibility is calculated using the 2SFCA method. The generalized form of 2SFCA can be written as Equation (2).

$$A_i^l = \sum_{j}^{n} \frac{S_j^l f(d_{ij})}{\sum_k^m P_k f(d_{kj})} \tag{2}$$

where S_j^l is the supply size (amount of physicians in this study) at candidate level-l facility location j, n is the amount of candidate facility locations, d_{ij} is the travel cost (e.g., travel time or distance) between each demand node and each candidate facility location, and f is a function that describes the distance friction effect. In this study, the Gaussian-based 2SFCA method is adopted to measure spatial accessibility, which is advocated by existing studies on measuring healthcare accessibility [16]. The model takes a Gaussian distance friction function that is suitable for hierarchical facilities, which can be expressed as Equation (3):

$$f(d_{ij}) = \begin{cases} \frac{e^{-1/2 \times (d_{ij}/D_l)^2} - e^{-1/2}}{1 - e^{-1/2}}, & d_{ij} \leq D_l \\ 0, & d_{ij} > D_l \end{cases} \tag{3}$$

where D_l is the catchment area size, i.e., the radius of service scope, of candidate facility at level l. Note that in the traditional spatial accessibility and maximal equality optimization studies, the catchment area is the same for all facilities. When applied to hierarchical facilities, however, this setting is inappropriate. Following existing studies on the spatial accessibility to hierarchical healthcare facilities [32], we assigned different catchment sizes for facilities at different levels. Generally, the catchment size is larger for facilities at a higher level. Furthermore, on the basis of the Gaussian function, a larger catchment size also means a weaker distance friction for higher levels, which is another important characteristic of hierarchical healthcare facilities [32]. Note that the HMAE model in this study intends to maximize the equality of accessibility to healthcare facilities for each level independently. The reason for this setting is that the current hierarchical healthcare system in Shenzhen is a multi-flow and nested hierarchical system, where facilities at various levels provide service to residents independently [16].

3.3. Implementation of the HMAE Model

The process of spatial optimization is to determine the optimal value of the decision variable that can optimize the objective function for each facility level (tertiary and secondary) independently. Figure 2 summarizes the procedures in this study. The decision variable of the maximal equality model is the supply size S_j at each candidate location. It can be zero, which means no facility is located at the location, or any positive value. Therefore, the selection of candidate locations is crucial for spatial optimization. In existing studies, there are two ways to select candidate locations. The first way is to set up candidate locations without consideration of existing facilities. The candidate locations can be the centroids or random locations within administrative or census units, or evenly distributed locations across the study area. The second way, by contrast, aims to rearrange the supply of existing facilities. In other words, the locations of existing facilities are used as candidate locations. The advantage of the second way is that the existing resources can be accounted for, making it more cost-efficient and realistic for policy decision making.

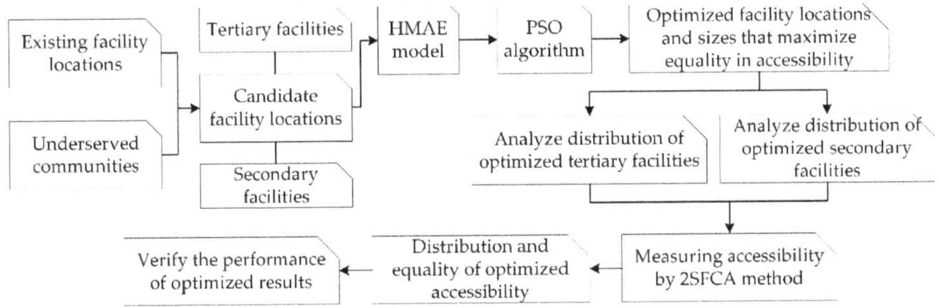

Figure 2. The framework of the procedures in this study.

In this study, the selection of candidate locations was based on the existing facility locations. Furthermore, considering that the existing facility locations might not be enough to provide coverage to all demand nodes within given catchment areas, we also accounted for the possibility that new facility locations may be needed. Despite the existing facility locations, we examined whether there are demand nodes that are located outside the catchment areas of all existing facilities. If yes, additional candidate locations were be selected from these underserved demand nodes. In sum, candidate locations for locating facilities consisted of two parts, i.e., existing facilities and underserved demand nodes. The selected candidate locations are described below.

This study introduces online map API to improve the estimation of travel time from patients to facilities. Specifically, the driving and transit navigation APIs provided by Baidu Map, a leading online map provider in China, were utilized to estimate travel time by driving or by public transit, respectively. Online map API can provide more accurate and reliable estimates of travel time on the basis of the frequently updated transport network, navigation rules, transit schedule, and traffic status [16,47].

A few studies have demonstrated that various transport modes should be considered in accessibility analysis such that heterogenous demand of different socio-economic groups can be reflected [16,34,48]. This study considered two transport modes, i.e., private car and public transit. The latter includes both regular buses and subway and inter-mode transfers. Following Tao and Cheng [19], travel times by the two modes were combined on the basis of modal shares. Modal shares of private car, bus, and subway at the district level were collected from the Shenzhen 2016 Travel Survey.

As for hierarchical healthcare facilities, the catchment sizes should vary across different levels. Following existing studies [19,21], we determined the catchment sizes on the basis of the exceptional breakpoint of the distribution of the travel time from each demand node to the closest existing facility. The threshold was determined so that most demand nodes were within the catchment areas of existing facilities. Only a few extreme demand nodes that were extremely far away from existing facilities were excluded from the catchment areas. As a result, the catchment sizes for tertiary and secondary facilities were 70 and 40 min of travel times, respectively (see Figure 3).

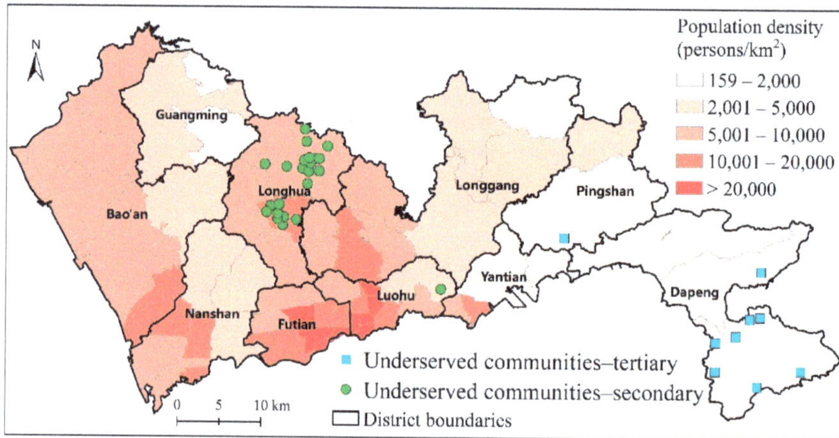

Figure 3. Underserved demand nodes of existing healthcare facilities.

In the next step, the travel times from each demand node to the closest existing tertiary and secondary facilities were calculated on the basis of the travel time matrix. If the minimum travel time between a demand node and tertiary facilities was larger than the catchment size, it could be considered to be an underserved demand node of tertiary facilities. The same procedure was executed for secondary facilities. The distribution of underserved demand nodes (communities) is shown in Figure 3.

The underserved demand nodes of tertiary facilities are concentrated in the east part of Shenzhen, one in Pingshan District, and the others in Dapeng District. Population density is relatively low in these areas. Therefore, one new candidate location was added in each district. Furthermore, even though all demand nodes in the west part of Shenzhen (i.e., Bao'an and Guangming districts) are covered by existing tertiary facilities, most of these demand nodes are quite far away from existing tertiary facilities. The closest tertiary facility is in the southernmost area of Bao'an District. Therefore, two new candidate locations were added in the northern Bao'an and Guangming Districts. The underserved demand nodes of secondary facilities are mainly concentrated in Longhua district, where moderate population density presents. There is another underserved demand node in Luohu District. Although Luohu is one of the central districts, population density in this subdistrict is relatively low, where the highest mountain, Wutong mountain, is located. Therefore, new candidate locations were only selected in Longhua District, on the basis of the distribution pattern of underserved demand nodes. Taken together, as shown in Figure 3, there are 23 and 39 candidate locations for tertiary facilities and secondary facilities, respectively. The average supply size of each facility was set as the same with existing facilities at each level. The total numbers of physicians at the two levels were found to be 9700 and 7100, respectively. Table 2 summarizes the setting of all parameters.

Table 2. Parameters at each level.

Facility Level	Catchment Size	Number of Underserved Communities	Number of Candidate Locations	Total Physicians
Secondary facilities	40 min	23	39	7100
Tertiary facilities	70 min	9	23	9700

Following existing studies [29,49], the HMAE model was solved by using the particle swarm optimization (PSO) algorithm developed by Kenned and Eberhart [50]. PSO specifies a fitness function to evaluate the performance of each possible solution, which is represented as the total accessibility difference calculated by Equation (1). It provides an efficient approach to solving optimization problems by simultaneously considering a wide range of possible solutions and moving towards the optimal solution in an evolutionary manner. Each solution is termed as a particle, which consists of the facility sizes at all candidate locations. In each iteration, the performance of current solutions is evaluated by the fitness function and compared with the previous iteration. If the current solution generates better performance, it would be retained and the solutions in the next iteration would be determined on the basis of the trend. These solutions are expected to converge to a global optimal solution after a certain number of iterations.

PSO was first introduced by Tao et al. [29] into the maximal equality model. PSO can be operationalized with a toolbox in MATLAB developed by Birge [51]. The implementation of PSO is required for set-up of a few parameters, among which range of X (i.e., the size of facility) and dimension of particles are related to the specific case. Dimension of particles was set as the number of candidate locations, i.e., 39 and 23 for tertiary and secondary facilities, respectively. Range of X is defined by upper bound and lower bound. The lower bound was set as 0. The upper bound was set as two times of the size of the largest existing facility, i.e., 1500 and 800 for tertiary and secondary facilities, respectively. Other parameters were determined on the basis of the work of Tao et al. [29] and the default parameters given by the manual of the toolbox.

4. Results

4.1. Optimal Distribution of Tertiary Healthcare Facilities

In addition to the existing 19 tertiary facilities, four new facility locations were added in areas that are quite far away from existing facilities. The newly added locations are located in Bao'an, Guangming, Pingshan, and Dapeng Districts. The optimal sizes of these candidate locations, both the existing facilities and newly added locations, were determined by using the hierarchical maximal equality model, aiming to minimize the variation in the spatial accessibility to facilities for all demand nodes.

The results are shown in Figure 4. The optimized facilities were classified into small-, middle-, and large-sized facilities (corresponding to facilities with 100–300, 300–800, or 800–1200 physicians, respectively) on the basis of the natural-breaks method. The ratios of three types of facilities were 48%, 39%, and 13%, respectively. Small- and middle-sized facilities are dominant. There are only three large-sized tertiary facilities, which are respectively located in Nanshan, Bao'an and Guangming Districts. By contrast, the tertiary facilities in Futian and Luohu Districts, which are regarded as the core of Shenzhen, are middle- or small-sized. However, a relatively large number of existing tertiary facilities are concentrated in Futian and Luohu Districts. The tertiary facilities in Longhua, Longgang, and Pingshan Districts, where the distribution of facilities is relatively dispersed, are mainly middle-sized. The only tertiary facility in Dapeng District, which is newly added, is small-sized, due to the low population density in Dapeng and the surrounding areas. Generally, the optimized distribution of tertiary healthcare facilities presents a pattern in that facilities in the central areas are densely distributed but small-to-middle-sized, while facilities in the peripheral areas are middle-to-large-sized but dispersedly distributed.

The differences between optimized sizes and actual sizes of existing tertiary facilities were also calculated. This can help determine which adjustments of existing facilities are needed to achieve the optimal distribution, which is useful for decision making. As shown in Figure 5, the differences were significant, indicating that large adjustments are needed to materialize healthcare accessibility equality. In other words, the distribution of existing tertiary facilities is poorly performed in terms of providing equal healthcare accessibility.

Existing tertiary facilities that require positive size adjustments are mainly located in Nanshan, Bao'an, Longgang, and Longhua districts, while most facilities in Luohu and Futian districts need to be cut down in size. The pattern reveals that to achieve equal accessibility, more healthcare resources need to be allocated in the peripheral areas. Note that in the optimized distribution, some downsized ("negative adjustment") facilities may be close to upsized ("positive adjustment") facilities, e.g., the example marked by a yellow box in Figure 5. In such cases, the positive and negative adjustments close to each other can be counteracted. By doing so, many costs of adjustments can be saved, but with only negligible impacts on resources distribution and healthcare accessibility.

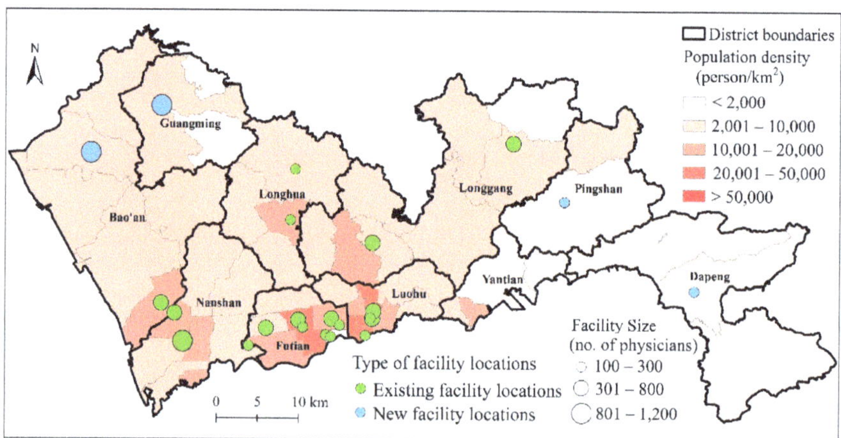

Figure 4. Optimized distribution of tertiary healthcare facilities.

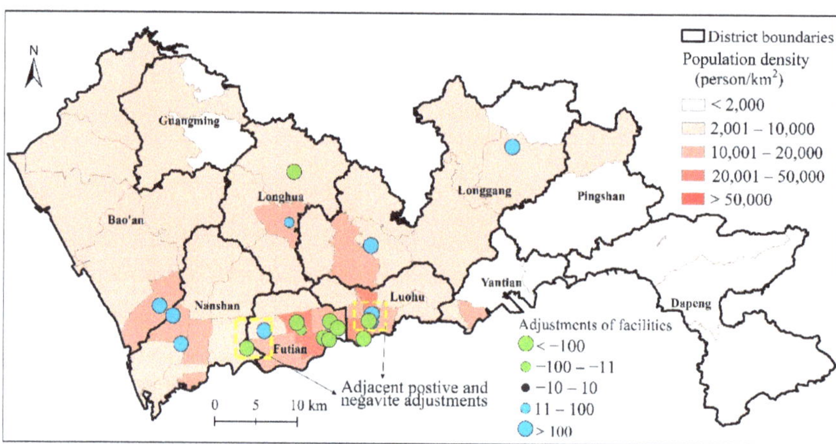

Figure 5. Size adjustments of existing tertiary healthcare facilities according to optimization.

4.2. Optimal Distribution of Secondary Healthcare Facilities

Similarly, the optimized secondary healthcare facilities are classified according to their sizes using the natural breaks method. The sizes of small-, middle-, and large-sized secondary facilities were less than 100, 100–300, and 300–600 physicians, respectively. The ratios of facility amount for three types were 46%, 36%, and 18%. The number of small-sized secondary facilities was the largest. As shown in Figure 6, the distribution of each type of secondary facilities is relatively even in most districts. In Pingshan, Dapeng, and Yantian districts, however, optimized secondary facilities were found to be relatively small in size. Large-sized secondary facilities with more than 300 physicians are relatively evenly distributed, which can cover moderate-to-high population density (higher than 10,000 persons/km²) areas within a relatively small distance.

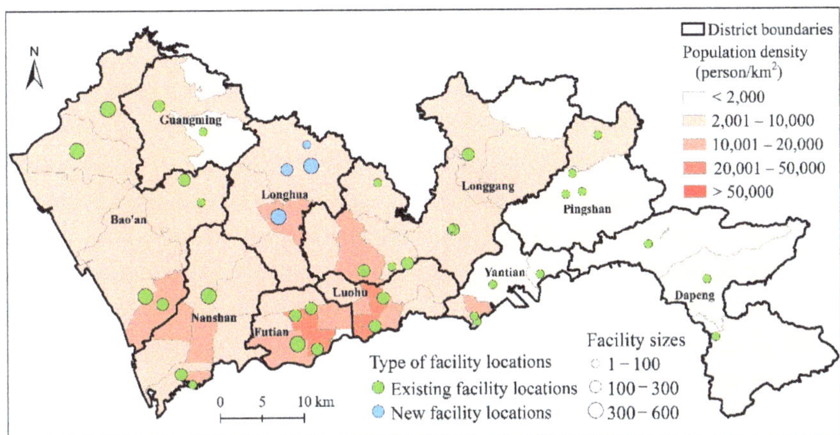

Figure 6. Optimized distribution of secondary healthcare facilities.

The needed size adjustments of existing secondary facilities to achieve equal accessibility are shown in Figure 7. Facilities that require positive size adjustments are mainly located in areas with relatively high population density. This indicates that the distribution of existing secondary facilities may fail to match the distribution of the demand, which can result in poor and unequal healthcare accessibility. There are also situations where negative and positive adjustment are close to each other, e.g., the areas marked by yellow boxes in Futian and southern Longgang districts. Counteracting these inverse size adjustments can make the optimized solution more economically feasible with negligible impacts on healthcare accessibility.

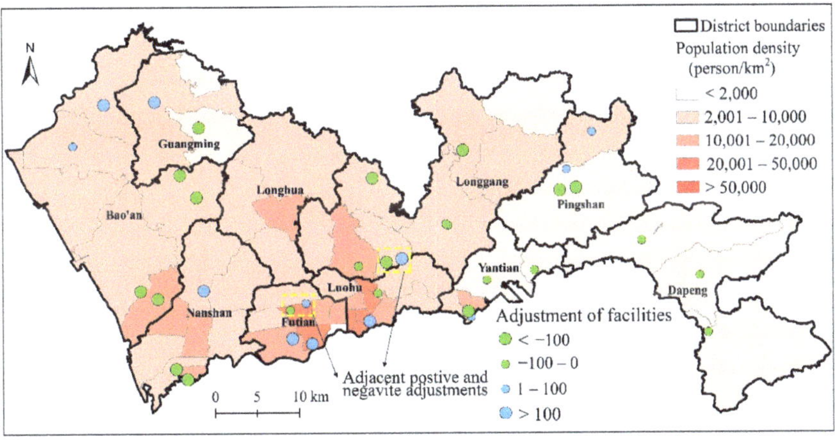

Figure 7. Size adjustments of existing secondary healthcare facilities according to optimization.

4.3. Examining the Improvement of Accessibility Equality

The disparities and distributions of the optimized as well as actual healthcare accessibility were further estimated and compared to examine whether and how the equality in accessibility is improved. The healthcare accessibility based on the actual distribution of healthcare facilities was estimated by using the Gaussian-based 2SFCA method with the same parameters as in the above optimization model. The disparity in accessibility was measured by coefficient of variation (CV), which ranges from 0 to 1, with a larger

CV representing larger disparity. CV was selected as the measure of disparity because it can make the measures at two levels comparable. In the calculation of CV, the standard deviation of accessibility is divided by the mean.

As shown in Table 3, the CVs of actual and optimized distributions of healthcare accessibility to tertiary facilities were 0.53 and 0.32, respectively. Similarly, the respective CVs for actual and optimized secondary facilities were 0.58 and 0.36. After optimization, the disparities in healthcare accessibility to tertiary and secondary facilities decreased by 40% and 38%, respectively. In other words, the optimization improved the equality in the spatial accessibility to the tertiary and secondary healthcare facilities by 40% and 38%, respectively.

Table 3. The coefficients of variation in actual and optimized healthcare accessibility.

Facility Level	Actual	Optimized	Improvement
Secondary facilities	0.58	0.36	38%
Tertiary facilities	0.53	0.32	40%

The accessibility was calculated first for discrete community locations, and then extrapolated into continuous distribution with the inverse distance weighted spatial interpolation method. As shown in Figure 8, after optimization, the healthcare accessibility to the tertiary facilities ranging from 0.0005 to 0.0010 is relatively evenly distributed in Shenzhen. The distribution of higher accessibility was found to be positively related to population density distribution. Low accessibility could only be observed in few marginal areas.

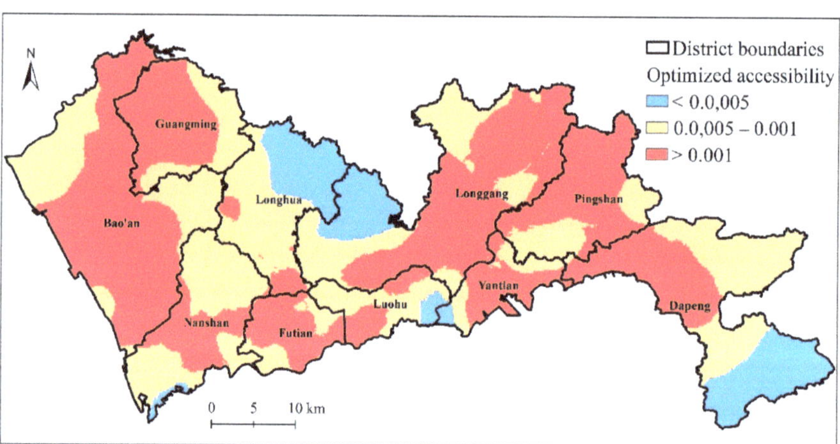

Figure 8. The distribution of optimized accessibility to tertiary healthcare facilities.

As shown in Figure 9, the optimized healthcare accessibility to the secondary facilities in most areas was found to range from 0.0004 to 0.008 in Shenzhen. However, the distributions of higher and lower accessibility were more dispersed than that of the tertiary facilities. This corresponds to the fact that the distribution of higher-level facilities is generally more concentrated. This proves that the optimization resulted in relatively equal healthcare accessibility to both the tertiary and secondary facilities.

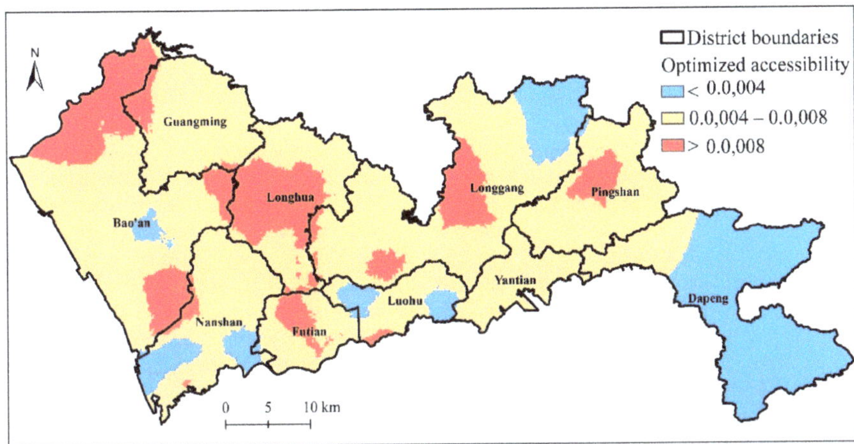

Figure 9. The distribution of optimized accessibility to secondary healthcare facilities.

5. Discussion

Rational distribution and equalization of healthcare services are critical for the improvement of health and well-being. Equality-oriented spatial optimization models can act as scientific tools for healthcare facilities planning and policymaking. This study developed a hierarchical maximal equality model that sought to maximize the equality in healthcare accessibility. The case study of Shenzhen proves that the model can significantly improve the equality in healthcare accessibility at each level as compared to the status quo. Considering the ubiquitous inequality in healthcare accessibility and the popularity of equalization of healthcare services as a key policy goal in different contexts, this study provides replicable procedures and methods for promoting the equalization of healthcare accessibility. Furthermore, the model can be applied or adapted for analyzing other hierarchical facilities (e.g., educational facilities and public parks) because the equal accessibility to various public services is a common public policy goal.

Compared to the original maximal accessibility equality model, the hierarchical maximal accessibility equality model developed in this study further incorporates hierarchical features of healthcare facilities. It specifies different catchment area sizes of, and distance friction effects for facilities at various levels. The empirical analyses demonstrate that the numbers and distribution of healthcare facilities at various levels are quite different. Existing studies have reported significant differences between healthcare accessibility at various levels [32–34]. The model developed in this study can better quantify the accessibility to hierarchical healthcare facilities across levels and optimize the equality of such accessibility. It highlights that the hierarchy structure of healthcare facilities should be carefully considered in the spatial optimization of public facilities.

Spatial optimization of hierarchical facilities is a classic and recurrent topic, as mentioned previously. A set of hierarchical location–allocation models have been developed in the past several decades [24,36,37]. These existing studies highlight the needs for taking into account the hierarchy structure of healthcare facilities in the optimization but fail to address the equality issue and comprehensively measure accessibility. This study contributes to the efforts in this respect by extending the maximal equality model into a hierarchical location–allocation model. Compared to existing hierarchical location–allocation models, it improves the measurement of accessibility, namely, interactions between demand and supply, by using a Gaussian-based 2SFCA method.

Furthermore, this study is one of the first studies that has attempted to account for both existing facility locations and newly added locations in the location–allocation analysis. The results reveal that not only existing facilities should be adjusted, but more new facilities need to be added in areas that are not well served by existing facilities. Considering that substantial fixed resources have been invested in existing facilities, such setting of candidate facility locations can better reflect the actual base of optimization or planning. Therefore, the procedures developed in this study can provide more feasible solutions of facility planning.

Our findings suggest a dispersal strategy to improve spatial equality in healthcare accessibility in Shenzhen by reallocating the existing healthcare resources and supplying extra resources. Generally, the sizes of existing facilities in central areas should be decreased, while the sizes of existing facilities should be increased, and more new facilities should be constructed in periphery areas. Considering that the space and land resources in central areas are in great shortage, the dispersal strategy is feasible to be put into implementation. However, it should be noted that our analyses are based on the population data (i.e., demand for healthcare resources) in 2010, and future growth of demands for healthcare resources were not considered. According to relevant plans in Shenzhen, peripheral areas are expected to experience larger population growth in the future. Therefore, the dispersal strategy can work even when the future population growth is taken into account. Note that the optimized healthcare facilities in the peripheral areas are middle-to-large-sized but dispersedly distributed. This suggests that the sizes of facilities in the peripheral areas should be expanded on one hand, and more new facilities may need to be built in these areas.

Despite the strengths of our study, there are also some limitations. First, the costs of the size adjustments of existing facilities are not considered in the optimization. As a result, in the optimized distribution of healthcare facilities, some facilities that need positive- or negative-size adjustments are close to each other. It is suggested that adjustments of existing facilities in such cases may be unnecessary and should not be implemented. By doing so, the costs of adjustments could be saved while the impacts on healthcare accessibility are negligible. In future study, such costs of size adjustments should be modelled into the optimization objectives or constraint conditions in a more normative way. Nevertheless, our optimization results can act as a scientific baseline for decision making. Second, although the analysis unit in this study (community) is already the smallest geographical area used by the local government, the analysis may still be faced with the modifiable area unit problem (MAUP), due to different areas and irregular shapes of the communities. More efforts are needed to examine whether and to what extent MAUP exists, as well as to figure out how to address MAUP in location–allocation modelling. Third, the accessibility to facilities and its equality was optimized independently for each level on the basis of the characteristics of the current hierarchical healthcare system in Shenzhen. However, this assumption may be inappropriate in some cases. In future works, efforts should be made to explore the interaction between various facility levels and to optimize the equality of the overall healthcare accessibility.

6. Conclusions

This study proposes a hierarchical maximal accessibility equality model for optimizing the locations of hierarchical healthcare facilities to improve the equality in accessibility to them. It extends the maximal accessibility equality model, which aims to minimize the variance of spatial accessibility to facilities by accounting for the hierarchical features of healthcare facilities. The minimal variance in spatial accessibility is pursued at each respective level. The Gaussian-based 2SFCA method is applied to measure the spatial accessibility to healthcare facilities at each level. The optimization model is solved by using the PSO algorithm. The empirical results demonstrate that healthcare facilities at each level need to be more dispersedly distributed to achieve maximal accessibility equality in Shenzhen. Compared to the current distribution, the accessibility equality of tertiary and secondary healthcare facilities in optimal solutions can be improved by 40% and 38%, respectively, which proves the validity of the proposed optimization model. Both newly added facilities and adjustments of existing facilities are needed to achieve equal healthcare accessibility. Furthermore, the optimization results are quite different for facilities at different levels, which highlights the importance of considering hierarchy structure in the optimization of healthcare facilities. The findings provide evidence-based suggestions for the policymaking in Shenzhen to improve the accessibility to hierarchical healthcare facilities. All in all, this study provides transferable methods for the equality-oriented spatial optimization of hierarchical facilities.

Author Contributions: Conceptualization, Z.T., Q.W. and W.H.; methodology, Z.T.; software, Z.T.; validation, Z.T. and W.H.; formal analysis, Z.T.; investigation, Z.T., Q.W. and W.H.; resources, Z.T., Q.W. and W.H.; data curation, Z.T.; writing—original draft preparation, Z.T. and W.H.; writing—review and editing, Z.T., Q.W. and W.H.; visualization, Z.T.; project administration, Z.T.; funding acquisition, Z.T. All authors have read and agreed to the published version of the manuscript.

Funding: This work was funded by the National Natural Science Foundation of China (grant number 42101189) and the Fundamental Research Funds for the Central Universities. The funding bodies had no direct role in the design of the study or the collection, analysis, and interpretation of data, or in writing the manuscript.

Institutional Review Board Statement: Not applicable.

Informed Consent Statement: Not applicable.

Data Availability Statement: The healthcare facility data can be obtained from the website of Shenzhen Municipal Health Commission (http://wjw.sz.gov.cn/bmfw/wycx/fwyl/yycx/index.html accessed on 26 October 2021). The travel time data were estimated using the public web API of Baidu Map (http://lbsyun.baidu.com/index.php?title=jspopular/guide/routeplan accessed on 26 October 2021). The Particle Swarm Optimization Toolbox is available at https://ww2.mathworks.cn/matlabcentral/fileexchange/7506-particle-swarm-optimization-toolbox accessed on 26 October 2021.

Acknowledgments: We would like to thank the editors and anonymous reviewers for their insightful suggestions.

Conflicts of Interest: The authors declare no conflict of interest.

References

1. Corburn, J. *Toward the Healthy City: People, Places, and the Politics of Urban Planning*; The MIT Press: Cambridge, MA, USA, 2009.
2. Fujita, M.; Sato, Y.; Nagashima, K.; Takahashi, S.; Hata, A. Impact of geographic accessibility on utilization of the annual health check-ups by income level in Japan: A multilevel analysis. *PLoS ONE* **2017**, *12*, e0177091.
3. Onega, T.; Duell, E.J.; Shi, X.; Wang, D.; Demidenko, E.; Goodman, D. Geographic access to cancer care in the U.S. *Cancer* **2008**, *112*, 909–918. [CrossRef]
4. Wang, F. Measurement, optimization and impact of healthcare accessibility: A methodological review. *Ann. Assoc. Am. Geogr.* **2012**, *102*, 1104–1112. [CrossRef]
5. Khan, A.A. An integrated approach to measuring potential spatial access to health care services. *Socioecon. Plan. Sci.* **1992**, *26*, 275–287. [CrossRef]
6. Wang, F.; Luo, W. Assessing spatial and nonspatial factors for healthcare access: Towards an integrated approach to defining health professional shortage areas. *Health Place* **2005**, *11*, 131–146. [CrossRef]
7. Hansen, W.G. How Accessibility Shapes Land Use. *J. Am. Inst. Plan.* **1959**, *25*, 73–76. [CrossRef]
8. Falchetta, G.; Hammad, A.; Shayegh, S. Planning universal accessibility to public health care in sub-Saharan Africa. *Proc. Natl. Acad. Sci. USA* **2020**, *117*, 31760–31769. [CrossRef]
9. Weiss, D.J.; Nelson, A.; Vargas-Ruiz, C.A.; Gligorić, K.; Bavadekar, S.; Gabrilovich, E.; Bertozzi-Villa, A.; Rozier, J.; Gibson, H.S.; Shekel, T.; et al. Global maps of travel time to healthcare facilities. *Nat. Med.* **2020**, *26*, 1835–1838. [CrossRef] [PubMed]
10. The State Council of the People's Republic of China. The Guiding Opinions on Promoting the Construction of Hierarchical Medical System. 2015. Available online: http://www.gov.cn/xinwen/2016-10/25/content_5124174.htm (accessed on 12 April 2020).
11. The State Council of the People's Republic of China. The Outline of the Healthy China 2030 Plan. 2016. Available online: http://www.gov.cn/zhengce/content/2015-09/11/content_10158.htm (accessed on 12 April 2020).
12. Kang, J.-Y.; Michels, A.; Lyu, F.; Wang, S.; Agbodo, N.; Freeman, V.L.; Wang, S. Rapidly measuring spatial accessibility of COVID-19 healthcare resources: A case study of Illinois, USA. *Int. J. Health Geogr.* **2020**, *19*, 36. [CrossRef]
13. Pereira, R.H.M.; Braga, C.K.V.; Servo, L.M.; Serra, B.; Amaral, P.; Gouveia, N.; Paez, A. Geographic access to COVID-19 healthcare in Brazil using a balanced float catchment area approach. *Soc. Sci. Med.* **2021**, *273*, 113773. [CrossRef]
14. McGrail, M.R.; Humphreys, J.S. Measuring spatial accessibility to primary health care services: Utilising dynamic catchment sizes. *Appl. Geogr.* **2014**, *54*, 182–188. [CrossRef]

15. Polzin, P.; Borges, J.; Coelho, A.N. An extended kernel density two-step floating catchment area method to analyze access to health care. *Environ. Plan. B Plan. Des.* **2014**, *41*, 717–735. [CrossRef]

16. Tao, Z.; Liu, Z.; Cheng, Y. Hierarchical two-step floating catchment area (2SFCA) method: Measuring the spatial accessibility to hierarchical healthcare facilities in Shenzhen, China. *Int. J. Equity Health* **2020**, *19*, 164. [CrossRef]

17. Dai, D. Racial/ethnic and socioeconomic disparities in urban green space accessibility: Where to intervene? *Landsc. Urban Plan.* **2011**, *102*, 234–244. [CrossRef]

18. Langford, M.; Higgs, G.; Fry, R. Multi-modal two-step floating catchment area analysis of primary health care accessibility. *Health Place* **2016**, *38*, 70–81. [CrossRef]

19. Tao, Z.; Cheng, Y. Modelling the spatial accessibility of the elderly to healthcare services in Beijing, China. *Environ. Plan. B Urban Anal. City Sci.* **2019**, *46*, 1132–1147. [CrossRef]

20. Neutens, T. Accessibility, equity and health care: Review and research directions for transport geographers. *J. Transp. Geogr.* **2015**, *43*, 14–27. [CrossRef]

21. Cheng, G.; Zeng, X.; Duan, L.; Lu, X.; Sun, H.; Jiang, T.; Li, Y. Spatial difference analysis for accessibility to high level hospitals based on travel time in Shenzhen, China. *Habitat Int.* **2016**, *53*, 485–494. [CrossRef]

22. Drezner, Z.; Hamacher, H.W. *Facility Location: Applications and Theory*; Springer: New York, NY, USA, 2002.

23. Tong, D.; Murray, A.T. Spatial Optimization in Geography. *Ann. Assoc. Am. Geogr.* **2012**, *102*, 1290–1309. [CrossRef]

24. Owen, S.H.; Daskin, M.S. Strategic facility location: A review. *Eur. J. Oper. Res.* **1998**, *111*, 423–447. [CrossRef]

25. Kontodimopoulos, N.; Nanos, P.; Niakas, D. Balancing efficiency of health services and equity of access in remote areas in Greece. *Health Policy* **2006**, *76*, 49–57. [CrossRef]

26. Church, R.L. Location modelling and GIS. In *Geographical Information Systems*; Longley, P.A., Goodchild, M., Maguire, D., Rhind, D., Eds.; John Wiley: New York, NY, USA, 1999; pp. 293–303.

27. Murray, A.T. Maximal Coverage Location Problem. *Int. Reg. Sci. Rev.* **2015**, *39*, 5–27. [CrossRef]

28. Wang, F.; Tang, Q. Planning toward equal accessibility to services: A quadratic programming approach. *Environ. Plan. B Plan. Des.* **2013**, *40*, 195–212. [CrossRef]

29. Tao, Z.; Cheng, Y.; Dai, T.; Rosenberg, M.W. Spatial optimization of residential care facility locations in Beijing, China: Maximum equity in accessibility. *Int. J. Health Geogr.* **2014**, *13*, 33. [CrossRef]

30. Luo, J.; Tian, L.; Luo, L.; Yi, H.; Wang, F. Two-Step Optimization for Spatial Accessibility Improvement: A Case Study of Health Care Planning in Rural China. *BioMed Res. Int.* **2017**, *2017*, 1–12. [CrossRef] [PubMed]

31. Dai, T.; Liu, Z.; Liao, C.; Cai, H. Toward Equal Opportunity of Primary Education: Introducing a Lottery into China's Proximity-Based Enrollment System. *Prof. Geogr.* **2019**, *71*, 210–220. [CrossRef]

32. Jin, M.; Liu, L.; Tong, D.; Gong, Y.; Liu, Y. Evaluating the Spatial Accessibility and Distribution Balance of Multi-Level Medical Service Facilities. *Int. J. Environ. Res. Public Health* **2019**, *16*, 1150. [CrossRef]

33. Zhang, S.; Song, X.; Wei, Y.; Deng, W. Spatial Equity of Multilevel Healthcare in the Metropolis of Chengdu, China: A New Assessment Approach. *Int. J. Environ. Res. Public Health* **2019**, *16*, 493. [CrossRef] [PubMed]

34. Ma, X.; Ren, F.; Du, Q.; Liu, P.; Li, L.; Xi, Y.; Jia, P. Incorporating multiple travel modes into a floating catchment area framework to analyse patterns of accessibility to hierarchical healthcare facilities. *J. Transp. Health* **2019**, *15*, 100675. [CrossRef]

35. Hodgson, M.J. Alternative Approaches to Hierarchical Location-Allocation Systems. *Geogr. Anal.* **1984**, *16*, 275–281. [CrossRef]

36. Şahin, G.; Süral, H. A review of hierarchical facility location models. *Comput. Oper. Res.* **2007**, *34*, 2310–2331. [CrossRef]

37. Farahani, R.Z.; Hekmatfar, M.; Fahimnia, B.; Kazemzadeh, N. Hierarchical facility location problem: Models, classifications, techniques, and applications. *Comput. Ind. Eng.* **2014**, *68*, 104–117. [CrossRef]

38. García-Palomares, J.C.; Gutiérrez, J.; Latorre, M. Optimizing the location of stations in bike-sharing programs: A GIS approach. *Appl. Geogr.* **2012**, *35*, 235–246. [CrossRef]

39. Drezner, T.; Drezner, Z. The gravity p-median model. *Eur. J. Oper. Res.* **2007**, *179*, 1239–1251. [CrossRef]

40. Narula, S.C. Hierarchical location-allocation problems: A classification scheme. *Eur. J. Oper. Res.* **1984**, *15*, 93–99. [CrossRef]

41. Espejo, L.G.A.; Galv, O.R.D.; Boffey, B. Dual-based heuristics for a hierarchical covering location problem. *Comput. Oper. Res.* **2003**, *30*, 165–180. [CrossRef]

42. Teixeira, J.C.; Antunes, A.P. A hierarchical location model for public facility planning. *Eur. J. Oper. Res.* **2008**, *185*, 92–104. [CrossRef]

43. Wang, F.; Fu, C.; Shi, X. Planning towards maximum equality in accessibility to NCI cancer centers in the U.S. In *Spatial Analysis in Health Geography*; Kanaroglou, P., Delmelle, E., Ghosh, D., Paez, A., Eds.; Ashgate Publishing: Farnham, UK, 2015; pp. 261–274.

44. Li, X.; Wang, F.; Yi, H. A two-step approach to planning new facilities towards equal accessibility. *Environ. Plan. B Urban Anal. City Sci.* **2017**, *44*, 994–1011. [CrossRef]

45. Shenzhen Municipal Health Commission. List of Healthcare Facilities in Shenzhen. 2020. Available online: http://wjw.sz.gov.cn/bmfw/wycx/fwyl/yycx/index.html (accessed on 9 October 2020).

46. Baidu Map. Development Document of Baidu Map Route Plan Web API. 2020. Available online: http://lbsyun.baidu.com/index.php?title=jspopular/guide/routeplan (accessed on 9 October 2020).

47. Wang, F.; Xu, Y. Estimating O-D travel time matrix by Google Maps API: Implementation, advantages, and implications. *Ann. GIS* **2011**, *17*, 199–209. [CrossRef]

48. Dony, C.C.; Delmelle, E.M.; Delmelle, E.C. Re-conceptualizing accessibility to parks in multi-modal cities: A Variable-width Floating Catchment Area (VFCA) method. *Landsc. Urban Plan.* **2015**, *143*, 90–99. [CrossRef]
49. Chu, H.-J.; Lin, B.-C.; Yu, M.-R.; Chan, T.-C. Minimizing Spatial Variability of Healthcare Spatial Accessibility—The Case of a Dengue Fever Outbreak. *Int. J. Environ. Res. Public Health* **2016**, *13*, 1235. [CrossRef] [PubMed]
50. Kennedy, J.; Eberhart, R.C. Particle swarm optimization. In Proceedings of the ICNN'95—International Conference on Neural Networks, Perth, WA, Australia, 27 November–1 December 1995.
51. Birge, B. Particle Swarm Optimization Toolbox. 2006. Available online: https://ww2.mathworks.cn/matlabcentral/fileexchange/7506-particle-swarm-optimization-toolbox (accessed on 5 October 2020).

Article

Incorporating a Topic Model into a Hypergraph Neural Network for Searching-Scenario Oriented Recommendations

Xin Huang * and Xiaojuan Liu

School of Government, Beijing Normal University, Beijing 100875, China; lxj_2007@bnu.edu.cn
* Correspondence: 11132020309@bnu.edu.cn

Abstract: The personalized recommendation system is a useful tool adopted by e-retailers to help consumers to find items in line with their preferences. Existing methods focus on learning user preferences from a user-item matrix or online reviews after purchasing, and they ignore the interactive features in the process of users' learning about product information through search queries before they make a purchase. To this end, this study develops a topic augmented hypergraph neural network framework to predict the user's purchase intention by connecting the latent topics embedded in a consumer's online queries to their click, purchase, and online review behavior, which aims at mining the connection information existing in the interaction graph domain. Meanwhile, in order to reduce the influence of text noise words by fusing topic information, we integrate the topic distribution and convolutional embedding to better represent each user and item, which can make up for the lack of topic information in traditional convolutional neural networks. Extensive empirical evaluations on real-world datasets demonstrate that the proposed framework improves the novelty of recommendation items as well as accuracy. From a managerial perspective, recommending diversified and novel items to consumers may increase the users' satisfaction, which is conducive to the sustainable development of e-commerce enterprises.

Keywords: personalized recommender system; online query sessions; user's preference modeling; topic model; hypergraph neural network

Citation: Huang, X.; Liu, X. Incorporating a Topic Model into a Hypergraph Neural Network for Searching-Scenario Oriented Recommendations. *Appl. Sci.* **2022**, *12*, 7387. https://doi.org/10.3390/app12157387

Academic Editor: Antonio Moreno

Received: 29 June 2022
Accepted: 21 July 2022
Published: 22 July 2022

Publisher's Note: MDPI stays neutral with regard to jurisdictional claims in published maps and institutional affiliations.

1. Introduction

As mobile Internet and information technology has achieved great technological progress, consumers can browse products and make purchase through mobile devices from anywhere at any time [1]. The rapid development of mobile e-commerce has intensified competition between e-commerce companies. E-commerce enterprises maintain their competitive advantage by implementing product differentiation strategies, offering consumers more products and discounts, as well as exploiting intelligent information filtering systems to assist online users in quickly finding products that are in line with their preferences [2]. Online retailers need to provide customers with targeted goods and services according to their different needs to avoid homogeneous competition [3]. The recommender system is a classical type of information filtering system that attempts to recommend products to users that can conform to their different hobbies and personal experience [4,5]. However, the traditional recommendation method only uses ratings to reflect the user's overall preference for items, but it is difficult to depict users' relative preferences for multiple dimensions of product features [6]. E-commerce platforms hope to help consumers to quickly find the right product that satisfies the heterogeneous needs of consumers and delivers their business ideas and product information to these potential consumers in a targeted manner [7,8]. E-commerce enterprises are committed to developing a more powerful personalized recommender system to enhance shopping experience of online consumers.

Although traditional personalized recommendation methods (such as collaborative filtering and content-based recommendation algorithms) are widely used, they all have their own shortcomings. Therefore, hybrid recommender systems are proposed to deal with these shortcomings by combining different recommendation algorithms. Recently, the most widely used hybrid recommender systems are based on collaborative filtering algorithms and content-based algorithms, while other types of combinations have also been developed. The main idea of collaborative filtering is to use the preferences of user groups with similar tastes to the target user to predict what the target user might like. The data sparsity problem and the cold start problem are considered as two key problems faced by collaborative filtering techniques [9–12]. The data sparsity problem seriously restricts the performance of collaborative filtering. For large business websites, due to the large number of products and users, the user rating products generally do not exceed 1% of the total number of products. The cold start problem usually occurs when new users arrive. As there is no user behavior data when a new user enters the system, it is difficult to make effective recommendations. The basic idea of content-based filtering is to recommend other items similar to the items that the user liked in the past. The content-based filtering technology relies on user portraits. Therefore, even if the database does not contain user interests, it will not affect the accuracy of the recommendation results [13]. However, the content-based filtering technique depends on the item's metadata. That is to say, the system needs rich item content descriptions and complete user portraits. Hence, users can only get recommendations similar to items in their own profile, hardly getting diversified options. One of the ways to build a hybrid recommender system is to independently apply collaborative filtering, content-based and other algorithms, for recommendations by combining the recommendation results of two or more systems and using the linear combination of prediction scores to make recommendations. Some hybrid recommender systems are content-based collaborative filtering algorithms. That is, the similarity of users is calculated through content-based profiles, rather than the information of products that are rated together. This can overcome the sparsity problem in collaborative filtering systems. Another hybrid recommendation mechanism is utilizing multiple independent recommendation algorithms, each of which generates its own recommendation results, and fusing these recommendation results in the mixing stage to generate the final recommendation result. It can be seen from the above analysis that the above recommendation techniques predict consumer purchase intention based on product ratings of what other similar users have purchased or what they themselves have purchased. User preferences characterized by these methods are commonly presented based on user ratings of 1 to 5, which can capture a user's overall evaluation of the product. However, ratings data is too simple to capture consumers' multi-dimensional and fine-grained evaluation of product attributes. Unlike sparse consumer purchasing data, consumers conduct extensive online search queries before making a purchasing decision. Take the laptop as an example, customers formulate queries like "best laptop for programming" that directly reflect their content preferences for product features. Thus, it is important to understand the navigation keywords associated with product features in users' online query sessions. It is critical for e-commerce platforms to extract consumer content preferences from online search sessions [14].

Despite the importance of inferring user preferences from online query sessions, few studies have focused on this area. Roscoe et al. [15] revealed that online search queries focused on superficial product features rather than key knowledge. Information search behavior is an important factor that is assessed to identify differences in consumers regarding their purchasing patterns and preferences [16]. Kim et al. [17] verified that there is a significant relationship between new product diffusion and internet search volume. Internet search volume is an important indicator for predicting new product demand. Liu and Toubia [18] suggested that marketers should focus their efforts on keywords and queries that reflect content preferences that are well aligned with the content they are trying to promote. Codignola et al. [19] found that these browsing data can be saved with cookies and can be used to show customers potentially suitable items. Although numerous studies have been conducted to empirically verify that online queries can explicitly express consumers' content preference or can be used to predict product demand, quantitative studies that can estimate content preferences from online queries in an interpretable manner are lacking. Therefore, it is managerially important for sustainable e-retailers to develop intelligent recommendations based on learning dynamic customer preference from online query sessions [20].

In this study, we develop a topic augmented hypergraph neural network (Topic-HGNN) framework, which uses the hypergraph structure to capture the complex multivariate relationship among users, query topics, items, and item features. Besides, we incorporate topic models into the hypergraph neural network to more finely depict user preferences and product characteristics. To this end, we specifically propose an Aggregated Latent Dirichlet Allocation model to jointly extract users' content preference topics from queries and webpages, and apply the Latent Dirichlet Allocation [21] model to extract product feature topics from online reviews, which is useful to enhance feature interaction interpretability. In detail, the proposed Topic-HGNN framework involves: (1) adopting a hypergraph to model the multivariate relationship among users, query topics, items, and item features and applying the dual-embedding mechanism to handle complex and high-order correlations; (2) applying hyperedge corruption to generate a user-query hypergraph and an item-feature hypergraph and utilizing the hyperedge convolution layer to obtain user embedding and item embedding; (3) developing an Aggregated Latent Dirichlet Allocation model to jointly extract users' content preference topics from queries and webpages and applying the Latent Dirichlet Allocation model to extract product feature topics from online reviews; (4) integrating topic distribution and convolutional embedding to represent each user and item; and (5) using multilayer perceptron to calculate the soft match score between query entities and item entities.

We summarize the main contributions in the paper as follows:

- Despite the importance of inferring user preferences from online query sessions, very little research has focused on this area. In this paper, we propose a sustainable recommender system architecture based on inferring users' preferences from online query sessions, which can more accurately predict user purchase intentions;
- We develop an Aggregated Latent Dirichlet Allocation (ALDA) model, a novel topic model that can simultaneously learn user query topics and topics of corresponding clicked webpages. The ALDA model treats the joint topic distribution of queries and webpages as the topic distribution of user preferences. The data sparsity of online query data is avoided by aggregating corresponding webpages to assist in learning users' content preferences;
- To handle the complex multivariate relationship among users, query topics, items, and item features, we design a topic augmented hypergraph neural network (Topic-HGNN) framework to more accurately represent each user and item by integrating the convolution information and the topic information. The Topic-HGNN framework can significantly improve the accuracy and the novelty of recommended items;
- Extensive tests verify that our approach can better capture consumers' multi-dimensional preferences for product attributes and can better predict consumers' purchase intentions.

We organize the rest of this paper as follows. Section 2 summarizes the related works. Section 3 describes the proposed recommender system in details. Section 4 presents the extensive experiments designed to evaluate the effectiveness and the efficiency of the proposed framework. Section 5 summarizes the paper.

2. Related Work

From the above analysis, this paper aims to develop a novel personalized recommendation system based on learning dynamic customer preference from online query sessions. Thus, in this section, we briefly summarize related works from the following three aspects: traditional recommendation systems, online query sessions, and recommendations based on graph learning.

2.1. Traditional Recommendation Systems

The recommendation system is a widely used information filtering tool to provide customers with product information and suggestions to help users to decide which products they should purchase. Bobadilla et al. [22] classified recommendation methods into three categories: collaborative filtering, content-based filtering, and hybrid recommender systems.

Collaborative filtering is one of the earliest and the most successful techniques used in recommender systems. It generally uses the preference of user groups with similar tastes to the target user to predict the target user's preference for a specific product. Generally speaking, there are two types of collaborative filtering techniques. The first is user-based collaborative filtering [23,24], and the second is item-based collaborative filtering [25,26]. User-based collaborative filtering mainly considers the similarity between users. It predicts the target user's rating for a particular item based on the ratings of items liked by similar users. The basic idea of item-based filtering is to calculate the similarity between items based on the historical preference data of all users, and then to recommend items similar to the user's favorite item to the target user. Currently, a large number of scholars focus on utilizing machine learning models to improve the performance of collaborative filtering technique. Matrix factorization [27], neural network [28], and graphic models [29] are commonly used in combination with collaborative filtering. The most difficult challenge faced by the collaborative filtering technique is the cold start problem when a new user arrives. Since the recommender system does not have any data of new users, it cannot effectively recommend items for new users. In addition, collaborative filtering cannot understand different scenarios, which is unable to capture the specific consumption purpose of users at a specific moment.

Content-based filtering works by evaluating the similarity between items that the user has not seen and items that the user has liked in the past. To generate meaningful recommendation results, content-based filtering uses different models to find similarities between items. It typically uses a vector space model (e.g., term frequency inverse document frequency) or a probabilistic model (naive Bayes classifier, decision tree, and neural network) to model relationships between different items [30–32]. Content-based filtering technology does not need to refer to other user portraits because other user portraits will not affect the final recommendations. Moreover, content-based filtering technology can still adjust the recommendation results in a very short period of time if the user profile changes. The main disadvantage of this technique is that it requires the system to have a deep understanding of the characteristics of the item. Since content-based filtering depends only on the user's past preferences for certain items, users can only get recommendations similar to items in their own profile, hardly getting diversified options.

Hybrid recommender systems combine multiple recommendation algorithms to avoid the problems of a single technique. Burke [33] distinguished hybrid recommender systems into three basic design ideas: monolithic, parallelized, and pipelined. The monolithic paradigm integrates multiple recommendation algorithms into the same algorithm system, and the integrated recommendation algorithm provides a unified recommendation service. The parallelized paradigm utilizes multiple independent recommendation algorithms, each of which generates its own recommendation results, and fuses these recommendation results in the mixing stage to generate the final recommendation result. In the pipelined paradigm, the recommendation result generated by one algorithm is given to another recommendation algorithm as input, and then the recommendation result is generated, which is input to the next recommendation algorithm, and so on.

With the rapid development of mobile commerce, more and more recommendation services occur in dynamically changing contexts, such as user location, access time, current traffic, and other surrounding environments. Traditional personalized recommendation technology is no longer enough to deal with the new impact caused by contextual factors [34,35]. Therefore, a current trend is to integrate and to apply contextual information in traditional recommendation systems to form a context-based recommendation system, so as to accurately and to efficiently provide information resources that not only conform to the current situation of the user but also satisfies the user's preference [36].

In summary, recent recommendation techniques predict the consumer's purchase intention based on product ratings of what other similar users have purchased or what they themselves have purchased. Unlike sparse consumer purchasing data, consumers conduct extensive online search queries before making a purchasing decision. Different from these studies, this paper tried to extract users' explicit content preferences from online query sessions to alleviate the problems mentioned above.

2.2. Online Query Sessions

Online query sessions contain a wealth of valuable information about users' hobbies, preferences and intensions. The content and the quantity of online search queries can be used to predict product or service demand in the era of big data [37]. Choi and Varian [38] showed how to predict near-term values of economic indicators, e.g., automobile sales, unemployment claims, travel destination planning, and consumer confidence, based on Google search data. Yang, Pan, and Song [39] utilized traditional econometric models to predict hotel demand and hotel occupancy in tourist destinations based on web query volumes. Roscoe et al. [15] debated how online search and the holistic stance of a web search toward a consumer product contributed to decision making, and they assessed decision making by combining analyses of online searches with robust choice modeling. Taking bottled water as an example, this approach revealed how different product attributes (e.g., type of product, type of packaging, and cost) affected users purchase intentions in different degrees. Tibau et al. [40] applied the Exploratory Search Knowledge-intensive Process Model to visualize search patterns and to identify best practices associated with users' decision-making processes. They identified four important characteristics of users' decision-making processes while searching online. Liu and Toubia [18] suggested that marketers should focus their efforts on keywords and queries that reflect content preferences that are well aligned with the content they are trying to promote. Codignola et al. [19] found that these browsing data can be saved with cookies and can be used to show customers potentially suitable items.

Although numerous studies have been done to empirically verify that online queries can explicitly express consumers' content preference or can be used to predict product demand, quantitative studies that can estimate content preferences from online queries in an interpretable manner are lacking. In this paper, we propose a novel Aggregated Latent Dirichlet Allocation (ALDA) topic model that can simultaneously learn the potential topics hidden in user's online search queries and the corresponding webpages. Since online query phrases data is sparse, the ALDA model aggregates click documents corresponding to user queries to assist in more accurately learning users' content preferences.

2.3. Recommendation Based on Graph Learning

Graph is becoming a core area of machine learning. Graph learning is widely used to understand the structure of social networks by predicting potential connections, detecting fraud, understanding consumer behavior, or making real-time recommendations. Graph neural network (GNN) techniques have been widely used in recommender systems because most of the information in recommender systems has a graph structure in nature and GNNs have excellent performance in learning graph structures [41,42]. He et al. [43] proposed a light graph convolution network (LightGCN) model that uses user-interacted item records to enhance user representation and interacted user records to enhance item representation. Multi-layer GNNs can simulate the information transfer process and efficiently establish higher-order connections. Li et al. [44] designed a novel feature interaction graph neural network (Fi-GNN) to model sophisticated feature interactions in a flexible and an explicit fashion, which provides good model explanations for click-through rate prediction. Chang et al. [45] proposed a new graph-based geographical latent representation (GGLR) that models geographic influences between the POIs and the transition patterns of user sequence behavior based on spatial and temporal features, which can capture highly non-linear geographical influences from complex user-POI networks.

The GNN methods mentioned above employ pairwise connections between data. However, data structures in real-world applications can go beyond pairwise connections and they can be even more complicated. Feng et al. [46] proposed a hypergraph neural network (HGNN) framework that can deal with complex data correlations by encoding high-order data correlation (beyond pairwise connections) using its degree-free hyperedges. Chen et al. [47] proposed a neural signed hypergraph to extract non-linear relationships among users, items, and features. He et al. [48] proposed a hypergraph click-through rate prediction framework (HyperCTR) that learns item representations based on multi-modal information interactions among users and items. However, existing research focuses on learning user interaction characteristics with products during and after purchase (e.g., purchase and online review), and it ignores the interactive features in the process of users' learning about product information through search queries before they make a purchase (e.g., product information search). However, user association with a product is a coherent process that should not be isolated into different nodes. Only sorting the user's process of searching-understanding-purchasing-using products and finding opportunity points from each stage can help the recommender system to better discover the potential needs of users. In this paper, we develop a hypergraph framework to handle the interaction behavior of consumers in the whole process of shopping (i.e., searching-understanding-purchasing-using).

3. Materials and Methods

In order to make effective recommendations to users, recommendation systems need to solve two problems. One is to predict consumers' product ratings, that is, recommending products with higher predicted scores to target consumers. The second is the interpretation of the recommendation results, that is, explaining the working mechanism of the recommendation system and the specific reasons for recommending a product to consumers in an appropriate way. Since the recommendation process is still a relatively mysterious process for most consumers, a reasonable explanation of the recommendation results is necessary to improve consumers' trust in the recommendation system, which greatly affects consumers' perception and acceptance of recommendation results. The existing recommendation algorithms generally directly rely on the users' overall rating score for products, and the obtained recommendation results are greatly affected by the sparsity of the rating matrix and the cold start problem. This study believes that this situation is mainly caused by the coarse information granularity of the user's product ratings. That is to say, it is impossible for any product to fully meet all the needs of users, and it is impossible for users to have the same degree of preference for all attributes of a product. The recommendation results generated by directly relying on the user's overall ratings cannot reflect the users' preferences for various attributes of the product, and it is difficult to explain the real reasons for the user's preference for the product.

As consumers are more likely to submit online search phrases to search engines to gather information before making an intended purchase decision. They enter keywords to explicitly express their preferences for product attributes. For example, customers formulate queries such as "best laptop for programming" that directly reflect their content preferences for product configurations. Interpreting consumers' search phrases renders a better understanding of their purchase intentions and preferences for product attributes, which is critical for developing an effective personalized recommendation system.

In this paper, we introduce a sustainable recommender system architecture based on fusing a topic model and a hypergraph neural network, which can deal with the interaction behavior of consumers in the whole process of shopping (i.e., searching-understanding-purchasing-using). Figure 1 shows the topic augmented hypergraph neural network (Topic-HGNN) framework for searching-scenario oriented recommendation. First, we adopt a hypergraph to model the multivariate relationship among users, query topics, items, and item features, which aims at mining the connection information existing in the interaction graph domain. Then, we utilize hyperedge corruption [47] to generate a user–query hypergraph and an item–feature hypergraph, and we utilize the hyperedge convolution layer [46] to obtain user embedding and item embedding. Meanwhile, in order to reduce the influence of text noise words by fusing topic information, we specially design an Aggregated Latent Dirichlet Allocation (ALDA) model to jointly extract users' content preference topics from queries and webpages and to apply Latent Dirichlet Allocation model to extract product feature topics from online reviews. Then, we integrate the topic distribution and convolutional embedding to represent each user and item, which can make up for the lack of topic information in traditional convolutional neural networks. Finally, we use multilayer perceptron to calculate the soft match score between query entities and item entities.

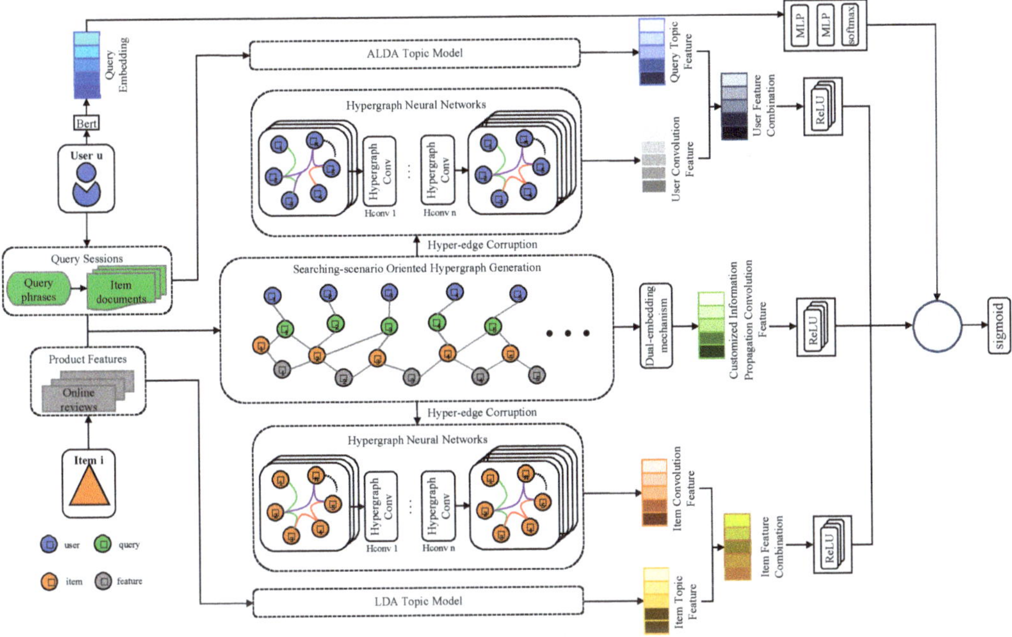

Figure 1. The proposed topic augmented hypergraph neural network (Topic-HGNN) framework for searching-scenario oriented recommendation.

3.1. Searching-Scenario Oriented Hypergraph Generation

Existing research focuses on learning user interaction characteristics with products during and after purchase (e.g., purchase and online review), and ignores the interactive features in the process of users' learning about product information through search queries before they make a purchase (e.g., product information search). However, user association with a product is a coherent process that should not be isolated into different nodes. Only sorting the user's process of searching-understanding-purchasing-using products and finding opportunity points from each stage can help the recommender system to better discover the potential needs of users. Thus, this work considers quaternary relationships between interacting entities (user, query topic, item, and item feature) and employs a hypergraph to model the interaction behavior of consumers in the whole process of shopping.

Let $V = \left\{ V_u, V_q, V_i, V_f \right\}$ denote the vertex set, where V_u represents user vertex, V_q is the query vertex sent by the user, V_i represents item vertices, and V_f is the product feature node extracted from the product online reviews. E represents the set of hyperedges e_j built from V. Each hyperedge "v_u-v_q-v_i-v_f" is a complete purchasing path for the user, which means that user u finds a product i that matches his preference for feature f through query q, and makes a purchase. Thus, $G = (V, E)$ represents a hypergraph, and a hypergraph G can be represented by a $|V| \times |E|$ incidence matrix H, with entries defined as:

$$h(v, e) = \begin{cases} 1, & v \in e \\ 0, & v \notin e \end{cases}$$

For a vertex $v \in V$, its degree is defined as $d(v) = \sum_{e \in E} w(e)h(v, e)$, where $w(e)$ represents the weight of the hyperedge e. For an hyperedge $e \in E$, its degree is defined as $\delta(e) = \sum_{v \in V} h(v, e)$. The degree matrices of vertex and hyperedge are represented by the diagonal matrices D_v and D_e, respectively.

3.2. Topic Feature Learning of User and Item

In this section, we introduce the Aggregated Latent Dirichlet Allocation (ALDA) model in detail. ALDA is a bag-of-word model that depicts the semantic relation between user preferences and their online query sessions. Instead of modeling the topic intensities in the query sessions and the topic intensities in the webpages hierarchically [49], the ALDA conjointly models the topic intensities in the query sessions and the topic intensities in the webpages into the same document layer. The data sparsity of online query data is avoided by aggregating corresponding webpages to assist in learning users' content preferences. Consumers' online shopping behavior is usually a learning process. First, users may enter inaccurate keywords to express their needs. Then, users enhance their understanding of products through browsing the search results and adjusting the input keywords. Consumers will repeat this learning process until finding the right product. That is to say, the topics of query keywords and the topic of search results are semantically related to each other. Liu and Toubia [49] assumed the topic intensities in webpages is affected by query keywords while ignoring that webpages can in turn affect the topic intensities in query keywords. Thus, we model the interactive relationship between queries and webpages in ALDA. The graphical representation of ALDA proposed in this paper is illustrated in Figure 2. The main notations in ALDA are listed in Table 1.

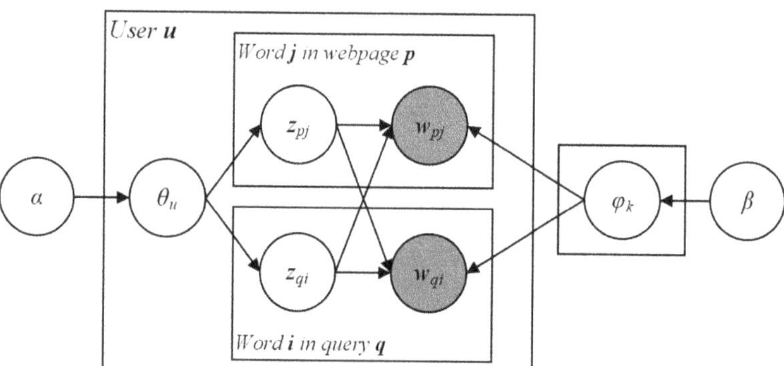

Figure 2. The graphical representation of ALDA model.

Table 1. Summary of the main notations.

Notations	Explanation
$u \in \{1, 2, \cdots, U\}$	Set of users
$q \in \{1, 2, \cdots, Q\}$	Set of user queries
$p \in \{1, 2, \cdots, P\}$	Set of webpages
θ_u	The vector of topic probabilities in users' preferences
α	Dirichlet prior distribution for θ_u
φ_k	The vector of word probabilities for topic k
β	Dirichlet prior distribution for φ_k
z_{qi}	Topic assignment of the ith word in query q
z_{pj}	Topic assignment of the jth word in webpage p
w_{qi}	The ith observed word in query q
w_{pj}	The jth observed word in webpage p

3.2.1. Model Description

First, we introduce the notations of the ALDA model. Supposing that there is a collection of U users in a particular e-commerce platform: $u \in \{1, 2, \cdots, U\}$. The user u entered different queries for a particular search domain: $q \in \{1, 2, \cdots, Q\}$. There are P webpages underlying a particular query q: $p \in \{1, 2, \cdots, P\}$. There are K topics that the user u is interested in: $k \in \{1, 2, \cdots, K\}$. There are V topic words in the vocabulary. w_{qi} represents the ith word in the query q. w_{pj} represents the jth word in the webpage p.

- θ_u denotes the topics probability distribution in user u's preferences.
- φ_k denotes the words probability distribution of the kth topic.
- α is the symmetric Dirichlet prior hyper-parameter for θ_u.
- β is the symmetric Dirichlet prior hyper-parameter for φ_k^q and φ_k^q.
- z_{qi} denotes the topic of the ith word in query q.
- z_{pj} denotes the topic of the jth word in webpage p.
- w_{qi} denotes the ith word in the query q.
- w_{pj} denotes the jth word in the webpage p.

Formally, the generative process of query sessions and webpages based on the ALDA model is described as follows:

Topics: We continue to work on the assumption proposed by Liu and Toubia [49]. Liu and Toubia [49] assumed that search query documents and webpage documents follow the same topic distributions. The topic intensities in the documents are reflected by the words displayed in the documents and each document has different topic intensities. Similar to an LDA, each topic $k \in \{1, 2, \cdots, K\}$ is represented as a topic-word distribution vector φ_k. The vector φ_k follows a *Dirichlet* distribution over V topic words in the vocabulary:

$$\varphi_k \sim Dirichlet(\beta)$$

Queries: To model the ith word w_{qi} observed in the query q, ALDA sequentially samples the topic distribution of the query q and the topic assignment of the ith word in the query q. The generation process of users' query online queries is as follows:

1   For each query q ($q \in \{1, 2, \cdots, Q\}$):

     1.1   Generate topic probabilities θ_u from a homogeneous Dirichlet distribution with parameter α: $\theta_u \sim Dirichlet(\alpha)$

2   For each topic k ($k \in \{1, 2, \cdots, K\}$):

     2.1   Generate φ_k independently from a homogeneous Dirichlet distribution with parameter β: $\varphi_k \sim Dirichlet(\beta)$

3   For each word w_{qi} in the query q:

     3.1   Choose a topic z_{qi} from the K topics with probabilities given by θ_u: $z_{qi} \sim Multinomial(\theta_u)$

    3.2   Choose a word w_{qi} from the dictionary with probabilities given by φ_k: $w_{qi} \sim Multinomial(\varphi_k)$

Webpages: To model the jth word w_{pj} observed in the webpage p, ALDA sequentially samples the topic distribution of the webpage p and the topic assignment of the jth word in the webpage p. The generation process of webpages related to online queries is as follows:

1   For each query p ($p \in \{1, 2, \cdots, P\}$):

    1.1   Generate topic probabilities θ_u from a homogeneous Dirichlet distribution with parameter α: $\theta_u \sim Dirichlet(\alpha)$

2   For each topic k ($k \in \{1, 2, \cdots, K\}$):

    2.1   Generate φ_k independently from a homogeneous Dirichlet distribution with parameter β: $\varphi_k \sim Dirichlet(\beta)$

3   For each word w_{pj} in the query p:

    3.1   Choose a topic z_{pj} from the K topics with probabilities given by θ_u: $z_{pj} \sim Multinomial(\theta_u)$

    3.2   Choose a word w_{pj} from the dictionary with probabilities given by φ_k: $w_{pj} \sim Multinomial(\varphi_k)$

3.2.2. Parameter Estimation

It is an intractable task to exactly estimate the parameters θ_u, φ_k. Similar to LDA, we use Gibbs sampling to approximately infer the parameters. First, we need to sample $P(z_{qi}|w_{qi}, w_{pj})$ and $P(z_{pj}|w_{qi}, w_{pj})$ to obtain the topic assignment z_{qi} in query documents and the topic assignment z_{pj} in webpage documents. Thus, the following conditional probability distribution is derived:

$$P(z_{qi} = k, z_{pj} = k | z_{-qi}, z_{-pj}, w_{qi}, w_{pj})$$

$$
\begin{aligned}
&P(z_{qi} = k, z_{pj} = k | z_{-qi}, z_{-pj}, w_{qi}, w_{pj}) \\
&\propto P(z_{qi} = k, z_{pj} = k, w_{qi} = t_1, w_{pj} = t_2 | z_{-qi}, z_{-pj}, w_{-qi}, w_{-pj}) \\
&= \int P(z_{qi} = k, z_{pj} = k, w_{qi} = t_1, w_{pj} = t_2, \theta_u, \theta_u, \varphi_k | z_{-qi}, z_{-pj}, w_{-qi}, w_{-pj}) d\theta_u d\theta_u d\varphi_k \\
&= \int P(z_{qi} = k, \theta_u | z_{-qi}, z_{-pj}, w_{-qi}, w_{-pj}) \cdot P(z_{pj} = k, \theta_u | z_{-qi}, z_{-pj}, w_{-qi}, w_{-pj}) \\
&\quad \cdot P\left(w_{qi} = t_1, \varphi_{k,w_{qi}} \Big| z_{-qi}, z_{-pj}, w_{-qi}, w_{-pj}\right) \\
&\quad \cdot P\left(w_{pj} = t_2, \varphi_{k,w_{pj}} \Big| z_{-qi}, z_{-pj}, w_{-qi}, w_{-pj}\right) d\theta_u d\theta_u d\varphi_{k,w_{qi}} d\varphi_{k,w_{pj}} \\
&= \int P(z_{qi} = k | \theta_u) P(\theta_u | z_{-qi}, z_{-pj}, w_{-qi}, w_{-pj}) \\
&\quad \cdot P(z_{pj} = k | \theta_u) P(\theta_u | z_{-qi}, z_{-pj}, w_{-qi}, w_{-pj}) \\
&\quad \cdot P\left(w_{qi} = t_1 \Big| \varphi_{k,w_{qi}}\right) P\left(\varphi_{k,w_{qi}} \Big| z_{-qi}, z_{-pj}, w_{-qi}, w_{-pj}\right) \\
&\quad \cdot P\left(w_{pj} = t_2 \Big| \varphi_{k,w_{pj}}\right) P\left(\varphi_{k,w_{pj}} \Big| z_{-qi}, z_{-pj}, w_{-qi}, w_{-pj}\right) d\theta_u d\theta_u d\varphi_{k,w_{qi}} d\varphi_{k,w_{pj}} \\
&= \int P(z_{qi} = k | \theta_u) Dir(\theta_u | n_{q,-qi} + \alpha) \cdot P(z_{pj} = k | \theta_u) Dir(\theta_u | n_{p,-pj} + \alpha) \\
&\quad \cdot P\left(w_{qi} = t_1 \Big| \varphi_{k,w_{qi}}\right) Dir\left(\varphi_{k,w_{qi}} \Big| n^{q}_{k,-qi} + \beta\right) \\
&\quad \cdot P\left(w_{pj} = t_2 \Big| \varphi_{k,w_{pj}}\right) Dir\left(\varphi_{k,w_{pj}} \Big| n^{p}_{k,-pj} + \beta\right) d\theta_u d\theta_u d\varphi_{k,w_{qi}} d\varphi_{k,w_{pj}} \\
&= \int \theta_u^{(k)} Dir\left(\theta_u \Big| n^{(k)}_{q,-qi} + \alpha\right) \cdot \theta_u^{(k)} Dir\left(\theta_u \Big| n^{(k)}_{p,-pj} + \alpha\right) \\
&\qquad \cdot \varphi_k^{(t_1)} Dir\left(\varphi_{k,w_{qi}} \Big| n^{q(t_1)}_{k,-qi} + \beta\right) \\
&\qquad \cdot \varphi_k^{(t_2)} Dir\left(\varphi_{k,w_{pj}} \Big| n^{p((t_2))}_{k,-pj} + \beta\right) d\theta_u d\theta_u d\varphi_{k,w_{qi}} d\varphi_{k,w_{pj}} \\
&= E\left(\theta_u^{(k)}\right)^2 \cdot E\left(\varphi_k^{(t_1)}\right) \cdot E\left(\varphi_k^{(t_2)}\right) \\
&= \frac{n^{(k)}_{q,-qi} + \alpha}{\sum_{k=1}^{K} n^{(k)}_{q,-qi} + K\alpha} \cdot \frac{n^{(k)}_{p,-pj} + \alpha}{\sum_{k=1}^{K} n^{(k)}_{p,-pj} + K\alpha} \cdot \frac{n^{q(t_1)}_{k,-qi} + \beta}{\sum_{v=1}^{V} n^{q(v)}_{k,-qi} + V\beta} \cdot \frac{n^{p((t_2))}_{k,-pj} + \beta}{\sum_{v=1}^{V} n^{p(v)}_{k,-pj} + V\beta}
\end{aligned}
\tag{1}
$$

Inside, $w_{qi} = t_1$ denotes the ith word in the query q is t_1. $w_{pj} = t_2$ denotes the jth word in the webpage p is t_2. z_{-qi} denotes the topic assignments to all words except the ith word in the query q. z_{-pj} denotes the topic assignments to all words except jth word in the webpage p. w_{-qi} denotes all words except the ith word in the query q. w_{-pj} denotes all words except the jth word in the webpage p. $n_{q,-qi}^{(k)}$ denotes the number of words generated by topic k in the query q excluding the ith word in the query q, $n_{p,-pj}^{(k)}$ denotes the number of words generated by topic k in the webpage p excluding the jth word in the webpage p, $n_{q,-qi} = \left(n_q^{(1)}, n_q^{(2)}, \cdots, n_q^{(k)} - 1, \cdots, n_q^{(K)} \right)$ denotes the number of words generated by topic k in the query q excluding the ith word, $n_{p,-pj} = \left(n_p^{(1)}, n_p^{(2)}, \cdots, n_p^{(k)} - 1, \cdots, n_p^{(K)} \right)$ denotes the number of words generated by topic k in the webpage p excluding the jth word. $n_{k,-qi}^{q(t)}$ denotes the number of times the word t is assigned to the topic k excluding the ith word in the query q, $n_{k,-qi}^q = \left(n_k^{q(1)}, n_k^{q(2)}, \cdots, n_k^{q(t)} - 1, \cdots, n_k^{q(V)} \right)$. $n_{k,-pj}^{p(t)}$ denotes the number of times the word t is assigned to the topic k excluding the jth word in the webpage p, $n_{k,-pj}^p = \left(n_k^{p(1)}, n_k^{p(2)}, \cdots, n_k^{p(t)} - 1, \cdots, n_k^{p(V)} \right)$.

Algorithm 1 summarizes the overall procedure of Gibbs sampling to estimate the parameters θ_u, φ_k. First, the assignments of topic to each word are initialized according to a uniform distribution. Then, the assignment of topics to each word will be updated by examining Equation (1). Finally, $n_q^{(k)}$, $n_p^{(k)}$, $n_k^{q(v)}$, $n_k^{p(v)}$ can be counted after a sufficient number of iterations. $n_q^{(k)}$ denotes the number of times the topic k occurs in the query q. $n_p^{(k)}$ denotes the number of times the topic k occurs in the webpage p. $n_k^{q(v)}$ denotes the number of times the word v is assigned as a query word to topic k. $n_k^{p(v)}$ denotes the number of times the word v is assigned as a webpage word to topic k.

Here, we only give the derivation of the parameter θ_u, the derivation of other parameters is the same.

$$
\begin{aligned}
P(\theta_u | n_u, \alpha) &= \frac{P(n_u | \theta_u) P(\theta_u | \alpha)}{\int P(n_u | \theta_u) P(\theta_u | \alpha) d\theta_u} \\
&= \frac{Mult(n_u | \theta_u) Dir(\theta_u | \alpha)}{\int Mult(n_u | \theta_u) Dir(\theta_u | \alpha) d\theta_u} \\
&= Dir(\theta_u | \alpha + n_u)
\end{aligned}
$$

Inside, $n_u = n_q + n_p$

The estimated value of each parameter is:

$$
E(\theta_u) = \left(\frac{n_q^{(1)} + n_p^{(1)} + \alpha}{\sum_{k=1}^{K} n_q^{(k)} + n_p^{(k)} + K\alpha}, \cdots, \frac{n_q^{(k)} + n_p^{(k)} + \alpha}{\sum_{k=1}^{K} n_q^{(k)} + n_p^{(k)} + K\alpha}, \cdots, \frac{n_q^{(K)} + n_p^{(K)} + \alpha}{\sum_{k=1}^{K} n_q^{(k)} + n_p^{(k)} + K\alpha} \right) \tag{2}
$$

$$
E(\varphi_k) = \left(\frac{n_k^{q(1)} + n_k^{p(1)} + \beta}{\sum_{v=1}^{V} n_k^{q(v)} + n_k^{p(v)} + V\beta}, \cdots, \frac{n_k^{q(v)} + n_k^{p(v)} + \beta}{\sum_{v=1}^{V} n_k^{q(v)} + n_k^{p(v)} + V\beta}, \cdots, \frac{n_k^{q(V)} + n_k^{p(V)} + \beta}{\sum_{v=1}^{V} n_k^{q(v)} + n_k^{p(v)} + V\beta} \right) \tag{3}
$$

Finally, the topic feature vector of each user can be expressed as θ_u.

Similarly, we can use LDA [21] to mine each product's topic feature vector from its online reviews: θ_i.

Algorithm 1: The Gibbs sampling for ALDA

Input : topic number K, vocabulary number V, document sets, α, β.
Output : θ_u, φ_k.

1. Initialization
Sample z_{qi}, z_{pj} according to the uniform distribution
$n_q^{(k)} = n_q^{(k)} + 1$, $n_q = n_q + 1$, $n_p^{(k)} = n_p^{(k)} + 1$, $n_p = n_p + 1$, $n_k^{q(t)} = n_k^{q(t)} + 1$, $n_k^q = n_k^q + 1$, $n_k^{p(t)} = n_k^{p(t)} + 1$, $n_k^p = n_k^p + 1$.
2. Gibbs sampling
For each query q and webpage p do:
    For each word w_{qi} in query q do:
        (1) $z_{qi} = k \rightarrow n_q^{(k)} = n_q^{(k)} - 1$, $n_q = n_q - 1$, $n_k^{q(t)} = n_k^{q(t)} + 1$, $n_k^q = n_k^q + 1$.
        (2) Sample $z_{qi} = \hat{k} \sim \mathrm{P}\left(z_{qi} = k \middle| z_{-qi}, w_{qi}\right)$ according to Equation (1)
        $n_q^{(\hat{k})} = n_q^{(\hat{k})} - 1$, $n_q = n_q - 1$, $n_{\hat{k}}^{q(t)} = n_{\hat{k}}^{q(t)} + 1$, $n_{\hat{k}}^q = n_{\hat{k}}^q + 1$.
    For each word w_{pj} in webpage p do:
        (1) $z_{pj} = k \rightarrow n_p^{(k)} = n_p^{(k)} - 1$, $n_p = n_p - 1$, $n_k^{p(t)} = n_k^{p(t)} + 1$, $n_k^p = n_k^p + 1$.
        (2) Sample $z_{pj} = \hat{k} \sim \mathrm{P}\left(z_{pj} = k \middle| z_{-pj}, w_{pj}\right)$ according to Equation (1)
        $n_p^{(\hat{k})} = n_p^{(\hat{k})} - 1$, $n_p = n_p - 1$, $n_{\hat{k}}^{p(t)} = n_{\hat{k}}^{p(t)} + 1$, $n_{\hat{k}}^p = n_{\hat{k}}^p + 1$.

3. Parameter estimation
Estimating θ_u, φ_k according to Equations (2) and (3)

3.3. Convolutional Feature Learning of User and Item

The searching-scenario oriented hypergraph obtains high-order correlations between data, while it contains heterogeneous vertices (i.e., user vertex, query vertex, item vertex, feature vertex). Thus, it is necessary to obtain not only high-order information between paths but also vertex-based semantic information within paths. Therefore, based on the searching-scenario oriented hypergraph, this paper utilizes a dual-embedding mechanism [47] and hyperedge convolution [46] to obtain high-order information between paths and vertex-based semantic information within paths, respectively.

3.3.1. Path Semantic Association Learning

A path contains any number of nodes, these nodes are of the same or different types, so the generated paths have different semantic information. In this paper, dual-embedding mechanism [47] is used to obtain semantic associations among consumers' online queries, their click, purchase, and online review behavior.

The semantic associations among consumers' online queries, their click, purchase, and online review behavior is illustrated as follows. Take the query "harry potter" for example. By using the searching-scenario oriented hypergraph, "harry potter" entered by different users can reach different items such as "harry potter PVC figure", "harry potter book", "harry potter magic wand" or "harry potter LEGO". Obviously, we can obtain more recommendation candidates for the query "harry potter" by using the searching-scenario oriented hypergraph. More importantly, the structural superiority of the searching-scenario oriented hypergraph gives the recommender system a chance to identify different semantic facets of the input search phrases. Similarly, the searching-scenario oriented hypergraph can leverage user behavior to mine related queries with different query phrases. For example, the query "python" entered by user A and the query "Data Analysis" entered by user B can reach the same book "Python for Data Analysis". We can infer from this example that consumer B who bought the book had a preference for using Python even though it was not explicitly expressed in his query. Query-item collaborative filtering greatly solves the item entity recall problem under sparse data.

Therefore, to augment semantic information propagation and training efficiency, we use second-order neighbor relations instead of first-order neighbor relations. To ensure the timeliness of recommended items, we use a strategy of 20% uniform sampling and 80% popularity-based sampling to sample node neighbors.

3.3.2. Convolutional Semantic Features Learning

Not only are there complex associations between paths but the vertices in paths also contain rich semantic information. This paper adopts hyperedge corruption [47] to cut the hyperedge into ordinary edges, which connect the user-query, query-item, and item-feature, respectively. Then, ordinary edges are used to generate association matrices, and the initial weights of the vertices are calculated to generate the hypergraph Laplacian matrix based on meta-path information. Then, this matrix is added to the hypergraph neural network [46] to learn the hyperedge convolution:

$$X^{(l+1)} = \sigma\left(D_v^{-\frac{1}{2}} HWD_e^{-1} H^T D_v^{-\frac{1}{2}} X^{(l)} \Theta^{(l)}\right) \tag{4}$$

where X, D_v, D_e, and Θ is the signal of the hypergraph at l layer, σ denotes the nonlinear activation function.

Therefore, the final convolutional feature can be obtained by connecting L layer features:

$$x_u = \left[X^0, X^1, \cdots, X^L\right] \tag{5}$$

Similarly, for the online reviews of each item i, the corresponding convolutional semantic feature x_i can be obtained through the hypergraph neural network.

3.4. Prediction

For each user u, the obtained convolutional semantic feature x_u and query topic feature θ_u are combined to represent the final user embedding X_u of user u:

$$X_u = x_u \oplus \theta_u \tag{6}$$

Similarly, the final feature X_i of each item i is:

$$X_i = x_i \oplus \theta_i \tag{7}$$

Since the number of words in each query is different, the dimension of the word vector matrix is inconsistent, which cannot be processed by the convolutional neural network. Therefore, this paper fixes the number of search phrases in each query as 32, that is, when the number of words is less than 32, it is filled with 0, and when the number of words is greater than 32, the first 32 words are taken. This paper uses BERT to pre-train all the obtained text content to obtain vectors of words X_q.

We want to integrate query embedding, user embeddings, item embedding, and high-order correlations to capture more complex connections. We utilize a deep architecture [48] to predict link relationships between users, queries, items, and features:

$$\hat{y} = \phi_L\left(\phi_{L-1}\left(\ldots \phi_1\left(\left[X_q; X_u; X_i\right]\right)\right)\right) \tag{8}$$

where $[;;;]$ concatenates the input vectors and $\{\phi_1, \phi_2, \ldots, \phi_L\}$ are non-linear layers with sigmoid as the active function.

We also take the widely used binary cross-entropy as the loss function:

$$\mathcal{L} = \sum y \log(\hat{y}) + (1 - y) \log(1 - \hat{y}) + \lambda \omega^2 \tag{9}$$

where ω is the learnable parameters set, λ is the regularization parameter.

4. Results

In order to test the improvement of the proposed Topic-HGNN framework, we conducted experiments based on different datasets obtained from real-world applications. The experiments were designed to verify two aspects of the proposed recommender framework: (1) the quality of topics in online query sessions identified by the ALDA model, and (2) the improvement of recommendation accuracy and novelty of the Topic-HGNN framework that connects the latent topics embedded in consumers' online queries to their click, purchase, and online review behavior.

All empirical evaluations in this paper were implemented on a Dell Precision T5820 workstation with Xeon W-2102 CPU, 8.00 GB RAM, and we chose to implement the program in the Python language.

4.1. Data Description

The public AOL query log dataset (http://www.gregsadetsky.com/_aol-data accessed on 18 September 2019) in the real word is used for experimental verification. This collection consists of 20 M web queries collected from 650 k users over three months in 2006. The data is sorted by anonymous user ID and sequentially arranged. The data set includes {AnonID, Query, QueryTime, ItemRank, ClickURL}. AnonID represents an anonymous user ID number. Query indicates the query issued by the user. QueryTime indicates the time at which the query was submitted for search. If the user clicked on a search result, the rank of the item on which they clicked is listed, and it is marked as ItemRank. If the user clicked on a search result, the domain portion of the URL in the clicked result is listed, which is marked as ClickURL.

We preprocessed the AOL query log dataset before conducting experiments. First, we successively removed query terms containing URL strings, query terms containing special characters, and query terms that did not contain click URLs. Then, we utilized "15 min interval" [50] to derive reasonable session breaks in online queries in order to better investigate the effectiveness of the ALDA model. Finally, we divided each user's search records into training sets and test sets with a ratio of 80%/20%. Part of the AOL query log dataset format is shown in Table 2.

The Retailrocket data (https://www.kaggle.com/retailrocket/ecommerce-dataset accessed on 18 September 2019) was collected from a real-world e-commerce site. The data includes 2,756,101 behavior records from 1,407,580 users, including 2,664,312 views, 69,332 cart additions, and 22,457 purchases.

Table 2. The example of the AOL query log dataset.

AnonID	Query	QueryTime	ItemRank	ClickURL
479	car decals	2006-03-03 23:20:12	4	http://www.decaljunky.com
479	car decals	2006-03-03 23:20:12	1	http://www.modernimage.net
479	car decals	2006-03-03 23:20:12	5	http://www.webdecal.com
479	car window decals	2006-03-03 23:24:05	9	http://www.customautotrim.com
479	car window sponsor decals	2006-03-03 23:27:17	3	http://www.streetglo.net
1020	slot machine tips	2006-04-18 12:43:46	1	http://www.slotadvisor.com
1020	slot machine tips	2006-04-18 12:43:46	4	http://www.thegamblersedge.com
1020	slot machine tips	2006-04-18 12:43:46	8	http://www.gambling.jaxworld.com
1020	slot machine tips	2006-04-18 13:06:52	11	http://www.licensed4fun.com

The entire dataset contains three files: behavioral data file, category relationship file, and item properties file. Each row of data describes the user's behavior on an item at a certain time.

4.2. Evaluation of the ALDA Model

In order to examine the quality of topics in online query sessions identified by the ALDA model proposed in our paper, five typical methods for inferring user preference distributions are selected as baseline methods.

- LDA is a generative probabilistic model in which each document is modeled as a finite mixture over an underlying set of topics and each topic is modeled as an infinite mixture over an underlying set of word distributions [21].
- Twitter-BTM aggregates user-based biterms to learn user specific topic distribution and incorporates a background topic to distinguish user's preference between background words and topical words [51].
- UCIT learns users' short-term and long-term preferences based on their followees' topic distributions, the content of current short texts, and the previously estimated distributions [52].
- HDLDA is a hierarchically dual latent Dirichlet allocation that assumes there is a semantic relation between search query documents and search result documents, and it quantitatively characterizes how consumers translate their content preferences into search queries [49].
- UATM infers topic intensities in user's preference by learning topic intensities in user's preference and topic intensities in followees' preference, which can efficiently alleviate the sparsity problem [53].

We use the AOL query log dataset in this section. By comparing the parameter settings of the above models, we set the hyperparameters $\alpha = 50/K$, $\beta = 0.01$, $\gamma = 0.5$.

4.2.1. Topic Coherence

Topic coherence is mainly used to measure whether the words within a topic are coherent. So, how can these words be considered coherent? If the words support each other, then the group of words is coherent. In other words, if you put words from multiple topics together and cluster them with a perfect cluster, then words from the same topic should be in the same category. *PMI* uses external text datasets to measure the coherence of a topic, which is a fair metric of evaluating the quality of topics extracted by each model. The PMI can be calculated by:

$$PMI(w_i, w_j) = \log \frac{p(w_i, w_j) + \epsilon}{p(w_i) \cdot p(w_j)}$$

where w_i and w_j are topic words, and ϵ is a random disturbance term. The larger the value of PMI, the better the coherence between topic words.

To further evaluate the PMI of randomly selected topics, Wikipedia articles downloaded from the official Wikipedia website were used as an auxiliary corpus. We selected the top 5, 10, and 20 words in each topic and calculated the average PMI score. Figure 3 shows the topic coherence results of selected topics learned by each topic discovery model. In the comparison of six models, it clearly shows that the PMI score of our ALDA model is significantly better than the other models. The results demonstrate that topics extracted by our ALDA are more coherent than other models. This is due to the fact that our ALDA conjointly models the topic intensities in the query sessions and the topic intensities in the webpages into the same document layer. The data sparsity of online query data is avoided by aggregating corresponding webpages to assist in learning users interested topics. Because Twitter-BTM and LDA can only model query documents and webpage documents separately, these two models perform worst. Twitter-BTM outperforms LDA because Twitter-BTM inherits BTM's excellent ability to deal with short texts. UCIT and UATM significantly outperforms Twitter-BTM and LDA. This is because UCIT and UATM not only extract topics from content generated by the user themself but also extracts topics from content generated by user clusters that are similar to them. HDLDA can generate more coherent topics than UATM, UCIT, Twitter-BTM, and LDA. This is because HDLDA models query the document and the webpage document in two hierarchical LDA processes. HDLDA can better capture the semantic relation between query and webpage.

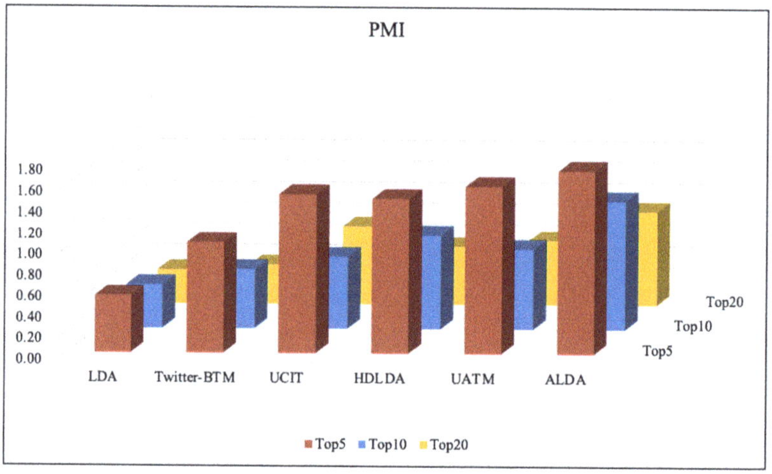

Figure 3. PMI score of each model on AOL query log dataset.

Unlike HDLDA, which models the topic intensities in the query sessions and the topic intensities in the webpages, our ALDA conjointly models the topic intensities in the query sessions and the topic intensities in the webpages into the same document layer. Thus, our ALDA obtained better results than HDLDA.

4.2.2. User's Preference Prediction

We utilize perplexity to compare the accuracy of predicting users' content preference drift estimated by these models. As perplexity in information theory is a measure that is often used to judge probability models or probability distribution prediction samples, we utilize perplexity to evaluate the effect of user's preference inferred by each model. The ability of perplexity is to predict words in new documents, which are not observed. The smaller the value of perplexity, the better the performance of the model in mining user's intention. *Perplexity* can be calculated as follows:

$$Perplexity_{portion}(\mathcal{M}) = \left(\sum_{d=1}^{D} \sum_{i=P+1}^{N_d} p(w_i | \mathcal{M}, w_{1:p}) \right)^{-\frac{1}{\Sigma_{d=1}^{D}(N_d - p)}}$$

where $\mathcal{M}$ is the set of model parameters learned from the training set, d represents the document, and N_d is the number of words in the document.

To make the experimental results more reliable, we sample the observed in the AOL dataset at different scales (from 10% to 90%). It can be seen from Figure 4 that the perplexity of each model gradually decreases with the expansion of the percentage of the observed data. This shows that each model gets better at predicting consumer preferences with the growth of the observed data. Compared with the other five models, the perplexity degree of our ALDA model is the smallest, from 1100 to 2500, which indicates that ALDA preforms best among the six models for identifying consumer interests. This is because ALDA models the interactive relationship between queries and webpages. In reality, a consumer's shopping process is actually a process of understanding and evaluating products. First, users may enter inaccurate keywords to express their needs. Then, users enhance their understanding of products through browsing the search results and adjusting the input keywords. Consumers will repeat this learning process until finding the right product.

That is to say, the topics of query keywords and the topic of search results are semantically related to each other. Thus, modeling this interaction between queries and webpages helps us to more accurately capture changes in consumer's interests and preferences. This is the fundamental reason our model is better than other models in identifying consumers' purchase intentions.

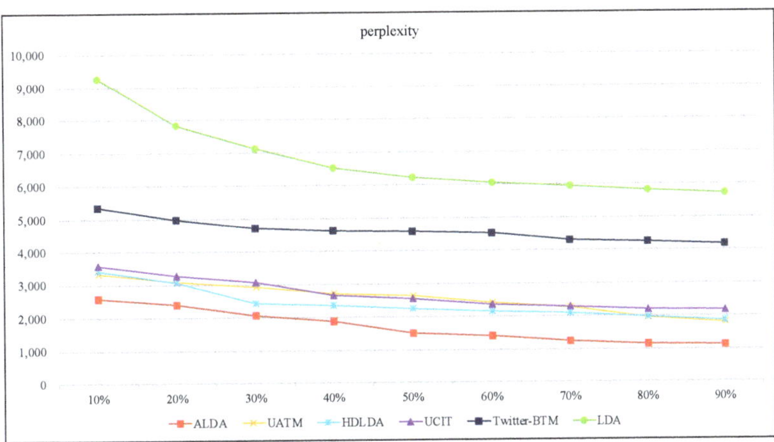

Figure 4. Comparison of user's preference inferring performance.

As LDA and Twitter-BTM do not model how the topics in search queries relate to the topics in the corresponding search results, they obtain the worst performance on understanding users' preference. Both UCIT and UATM learn the topic distributions in the user's content and followees' content, which enables extensive mining and understanding of user's preference and intention, and the experimental results also confirm that UCIT and UATM significantly perform better than LDA and Twitter-BTM. HDLDA models query the document and the webpage document in two hierarchical LDA processes, and they assume that the query document is semantically related to the webpage document, which contributes to a slight lead over UCIT and UATM in understanding the user's interest. Although HDLDA produces good results, it performs worse than ALDA. This is due to HDLDA failing to capture the interactive relationship between queries and webpages. In summary, our ALDA model always outperforms the other comparison models on predicting consumers' purchase intentions.

4.3. Evaluation of Recommendation Results

The proposed Topic-HGNN framework incorporates the topic model into a hypergraph neural network for enhancing user and item embedding representation. Five typical topic model-based recommendation techniques and two state of art neural network-based recommendation methods are selected as baselines.

In order to examine whether the user and the item feature identified by the Topic-HGNN can achieve better personalized recommendations, we utilized precision and diversification to evaluate the recommendation results in detail. The experiment was conducted on the Retailrocket dataset.

- CTR provides an interpretable latent structure for users and items by combining the merits of traditional collaborative filtering and probabilistic topic modeling [54].
- SVD-LDA improves SVD-based recommendations for items with textual content with topic modeling of this content [55].
- CoAWILDA relies on an adaptive online Latent Dirichlet Allocation to model newly available items arriving as a document stream and incremental matrix factorization for collaborative filtering [56].

- AR-LDA uses topic modeling and sequential association rule mining to capture the preference of the user's product changes over time [57].
- EUU-CF extracts topics in Wikipedia by using the LDA model and then uses the topics on user browsing history to extract user preferences [58].
- Graph-CNN is a graph convolutional neural network-based approach to recommend products to users by analyzing their previous interactions [42].
- HyperCTR learns item representations based on multi-modal information interactions among users and items [48].

4.3.1. Precision of Recommendation Results

We adopt two commonly used metrics, *Precision* and *Recall*, to evaluate the accuracy of recommendation results obtained by each recommender method. Precision and Recall are defined as:

$$Precision = \frac{\sum_{u \in U} |R(u) \cap T(u)|}{|R(u)|}$$
$$Recall = \frac{\sum_{u \in U} |R(u) \cap T(u)|}{|T(u)|}$$

where $R(u)$ denotes the recommendation list based on the training dataset, and $T(u)$ denotes the recommendation list based on the test dataset.

To evaluate the accuracy of recommendation results obtained by each recommender technique, we set the number of recommendations from top10 to top100.

Figure 5 shows the comparison of the accuracy of recommendation results generated by each recommender technique.

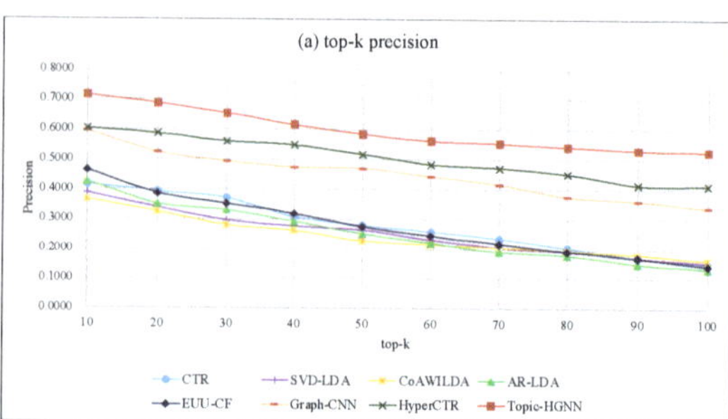
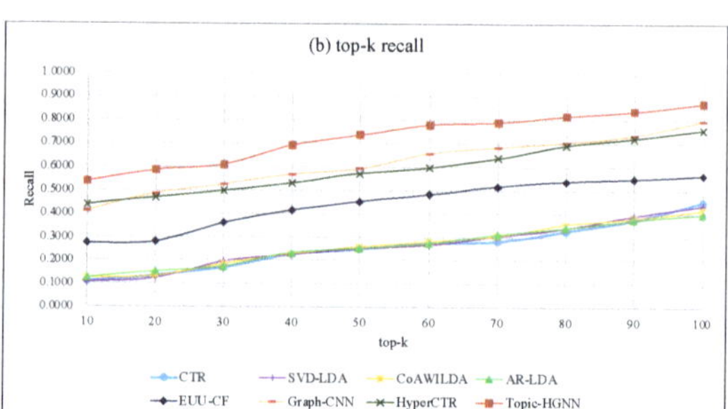

Figure 5. Comparison of recommendation result accuracy: (**a**) precision of top-k items; (**b**) recall of top-k items.

We can observe that the accuracy of recommendation results generated by topic-based methods CTR, SVD-LDA, CoAWILDA, AR-LDA, and EUU-CF are very close to each other and are significantly worse than Graph-CNN, HyperCTR, and Topic-HGNN. This is because topic-based methods focus on improving recommendations for items with textual content. They infer the user's interest based on the user's purchase behavior, which is difficult to refine user preferences for different product attributes and capture high-order correlations between users and items. Different from topic-based recommendation models, Graph-CNN, HyperCTR, and Topic-HGNN infer the user's preference from rich user-product interaction information. Although Graph-CNN and HyperCT also produces good accurate recommendations, it performs worse than Topic-HGNN. This is due to the Graph-CNN and HyperCT only focusing on learning user interaction characteristics with products during and after purchase (e.g., purchase and online review) and ignoring the interactive features in the process of users' learning about product information through search queries before they make a purchase (e.g., product information search). However, user association with a product is a coherent process that should not be isolated into different nodes. Our Topic-HGNN integrates a topic model and a hypergraph neural network, which can deal with the interaction behavior of consumers in the whole process of shopping (i.e., searching-understanding-purchasing-using). Besides, the Topic-HGNN obtains the convolutional semantic features of users and items, and uses the topic model to obtain the corresponding topic features. The result shows that incorporating the topic information from users and items into a convolutional neural network can effectively represent user preferences and item features, which can significantly improve the accuracy of prediction scores. The result also demonstrates the structural superiority of the searching-scenario oriented hypergraph, which gives the recommender system a chance to identify different semantic facets of the input search phrases.

4.3.2. Novelty of Recommendation Results

Only verifying the accuracy of model recommendation results is not enough to explain the personalized effect of a recommendation model. As the collaborative filtering only depends on the user's past purchase behavior, users can only get recommendations similar to items in their own profile and hardly get diversified options. So, experiments are further designed to verify the ability of the recommendation model to discover novel items to the target user. We adopt the novelty metric [59] to measure the ability of recommendation model to find novel items. The lower the Novelty is, the more novel products are recommended. *Novelty* is defined as:

$$novelty = \frac{1}{mk} \sum_{u=1}^{m} \sum_{i \in L_u} d_i$$

where L_u is the top-k list of a user u, m is the number of users, and d_i is the degree of item i, i.e., the number of users that rated the item i.

We set the number of recommendation items to 10, and experimented on the Retailrocket datasets 32 times each. A smaller novelty value indicated that the recommendation items were more novel.

Figure 6 shows the comparison of the novelty of recommendation results generated by each recommender technique. We can observe that the novelty of recommendation results generated by CTR, SVD-LDA, CoAWILDA, EUU-CF, and AR-LDA are very close to each other and are significantly worse than Graph-CNN, Hyper-CTR, and Topic-HGNN. This is because CTR, SVD-LDA, CoAWILDA, EUU-CF, and AR-LDA infer the user's interest based on the user's historical purchase behavior, which is difficult to discover new products for consumers. This result demonstrates that the topic-based method is significantly worse than the graph-based method. The Topic-based method regards the interaction between users and products as a matrix, and it focuses on mining linear correlation and low-rank information. However, graph-based methods focus on mining interaction information and high-order relation in the graph. Compared with the matrix, the graph can describe more information, such as the link to describe the connection between adjacent vertices, the overall connection between all vertices in the graph, and the link density to describe the community structure in the graph. The graph has a powerful representation ability and the effect of the graph-based method is significantly better than that of the traditional recommendation algorithm.

Our Topic-HGNN is significantly better than Graph-CNN, Hyper-CTR, which demonstrates that Topic-HGNN can identify different semantic facets of input search phrases. Topic-HGNN can obtain semantic associations among consumers' online queries, their click, purchase, and online review behavior that are better than Graph-CNN, Hyper-CTR. Topic-HGNN simultaneously considers heterogeneous interactions and homogeneous interactions in the user purchasing paths, which can better utilize the deep connection information contained in the interactive graph domain, and it is not limited to the observed links.

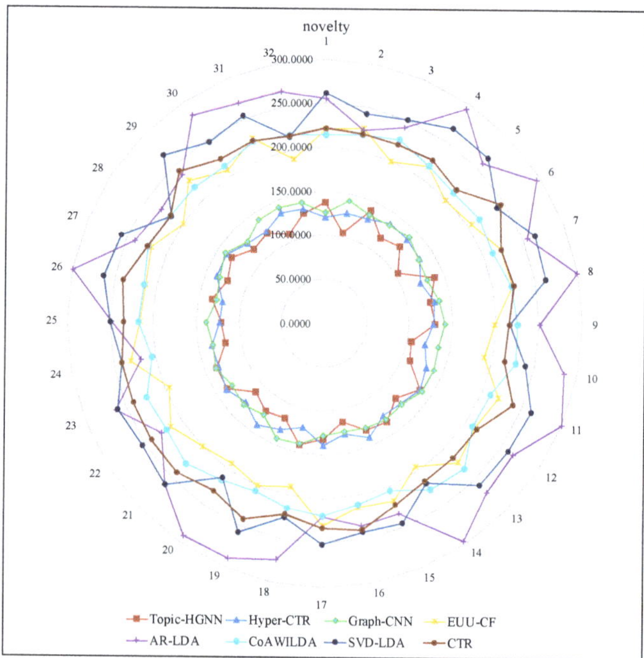

Figure 6. Comparison of recommendation result novelty.

In summary, our Topic-HGNN could improve the novelty of recommendation items without sacrificing accuracy.

4.3.3. Efficiency of Topic-HGNN

The running time and the memory consumption of each method under different query search volumes on Retailrock dataset is shown in Tables 3 and 4. We set the number of recommendation results as 10. From Table 3, it can be seen that the recommendation framework based on a topic model is significantly better than the recommendation framework based on graph learning in terms of running time. Although the recommendation framework based on a topic model is approximately 15% more efficient than the recommendation framework based on graph learning, the quality of the results identified by the recommendation framework based on graph learning on the accuracy, recall, and novelty indicators improved by 53%, 51%, and 46%. This also demonstrates that the method based on graph learning can significantly improve the quality of recommendation results at the expense of a small amount of operating efficiency. Among the three graph-learning-based methods, the running time of our model is slightly higher since our method models the quaternary higher-order relationship among consumers, queries, items, and features. Thus, Topic-HGNN is significantly superior to that of Hyper-CTR and Graph-CNN, when sacrificing a relatively low efficiency.

Table 3. The running time of each method under different query search volumes on Retailrock dataset (the number of recommendation results is 10).

Method	Running Time (10^3 Queries)	Running Time (10^4 Queries)	Running Time (10^5 Queries)
CTR	12.12 ms	2095.54 ms	49,514.16 ms
SVD-LDA	11.87 ms	2294.63 ms	48,510.53 ms
CoAWILDA	11.65 ms	3220.22 ms	46,767.78 ms
AR-LDA	8.02 ms	3076.21 ms	38,881.03 ms
EUU-CF	8.54 ms	3085.96 ms	48,736.47 ms
Graph-CNN	15.57 ms	5014.59 ms	58,294.41 ms
Hyper-CTR	19.56 ms	4963.22 ms	59,324.57 ms
Topic-HGNN	19.67 ms	5038.40 ms	58,290.89 ms

Table 4. The memory consumption of each method under different query search volumes on Retailrock dataset (the number of recommendation results is 10).

Method	Memory Consumption (103 Queries)	Memory Consumption (104 Queries)	Memory Consumption (105 Queries)
CTR	83 MB	347 MB	970 MB
SVD-LDA	89 MB	385 MB	1102 MB
CoAWILDA	96 MB	403 MB	1165 MB
AR-LDA	77 MB	311 MB	928 MB
EUU-CF	79 MB	284 MB	944 MB
Graph-CNN	882 MB	1509 MB	3259 MB
Hyper-CTR	926 MB	1647 MB	3895 MB
Topic-HGNN	974 MB	1802 MB	3971 MB

As can be seen from Table 4, the Topic-HGNN framework does not consume additional memory compared to other graph-based learning methods. This is because the Topic-HGNN is decomposed by hyperedge corruption, importing batches of vertices and hyperedges each time to relieve memory pressure. Therefore, in summary, the Topic-HGNN proposed in this work can produce better recommendation results, while being almost as effective as other graph-based methods.

5. Conclusions

Personalized product recommendation systems are a useful tool adopted by e-retailers to help consumers find items in line with their preferences. Existing research focuses on learning user interaction characteristics with products during and after purchase (e.g., purchase and online review), while ignoring the interactive features in the process of users' learning about product information through search queries before they make a purchase (e.g., product information search). However, users' association with a product is a coherent process that should not be isolated into different nodes. Only sorting the user's process of searching-understanding-purchasing-using products and finding opportunity points from each stage can help the recommender system to better discover the potential needs of users. To this end, we develop a topic augmented hypergraph neural network framework to predict users' purchase intentions by connecting the latent topics embedded in consumers' online queries to their click, purchase, and online review behavior. First, we adopt a hypergraph to model the multivariate relationship among users, query topics, items, and item features, which aims at mining the connection information existing in the interaction graph domain. Then, we utilize the hyperedge corruption to generate a user-query hypergraph and an item-feature hypergraph and utilize the hyperedge convolution layer to obtain user embedding and item embedding. Meanwhile, in order to reduce the influence of text noise words by fusing topic information, we specially design an Aggregated Latent Dirichlet Allocation (ALDA) model to jointly extract users' content preference topics from queries and webpages and apply Latent Dirichlet Allocation model to extract product feature topics from online reviews. Then, we integrate the topic distribution and convolutional embedding to represent each user and item, which can make up for the lack of topic information in traditional convolutional neural networks. Finally, we use multilayer perceptron to calculate the soft match score between query entities and item entities. Extensive empirical evaluations on real-world datasets demonstrate that the proposed framework could improve the novelty of recommendation items without sacrificing accuracy. From the managerial perspective, recommending diversified and novel items to consumers may increase the user's satisfaction, which is conducive to the sustainable development of e-commerce enterprises.

With the rapid development of mobile commerce, more and more recommendation services occur in dynamically changing contexts, such as user location, access time, current traffic, and other surrounding environments. Traditional personalized recommendation technology is no longer enough to deal with the new impact caused by contextual factors. Therefore, our future work will focus on integrating and applying contextual information into the hypergraph framework, which aims at combining context development diagram and user behavior prediction to form a unified and concise context-based recommendation model. In this work, we assumed that search query documents and webpage documents follow the same topic distributions. In reality, search query documents and webpage documents sometimes didn't follow the same topic distributions. Thus, examining the impact in the results when search query documents and webpage documents did not follow the same topic distribution is also a future research topic.

Author Contributions: Conceptualization, X.H. and X.L.; methodology, X.H.; software, X.H.; validation, X.L.; formal analysis, X.L.; investigation, X.H. and X.L.; resources, X.H.; data curation, X.H.; writing—original draft preparation, X.H.; writing—review and editing, X.L.; visualization, X.H.; supervision, X.L.; project administration, X.L.; funding acquisition, X.H. and X.L. All authors have read and agreed to the published version of the manuscript.

Funding: This work is supported by the Fundamental Research Funds for the Central Universities under Grant No. 310422121.

Institutional Review Board Statement: Not applicable.

Informed Consent Statement: Not applicable.

Data Availability Statement: The data used in this work can be found at http://www.gregsadetsky.com/_aol-data and https://www.kaggle.com/retailrocket/ecommerce-dataset accessed on 18 September 2019.

Acknowledgments: We would like to thank the editors and the anonymous reviewers for their insightful suggestions.

Conflicts of Interest: The authors declare no conflict of interest.

References

1. Pantano, E.; Priporas, C.-V. The effect of mobile retailing on consumers' purchasing experiences: A dynamic perspective. *Comput. Hum. Behav.* **2016**, *61*, 548–555. [CrossRef]
2. Hussien, F.T.A.; Rahma, A.M.S.; Abdulwahab, H.B. An E-Commerce Recommendation System Based on Dynamic Analysis of Customer Behavior. *Sustainability* **2021**, *13*, 10786. [CrossRef]
3. Patten, E.; Ozuem, W.; Howell, K.; Lancaster, G. Minding the competition: The drivers for multichannel service quality in fashion retailing. *J. Retail. Consum. Serv.* **2020**, *53*, 101974. [CrossRef]
4. Lee, D.; Hosanagar, K. How Do Recommender Systems Affect Sales Diversity? A Cross-Category Investigation via Randomized Field Experiment. *Inf. Syst. Res.* **2019**, *30*, 239–259. [CrossRef]

5. Jesse, M.; Jannach, D. Digital nudging with recommender systems: Survey and future directions. *Comput. Hum. Behav. Rep.* **2021**, *3*, 100052. [CrossRef]
6. Archak, N.; Ghose, A.; Ipeirotis, P.G. Deriving the Pricing Power of Product Features by Mining Consumer Reviews. *Manag. Sci.* **2011**, *57*, 1485–1509. [CrossRef]
7. Xie, K.; Wu, Y.; Xiao, J.; Hu, Q. Value co-creation between firms and customers: The role of big data-based cooperative assets. *Inf. Manag.* **2016**, *53*, 1034–1048. [CrossRef]
8. Ebrahimi, P.; Hamza, K.A.; Gorgenyi-Hegyes, E.; Zarea, H.; Fekete-Farkas, M. Consumer Knowledge Sharing Behavior and Consumer Purchase Behavior: Evidence from E-Commerce and Online Retail in Hungary. *Sustainability* **2021**, *13*, 10375. [CrossRef]
9. Koren, Y.; Rendle, S.; Bell, R. Advances in Collaborative Filtering. In *Recommender Systems Handbook*; Springer: New York, NY, USA, 2021; pp. 91–142. [CrossRef]
10. Khojamli, H.; Razmara, J. Survey of similarity functions on neighborhood-based collaborative filtering. *Expert Syst. Appl.* **2021**, *185*, 115482. [CrossRef]
11. Lee, K.; Hwangbo, Y.; Jeong, B.; Yoo, J.; Park, K. Extrapolative Collaborative Filtering Recommendation System with Word2Vec for Purchased Product for SMEs. *Sustainability* **2021**, *13*, 7156. [CrossRef]
12. Lika, B.; Kolomvatsos, K.; Hadjiefthymiades, S. Facing the cold start problem in recommender systems. *Expert Syst. Appl.* **2014**, *41*, 2065–2073. [CrossRef]
13. Son, J.; Kim, S.B. Content-based filtering for recommendation systems using multiattribute networks. *Expert Syst. Appl.* **2017**, *89*, 404–412. [CrossRef]
14. Humphreys, A.; Isaac, M.S.; Wang, R.J.-H. Construal Matching in Online Search: Applying Text Analysis to Illuminate the Consumer Decision Journey. *J. Mark. Res.* **2021**, *58*, 1101–1119. [CrossRef]
15. Roscoe, R.D.; Grebitus, C.; O'Brian, J.; Johnson, A.C.; Kula, I. Online information search and decision making: Effects of web search stance. *Comput. Hum. Behav.* **2016**, *56*, 103–118. [CrossRef]
16. Park, J.; Kim, R.B. A new approach to segmenting multichannel shoppers in Korea and the US. *J. Retail. Consum. Serv.* **2018**, *45*, 163–178. [CrossRef]
17. Kim, D.; Woo, J.; Shin, J.; Lee, J.; Kim, Y. Can search engine data improve accuracy of demand forecasting for new products? Evidence from automotive market. *Ind. Manag. Data Syst.* **2019**, *119*, 1089–1103. [CrossRef]
18. Liu, J.; Toubia, O. Search query formation by strategic consumers. *Quant. Mark. Econ.* **2020**, *18*, 155–194. [CrossRef]
19. Codignola, F.; Capatina, A.; Lichy, J.; Yamazaki, K. Customer information search in the context of e-commerce: A cross-cultural analysis. *Eur. J. Int. Manag.* **2021**, *16*, 28–59. [CrossRef]
20. Liu, Z.; Chen, H.; Sun, F.; Xie, X.; Gao, J.; Ding, B.; Shen, Y. Intent preference decoupling for user representation on online recommender system. In Proceedings of the Twenty-Ninth International Conference on International Joint Conferences on Artificial Intelligence, Yokohama, Japan, 7–15 January 2021; pp. 2575–2582.
21. Blei, D.M.; Ng, A.Y.; Jordan, M.I. Latent dirichlet allocation. *J. Mach. Learn. Res.* **2003**, *3*, 993–1022.
22. Bobadilla, J.; Ortega, F.; Hernando, A.; Gutiérrez, A. Recommender systems survey. *Knowl.-Based Syst.* **2013**, *46*, 109–132. [CrossRef]
23. Bellogín, A.; Castells, P.; Cantador, I. Neighbor selection and weighting in user-based collaborative filtering: A performance prediction approach. *ACM Trans. Web* **2014**, *8*, 1–30. [CrossRef]
24. Zhang, Z.; Kudo, Y.; Murai, T. Neighbor selection for user-based collaborative filtering using covering-based rough sets. *Ann. Oper. Res.* **2016**, *256*, 359–374. [CrossRef]
25. Jiang, J.; Lu, J.; Zhang, G.; Long, G. Scaling-up item-based collaborative filtering recommendation algorithm based on hadoop. In Proceedings of the 2011 IEEE World Congress on Services, Washington, DC, USA, 4–9 July 2011; pp. 490–497.
26. Xue, F.; He, X.; Wang, X.; Xu, J.; Liu, K.; Hong, R. Deep Item-based Collaborative Filtering for Top-N Recommendation. *ACM Trans. Inf. Syst.* **2019**, *37*, 1–25. [CrossRef]
27. Ortega, F.; Hernando, A.; Bobadilla, J.; Kang, J.H. Recommending items to group of users using Matrix Factorization based Collaborative Filtering. *Inf. Sci.* **2016**, *345*, 313–324. [CrossRef]
28. He, X.; Liao, L.; Zhang, H.; Nie, L.; Hu, X.; Chua, T.S. Neural collaborative filtering. In Proceedings of the 26th International Conference on World Wide Web, Perth, Australia, 3–7 May 2017; pp. 173–182.
29. Thakur, S.; Sing, J. Online product prediction and recommendation using probability graphical model and collaborative filtering: A new approach. In Proceedings of the 2011 IEEE Recent Advances in Intelligent Computational Systems, Trivandrum, India, 22–24 September 2011; pp. 151–156.
30. Hu, Y.; Guo, C.; Ngai, E.W.; Liu, M.; Chen, S. A scalable intelligent non-content-based spam-filtering framework. *Expert Syst. Appl.* **2010**, *37*, 8557–8565. [CrossRef]
31. Philip, S.; Shola, P.; Ovye, A. Application of Content-Based Approach in Research Paper Recommendation System for a Digital Library. *Int. J. Adv. Comput. Sci. Appl.* **2014**, *5*, 37–40. [CrossRef]
32. Shahi, T.B.; Yadav, A. Mobile SMS spam filtering for Nepali text using naïve bayesian and support vector machine. *Int. J. Intell. Sci.* **2014**, *4*, 24–28. [CrossRef]
33. Burke, R. Hybrid Recommender Systems: Survey and Experiments. *User Model. User-Adapt. Interact.* **2002**, *12*, 331–370. [CrossRef]
34. Annunziata, G.; Colace, F.; De Santo, M.; Lemma, S.; Lombardi, M. ApPoggiomarino: A Context Aware App for e-Citizenship. In Proceedings of the 18th International Conference on Enterprise Information Systems (ICEIS (2)), Rome, Italy, 25–28 April 2016; pp. 273–281.

35. Colace, F.; Lemma, S.; Lombardi, M.; Pascale, F. A Context Aware Approach for Promoting Tourism Events: The Case of Artist's Lights in Salerno. In Proceedings of the 19th International Conference on Enterprise Information Systems (ICEIS (2)), Porto, Portugal, 26–29 April 2017; pp. 752–759.
36. Ricci, F.; Shapira, B.; Rokach, L. Recommender systems: Introduction and challenges. In *Recommender Systems Handbook*; Springer: Boston, MA, USA, 2015; pp. 1–34.
37. Li, X.; Pan, B.; Law, R.; Huang, X. Forecasting tourism demand with composite search index. *Tour. Manag.* **2017**, *59*, 57–66. [CrossRef]
38. Choi, H.; Varian, H. Predicting the Present with Google Trends. *Econ. Rec.* **2012**, *88*, 2–9. [CrossRef]
39. Yang, Y.; Pan, B.; Song, H. Predicting Hotel Demand Using Destination Marketing Organization's Web Traffic Data. *J. Travel Res.* **2014**, *53*, 433–447. [CrossRef]
40. Tibau, M.; WM Siqueira, S.; Pereira Nunes, B.; Bortoluzzi, M.; Marenzi, I.; Kemkes, P. Investigating users' decision-making process while searching online and their shortcuts towards understanding. In *International Conference on Web-Based Learning*; Springer: Berlin/Heidelberg, Germany, 2018; pp. 54–64.
41. Wu, Z.; Pan, S.; Chen, F.; Long, G.; Zhang, C.; Philip, S.Y. A comprehensive survey on graph neural networks. *IEEE Trans. Neural Netw. Learn. Syst.* **2020**, *32*, 4–24. [CrossRef] [PubMed]
42. Shafqat, W.; Byun, Y.-C. Enabling "Untact" Culture via Online Product Recommendations: An Optimized Graph-CNN based Approach. *Appl. Sci.* **2020**, *10*, 5445. [CrossRef]
43. He, X.; Deng, K.; Wang, X.; Li, Y.; Zhang, Y.; Wang, M. Lightgcn: Simplifying and powering graph convolution network for recommendation. In Proceedings of the 43rd International ACM SIGIR Conference on Research and Development in Information Retrieval, Xi'an, China, 25–30 July 2020; pp. 639–648.
44. Li, Z.; Cui, Z.; Wu, S.; Zhang, X.; Wang, L. Fi-gnn: Modeling feature interactions via graph neural networks for ctr prediction. In Proceedings of the 28th ACM International Conference on Information and Knowledge Management, Beijing, China, 3–7 November 2019; pp. 539–548.
45. Chang, B.; Jang, G.; Kim, S.; Kang, J. Learning graph-based geographical latent representation for point-of-interest recommendation. In Proceedings of the 29th ACM International Conference on Information & Knowledge Management, Virtual, 19–23 October 2020; pp. 135–144.
46. Feng, Y.; You, H.; Zhang, Z.; Ji, R.; Gao, Y. Hypergraph neural networks. In Proceedings of the AAAI Conference on Artificial Intelligence, Honolulu, HI, USA, 27 January–1 February 2019; Volume 33, pp. 3558–3565.
47. Chen, X.; Xiong, K.; Zhang, Y.; Xia, L.; Yin, D.; Huang, J.X. Neural Feature-aware Recommendation with Signed Hypergraph Convolutional Network. *ACM Trans. Inf. Syst.* **2020**, *39*, 1–22. [CrossRef]
48. He, L.; Chen, H.; Wang, D.; Jameel, S.; Yu, P.; Xu, G. Click-Through Rate Prediction with Multi-Modal Hypergraphs. In Proceedings of the 30th ACM International Conference on Information & Knowledge Management, Gold Coast, Australia, 1–5 November 2021; pp. 690–699.
49. Liu, J.; Toubia, O. A Semantic Approach for Estimating Consumer Content Preferences from Online Search Queries. *Mark. Sci.* **2018**, *37*, 930–952. [CrossRef]
50. He, D.; Göker, A. Detecting session boundaries from web user logs. In Proceedings of the BCS-IRSG 22nd annual Colloquium on Information Retrieval Research, Lisbon, Portugal, 14–17 April 2020; pp. 57–66.
51. Chen, W.; Wang, J.; Zhang, Y.; Yan, H.; Li, X. User based aggregation for biterm topic model. In Proceedings of the 53rd Annual Meeting of the Association for Computational Linguistics and the 7th International Joint Conference on Natural Language Processing (Volume 2: Short Papers), Beijing, China, 26–31 July 2015; Association for Computational Linguistics: Stroudsburg, PA, USA, 2015; pp. 489–494.
52. Liang, S.; Yilmaz, E.; Kanoulas, E. Collaboratively Tracking Interests for User Clustering in Streams of Short Texts. *IEEE Trans. Knowl. Data Eng.* **2018**, *31*, 257–272. [CrossRef]
53. Shi, L.; Song, G.; Cheng, G.; Liu, X. A user-based aggregation topic model for understanding user's preference and intention in social network. *Neurocomputing* **2020**, *413*, 1–13. [CrossRef]
54. Wang, C.; Blei, D.M. Collaborative topic modeling for recommending scientific articles. In Proceedings of the 17th ACM SIGKDD International Conference on Knowledge Discovery and Data Mining, San Diego, CA, USA, 21–24 August 2011; pp. 448–456.
55. Nikolenko, S. SVD-LDA: Topic modeling for full-text recommender systems. In *Mexican International Conference on Artificial Intelligence*; Springer: Cham, Switzerland, 2015; pp. 67–79.
56. Al-Ghossein, M.; Murena, P.A.; Abdessalem, T.; Barré, A.; Cornuéjols, A. Adaptive collaborative topic modeling for online recommendation. In Proceedings of the 12th ACM Conference on Recommender Systems, Vancouver, BC, Canada, 2–7 October 2018; pp. 338–346.
57. Kang, S.Y.; Kim, J.K.; Choi, I.Y.; Kang, C.D. A Topic Modeling-based Recommender System Considering Changes in User Preferences. *J. Intell. Inf. Syst.* **2020**, *26*, 43–56.
58. Rajendran, D.P.D.; Sundarraj, R.P. Using topic models with browsing history in hybrid collaborative filtering recommender system: Experiments with user ratings. *Int. J. Inf. Manag. Data Insights* **2021**, *1*, 100027. [CrossRef]
59. Lü, L.; Medo, M.; Yeung, C.H.; Zhang, Y.C.; Zhang, Z.K.; Zhou, T. Recommender systems. *Phys. Rep.* **2012**, *519*, 1–49. [CrossRef]

Article

An Alternative Globalization Barometer for Investigating the Trend of Globalization

Sha Sun [1], Haiyue Xu [2], Minsong He [3], Yao Xiao [1] and Huayong Niu [1,*]

[1] International Business School, Beijing Foreign Studies University, Beijing 100089, China
[2] School of Russian Language, Beijing Foreign Studies University, Beijing 100089, China
[3] Faculty of Languages and Literatures, Ludwig Maximilian University of Munich, 80539 Munich, Germany
* Correspondence: niuhuayong@bfsu.edu.cn; Tel.: +86-10-8881-6347

Abstract: Analyzing, evaluating, and predicting the trend of globalization are highly valuable endeavors. However, existing literature lacks a quantifiable metric for objective evaluation. To fill the gap, we first compiled a Globalization Index based on existing globalization indices and using the CRITIC weighting method. Second, we constructed the Globalization Barometer and a trend term for trend analysis using the HP filtering method. Third, we conducted time-series predictions for globalization trajectory by applying the Random Forest model. Our results indicate that: (1) The *de facto* and *de jure* globalization both displayed a gradually upward trend over time; (2) the 2008 financial crisis and the 2020 COVID-19 pandemic negatively impacted globalization and served as turning points; (3) on a positive note, COVID-19 has narrowed the gap in both *de facto* and *de jure* globalization. This is due to the fact that the shocks were uneven, with economies that participated more in globalization weathering the brunt of the impact, while economies that participated less experiencing little changes; (4) the *de facto* and *de jure* globalization are predicted to remain on an upward trend for the subsequent 5 years. This research provides essential references for assessing and predicting globalization trends.

Keywords: globalization; COVID-19 pandemic; globalization barometer; trend analysis; trend forecasting

Citation: Sun, S.; Xu, H.; He, M.; Xiao, Y.; Niu, H. An Alternative Globalization Barometer for Investigating the Trend of Globalization. *Appl. Sci.* **2022**, *12*, 7896. https://doi.org/10.3390/app12157896

Academic Editor: Jianbo Gao

Received: 6 July 2022
Accepted: 4 August 2022
Published: 6 August 2022

Publisher's Note: MDPI stays neutral with regard to jurisdictional claims in published maps and institutional affiliations.

1. Introduction

Globalization is an impactful force for countries and regions all over the world. The trend of globalization influences multiple stakeholders, not least of which include actors in the economic, social, and even political fields. Most stakeholders need to base and readjust their strategies on how globalization will proceed.

Despite its importance, there is no consensus on assessing the trend of globalization in the existing literature. Some scholars believe that globalization is irreversible and it will continue to move forward. They argue that the globalization variables are more resilient than most people expected [1]. The increasing mobility of people, information, and technology worldwide has reduced the possibility of deglobalization [2]. Among them, positive globalization trends are especially reflected in increasing global exchanges of services and data [3]. Moreover, based on the fact that the world remains highly collaborative during the coronavirus outbreak, globalization will not end as a result of the pandemic [4]. Some foresee a slowdown in globalization in the near and long-term future, characterized by the concept of "slowbalization" [5]. These ideas predate the COVID-19 outbreak, such as the likely deceleration of globalization suggested by Bordo [6]. However, slowbalization is not a uniform trend. It includes a recession in economic globalization and a boom in information globalization [7]. Others hold the view that globalization is suffering a downturn. The current deglobalization is partially triggered by the pandemic exposing the underlying fragility in globalization [8,9]. This crisis has spurred the pre-pandemic globalization skeptics [10], with economic and social factors further accelerating this skepticism worldwide [11]. Scholars conclude that, due to various anti-globalization factors, including inter-country inequalities, populism, protectionism, and unilateralism, a greater globalization process is difficult to achieve in the current world economy [12]. Populism, in particular, is heavily impactful. It threatens not only economic but also social and political globalization [13,14]. This rise of populism is fueled by the backlash against neoliberal constitutionalism [15]. Some scholars have refrained from defining a fixed standpoint, as it remains uncertain whether the crisis triggered by the COVID-19 pandemic marks the end of globalization [16]. However, they have pointed out crucial factors that may influence future trends. Digitalization, for instance, projects both centrifugal and centripetal forces on globalization [17], while the pandemic has transformative effects that paint a new image of a post-Covid era global market [18]. While recent studies have made important progress in evaluating and projecting how connections in the world will develop moving forward, they are mainly qualitative studies and therefore by nature, are prone to subjective judgements.

Another related strand of literature is the study of globalization indices. Traditionally, sociologists, economists, and others worked on different dimensions of globalization [19]. However, globalization by definition is a multifaceted concept that includes economic, social, and political aspects [20]. Therefore, to measure globalization in a more comprehensive way, most of the existing globalization indices have adopted an interdisciplinary approach, i.e., a composite index of globalization. The A.T. Kearney/Foreign Policy Globalization Index [21] was the first systematic measure of globalization, which measured and ranked 62 countries worldwide on four dimensions: Economic integration, personal contact, technological connectivity, and political engagement. Noteworthy indices include: The KOF Globalization Index [22] uses 43 indicators in the economic, social, and political dimensions and covers data pooled from 203 economies between 1970 and 2018. The CSGR Globalization Index [23] applies 16 indicators along the economic, social, and political dimensions, covering data from 119 countries and regions from 1982 to 2004. The Maastricht Globalization Index [24] measures the level of globalization in 117 countries in 2002, 2008, and 2012 presenting five dimensions: Political, economic, social and cultural, technological, and environmental. The DHL Global Connectedness Index [1] measures the depth and breadth of global connectivity of 140 economies between 2005 and 2020, using 12 indicators along four dimensions: Trade, capital, information, and people. The prior research showed its merits in providing the basic quantitative framework and methodology for constructing a globalization index. However, their scope is rather limited to presenting globalization in the past and present rather than trend analysis and forecasting.

In this case, an approach which quantifies the trend of globalization will be useful to provide forward-looking analyses, especially given that it is the changes of globalization than globalization per se that fulfills the greatest need. Current literature in economics and other fields used barometers to assess the trend of specific variables [25,26]. A barometer is a composite indicator designed for assessing the trend of growth and discovering turning points [25,26]. This implies that the construction of a barometer would serve as a suitable way to quantify globalization trends, which is rarely covered in the existing literature. The barometer can deliver fact-based, future-oriented solutions for industrial activities, provide theoretical instruments for academic purposes, and serve as a window for the public to monitor the current state of global interconnectedness.

In this paper, we constructed the Globalization Barometer and provided a trend analysis of globalization. Our research pooled data from 142 economies from 2000 to 2020, spanning economic, social, and political dimensions. The CRITIC method was used to assign indicator weights. The HP filter was used to implement the trend analysis. Finally, the Random Forest model was used to conduct time-series predictions for globalization trajectory.

The rest of the article is organized as follows. Section 2 illustrates the data and methods. Section 3 presents the results. Sections 4 and 5 are intended for discussion and conclusion.

2. Materials and Methods

2.1. Methodology Procedures

As presented in Figure 1, the assessment and prediction of globalization trend involve three steps:

(1) The compilation of the Globalization Index. Initially, we compiled our globalization indicator framework by making revisions and amendments to existing globalization indices. Accordingly, relevant data were collected, imputed and normalized. Then, we used the CRITIC method to assign weights and aggregated the indicators to the globalization index.

(2) The construction of the Globalization Barometer. We applied the HP filtering method to decompose the globalization index into two parts: The trend term and the deviation term, of which the latter is used to construct the Globalization Barometer.

(3) Time-series prediction of globalization trajectory. The Random Forest model is used to predict the subsequent periods of globalization level.

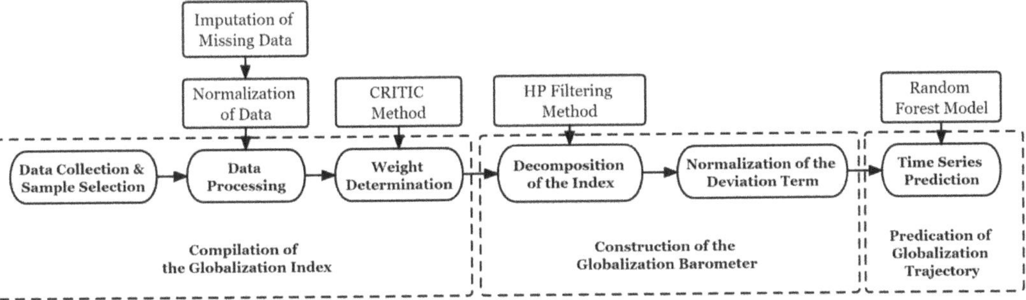

Figure 1. Methodology procedures.

2.2. Compilation Framework of the Globalization Index

A clear and universally recognized definition of globalization is necessary for compiling a globalization index. Based on emblematic articles in the field of globalization, including Sklair [27,28], Stoudmann and Al-Rodhan [29], and Scholte [30], globalization is defined as the global increase in connection, interdependence, and convergence of all economies in the economic, social, and political fields. In this work, we define globalization in three sub-dimensions: Economic globalization is reflected in the cross-border flows of products and services and the allocation of production factors on a global scale. Social globalization covers the migration of people and the transmission of information, accompanied by the convergence of cultures and exchanges in science and technology. Political globalization denotes intergovernmental cooperation and collaboration within the framework of international organizations.

Existing globalization indices have provided an adequate structure to quantify the level of globalization. Following Gygli et al. [22], we introduce two dimensions, namely, *de facto* and *de jure* globalization. *De facto* globalization refers to the extent to which a country's participation in globalization has been achieved, while *de jure* globalization is defined as the decisions, policies, institutions, and other proactive factors that the country has put in place to make its participation in globalization more possible.

Based on prior indices, our index also made several adjustments and improvements as described below.

(1) We measured political globalization more broadly. As an expansion of existing indices [22–24], the political dimension is determined by two sub-dimensions: International cooperation and global governance. International cooperation measures the degree of intergovernmental coordination and communication, while global gov-

ernance evaluates the participation level of each government in the framework of international organizations. Of note, this paper differs from existing indices in the identification of international organizations. Existing indices mostly use all inter-governmental organizations (IGOs) and non-governmental organizations (NGOs). This approach assumes that all IGOs and NGOs contribute equally to the process of globalization, neglecting the individual difference in capacity. In our context, the international organization mainly refers to the UN and its affiliated institutions, since we propose that UN plays a crucial role in the formation of modern globalization and thus should lay emphasis on its work.

(2) We measured social globalization more impartially. Indicators, such as McDonald's restaurants and IKEA stores, have long been criticized for measuring Americanization or Westernization rather than cultural globalization, as pointed out in the founding paper of the KOF index [22]. Accordingly, we removed these indicators and used trade in printed goods and international trademarks instead to evaluate *de facto* cultural globalization.

Another adjustment reflects in the inclusion of decentralized indicators. Some new indicators, such as the language popularity index, are added to our index system. Rather than measuring the average proficiency of English in a given country, the language popularity index measures the number of foreign nationals speaking the majority language of that country (the most spoken language among all official languages of that country) as a percentage of those speaking that language worldwide. Language functions as the medium of culture [31], opens more possibilities for cultural transmission and exchange, and thus could be included in cultural globalization. Additionally, the social tolerance index, which refers to the extent of recognition and acceptance of differences, and willingness to grant equal rights [32], was included to address the importance of mutual respect and appreciation of different cultures.

(3) In addition to the adjustments in the indicators, there are some amendments to the measurements. Previous studies have shown that absolute indicators are prone to the impact of scale [22]. Therefore, the indicators chosen in this paper are mostly relative indicators. This method ensures all components are statistically comparable.

In conclusion, this paper constructed a *de facto* and *de jure* globalization index system consisting of three primary, eight secondary, and twenty-eight tertiary indicators as demonstrated in Table 1. See Table A1 in Appendix A for measurement of all indicators.

Table 1. The globalization index: Variables description.

Primary Indicators	Secondary Indicators		Tertiary Indicators	Source
Economic	Trade	*de facto*	Trade in products	World Bank WDI
			Trade in services	World Bank WDI
		de jure	Tariffs	World Bank WDI
			Trade agreements	DESTA
	Financial	*de facto*	Foreign direct investment	IMF IIP
			Portfolio investment	IMF IIP
			International income payments	IMF BoP
		de jure	Capital account openness	Knoema
			International investment agreements	Investment Policy Hub

Table 1. *Cont.*

Primary Indicators	Secondary Indicators		Tertiary Indicators	Source
Social	Informational	*de facto*	Used internet bandwidth	ITU ICT-Eye
		de jure	Internet access	ITU ICT-Eye
	Interpersonal	*de facto*	International tourism	World Bank WDI
			International students	World Bank WDI
			Migration	World Bank WDI
		de jure	Freedom to visit	DEMIG VISA
			International airports	CIA World Factbook *
	Technological	*de facto*	International patents	World Bank WDI
			High technology exports	World Bank WDI
		de jure	Global innovation index	GII
	Cultural	*de facto*	Trade in printed goods	UN Comtrade
			International trademarks	WIPO IP Portal
		de jure	Social tolerance index	World Value Survey
			Language popularity index	Ethnologue *
Political	International Cooperation	*de facto*	Foreign affairs agencies	Lowy Global Diplomacy
		de jure	International organizations	CIA World Factbook *
			International treaties	UN Treaty Collection *
	Global governance	*de facto*	Speech contribution in UN	UN Digital Library *
		de jure	UN peacekeeping contribution	UN Peacekeeping *

* Indicators manually collected and calibrated from accessible databases.

2.3. Data

2.3.1. Data Collection and Sample Selection

In this work, we covered diverse data from the economic, social, and political fields. Our primary sources of data are obtained from databases of the UN, the World Bank, and the IMF. In addition, our work collected data from trustworthy sources, such as the CIA, the World Value Survey, and Ethnologue. In total, all 19 different databases were consulted (Table 1).

The raw dataset covers 217 economies with 28 variables (indicators) over the timespan (year 2000 to 2020). However, it suffers from a severe missing observation problem, which calls for sample selection. We decide whether to retain a sample following two principles: (1) Data coverage ratio should be improved; (2) the structure of the sample should be in proportion to the raw dataset from a geographic and economic perspective. With careful consideration, we narrowed our sample to 142 economies. See Table A2 in Appendix A for the structure of the sample compared with the raw dataset.

2.3.2. Data Processing

Data processing involves imputation of missing data and data normalization. Following the practice of existing globalization indices [22–24], we imputed the missing data within a series using linear interpolation and extrapolation. In addition, the values of the indicators themselves are not comparable due to differences in the scale and units of the indicators. Therefore, we normalized the data using the max-min method.

2.4. Methods

2.4.1. The CRITIC Method

Two major weighting systems are used in related studies: The objective and subjective weighting methods [33]. The former is dataset-driven, while the latter is expert-driven. In order to eliminate subjective biases, our work employs the CRITIC method to aggregate our indicators. This method is based on evaluating the comparative strength between indicators and the deviation of indicators to determine the weights of indicators.

Consider a normalized dataset represented by matrix M:

$$M = \begin{bmatrix} x_{11} & x_{12} & \cdots & x_{1n} \\ x_{21} & x_{22} & \cdots & x_{2n} \\ \vdots & \vdots & \ddots & \vdots \\ x_{m1} & x_{m2} & \cdots & x_{mn} \end{bmatrix} \tag{1}$$

where m denotes the number of observations, and n denotes the number of indicators under the same category.

While determining the weights of a specific indicator, both standard deviation and its correlation between indicators are considered. In this regard, the weights are obtained as follows:

$$w_j = \frac{C_j}{\sum_{j=1}^{n} C_j} \tag{2}$$

where C_j denotes the quantity of information j-th indicator contains, which is determined as:

$$C_j = \sigma_j \cdot \sum_{k=1}^{n} \left(1 - r_{kj}\right) \tag{3}$$

where σ_j represents the standard deviation, and r_{kj} represents the correlation coefficient between indicator k and j.

2.4.2. HP Filter

Inspired by the Global Trade Barometer issued by World Trade Organization (WTO) [28], our work uses the Hodrick-Prescott (HP) filter method to formulate the Globalization Barometer. The HP filter, first proposed by Hodrick and Prescott [34], is a commonly used data-smoothing tool in macroeconomics. We use the HP filter method to decompose a series into trend and cyclical components. The original series can be represented as the following function:

$$y_t = \tau_t + c_t \tag{4}$$

where y_t is the original series, τ_t is the trend component for the long-term path, and c_t is the cyclical component which denotes short-run dynamics. The deviation is the actual level of output from the long-term trend. Therefore, the HP filter is defined as the following optimization function:

$$min_{\{\tau_t\}} \left\{ \sum_{T=1}^{T} (y_t - \tau_t)^2 + \lambda \sum_{t=2}^{T-1} [(\tau_{t+1} - \tau_t) - (\tau_t - \tau_{t-1})]^2 \right\} \tag{5}$$

where the first term is interpreted as the sum of the squared differences between actual globalization and trend, and the second term is a second-order difference equation that exists for the trend multiplied by the smoothing parameter λ. This parameter will determine the amount of volatility associated with a trend, namely, the higher λ that is used, the smaller the volatility will be. Following Backus and Kehoe [35], the value of the smoothing parameter λ is assigned to a value of 100 for annual data in this paper.

2.4.3. Random Forest Model

Machine learning is the learning process of analyzing data automatically to obtain a model from the data and using the model to make predictions about unknown data. Random Forest is an integrated algorithm that reduces the variance of a model by combining multiple decision trees, correcting the habit of a single decision tree to over-fit its training set. Among various machine learning algorithms, Random Forest generally has better generalization performance [36]. It is resilient to outliers in the dataset and does not require considerable parameter-tuning. In 2001, Breiman [37,38] improved the Random Forest model, which not only simplified the computational effort and improved accuracy, but also better predicted small sample sizes and unbalanced datasets.

In this work, 5-fold cross-validation and "out-of-bag" (OOB) method are used to evaluate the model performance. K-fold cross-validation is a resampling procedure for evaluating machine learning models on the limited sample [39]. We validate our results by randomly partitioning data into k mutually exclusive subsets. One set is used for validation, the other k-1 sets are used for training. The validation process is repeated for k times. K is usually chosen to be 5 or 10, but there is no formal rule. We choose k = 5 in this paper. The OOB method is another evaluation method, in which OOB observations are used to create training samples [40].

3. Results

3.1. Globalization Barometers and Trend Terms on World Average

Drawing on the idea of the Global Trade Barometer of WTO [28], this paper compiles the Globalization Barometer with *de facto* and *de jure* dimensions. The steps are as follows: First, the HP filter is applied to the Globalization Index measured in Section 2, and the trend term is extracted; next, the index value is subtracted from the trend term to obtain the deviation term; finally, the deviation term is normalized and added by 100, and compared against the barometer standard interval. The standard barometer interval is set by taking 100 as the baseline, and the intervals (99, 101), (101, +∞), and (−∞, 99) are defined as "in trend" (yellow), "above trend" (green), and "below trend" (red), respectively.

Figure 2 shows the *de facto* and *de jure* Globalization Barometer, as well as the trend components. In terms of *de facto* globalization, the trend term is generally upward. During the past two decades, two global events, i.e., the 2008 financial crisis and the COVID-19 outbreak in 2020, influenced its progress, as shown by the grey areas. The financial crisis in 2008 caused a significant dip in *de facto* globalization. The *de facto* Globalization Barometer soared to 101.58 on the eve of the 2008 crisis, and then fell to the bottom of 98.84 when the crisis started to spread, turning from "above trend" to "below trend". Since then, *de facto* globalization resumed its positive development until the pandemic hit the world in 2020, when *de facto* globalization suffered an even greater fluctuation. As shown in Figure 2b, a significant "above trend" could be observed in 2019, before a sharp drop to "below trend" in 2020, when the pandemic began to exert its negative influence on the world economy as well as other aspects of human lives. Moreover, the pandemic even caused the trend component to decrease. This is partly a reflection of the fragility and relative instability of globalization, where global events at the economic, social or political level could destabilize the level of *de facto* globalization. Overall, however, a positive trend is evident in *de facto* globalization.

The *de jure* globalization shows a relatively flat upward trend in general, which is also impacted by the 2008 crisis and the 2020 pandemic. The 2008 crisis had a lagging effect on the *de jure* globalization. Since 2010, there has been a general slowdown in the growth of *de jure* globalization, even if it remained positive until 2020. The *de jure* Globalization Barometer fell from 101.86 (above trend) in 2008 to 98.71 (below trend) in 2011. From 2018 onwards, *de jure* globalization remained at the previous level until 2020 when *de jure* globalization dipped following the outbreak of the COVID-19 pandemic. The trend component eventually fell, and the barometer dropped to 98.41 (below trend). The above results further illustrate the continuing negative impact of the crisis on *de jure* globalization.

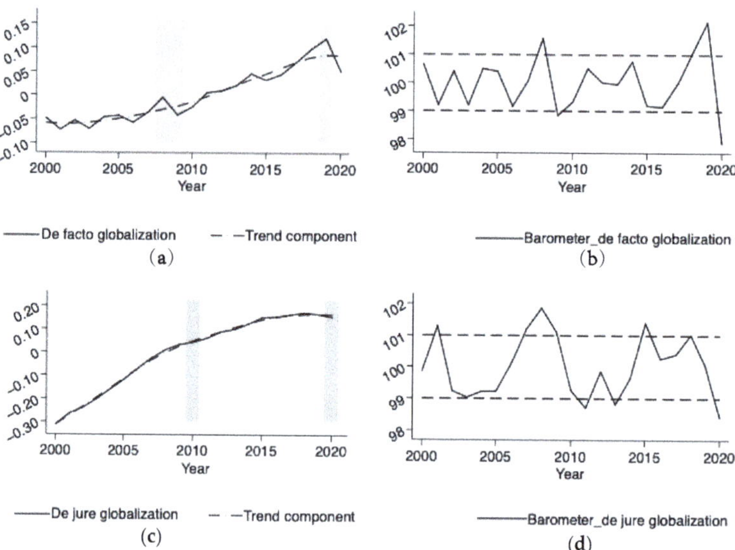

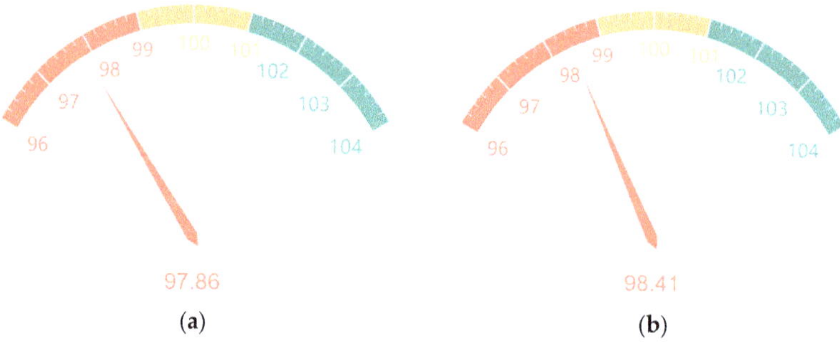

Figure 2. Barometers and trend terms of *de facto* and *de jure* globalization. (**a**) *De facto* globalization trend term. (**b**) *De facto* globalization barometer. (**c**) *De jure* globalization trend term. (**d**) *De jure* globalization barometer.

Figure 3 shows the *de facto* and *de jure* Globalization Barometer for 2020. In 2020, the *de facto* Globalization Barometer of 97.86 was "below trend". The *de jure* Globalization Barometer was 98.41, also "below trend". The results show that the pandemic in 2020 has challenged the development of globalization in the world and has dealt a huge blow to the globalization process.

Figure 3. Globalization barometer. (**a**) *De facto* globalization barometer. (**b**) *De jure* globalization barometer.

3.2. Spatial Variations

For further analysis, we have systematically analyzed the level of *de facto* and *de jure* globalization and barometers for all sample economies. The spatial variations of *de facto* and *de jure* globalization and their barometers in 2020 are illustrated in Figure 4. We observe that the variations of *de facto* and *de jure* globalization are characterized by regional spatial agglomeration.

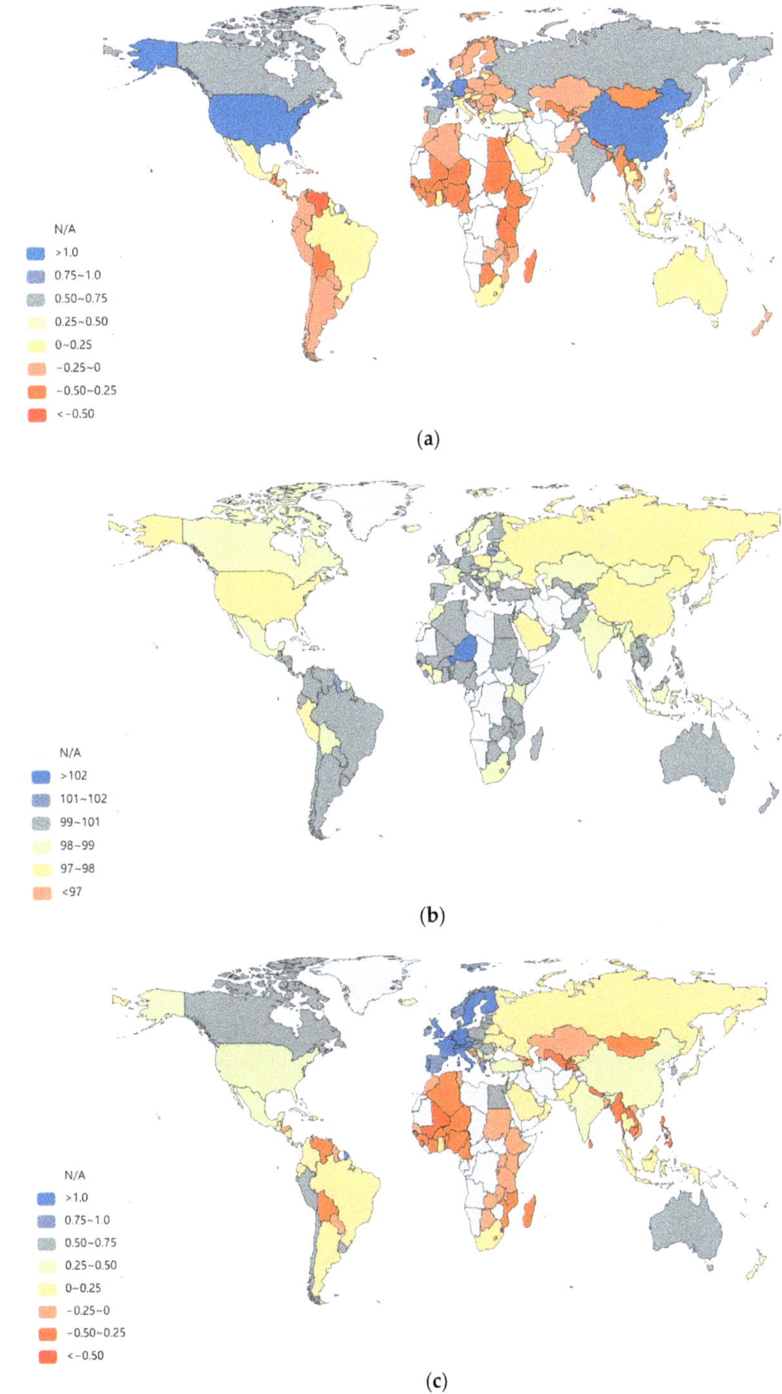

Figure 4. *Cont.*

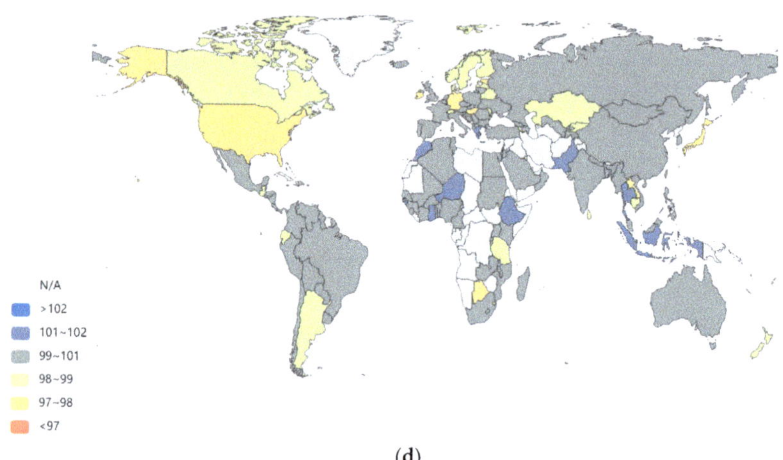

(**d**)

Figure 4. Spatial variation of the globalization index and globalization barometer in 2020. (**a**) Spatial variations of *de facto* globalization. (**b**) Spatial variations of *de facto* globalization barometer. (**c**) Spatial variations of *de jure* globalization. (**d**) Spatial variations of *de jure* globalization barometer.

For *de facto* globalization and its barometer in 2020, we can observe the following, as illustrated in Figure 4a,b:

(1) Economies with higher *de facto* globalization tend to cluster in the North America, North-Eastern Asia, and Europe. For instance, the *de facto* globalization of the United Kingdom, the United States, Germany, China, and France were higher than 1.0, which indicates that their participation level surpassed the global average by at least one standard deviation.

(2) Comparatively speaking, South America, Central Asia, Central, and Eastern Europe, and Africa were less involved. For instance, the *de facto* globalization of Uganda, Nepal, Venezuela, and Sri Lanka were less than −0.50.

(3) The Barometers of different regions and countries were also differentially impacted by the pandemic. Interestingly, we have found a greater impact in economies with more involvement in globalization. Economies with traditionally high *de facto* globalization values, such as China, Russia, and the US, ranked lower in the barometer, scoring 97.74, 97.55, and 97.53, respectively. Meanwhile, countries with lower *de facto* globalization were less exposed to shocks, especially those in South America, Central Asia, and Africa as well as Oceania, with economies sitting at or above trend. These results indicate that the impact of the pandemic on globalization is not uniform across the globe, serving as an equalizer of sorts, leveling the differences in globalization participation in the post-COVID world.

Second, let us focus on the *de jure* globalization and its barometer in 2020. From Figure 4c,d, we can observe:

(1) Economies in Western Europe, Northern Europe, and North America tend to have higher *de jure* globalization scores, with the *de jure* globalization of France, Germany, Netherlands, Belgium, and the United Kingdom higher than 1.0.

(2) Certain countries or regions showed lower *de jure* participation, especially those in South America, Central Asia, and Africa, with the Philippines, Myanmar, Fuji, Sri Lanka, and Uzbekistan scoring less than −0.50.

(3) Of note, the *de jure* globalization of most countries has not been significantly hampered by the pandemic. The proportion of countries with the barometer of *de jure* globalization above or in trend was about 74.6%. Only North America and a few countries in Asia and Europe were below trend, including Germany, Japan, and the

United States, scoring 97.86, 97.77, and 97.55 in 2020, respectively. Compared with *de facto* globalization, *de jure* globalization was more stable.

3.3. Driving Forces

Figure 5 presents the three-dimensional forces, i.e., economic, social, and political factors, of the *de facto* and *de jure* Globalization Barometer for 2020.

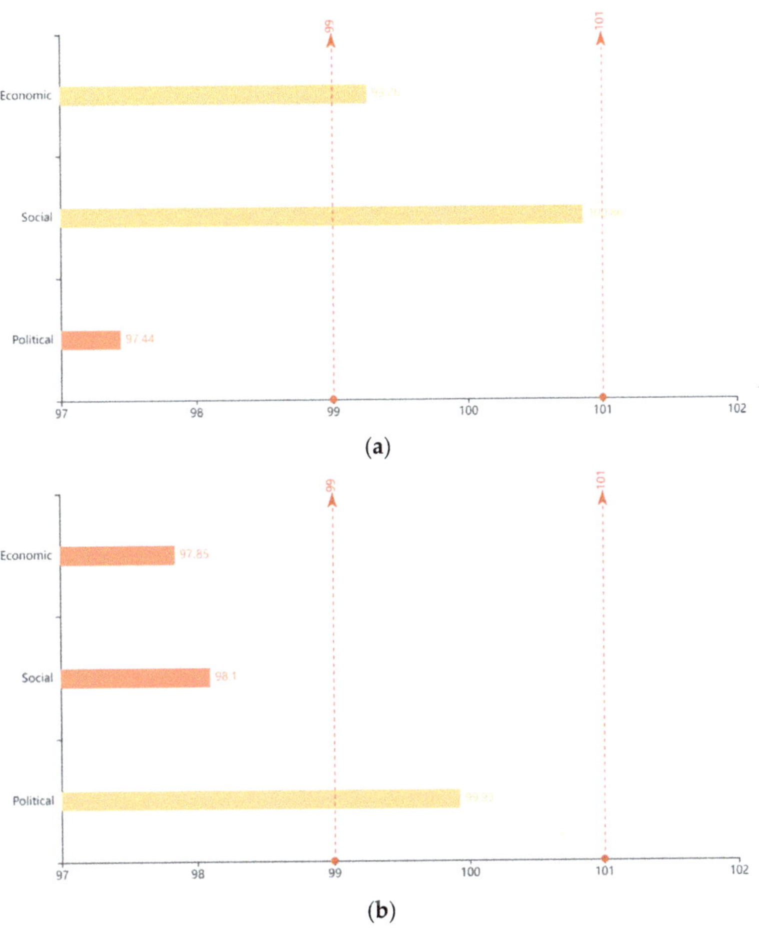

Figure 5. Three-dimensional forces of globalization barometer. (**a**) Three-dimensional forces of *de facto* globalization barometer. (**b**) Three-dimensional forces of *de jure* globalization barometer.

3.3.1. Driving Forces of de Facto Globalization Barometer

In 2020, the *de facto* economic Globalization Barometer was in trend (99.26). The *de facto* social Globalization Barometer was also in trend (100.86), while the *de facto* political Globalization Barometer was below trend (97.44).

First, the *de facto* economic globalization displayed remarkable buoyancy in the face of the significant shocks caused by the COVID-19 outbreak. In response to the virus, economies around the world implemented lockdowns and restrictions, generally disrupting international trade. However, after a brief negative depression, global trade quickly turned positive in the second half of 2020, as suggested by the DHL connectedness index. Mirroring this trend was the global financial markets, which responded negatively with increased uncertainty and systematic risks during the pandemic, likely exacerbated by the loss of confidence from investors and other actors responding to more pessimistic forecasts. Yet, a general recovery in the global trade was thought to have contained the panic in the global financial markets, contributing to more stability and positive prospects regarding international capital flows.

Second, *de facto* social globalization is characterized by structural differences. Restrictions on international travel greatly disrupted cross-border human exchange, leading to a fall in interpersonal globalization. The main means of international transportation, air transport services, including the airline industry, suffered a great blow. However, it is also worth mentioning that informational globalization boomed during the pandemic, to some extent mitigating social impacts. Limitations on public gatherings and activities led to a significant increase in online activities and far more frequent internet use compared with pre-pandemic times, which fostered thriving online communities. The exchanges of information became considerably more efficient thanks to the internet, and informational globalization acted as a major counter-weight against the negative impact of interpersonal globalization. Therefore, *de facto* social globalization maintained a mildly steady trend in 2020.

Third, as for political globalization, both international cooperation and governance displayed a downward trend in the post-pandemic period. A relatively lower speech contribution in UN conferences, which brings more uncertainty to political globalization, may provide one of the reasonable explanations for the trend.

Overall, there are well-grounded reasons to expect the optimistic development of economic and social globalization in the short-term, whereas more attention needs to be paid to the dynamics of *de facto* political globalization.

3.3.2. Driving Forces of de Jure Globalization

In 2020, the *de jure* economic Globalization Barometer is falling at 97.85, far below trend; the *de jure* social Globalization Barometer is 98.1, also "below trend"; the *de jure* political Globalization Barometer is 99.93, in trend.

First, before the pandemic, populism and protectionism actions by several economies have stirred up hostile emotions in international trade. These emotions were exacerbated by the 2020 global health crisis and its concomitant economic fallout, as economies trying to establish new trade agreements diverted strategies to cope with the ongoing pandemic. Consequently, fewer trade agreements have been signed since the onset of the pandemic. According to the United Nations Treaty Collection, 356 trade agreements were signed in 2020, six less than in 2019; 437 trade agreements entered into force in 2020, 51 less than in 2019.

Second, when considering *de jure* social globalization, the clear inconvenience caused by restrictive measures discouraged international human exchange, especially cross-border travel. The construction of new airports also slowed down due to a decrease in international traffic demand, following a decrease in visas issued due to stricter border controls and other relevant COVID-19 countermeasures. An inactive trend in interpersonal infrastructure was one of the major factors that explained the falling trend of *de jure* social globalization.

Third, from the perspective of *de jure* political globalization, international organizations, especially World Health Organization, have been working around the clock to mitigate the global impact of the pandemic with programs, such as the COVAX initiative for worldwide vaccine distribution. Furthermore, relaxed restrictions related to the pandemic lead to a recovery of some UN peacekeeping missions around the world, leading to higher participation in peacekeeping missions and indicating that political globalization is regaining its strengths.

3.4. Forecasting Globalization Trends

Following Petukhova et al. [41], we use the Random Forest model to predict the time-series data. The *de facto* and *de jure* globalization, as well as the economic, social, and political dimensions, for the subsequent 5 years after 2020 are predicted and analyzed in this section.

In this paper, eight Random Forest models were built based on eight dimensions, i.e., the *de facto* and *de jure* globalization and the three subdimensions of each. For example, the dependent variable is the *de facto* globalization for each economy in period t, and the independent variables would be set to be the *de facto* globalization for each economy in period t-*n* to t-1. To expand the sample size, we chose $n = 10$. Since our sample spans from year 2000 to 2020, period t would be from year 2010 to 2020. In this way, we use data of each dimension as dependent variables, and the independent variables are set as the previous 10-year data for the corresponding dimension.

We used 5-fold cross-validation to find the set of parameters with the best prediction results and chose parameters with the number of subtrees of 100, a minimum number of samples required for internal node subdivision of 2, a minimum number of samples for leaf nodes of 1, and the maximum number of features used for a single decision tree is $\sqrt{N}$ (N is the total number of features). The performance of the model was evaluated by the coefficient of determination (R^2), root mean squared error (RMSE), and mean absolute error (MAE) via 5-fold cross-validation (CV) and the "out-of-bag" (OOB) method. The R^2 are all above 84%, the RMSE and MAE are small for all eight models. The model is generalized well, with no overfitting occurring. The detailed model evaluation results are shown in Table 2.

Table 2. Model evaluation results.

Dimension	5-Fold Cross-Validation			Out-of-Bag		
	R^2	RMSE	MAE	R^2	RMSE	MAE
De facto globalization	0.9502	0.0129	0.0628	0.9508	0.0126	0.0620
De facto economic globalization	0.9750	0.0248	0.0630	0.9818	0.0190	0.0551
De facto social globalization	0.8731	0.0608	0.0697	0.8486	0.0490	0.0649
De facto political globalization	0.9400	0.0469	0.1344	0.9390	0.0462	0.1335
De jure globalization	0.9864	0.0038	0.0366	0.9865	0.0037	0.0357
De jure economic globalization	0.9896	0.0088	0.0576	0.9899	0.0085	0.0560
De jure social globalization	0.9885	0.0035	0.0332	0.9903	0.0029	0.0310
De jure political globalization	0.9459	0.0257	0.0668	0.9547	0.0228	0.0624

After training the model, we used the 10-year data from 2011 to 2020 to forecast the result of 2021. Then, with the predicted data of 2021 as one of the features, we used the data from 2012 to 2021 to forecast the result in 2022, followed by four more periods of prediction from 2022 to 2025 using the same method.

Model Results

Figure 6 shows prediction results for globalization trajectory using the Random Forest model. Both *de facto* and *de jure* globalization are projected to exhibit a steady upward trend in the next five periods after 2020.

The projected *de facto* globalization for 2021 to 2025 is as follows: A significant rebound is expected for 2021, describing a "V" curve. After the rebound, *de facto* globalization is highly likely to follow a steady upward trajectory for the remainder of the 5 years; *de facto* economic globalization in the same 5-year period would experience a large-scale rebound; *de facto* social globalization would decrease in 2021, followed by a steady, gradual recovery; while *de facto* political globalization has the potential to experience a large-scale rebound in 2021, followed by a steadily increasing trend.

The *de jure* globalization trend in general may remain relatively stable. For the next 5 years, *de jure* globalization is projected to maintain steady yet slow levels of upward momentum. *De jure* economic globalization keeps pace with overall globalization, i.e., a rebound to form a "V" curve, tapering into a slow increase; *de jure* social globalization maintains a slowly upward pace for this period; while *de jure* political globalization is expected to maintain its previous level for 2021, and increases after 2022.

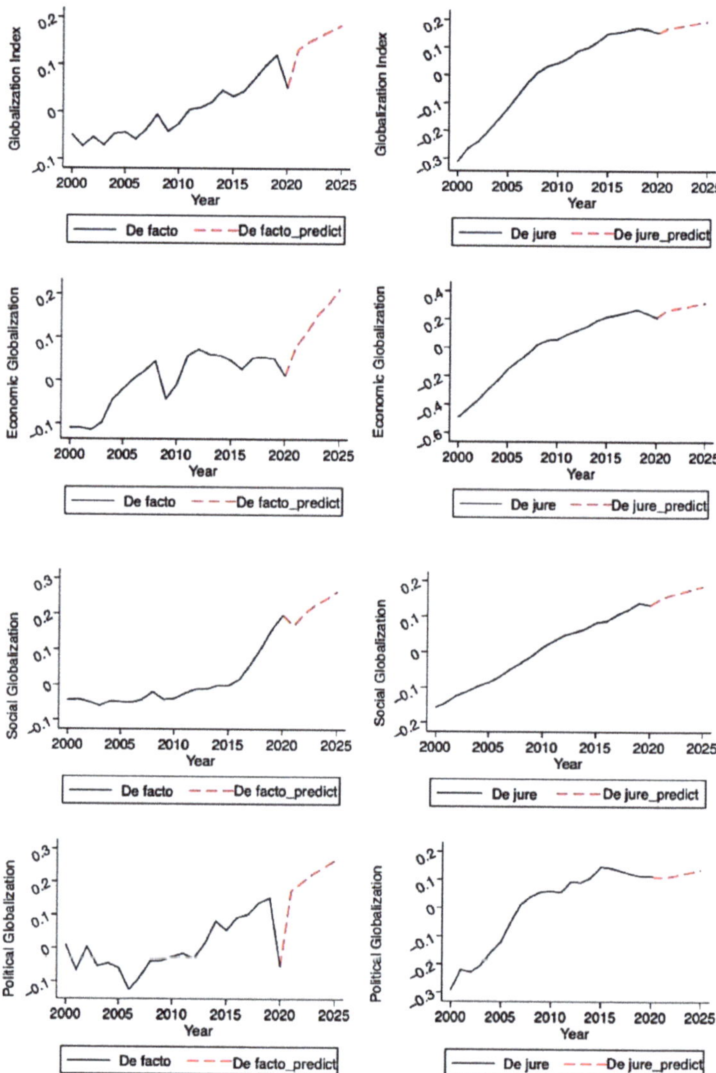

Figure 6. Forecasting of the globalization trends.

4. Discussion

The globalization trend impacts the economic, social, and political development of all countries around the world, making it a highly relevant topic of discussion. Current literature is limited to qualitative judgements or point-quantifications, leaving the broader stroke issue of trend largely untouched. In order to fill the gap, we developed the Globalization Barometer to evaluate the trend of globalization in more depth.

We constructed a Globalization Index based on existing indices and used the HP filter method to decompose the trend and deviation terms, of which the latter is used to construct the Globalization Barometer. Our results indicate that since 2000, *de facto* and *de jure* globalization generally maintained an upward trajectory. The 2008 financial crisis and the 2020 COVID-19 pandemic have negatively impacted globalization trends, serving as turning points on the curve. From a world average perspective, *de facto* globalization saw a significant setback post the 2008 financial crisis, followed by a rapid rebound until the pandemic in 2020. *De jure* globalization generally maintained steady growth but also experienced the shocks of the two crises, including the 2008 slowdown in growth and the dip in trend in 2020. Our results are intuitive and support previous observations [5,7].

Although the 2020 pandemic has exerted a negative impact on globalization, it is well-worth noting the unexpected positive effect of reducing the gap in *de facto* and *de jure* globalization. Further analysis revealed an uneven distribution of globalization across the world. We find that economies in North America, Northeast Asia, and Europe feature high *de facto* globalization; on the other hand, economies with high *de jure* globalization are generally concentrated in Western Europe, Northern Europe, and North America. However, the spatial variation of globalization barometer shows that the impact of the pandemic is not evenly felt around the world. Economies with higher scores in *de facto* and *de jure* globalization experience greater shocks, while economies with lower scores remain relatively untouched. Our results complement current literature on globalization trends and the pandemic's impacts and threats [8,9] while also adding to the current debate the argument that the impact of globalization mainly was felt in a few economies and regions and may help form a more balanced post-COVID global paradigm.

By deconstructing the driving forces of globalization, our research shows uneven levels of globalization across the economic, social, and political dimensions. *De facto* globalization grew below trend with economic and social globalization on trend, and political globalization below trend. *De jure* globalization turned out to be below trend as well for this year, due mainly to economic and social globalization's downturn compared with political globalization. Our results indicate that albeit the fact that the current development of globalization in general failed to meet our expectations, there are still positive dimensions in globalization, and may serve as the future driving forces of stable and positive development.

Finally, we have also used the Random Forest model to conduct time-series predictions for the years between 2021 and 2025. For the subsequent 5 years, *de facto* and *de jure* globalization will likely maintain an upward trajectory, thereby providing quantitative, machine learning-backed response to current qualitative research on future globalization trends.

5. Conclusions

In summary, our research has confirmed and complemented existing studies on globalization trends in the following two aspects. Approach-wise, our research adopted appropriate methods to quantify globalization trends. Although these methods, including the HP filter and Random Forest, are not novel in the scientific community, we are one of the early adopters of these methods in the research of globalization. Conclusion-wise, our research can adequately respond to the theoretical debates on the direction of post-pandemic globalization trend within the greater sphere of sociology and international relations. For instance, Contractor [2] believes the pandemic has a short-term impact from which the world will recover soon, while Ciravegna and Michailova [12] believe that the pandemic "will have significant long-lasting effects on globalization." Additionally, even fewer researchers conduct detailed and in-depth discussions on the globalization trend due to the lack of quantifiable metrics. Our paper fills in the theoretical gap and comes to an unexpectedly interesting conclusion that the pandemic has decreased the uneven distribution of globalization.

Other than academic research, globalization quantification can be used by business analysis, mass and specialized media, and public policy. Stakeholders can make informed predictions and decisions on globalization trends using the Globalization Barometer. Companies can adjust how they deploy their regional investment strategy and transnational operations. Media can use the barometer to provide their audience with a more neutral and accurate image of globalization. In addition, policymakers can reference this research to set foreign policy and international relations decisions.

Considering the complicated nature of globalization development, quantifiable trend analysis will continue to pose a challenge for academia. Our paper is an exploratory attempt at quantifying globalization and is far from perfect. For the future, we will focus on: (1) The barometer's application and validation using more variables and time-series; (2) more detailed study on national or regional globalization trends that were excluded in this paper due to space constraints; and (3) a more in-depth study on the driving factors of globalization to better aid globalization development trend analysis.

Author Contributions: Conceptualization, S.S. and H.N.; methodology, S.S., H.X. and H.N.; software, H.X. and Y.X.; validation, S.S., H.X. and Y.X.; formal analysis, S.S.; investigation, S.S. and H.X.; resources, S.S. and H.X.; data curation, S.S. and H.X.; writing—original draft preparation, S.S., M.H., H.X. and Y.X.; writing—review and editing, S.S., M.H., H.X., Y.X. and H.N.; visualization, H.X. and Y.X.; supervision, S.S. and H.N.; project administration, S.S. and H.N.; funding acquisition, S.S. and H.N. All authors have read and agreed to the published version of the manuscript.

Funding: This research was funded by Beijing Foreign Studies University First-class Discipline Development Research Project under Grant No. YY19ZZB012, by Beijing Foreign Studies University's "COVID-19 Epidemic" First-class Special Research Project under Grant No. SYL2020ZX015, by Incoming Junior Faculty Research Program under Grant No. 2018QD013 and by Beijing Foreign Studies University First-Class Major Signature Research Project under Grant No. 2022SYLZD001.

Institutional Review Board Statement: Not applicable.

Informed Consent Statement: Not applicable.

Data Availability Statement: Data available in a publicly accessible repository.

Acknowledgments: The authors thank the editors and anonymous reviewers for their insightful suggestions.

Conflicts of Interest: The authors declare no conflict of interest.

Appendix A

Table A1. Measurements of all indicators in the globalization index.

Primary Indicators	Secondary Indicators		Tertiary Indicators	Measurements
Economic	Trade	*de facto*	Trade in products	goods imports and exports (% of GDP)
			Trade in services	service imports and exports (% of GDP)
		de jure	Tariffs	the unweighted mean of custom duties
			Trade agreements	the number of trade agreements
	Financial	*de facto*	Foreign direct investment (FDI)	the inbound and outbound flows of FDI (% of GDP)
			Portfolio investment (PI)	the inbound and outbound flows of PI (% of GDP)
			International income payments (IIP)	the asset and liability of IIP (% of GDP)
		de jure	Capital account openness	Chinn-Ito index
			International investment agreements	the total number of Bilateral Investment Treaties (BIT) and Treaties with Investment Provisions (TIP)

Table A1. *Cont.*

Primary Indicators	Secondary Indicators		Tertiary Indicators	Measurements
Social	Informational	*de facto*	Used internet bandwidth	international bandwidth measured in Mbit/s
		de jure	Internet access	the total of individuals using the Internet (% of population)
	Interpersonal	*de facto*	International tourism	the inbound and outbound tourists (% of population)
			International students	the inbound and outbound tourists (% of population)
			Migration	the immigrants and emigrants (% of population)
		de jure	Freedom to visit	the number of visa-free countries or regions
			International airports	the number of international airports (% of population)
	Technological	*de facto*	International patents	the nonresident-applied patents (% of total)
			High technology exports	the high-tech exports (% of manufactured exports)
		de jure	Global innovation index	Global innovation index
	Cultural	*de facto*	Trade in printed goods	the imports and exports in printed goods HS Code 49 (% of GDP)
			International trademarks	the nonresident-applied trademarks (% of total trademarks)
		de jure	Social tolerance index	Social tolerance index
			Language popularity index	foreign nationals speaking the majority language of that country (% of total speakers)
Political	International Cooperation	*de facto*	Foreign affairs agencies	the sum of embassies, consulates, permanent missions, and other representations
		de jure	International organizations	the number of international organizations
			International treaties	the number of international treaties signed after 1945 and ratified by the legislative organization
	Global governance	*de facto*	Speech contribution in UN	the total speech number made in UN
		de jure	UN peacekeeping contribution	the number of peacekeeping personnel

Table A2. The structure of the original and selected dataset.

	Categories	Original		Selected	
		Number	Percentage	Number	Percentage
Geographic Location	Europe and Central Asia	58	26.73%	46	32.39%
	Sub-Saharan Africa	48	22.12%	33	23.24%
	Latin America and Caribbean	42	19.35%	25	17.61%
	East Asia and Pacific	37	17.05%	19	13.38%
	Middle East and North Africa	21	9.68%	12	8.45%
	South Asia	8	3.69%	5	3.52%
	North America	3	1.38%	2	1.41%
Income Group *	High income	79	36.57%	65	45.77%
	Lower middle income	55	25.46%	39	27.46%
	Upper middle income	55	25.46%	33	23.24%
	Low income	27	12.50%	5	3.52%

* Venezuela, RB was not assigned to any income group when data were collected.

References

1. Altman, S.A.; Bastian, P. *DHL Global Connectedness Index 2020—The State of Globalization in a Distancing World*; Deutsche Post DHL Group: Bonn, Germany, 2020; 104p.
2. Contractor, F.J. The world economy will need even more globalization in the post-pandemic 2021 decade. *J. Int. Bus. Stud.* **2021**, *53*, 156–171. [CrossRef] [PubMed]
3. Straubhaar, T. Nicht das Ende, sondern der Anfang einer neuen Globalisierung. *Wirtschaftsdienst* **2021**, *101*, 841–844. [CrossRef]
4. Coronavirus Won't Kill Globalization. But It Will Look Different after the Pandemic. Available online: https://time.com/583875 1/globalization-coronavirus/ (accessed on 27 June 2022).
5. Wang, Z.; Sun, Z. From Globalization to Regionalization: The United States, China, and the Post-COVID-19 World Economic Order. *J. Chin. Political Sci.* **2021**, *26*, 69–87. [CrossRef] [PubMed]
6. Bordo, M.D. *The Second Era of Globalization is Not Yet Over: An Historical Perspective*; Working Paper Series No. 23786; National Bureau of Economic Research: Cambridge, MA, USA, 2017. [CrossRef]
7. Titievskaia, J.; Kononenko, V.; Navarra, C.; Stamegna, C.; Zumer, K. *Slowing down or Changing Track? Understanding the Dynamics of 'Slowbalization'*; PE 659.383; European Parliamentary Research Service: Brussels, Belgium, 2020.
8. Witt, M.A.; Li, P.P.; Välikangas, L.; Lewin, A.Y. De-globalization and Decoupling: Game Changing Consequences? *Manag. Organ. Rev.* **2021**, *17*, 6–15. [CrossRef]
9. Madhok, A. Globalization, de-globalization, and re-globalization: Some historical context and the impact of the COVID pandemic. *BRQ Bus. Res. Q.* **2021**, *24*, 199–203. [CrossRef]
10. Delios, A.; Perchthold, G.; Capri, A. Cohesion, COVID-19 and contemporary challenges to globalization. *J. World Bus.* **2021**, *56*, 101197. [CrossRef]
11. Garg, S.; Sushil. Determinants of deglobalization: A hierarchical model to explore their interrelations as a conduit to policy. *J. Policy Model.* **2021**, *43*, 433–447. [CrossRef]
12. Ciravegna, L.; Michailova, S. Why the world economy needs, but will not get, more globalization in the post-COVID-19 decade. *J. Int. Bus. Stud.* **2022**, *53*, 172–186. [CrossRef]
13. Broz, J.L.; Frieden, J.; Weymouth, S. Populism in Place: The Economic Geography of the Globalization Backlash. *Int. Organ.* **2021**, *75*, 464–494. [CrossRef]
14. Walter, S. The Backlash Against Globalization. *Annu. Rev. Political Sci.* **2021**, *24*, 421–442. [CrossRef]
15. Slobodian, Q. The Backlash Against Neoliberal Globalization from Above: Elite Origins of the Crisis of the New Constitutionalism. *Theory Cult. Soc.* **2021**, *38*, 51–69. [CrossRef]
16. Hameiri, S. COVID-19: Is this the end of globalization? *Int. J.* **2021**, *76*, 30–41. [CrossRef]
17. Autio, E.; Mudambi, R.; Yoo, Y. Digitalization and globalization in a turbulent world: Centrifugal and centripetal forces. *Glob. Strategy J.* **2021**, *11*, 3–16. [CrossRef]
18. McNamara, K.R.; Newman, A.L. The Big Reveal: COVID-19 and Globalization's Great Transformations. *Int. Organ.* **2020**, *74*, E59–E77. [CrossRef]
19. Martens, P.; Raza, M. Globalisation in the 21st Century: Measuring Regional Changes in Multiple Domains. *Integr. Assess. J.* **2009**, *9*, 1–18.
20. Potrafke, N. The Evidence on Globalisation. *World Econ.* **2015**, *38*, 509–552. [CrossRef]
21. Kearney, A.T. Measuring globalization. *Foreign Policy* **2001**, *122*, 56–65.
22. Gygli, S.; Haelg, F.; Potrafke, N.; Sturm, J.E. The KOF Globalization Index-revisited. *Rev. Int. Organ.* **2020**, *14*, 543–574. [CrossRef]
23. Lockwood, B.; Redoano, M. *The CSGR Globalization Index: An Introductory Guide*; CSGR Working Paper; Centre for the Study of Globalisation and Regionalisation: Coventry, UK, 2005.
24. Figge, L.; Martens, P. Globalization Continues: The Maastricht Globalization Index Revisited and Updated. *Globalizations* **2014**, *11*, 875–893. [CrossRef]
25. WTO Trade Barometers. Available online: https://www.wto.org/english/res_e/statis_e/wtoi_e.htm#:~{}:text=WTO%20trade% 20barometers%20The%20WTO%20has%20developed%20a,three%20months%20ahead%20of%20merchandise%20trade%20 volume%20statistics (accessed on 15 June 2021).
26. Abberger, K.; Graff, M.; Müller, O.; Sturm, J.-E. Composite Global Indicators from Survey Data: The Global Economic Barometers. *Rev. World Econ.* **2022**, *158*, 917–945. [CrossRef]
27. Sklair, L. Competing Conceptions of globalization. *J. World-Syst. Res.* **1999**, *5*, 143–163. [CrossRef]
28. Sklair, L. The Emancipatory Potential of Generic Globalization. *Globalizations* **2009**, *6*, 525–539. [CrossRef]
29. Stoudmann, A.G.; Al-Rodhan, N.R.F. *Definitions of Globalization: A Comprehensive Overview and a Proposed Definition*; GCSP Working Papers; Geneva Centre for Security Policy: Geneva, Switzerland, 2006; 21p.
30. Scholte, J.A. Defining Globalisation. *World Econ.* **2008**, *31*, 1471–1502. [CrossRef]
31. Tong, H.-K.; Lin, H.-C. Cultural Identity and Language: A Proposed Framework for Cultural Globalization and Glocalization. *J. Multiling. Multicult. Dev.* **2011**, *32*, 55–69. [CrossRef]
32. Zanakis, S.; Newburry, W.; Taras, V. Global Social Tolerance Index and multi-method country rankings sensitivity. *J. Int. Bus. Stud.* **2016**, *47*, 480–497. [CrossRef]
33. Diakoulaki, D.; Mavrotas, G.; Papayannakis, L. Determining Objective Weights in Multiple Criteria Problems: The Critic Method. *Comput. Oper. Res.* **1995**, *22*, 763–770. [CrossRef]

34. Hodrick, R.J.; Prescott, E.C. Postwar U.S. business cycles: An empirical investigation. *J. Money Credit Bank.* **1997**, *29*, 1–16. [CrossRef]
35. Backus, D.; Kehoe, P. International evidence on the historical properties of business cycles. *Am. Econ. Rev.* **1992**, *82*, 864–888. [CrossRef]
36. Fernández-Delgado, M.; Cernadas, E.; Barro, S.; Amorim, D. Do we need hundreds of classifiers to solve real world classification problems? *J. Mach. Learn. Res.* **2014**, *15*, 3133–3181. [CrossRef]
37. Breiman, L. Random Forests. *Mach. Learn.* **2001**, *45*, 5–32. [CrossRef]
38. Breiman, L. Statistical Modeling: The Two Cultures. *Stat. Sci.* **2001**, *16*, 199–231. [CrossRef]
39. Stuart, R.; Peter, N. *Artificial Intelligence: A Modern Approach*, 3rd ed.; Pearson: London, UK, 2009; p. 708.
40. Gareth, J.; Daniela, W.; Trevor, H.; Rob, T. *An Introduction to Statistical Learning*, 1st ed.; Springer: Berlin, Germany, 2013; pp. 316–321.
41. Petukhova, T.; Ojkic, D.; McEwen, B.; Deardon, R.; Poljak, Z. Assessment of autoregressive integrated moving average (ARIMA), generalized linear autoregressive moving average (GLARMA), and Random Forest (RF) time series regression models for predicting influenza A virus frequency in swine in Ontario, Canada. *PLoS ONE* **2018**, *13*, e0198313. [CrossRef] [PubMed]

Article

Normality in the Distribution of Revealed Comparative Advantage Index for International Trade and Economic Complexity

Bin Liu [1] and Jianbo Gao [2,3,*]

[1] Business School, Guangxi University, Nanning 530004, China; binliu.ssxs@gmail.com
[2] Center for Geodata and Analysis, Faculty of Geographical Science, Beijing Normal University, Beijing 100087, China
[3] Institute of Automation, Chinese Academy of Sciences, Beijing 100087, China
* Correspondence: jbgao.pmb@gmail.com; Tel.: +86-185-7787-0637

Abstract: The Revealed Comparative Advantage (RCA) index is an important metric for evaluating competitiveness of a country in exporting certain commodity. While it is desirable to have a normally distributed RCA index, the opposite is often found in empirical studies, and efforts for developing alternative indices of the RCA index have not been very successful. This motivates us to ask a more fundamental question: what is the significance of a normally distributed RCA index? To answer this question, we have defined a quantity called the Deviation from Gaussianity (DfG) based on the KS test, which quantifies the deviation of the distribution of a country's RCA index from normality. By systematically analyzing the distribution characteristics of RCA index for each country from 1991 to 2019, we find that DfG is strongly negatively correlated with the logarithm of GDP and the Economic Complexity Index (ECI). In particular, correlation between DfG and GDP is stronger than that between ECI and GDP since 2008. These results suggest that DfG may serve as a new excellent index to quantify the economic complexity and economic performance of a country.

Keywords: RCA index; economic complexity; Gaussian distribution; economic development

Citation: Liu, B.; Gao, J. Normality in the Distribution of Revealed Comparative Advantage Index for International Trade and Economic Complexity. *Appl. Sci.* **2022**, *12*, 1125. https://doi.org/10.3390/app12031125

Academic Editor: Elisa Quintarelli

Received: 7 December 2021
Accepted: 20 January 2022
Published: 21 January 2022

Publisher's Note: MDPI stays neutral with regard to jurisdictional claims in published maps and institutional affiliations.

1. Introduction

The revealed comparative advantage (RCA) index, also called Balassa index as it was first proposed by Balassa in 1965 [1], is an important metric for quantifying the relative strength of a country in producing a product vis-à-vis its trading partners. While it has been widely used in empirical studies, the RCA index has also been further studied theoretically. Those works mainly focus on the statistical features of the RCA index, and can be roughly classified into two groups. One group is somewhat traditional, with emphasis on clarifying the statistical characteristics of the RCA index across sectors or countries. In many applications, it is desirable to have a normally distributed RCA index, so that it can reliably measure a country's revealed comparative advantage [2]. However, in the majority of empirical studies, a non-Gaussian distribution of the RCA index has been observed. The non-Gaussianity has made the RCA index to suffer from many disturbing properties such as unstable distribution and poor ordinal ranking property [3], the unstable mean [4,5], asymmetric distributional shape [2], and skewness and variable upper bound [6,7]. These features of the RCA index have made its interpretation difficult [3,4,8–10], and thus have motivated a lot of researchers to develop alternative indices of the RCA index so that the new indices can be more normally distributed [3–5,11–15]. These efforts are not very successful, however. To understand why the RCA index and its alternatives may not follow Gaussian distributions, Liu and Gao systematically analyzed the distribution characteristics of the RCA index cross sectors and countries [16]. They find that the RCA index in the majority of the situations cannot be normally distributed, since it is the ratio of two distributions, one following an exponentially truncated Zipf–Mandelbrot's law, the other being a permutation of the truncated Zipf–Mandelbrot's law [16]. Only occasionally can a normally distributed RCA index be observed—it may emerge with about 1% chance. The significance of a normally distributed RCA index has not yet been explained, however.

The other group of the work on the theoretical aspects of the RCA index mainly employs matrix and complex network theory by constructing the country–product bipartite network, where countries are connected to the products they export. The bipartite network is an 0–1 adjacency matrix constructed according to the value of the RCA index (the element is 1 if the corresponding RCA $\geq$ 1 and 0 otherwise). By developing the Method of Reflections to interpret an export bipartite network, Hidalgo and Hausmann proposed the Economic Complexity Index (ECI) and Product Complexity Index (PCI) [17,18]. Hidalgo and Hausmann's approach has been proven to be equivalent to a spectral clustering algorithm that partitions a similarity graph into two parts [19]. Although, the ECI may offer a good description of global macroeconomic relations, technological trends, and growth dynamics [20], and could be used to measure the gap in the economic development between countries [21], the approach suffers from a number of conceptual and practical problems [22–26]. To overcome these problems, the Fitness Index (FI) and some other variants of the ECI have been developed [22–24,27,28]. The ECI and its variants have been widely used to study the impact of economic structures on economic development [18,20,29–42]. Fundamentally speaking, however, the FI and the other new variants of the ECI are not very different from the ECI, since the ECI and FI (or log FI) are strongly positively correlated [37–41], and both metrics have almost the same skill in predicting economic growth [42]. This raises an important question as to which of the neglected aspects of the RCA index by the network based approach should be reinstated so that characterization of economic complexity can be fundamentally improved.

In this article, we attempt to answer both the above questions: why a normally distributed RCA index is important and how to better quantify economic complexity. In doing so, we will find a bridge connecting the two groups, one more traditional, the other based on the network approach. Concretely, we will define a quantify called the Deviation from Gaussianity (DfG) based on the KS test, which measures the deviation of the distribution of a country's RCA index from normality. Then, we will systematically analyze the distribution characteristics of RCA index for each country from 1991 to 2019, and examine the relationship between the DfG and economic development and economic complexity.

The remainder of the paper is organized as follows: Section 2 describes Materials and methods, Section 3 presents the main results, and Section 4 contains conclusion and discussion.

2. Materials and Methods

2.1. Materials

In this work, we analyze international commodity trade data with products disaggregated according to the COMTRADE Harmonized System at the four-digital level (abbreviated as HS4). The data covered 29 years from 1991 to 2019, and were downloaded from UNComtrade database (International Trade Statistics Database: https://comtrade.un.org/ accessed on 5 August 2021).

2.2. Methods

2.2.1. RCA Index

The RCA index is defined as

$$\text{RCA}^{k}_{(ix)} = \frac{X^{k}_{(i)} / X_{(i)}}{X^{k}_{(w)} / X_{(w)}} = \frac{p^{k}_{(ix)}}{p^{k}_{(wx)}}, \tag{1}$$

where X (or x) denotes export, i denotes country, while w denotes world, k denotes product. For example, $X^{k}_{(i)}$ represents country i's export of product k, $X_{(i)}$ denotes country i's total export, and $X^{k}_{(i)} / X_{(i)}$ is the export share of country i in product k. Being a probability, it can also be expressed as $p^k_{(ix)}$, and $\sum_{k=1}^{N_c} p^k_{(ix)} = 1$, where N_c represents the number of products in a country.

2.2.2. Economic Complexity Index

The Economic Complexity Index (ECI) was developed by Hidalgo and Hausmann in 2009 [18]. The algorithm for computing it is as follows. Consider a country-product bipartite network represented by a matrix with elements M_{cp} defined as 1 or 0, depending on whether the corresponding RCA ≥ 1 or RCA < 1. Summing up rows and columns of the matrix, one obtains $k_{c,0} = \sum_p M_{cp}$, $k_{p,0} = \sum_c M_{cp}$, which represent, respectively, the observed the number of products exported by some country, and the number of countries exporting some product. The ECI is obtained by an iteration algorithm,

$$k_{c,N} = \frac{1}{k_{c,0}} \sum_p M_{cp} k_{p,N-1}, \tag{2}$$

$$k_{p,N} = \frac{1}{k_{p,0}} \sum_p M_{cp} k_{c,N-1}, \tag{3}$$

where $N \geq 2$ is the number of iterations. Collecting $k_{c,N}$, $c = 1, \cdots, C_n$, where C_n is the total number of countries with data, we then obtain ECI as

$$\mathrm{ECI}_{c^*} = \frac{k_{c^*,N} - \mathrm{mean}\{k_{c,N}\}}{\mathrm{stdev}\{k_{c,N}\}}, \tag{4}$$

where c^* denotes a country of interest, and mean and stdev are performed over all the countries with data. It is thought that the larger the ECI, the higher the economic complexity.

2.2.3. Deviation from Gaussianity Based on KS Test

The KS test (Kolmogorov–Smirnov test or K-S test) is one of the most useful and general nonparametric methods. The one-sample KS test can be used to compare a sample with a reference probability distribution. In this paper, we define the Deviation from Gaussianity (DfG) based on one-sample KS test. The algorithm is as follows:

$$F_n(x) = \frac{1}{n} \sum_{n=1}^{N} I_{[-\infty,x]}(X_i), \tag{5}$$

where $I_{[-\infty,x]}(X_i)$ is the indicator function, which is equal to 1 if $X_i < x$ and 0 otherwise. The Kolmogorov–Smirnov statistic for a given cumulative distribution function $F(x)$ is

$$D_n = sup_x |F_n(x) - F(x)|, \tag{6}$$

where sup_x is the supremum of the set of distances. We define the divergence of DfG in the distribution of RCA index as follows:

$$\mathrm{DfG} = D_n - CV, \tag{7}$$

where CV is the critical value of KS test. A negative DfG indicates Gaussian distribution of RCA index, while a positive DfG indicates rejection of the Gaussian distribution—the more positive DfG, the larger the deviation from Gaussianity [16].

2.2.4. Pooled OLS and Panel VAR

In this article, we will also employ regression analysis to further explore the connections among DfG, ECI, and economic development. Considering that our data may be considered panel data, we will employ two regression models—pooled Ordinary Least

Square (OLS) and panel Vector Autoregression (VAR) models. The general econometric model for panel data is as follows [43,44]:

$$Y_{it} = \alpha_i + \vec{\beta}_i \cdot \vec{X}_{it} + \mu_{it}, \tag{8}$$

where, $i = 1, 2, \ldots, N$, $t = 1, 2, \ldots, T$, and N and T are the number of individual countries and total time (in year), respectively. Y_{it} is the dependent variable, $\vec{X}_{it}$ is the independent variables (column vector), α_i and $\vec{\beta}_i$ are parameters (the latter a row vector with dimension matched to the column vector $\vec{X}_{it}$ so that the inner product is defined), and μ_{it} is the error term. As our purpose in this research is to find (and design) effective measures for quantifying economic complexity, we first assume that α_i and $\vec{\beta}_i$ are constant for all countries and time. This scenario is called the pooled OLS model, which is equivalent to the simple OLS model performed on panel data. The concrete equation used here is as follows:

$$\ln \mathrm{GDP}_{it} = \alpha + \beta_1 \mathrm{DfG}_{it} + \beta_2 \mathrm{ECI}_{it} + \mu_{it}, \tag{9}$$

We also use a panel-data VAR methodology. This technique combines the traditional VAR approach, which treats all the variables in the system as endogenous, with the panel-data approach, which allows for unobserved individual heterogeneity [45,46]. We employ a first-order panel VAR model as follows:

$$z_{i,t} = \Gamma_0 + \Gamma_1 z_{i,t-1} + \mu_t, \tag{10}$$

where i represents the country in the panel-data, $z_{i,t}$ is a three-variable vector $(\ln \mathrm{GDP}, \mathrm{DfG}, \mathrm{ECI})$, Γ_1 is a 3×3 matrix of coefficients, Γ_0 is a vector of individual effects. The stationarity of the three variables will be examined by using the LLC test [47] before we employ the PVAR model. Moreover, we can explore the statistical causality between the three variables based on the PVAR model.

3. Results

3.1. DfG and Economic Growth

There are two types of distributions for the RCA index. One is the distribution of the RCA index for all the sectors/products of an economy or a country. The other is the distribution of the RCA for all countries in the world given a sector/product. In this article, we focus on the former. Since the RCA index is the ratio of two probabilities, it is useful to first understand the distributions of the two probabilities. It turns out that both the numerator and the denominator defining the RCA index ($p_{(ix)}^k$ and $p_{(wx)}^k$) basically follow exponentially truncated Zipf–Mandelbrot's law, given by:

$$p(k) \sim (k + p)^{-\alpha} e^{-\beta k^{\gamma}}, k > k^*, \tag{11}$$

where p, α, β, and γ are parameters. The exponential truncation can be naturally expected due to finiteness of the data.

To better understand deviations from normality in the distribution of RCA index for different countries, we use Japan and Germany as two examples. Figure 1 shows the distribution features of the two parts of RCA index and the probability distribution function (PDF) of the RCA index for Japan and Germany under the HS4 scheme in 2018. Obviously, the $p_{(wx)}$ in Figure 1a,b follows exponentially truncated Zipf–Mandelbrot's law. If the $p_{(ix)}$ in Figure 1a,b are also arranged in descending order, they will also follow exponentially truncated Zipf–Mandelbrot's law (but possibly with different parameters). Interestingly, by comparing the layout of $p_{(ix)}$ (red diamonds) around $p_{(wx)}$ (blue circles) in Figure 1a,b, we can observe that the $p_{(ix)}$ of Germany is more concentrated around $p_{(wx)}$ than Japan's. This highlights that Germany's export share of most products relative to its total exports is closer to the world average level than Japan's.

Next, we discuss how the differences between Figure 1a,b results in the differences in the distribution of the RCA index shown Figure 1c,d. Clearly, the PDFs for the RCA index of Germany and Japan are very different. Concretely, the PDF of Japan's RCA index has more asymmetry, stronger skewness, and longer tail than that of Germany's. This suggests that the PDF of Germany's RCA index should be closer to a Gaussian distribution than Japan's. To better quantify how the PDF of a country's RCA index deviates from normality, we employ DfG we have defined earlier. The DfG for Germany and Japan is 0.088 and 0.266 in 2018, respectively. According to the nature of DfG—the more positive DfG, the larger the deviation from Gaussianity, one can conclude that the PDF of Germany's RCA index is indeed closer to a normal distribution than that of Japan's, just as one has anticipated from Figure 1.

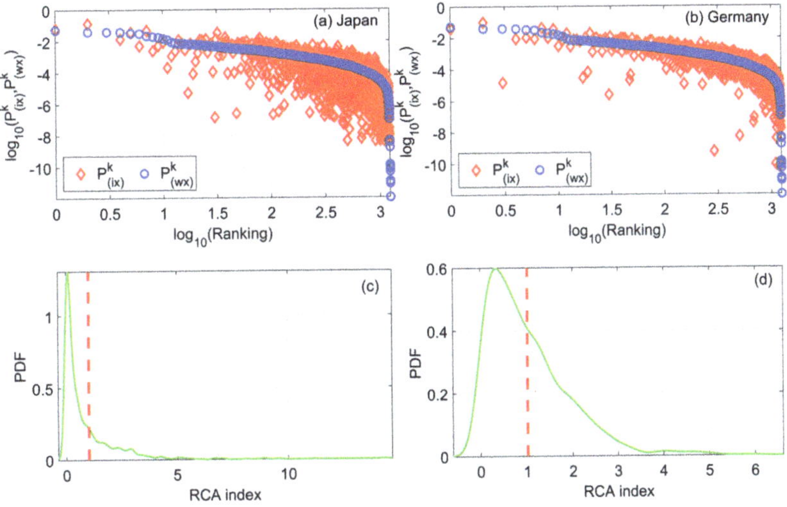

Figure 1. The distribution features of the two parts of RCA index (**a**,**b**) and the Probability Distribution Function (PDF) of RCA index (**c**,**d**) for Germany and Japan under the HS4 scheme in 2018.

It is interesting to examine the spatiotemporal evolution of the DfG of all the economies in the world. For this purpose, we have systematically computed DfG for all the economies in the world from 1991 to 2019. The spatial variations of the DfG in 1998, 2008 and 2018 are illustrated in Figure 2. We observe that the variations of DfG are characterized by spatiotemporal heterogeneity and regional spatial agglomeration.

First, let us focus on the spatiotemporal heterogeneity. From Figure 2a, we can observe: (1) only the DfG of USA and Germany was less than 0.1, followed by France and Italy, (2) only a few countries (such as China, South Korea, Japan, etc.) had DfG between 0.2 and 0.3, and (3) the DfG of most countries was greater than 0.3, especially in Africa, South America, Southern and Western Asia, and Eastern Europe. By 2008, which is shown in Figure 2b, the spatial variation of DfG had undergone some changes. Now Germany is the only economy with DfG < 0.1, indicating that Germany is the only country with the PDF of its RCA index to be very close to a normal distribution. The decrease in China's DfG was significant. In contrast, the DfG in some countries has become larger, such as USA, France, Australia, Egypt, etc. The DfG in most other countries and regions did not change much though, especially in Africa and South America. The major changes in DfG can at least be partially be attributed to the global financial crisis in 2008. Interestingly, by 2018, as shown in Figure 2c, the DfG in India and Vietnam had decreased significantly. This clearly reflected transfer of many production activities to India and Vietnam in recent years. Overall, compared with 2008, the pattern of the spatial variation of DfG for most countries in the world in 2018 did not change significantly. This suggests that the negative impact of the 2008 global financial crisis has been quite long-lasting.

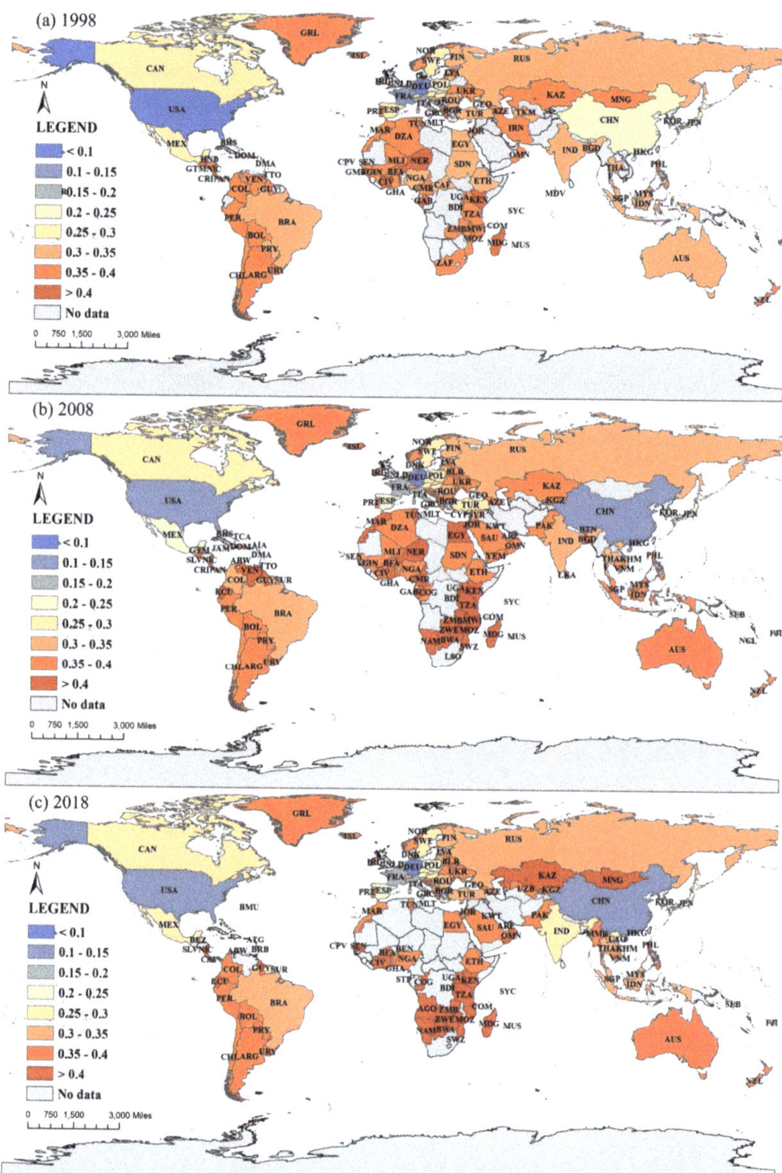

Figure 2. Spatial variations of the DfG in 1998, 2008 and 2018 under the HS4 scheme.

Second, let us focus on the regional spatial aggregation phenomena in Figure 2. That is, countries with smaller DfG are mainly concentrated in North America, Western Europe and Eastern Asia, while countries with larger DfG are mainly concentrated in Africa, South America, Western and Southern Asia. It is worth paying attention to the Eastern Asia represented by China, Japan and South Korea. In 1998, the DfG in this region was larger than USA and Germany. By 2008, this gap had shrunk substantially, and by 2018, the level of DfG in this region was already comparable to that in North America and Western Europe. By now, we can conclude that this aggregated region with smaller DfG represented by China, Japan and South Korea has been well formed. It is worth noting that these three areas with fairly small DfG are very consistent with the description of "The world seems to have three interconnected production hubs for the extensive trade in parts and components" in the "Global Value Chain Development Report 2017—Measuring and Analyzing the Impact of GVCs on Economic Development".

The pattern of DfG's spatial variation suggests that DfG may be indicative of a country's economic performance. To check this idea, we have examined the relationship between DfG and GDP (current dollars) from 1991 to 2019. The result is shown in Figure 3. We observe that DfG and the logarithm of GDP is strongly negatively correlated. This means that the larger the economic scale of a country, the smaller its DfG. In other words, the larger GDP a country has, the easier for the country to have the distribution of its RCA index to converge to a Gaussian distribution. This observation suggests that the level of specialization and division of labor is connected to the deviations from normality in the distribution of a country's RCA index. Generally, the bigger a market (as characterized by GDP) is, the more its participants can specialize and the deeper the division of labor in the market can be achieved.

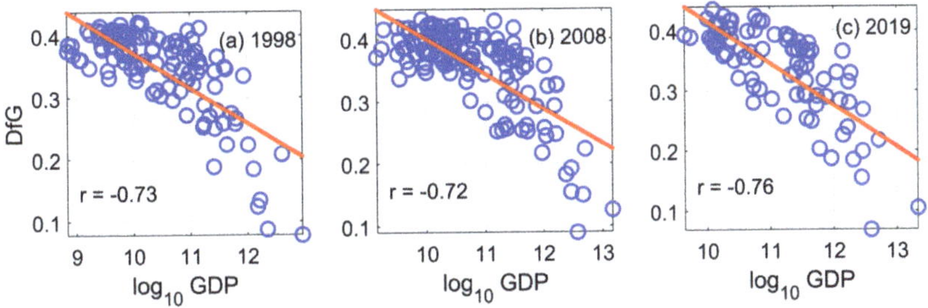

Figure 3. Regression analysis showing correlation between DfG and the logarithm of GDP in 1998, 2008 and 2019.

Finally, let us turn to discuss the dynamic evolution of the DfG for a few more or less arbitrarily chosen countries, including China, India, Australia and Zambia. The results are shown in Figure 4. We observe that China's DfG bascially monotonically decreases in most of the time. India has similar behavior, especially after 1999. In contrast, Australia's DfG has largely been increasing most of the time, while the DfG for Zambia has been fluctuating. Considering that DfG is highly negatively correlated with the logarithm of GDP, we have good reason to conclude that DfG characterizes the trade as well as economic structure of a country to some degree. Therefore, we can associate the temporal variation of DfG for a country with the temporal evolution of its trade and economic structure, as a result of its effort in maintaining competitiveness in the world economy. In short, in general, DfG of a country must be expected to vary with time with trends, instead of being stationary.

3.2. DfG and Economic Complexity

Considering that the level of DfG is closely related to specialization and division of labor, it is necessary to examine the connection between DfG and economic complexity. Figure 5a–c show correlations between DfG and ECI in 1998, 2008 and 2019, respectively. Clearly, we observe that the DfG is very strongly negatively correlated with the ECI. This suggests that the higher level of economic complexity, the smaller the DfG. In other words, the higher level of economic complexity, the closer a country's RCA index to a normal distribution. Therefore, relationships between the DfG and economic development and economic complexity reflect that a closer a country's RCA index to a normal distribution, the higher degree of economic complexity and better economic performance of a country.

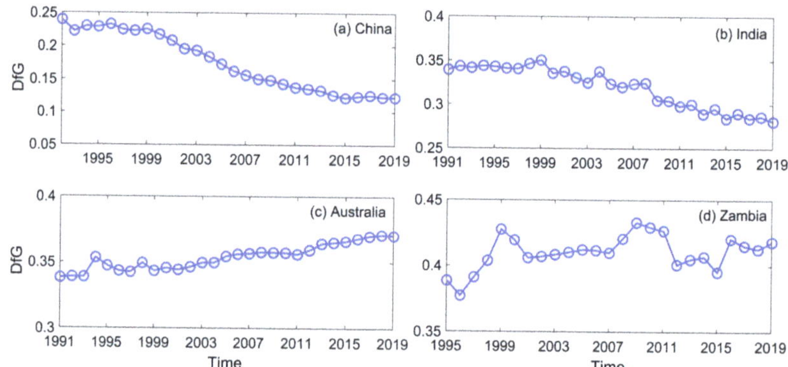

Figure 4. Dynamic evolution of the DfG for China, India, Australia and Zambia from 1991 to 2019 under the HS4 scheme.

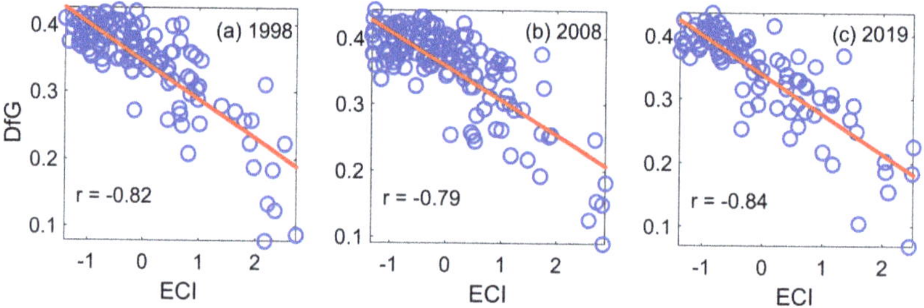

Figure 5. Regression analysis showing correlation between DfG and ECI in 1998, 2008 and 2019.

It is interesting to compare the Pearson correlation coefficient between DfG and the logarithm of GDP and that between ECI and the logarithm of GDP. Since the correlation coefficient for the former is negative but positive for the latter, it is more convenient to use the Pearson correlation coefficient between DfG and the logarithm of GDP in absolute value. The result for the comparison is shown in Figure 6, where the red curve denotes the absolute value of the correlation coefficient between DfG and the logarithm of GDP, and the blue curve is for the correlation coefficient between ECI and the logarithm of GDP. We observe that before the global financial crisis of 2008, the correlation coefficients between DfG and the logarithm of GDP, and between ECI and the logarithm of GDP, are comparable. However, after the global financial crisis, the correlation coefficients between DfG and the logarithm of GDP are persistently larger than those between ECI and the logarithm of GDP. The significance of this feature for designing better indicators of economic complexity will be further discussed in the last section.

Out of curiosity, we have examined whether DfG using import data is still strongly negatively correlated with the logarithm of GDP. The answer is positive. In fact, the correlation coefficient using import data is basically identical to that using export data. This interesting property however, is not shared by ECI—when using import data, whether we focus on adjacency matrices based on RCA $\geq$ 1 or RCA $<$ 1, the computed "ECI" essentially has no correlation with the logarithm of GDP. This signifies that RCA $\geq$ 1 or RCA $<$ 1 based on import data cannot be interpreted as that based on export data to have comparative advantage or disadvantage.

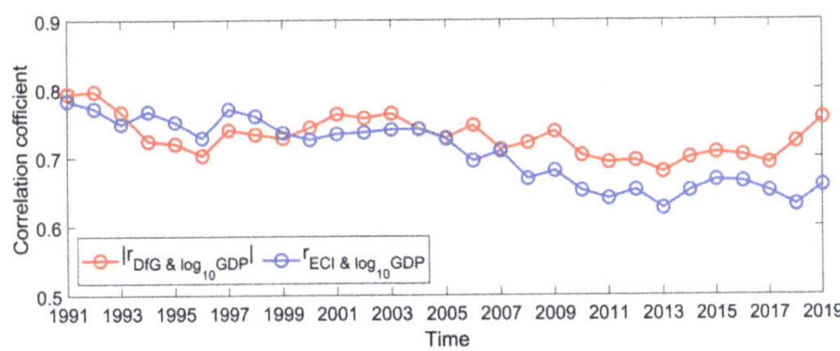

Figure 6. Variation of the Pearson correlation with time, where the red and the blue curves are for the absolute value of the correlation coefficient between DfG and the logarithm of GDP, and the correlation coefficient between ECI and the logarithm of GDP.

3.3. Regression and Causality Analysis

To understand more deeply the connection between DfG and economic development, we have employed the Pooled OLS model. The results are summarized in Table 1. Here, we select 60 countries which have continuous data from 1996 to 2019. We thus have a total of 1440 observations. We have first run a pooled OLS regression for the whole period. The results are shown in columns 1 to 3 of Table 1, where the 1st column is for the model with only DfG considered, the 2nd column for the results with only ECI considered, and the 3rd column for both DfG and ECI considered. We call these models 1–3. We observe that the regression coefficients for models 1–3 are significant at the 1% level. By comparing the columns 1 and 2, we find that DfG can explain 57.3 percent of the variance in GDP, while ECI accounts for 45.7 percent, as shown by the R^2 of the regression. This suggests that the explanatory power of DfG on GDP is stronger than that of ECI. After both DfG and ECI are considered, the model explains 58.7 percent of the variance in GDP, which is slightly better than model 1.

Table 1. Regression results for GDP, DfG and ECI.

Variables	Model 1–3 (1996–2019)			Model 4–6 (1996–2007)			Model 7–9 (2008–2019)		
	ln GDP	ln GDP	ln GDP	ln GDP	ln GDP	ln GDP	ln GDP	ln GDP	ln GDP
DfG	−18.97 ***[1] (−43.95)		−15.04 *** (−21.29)	−19.78 *** (−33.74)		−13.65 *** (−14.34)	−18.48 *** (−33.05)		−16.88 *** (−14.34)
ECI		1.26 *** (34.78)	0.37 *** (6.95)		1.43 *** (29.2)	0.58 *** (7.98)		1.11 *** (22.73)	0.14 ** (2.16)
Constant	31.94 *** (221.73)	25.16 *** (593.36)	30.48 *** (120.6)	31.75 *** (163.24)	24.67 *** (442.62)	29.49 *** (86.8)	32.22 *** (171.91)	25.64 *** (437.45)	31.64 *** (96.43)
Observations	1440	1440	1440	720	720	720	720	720	720
Adjusted R^2	0.573	0.457	0.587	0.623	0.543	0.645	0.603	0.418	0.605
F-Statistics	1931.66	1209.56	1021.59	1138.19	852.59	650.51	1092.18	516.62	551.22
Prob > F	0.000	0.000	0.000	0.000	0.000	0.000	0.000	0.000	0.000

[1] **Note:** *** $p < 0.01$, ** $p < 0.05$.

Considering that DfG has a higher correlation with GDP than ECI since the global financial crisis of 2008, we have also divided the whole time period into two, one from the year 1996 to 2007, the other from 2008 to 2019. The results are shown in the columns 4–6 and 7–9 of Table 1, for the models 1–3 explained earlier. By comparing the results of regression models for these two groups, we find: (1) in both time periods DfG has a stronger explanatory power on the variance of GDP than ECI, (2) the explanatory power of DfG and ECI coombined on the variance of GDP in first group is stronger that that of the second group. It is worth noting that ECI does not significantly improve the explanatory power of the model on the variance of GDP in these three scenarios of regression models, especially in the period after the global financial crisis of 2008. Therefore, DfG better explains the variance in GDP than ECI.

We have also performed a panel VAR analysis. LLC test indicates that the three variables with one period lag and trend are stationary. This allows us to estimate the coefficients of the system described by Equation (9) after the individual effects removed. Robustness test shows that the PVAR model is reasonable, as shown in Figure 7. Table 2 shows the results of the model with three variables, from the columns of which we find that the impact of ln GDP with one period lag on ln GDP, DfG and ECI are significant for all three different panel VARs, the impact of DfG with one period lag on DfG and ECI are significant, and the impact of ECI with one period lag on DfG and ECI are significant. However, impacts of DfG and ECI with one period lag on ln GDP are not significant. On the other hand, impact of DfG with one period lag on DfG is positive but negative on ECI, while the impacts of ECI with one period lag on both DfG and ECI are positive.

Table 2. Main results of a three-variables panel VAR.

Response of	Response to		
	ln GDP$_{(t-1)}$	DfG$_{(t-1)}$	ECI$_{(t-1)}$
ln GDP$_{(t)}$	0.961 ***[1]	−0.313	0.027
	(130.16)	(−1.40)	(0.6)
DfG$_{(t)}$	−0.008 ***	0.734 ***	0.027 ***
	(−7.96)	(24.23)	(4.41)
ECI$_{(t)}$	0.292 ***	−4.864 ***	0.874 ***
	(15.10)	(−8.31)	(7.43)
Observations	1440		
N countries	60		

[1] **Note:** *** $p < 0.01$.

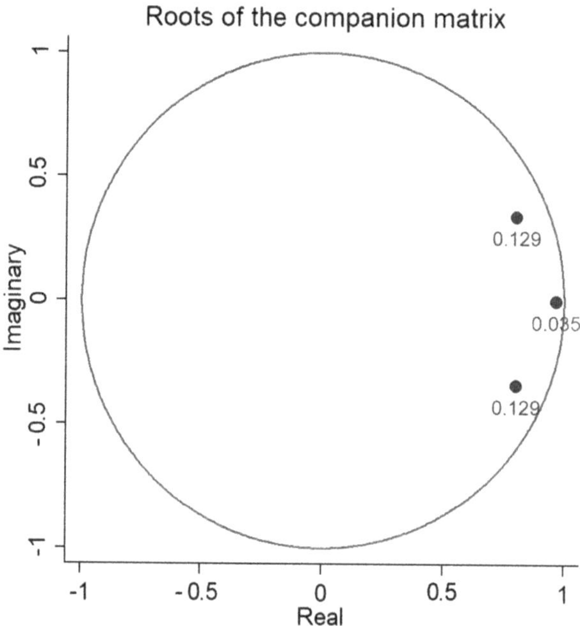

Figure 7. Robustness test of Panel VAR.

Finally, we have examined the statistical causality among the three variables based on PVAR by using panel Granger causality Wald test. The results are shown in Table 3. We observe that the ln GDP is not the Granger cause of DfG and ECI at the 5% level, while the DfG and ECI are the Granger cause of ln GDP at the 1% level. This result is as anticipated.

Table 3. Granger causality Wald tests for Panel VAR.

Hypothesis	chi2	df	Prob > chi2
ln GDP does not Granger cause DfG	1.971	1	0.160
ln GDP does not Granger cause ECI	0.365	1	0.546
DfG does not Granger cause ln GDP	63.338	1	0.000 ***[1]
DfG does not Granger cause ECI	19.434	1	0.000 ***
ECI does not Granger cause ln GDP	227.91	1	0.000 ***
ECI does not Granger cause DfG	69.104	1	0.000 ***

[1] **Note:** *** $p < 0.01$.

4. Discussion

Understanding the difference in economic development among countries or regions is a long-standing issue in economics. A crucial perspective to shed light on the issue is to evaluate competitiveness of a country in international trade as characterized by the RCA index. Although it is desirable to have a normally distributed RCA, empirical studies have often found the opposite. This discrepancy has stimulated a lot of researchers to develop alternative indices of the RCA index so that their distributions would be closer to Gaussian distributions. Yet, those efforts are not very successful. This calls for a deeper understanding of the significance of a normally distributed RCA index.

To gain insights into this issue, we have defined a quantity, DfG, based on the KS test, which quantifies the deviation of the distribution of a country's RCA index from normality. We have found that the variations of DfG are characterized by spatiotemporal heterogeneity and regional spatial agglomeration. The spatiotemporal heterogeneity of the DfG refers to the significant differences in many countries' DfG and their dynamic evolution. Regional spatial agglomeration of the DfG refers to that countries with smaller DfG are mainly concentrated in North America (represented by USA), Western Europe (represented by Germany), and Eastern Asia (represented by China, Japan and South Korea). Interestingly, these three areas are very consistent with the description of "The world seems to have three interconnected production hubs for the extensive trade in parts and components" in the "Global Value Chain Development Report 2017—Measuring and Analyzing the Impact of GVCs on Economic Development". It suggests that the DfG has some connections with the development of GVCs. On the other hand, countries with larger DfG are mainly concentrated in Africa, South America, Western and Southern Asia.

The pattern of DfG's spatial variation suggests that the DfG can act as a good indicator of a country's economic performance. This is indeed so, as DfG is found to be strongly negatively related with both the logarithm of GDP and the ECI. Therefore, the closer the distribution of a country's RCA index to a normal distribution, the higher degree of economic complexity and better economic performance of the country. This highlights the optimality of a country's export when its RCA index follows a normal distribution, and provides a new perspective to understand the difference in economic development among countries or regions. Furthermore, we have found that the correlation coefficients between DfG and the logarithm of GDP are persistently larger than those between ECI and the logarithm of GDP after the 2008 global financial crisis. This is further corroborated by regression analysis which shows that DfG better explains the variance in GDP than ECI. Further Granger causality analysis shows that DfG and ECI are the Granger cause of ln GDP, but not the vice versa. It is worth emphasizing that Gaussianity is not a cause, it is more a consequence indicating economic development.

The last feature, that DfG is more strongly correlated with GDP than ECI, suggests an interesting way to improve characterization of economic complexity of a country. For this purpose, we need to first understand the meaning of the correlation between ECI and the logarithm of GDP. This is due to the strong correlation between export and GDP—ECI amounts to retaining only products with RCA equal to or greater than 1 and approximating the amount of export by counting the number of products with RCA $\geq$ 1. Our observation that after the 2008 global financial crisis, the correlation between DfG and GDP is stronger than that between ECI and GDP, can at least be partially attributed to the enhancement of the global participation in production chains, or simply, greater participation in global value chains (GVCs). Therefore, simply focusing on RCA $\geq$ 1, which has been used in designing ECI and its variants, is no longer sufficient. In other words, information contained in products with RCA $<$ 1 can no longer be simply discarded. Therefore, in future, it would be extremely interesting to develop a new economic complexity index by using DfG alone, or by combining DfG and ECI (or its variants).

Author Contributions: Conceptualization, J.G. and B.L.; methodology, J.G. and B.L.; software, B.L.; validation, J.G. and B.L.; formal analysis, J.G. and B.L.; investigation, J.G. and B.L.; resources, B.L.; data curation, B.L.; writing—original draft preparation, J.G. and B.L.; writing—review and editing, J.G. and B.L.; visualization, B.L.; supervision, J.G.; project administration, J.G.; funding acquisition, J.G. All authors have read and agreed to the published version of the manuscript.

Funding: This research was supported by the National Key Research and Development Program of China under Grant No. 2019AAA0103402, by the Fundamental Research Funds for the Central Universities, and by Innovation Project of Guangxi Graduate Education under Grant No. YCBZ2017017. It is also supported by the National Natural Science Foundation of China under Grant No. 41671532. One of the authors (JG) also benefited tremendously from participating the long program on culture analytics organized by the Institute for Pure and Applied 746 Mathematics (IPAM) at UCLA, which was supported by the National Science Foundation.

Institutional Review Board Statement: Not applicable.

Informed Consent Statement: Not applicable.

Data Availability Statement: Data available in a publicly accessible repository.

Acknowledgments: The authors thank Xin He and Qiyue Hu for help plotting some of the figures presented in the paper.

Conflicts of Interest: The authors declare no conflict of interest.

References

1. Balassa, B. Trade liberalization and 'revealed' comparative advantage. *Manch. Sch. Econ. Soc. Stud.* **1965**, *32*, 99–123. [CrossRef]
2. Yeats, A.J. On the Appropriate Interpretation of the Revealed Comparative Advantage Index: Implications of a Methodology based on Industry Sector Analysis. *Weltwirtsch. Arch.* **1985**, *121*, 61–73. [CrossRef]
3. Leromain, E.; Orefice, G. New revealed comparative advantage index: Dataset and empirical distribution. *Int. Econ.* **2014**, *139*, 48–70. [CrossRef]
4. Hoen, A.R.; Oosterhaven, J. On the measurement of comparative advantage. *Ann. Reg. Sci.* **2006**, *40*, 677–691. [CrossRef]
5. Yu, R.; Cai, J.; Leung, P. The normalized revealed comparative advantage index. *Ann. Reg. Sci.* **2009**, *43*, 267–282. [CrossRef]
6. Hinloopen, J.; Marrewijk, C.V. On the empirical distribution of the Balassa index. *Weltwirtsch. Arch.* **2001**, *137*, 1–35. [CrossRef]
7. De Benedictis, L.; Tamberi, M. Overall specialization empirics: Techniques and applications. *Open Econ. Rev.* **2004**, *15*, 323–346. [CrossRef]
8. Bowen, H.P. On the Theoretical Interpretation of Indices of Trade Intensity and Revealed Comparative Advantage. *Weltwirtsch. Arch.* **1983**, *119*, 464–472. [CrossRef]
9. Laursen, K. Revealed comparative advantage and the alternatives as measures of international specialization. *Eurasian Bus. Rev.* **2015**, *5*, 99–115. [CrossRef]
10. Deb, K.; Sengupta, B. On empirical distribution of RCA Indices. *IIM Kozhikode Soc. Manag. Rev.* **2017**, *6*, 23–41. [CrossRef]
11. Vollrath, T.L. A Theoretical Evaluation of Alternative Trade Intensity Measures of Revealed Comparative Advantage. *Weltwirtsch. Arch.* **1991**, *127*, 265–280. [CrossRef]
12. Dalum, B.; Laursen, K.; Villumsen, G. Structural change in OECD export specialisation patterns: De-specialisation and 'stickiness'. *Int. Rev. Appl. Econ.* **1998**, *12*, 423–443. [CrossRef]
13. Proudman, J.; Redding, S. Evolving patterns of international trade. *Rev. Int. Econ.* **2000**, *8*, 373–396. [CrossRef]
14. Amador, J.; Cabral, S.; Maria, J.R. A simple cross-country index of trade specialization. *Open Econ. Rev.* **2011**, *22*, 447–461. [CrossRef]
15. Jenny, P.D.B.; Rémi, S. A New Class of Revealed Comparative Advantage Indexes. *Open Econ. Rev.* **2021**, 1–27. [CrossRef]
16. Liu, B.; Gao, J.B. Understanding the non-Gaussian distribution of revealed comparative advantage index and its alternatives. *Int. Econ.* **2019**, *158*, 1–11. [CrossRef]

17. Hidalgo, C.A.; Klinger, B.; Barabási, A.L.; Hausmann, R. The product space conditions the development of nations. *Science* **2007**, *317*, 482–487. [CrossRef] [PubMed]
18. Hidalgo, C.; Hausmann, R. The building blocks of economic complexity. *Proc. Natl. Acad. Sci. USA* **2009**, *106*, 10570–10575. [CrossRef] [PubMed]
19. Mealy, P.; Farmer, J.D.; Teytelboym, A. Interpreting economic complexity. *Sci. Adv.* **2019**, *5*, 1–8. [CrossRef] [PubMed]
20. Andrea, T.; Mazzilli, D.; Pietronero, L. A dynamical systems approach to gross domestic product forecasting. *Nat. Phys.* **2018**, *14*, 861–865.
21. Gao, J.; Zhang, Y.C.; Zhou, T. Computational socioeconomics. *Phys. Rep.* **2019**, *817*, 1–104. [CrossRef]
22. Tacchella, A.; Cristelli, M.; Caldarelli, G.; Gabrielli, A.; Pietronero, L. A new metrics for countries' fitness and products' complexity. *Sci. Rep.* **2012**, *2*, 723. [CrossRef] [PubMed]
23. Tacchella, A.; Cristelli, M.; Caldarelli, G.; Gabrielli, A.; Pietronero, L. Economic complexity: Conceptual grounding of a new metrics for global competitiveness. *J. Econ. Dynam. Control* **2013**, *37*, 1683–1691. [CrossRef]
24. Caldarelli, G.; Cristelli, M.; Gabrielli, A.; Pietronero, L.; Scala, A.; Tacchella, A. A network analysis of countries' export flows: Firm grounds for the building blocks of the economy. *PLoS ONE* **2012**, *7*, e47278. [CrossRef] [PubMed]
25. Cristelli, M.; Gabrielli, A.; Tacchella, A.; Caldarelli, G.; Pietronero, L. Measuring the intangibles: A metrics for the economic complexity of countries and products. *PLoS ONE* **2013**, *8*, e70726. [CrossRef]
26. Battiston, F.; Cristelli, M.; Tacchella, A.; Pietronero, L. How metrics for economic complexity are affected by noise. *Complex. Econ.* **2014**, *3*, 1–22.
27. Mariani, M.S.; Vidmer, A.; Medo, M.; Zhang, Y.C. Measuring economic complexity of countries and products: Which metric to use? *Eur. Phys. J. B* **2015**, *88*, 293. [CrossRef]
28. Wu, R.J.; Shi, G.Y.; Zhang, Y.C.; SebastianMariani, M. The mathematics of non-linear metrics for nested networks. *Phys. A* **2016**, *460*, 254–269. [CrossRef]
29. Felipe, J.; Kumar, U.; Abdon, A.; Bacate, M. Product complexity and economic development. *Struct. Chang. Econ. Dyn.* **2012**, *23*, 36–68. [CrossRef]
30. Poncet, S.; Waldemar, F.S.D. Export upgrading and growth: The prerequisite of domestic embeddedness. *World Dev.* **2013**, *51*, 104–118. [CrossRef]
31. Hausmann, R.; Hidalgo, C. *The Atlas of Economic Complexity: Mapping Paths to Prosperity*; MIT Press: Cambridge, MA, USA, 2014.
32. Zhu, S.; Yu, C.; He, C. Export structures, income inequality and urban-rural divide in China. *Appl. Geogr.* **2020**, *115*, 102150. [CrossRef]
33. Neagu, O.; Teodoru, M.C. The relationship between economic complexity, energy consumption structure and greenhouse gas emission: Heterogeneous panel evidence from the EU countries. *Sustainability* **2019**, *11*, 497. [CrossRef]
34. Romero, J.P.; Gramkow, C. Economic complexity and greenhouse gas emissions. *World Dev.* **2021**, *139*, 105317. [CrossRef]
35. Lapatinas, A. Economic complexity and human development: A note. *Econ. Bull.* **2016**, *366*, 1441–1452.
36. Vu, T.V. Economic complexity and health outcomes: A global perspective. *Soc. Sci. Med.* **2020**, *265*, 113480. [CrossRef] [PubMed]
37. Hartmann, D.; Guevara, M.R.; Jara-Figueroa, C.; Manuel, A.; Hidalgo, C.A. Linking economic complexity, institutions, and income inequality. *World Dev.* **2017**, *93*, 75–93. [CrossRef]
38. Chavez, J.C.; Marco, T.M.; Manuel, G.Z. Economic complexity and regional growth performance: Evidence from the Mexican Economy. *Rev. Reg. Stud.* **2017**, *47*, 201–219. [CrossRef]
39. Gao, J.; Zhou, T. Quantifying China's regional economic complexity. *Phys. A Stat. Mech. Its Appl.* **2018**, *492*, 1591–1603. [CrossRef]
40. Fritz, B.S.; Robert, A.M. The economic complexity of US metropolitan areas. *Reg. Stud.* **2021**, 1–12. [CrossRef]
41. Hidalgo, C.A. Economic complexity theory and applications. *Nat. Rev. Phys.* **2021**, *3*, 92–113. [CrossRef]
42. Stojkoski, V.; Utkovski, Z.; Kocarev, L. The impact of services on economic complexity: Service sophistication as route for economic growth. *PLoS ONE* **2016**, *11*, e0161633. [CrossRef] [PubMed]
43. Wooldridge, J.M. Selection corrections for panel data models under conditional mean independence assumptions. *J. Econom.* **1995**, *68*, 115–132. [CrossRef]
44. Wooldridge, J.M. *Econometric Analysis of Cross Section and Panel Data*; MIT Press: Cambridge, MA, USA, 2010.
45. Love, I.; Zicchino, L. Financial development and dynamic investment behavior: Evidence from panel VAR. *Q. Rev. Econ. Financ.* **2006**, *46*, 190–210. [CrossRef]
46. Koutsomanoli-Filippaki, A.; Mamatzakis, E. Performance and Merton-type default risk of listed banks in the EU: A panel VAR approach. *J. Bank. Financ.* **2009**, *33*, 2050–2061. [CrossRef]
47. Levin, A.; Lin, C.F.; Chu, C.S.J. Unit root tests in panel data: Asymptotic and finite-sample properties. *J. Econom.* **2002**, *108*, 1–24. [CrossRef]

applied
sciences

Essay

Formal Matters on the Topic of Risk Mitigation: A Mathematical Perspective

Giuseppe Bilotta *, Annalisa Cappello and Gaetana Ganci

Osservatorio Etneo, Istituto Nazionale di Geofisica e Vulcanologia, Piazza Roma, 2, 95125 Catania, Italy
* Correspondence: giuseppe.bilotta@ingv.it

Abstract: How (in)formal should the classic expression describing risk as the product of hazard, exposure, and vulnerability be considered? What would be the most complete way to describe the process of risk mitigation? These are the questions we try to answer here, using a formal, mathematically sound yet abstract description of hazard, exposure, vulnerability, and risk. We highlight the elements that can be affected for the purpose of mitigation and show how this can improve the quantitative assessment of the procedural aspects of risk mitigation, both long- and short-term, down to the timescale of emergency response.

Keywords: risk; vulnerability; exposure; hazard; mitigation

1. Introduction

The groundwork for a clarifying definition of the concepts of risk and hazard, and their relation, was first carried out in an international setting in [1], wherein *risk* is defined as the possibility of loss (whose type and cause are further specified by attributes, such as seismic risk for the chance of loss caused by earthquakes), and (natural) *hazard* is defined as «*the state of risk due to the possibility of occurrence of a*» (natural) «*disaster*». The document also includes a brief section concerning protection and insurance, arguably the first mention of the need of a systematic approach to risk mitigation.

Risk Management in Formulas

The current commonly (at least in the field of natural hazards) accepted informal expression of risk as the product of hazard, exposure (or value), and vulnerability is due to [2]:

$$\text{Risk} = \text{Hazard} \times \text{Exposure} \times \text{Vulnerability} \tag{1}$$

where Hazard indicates the probability of occurrence of the event (e.g., lava flow inundation, earthquake, etc.), Exposure is a quantification of (the value of) the people, systems, and property potentially subject to the hazardous phenomenon (in fact, Ref. [2] explicitly uses the term Value rather than Exposure), and Vulnerability is a quantification of the effective relative impact of the event, expressed as a percentage, with $V = 1$ indicating total loss (100%) and $V = 0$ indicating total resilience. Contextually, the author also presents some key elements for risk management (in the context of volcanic hazard, but of general applicability), such as land-use planning ("zoning") to reduce exposure, and preparedness (including monitoring, early warning systems and response planning) to reduce vulnerability.

We note that Equation (1) is essentially *qualitative* in nature, rather than quantitative, as pointed out e.g., by [3]. Other functional relationships are available in the literature, especially in engineering contexts. For example, Ref. [4], and more recently [5], prefer an even more informal

$$\text{Risk} = \text{Uncertainty} + \text{Damage}. \tag{2}$$

Citation: Bilotta, G.; Cappello, A.; Ganci, G. Formal Matters on the Topic of Risk Mitigation: A Mathematical Perspective. *Appl. Sci.* **2023**, *13*, 265. https://doi.org/10.3390/app13010265

Academic Editors: Andrea L. Rizzo, Na Zhao and Hua Zhong

Received: 29 November 2022
Revised: 19 December 2022
Accepted: 20 December 2022
Published: 26 December 2022

When relating risk to hazard, Ref. [4] provides a different, but still informal relationship:

$$\text{Risk} = \frac{\text{Hazard}}{\text{Safeguards}} \tag{3}$$

that provides an explicit indication of the possibility of intervention to reduce risk. This is in contrast to [2] whose formula (1) provides no explicit mention of the possible quantification of mitigation efforts, even though the paper presents several approaches to risk management.

In the field of natural hazards, an extension of (1) that includes an explicit dependency on mitigation is given by [3]:

$$\text{Risk} = \text{Hazard} \times \text{Exposure} \times (\text{Vulnerability} - \text{Risk mitigation efforts}), \tag{4}$$

and more recently in [6] that provides a different formulation for mitigated risk as:

$$\text{Risk} = \frac{\text{Hazard} \times \text{Exposure} \times \text{Vulnerability}}{\text{Mitigation measures}}. \tag{5}$$

Equations (4) and (5) are also intended to be informal and qualitative rather than quantitative, although, like (1), they *can* be used in a more quantitative sense (see also Section 2.4). For example, (5) could be used to compute an a posteriori value for the *efficiency* of the given mitigation measures as $M = R_U / R_M$, where M represents the efficiency of the mitigation measures, R_U the unmitigated risk, and R_M the mitigated risk.

Arguably, (4) and (5) illustrate a philosophical difference in the approach to risk mitigation: while [3] focuses exclusively on reducing vulnerability, [6] applies mitigation to risk *as a whole*, and it is thus closer to the arguments brought forth in [2], which include the exposure-related land-use planning as a risk management feature. In relation to the other formulas seen so far, it may also be considered a more detailed version of (3), and a more general version of (4), even though the latter is not clear, due to the different choice of mathematical operators used to indicate the influence of mitigation measures.

Considering their qualitative, informal nature, the functional difference between (4) and (5) is actually largely inessential. It is only when aiming at a more rigorous and quantitative assessment of risk and its mitigation that the specifics of the mathematical formulation become relevant—an aspect that so far has received more attention in engineering [4,5,7,8] than in natural hazards [9–11].

In engineering, this quantifying effort is required to manage the multi-objective problem of minimizing both the risk and the costs associated with the mitigation [8]. In this sense, risk takes the form of an expected loss of value in a strictly probabilistic sense, and may be more in general treated not as a single value, but as a formal collection of all the elements that contribute to its assessment. For example, Ref. [4] defines risk as a set of triplets that describe all of the known possible, mutually exclusive scenarios, their probability, and their outcomes; no single value is associated with risk.

In an effort to try and bridge the gap between the more qualitative formulation of risk (1) presented by [2] and common in natural hazards, and the more formal approaches to the quantification of risk and its mitigation common in engineering, we present here a detailed mathematical approach to the quantification of risk assessment and mitigation.

While the description will be kept as abstract and generic as possible, much of it can be seen simply as a formalization of common practices [11,12]. In addition, our formalism will make an effort to bring out the explicit dependency of risk (and its components) on several variables, highlighting the distinction between decision variables (i.e., quantities that can be influenced by decision-makers and other stakeholders) and other input variables [8], which will be crucial to our discussion about the quantitative approach to risk mitigation. The focus will be specifically on (1) and will not directly touch on the mathematical aspects of hazard and risk assessment (including details about modeling and quantification) that have been extensively discussed in the literature [3,9–11].

2. Formalizing Risk

To formalize the risk assessment Equation (1), we must first define hazard, exposure, vulnerability, and risk in a mathematically rigorous sense, while preserving the spirit and, as far as possible, the actual functional relationship of (1).

To this end, consider a two-dimensional set $\Omega \subset \mathbb{R}^2$ that represents our *area of study* in some appropriate reference system, e.g., a two-dimensional section of the Earth surface with a specific choice of coordinate system. such as, for example, EPSG:32633 (WGS-84 spheroid with UTM projection, zone 33 north) to study Mt Etna [13] and references within, or EPSG:32740 (WGS-84 spheroid with UTM projection, zone 40 south) for Piton de la Fournaise [14].

(Our formalization is actually independent from the dimensionality of the problem: we could just as well consider $\Omega \subset \mathbb{R}^3$ and reason in three dimensions, e.g., for the risk associated with hazards in industrial complexes, taking into account the three-dimensionality of the distribution of people and other exposed elements. Time as a parameter could be included in a similar fashion.)

2.1. Hazard

Assume for simplicity that we are looking at the case of a single hazard expressed as the probability of occurrence of a dangerous event hitting a specific area, with no intensity information. Formally, this translates to a *pointwise hazard* probability density function $h : \Omega \to [0, 1]$ such that, for any *area of interest* $A \subseteq \Omega$, the probability of the hazard affecting the area A is

$$H(A) = \int_A h(x, y)\mathrm{d}x\mathrm{d}y.$$

Hazard may depend on the location (coordinates) directly, or implicitly through some other spatial property that can be affected by human action (e.g., many geophysical flows may be affected by building ditches and barriers).

We can make this dependency explicitly by writing $h : \mathcal{T} \times \Omega \to [0, 1]$, where $\mathcal{T}$ is a family of functions T defined in Ω and with values in some appropriate codomain $\mathcal{D}_T$. As a practical example, T might be a mathematical description of the topography of Ω, and $h(T, x, y)$ is the pointwise hazard associated with a geophysical flow whose behavior depends on the given topography. In this case, we would have $\mathcal{D}_T = \mathbb{R}$, i.e., the set of real numbers describing the pointwise altitude a.s.l. of the area of study.

Note that T may have "long range" effects, in the sense that a change in the value of T at some point (x_0, y_0) may affect the hazard in points $(x, y) \neq (x_0, y_0)$: for example, building an embankment is a local alteration of the topography that can reduce hazard in all points downstream of the structure. For this reason, h must depend explicitly on T as a function, rather than simply as $h(x, y) = h(T(x, y), x, y)$.

In general, man-made structures have an influence on hazard even if that is not their intent. For example, buildings and roads can influence geophysical flows, and coastal/river structures can influence flooding hazard. While the influence of such structures could be incorporated in T, to simplify notation, we will separate this into an additional dependency of hazard on some $B \in \mathcal{B}$ that will in turn depend on exposure-related elements that will be presented momentarily.

2.2. Exposure

In a similar fashion to hazard, we can define a *pointwise exposure* $e : \Omega \to [0, +\infty]$ such that, for any area of interest $A \subseteq \Omega$, its exposure value is defined as

$$E(A) = \int_A e(x, y)\mathrm{d}x\mathrm{d}y.$$

In general, however, exposure does not depend directly on the coordinates themselves, but rather on the distribution of elements at risk, such as population, land use, presence of buildings or infrastructure, etc. As carried out with hazard, it is thus better to write

$e : \mathcal{P} \times \Omega \to [0, +\infty[$, where $\mathcal{P}$ is the family of functions P (defined in Ω with an appropriate codomain $\mathcal{D}_\mathcal{P}$) that describe mathematically the distribution of the key exposed elements. It should be assumed that the elements of $\mathcal{P}$ have some kind of constraints (e.g., if P represents population distribution, we can assume that the total population $\mathbf{P} = \int_\Omega P(x, y)\mathrm{d}x\mathrm{d}y$ is independent of the choice of $P \in \mathcal{P}$).

Moreover, as noted before, the choice of P can have an influence on hazard too, inasmuch as the associated infrastructure impacts the evolution of the hazardous phenomenon. For a given $P \in \mathcal{P}$, we can thus define a set $\mathcal{B}(P)$ of functions B in Ω that represents the man-made structures supporting the exposed element distribution P and affecting the hazard.

Obviously, exposure itself depends on these elements too. The full function signature for e and h is thus $h : \mathcal{T} \times \mathcal{H}(\mathcal{P}) \times \Omega \to [0, 1]$, $e : \mathcal{P} \times \mathcal{H}(\mathcal{P}) \times \Omega \to [0, +\infty[$. Then, the hazard H and the exposed value E can be computed respectively as

$$H(A, T, P, B) = \int_A h(T, P, B, x, y)\mathrm{d}x\mathrm{d}y, \qquad E(A, P, B) = \int_A e(P, B, x, y)\mathrm{d}x\mathrm{d}y \qquad (6)$$

with $A \subseteq \Omega$ the area, $T \in \mathcal{T}$ any natural or man-made elements that influences hazard, but not exposure, $P \in \mathcal{P}$ the distribution of exposed elements with no impact on hazard, and $B \in \mathcal{B}(P)$ the distribution of P-dependent exposed elements that influence hazard.

2.3. Vulnerability

As implemented for hazard and exposure, the *pointwise vulnerability* can also be defined as a function $v : \Omega \to [0, 1]$. Vulnerability, though, does not depend only on the coordinates, but also on the resilience of the individual exposed elements, as well as on their interactions with the other exposed elements.

Consider the example of seismic hazard: the vulnerability of a building depends on the ground properties (sand vs. rock) of the location where it was built, on the resilience of the building to the shaking, but also on the presence of other surrounding buildings that could affect it by pounding due to their oscillation during an earthquake [15]. Finally, the vulnerability may also depend on the same environmental factors that also affect hazard (e.g., a rampart may reduce the vulnerability of a building, and also divert a geophysical flow, affecting the hazard).

If we denote by $\mathcal{Q}(P, B)$ the family of functions that describe the resilience of the individual exposed elements described by $P \in \mathcal{P}$ and $B \in \mathcal{B}(P)$, then v depends both on P directly, but also through $B \in \mathcal{B}(P)$, and through some $Q \in \mathcal{Q}(P, B)$, making the signature of the vulnerability function

$$v : \mathcal{P} \times \mathcal{B}(\mathcal{P}) \times \mathcal{Q}(\mathcal{P}, \mathcal{B}(\mathcal{P})) \times \Omega \to [0, 1].$$

2.4. Risk

With all the components of risk defined, the *pointwise risk* can be defined as the formally correct application of (1):

$$r(T, P, B, Q, x, y) = h(T, P, B, x, y) \cdot e(P, B, x, y) \cdot v(P, B, Q, x, y) \qquad (7)$$

where $T \in \mathcal{T}$, $P \in \mathcal{P}$, $B \in \mathcal{B}(P)$ are defined as in (6), and $Q \in \mathcal{Q}(P, B)$ is the resilience of the individual exposed element.

The risk associated with a specific area $A \subseteq \Omega$ can then be obtained by integration:

$$R(A, T, P, B, Q) = \int_A r(T, P, B, Q, x, y)\mathrm{d}x\mathrm{d}y =$$
$$= \int_A h(T, P, B, x, y)e(P, B, x, y)v(P, B, Q, x, y)\mathrm{d}x\mathrm{d}y, \qquad (8)$$

and the total risk over the entire domain is thus $R(\Omega, T, P, B, Q)$.

The relationship between the input variables and hazard, exposure, vulnerability, and risk is illustrated in Figure 1.

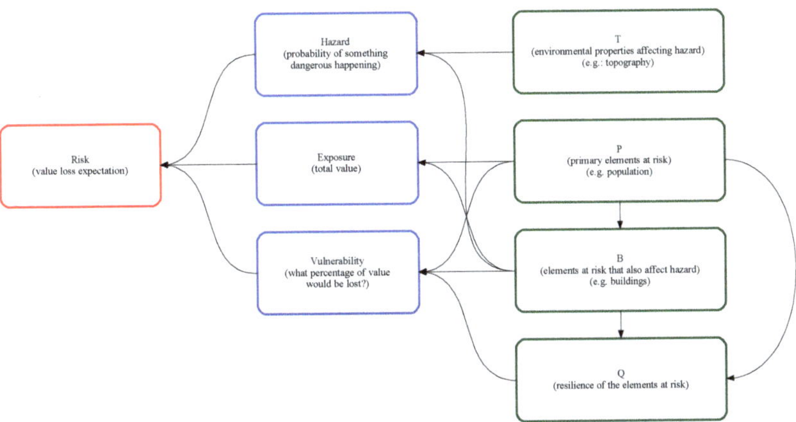

Figure 1. The functional dependencies between risk, vulnerability, exposure, hazard, and their respective input variables. Arrows point towards the dependent (i.e., $X \rightarrow Y$ indicates that Y depends on X).

We observe that (7) matches (1), as intended, and [2] reasoning can be applied to our definition of risk on each subset of Ω that has a (spatially) uniform hazard, exposure and vulnerability, making (1) a zero-order (piecewise constant) approximation of the more complete (8). Equation (8) on the other hand can also be interpreted as describing risk as the expected (loss of) value for the random variable $e \cdot v$ with probability density h, in line with the approach used to quantify hazard in engineering [8]. Indeed, (8) is already frequently used in its discrete form (and without explicit mention of T, P, B, Q in the literature (see, e.g., [11,12]).

3. Formalizing Risk Mitigation

When discussing risk mitigation, we should consider the h, e, v (and *a fortiori r*) functions to be *fixed*: they are the (mathematical or numerical) models that describe how to compute the hazard, exposure, and vulnerability (and risk) given the appropriate input data. For example, the values of h may be obtained using deterministic physical-mathematical models of the phenomenon, and e may be computed from well-established criteria that assign value to human resources present in the area.

Given the unchanging nature of h, e, v, to mitigate the risk, we must operate on the *input data* to these functions. Thus, while Equation (8) for risk may not seem particularly innovative, the explicitation of the dependency on the choice of T, P, B, Q in their respective sets is essential for the formalization of risk mitigation: these are the model inputs on which decision-makers have influence, i.e., the *decision variables* [8].

To see how the choice of these functions maps to risk mitigation efforts, consider, for example, that to reduce risk, we could strengthen the buildings to make them less vulnerable to earthquakes [16], which corresponds to choosing a different Q, raise the river banks to reduce flooding hazard [17] (resulting in a different T), and displace population [18–20] or reconsider land-use patterns [21] so that high-exposure elements are moved to low-hazard areas (equivalent to a different choice of P, with a possible indirect effect on B and Q).

More formally, assume we have a given $T_0 \in \mathcal{T}, P_0 \in \mathcal{P}, B_0 \in \mathcal{B}(P_0), Q_0 \in \mathcal{Q}(P_0, B_0)$ and a corresponding (pointwise) risk $r(T_0, P_0, B_0, Q_0, x, y)$ and total risk $R(\Omega, T_0, P_0, B_0, Q_0) = \int_\Omega r(T_0, P_0, B_0, Q_0, x, y) dx dy$. To mitigate risk, we need to find $T_1 \in \mathcal{T}, P_1 \in \mathcal{P}, B_1 \in \mathcal{B}(P_1)$,

$Q_1 \in \mathcal{Q}(P_1, B_1)$ such that the total risk $R(\Omega, T_1, P_1, B_1, Q_1) = \int_\Omega r(T_1, P_1, B_1, Q_1, x, y)\,dx\,dy$ satisfies:

$$R(\Omega, T_1, P_1, B_1, Q_1) < R(\Omega, T_0, P_0, B_0, Q_0).$$

In practice, the function quartet $C_0 = (T_0, P_0, B_0, Q_0)$ is a functional representation of the current situation, and the function quartet $C_1 = (T_1, P_1, B_1, Q_1)$ would be the functional representation of a distribution of resources that leads to a lower overall risk than the current situation.

3.1. Mitigation as a Minimization Process

From a mathematical perspective, risk mitigation can be considered a minimization process: given the set $\mathcal{C}$ of all possible configurations

$$\mathcal{C} = \{(T, P, B, Q) : T \in \mathcal{T}, P \in \mathcal{P}, B \in \mathcal{B}(P), Q \in \mathcal{Q}(P, B)\},$$

we might be interested in finding, for example, the lowest risk conceivable in the region

$$\inf_{C \in \mathcal{C}} R(\Omega, C),$$

and whether or not this can actually be achieved, i.e., if there exists $\bar{C} \in \mathcal{C}$ such that $R(\Omega, \bar{C}) = \inf_{C \in \mathcal{C}} R(\Omega, C)$ (note that, in this case, the infimum is an actual minimum in the mathematical sense). This can be important to determine the optimal land-use planning in a "virgin" territory, but also to determine what can be expected "at best" by any risk mitigation process.

When given an initial configuration $C_0 \in \mathcal{C}$, risk mitigation would imply studying the subset of configurations with lower risk:

$$\mathcal{C}_{<0} = \{C \in \mathcal{C} : R(\Omega, C) < R(\Omega, C_0)\}$$

and possibly look for some $C \in \mathcal{C}_{<0}$ that is "optimal" in some mathematical sense (possibly in relation to C_0 itself, as we shall see momentarily).

3.2. Cost Functions

In practice, risk mitigation has a cost: levees must be raised, ditches must be dug, buildings must be reinforced, and infrastructure needs to be changed to accommodate for the redistribution of population, etc.

Mathematically, this can be taken into account by associating a *cost function* κ to each pair of configurations C_0, C_1, with $\kappa(C_0, C_1) \geq 0$ modeling the cost of migration from configuration C_0 to configuration C_1.

In risk mitigation, it is therefore in general appropriate to look for new configurations such that the cost of migration from the previous to the new configuration is less than the difference in risk, i.e., for configurations in the set

$$\mathcal{C}^\kappa_{<0} = \{C \in \mathcal{C} : R(\Omega, C) < R(\Omega, C_0) + \kappa(C_0, C)\}$$

for some cost function κ. This is a way to express mathematically the idea that the cost of reducing risk should not be higher than the value saved by reducing the risk.

In this case, one could consider the optimality of a lower risk configuration $C \in \mathcal{C}^\kappa_{<0}$ for example as the "most bang for the buck", i.e., a configuration that minimizes both $R(\Omega, \cdot)$ and $\kappa(C_0, \cdot)$. This translates to a multi-objective optimization problem and the study of Pareto-optimal configurations [8,17].

The previous formulation is formally complete if the cost of migration from C_0 to C is unique. This, however, is not the case in general: for example, the same change in single-building resilience (from Q_0 to Q) may be achieved with different engineering efforts, each with a different cost.

One possible approach to simplify this is to consider as $\kappa(C_0, C)$ the *minimum* cost necessary to enact the change in configuration (or at least the infimum of the costs, if the minimum does not exist). This is sufficient to make the cost function unique but may result in unrealistic results in the estimation of $C_{<0}^{\kappa}$, since, in practice, the costs will have a probability of being higher than the estimated lower bound κ, possibly resulting in a risk mitigation process that is in practice more expensive than the expected savings in value loss.

A more complete way to approach the multiplicity of the cost function for each pair of configurations is to take inspiration from homotopies. Mathematically, we define a *transition* from configuration C_0 to configuration C_1 as a function $M_{C_0,C_1} : [0,1] \to \mathcal{C}$ such that $M_{C_0,C_1}(0) = C_0$ and $M_{C_0,C_1}(1) = C_1$. The mitigation cost is then associated not with the *endpoints* of the transition C_0, C_1, but with the specific transition, i.e., not $\kappa(C_0, C_1)$, but $\kappa(M_{C_0,C_1})$.

If we indicate by $\mathcal{M}(C_0, C_1)$ the set of possible transitions from C_0 to C_1, risk mitigation as an optimization problem translates then to the problem of finding $C \in \mathcal{C}$ such that there exists $M_{C_0,C} \in \mathcal{M}(C_0, C)$ such that $R(\Omega, C) < R(\Omega, C_0) + \kappa(M_{C_0,C})$. Of course, while this description is more accurate and complete, it significantly increases the search space of the problem.

3.3. Transitions and Emergency Response

By ensuring that $M_{C_0,C_1}(t) \in \mathcal{C} \; \forall t \in [0,1]$, we are acknowledging the fact that each intermediate stage of the transition is a configuration in and on itself, potentially with its own associated risk assessment. While this may not seem to be particularly relevant for long-term risk assessment (unless the material time to complete the transition is comparable with the expected occurrence timescale of the hazardous events), the significance of this formulation becomes evident when considering its application to emergency responses.

Consider a small-scale example such as a single building and its fire hazard. During an emergency (one or more fires have started), the response might include an evacuation plan that, in our formulation, maps to a *transition* from an initial configuration C_0 (in which people are distributed e.g., to their habitual workplaces within the building) to a new configuration C_1 in which no people are left in the building. The transition itself will involve, at every instant in time, a new distribution of people in the building, along the established evacuation routes. However, the choice of the evacuation routes (i.e., the choice of transition M_{C_0,C_1}) has an impact on the inherent risk associated *with the transition itself*, due to the different distribution of people along them at each moment during the evacuation.

Note also that, in such a case, the pointwise risk function r itself would not be fixed, so a more sophisticated formulation that takes this into account would be necessary to complete the mathematical formulation necessary for the design of the emergency response.

4. Discussion

The assessment of risk involves determining the probability of a hazard occurring and estimating the consequences through the quantification of exposure and vulnerability, while mitigation refers to any action aiming at reducing the risk, and includes prevention, preparedness, and response. Prevention is focused on a conscientious land-use planning in order to reduce exposure. Preparedness includes all strategies to better understand the hazardous phenomenon in order to limit its impact, like the development of monitoring and early warning systems. Response consists of the design of actions to contain the threat and for the possible evacuation, reducing vulnerability.

The formalization of risk assessment introduced here does not invalidate the more informal approaches normally adopted for natural hazards, but extends them in such a way that the informal approach can be formally recognized as a numerical approximation stemming from epistemic limits [22] or the need to compromise between accuracy and computational complexity.

For example, the digital elevation and surface models used as input to numerical fluid dynamics computational models typically employed in hazard and risk assessment for geophysical flows (such as floods [23], landslides [24], pyroclastic density currents [25], or lava flows [26]) are piece-wise constant approximations of the reality, whose higher or lower horizontal resolution and vertical accuracy can influence the models' output used in hazard assessment [27]. Hazard maps for these phenomena are typically assembled from a large number of such simulations: the choice of the combinations of initial conditions, source location, geometry, etc. provide a discretized approximation of all the possible scenarios that may impact the region of interest [28] whose combination is a numerical approximation of the integral form (8).

We expect that this kind of insight may be useful in the selection of the representative scenarios for the given problem space, with an eye on well-established numerical integration schemes that may provide higher accuracy or lower computational loads, such as the Clenshaw–Curtis [29] or Gauss–Kronrod [30] quadrature formulas rather than the simpler rectangle formula typically adopted when choosing scenarios on a regular distributed grid.

The main benefit of the formalization proposed here, however, is in the more refined functional dependency proposed between the components of risk and the underlying decision variables on which policy makers should act to improve prevention, preparedness, and response.

The classification of these variables (environmental properties, elements at risk without direct influence on hazard, elements at risk with an influence on hazard, and resilience of the elements at risk) can provide insights on the extent to which each of them impacts the final risk assessment, and thus guide the decision-making process in risk management and mitigation.

For example, the category described by the family $\mathcal{B}$ of at-risk elements with an impact on hazard (such as buildings and roads in the case of lava flows hazard) presents the unique property of influencing risk assessment through all three of its components (hazard, exposure, and vulnerability), giving it a potentially higher priority over variables in other categories. This is particularly important in land-use and zoning plans (prevention), for which policies should focus not only on prioritizing construction in low-hazard areas, but also on favoring designs with higher resilience (preparedness), as well as ensuring that the associated infrastructure does not amplify hazard itself in the areas of interest (notoriously, for example, roads can become a preferential course for geophysical flows, directing them towards densely populated areas that would otherwise be less threatened by these hazardous phenomena).

Such policies must of course take into account a cost–benefit analysis. While these may be considered statically in the planning stages, risk management for existing distributions of elements (population, infrastructure, etc) must be considered in a dynamic sense. For the scientific community, this implies not only that no risk mitigation result should be considered complete without an indication of the practical means by which the mitigating results may be achieved, including an associated estimate of the possible costs, but also that the design of strategies to minimize such costs should be considered valuable results in their own right.

As an example, in the response to the hazard associated with geophysical flows, this means that the optimality in the design of barriers (or other diversion mechanisms) and their placement should take into account not only the effectiveness of the obstruction/ diversion *per se*, but also the costs of construction and deployment. More importantly, it also means that additional research opportunities can be found in the reduction of such costs, for example by devising deployment strategies that minimize both storage and transport (a classic application of domination problems from graph theory).

For decision makers, three action points should thus be encouraged, to help the scientific community in providing more effective results: (i) advertise the costs of current strategies for risk management and mitigation; (ii) minimize administrative costs of current and future strategies; and (iii) foster interdisciplinary collaboration between natural sciences, engineering, and mathematics to maximize the usefulness of research products.

5. Conclusions

The informal expression commonly used for risk assessment in the context of natural hazards can be formalized in a way that exposes more clearly the relation between the fundamental building blocks of hazard, exposure, and vulnerability, while still reducing to the informal expression with the appropriate simplifications. For the sake of brevity, we have shown here the formalization for the case of a single risk with no intensity information, but the same approach can be used by including intensity and multiple interacting risks, at the cost of an even higher complexity in the functional dependencies between the components of the final formulation.

While the variables contributing to hazard, exposure, and vulnerability have been presented here in the most abstract and general form, any expert in the field will be able to match easily the data and models they operate with, and the corresponding sets and functions discussed in this perspective. This correspondence will help identify the decision variables (*«what can be acted upon to decrease risk?»*) and their weight in the formulation (*«how effective will it be to act on this variable to reduce risk?»*). The actual extent to which the mathematical formulation presented here can be used depends then on the accuracy and completeness of the data and information available.

A key novelty of our perspective is the choice to view risk mitigation not only (or primarily) in terms of its final effect (the scaling or reduction factor previously discussed in the literature), but as a dynamic process, whose duration and costs have a distinct influence on the final results. While this increases the complexity of evaluation of risk mitigation measures, it shifts the attention towards a more realistic, and thus hopefully more useful, approach, where answering the question *«how do we get there?»* is of equal, if not higher, importance as *«where do we want to get?»*

Author Contributions: G.B. defined the original idea, and A.C. and G.G. contributed to its development. All authors contributed to the preparation of the manuscript. All authors have read and agreed to the published version of the manuscript.

Funding: This work was supported by the research project "SHIELD—Optimization strategies for lava flow risk reduction at Etna volcano" (Bando di Ricerca Libera 2019 of INGV).

Institutional Review Board Statement: Not applicable.

Informed Consent Statement: Not applicable.

Data Availability Statement: Not applicable.

Conflicts of Interest: The authors declare no conflict of interest

References

1. Unesco. Report of Consultative Meeting of Experts on the Statistical Study of Natural Hazards and Their Consequences. 1972, Document SC/WM/500. Available online: https://unesdoc.unesco.org/ark:/48223/pf0000001657 (accessed on 15 October 2022).
2. Fournier d'Albe, E.M. Objectives of volcanic monitoring and prediction. *J. Geol. Soc.* **1979**, *136*, 321–326.
3. Lockwood, J.P.; Hazlett, R.W. *Volcanoes: Global Perspectives*; Wiley–Blackwell: Chichester, UK, 2010.
4. Kaplan, S.; Garrick, B. On The Quantitative Definition of Risk. *Risk Anal.* **1981**, *1*, 11–27. [CrossRef]
5. Singh, V.P.; Jain, S.K.; Tyagi, A. *Risk and Reliability Analysis*; ASCE Press: Reston, VA, USA, 2007.
6. Martí Molist, J. *Assessing Volcanic Hazard: A Review*; Oxford Handbooks; Oxford University Press: Oxford, UK, 2017.
7. Lowrance, W.W. *Of Acceptable Risk: Science and the Determination of Safety*; W. Kaufmann: Los Altos, CA, USA 1976.
8. Haimes, Y.Y. *Risk Modeling, Assessment, and Management*; John Wiley & Sons, Ltd.: Hoboken, NJ, USA, 2004.
9. Woo, G. *The Mathematics of Natural Catastrophes*; Imperial College Press: London, UK, 1999. [CrossRef]
10. Taylor, C.; VanMarcke, E. (Eds.) *Acceptable Risk Processes: Lifelines and Natural Hazards*; American Society of Civil Engineers: Technical Council on Lifeline Earthquake Engineering; ASCE Press: Reston, VA, USA, 2002.
11. Uddin, N.; Ang, A.H. (Eds.) *Quantitative Risk Assessing (QRA) for Natural Hazards*; ASCE Council on Disaster Risk Management; ASCE Press: Reston, VA, USA, 2011. [CrossRef]
12. Kiremidjan, A.; Stergio, E.; Lee, R. Quantitative Earthquake Risk Assessment. In *Quantitative Risk Assessing (QRA) for Natural Hazards*; Uddin, N., Ang, A.H., Eds.; ASCE Press: Reston, VA, USA, 2011.
13. Cappello, A.; Ganci, G.; Bilotta, G.; Corradino, C.; Hérault, A.; Del Negro, C. Changing Eruptive Styles at the South-East Crater of Mount Etna: Implications for Assessing Lava Flow Hazards. *Front. Earth Sci.* **2019**, *7*, 213. [CrossRef]

14. Chevrel, M.O.; Favalli, M.; Villeneuve, N.; Harris, A.J.L.; Fornaciai, A.; Richter, N.; Derrien, A.; Boissier, P.; Di Muro, A.; Peltier, A. Lava flow hazard map of Piton de la Fournaise volcano. *Nat. Hazards Earth Syst. Sci. Discuss.* **2020**, *2020*, 1–33. [CrossRef]
15. Anagnostopoulos, S.A. Pounding of buildings in series during earthquakes. *Earthq. Eng. Struct. Dyn.* **1988**, *16*, 443–456.
16. Frascadore, R.; Ludovico, M.D.; Prota, A.; Verderame, G.M.; Manfredi, G.; Dolce, M.; Cosenza, E. Local Strengthening of Reinforced Concrete Structures as a Strategy for Seismic Risk Mitigation at Regional Scale. *Earthq. Spectra* **2015**, *31*, 1083–1102.
17. Ang, A.H. An application of quantitative risk assessment in infrastructures engineering. In *Quantitative Risk Assessing (QRA) for Natural Hazards*; Uddin, N., Ang, A.H., Eds.; ASCE Press: Reston, VA, USA, 2011.
18. Anonymous. Prediction of the Haicheng earthquake. *Eos Trans. Am. Geophys. Union* **1977**, *58*, 236–272. [CrossRef]
19. Wang, K.; Chen, Q.F.; Sun, S.; Wang, A. Predicting the 1975 Haicheng Earthquake. *Bull. Seismol. Soc. Am.* **2006**, *96*, 757–795.
20. Wang, H.; Mostafizi, A.; Cramer, L.A.; Cox, D.; Park, H. An agent-based model of a multimodal near-field tsunami evacuation: Decision-making and life safety. *Transp. Res. Part C Emerg. Technol.* **2016**, *64*, 86–100. [CrossRef]
21. Banba, M.; Shaw, R. (Eds.) *Land Use Management in Disaster Risk Reduction: Practice and Cases from a Global Perspective*; Springer: Tokyo, Japan, 2017. [CrossRef]
22. Beven, K.J.; Almeida, S.; Aspinall, W.P.; Bates, P.D.; Blazkova, S.; Borgomeo, E.; Freer, J.; Goda, K.; Hall, J.W.; Phillips, J.C.; et al. Epistemic uncertainties and natural hazard risk assessment—Part 1: A review of different natural hazard areas. *Nat. Hazards Earth Syst. Sci.* **2018**, *18*, 2741–2768. [CrossRef]
23. Mosquera-Machado, S.; Ahmad, S. Flood hazard assessment of Atrato River in Colombia. *Water Resour. Manag.* **2007**, *21*, 591–609. [CrossRef]
24. Hürlimann, M.; Copons, R.; Altimir, J. Detailed debris flow hazard assessment in Andorra: A multidisciplinary approach. *Geomorphology* **2006**, *78*, 359–372. [CrossRef]
25. Clarke, B.; Tierz, P.; Calder, E.; Yirgu, G. Probabilistic Volcanic Hazard Assessment for Pyroclastic Density Currents from Pumice Cone Eruptions at Aluto Volcano, Ethiopia. *Front. Earth Sci.* **2020**, *8*, 348. [CrossRef]
26. Del Negro, C.; Cappello, A.; Bilotta, G.; Ganci, G.; Hérault, A.; Zago, V. Living at the edge of an active volcano: Risk from lava flows on Mt. Etna. *GSA Bull.* **2019**, *132*, 1615–1625.
27. Bilotta, G.; Cappello, A.; Hérault, A.; Del Negro, C. Influence of topographic data uncertainties and model resolution on the numerical simulation of lava flows. *Environ. Model. Softw.* **2019**, *112*, 1–15. [CrossRef]
28. Cappello, A.; Vicari, A.; Del Negro, C. Retrospective validation of a lava-flow hazard map for Mount Etna volcano. *Ann. Geophys.* **2011** . [CrossRef]
29. Gentleman, W.M. Implementing Clenshaw–Curtis Quadrature, I Methodology and Experience. *Commun. ACM* **1972**, *15*, 337–342. [CrossRef]
30. Caliò, F.; Gautschi, W.; Marchetti, E. On Computing Gauss–Kronrod Quadrature Formulae. *Math. Comput.* **1986**, *47*, 639–650. [CrossRef]

Review

Complex Systems, Emergence, and Multiscale Analysis: A Tutorial and Brief Survey

Jianbo Gao [1,2,*] and Bo Xu [2]

1 Center for Geodata and Analysis, Faculty of Geographical Science, Beijing Normal University, Beijing 100875, China
2 Institute of Automation, Chinese Academy of Sciences, Beijing 100190, China; xubo@ia.ac.cn
* Correspondence: jbgao.pmb@bnu.edu.cn

Abstract: Mankind has long been fascinated by emergence in complex systems. With the rapidly accumulating big data in almost every branch of science, engineering, and society, a golden age for the study of complex systems and emergence has arisen. Among the many values of big data are to detect changes in system dynamics and to help science to extend its reach, and most desirably, to possibly uncover new fundamental laws. Unfortunately, these goals are hard to achieve using black-box machine-learning based approaches for big data analysis. Especially, when systems are not functioning properly, their dynamics must be highly nonlinear, and as long as abnormal behaviors occur rarely, relevant data for abnormal behaviors cannot be expected to be abundant enough to be adequately tackled by machine-learning based approaches. To better cope with these situations, we advocate to synergistically use mainstream machine learning based approaches and multiscale approaches from complexity science. The latter are very useful for finding key parameters characterizing the evolution of a dynamical system, including malfunctioning of the system. One of the many uses of such parameters is to design simpler but more accurate unsupervised machine learning schemes. To illustrate the ideas, we will first provide a tutorial introduction to complex systems and emergence, then we present two multiscale approaches. One is based on adaptive filtering, which is excellent at trend analysis, noise reduction, and (multi)fractal analysis. The other originates from chaos theory and can unify the major complexity measures that have been developed in recent decades. To make the ideas and methods better accessed by a wider audience, the paper is designed as a tutorial survey, emphasizing the connections among the different concepts from complexity science. Many original discussions, arguments, and results pertinent to real-world applications are also presented so that readers can be best stimulated to apply and further develop the ideas and methods covered in the article to solve their own problems. This article is purported both as a tutorial and a survey. It can be used as course material, including summer extensive training courses. When the material is used for teaching purposes, it will be beneficial to motivate students to have hands-on experiences with the many methods discussed in the paper. Instructors as well as readers interested in the computer analysis programs are welcome to contact the corresponding author.

Keywords: complexity; emergence; chaos; fractal; power-law; multiscale analysis; social complexity

Citation: Gao, J.; Xu, B. Complex Systems, Emergence, and Multiscale Analysis: A Tutorial and Brief Survey. *Appl. Sci.* **2021**, *11*, 5736. https://doi.org/10.3390/app11125736

Academic Editor: Itzhak Katra

Received: 12 May 2021
Accepted: 16 June 2021
Published: 21 June 2021

Publisher's Note: MDPI stays neutral with regard to jurisdictional claims in published maps and institutional affiliations.

1. Introduction

The ever increasing amount of big data in science, engineering, and society, including meteorological, hydrological, ecological, environmental, as well as various kinds of biomedical, manufacturing, e-commerce, and government management data, has fueled enormous optimism among researchers, entrepreneurs, government officials, the media, and the general public [1,2]. It is now hoped that by recording and analyzing the errors of all the components of a sophisticated machine, one can quickly diagnose and then fix its malfunctioning. When one is sick, one hopes that in the near future, with all the increasingly detailed data about oneself, including genomic, cellular, clinical, psychological, and environmental data, one may promptly get optimized treatment. One also hopes to identify the most promising stocks by collecting and analyzing all the relevant economic data and then investing on them.

Such optimism is not entirely unfounded, as big data indeed has brought some pleasant surprises to science and society. For example, a good online shopping system can quickly and fairly accurately infer what an online shopper is looking for by analyzing the shopper's online behavior in real time. By analyzing the tweets about major natural disasters, key information of disasters can be accurately obtained [3]. Google Flu Trends did an impressive job in predicting the 2008 influenza [4].

While the big data showcase does not stop at the above successful examples, it is important that one is not carried away by those successes. In fact, many more not so successful cases also exist. For example, right after 2008, Google Flu Trends over-predicted influenza outbreaks, and by 2012, the error was by as much as a factor of two [5], which then prompted Google to give up the predictor. The box office price of the film "Golden Times", which was first released in China during the National Holiday, 1 October 2014, was only slightly more than 40 million, while Baidu, the leading Chinese web services company, predicted it to be about 200–230 million. The poor prediction by Baidu made a reviewer of the film to lament that big data may not be dependable [6]. Of course, we have to add the failed prediction of the Trump presidency in 2016 by many predictors, whose implications to the Americas, and even the world's politics, are almost unfathomable.

Among the most important values of big data analysis are to detect changes in system dynamics (e.g., detect and understand abnormal behaviors) and to help science to extend its reach (and most desirably, to possibly uncover new fundamental laws). This includes timely diagnosis and treatment of various kinds of diseases in health care, proper prediction of regime changes in weather and climate patterns, timely forewarning of natural disasters, and timely detection and fixing of malfunctioning of various kinds of devices, infrastructure, and software in the field of operation and maintenance [7–10], among many others. Understandably, abnormal behaviors cannot be expected to occur frequently, and thus the relevant data may not be so abundant that direct application of machine-learning based approaches will always be very rewarding. In those situations, the systems often generate data with complex characteristics including long-range spatial–temporal correlations, extreme variations (sometimes caused by small disturbances), time-varying mean and variance, and multiscale analysis (i.e., different behavior depending on the scales at which the data are examined). Such situations have been increasingly manifesting themselves in science, engineering, and society. To adequately cope with these situations, it is often beneficial to resort to complexity science to analyze the relevant data. In fact, when dealing with such highly challenging situations, many analyses using machine-learning based approaches may be considered pre-processing of the data or the first step that can facilitate further application of complexity-based approaches, or as post-processing of the features obtained through multiscale analysis. An excellent article along this line (more precisely, study of segmental organization of the human genome by combining complexity with machine learning approaches) has recently been reported by Karakatsanis et al. [11]. In short, the complex behaviors in nature, science, engineering, and society must be infinite. To help one to peek into the infinity of the complex behaviors, going beyond statistical analysis and machine-learning by resorting to the type of mathematics that embodies an element of infinity will often be beneficial.

At this point, it is important to pause for a moment to discuss a peculiar phenomenon: while many consider complexity science to be very useful, some others doubt its relevance to reality. Why is this so? The basic reason is that in complexity research, conceptual thinking, simulational study, and applications have not been well connected. For example, Science magazine dedicated the April 1999 issue to Complex Systems. A number of leading experts in their respective fields, including chemistry, physics, economics, ecology, and biology, expressed their views on the relevance/importance of complexity science in their fields. While the special issue is influential in making some concepts of complex systems known to a wider research community and even the general public, it does little in teaching readers how to solve real-world problems. This may have contributed to the waning of enthusiasm in complexity science research in the subsequent years, as most readers cannot see how complexity science can help solve their problems. Fortunately, the tide appears to have been reversed (please see recent reviews on complexity theory and leadership practice [12] and health [13]).

The purpose of this article is to convey how the many concepts in complexity science can be effectively applied to help one formulate stimulating problems pertinent to the data and the underlying system. We will particularly focus on multiscale approaches. They are the key to find scaling laws from the data. With the scaling laws, we can then find defining parameters/properties of the data and eliminate spurious causal relations in the data. The latter can help to shed some light on a new generation of AI, which is based on correlation/causality rather than pure probabilistic thinking [14]. To better serve our goal, we will discuss various kinds of applications right after a concept/method is introduced. Our goal here is to fully arouse readers' interest in the materials covered, and to equip them with a set of widely applicable concepts and methods to help solve their own interesting problems.

2. Basics of Complex Systems and Emergence

2.1. Complex Systems and Emergence: Working Definitions

To better understand which systems can be considered complex, we first explain how complexity is quantified. There are two major types of measures. One is called Deterministic complexity, which increases with the degree of randomness. See Figure 1 (left). Widely used measures in this category include Shannon entropy [15], Kolmogorov–Sinai (KS) entropy [16,17], Kolmogorov–Chaitin complexity [18–20], and the Lempel–Ziv (LZ) complexity [21]. The other is called *Structural complexity*. Here, the measure attains a maximal value for an intermediate level of randomness. See Figure 1 (right).

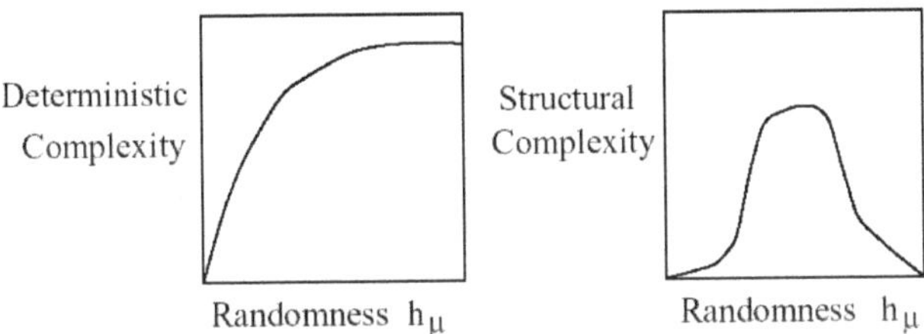

Figure 1. Deterministic vs. structural complexity.

Let us now examine the main features of a complex system. It is often thought that a complex system must consist of many interconnecting components or parts. The individual components together with their dynamics could be quite simple. The system as a whole, however, must exhibit complex dynamics. Note that with this view, a pendulum with chaotic behavior is no longer considered a complex system. In addition, note that some researchers (e.g., Kastens et al. [22]) advocate to assign a complex system with many more quantifiable features, such as feedback loops, multiple inputs and multiple outputs, non-Gaussian distributions of the outputs, nonlinear interactions, multiple stable states, fractal and chaotic behaviors, self-organized criticality, hierarchy, and so on. Our view is that it is extremely rare for a single system to simultaneously possess so many distinguished properties at the same time. Therefore, simpler definitions that give more room and freedom to think and work could be more beneficial.

Complex systems often defy pure statistical analysis. To illustrate the idea, let us discuss an author (JB)'s personal experience. JB worked at Guangxi University in Nanning for a few years. The campus was full of natural wonders, with flowers blossoming and many kinds of tropical and subtropical fruits dangling on trees all year long. Thus, JB and many of his friends truly enjoyed the campus. Approximately 100,000 people, including University employees and students, lived on campus. JB used to buy vegetables and meat at a farmer's market in the east campus of the University. Although the farmer's market was a bit shabby, it was in a convenient location and was visited by a lot of customers everyday. In the market, there was a pork meat seller who normally would sell out all the meat within 2.5 h before 11 am in the morning. Around October 2017, the market was relocated to a new place about 7 min walk from the original site. Surprisingly, the number of customers to the market dropped considerably. As a result, the pork meat seller would still be selling meat around 1–2 pm. After that, the seller had to take the meat to some fast food restaurants, as otherwise the pork, not refrigerated, would become spoiled and smelly. Surely, quite a few fruit and vegetable sellers eventually gave up. Such dramatic drop in customer number is very difficult to predict with statistical models, however sophisticated they are. One can readily see that to truly understand the phenomenon, one has to systematically analyze the dynamics of the customer behavior by considering diverse factors such as the variety, cost, and freshness of food; convenience of the market; competitors of the market; and customer psychology.

Next, let us consider emergence in complex systems. Emergence is a bulk property of the system involving many of the interacting components of the system [23,24]. As a result, its scale usually is much larger than that of the individual component. Outstanding examples of emergence include the spiral galaxy [25], the great red spot of Jupiter [26], hurricanes, tornadoes, phase transitions and critical phenomena [27], bird flocking [28,29], fish schooling [30–33], sand dunes [34], mass parades or protests, and bursts of anger (where many neurons in certain regions of the brain fire synchronously). Less frequently mentioned examples of emergence that are of tremendous significance to our society include the many innovations in technology, including Internet-enabled platform economy, where large numbers of sellers and buyers interact and transact through the platform. Among the important and fascinating questions concerning such platform-enabled emergent behaviors are to identify the conditions under which such services will become attractive and widely adopted, and to quantify the generic statistical properties underlying such services.

Often it is thought that for a system to exhibit an emergent behavior, it must have a hierarchical structure. This thinking is, however, not quite consistent with the fact that simple models with local interaction rules may simulate certain emergent behaviors quite well, including bird flocking and fish schooling [28–33].

We now consider *Complex giant systems*, a notion that has been widely discussed in many fields in China, including physics, mathematics, philosophy, and humanities. As fluid motions including turbulence are considered not to belong to such systems, social systems become the prototypical model here. While a big social system is certainly a giant system, as it contains so many individuals and their interactions, it is not necessarily a complex system. For example, in an autocratic state where governance is strictly hierarchical, from top to bottom, and all means of feedback, such as election, parade protests, and so on, are prohibited, the social dynamics of a specific layer are only directionally connected to its nearest upper and lower layers (driven and driving, respectively). This is the consequence of lacking a persistent negative feedback loop in the society. As a result, the complexities of such societies cannot be considered very high, as those societies do not possess well-developed dynamics that have to be enabled by feedback loops. In particular, they lack many emergent behaviors that a democratic society has, such as parade protests instigated by explosions in public opinion.

In the study of complex systems, different researchers may have different emphasis [35,36]. One school focuses on the mathematics and mechanics of complex systems. Here, one is mainly concerned about rigorous mathematical analysis of the system under study, most desirably starting from fundamental governing equations of the system, and using mechanics (quantum, classical, and statistical) to analyze the system. While in principle a living organism (e.g., the human body) may be modeled by a large set of differential equations with a lot of controlling parameters, with the values of the parameters indicating healthy or diseased states, this may not be achieved in the near future. To better exploit the unprecedented opportunities provided by the explosion of data in all areas of science, technology, and society, in this article we adopt a data-driven approach to study complex systems. Among the many techniques to analyze data is distribution analysis. As the power law is a distribution with many interesting properties that are not shared by most commonly used distributions in conventional statistical analysis, in the next subsection we will discuss the power law and the related heavy-tailed distributions.

2.2. Power Law and Heavy-Tailed Distributions

In contrast to Gaussian, exponential, and other thin-tailed distributions that have a well-defined scale, a power law distribution does not have a scale. It has been observed in various kinds of physical, biological, technological, and and social systems. Well-known examples include the distribution of word frequency, web hits, citations of scientific papers, telephone calls, copies of books sold, diameter of moon craters, intensity of solar flares, intensity of wars, magnitude of earthquakes, wealth of the richest people, and population of cities [37].

A power law distribution can be expressed by its probability density function (PDF) [38]

$$f(x) \sim x^{-\alpha-1}, \quad x \to \infty, \tag{1}$$

or equivalently by the complementary cumulative distribution function (CCDF) [38]

$$P[X \geq x] \sim x^{-\alpha}, \quad x \to \infty. \tag{2}$$

Notice here the emphasis that $x \to \infty$. An interesting property of the power law distribution is that for a given α, its moments with order higher than α do not exist. Therefore, when $0 < \alpha < 2$, the variance and all moments higher than the second order do not exist, and when $0 < \alpha \leq 1$, even the mean is infinite. When the power law relation extends to the entire range of the allowable x, we have the Pareto distribution [39]:

$$P[X \geq x] = \left(\frac{b}{x}\right)^{\alpha}, \quad x \geq b > 0, \ \alpha > 0, \tag{3}$$

Here, α is the shape parameter, and b the location parameter. In the discrete case, the Pareto distribution is called the Zipf distribution, which provides an excellent description between the frequency of any word in a corpus of natural language and its rank in the frequency table.

Somewhat related to the Zipf distribution is another distribution called Benford's law [40], which is about the probability of occurrence of leading digits $d \in \{1, 2, \cdots, 9\}$,

$$P(d) = \log_{10}(d+1) - \log_{10}(d) = \log_{10}\left(1 + \frac{1}{d}\right) \tag{4}$$

A good mechanism for explaining the uneven distributions stipulated by Benford's law has been proposed in [41].

Benford's law has been used for evaluating possible fraud in accounting data [42], legal status [43], election data [44–46], macroeconomic data [47], price data [48], etc. From Equation (4), we observe that beyond the small digits, the probability approximately approaches the Zipf distribution with $\alpha = 1$,

$$P(d) = \log_{10}\left(1 + \frac{1}{d}\right) \sim d^{-1}/\ln(10), \quad d = 3, 4, \cdots, 9 \tag{5}$$

2.2.1. Pareto Principle or the 80/20 Rule

The 80/20 rule or the Pareto principle was first put foreword by the Italian economist Vilfredo Pareto in 1896: approximately 80% of the land in Italy was owned by 20% of the population. The rule later more generally applies, as approximately 80% of the wealth in a society is owned by 20% of the population. It can be derived from the Pareto distribution with a specific parameter α. To see this, we can demonstrate as follows.

Suppose in a society the number of people with wealth at least x follows a power law:

$$N(X \geq x) = Ax^{-\alpha} \tag{6}$$

where A is some coefficient. If the minimal wealth of a person is x_0, then the total number of people in the society can be denoted as $N(X \geq x_0)$, and

$$N(X \geq x_0) = Ax_0^{-\alpha} \tag{7}$$

Their ratio gives the percentage of rich people with wealth at least x and is equal to

$$\left(\frac{x}{x_0}\right)^{-\alpha} \tag{8}$$

The proability density function for a person to have wealth of x is

$$f(x) = \alpha x^{-\alpha-1} \tag{9}$$

Thus, the society's total wealth is

$$\int_{x_0}^{\infty} \alpha x^{-\alpha-1} x\, dx \tag{10}$$

and the total wealth of rich people with at least wealth x is given by

$$\int_{x}^{\infty} \alpha x^{-\alpha-1} x\, dx \tag{11}$$

Note these two integrals are from x_0 to ∞ and x to ∞, respectively. The ratio between the latter and the former is given by

$$\left(\frac{x}{x_0}\right)^{1-\alpha} \tag{12}$$

Solving for α by letting the ratios given by Equations (8) and (12) to be 0.2 and 0.8, respectively, we find

$$\alpha = \ln 5 / \ln 4 \approx 1.16 \tag{13}$$

As a non-wealthy person might not be in a good mood or even become cynical when hearing about the 80/20 rule, it is good to be reminded of one of two insights offered by Will Durant and Ariel Durant, the famed authors of the prominent history book *The Story of Civilization*: "For in modern states the men who can manage men manage the men who can manage only things; and the men who can manage money manage all [49]. ... As everywhere, the majority of abilities was contained in a minority of men, and led to a concentration of wealth" [50] The lesson here is that whatever one does, if one does not want to be one of the 80% of the people, then one cannot be a follower; instead, one has to strive to do new things, as only in those situations, can one have 80% rewards with 20% efforts.

2.2.2. Simulation and Parameter Estimation

To simulate a Pareto distributed random variable U, we can associate U with an outcome of a random experiment. The same outcome may also be represented by the value of another random variable X. The probability of an event of the experiment is then either $dF_U(u) = f_U(u)du$ or $dF_X(x) = f_X(x)dx$, where $F_U(u)$ and $F_X(x)$ are the cumulative distribution functions (CDFs) for the U and X, while $f_U(u)$ and $f_X(x)$ are the PDFs. Then we have

$$\int_a^X dF_X(x) = \int_0^U du. \tag{14}$$

Since $F_X(x)$ is monotonically nondecreasing, its inverse function exists. We then have

$$X = F_X^{-1}(U). \tag{15}$$

Now suppose U is a uniform $[0, 1]$ random variable, while X is a Pareto random variable, then

$$X = bU^{-\frac{1}{\alpha}}. \tag{16}$$

The most important parameter of the Pareto distribution is the exponent α. To estimate it, we only need to notice that $\ln P[X \geq x]$ vs. $\ln x$ is a linear function, with the slope being $-\alpha$. When estimating α from a finite set of data points, it is important to first take the logarithm of x, then estimate the CCDF for $\ln x$, and finally check if the logarithm of CCDF has a linear relation with $\ln x$. If one straightforwardly estimates a PDF or CCDF for the original data, then take log-log of both axes to estimate α, one will often get a very inaccurate or even wrong estimation. The reason is many of the small intervals used for counting the number of data points x falling within them will be empty.

2.2.3. Reasons Why the Power Law Is Favored in Modeling

Two reasons make the power law extremely important in complexity science. One reason is that it embodies the notion of self-similarity, and thus is the natural mathematical tool for describing fractal phenomena. The other reason is that it often signifies great risk, due to infinite variance or even mean. To understand the first reason, imagine a large room with a lot of balls flying around. See Figure 2.

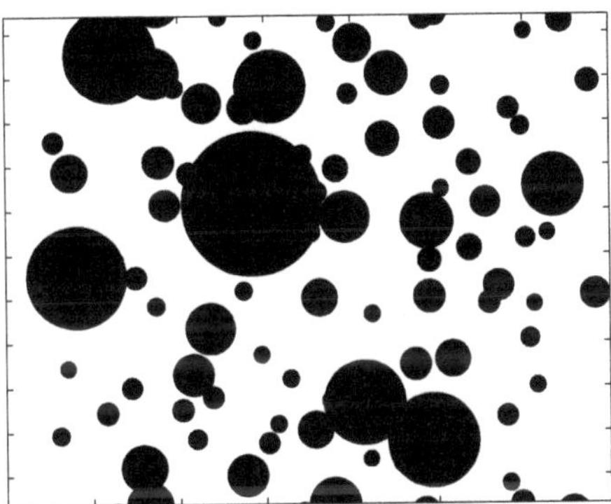

Figure 2. Pareto-distributed balls, where $\alpha = 1.8$.

Assume the size of the balls follows a power law distribution,

$$p(r) \sim r^{-\alpha}. \tag{17}$$

When we observe the balls with our naked eyes, we normally will only pay more attention to the balls of certain size ranges—large balls will block our vision, and very small balls cannot be seen. Now assume that our eyes are comfortable with the scales r_0, $2r_0$, $r_0/2$, etc. Our perception is determined by the relevant abundance or the ratio of the balls of sizes $2r_0$, r_0, and $r_0/2$:

$$p(2r_0)/p(r_0) = p(r_0)/p(r_0/2) = 2^{-\alpha}. \tag{18}$$

It is independent of r_0. Now suppose we view the balls through a microscope with a magnifying power of 100, so now our eyes will be focusing on the balls with scales $2r_0/100$, $r_0/100$, $r_0/200$, etc. The ratio of the balls on those scales will again be independent of the scale $r_0/100$. A perception independent of the scale is the essence of self-similarity.

The second reason that the power law is associated with higher risks is easier to understand, since a power law distribution has infinite variance when $0 < \alpha < 2$ and even infinite mean when $0 < \alpha \leq 1$. Here, on one hand, one has to have some awe with the power law, as otherwise the cost could be tremendous. For example, during financial crises or economic downturns, the loss of the listed companies follows a power law distribution that is even heavier than the distribution of the gains of all profitable companies [51,52]. As further examples, the size of forest fires and volcanic eruptions also follow power law distributions (see Figures 3 and 4), which has obvious implications for fire fighting or observation of volcanoes—going too close to the sites could easily lead to casualties. However, on the other hand, one also has to be mindful that having infinite variance or mean is not always associated with the severity of natural disasters. An important counterexample is flooding, as it has been found that stream flow of rivers in dry seasons (especially in desert areas) is better described by power law distributions, while that in wet seasons is better described by log-normal distributions [53]. In deserts, surely flooding does not constitute a major risk.

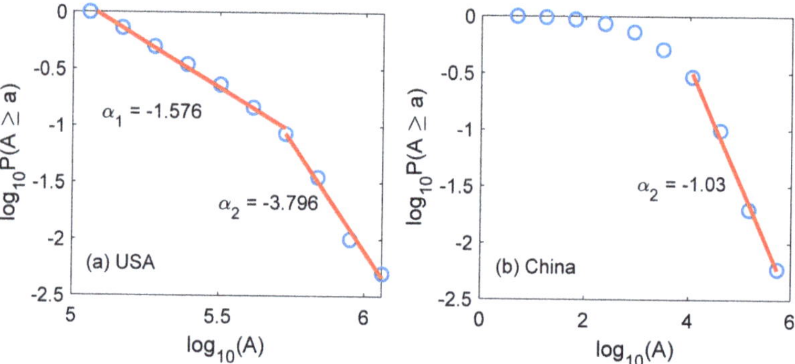

Figure 3. Complementary cumulative distribution function (CCDF) for the forest fires in USA and China, where the size of a fire is measured by its area A. The data for USA are the sizes of individual fires from 1997 to 2018, while those for China are the total annual size of forest fires in the 30 provinces from 1998 to 2017.

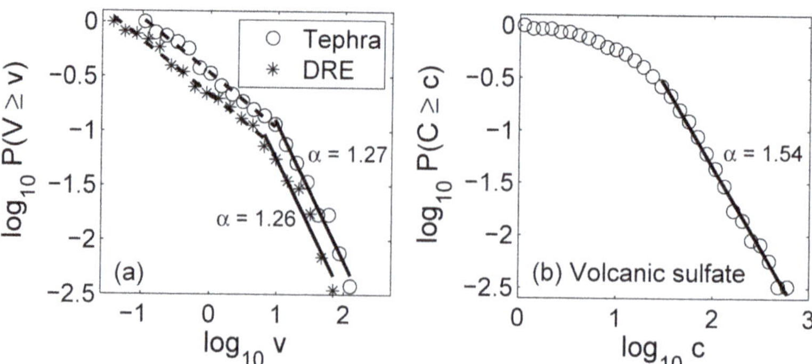

Figure 4. Complementary cumulative distribution function (CCDF) for the products of volcanic eruptions in the Holocene: (**a**) tephra volume (km^3) and dense rock equivalent (DRE) (km^3), and (**b**) volcanic sulfate (data were from [54]).

2.2.4. Mechanisms for Power Laws

The prevalence of power laws calls for development of models to explain the mechanism. Various models have been proposed, including Tsallis non-extensive statistics [55–57]. For a systematic discussion, we refer to Chapter 11 of [38]. Here, we note two of them, which appear to be relevant to many different scenarios and thus may better stimulate readers to readily find mechanisms when they find power laws from their data. One model is related to spatial heterogeneity and resource allocation (or availability). It is provided by the model that superposition of exponential distributions with different parameters can give rise to power law distributions. The other reflects the underlying local dynamics of the problem to some degree, and thus is in some sense more thought-provoking. The most well-known example of this class is perhaps the scale-free power law network model [58]. Another example is related to social segregation and crimes in a society: distributions of the ratio between sex offenders and the total population in the states of Ohio and New York in the USA follow power laws, as shown in Figure 5 [59]. While intuitively this must be driven by crimes (more concretely, sexual offenses) and instigated by laws preventing crimes, so far, however, a concrete model is still lacking. Such a model is surely worth developing in the future.

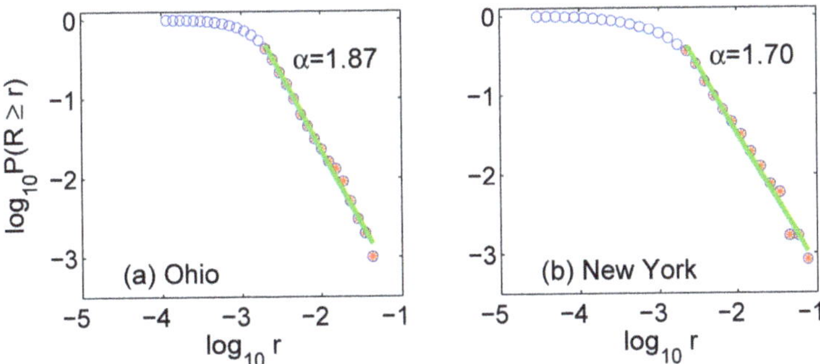

Figure 5. Distribution for the ratio between sex offenders and the total population in (a) Ohio and (b) New York (adapted from [59]).

2.3. Essentials of Chaos Theory

Many readers can easily recall observing a sinusoidal signal with an oscilloscope. Assume we are examining some production line through monitoring of some signal. An aperiodic, highly irregular time series pops up. Is the signal simply some kind of noise? Very unlikely, since our system is deterministic. Can a seemingly random signal come from a deterministic system which can be described by only a few variables instead of a random system with infinite numbers of degrees of freedom? Yes, a chaotic system can do that! Not only so, many universal behaviors behind chaos have been uncovered. These findings have fundamental, far-reaching implications in science and engineering, and thus chaos theory, relativity, and quantum mechanics are considered the three most revolutionary scientific theories of the twentieth century.

To facilitate understanding of the essentials of chaos theory, in this section, we first explain the notion of phase space and transformation, then we present the basic properties of chaos. To satisfy curious minds, we will also give a flavor of analytical thinking. Finally, we explain how to reconstruct a proper phase space from a single variable (scalar time series) and estimate the few basic metrics (called invariants) that characterize a chaotic system.

2.3.1. Phase Space and Transformation

Phase space is the arena for the evolution of a dynamical system to unfold. It is spanned by all the variables needed to fully characterize the evolution of the system. To help one to better understand the idea, let us start from a system characterized by only two state variables, X_1 and X_2. Monitoring the system often amounts to examining the waveforms of $X_1(t)$ and $X_2(t)$. One may instead try to examine the trajectory defined by $(X_1(t), X_2(t))$, where t now is treated as an implicit parameter. The space spanned by X_1 and X_2 is the phase space (or state space) we are discussing. They could be position and velocity, for example. Employing phase space facilitates one to study the dynamics of a complicated system with a geometrical viewpoint. For some dynamical systems, irrespective of initial conditions, the trajectory eventually approaches a single point; this is called a globally stable fixed point solution. Of course, the situation could be more complicated. For example, the trajectory may converge to a closed loop, again irrespective of where the trajectory starts. This is called a globally stable limit cycle. The discrete counter part of a limit cycle is a periodic motion with certain period (say N): the corresponding attractor consists of N points, and the trajectory amounts to hopping among the N points with a definite order.

To be more familiar with the concept of phase space, it is useful to examine certain experience in daily life. To illustrate the idea, suppose we were going to a meeting by a taxi. On our way, there was a traffic jam, and the taxi got stuck. Afraid of being late, we decided to call the organizer. How would we describe our situation? Usually, we would tell the organizer where we got stuck and how quickly or slowly the taxi was moving. In other words, we actually have been using the concept of phase space as part of our daily language.

Although the concept of phase space is among the most basic in dynamical systems theory, its usefulness in geographical science has yet to be seriously explored [60]. To accelerate the coming of a time that phase space becomes as basic in geographical science as in complexity science, it is helpful to discuss two potential applications of phase space in geographical science. One application is top-down, that is, to systematically think about how many independent variables are needed to fully characterize an interesting and important problem in geographical science, and how each variable can be measured. The other application is bottom-up. It is easiest to illustrate the idea by using some variables in the World Value Survey (WVS, accessed on 17 April 2021, http://www.worldvaluessurvey. org/wvs.jsp) as an example. WVS is an interesting project that explores values and beliefs of people around the globe, how the values and beliefs evolve with time, and what social and political implications they may have. Since 1981, researchers have conducted representative national surveys in almost 100 countries. During the survey, a lot of variables have been deduced. We show here that phase space offers a convenient geometrical way to visualize the data and identify co-variations of the variables. For this purpose, we choose a variable that gives three levels of religious participation for people in the nations surveyed. The other variable we choose is happiness, which is given in four levels. How are the two variables related? How different are people in different countries in terms of these two variables? To gain insights into these interesting questions, we can form a phase space spanned by these two variables. The format of the survey data determines that people surveyed in a nation will belong to one of the 12 different categories. To fully utilize the notion of space, we can associate each category with a box. Instead of putting every person belonging to that category at one single point (e.g., the center of the box), we can generate two uniformly distributed random variables as the coordinate of the person in the corresponding box. Please see Figure 6. With such a visualization, one can immediately see the abundance of each category. When WVS data of different waves (times) are used, one can then examine variation of the percentage of people in each category over time for a nation, compare among different nations, deduce functional relationships between these two variables, and classify nations in the world into different clusters. Note Figure 6 may be called phase space ensemble based visualization, where an ensemble amounts to a participant in the survey.

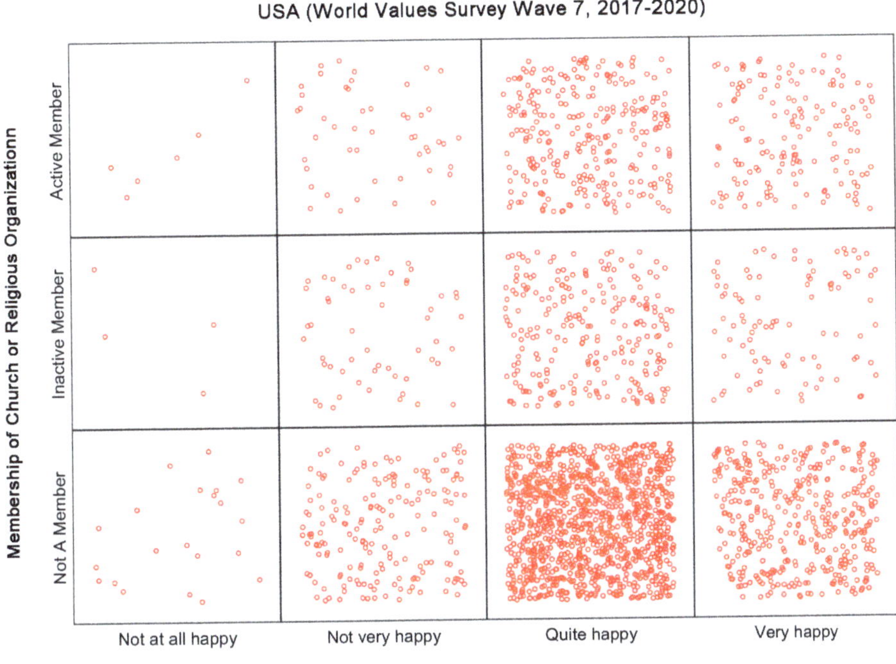

Figure 6. Phase space diagram of religious participation vs. happiness for the USA based on wave 7 of the World Value Survey Data.

Next, let us consider transformations in phase space. A good way to grasp the idea is to imagine the following situation: on a very weedy day, a little boy went outside with a sheet of paper in his hand. He grabbed a handful of sand and put it on the paper. Then he released the paper in the air. How would the sand be swept across the sky? One could even think that originally the boy had arranged the sand to resemble the face of a person. How would the face be twisted by the wind? To make this discussion more concrete, we can consider how a unit circle is transformed by the Henon map [61]:

$$\begin{aligned}
x_{n+1} &= 1 - ax_n^2 + y_n, \\
y_{n+1} &= bx_n,
\end{aligned} \tag{19}$$

where $a = 1.4, b = 0.3$. Figure 7 shows the successive (from left to right and top to bottom) images of the unit circle after $n = 1, \cdots, 5$ iterations. Note that the fifth image is basically the Henon attractor one can find in textbooks, journal papers, or certain web sites. It is usually obtained by choosing an arbitrary initial condition and iterate the Henon map long enough. If the trajectory does not diverge, then after removing the transient points (which are the first few points here), the remaining trajectory (not connected by lines) will be very similar to the fifth image shown here. In our ensemble scenario, we observe that just after one iteration, the unit circle is already changed to a very different shape, and by the fourth iteration, the shape of the image is already very similar to the Henon attractor. By now, one could easily understand that the Henon attractor can either be readily obtained from an arbitrarily shaped phase space region (discarding initial conditions which lead to the divergence of the iterations) or by iterating a single arbitrary initial condition many times. The equivalence of the two approaches, one based on the evolution of ensembles in the phase space, the other based on long-time iterations, is a clear manifestation of the ergodic property of the Henon map (and more generally, chaotic systems).

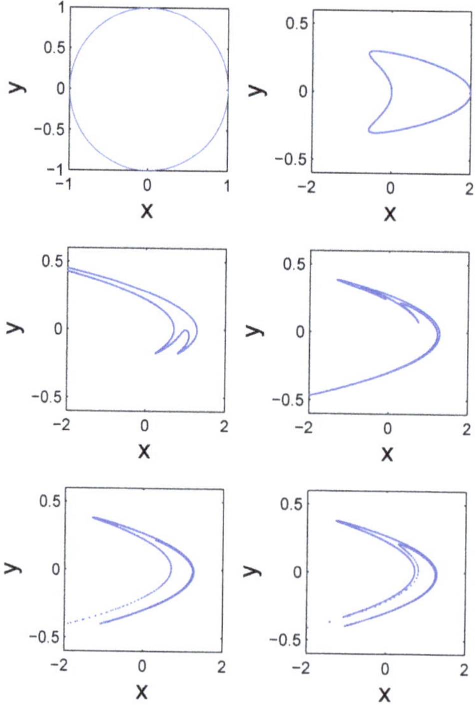

Figure 7. Successive transformation of a unit circle by the Henon map. The unit circle is represented by 36,000 points with equal arc spacing. These points are then taken as initial conditions for the Henon map. Successive ($i = 1, 2, \ldots, 5$) images of the unit circle (discarding initial conditions which lead to divergence of the iterations) are shown from left to right and top to bottom in the figure.

To enhance our understanding of the materials discussed so far, let us visually observe how chaos manifests itself in the chaotic Lorenz system:

$$
\begin{aligned}
dx/dt &= -16(x - y), \\
dy/dt &= -xz + 45.92x - y, \\
dz/dt &= xy - 4z.
\end{aligned}
\tag{20}
$$

For this purpose, let us arbitrarily choose an initial condition, $(-17.3432, -24.5966, 40.1096)$, perturb it 2500 times using standard Gaussian random variables with very small variance, and monitor the evolution of all those points. These initial conditions are shown in Figure 8 as a magenta block centered at our chosen initial condition. After 2 units of time, these initial conditions spread to the points labeled as red in the Figure. After another 2 units of time, the red points further evolve to the points labeled as green. Two more units of time later, the green points become the blue points. By that time, the shape of the points already resembles the chaotic Lorenz attractor we usually see in books, papers, and on the Internet.

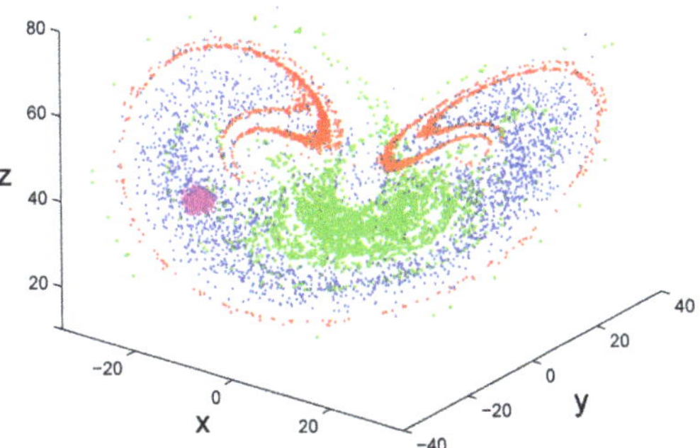

Figure 8. Evolution of point clouds in the chaotic Lorenz system: magenta, red, green, and blue correspond to $t = 0, 2, 4, 6$, respectively.

2.3.2. Defining Properties of Chaotic Systems

The most important property of chaos is sensitive dependence on initial conditions. It means that a very small difference in the initial condition may lead to a completely different trajectory. To appreciate this property, one may imagine a butterfly flapping its wings sometime on a day in the Amazon rain forest. This contributes to a minor change in the global air currents. If the motion on that day is chaotic, then sunny weather in some city, say Ney York, could have been replaced by a rainy weather not long after the flapping of the butterfly's wings. One may contrast this feature with a the traditional view, largely drawn from the study of linear systems, that small disturbances only produce proportional effects. Under the latter scenario, in order for the motion of the system to be random, the number of degrees of freedom has to be infinite.

Being the most important property of chaos, sensitive dependence on initial conditions has to be quantified. This is achieved by equating this property with an exponential divergence of nearby trajectories in the phase space. Let $d(0)$ be the small distance between two arbitrary trajectories at time 0, and let $d(t)$ be the distance between them at time t. Then, for true low-dimensional deterministic chaos, we have

$$d(t) \sim d(0)e^{\lambda_1 t} \tag{21}$$

where λ_1 is called the largest positive Lyapunov exponent. This property of sensitive dependence on initial conditions of chaos can be conveniently illustrated by the chaotic Logistic map:

$$x_{n+1} = \mu x_n(1 - x_n), \tag{22}$$

where $\mu = 4$. We can generate, for example, 100 initial conditions by using uniformly distributed random numbers, and iterate the Logistic map to get 100 trajectories. We then perturb each of the initial conditions by a small error of 10^{-4} and regenerate the 100 trajectories. The evolution of the errors between the original and the perturbed trajectories is shown in Figure 9. Clearly we observe that the logarithm of the errors first increases with time linearly to about a time of $n = 25$, then is saturated. Linear growth in a logarithmic scale amounts to exponential growth. By visual inspection, we can identify that λ_1 here is close to 0.7 (more precisely, $\ln 2$, which will be explained shortly). That errors very soon saturate is due to the fact that x defined by the logistic map is in the unit internal, as is the absolute value of the errors.

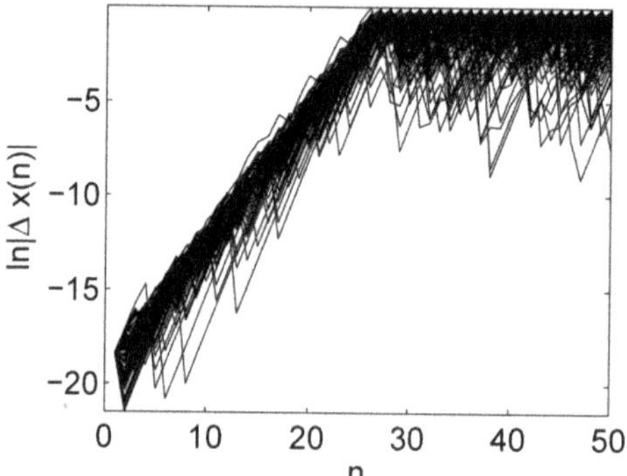

Figure 9. Error growth in the logistic map.

The largest positive Lyapunov exponent for the Henon map and the chaotic Lorenz system we discussed in Section 2.3.1 can also be conveniently computed based on time series data. This will be discussed shortly.

The trajectories of a chaotic attractor are bounded in the phase space. This is another fundamental property of the chaotic attractor. The ceaseless stretching due to exponential divergence of nearby trajectories, and folding from time to time due to boundedness of the attractor, make the chaotic attractor a fractal, characterized by

$$N(\epsilon) \sim \epsilon^{-D}, \ \epsilon \to 0 , \tag{23}$$

where $N(\epsilon)$ represents the (minimal) number of boxes, with linear length not larger than ϵ, needed to completely cover the attractor in the phase space. D is called the box-counting dimension of the attractor. Typically, it is a nonintegral number. For the chaotic Henon and Lorenz attractor, D is 1.2 and 2.05, respectively.

2.3.3. A Taste of Analysis

In order to better understand the key concept of chaotic dynamics, the sensitive dependence on initial conditions, let us engage in some analytic analysis. In practice, if one can identify from the problem a transformation similar to the following map, then one can be more than excited,

$$x_{n+1} = 2x_n \ \text{mod} \ 1, \tag{24}$$

This is a map on the unit interval, where x is positive, and mod 1 means that only the fractional part of $2x_n$ is retained as x_{n+1}. The map can also be written as

$$x_{n+1} = \begin{cases} 2x_n, & 0 \le x_n < 1/2 \\ 2x_n - 1, & 1/2 \le x_n < 1, \end{cases} \tag{25}$$

This map in fact acts as a Bernoulli shift [62], or binary shift, since if we represent an initial condition x_0 in binary form

$$x_0 = 0.a_1 a_2 a_3 \cdots = \sum_{j=1}^{\infty} 2^{-j} a_j, \tag{26}$$

then

$$x_1 = 0.a_2 a_3 a_4 \cdots ,$$

$$x_2 = 0.a_3 a_4 a_5 \cdots ,$$

and so on, where each of the digits a_j is either 1 or 0. Now it is clear that when x_0 is a rational number, the trajectory is periodic. In fact, we can easily find cycles of any length.

For example, if x_0 is a 3-bit repeating sequence, such as $x_0 = 0.001001001\cdots$, then the trajectory is periodic with period 3. Since there are infinitely more irrational numbers than rational numbers in $[0, 1)$, an arbitrary initial condition x_0 will be an irrational number with probability 1, and will almost surely generate an aperiodic, chaotic trajectory. Since after each iteration the map shifts one bit, a digit that is initially very unimportant, say the 80th digit (corresponding to $2^{-80} \approx 10^{-24}$), becomes the first and the most important digit after 80 iterations. This is a vivid example that a small change in the initial condition makes a profound change in x_n. Clearly, the largest Lyapunov exponent λ_1 here is $\ln 2$.

Next, let us re-consider the logistic map with $\mu = 4$. If we make a transformation,

$$x_n = \sin^2(2\pi y_n) \tag{27}$$

then the logistic map becomes the Bernoulli shift map discussed above. Therefore, the largest Lyapunov exponent λ_1 for the logistic map with $\mu = 4$ is also $\ln 2$, as we already mentioned.

Now that we have gained some understanding by considering simple model systems, we can discuss how to characterize general chaotic systems. For a chaotic dynamical system with dimensions higher than 1, first we need to realize that exponential divergence can occur in more than one direction, and possibly in many directions. That means we have multiple positive Lyapunov exponents. We denote them by λ^+, among them, the largest one is usually denoted as λ_1. How are these Lyapunov exponents related to the rate of creation of new information, or in other words, loss of prior knowledge, in the system? To find the answer, we may partition the phase space into boxes of size ϵ, compute the probability p_i that the trajectory visits box i, and finally calculate the Shannon entropy $I = -\sum p_i \ln p_i$. For many systems, when $\epsilon \to 0$, information increases with time linearly [63]

$$I(\epsilon, t) = I_0 + Kt \tag{28}$$

Here, I_0 is the initial entropy, and K is the celebrated Kolmogorov–Sinai (K-S) entropy [16,17]. Now let us consider the situation that all the initial conditions of the system are confined in a small region in the phase space. In this case, the initial probability in the chosen small region is 1, and 0 in all other regions. Therefore, $I_0 = 0$. For a chaotic system, because of the exponential divergence, the number of phase space regions visited by the system after a time of T is $N \propto e^{(\sum \lambda^+)T}$, where λ^+ are the positive Lyapunov exponents we have already explained. If all these regions are visited by the trajectories with equal probability, then $p_i(T) \sim 1/N$, and the information function becomes

$$I(T) = -\sum_{i=1}^{N} p_i(T) \ln p_i(T) = \left(\sum \lambda^+\right)T \tag{29}$$

We thus have $K = \sum \lambda^+$. In general, if these phase space regions are not visited equally likely, then

$$K \leq \sum \lambda^+ \tag{30}$$

Grassberger and Procaccia suggest that equality usually holds [64].

2.3.4. Bifurcations, Routes to Chaos, and Universality

In practice, whenever one has a dynamical system model described by discrete maps or differential equations, then the first thing one needs to consider is if the model has a unique fixed point solution, and if yes, if the solution is locally or globally stable. If the model contains some controlling parameter(s), then one also has to consider if the qualitative feature of the solution changes with the parameter(s), and if yes, find out what kind of changes they are. One can also think if any features of the system are shared by systems in other fields. The last point is the universality issue. These considerations make it clear that studies of bifurcations, routes to chaos, and universality are of fundamental importance to the study of dynamical systems.

Fixed point solutions are one of the the limiting behaviors of dynamical systems. It turns out the limiting behaviors of dynamical systems are very rich. In order of increasing complexity, they are fixed points, limit cycles, torus, chaos, turbulence, and random motions [38]. Fixed points correspond to motions without any change; limit cycles correspond to periodic motions. We have already mentioned these two in the beginning of this section. Torus corresponds to quasi-periodic motions, i.e., the motion is characterized by two or more independent frequencies. Periodic and quasi-periodic motions may be associated with crystals and quasi-crystals, finding of the latter won Professor Daniel Shechtman a Nobel Prize in Chemistry in 2011. Fixed points, limit cycles, and torus all belong to regular motions.

Since chaotic and regular motions appear almost everywhere, we should ask if a chaotic motion may arise from a regular motion, and vice versa. Interestingly, the answer can be found by studying bifurcations and routes to chaos in dynamical systems. Here, it is critical to realize that the qualitative behaviors of the dynamics of a system may change when one or more controlling parameters are changed. The parameter values that cause such qualitative changes are called bifurcation points.

To better understand the notion of transitioning from one state to another, let us briefly consider the anti-globalization movement. As often reported in the media, anti-globalization activities are often accompanied with grandeur and truly praiseworthy ideals such as better democratic representation, advancement of human rights, fair trade, and sustainable development. However, this is only part of the story. The more fundamental cause of the anti-globalization movement is the flipping of power ranking among the participating countries—a country afraid of losing competitive edges or even being demoted to a lower position in the power ranking would attribute that to unfair trade, infringement of intellectual property rights, etc. While these concerns are not entirely unfounded, one has to realize that reward to countries participating in economic globalization cannot be linearly proportional to their ranking. As a result, rearrangement of the power ranking surely will occur. Here, the basic parameter controlling the transition from globalization to anti-globalization is associated with the rearrangement of the (relative) power ranking among the participating countries.

To understand bifurcations, let us analyze the logistic map described by Equation (22) again. Let us set $\mu = 2$ and iterate the map starting with an initial condition $x_0 = 0.3$. With simple calculations, we can easily find that x_n soon equals 0.5 after a few iterations. If we choose $x_0 = 0.5$, then $x_1 = x_2 = \cdots = 0.5$. This means that 0.5 is a stable, fixed-point solution. While it is easy to prove this statement rigorously [38], here, let us resort to simulations: For any μ, where $\mu \in [2, 4]$, we choose an arbitrary initial value of x_0, and iterate Equation (22). After discarding the initial iterations so that the solution of the map has stabilized, we retain a large number (say, 100) of the value of the iterations, and form a scatter plot of those values with μ. When the map has a globally attracting fixed-point solution, then the recorded values of x_n will all become the same since the transients have been discarded. In this situation, one only observes a single point with the horizontal axis being the chosen μ and the vertical axis being the converged value of x_n. For a periodic solution with period m, one can observe m distinct points on the vertical axis. When the motion becomes chaotic, one observes on the vertical axis as many distinct points as one records (100 in our example). Figure 10a shows the bifurcation diagram for the logistic map—the interesting structure is the celebrated period-doubling bifurcation to chaos.

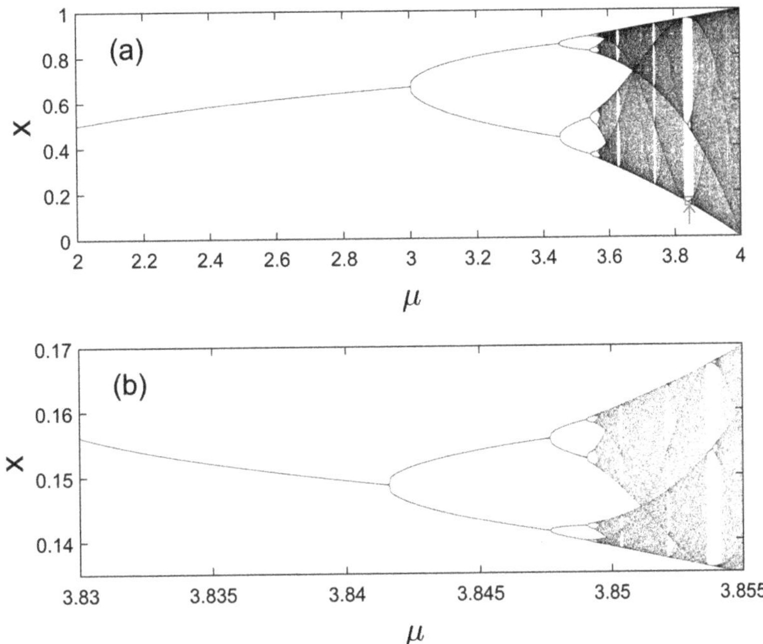

Figure 10. Bifurcation diagram for the logistic map; (**b**) is an enlargement of the little rectangular box indicated by the arrow in (**a**).

Figure 10a embodies more structures than one could comprehend by a simple glance. For example, if one enlarges Figure 10a the small rectangular region containing the period-3 window, then one obtains Figure 10b. We have again observed a period-doubling route to chaos! (To truly understanding the presentations here, it is beneficial for readers new to chaos theory to write a simple program to reproduce Figure 10a,b).

Having been observed in many diverse fields, period-doubling bifurcation to chaos is one of the most studied and most celebrated routes to chaos [65]. To better comprehend this universality, it is worth noting that it also underlies the bifurcations in the Henon map (see Figure 11) and the Lorenz system. In fact, the notion of universality can be quantified for the period-doubling bifurcation to chaos, through the Feigenbaum constant defined by

$$\delta = \lim_{k \to \infty} \frac{\mu_k - \mu_{k-1}}{\mu_{k+1} - \mu_k} = 4.669201\cdots. \tag{31}$$

Other routes to chaos also exist. They include the well-known quasi-periodicity route [66] and the intermittency route [67]. The former refers to when a controlling parameter is changed, the motion of the system changes from a periodic motion with one basic frequency, a quasi-periodic motion with two or more basic frequencies, to chaotic motions. This route has been observed in many mechanical and physical systems, including fluid systems. A bit surprisingly, this route has also manifested itself in the Internet transport dynamics (concretely, a variable amounting to the round-trip time of a message transmitting through the Internet can change from periodic and quasi-periodic motion to chaos when the congestion level increases [68]). The third classic route to chaos, intermittency, refers to the behavior that the motion of the system alters between smooth and chaotic modes, again when a controlling parameter is changed. This route to chaos is very relevant to many nonstationary phenomena in nature, including river flow dynamics, which are very different in wet and dry seasons.

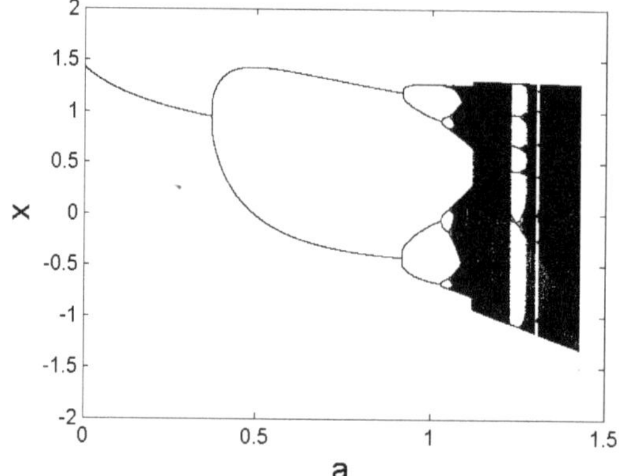

Figure 11. Bifurcation diagram for the Henon map.

2.3.5. Chaotic Time Series Analysis

In this big data era, data of all kinds, including time series data, have been accumulating explosively. Many techniques developed in the context of chaotic time series analysis will be of tremendous value for the analysis of all kinds of complex time series data whenever linear approaches are not sufficient. Below, we explain briefly but systematically all the main components of chaotic time series analysis.

A. Optimal embedding

Often, a complicated dynamical system described by $d\vec{U}/dt = f(\vec{U})$ lives in a high-dimensional phase space, where $\vec{U}$ is a vector. In many situations, we may only be able to access a single variable, say x, instead of many components of $\vec{U}$. In the simplest case, x is just a component of $\vec{U}$, say U_1. In general, x may be a function of $\vec{U}$. From $x(t)$, how much can we deduce the behavior of the dynamical system? The answer is a lot can be learned from x, thanks to the Takens embedding theorem. The basic procedure is to construct vectors according to the following equation [69–71],

$$V_i = [x(i), x(i+L), ..., x(i+(m-1)L)], \tag{32}$$

where m is the embedding dimension and L the delay time. More explicitly, we have

$$V_1 = [x(t_1), x(t_1+\tau), x(t_1+2\tau), ..., x(t_1+(m-1)\tau],$$
$$V_2 = [x(t_2), x(t_2+\tau), x(t_2+2\tau), ..., x(t_2+(m-1)\tau],$$
$$\vdots$$
$$V_j = [x(t_j), x(t_j+\tau), x(t_j+2\tau), ..., x(t_j+(m-1)\tau], \tag{33}$$
$$\vdots$$

where $t_{i+1} - t_i = \Delta t$ and $\tau = L\Delta t$. We thus obtain a discrete dynamical system (i.e., a map),

$$V_{n+1} = M(V_n). \tag{34}$$

If the original dynamical system has an attractor with a boxing counting dimension D defined by Equation (23), then so long as $m > 2D$, topologically the dynamics of the original system described by $d\vec{U}/dt = f(\vec{U})$ are equivalent to that described by Equation (34). In this case, the procedure using the delay coordinates is called an embedding. In proving this theorem, two properties of differential equations play key roles: (1) for any initial condition, a set of ODEs has a unique solution, and this ensures that trajectories corresponding to different initial conditions in the phase space do not intersect in the phase space; (2) a trajectory corresponding to a specific initial condition does not self-intersect in the phase space; when m is sufficiently large, self-intersection will be fully eliminated.

In practical applications, m and L have to be determined according to some optimization procedure. To appreciate the issue, let us consider the harmonic oscillator described below, which is among the simplest dynamical systems:

$$\frac{d^2 x}{dt^2} = -\omega x. \tag{35}$$

Of course, we can also write it as

$$\frac{dx}{dt} = y, \quad \frac{dy}{dt} = -\omega x, \tag{36}$$

The general solution is

$$x(t) = A\cos(\omega t + \phi_0), \quad y(t) = A\sin(\omega t + \phi_0). \tag{37}$$

Here, the phase space is a 2D plane with coordinates x and y. Now consider the case that we can only measure $x(t)$. Using the embedding procedure with $m = 2$, we obtain $V(t) = [x(t), x(t + \tau)]$. Figure 12 shows embeddings with $\tau = T/40$, $T/8$, $T/4$, where $T = 2\pi/\omega$ is the period of the oscillation. When $\tau = T/4$, the difference between the two components, $x(t)$ and $x(t + \tau)$, in terms of angle is $\pi/2$. With this angle difference, the cosine function becomes the sine function. That is, $x(t + \tau)$ becomes $y(t)$). Therefore, the reconstructed dynamical system is the same as the original one. In this simple example, the minimal embedding dimension m is 2, and the optimal delay time L is $1/4$ of the period. The consequence of using this optimal delay time is that the motion in the reconstructed phase plane is the most uniform—the phase velocity is the same everywhere in the case of Figure 12c, but not in those of Figure 12a,b.

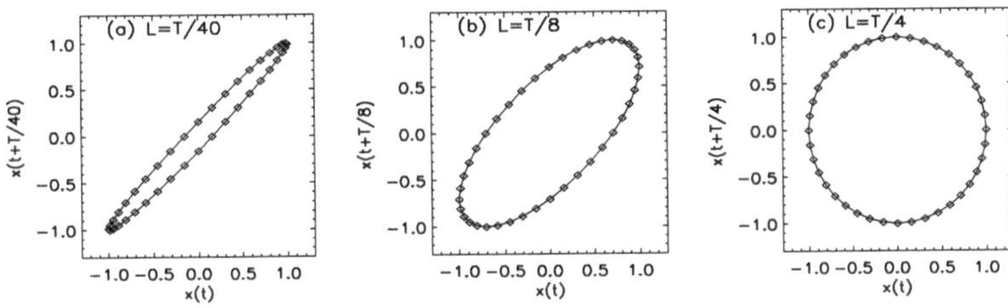

Figure 12. Embedding of the harmonic oscillator.

Since the 1980s, a number of excellent methods have been proposed to optimally determine m and τ. Below we describe two approaches, which have been extensively tested and are very systematic.

(1) **False nearest-neighbor method**: This is a geometrical method. Consider the situation in which an m_0-dimensional delay reconstruction is embedded but an $(m_0 - 1)$-dimensional reconstruction is not. Passing from $m_0 - 1$ to m_0, self-intersection in the reconstructed trajectory is eliminated. This feature can be quantified by the sharp decrease in the number of nearest neighbors when m is increased from $m_0 - 1$ to m_0. Therefore, the optimal value of m is m_0. More precisely, for each reconstructed vector $V_i^{(m)} = [x(t_i), x(t_i + \tau), x(t_i + 2\tau), \cdots, x(t_i + (m-1)\tau)]$, its nearest neighbor $V_j^{(m)}$ is found (to ensure unambiguity, here the superscript (m) is used to emphasize that this is an m-dimensional reconstruction). If m is not large enough, then $V_j^{(m)}$ may be a false neighbor of $V_i^{(m)}$ (something like both the north and south poles are mapped to the center of the equator, or multiple different objects have the same shadow). If embedding can be achieved by increasing m by 1, then the embedding vectors become

$$V_i^{(m+1)} = [x(t_i), x(t_i + \tau), x(t_i + 2\tau), \cdots, x(t_i + (m-1)\tau, x(t_i + m\tau)] = [V_i^{(m)}, x(t_i + m\tau)]$$ and $V_j^{(m+1)} = [V_j^{(m)}, x(t_j + m\tau)]$, and they will no longer be close neighbors. Instead, they will be far apart. The criterion for optimal embedding is then

$$R_f = \frac{|x(t_i + m\tau) - x(t_j + m\tau)|}{||V_i^{(m)} - V_j^{(m)}||} > R_T, \tag{38}$$

where R_T is a heuristic threshold value. Abarbanel [72] recommends $R_T = 15$. After m is determined, τ can be obtained by minimizing R_f.

While this method is intuitively appealing, it should be pointed out that it works less effectively in the noisy case. Partly, this is because nearest neighbors may not be well defined when data have noise.

(2) **Time-dependent exponent curves**: This is a dynamical method developed by Gao and Zheng [73,74]. The basic idea is that false neighbors will fly apart rapidly if we follow them on the trajectory. Denote the reconstructed trajectory by $V_1^{(m)}, V_2^{(m)}, \cdots$. If $V_i^{(m)}$ and $V_j^{(m)}$ are false neighbors, then it is unlikely that points $V_{i+k}^{(m)}, V_{j+k}^{(m)}$, where k is the evolution time, will remain close neighbors. That is, the distance between $V_{i+k}^{(m)}$ and $V_{j+k}^{(m)}$ will be much larger than that between $V_i^{(m)}$ and $V_j^{(m)}$ if the delay reconstruction is not an embedding. The metric recommended by Gao and Zheng is

$$\Lambda(m, L, k) = \left\langle \ln \left(\frac{||V_{i+k} - V_{j+k}||}{||V_i - V_j||} \right) \right\rangle. \tag{39}$$

Here, for simplicity, the superscript (m) in the reconstructed vectors is no longer indicated. The angle brackets denote the average of all possible (V_i, V_j) pairs satisfying the condition

$$\epsilon_i \leq ||V_i - V_j|| \leq \epsilon_i + \Delta\epsilon_i, \quad i = 1, 2, 3, \cdots, \tag{40}$$

where ϵ_i and $\Delta\epsilon_i$ are more or less arbitrarily chosen small distances. Geometrically speaking, Equation (40) defines a shell, with ϵ_i being the diameter of the shell and $\Delta\epsilon_i$ the thickness of the shell. When $\epsilon_k = 0$, the shell becomes a ball; in particular, if the embedding dimension m is 2, then the ball is a circle. Note that the computation is carried out for a series of shells, $i = 1, 2, 3, \cdots$, and $\Delta\epsilon_i$ may depend on the index i. With this approach, the effect of noise can be greatly suppressed.

As a rule of thumb, Gao and Zheng find that for a fixed small k, the minimal m is such that when further increasing m, $\Lambda(m, L, k)$ no longer decreases significantly. After m is determined, L can be chosen by minimizing $\Lambda(m, L, k)$.

Now that we have determined an optimal embedding, we can discuss how to estimate the largest positive Lyapunov exponent, dimension, and Kolmogorov entropy of chaotic attractors.

B. Estimation of the largest positive Lyapunov exponent

A number of algorithms for estimating the Lyapunov exponents have been developed. A classic method is Wolf et al.'s algorithm [75]. The basic idea is to select a fiducial trajectory and monitor how the deviation from it grows with time. Let the distance between the two trajectories at time t_i and t_{i+1} be d_i' and d_{i+1}. The rate of the exponential divergence over this time period is given by

$$\frac{\ln(d_{i+1}/d_i')}{t_{i+1} - t_i}.$$

To ensure exponential divergence, the distance between the two trajectories has to be always small. Therefore, when d_{i+1} exceeds a certain chosen threshold value, something has to be done: a new point in the direction of the vector of d_{i+1} is used so that d'_{i+1} is very small compared to the size of the attractor. This procedure is called normalization. After n repetitions of the procedure, we obtain

$$\lambda_1 = \sum_{i=1}^{n-1} \left[\frac{t_{i+1} - t_i}{\sum_{i=1}^{n-1}(t_{i+1} - t_i)} \right] \left[\frac{\ln(d_{i+1}/d_i')}{t_{i+1} - t_i} \right] = \frac{\sum_{i=1}^{n-1} \ln(d_{i+1}/d_i')}{t_n - t_1}. \tag{41}$$

Note the normalization procedure is where the novelty of the algorithm lies. The necessity of this step can be best understood by resorting to Figure 9: The computation from t_i to t_{i+1} amounts to one curve in Figure 9—when error saturates, a new round of computation has to begin; renormalization along the direction of the latest vector ensures that the evaluation of the largest positive Lyapunov exponent is along the most unstable dynamics of the data. This is especially important for high-dimensional cases, where there are multiple unstable directions (and therefore multiple positive Lyapunov exponents).

Unfortunately, the Wolf's algorithm suffers from two serious problems. One is that it does not and cannot tell how to determine a threshold value suitable for the normalization procedure. The other is even more serious: it assumes but does not test exponential divergence. As a consequence of the second problem, a positive λ_1 could arise from any type of noisy data, including independent identically distributed (IID) random variables, as long as all the distances used in the computation are small. Therefore, the approach can often interpret a noisy process as a chaotic motion. To see why this is so, consider the case that d_i' is small. At the next time, d_{i+1} usually will be larger than d_i'. This may be called that evolution would move d_i' to the most probable spacing. In the case of fully random sequence and without embedding, this "evolution" will be completed in just one time step; when embedding is used, embedding vectors automatically incorporate correlations, and this "evolution" will be completed in m time steps, where m is the embedding dimension. In both situations, d_{i+1}, being in the middle step evolving from d_i', typically will be larger than d_i'; consequently, a quantity computed using Equation (41) will be positive.

While a positive λ_1 is more likely to be produced by Wolf's algorithm, it should also be noted that certain implementations of the algorithm, such as that based on neural networks, may have to choose an initial spacing of d_i' larger than the most probable spacing, so that the computation can return a nonempty result—this is more so when noise is stronger. In that case, λ_1 estimated will be negative, enticing one to interpret the data under investigation to be non-chaotic when the data contain more noise. Of course, this interpretation is also incorrect since, in principle, entropy for noisy systems is infinite, but not negative (for more details on this issue, we refer to [76]).

To overcome the problems with Wolf's algorithm, a number of methods have been proposed. One algorithm is independently developed by Rosenstein et al. [77] and Kantz [78]. Another algorithm is developed by Gao and Zheng [73,74,79], published at about the same time. We first describe the former.

With the method of Rosenstein et al. [77] and Kantz [78], one first chooses a reference point and finds its ϵ-neighbors V_j. One then follows the evolution of all these points and computes an average distance after a certain time. Finally, one chooses many reference points and takes another average. Following the notation of Equation (39), these steps can be described by

$$\Lambda(k) = \left\langle \ln \left\langle \|V_{i+k} - V_{j+k}\| \right\rangle_{average\ over\ j} \right\rangle_{average\ over\ i}, \tag{42}$$

where V_i is a reference point and V_j are neighbors to V_i, satisfying the condition $\|V_i - V_j\| < \epsilon$. If $\Lambda(k) \sim k$ for a certain intermediate range of k, then the slope is the largest Lyapunov exponent. This is the most fundamental part of the algorithm: it explicitly tests whether the dynamics of the data possess exponential divergence or not.

While in principle this method can distinguish chaos from noise, with finite noisy data it may not function as desired. One of the major reasons is that in order for the *average over j* to be well defined, ϵ has to be small. In fact, sometimes the ϵ-neighborhood of V_i is replaced by the nearest neighbor of V_j. For this reason, the method cannot handle short, noisy time series well.

Gao and Zheng's algorithm [73,74,79] contains three basic ingredients: Equations (39) and (40), and the condition

$$|i - j| > w. \tag{43}$$

Equation (39) plays the same role as but is simpler than Equation (42), since it eliminates the necessity of performing two rounds of averages. More important are the conditions specified by two Inequalities (40) and (43). The condition specifying the series of shells makes the method a direct test for deterministic chaos, which will be explained momentarily. The condition specified by Inequality (43) ensures that tangential motions corresponding to the condition that V_i and V_j follow each other along the orbit are removed. Tangential motions contribute a Lyapunov exponent of zero and, hence, severely underestimate the positive Lyapunov exponent. An example is exhibited in Figure 13. We find that when $w = 1$, the slope of the curve severely underestimates the largest positive Lyapunov exponent, while $w = 54$ solves the problem. In practice, w can be chosen to be larger than one orbital time, when orbital times are defined in the dynamical system (Lorenz and Rossler attractor are such systems). If an orbital time cannot be defined, it can be more or less arbitrarily set to be a large integer if the dataset is not too small.

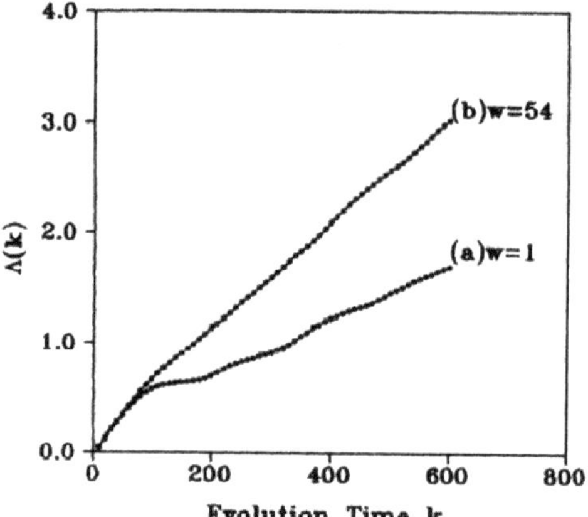

Figure 13. $\Lambda(k)$ vs. k curves for the Lorenz system. When $w = 1$, the slope of the curve severely underestimates the largest Lyapunov exponent. When w is increased to 54, the slope correctly estimates the largest Lyapunov exponent (reproduced from [74]).

To see how the condition specifying the series of shells gives rise to a direct test for deterministic chaos, we can compare the behavior of the time-dependent exponent curves for truly chaotic data and independent, identically distributed random variables. The basic results are illustrated in Figure 14. We observe that for true chaotic signals, the time-dependent exponent curves from different shells not only grow linearly for some intermediate range of the evolution time k, but form a common envelope. As one expects, the slope of the common envelope gives an accurate estimation of the largest positive Lyapunov exponent. Such a common envelope does not exist for IID random variables. In fact, the behavior of the IID random variables vividly illustrates the problems with Wolf's algorithm: $\Lambda(k)/k\delta t$ amounts to the largest positive Lyapunov exponent; the very fact that it critically depends on k and the size of the shells is a clear manifestation that the data under study are random.

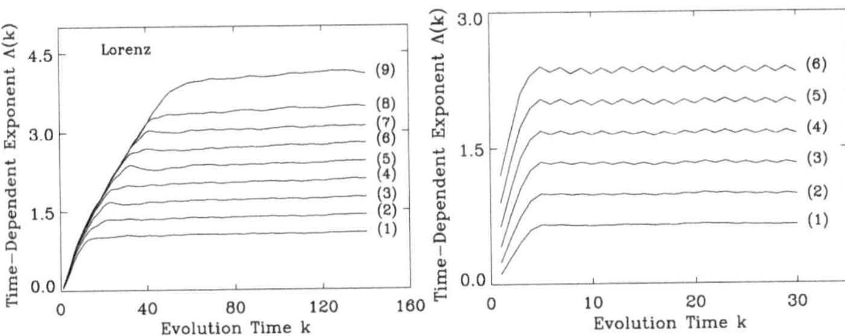

Figure 14. Time-dependent exponent curves for the chaotic Lorenz data (**left**) and IID random variables (**right**), where the curves, from bottom up, correspond to shells $(2^{-(i+1)}/2, 2^{-i/2})$, $1 = 1, 2, \cdots, 9$) (adapted from [74]).

As one can anticipate, when a chaotic signal is contaminated by noise, the common envelope will gradually disappear with an increasing amount of noise. In general, this is true for both measurement noise and dynamical noise, where measurement noise is the noise superimposed onto a signal during a measurement process, while dynamical noise is a noise that actively participates in the dynamics of the system (i.e., appears in the basic equation(s) of the dynamical system). When a system dynamic is oscillatory and characterized by a limit cycle, with dynamical noise, in certain situations, a stochastic oscillator will arise, with the frequency of the oscillation still close to that of the original limit cycle, but the amplitude differs from that of the original limit cycle considerably. In a phase space, it is characterized by a diffused limit cycle. An example is shown in Figure 15 (left) for essential tremor [80]. Such behavior has also been observed for Parkinsonian tremor [80], fluid dynamics in wakes behind circular cylinders in low Reynolds numbers and semiconductor lasers [81,82], and atomic force microscopy [83]. As chemical reactions are often oscillatory, one can also anticipate that stochastic oscillations are abundant in chemical reactions. Are stochastic oscillators also characterized by exponential divergence in the phase space, just as true chaos? Often, this is not the case. Instead, they are characterized by diffusional processes characterized by

$$\ln \| V_{i+k} - V_{j+k} \| = \ln \| V_i - V_j \| + \Lambda(k) \sim \ln k^{\alpha} \tag{44}$$

where the parameter α signifies what kind of diffusion the dynamic executes: the dynamic is called sub-diffusion, normal diffusion, and super-diffusion when $0 < \alpha < 1/2$, $\alpha \approx 1/2$, and $1/2 < \alpha$, respectively. In the case of tremors, the dynamics basically are normal diffusions [80]. Typical $\Lambda(k)$ curves for normal diffusions are of the shape shown in Figure 15 (right), which are also true for the fluid dynamics in wakes behind circular cylinders in low Reynolds numbers [81,82]. Other types of diffusions, although rarer, are also possible. We will return to this issue later when we consider chaos communications.

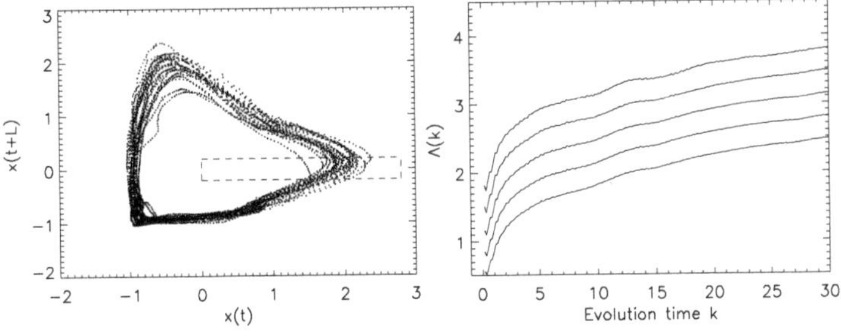

Figure 15. 2D phase diagram for essential tremor data (**left**) and time-dependent exponent curves (**right**), where the curves, from bottom up, correspond to shells $(2^{-(i+1)}/2, 2^{-i/2}, 1 = 1, 2, \cdots, 9)$ (adapted from [80]).

C. Estimation of fractal dimension and Kolmogorov entropy

There is an elegant algorithm, the Grassberger–Procaccia algorithm [64,84], that takes care of both. To fully understand the algorithm, we first extend the box-counting dimension defined in Equation (23). Recall that when we defined the box-counting or capacity dimension of a chaotic attractor, we partitioned the phase space where the attractor locates into many small regions called cells or boxes of linear size ϵ, and we counted the number of non-empty cells or boxes. We can monitor the non-empty boxes more precisely by counting how many points of the attractor have fallen into each of them. We can then assign a probability p_i to the ith cell that is not empty. The simplest way to compute p_i is by using n_i/N, where n_i is the number of points that fall within the ith cell, and N is the total number of points. Then

$$D_q = \frac{1}{q-1} \lim_{\epsilon \to 0} \left(\frac{\log \sum_{i=1}^{n} p_i^q}{\log \epsilon} \right), \tag{45}$$

where n is the total number of nonempty cells, and q is real. Generally speaking, D_q is a nonincreasing function of q. D_0 is the very box-counting or capacity dimension we have already discussed, since $\sum_{i=1}^{n} p_i^q = n$. D_1 gives the information dimension D_I,

$$D_I = \lim_{\epsilon \to 0} \frac{\sum_{i=1}^{n} p_i \log p_i}{\log \epsilon}. \tag{46}$$

Typically, D_I is equivalent to the pointwise dimension α defined as

$$p(l) \sim l^{\alpha}, \quad l \to 0, \tag{47}$$

where $p(l)$ is the measure (i.e., probability) for the trajectory to fall within a neighborhood of size l centered at a reference point. D_2 is called the correlation dimension. It is what the Grassberger–Procaccia algorithm calculates. It involves computing the correlation integral

$$C(\epsilon) = \lim_{N \to \infty} \frac{1}{N^2} \sum_{i,j=1}^{N} H(\epsilon - ||V_i - V_j||), \tag{48}$$

where V_i and V_j are the embedding vectors, $H(y)$ is the Heaviside function, which is 1 if $y \geq 0$ and 0 if $y < 0$. N is the number of points randomly chosen from the reconstructed vectors. The term involving the Heaviside function amounts to counting the number of points falling within a cell of radius ϵ that is centered around V_i. Therefore, $C(\epsilon)$ estimates the average fraction of points within a distance of ϵ. One then checks the following scaling behavior:

$$C(\epsilon) \sim \epsilon^{D_2}, \quad as \ \epsilon \to 0. \tag{49}$$

When calculating the correlation integral, one may compute pairwise distances, excluding points V_i and V_j that are too close in time (i.e., i and j are too close). A rule of thumb suggested by Theiler [85] is to remove the decorrelation time, which is equivalent to Inequality (43). This issue is best understood dynamically [74]: when V_i and V_j are close in time, they may be on the same orbit. The dimension corresponding to such tangential motion is 1, while the Lyapunov exponent is 0. Without removing them, the correlation dimension will be underestimated.

Next we consider entropy. First, let us precisely define the KS entropy. To be general, we consider a high dimensional dynamical system with F degrees of freedom. We partition the F-dimensional phase space into boxes of size ϵ^F. Assume the system has an attractor in the phase space. Let us focus on a transient-free trajectory $\vec{x}(t)$. Concretely, let us monitor the the state of the system at times $\tau, 2\tau, 3\tau, \cdots$. Let $p(i_1, i_2, \cdots, i_d)$ be the joint probability that the trajectory is in box i_1 at time τ, in box i_2 at time 2τ, $\cdots$, and in box i_d at time $d\tau$. The KS entropy is then

$$K = -\lim_{\tau \to 0} \lim_{\epsilon \to 0} \lim_{d \to \infty} \frac{1}{d\tau} \sum_{i_1, \cdots, i_d} p(i_1, \cdots, i_d) \ln p(i_1, \cdots, i_d) \, . \tag{50}$$

where K characterizes the rate of creation of entropy. To see this, we can start from the block entropy:

$$H_d(\epsilon, \tau) = -\sum_{i_1, \cdots, i_d} p(i_1, \cdots, i_d) \ln p(i_1, \cdots, i_d). \tag{51}$$

It is on the order of $d\tau K$. The difference between $H_{d+1}(\epsilon, \tau)$ and $H_d(\epsilon, \tau)$ gives the rate:

$$h_d(\epsilon, \tau) = \frac{1}{\tau}[H_{d+1}(\epsilon, \tau) - H_d(\epsilon, \tau)]. \tag{52}$$

Let

$$h(\epsilon, \tau) = \lim_{d \to \infty} h_d(\epsilon, \tau). \tag{53}$$

Taking proper limits in Equation (53), we obtain the KS entropy:

$$K = \lim_{\tau \to 0} \lim_{\epsilon \to 0} h(\epsilon, \tau) = \lim_{\tau \to 0} \lim_{\epsilon \to 0} \lim_{d \to \infty} \frac{1}{\tau}[H_{d+1}(\epsilon, \tau) - H_d(\epsilon, \tau)]. \tag{54}$$

The KS entropy can be generalized to the order-q Renyi entropies:

$$K_q = -\lim_{\tau \to 0} \lim_{\epsilon \to 0} \lim_{d \to \infty} \frac{1}{d\tau} \frac{1}{q-1} \ln \sum_{i_1, \cdots, i_d} p^q(i_1, \cdots, i_d). \tag{55}$$

When $q \to 1$, $K_q \to K$. Like the correlation dimension, the correlation entropy K_2 can be computed by the Grassberger–Procaccia algorithm by the following equation:

$$C_m(\epsilon) \sim \epsilon^{D_2} e^{-m\tau K_2}, \tag{56}$$

where $\tau = L\delta t$ is the actual delay time. The above equation can also be expressed as

$$K_2 = \lim_{\tau \to 0} \lim_{\epsilon \to 0} \lim_{m \to \infty} \frac{1}{\tau}[\ln C_m(\epsilon) - \ln C_{m+1}(\epsilon)]. \tag{57}$$

Although the above equations involve taking limits, in practice, data are of finite length, and one really looks for power-law scaling behaviors between $C_m(\epsilon)$ and ϵ when m is changed. When power law relations hold, in log-log scale, one should observe a series of curves, which are straight over a significant range of ϵ, and the curves for smaller embedding dimension m lie above those for larger m. In certain applications, one may just fix ϵ to some small value ϵ^*, say 10% or 15% of the standard deviation of the original time series, then compute $K_2(\epsilon^*)$. This $K_2(\epsilon^*)$ is called sample entropy, which has been widely used in various kinds of physiological data analyses. Sample entropy can also be computed for filtered data. When the filter is simply the moving average, which is the simplest ever known, the resulting series of entropies corresponding to different parameters for the moving average is called multiscale entropy. For more details, we refer to [86].

Before ending this subsection, we note a simple but very interesting and useful technique for testing nonlinearity. It is called the surrogate data approach [87,88]. The basic idea is to examine whether the original time series is distinctly different from a random time series sharing some basic properties of the original time series, such as the distribution or the power-spectral density. In the former case, the random time series can be readily obtained by simply shuffling the original time series. In the latter case, one can randomize the phase of the Fourier transform of the original time series and take the inverse transform.

2.3.6. Chaos-Based Communications and Effect of Noise on Dynamical Systems

Among the most promising applications of chaos theory is the exploitation of the short-term deterministic and long-term unpredictable aspects of chaotic behavior for the development of chaos-based communication systems. The actual research in this area goes in two directions. One, started in the early 1990s, is chaos-based secure communications [89]. The other, which is more recent, is to use chaos to rapidly generate random bits in physical devices, for a range of applications in cryptography and secure communication [90–99]. The potential of each direction is dictated by the role of noise played in the corresponding dynamical systems, which we will explain here.

In chaos-based secure communications, the most extensively studied is the scheme exploiting synchronization of chaos in two similar and coupled nonlinear systems [100–111]. The unpredictable behavior of chaos provides a means of security since chaotic signals are hard to decode by a third party (called an eavesdropper). The chaotic signal is used as a carrier to mask a message in the time or frequency domain. The synchronization of a chaotic receiver with the chaotic emitter is then used to retrieve the message. In mathematical notation,

- an emitter generates a chaotic signal $x(t)$,
- a message signal $s(t)$ is superimposed onto $x(t)$,
- the signal $r(t) = x(t) + s(t) + n(t)$ is sent to the receiver through the communication channel,
- a receiver is synchronized to the emitter so that $y(t) = x(t)$,
- signal $s(t)$ is retrieved at the receiver by taking the difference between $r(t)$ and $y(t)$.

Secure chaos communication was first realized in nonlinear electronic circuits [89]. In order to provide higher-speed encryption and be compatible with optical communication networks [112], later efforts have been focused on optical systems. Among the many optical systems studied in the field, the study of chaotic semiconductor diode lasers has been most fruitful. This type of laser, which is the preferred light source in telecommunications, has been an ideal test bed for many fundamental problems in nonlinear dynamics. The state-of-the art cryptosystems using diode lasers are now able to transmit Gb/s messages through a commercial fiber network of size 100 km [113].

The success of secure chaos communications depends on the realization of synchronization in two chaotic systems. While synchronization of periodic oscillators has been well-known since Huygens offered a mechanism in the seventeenth century, synchronization of chaotic systems was quite a surprise initially, since most researchers thought the exponential divergence in chaotic systems would prevent two chaotic systems from synchronizing. Amazingly, chaos synchronization can be proven analytically and demonstrated in laboratory experiments. To see the idea, let us consider two diffusively coupled dynamical systems,

$$
\begin{aligned}
x' &= F(x) + \alpha(y - x)x' = F(x) + \alpha(y - x) \\
y' &= F(y) + \alpha(x - y)y' = F(y) + \alpha(x - y)
\end{aligned}
\tag{58}
$$

Here, x and y are both vectors, $x' = F(x)$ is a chaotic system, and α is the parameter that couples the system x and y. An invariant subspace of the coupled system is given by $x(t) = y(t)$. If this subspace is locally attractive, then the two systems can synchronize perfectly. The role of $\alpha > 0$ is to suppress the divergence between the x and the y systems: in general, the larger the α, the easier the synchronization. To find the critical α, let us focus on $v = x - y$. Assuming v to be small, we can then use Taylor series expansion. Further assuming that higher order nonlinearities can be neglected, we obtain a linear differential equation

$$
v' = DF(x(t))v - 2\alpha v
\tag{59}
$$

Here, $DF(x(t))$ is the Jacobian of the vector field along the solution. When $\alpha = 0$, we have

$$u' = DF(x(t))u, \tag{60}$$

since the dynamics are chaotic, we have

$$\|u(t)\| \leq \|u(0)\|e^{\lambda_1 t}, \tag{61}$$

where λ_1 denotes the largest positive Lyapunov exponent of the isolated system. Now letting

$$v = ue^{-2\alpha t}, \tag{62}$$

we obtain

$$\|v(t)\| \leq \|u(0)\|e^{(-2\alpha + \lambda_1)t} \tag{63}$$

therefore, the critical coupling strength is

$$\alpha_c = \lambda_1/2. \tag{64}$$

In general, when $\alpha > \alpha_c$, and higher-order nonlinear terms in the Taylor series expansion can indeed be ignored, then the coupled system will exhibit complete synchronization. In building chaotic secure communication systems, the coupling is usually unidirectional, and the two systems are called drive and response (or master and slave) systems—in the example discussed here, if the term $\alpha(y - x)$ is dropped in the x system, then the x system is the drive system, and the y system is the response system.

To better understand the potential of chaotic secure communications, it is important to examine the effect of noise on dynamical systems. There are two types of noise, one is measurement noise. In chaotic secure communications, the channel noise is a type of measurement noise. The other type of noise is dynamical noise. It is in the equations governing the dynamics of the system. The channel noise becomes part of the dynamical noise for the response system (which can have additional dynamical noise sources). For two chaotic systems to synchronize, dynamical noise in the response system has to be small. This means the signal $s(t)$ has to be small compared with the chaotic signal $x(t)$. As a consequence, power consumption in chaotic secure communications is larger than traditional communication systems. This may be considered a cost for achieving better security.

Although in most situations noise is detrimental in chaotic secure communications, there are a few fortunate situations where noise is beneficial. This is enabled by an interesting phenomenon, the noise-induced chaos. The existence of the phenomenon can be demonstrated via a driven nonlinear oscillator [114], or the noisy logistic map [115], or other systems [116,117]. A mechanism for the phenomenon has also been developed [82,118]. The phenomenon is still a hot topic today, see for example [119,120].

Here we explain the basic properties of and the mechanism for noise-induced chaos via the noisy logistic map:

$$x_{n+1} = \mu x_n(1 - x_n) + P_n, \quad 0 < x_n < 1, \tag{65}$$

Here, μ is the bifurcation parameter, and P_n is a zero-mean Gaussian random variable with standard deviation σ. When $P_n = 0$, the map generates periodic orbits with periods 8, 6, 5, and 3 at parameter values $\mu = 3.55, 3.63, 3.74$, and 3.83, respectively. The period-8 motion at $\mu = 3.55$ is on the main 2^n cascade, and the period-3 motion at $\mu = 3.83$ is on the period(3)-doubling cascade (see Figure 10). For the case of $\mu = 3.55$, with a fairly large noise of $\sigma = 0.01$, the noisy trajectory is still very similar to the clean period-8 trajectory, as one can clearly see from Figure 16a. The case of $\mu = 3.74$ is very different. With σ as small as 0.0003, the noisy trajectory is already completely different from the original clean period-5 trajectory, as shown in Figure 16b. In fact, this noisy trajectory is chaotic, as shown by the time-dependent exponent curves shown in Figure 17c. In contrast, the noisy dynamics at $\mu = 3.55$ are definitely not chaotic, as shown by Figure 17a. The noisy dynamics at $\mu = 3.63$ and 3.83 are also chaos-like, though not as well defined as at $\mu = 3.74$. The mechanism for noise-induced chaos can be found by examining how a small amount of noise affects the dynamics. In general, the noisy dynamics when noise is very small is a diffusion characterized by Equation (44). The normal diffusion with $\alpha \approx 0.5$ corresponds to Brownian motions around the periodic orbit (or limit cycle), which is clear from Figure 16a. The case of super diffusion with $\alpha > 0.5$ is the very condition for noise-induced chaos to occur. This is shown in Figure 18 and can be readily understood as follows: chaos, which amounts to exponential divergence, can be more easily approached through larger α, especially when α is larger than 1, for a tiny amount of noise.

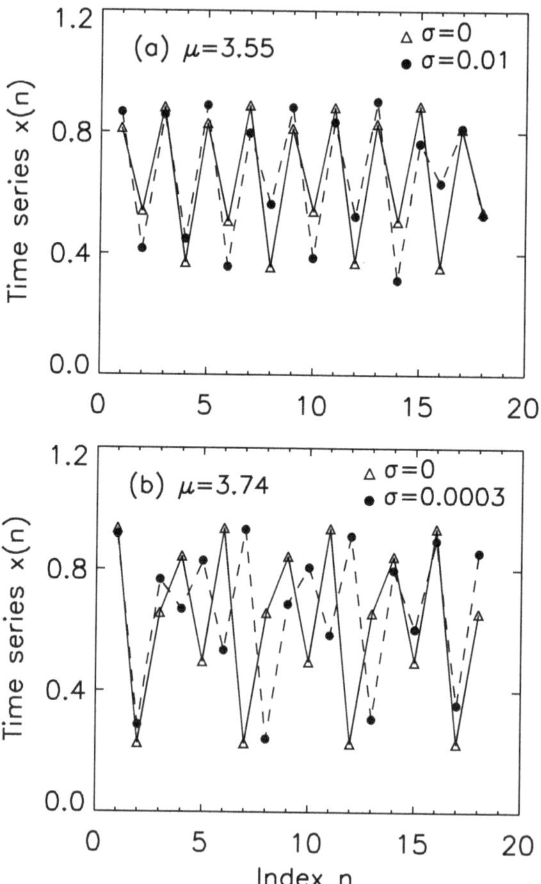

Figure 16. Clean (open triangles) and noisy (filled circles) trajectories for (**a**) $\mu = 3.55$ and (**b**) $\mu = 3.74$ (reproduced from [118]).

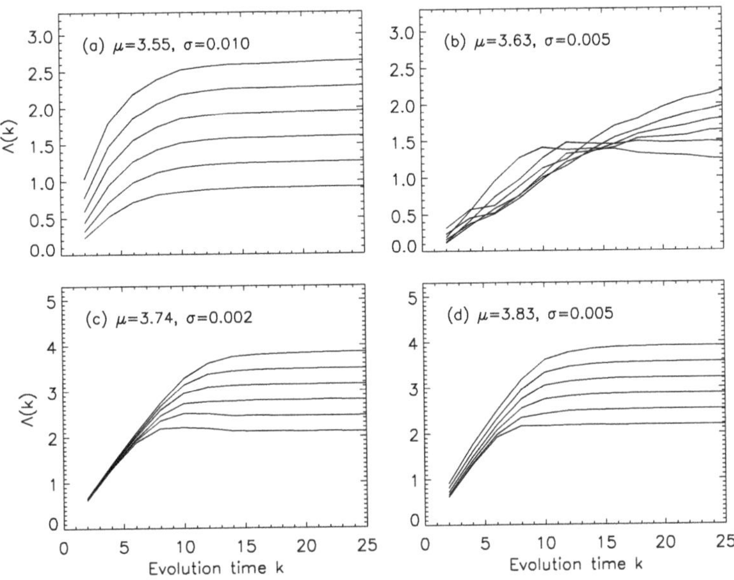

Figure 17. Time-dependent exponent curves for the noisy Logistic map: (**a**) $\mu = 3.55$ and $\sigma = 0.01$; (**b**) $\mu = 3.63$ and $\sigma = 0.005$; (**c**) $\mu = 3.74$ and $\sigma = 0.002$; and (**d**) $\mu = 3.83$ and $\sigma = 0.005$. Six curves, from the bottom up, correspond to shells $(2^{-(i+1)/2}, 2^{-i/2})$ with $i = 7, 8, 9, 10, 11$, and 12 (reproduced from [118]).

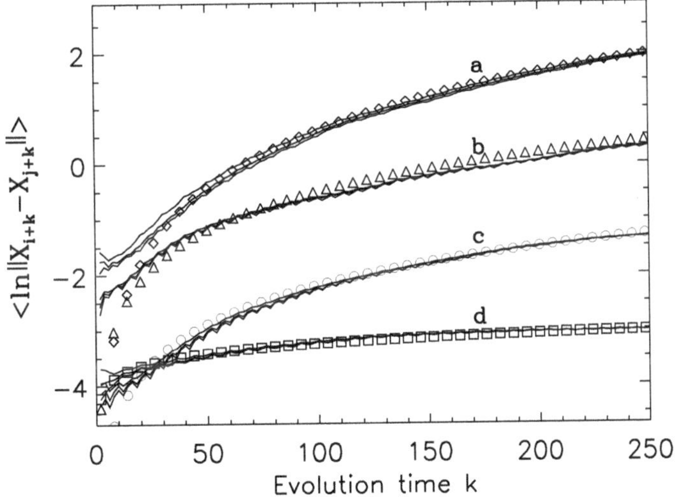

Figure 18. Logarithmic displacement curves illustrating the mechanism for noise-induced chaos. Each group actually consists of three curves, corresponding to shells $(2^{-(i+1)/2}, 2^{-i/2})$ with $i = 12, 13$, and 14. They basically collapse on each other. The parameters for the four groups are (**a**) $\mu = 3.74$ and $\sigma = 0.0003$; (**b**) $\mu = 3.83$ and $\sigma = 0.001$; (**c**) $\mu = 3.63$ and $\sigma = 0.0003$; and (**d**) $\mu = 3.55$ and $\sigma = 0.0005$. To separate these four groups (**a–d**) of curves from each other, they are shifted by 2, 1, -0.5, and -0.2 units, respectively, where a positive number indicates shifting upward, and a negative number indicates shifting downward. All four groups of curves are well modeled by $\ln k^{\alpha}$ with $\alpha = 1.5, 1.0, 1.0$, and 0.25 (reproduced from [118]).

Let us now come back to chaotic secure communications. Although noise-induced chaos can help with chaos synchronization, and thus chaos communication, the noise level has to be small. Otherwise, chaotic systems will desynchronize, and we will not be able to have any kind of communication at all [82].

In the beginning of this subsection, we have mentioned that recently there is a strong interest in using chaos to rapidly generate random bits in physical devices, for use in cryptography and secure communication. For this purpose, noise is always beneficial. The key here is to test whether a generated sequence of 0's and 1's is truly random. The usual tests for randomness, such as the widely used Statistical Test Suite for random number generator of NIST SP 800-22, basically test whether the distributions of 0's and 1's in the entire and the sub-sequences, as well as recurrences of certain patterns, are consistent with certain random distributions. The degree of divergence of nearby trajectories characterized by the time-dependent exponent curves offer additional information [109]. This is best understood by referring to Figure 17: the noise-induced chaos at $\mu = 3.74$ and 3.83 is more suitable to be used as fast physical random bit generator than at $\mu = 3.63$. The normal diffusion-like process at $\mu = 3.55$ will not pass the randomness test of NIST SP 800-22 since the dynamics are periodic-like.

Finally, as a side comment, we note that the pioneering works on chaos synchronization [100–111] are not cited evenly. Rather, some were only cited a few times, while the largest citation goes to [100], which is over 12,000 times. To better appreciate this somewhat astonishing behavior, we have listed these works in the reference not chronologically, but in descending order of the citations. The actual number of citations is shown in Figure 19, where the rank k from 1 to 12 denote references from [100–111]. Interestingly, the number of citations decays exponentially. This is in stark contrast with the behavior of the large-scale citation network mentioned earlier, which is a power law. This simple analysis has an interesting implication to using citation as a critical measure of the significance of scientific works. The analysis presented here clearly suggests that such a practice should not be taken too seriously, at least not taking citation as the sole measure of the significance of scientific works. In addition, there is an interesting lesson here: to enhance citations of one's work, it is important to get further involvement in the later development of a subject, after producing some pioneering work. For example, Dr. Pecora and Carrol have been actively involved in fostering the development of chaotic secure communications. Finally, there is an interesting question from this simple analysis: Can we develop a model to reconcile the exponential decay of citation to pioneering works with general power law decay of citations?

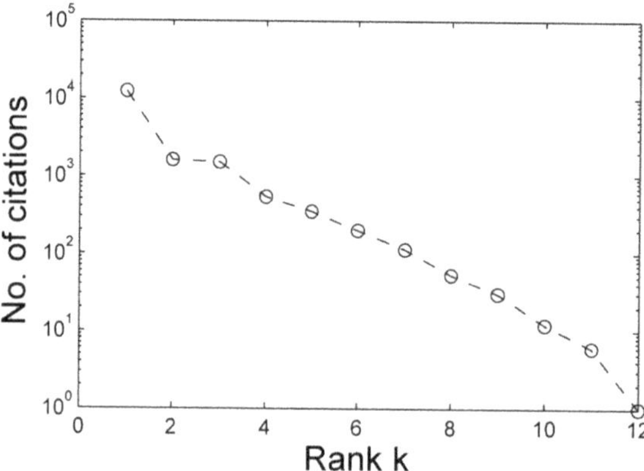

Figure 19. Number of citations of pioneering papers on chaos synchronization (data collected in March 2019).

2.4. Basics of Random Fractal Theory

In practice, many problems contain random elements. Random fractal theory is of crucial importance for finding structures and regularities in the random data, especially when the data involve a wide range of spatial and/or temporal scales (i.e., cover a long period of time or a wide extent of space).

Chaos and fractal theories are often discussed together and thought to be the same thing. This is a harmful perception because the part of fractal theory that is most useful for signal processing is the random fractal theory, whose foundation is fundamentally different from that of chaos theory. Chaos theory mainly studies irregular behaviors in nonlinear dynamical systems with only a few degrees of freedom. Here, noise or intrinsic randomness only has a minor role. Random fractal theory concerns systems that are inherently random. When equating chaos theory with fractal theory, one then will fail to fully understand the differences in the mathematics of the two theories, and fail to fully appreciate important issues such as distinguishing chaos from noise—a newcomer tackling the issue would think it sufficient to distinguish chaos from simple white noise. Unfortunately, this is not the case. Only if we can distinguish chaos from all known models of random processes can we say we can distinguish chaos from noise.

Below, we first discuss the basics of fractal theory, then we focus on random fractal theory. We will resume discussion of distinguishing chaos from noise in Section 2.5.

2.4.1. Introduction to Fractal Theory

Euclidean geometry studies simple shapes, including lines, planes, triangles, squares, cones, spheres, and so on. All these shapes are regular. Every one has seen clouds, mountains, and other complex shapes in nature. How well can those complex shapes be modeled by circles, spheres, cones, or other regular shapes? Very badly! When thinking along this line, Mandelbrot has created a new field, the fractal geometry [121].

Let us first try to understand fractal intuitively. The key here is self-similarity, which means that part of an object, when magnified, is similar to the whole. More concretely, whether we magnify the object by 10 times or 100 times, we always observe similar objects.

When discussing power laws, we have emphasized that a power law embodies self-similarity (please see Figure 2). Therefore, power law relations are natural mathematical tools to characterize fractals. When plotted in double-logarithmic scale, power laws become linear relations. To better appreciate the significance of power laws, imagine hiking on a mountain trail. Unlike many manmade trails with hundreds of stairs in the mountains of China, we assume the trail we are walking up is wild and irregular. How can we measure the distance we have walked? Let us measure the total distance by our step size. Denote it by ϵ. Note ϵ could be different for different hikers—one who rides a horse has a huge step size, while a little baby surely only has a tiny step size. The distance we have walked up is then

$$L = N(\epsilon) \cdot \epsilon, \tag{66}$$

where $N(\epsilon)$ is the number of intervals walked. Amazingly, $N(\epsilon)$ scales with ϵ as a power law, just as described by Equation (23), where $1 < D < 2$ is not an integer. Such a nonintegral D is the celebrated fractal dimension of the hiking trail.

What is the meaning of a nonintegral D? To find the answer, we start from the measurement of certain length, area, or volume. The basics of calculus teach us that we can measure a curve, a surface area, or a volume using very small intervals, squares, or cubes by properly covering the object we are interested to measure. Take the unit length, unit area, or unit volume as the unit of measurement, with linear size ϵ. Now suppose we measure the length of a straight line with length 1. What is the minimal number of intervals of length ϵ needed to fully cover this unit length? We need at least $N(\epsilon) \sim \epsilon^{-1}$ intervals. Extending to 2D and 3D, when covering an area or volume by boxes with linear length ϵ, we need at least $N(\epsilon) \sim \epsilon^{-2}$ squares to cover the area, and $N(\epsilon) \sim \epsilon^{-3}$ cubes to cover the volume. The D in $N(\epsilon) \sim \epsilon^{-D}$ here is called the topological dimension, which is 1 for a line, 2 for an area, and 3 for a volume. For M isolated points, the scaling law becomes $N(\epsilon) = M\epsilon^{-D}$, with $D = 0$. Therefore, the topological dimension D for isolated points is zero. We thus see that when we call a point, a line, an area, and a volume $0 - D$, $1 - D, 2 - D$, and $3 - D$ objects, we are talking about their topological dimensions.

We can now discuss the consequence of $1 < D < 2$ for an irregular mountain trail. Combining Equations (23) and (66), we have

$$L = \epsilon^{1-D},\tag{67}$$

Therefore, when ϵ becomes smaller, L becomes larger. In fact, when $\epsilon \to 0$, $L \to \infty$. This property was actually first found by Lewis Richardson, a mathematician, meteorologist, and pacifist who devoted himself in his later years to the study of the causes of wars and ways to prevent them. However, we have to wait for Mandelbrot to find the quantitative power law relation described by Equation (23), to create the new field of fractal.

With Equation (67), we can actually deduce more by using some concrete numbers. For example, let us take $D = 1.25$, and imagine a race between a hare and a tortoise. Taking into account the physical difference between a hare and a tortoise, it it reasonable to assume that the step size of the hare is 16 times that of the tortoise. Then we have

$$L_{\text{hare}} = \frac{1}{2}L_{\text{tortoise}}\tag{68}$$

The tortoise has to crawl twice the distance that the hare runs! Based on this simple calculation, we now understand when we walk along a wild trail, get tired, slow down, we are actually shrinking our step sizes, so we will be walking out a longer trail!

Next, we consider the Cantor set, one of the prototypical fractal objects, so that we can appreciate the concept of fractal dimension better.

The standard Cantor set is obtained by first partitioning a line segment into three equal parts and deleting the middle one. This step is then repeated, deleting the middle third of each remaining segment iteratively. See Figure 20a. Note that such a process can be related to the iteration of a nonlinear discrete map. The removed middle thirds can be related to the intervals that make the map diverge to infinity, while the remaining structures are linked to the invariant points of the map. At the limiting stage, $n \to \infty$, the Cantor set consists of infinitely many isolated points. Consistent with isolated point(s) having dimension 0, the topological dimension here is 0. The length of the total segments removed is

$$\frac{1}{3} + 2 \times \left(\frac{1}{3}\right)^2 + 4 \times \left(\frac{1}{3}\right)^3 + \cdots = \frac{1}{2} \times \frac{2}{3} + \frac{1}{2} \times \left(\frac{2}{3}\right)^2 + \frac{1}{2} \times \left(\frac{2}{3}\right)^3 + \cdots = 1\tag{69}$$

Therefore, the entire unit interval has been removed! Is the fractal dimension here the same as the topological dimension, which is 0?

To find out, let us focus on stage n. One needs $N(\epsilon) = 2^n$ boxes of length $\epsilon = \left(\frac{1}{3}\right)^n$ to cover the set. Hence, the fractal dimension for the Cantor set is

$$D = -\ln N(\epsilon) / \ln \epsilon = \ln 2 / \ln 3.\tag{70}$$

It is not zero!

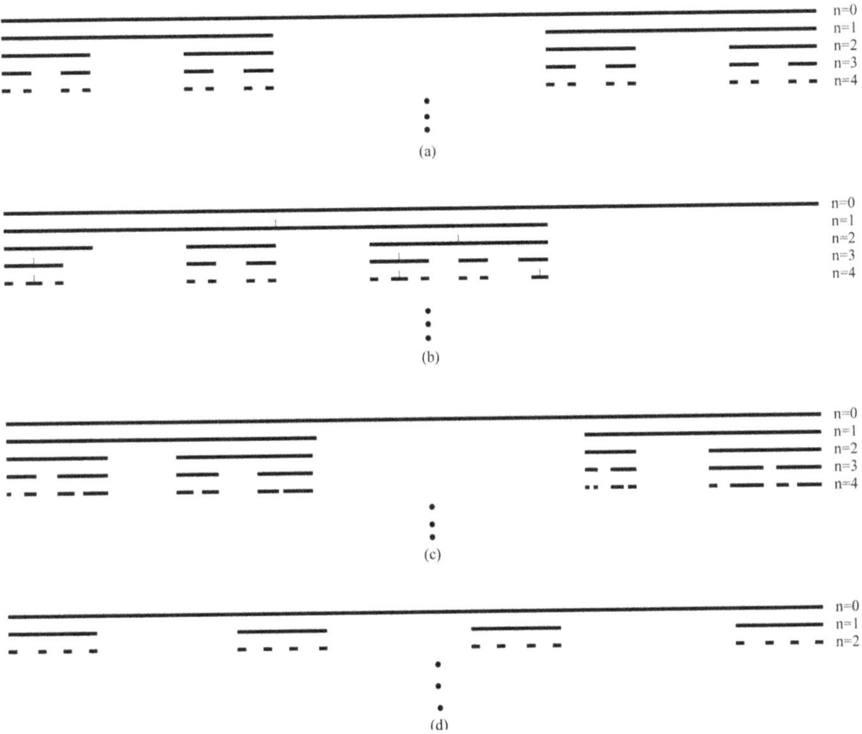

Figure 20. Standard Cantor set (**a**) and its variants (**b–d**). See the text for details.

The fractal dimension of the Cantor set can also be computed by employing the self-similar feature. Denote the number of intervals needed to cover the Cantor set at a certain stage with scale ϵ by $N(\epsilon)$. When the scale is reduced by 3, $N(\epsilon/3)$ is doubled. Since $N(\epsilon/3)/N(\epsilon) = 3^D = 2$, one immediately gets $D = \ln 2/\ln 3$.

To better transit to random fractals, we note that the standard Cantor set can be made random. One way is to divide each interval into three equal parts and randomly delete one of them (see Figure 20b). An alternative way of obtaining a random Cantor set is to delete a middle interval of random length at each stage (Figure 20c). Clearly, case (b) also has dimension $D = \ln 2/\ln 3$. When certain regulation is imposed on the length distribution for the subintervals in case (c), the fractal dimension can also be readily computed. One way of imposing such a regulation is to require that the ratio of the subinterval and its immediate parent interval follows some distribution that is stage-independent. Such a regulation is essentially a multifractal construction, which we will discuss soon.

The above discussion suggests that two different geometrical fractals may have the same fractal dimension. To further appreciate this point, we have shown in Figure 20d a different type of regular Cantor set. It is obtained by retaining four equally spaced segments whose length is 1/9th of the preceding segment. Denote the number of segments at a certain stage with length scale ϵ by $N(\epsilon)$. When the scale is reduced by 9, $N(\epsilon/9)$ is quadrupled. Here, D is again $\ln 2/\ln 3$, since $N(\epsilon/9)/N(\epsilon) = 9^D = 4$.

Based on the above discussions, one can readily realize that fractal curves and surfaces are more space filling. This property is beneficial in biological evolution. As a result, fractal forms are abundant in biology. Instead of giving actual examples here, we will refer readers to reference [122] for a menagerie of fractal forms in living things. This more space-filling property of fractals has also been exploited to design fractal antennas by maximizing the effective length or perimeter of the material that receives or transmits electromagnetic radiation. Fractal antennas are excellent for wideband and multiband applications [123].

2.4.2. Overview of Random Fractal Theory

Gaussian white noise is the most extensively studied noise in engineering. In complex systems, however, the temporal or spatial fluctuations often cannot be modeled by Gaussian white noise. Rather, they are characterized by a power law decaying spectrum in the frequency domain, denoted as $1/f^\alpha$ noise [38]. Its dimensionality cannot be reduced by popular methods such as principle component analysis [124]. Interesting examples of such processes include genetics [125–129], human cognition [130] and coordination [131], posture [132], vision [133,134], physiological signals [80,135–143], neuronal firing [144,145], urban traffic [146], tree rings [147], global terrorism [148], human response to natural and social phenomena [149], foreign exchange rate [76], and the distribution of prime numbers [150].

Basic Definitions and Equations

Denote a covariance stationary stochastic process as $X = \{X_t : t = 0, 1, 2, \dots \}$. Its mean is μ, variance is σ^2, and autocorrelation function $r(k), k \geq 0$ has the following form

$$r(k) \sim k^{2H-2}, \ \ as \ k \to \infty, \tag{71}$$

where H is a parameter called the Hurst exponent. It is in the unit interval, $0 < H < 1$. The exponent α for the spectra of the process, $1/f^\alpha$, is related to H by a simple equation,

$$\alpha = 2H - 1. \tag{72}$$

When $0 < H < 1/2$, the process is said to have anti-persistent correlations; when $H = 1/2$, the process is memoryless or only has short memory, when $1/2 < H < 1$, the process is said to have persistent long-range correlations. In this case, it is easy to prove $\sum_k r(k) = \infty$. This is why the process is said to have long-range correlation [38].

Let us now smooth the process X to obtain a time series $X^{(m)} = \{X_t^{(m)} : t = 1, 2, 3, \dots \}$, $m = 1, 2, 3, \dots$, where

$$X_t^{(m)} = (X_{tm-m+1} + \cdots + X_{tm})/m, \ t \geq 1. \tag{73}$$

The smoothing is carried out in a non-overlapping fashion; therefore, the length of $\{X_t^{(m)}\}$ is the largest integer that is not larger than N/m, where N is the length of $\{X_t\}$. Is there a relation between the variance of $X_t^{(m)}$, which is denoted by $V_m = var(X^{(m)})$, and that of the original process, which is denoted by σ^2? It is given by

$$var(X^{(m)}) = \sigma^2 m^{2H-2} \tag{74}$$

Equation (74) is often called the variance–time relation. It is fundamental and can help us understand the "little smoothing" phenomenon: when $H = 0.5$, $var(X^{(m)}) = 10^{-2}\sigma^2$ when $m = 100$. When $H = 0.75$, for $var(X^{(m)})$ to drop as much, m has to be 10,000. However, if $H = 0.25$, then $var(X^{(m)}) \approx 10^{-2}\sigma^2$ when $m \approx 23$. Therefore, when H increases to 1, smoothing has little effect in reducing the variance of the process.

As we have mentioned, the power spectral density (PSD) for X is

$$S_X(f) \sim f^{-\alpha} = f^{-(2H-1)}. \tag{75}$$

The integration of the X process, called the random walk process,

$$y_k = \sum_{i=1}^{k}(X_i - \overline{X}), \tag{76}$$

where $\overline{X}$ is the mean of X, has a PSD

$$S_Y(f) \sim f^{-\alpha-2} = f^{-(2H+1)}. \tag{77}$$

It is easy to see that the following relation is equivalent to Equation (74)

$$\left\langle |y(i+m) - y(i)|^2 \right\rangle \sim m^{2H}, \tag{78}$$

where the angle brackets denote averaging over i. Equation (78) is often called fluctuation analysis (FA). The superiority of Equation (78) over Equation (74) is that it can be readily generalized to a multifractal formulation.

The Fractional Brownian Motion (fBm) Process

The fBm process is the prototypical random walk model for $1/f^\alpha$ process [121]. It is a zero-mean Gaussian process, with stationary increments and variance

$$E[(B_H(t))^2] = t^{2H} \tag{79}$$

and covariance:

$$E[B_H(s)B_H(t)] = \frac{1}{2}\{s^{2H} + t^{2H} - |s-t|^{2H}\} \tag{80}$$

where H is the Hurst parameter. The increment process of the fBm, $X_i = B_H((i+1)\Delta t) - B_H(i\Delta t)$, $i \geq 1$, where Δt amounts to a sampling time, is called the fractional Gaussian noise (fGn). It is a zero-mean stationary Gaussian time series, with autocorrelation function:

$$\gamma(k) = E(X_i X_{i+k})/E(X_i^2) = \frac{1}{2}\left\{(k+1)^{2H} - 2k^{2H} + |k-1|^{2H}\right\}, \ k \geq 0 \tag{81}$$

Note $\gamma(k)$ is independent of Δt. In particular, $\gamma(1) = \frac{1}{2}\left(2^{2H} - 2\right)$. It is positive when $1/2 < H < 1$, and negative when $0 < H < 1/2$. When $k \to \infty$, $\gamma(k) \sim k^{2H-2}$, and we reproduce Equation (71).

Structure Function Based Multifractal Analysis

Since the Hurst parameter H is the defining parameter of random fractals, it is certainly of critical importance to estimate H. To facilitate estimation of H, it is most convenient to use the random walk process y, defined by Equation (76), and consider the following multifractal formulation:

$$F^{(q)}(m) = \left\langle |y(i+m) - y(i)|^q \right\rangle^{1/q} \sim m^{H(q)}, \tag{82}$$

where q is real-valued. The average is taken over all possible pairs of $(y(i+m), y(i))$. Note that $q > 0$ highlights large absolute value of $|y(i+m) - y(i)|$, while $q < 0$ highlights small absolute value of $|y(i+m) - y(i)|$ (to understand better, it is beneficial to take $q = 10$ and $max_i |y(i+m) - y(i)| = 100$, and $q = -10$ and $min_i |y(i+m) - y(i)| = 1/100$). $H(q)$ is a non-decreasing function of q. When $H(q)$ is a constant, the process is called a monofractal; otherwise, it is a multifractal.

Note that when $q = 2$, Equation (82) reduces to Equation (78), the FA, and $H(2) = H$. It can be readily proven that FA is equivalent to many other methods for estimating H, including the variance–time relation, the Fano factor analysis, and a few others [38,151] (the H value estimated by the R/S statistic is equivalent to $H(1)$). While all these methods are important, they have a limitation in that the largest H estimated by them is 1. Many processes, including auto-regressive processes, ON/OFF models, Levy walks, and processes with trends, have $H > 1$ on some time scale range. To accurately estimate those exponents, one has to use other methods, such as detrended fluctuation analysis (DFA) [152] and wavelet multi-resolution analysis [153]. In Section 3, we will present an improvement of DFA, adaptive fractal analysis (AFA) [149,154–157].

Singular Measure Based Multifractal Analysis

There is an alternative multifractal formalism to the structure-function based technique. It is based on probabilities and the thermodynamic formulation. The basic idea is to consider the scaling behaviors for the qth moments of the measure μ [38,153]:

$$Z(q,\epsilon) = \sum_{i=1}^{N(\epsilon)} \mu_i^q(\epsilon) \sim \epsilon^{\tau(q)}, \ \epsilon \to 0 \tag{83}$$

where $N(\epsilon)$ is the minimal number of boxes of linear size ϵ that are used to cover the support of the measure μ. The spectrum of the generalized dimensions D_q is defined by

$$D_q = \frac{\tau(q)}{q-1}, \tag{84}$$

Comparing with our discussions on the D_q spectrum for chaotic systems, we readily see that D_0 is the capacity (or box-counting) dimension, and D_1 is the information dimension. Just as the $H(q)$ spectrum, $D(q)$ is a non-decreasing function of q. When $D(q)$ is constant in q, the measure is called monofractal; otherwise, it is called multifractal.

There is another interesting way to characterize the properties of the measure. It is by the singular spectrum $f(\alpha)$, where α is called the pointwise dimension. The basic equation connecting the two characterizations is the Legendre transform,

$$q = df(\alpha)/d\alpha, \quad \tau(q) = q\alpha - f(\alpha). \tag{85}$$

Combining Equations (84) and (85), we have

$$D_q = \frac{1}{q-1}[q\alpha(q) - f(\alpha(q))]. \tag{86}$$

We thus see that D_q and $f(\alpha)$ provide the same amount of information.

The Random Cascade Model

In the study of multifractals, it is important to have a constructive model. This is provided by the random cascade model. It is among the most powerful models to understand the intermittency phenomenon of turbulence [158–162]. Here, we will use the notations developed for modeling Internet traffic and geophysical data [38,163–165] to present the model.

Consider a unit mass unevenly distributed on a unit interval. Let us divide the unit interval into two parts: call them the left and the right segments. By doing so, we have also partitioned the mass into two fractions, r and $1 - r$, which are on the left and right segments correspondingly. In general, the multiplier r is a random variable, having a probability density function (PDF) $P(r), 0 \leq r \leq 1$. Always with this rule we can further partition each new subinterval and the weight attached to it into two parts, ad infinitum. Figure 21 shows the procedure schematically. To facilitate mathematical analysis, the multiplier r has been rewritten as r_{ij}, where i indicates the stage number and j indicates the positions of a weight on that stage (we only use odd numbers, leaving even numbers for $1 - r_{ij}$). For many types of data analysis, it is important to explicitly introduce the notion of scale. This is provided by the interval length, which is 2^{-i} at stage i. Assuming bilateral symmetry, then we have to require that $P(r)$ is symmetric about $r = 1/2$. Let $P(r)$ have successive moments

$\mu_1, \mu_2, \cdots$. Hence, r_{ij} and $1 - r_{ij}$ both have marginal distribution $P(r)$. The weights at the stage N, $\{w_n, n = 1, ..., 2^N\}$, can be expressed as

$$w_n = u_1 u_2 \cdots u_N, \tag{87}$$

where $u_l, l = 1, \cdots, N$, are either r_{ij} or $1 - r_{ij}$. Thus, $\{u_i, i \geq 1\}$ are IID random variables all having PDF $P(r)$.

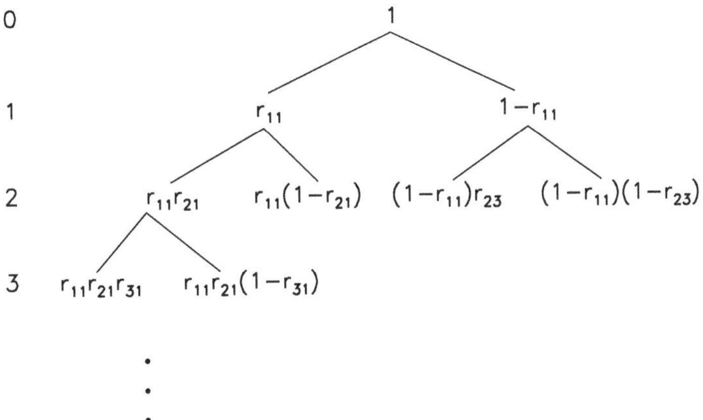

Figure 21. Schematic showing how a multiplicative multifractal is constructed.

The cascade model has many interesting properties. We list a few here:

- The weights at stage N are log-normally distributed. To see this, one can take logarithm on both sides of Equation (87), then the multiplication becomes summation, and one can use the central limit theorem.
- We can readily derive that

$$\tau(q) = -\ln(2\mu_q)/\ln 2. \tag{88}$$

- We can also derive that

$$H(q) \sim -\frac{1}{q}\ln \mu_q/\ln 2, \tag{89}$$

and

$$\tau(q) = qH(q) - 1. \tag{90}$$

We now illustrate Equations (88) and (89) using an example, the random binomial model, whose $P(r)$ is

$$P(r) = [\delta(r - p) + \delta(r - (1 - p))]/2 \tag{91}$$

where δ denotes the Dirac function. Therefore, $P(r = p) = P(r = 1 - p) = 1/2$. Here, the qth moment $\mu_q = [p^q + (1 - p)^q]/2$. We thus find

$$\tau(q) = -\ln[p^q + (1 - p)^q]/\ln 2 \tag{92}$$

and

$$H(q) = \frac{1}{q}\{1 - \ln[p^q + (1 - p)^q]/\ln 2\} \tag{93}$$

Clearly, $H(q)$ is a non-decreasing function of q. Without loss of generality, we may take $p \leq 1//2$. When $q \to -\infty$, H converges to its tight upper bound of $-\ln p/\ln 2 > 1$. When $q \to \infty$, H converges to its tight lower bound of $-\ln(1 - p)/\ln 2 < 1$.

In the cascade model, many different functional shapes for $P(r)$ can be used [163,164], and the model can simulate a random function with very high accuracy. Two examples are shown in Figure 22 for sea-clutter amplitude data [38,166]. The model can also be readily generalized to the high-dimensional case. The case for 2D is shown in Figure 23.

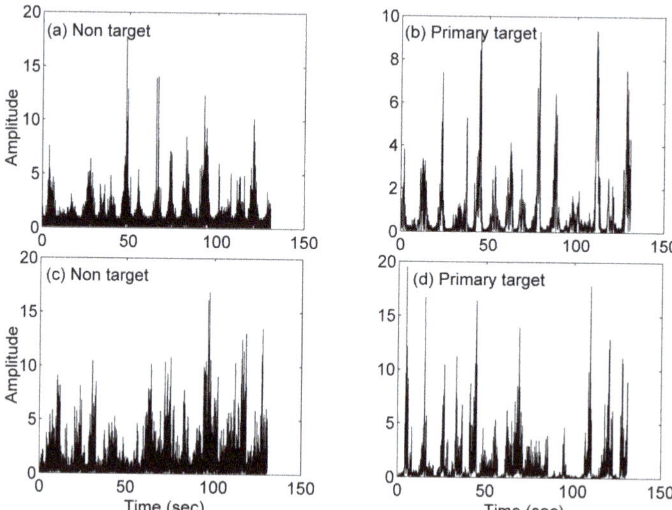

Figure 22. Sea clutter amplitude data: (**a**) is the original data without target, (**b**) is the original data with a primary target, and (**c**,**d**) are the modeled data.

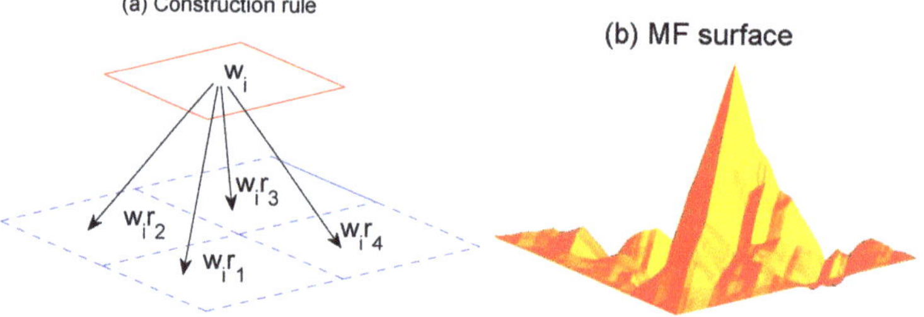

Figure 23. Construction of 2D multiplicative multifractals: (**a**) schematic rule, (**b**) an example.

2.5. Going from Distinguishing Chaos from Noise to Fully Understanding the System Dynamics

A long-standing problem in time series analysis, which is still of interest today, is to distinguish chaos from noise. This problem naturally arises when one wishes to understand whether certain complex behaviors in physics, finance, life sciences, ecology, and other fields, are of deterministic origin, or genuinely random. An unambiguous answer to the question can greatly help one to choose a proper model to study the behavior one wishes to understand. For a long time, however, when one computes a nonintegral fractal dimension, or a positive largest Lyapunov exponent, or a finite Kolmogorov entropy from a time series, one would think the time series is chaotic. In many applications, many researchers are still assuming so! Is this a sound assumption? Unfortunately, it is not. As one can expect, the most convincing counter-example would be the one that a genuinely random time series is interpreted as deterministic chaos by this assumption. It turns out that all $1/f^{\alpha}$ random processes can be proven to have non-integral fractal dimensions of $1/H$ [38], and finite Kolmogorov entropies [167,168], and thus may be misclassified as chaos. Because of this problem, it is desirable that whenever one studies chaos in observational data, one explicitly tests whether the data truly have the signature of chaos, the exponential divergence. In Section 4, we will discuss the scale-dependent Lyapunov exponent (SDLE), which generalizes the notion of the Lyapunov exponent. We will see there that SDLE can readily solve the problem.

Nowadays, efforts are still being made to develop innovative methods to distinguish chaos from noise. In our view, it is more important to find the defining parameters of the complex time series that one studies. In particular, one has to ask: If the time series is truly chaotic, what is the exponential growth rate? If the time series is random, what type of randomness is it? Only if we can unambiguously answer these fundamental questions can we truly understand the system under study. Clearly, this is more than simply trying to distinguish chaos from noise. In doing so, one will find that chaos and random fractals may both play significant roles in one's problem: chaos and random fractal may be manifested on different scales. This is the essence of multiscale phenomena: signals may exhibit different quantifiable features on different scales. Therefore, to best characterize a complex system, we need to use a number of tools synergistically. With this rationale, another fundamental question arises: what are the relations among the different complexity measures?

We have already introduced a number of different complexity measures, including the largest positive Lyapunov exponent, fractal dimension, generalized dimension spectrum, Kolmogorov–Sinai entropy, correlation dimension, correlation entropy, sample entropy, and multiscale entropy. Before discussing the connections among these complexity measures, we explain a few more measures, including the Lempel-Ziv (LZ) and the Kolmogorov–Chaitin complexity.

The LZ complexity is asymptotically equivalent to the Shannon entropy. The algorithm for computing the LZ complexity can be efficiently implemented and executed, and thus the LZ complexity and its many derivatives have found wide applications—the value of the LZ complexity of a numerical, text, or image file may be equated to the size of their compressed files using the commonly used compression schemes. To compute the LZ complexity for a time series, it is important to consider the effect of the finite length of the data. For more details, we refer to [169].

The Kolmogorov–Chaitin complexity is also called descriptive complexity, Kolmogorov complexity, algorithmic complexity, algorithmic entropy, and program-size complexity. It is a key measure in algorithmic information theory. The Kolmogorov–Chaitin complexity of a string of numbers or a text file is the length of the shortest computer program that generates the the string of numbers or the text file. Therefore, it measures the computational resources needed for specifying an object. To make the above discussions concrete, one can think of a completely random string. It is impossible to compress the string into a program with length shorter than the length of the string itself; the simplest program is to just read out the string. Although the lower bound for the Kolmogorov–Chaitin complexity of an object is difficult to obtain [20], the upper bound is easy to get, which are just the Shannon entropy or the LZ complexity. For dynamical systems and Markov information sources, this upper bound can almost surely be achieved [170].

Next, we explain a widely used entropy measure, the approximate entropy. The approximate entropy amounts to taking $q = 1$ in Equation (55) at a fixed scale ϵ and two small embedding dimensions (say m_0 and $m_0 + 1$) instead of taking the limits of $\lim_{\epsilon \to 0}$ and $\lim_{m \to \infty}$. While it is closely related to the sample entropy, it is not as effective as the sample entropy in resolving the scaling behavior. This is part of the reason that multiscale entropy is built op top of the sample entropy. For more details, we refer to [171].

Finally, we explain the permutation entropy (PE) [172]. Due to its simplicity, it has found numerous applications in time series analysis. Here, we describe PE following the notations of [173].

We start from an m-dimensional embedding vector, $X_i = [x(i), x(i + L), \cdots, x(i + (m - 1)L)]$. Let us sort the elements of the vector in ascending order, $[x(i + (j_1 - 1)L) \leq x(i + (j_2 - 1)L) \leq \cdots \leq x(i + (j_m - 1)L)$. When an equality occurs, e.g., $x(i + (j_{i1} - 1)L) = x(i + (j_{i2} - 1)L)$, we choose their natural order, i.e., if $j_{i1} < j_{i2}$, then

$x(i + (j_{i1} - 1)L) \leq x(i + (j_{i2} - 1)L)$. This way, the vector X_i is mapped onto a sequence of numbers, $(j_1, j_2, \cdots, j_m)$. Permutating it, we see that there are a total of $m!$ distinct combinations of $(j_1, j_2, \cdots, j_m)$. Each permutation can be considered as an m-dimensional symbol. Therefore, the reconstructed trajectory in the m-dimensional space is mapped to a m-dimensional symbol sequence. Let $P_1, P_2, \cdots, P_K$ be the probability for the $K \leq m!$ distinct symbols. The PE, denoted by E_p, for the time series $\{x(i), i = 1, 2, \cdots\}$ is defined as

$$E_p(m) = -\sum_{j=1}^{K} P_j \ln P_j. \tag{94}$$

The maximum of $E_P(m)$ is $\ln(m!)$ when $P_j = 1/(m!)$. It is convenient to normalize it to obtain

$$0 \leq E_p = E_p(m)/\ln(m!) \leq 1. \tag{95}$$

E_p essentially measures the randomness of the time series under study: with the passing of time, if data measured from a system become more regular, then E_p of the corresponding data becomes smaller. This statement suggests that if one wishes to detect dynamical changes in a system, one can partition a time series into short windows, compute PE for each window, and examine how PE changes with the window [173].

The construction of PE may be considered a generalization of symbolic dynamics of dynamical systems for finite data, recalling that the essence of symbolic dynamics is to map a trajectory in certain space to a few subspaces, such as a trajectory defined in the unit interval $[0\ 1]$ to two sub-intervals, $[0\ 1/2)$ and $[1/2\ 1]$. The usefulness of symbolic dynamics is a strong hint that PE is often very useful for analyzing complex time series.

While the connections among some of the complexity measures discussed here are obvious, a more comprehensive answer also exists. This, however, has to wait until we introduce a new complexity measure, SDLE, in Section 4.

3. Adaptive Detrending, Denoising, Multiscale Decomposition, and Fractal Analysis

Observational data may manifest both ordered and disordered behavior. To fully characterize a complex signal, it is desirable to synergistically use chaos and random fractal theory [38]. However, this goal is not easy to achieve, since a measured data set often contains noise and may also be nonstationary. This makes detecting chaos very difficult. On the other hand, many phenomena contain a rhythmic activity, such as diurnal cycle. This makes fractal analysis difficult since the essence of a fractal is scale-free. To tackle these problems, frequency-domain filtering and wavelet analysis have been widely used to filter away the undesired features in the data. With the rapid accumulation of complex data in all branches of science and engineering, it is important to have better approaches to solve these problems. In this section, we discuss an adaptive algorithm, which has a number of interesting properties: (1) it can accurately determine a trend in the signal; depending on the purpose of applications, one may treat the trend and associated nonstationarities as noise, and remove them, or retain them, as the signals one wishes to further study (such as the global warming trend); (2) it is more superior in reducing noise in the signals than linear filters, wavelet methods, and chaos-based methods; (3) it can conveniently decompose a complex signal into many functions of different frequency; (4) it is excellent in obtaining fractal properties from the data, especially when the data contain a strong and nonlinear trend. The method has been successfully applied to study traffic flow [146,174], various kinds of geophysical data including soil temperature, soil moisture, air temperature, and wind speed [175–178], tree rings [147], variation of electricity consumption with time [179], single neuron firing [145], clinical scalp EEG [180], ngram usage [149], quantum modeling of exciton diffusion in light harvesting systems [181], sentiments in novels [182,183], newspaper advertisements [184], textual cultural heritage [185], and global terrorism [148]. The method will be very useful for analyzing various kinds of geophysical time series that have been rapidly accumulating in recent years. Here, we will only present the key elements of the method; a concrete example of combining this method with a machine-learning method (random forest) for distinguishing epileptiform discharges from normal electroencephalograms can be found in Li et al. [180].

3.1. Adaptive Detrending, Denoising, and Multiscale Decomposition

The method is based on adaptive filtering [149,155,156,186]. It works this way: first we partition a time series into many segments. Let the length of each segment be $w = 2n + 1$ points, and neighboring segments overlap by $n + 1$ points. As we will see later, using segments with length containing odd number of sample points ensures symmetry. This operation also introduces a time scale $\frac{w+1}{2}\tau = (n+1)\tau$, where τ is the sampling time. For each segment, whose sample points represent a small portion of the curve we are studying, we assume the curve can be approximated by its Taylor series expansion very well. This suggests us to fit the segment by a polynomial of order M. Minimizing the error, the obtained polynomial fitting becomes the best local fitting. Here, an important parameter is the polynomial order M. When $M = 0$, the fitting is piece-wise constant. When $M = 1$, the fitting is locally linear (not necessarily also globally linear). Let $y^{(i)}(l_1)$, $y^{(i+1)}(l_2)$, $l_1, l_2 = 1, \cdots, 2n + 1$ be the fitted polynomial for the i-th and $(i + 1)$-th segments. The fitting for the overlapped part of the two adjacent segments can be obtained by properly combining these two polynomials:

$$y^{(c)}(l) = w_1 y^{(i)}(l + n) + w_2 y^{(i+1)}(l), \quad l = 1, 2, \cdots, n + 1 \tag{96}$$

The two weights, $w_1 = \left(1 - \frac{l-1}{n}\right), w_2 = \frac{l-1}{n}$, can be written as $(1 - d_j/n), j = 1, 2$, where d_j are the distances of the point from the centers of the two fitted polynomials. Therefore, the weights decrease linearly with the distance from the center of the segment. The weighting ensures symmetry. The scheme ensures that the overall fitted curve is continuous everywhere, has the right- or left-derivatives at the boundary, and is differentiable at non-boundary points.

The adaptive filter can readily determine any kind of trend from the data. An example for determining the trend from the global annual sea surface temperature (SST) data is shown in Figure 24a, where the blue straight line is the global linear fit, the black curve is the global second-order polynomial fit, and the red curve is the adaptive trend with a window size about the half of the total data length. It is amazing that with such a large window size, not only the global warming trend but also the local brief cooling periods are clearly shown. In fact, the residual noise (i.e., the difference between the fitting and the original data) shown in Figure 24b with these fits is comparable to that obtained by empirical mode decomposition (EMD) [187]. Since EMD involves dyadic decomposition, while the window size used by the adaptive method is continuous, the adaptive filtering is more flexible and can be accurate.

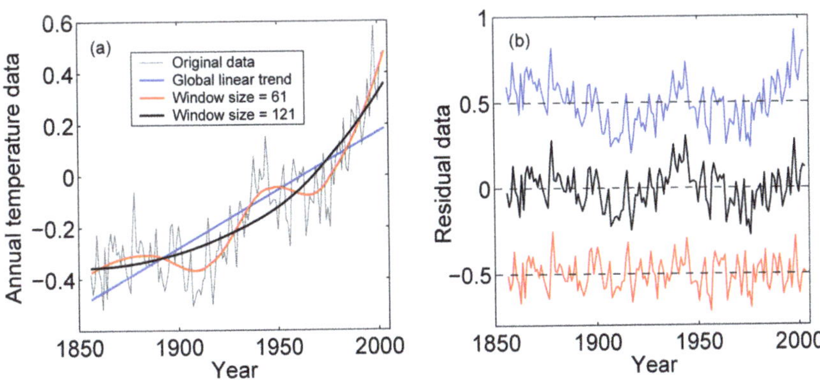

Figure 24. Analysis of the annual sea surface temperature (SST) data: (**a**) the original data and trend signals of different resolutions, (**b**) the residual signals.

Note that whether the trend is considered noise or the desired signal depends on one's purpose. When the trend is considered noise, the approach is a high-pass filter. When the trend signal is considered the desired signal, then the approach is a low-pass filter. We can also take two window sizes and determine two trend signals. If we take the difference between them, then the approach becomes a band-pass filter. More generally, if we use a series of window sizes, $w_1 = 2n_1 + 1 < w_2 = 2n_2 + 1 < w_3 = 2n_3 + 1 < \cdots$ and get the corresponding trend signals. The difference between the two trend signals of window sizes $w_i = 2n_i + 1$ and $w_j = 2n_j + 1$ is called a band-limited signal, with cutoff frequencies $1/(n_i\tau)$ and $1/(n_j\tau)$, where τ is the sampling time. These signals are called intrinsically band limited functions (IBFs) [154]. For an interesting application of the scheme (removing an ECG component from an EEG measurement for the study of apnea), we refer to [156].

The adaptive filter discussed here is more effective than linear filters, the wavelet method, and chaos-based approaches in reducing noise [155,156]. To see this, we have shown a comparison of these methods in Figure 25 for reducing measurement noise in the chaotic Lorenz system. The residual noise, characterized by the root-mean-square error (RMSE), is the smallest for the adaptive filter [154].

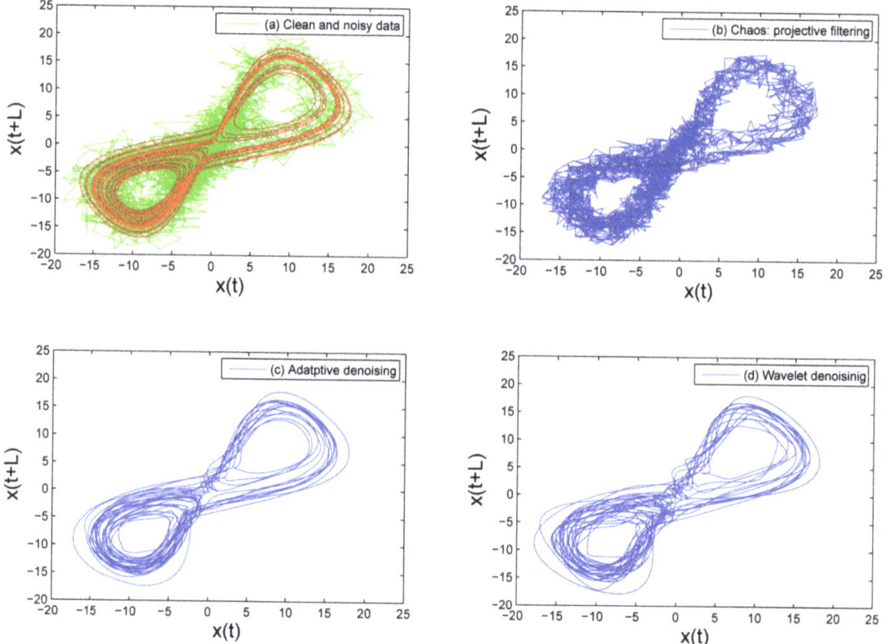

Figure 25. Denoising of the chaotic Lorenz signal: (**a**) phase diagrams constructed from the the clean and the noisy signal, which are marked as green and red, respectively; (**b**) the filtered signal obtained by a chaos-based approach; (**c**) the filtered signal obtained by the adaptive algorithm; and (**d**) the filtered signal obtained by a wavelet method.

To better appreciate the above discussed properties of the filter, let us consider a power load time series measured at a power plant in Guilin during a long time period (from 1 January 2005 to 29 April 2010). Guilin is a very well-known tourism city, with the saying "Guilin's landscape is the most uniquely beautiful in the world". Power load time series may be equated to electricity consumption in a city. Interesting questions one can ask include whether electricity consumption may be correlated with climate variations. The raw load time series from Guilin is shown in Figure 26a as the blue curve. Here, the sampling time is 15 min. We observe that the data are very irregular and non-stationary, reflecting that the city's businesses and population must have been changing a lot during the period the data were collected. The trend signal for the raw data is shown in Figure 26a as the red curve; it is obtained by using a window size of 699 sample points. The order of the polynomial used for fitting is 2 (if the raw signal is very spiky, then higher-order polynomials are recommended).

To facilitate further discussion, we denote the raw data by $x(t)$, and the trend signal by $trend(t)$. Then we have

$$x_{detrended} = x(t) - trend(t) \tag{97}$$

In order to see the details of $x_{detrended}$, Figure 26b shows a small segment of it as the blue curve. We observe a diurnal cycle in the data. This is reasonable since electricity consumption in daytime and during night has to be quite different. The signal does not have a fixed amplitude though, as it is still quite noisy. This noise, which is high frequency, can also be removed by applying the adaptive filter again, with a small window size. The trend thus determined will better represent the diurnal cycle. It is a band-pass signal. An example of this signal is shown in Figure 26b as the red curve, where we used a window size of 9 and a polynomial of order 2. From this signal, we can construct a phase diagram with delayed coordinates. This is shown in Figure 26c. A limit cycle-like structure does emerge.

We can further analyze the oscillatory feature of the trend signal by computing power-spectral density (PSD) from the data. The result is shown Figure 26d, where the blue, red, and green curves are for the raw, detrended, and the band-passed signals, respectively. The PSD curves show very sharp spectral peaks at frequency of 1 day^{-1} and its harmonics. Note the blue curves are basically covered by the other two colors, except at the very low frequency (i.e., close to 0 Hz). This is due to the red trend signal shown in Figure 26a.

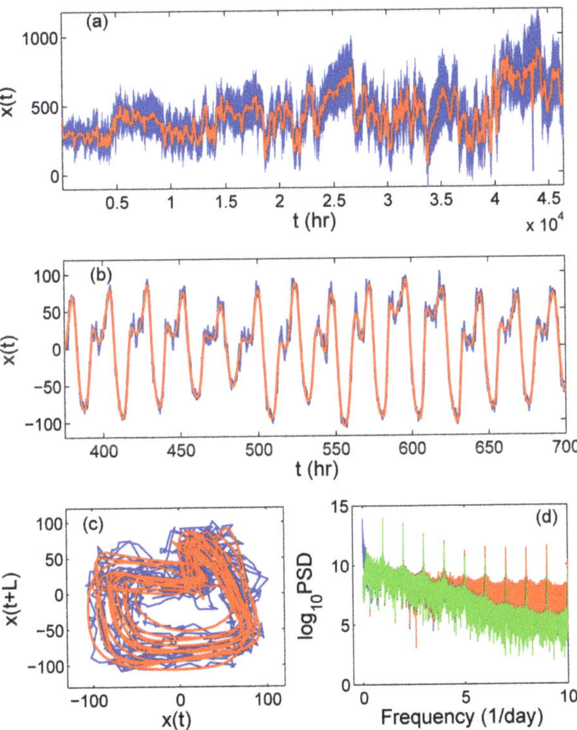

Figure 26. Electricity consumption analysis: (**a**) raw data (blue) and the trend signal (red); (**b**) enlargement of the high-frequency load data showing the diurnal cycle (blue) and its filtered band-pass data (red); (**c**) 2D phase diagrams constructed from the data shown in (**b**); (**d**) PSD for the raw, detrended, and denoised data, which are marked by blue, red, and green, respectively.

3.2. Adaptive Fractal Analysis (AFA)

In the past three decades, many efforts have been made to estimate H, the most important parameter for random fractals. As a result, many excellent methods for estimating H have been proposed. Among them is the celebrated detrended fluctuation analysis (DFA) [151,152]. It works as follows: To analyze a time series, $x_1, x_2, x_3, \cdots$, one first determines its mean $\bar{x}$, then constructs a random walk process using Equation (76). By doing so, one has assumed that the data are like a noise process. One then partitions the random walk into non-overlapping segments of length l (therefore, the number of distinct segments is not larger than N/l, where N is the length of the time series). One furthers determines the local trend in each segment by using the best linear or polynomial fitting. This procedure is schematically shown in Figure 27, where a short EEG signal is used as an example. Finally, one obtains the difference between the original "walk" and the local trend. Denote it by $u(n)$. H is then estimated by

$$F_d(l) = \left\langle \sum_{i=1}^{l} u_l(i)^2 \right\rangle^{1/2} \sim l^H \tag{98}$$

where the angle brackets is a short-hand notation for averages over all the segments.

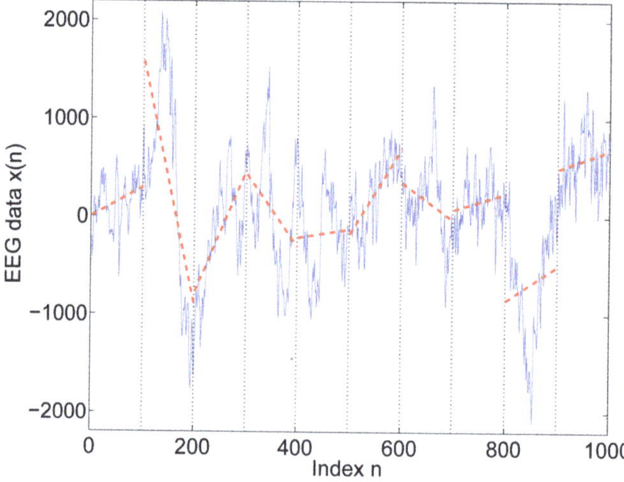

Figure 27. Schematic of DFA.

Although DFA is very good in many applications, when a signal has a strong nonlinear trend, such as an oscillatory component or a rhythmic activity, there may exist large discontinuities in adjacent segments of DFA (see Figure 27). These discontinuities can cause big problems. This problem can be readily solved by the adaptive fractal analysis (AFA) [149,154,157]. The difference with DFA is that we now have a globally, not only continuous but also almost everywhere, differentiable trend [155,156]. Denote it by $v(i)$. The difference between the original random walk process $u(i)$ and $v(i)$ can be used to accurately estimate H. The formula is given by [154]

$$F(w) = \left[\frac{1}{N} \sum_{i=1}^{N} (u(i) - v(i))^2\right]^{1/2} \sim w^H. \tag{99}$$

Generalizing to a multifractal analysis, we obtain:

$$F^{(q)}(w) = \left[\frac{1}{N} \sum_{i=1}^{N} |u(i) - v(i)|^q\right]^{1/q} \sim w^{H(q)} \tag{100}$$

where q is a real number. Just as we have discussed earlier, positive q values highlight large values in $|u(i) - v(i)|$, and negative q values highlight small values in $|u(i) - v(i)|$.

Equation (99) can readily be extended to long-range cross-correlation analysis [188] between two series: $x(i), i = 1 \cdots, n$ and $y(i), i = 1 \cdots, n$. Denote their trend signals corresponding to window size w by $\text{trend\_}x^{(w)}(i), i = 1 \cdots, n$ and $\text{trend\_}y^{(w)}(i)$, $i = 1 \cdots, n$, respectively. Then we have

$$F_{xy}(w) = \left[\frac{1}{N} \sum_{i=1}^{N} (x(i) - \text{trend\_}x^{(w)}(i)) \times (y(i) - \text{trend\_}y^{(w)}(i)) \right]^{1/2} \sim w^{H_{xy}}. \tag{101}$$

Following the generalization from Equations (99)–(100), Equation (101) can also be readily extended to multifractal analysis.

Let us now examine the fractal behavior of the power load data of Figure 26a using AFA. To cope with the nonstationary of the data, we partition the data into short windows, then we estimate H for each window. Recalling that the data are sampled 96 times a day, we choose the window size to be one month, containing $96 \times 30 = 2880$ sample points. To improve the resolution of the variation of H, the adjacent windows overlap by half of the window length. Figure 28a shows an example of the scaling analysis using AFA, for an arbitrarily window. The curve is linear for scale up to $w = 2^7$ sample points. It is a little longer than a day. H can be estimated as the slope of the linear portion of the curve. The temporal variation of H is plotted in Figure 28b as the red curve. Interestingly, it has a seasonal variation. To check whether this variation may be correlated with the yearly temperature variation, we have also shown in Figure 28b a curve in black reflecting the temperature variation. To facilitate comparison of the two variables, H and the temperature T, in the same plot, T is transformed to T' according to the following equation,

$$T' = T/100 + 0.5. \tag{102}$$

Interestingly, the local maxima of the $H(t)$ curve correspond to the seasonal minima of the curve for the temperature. This suggests that the power load data are characterized by stronger, persistent, long-range correlations during winter.

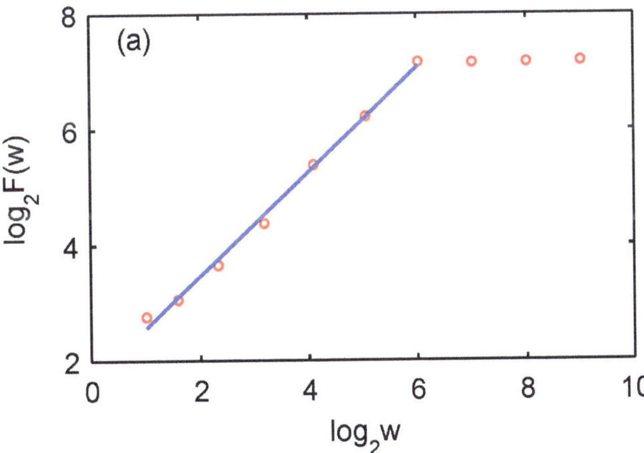

Figure 28. *Cont.*

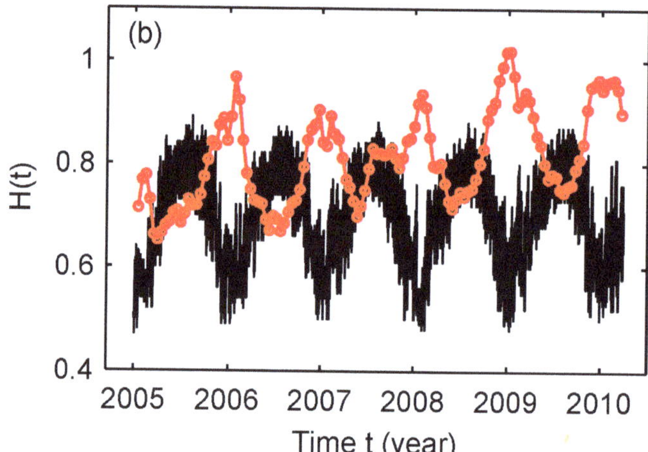

Figure 28. AFA of power load data: (**a**) an example of $\log_2 F(w)$ vs. $\log_2 w$ for the load data of an arbitrarily chosen day, (**b**) temporal variation of the Hurst parameter (red) and the rescaled temperature (black).

4. Multiscale Analysis with the Scale-Dependent Lyapunov Exponent (SDLE)

SDLE is developed for better distinguishing chaos from noise and for better characterizing complex data, especially through obtaining the defining parameters of the data [38,189]. SDLE is closely related to two other methods, the time-dependent exponent curves [73,74,79,81] and the finite size Lyapunov exponent [190–192]. SDLE was first introduced in [38,189], and has been further developed in [193,194] and applied to characterize EEG [143], HRV [195,196], financial time series [76], Earth's geodynamo [197], precipitation dynamics [198], sea clutter [199], THz imagery [200], and evaluate randomness [99]. As with the presentation of AFA, here, we will only present the key elements of the method; a concrete example of combining this method with a machine-learning approach (random forest) for distinguishing epileptiform discharges from normal electroencephalograms can be found in Li et al. [201].

SDLE is based on the evolution of vectors in a high-dimensional phase space. If initially the data are a time series, then one needs to obtain a suitable phase space using delay coordinates, as explained before. If the original data are a scalar random process, then the main advantage of the embedding procedure is to obtain a self-similar vector process from the original self-affine process. This is because x and t have different units and therefore have to be scaled differently in order for them to look "alike". All the components of a vector are of the same nature, and therefore can be stretched or shrunk with the same fashion. Consequentially, whenever a truly random time series is analyzed, the specific value of the embedding dimension m is not important. Often ensuring $m > 1$ is sufficient. After a phase space is obtained, one can examine the evolution of an ensemble of trajectories. Denote the initial distance between two nearby trajectories by ϵ_0. We further denote their *average distance* at time t by ϵ_t, and that at $t + \Delta t$ by $\epsilon_{t+\Delta t}$. A schematic showing how a small distance between two nearby trajectories grows with time is shown in Figure 29. With this setting, we can examine the relation between ϵ_t and $\epsilon_{t+\Delta t}$, where Δt is assumed to be small. When $\Delta t \to 0$, we have,

$$\epsilon_{t+\Delta t} = \epsilon_t e^{\lambda(\epsilon_t)\Delta t}, \tag{103}$$

where $\lambda(\epsilon_t)$ is the SDLE given by

$$\lambda(\epsilon_t) = \frac{\ln \epsilon_{t+\Delta t} - \ln \epsilon_t}{\Delta t}. \tag{104}$$

Equivalently, we can write,

$$\frac{d\epsilon_t}{dt} = \lambda(\epsilon_t)\epsilon_t. \tag{105}$$

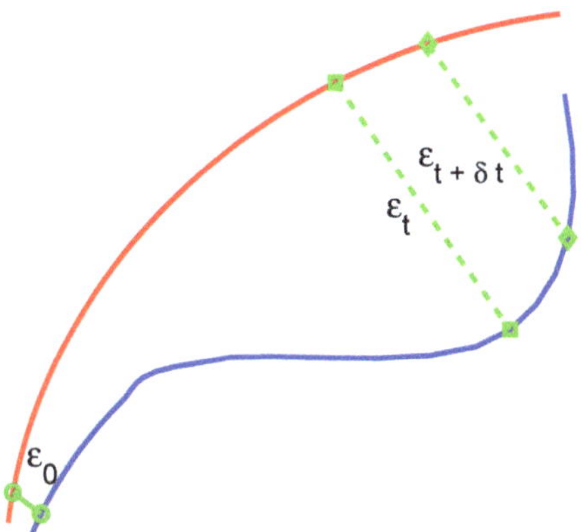

Figure 29. A schematic showing how a small distance between two nearby trajectories grows with time.

Now that we have introduced SDLE, we can better understand the classic algorithm for computing the largest Lyapunov exponent λ_1 discussed earlier [75]. That algorithm assumes $\epsilon_t \sim \epsilon_0 e^{\lambda_1 t}$ and then through averaging estimates λ_1 by $(\ln \epsilon_t - \ln \epsilon_0)/t$. This assumption may not even hold for true chaotic signals. This is reminded in the detail of the schematic plot shown in Figure 29 —$\epsilon_{t+\delta t}$ may be smaller than ϵ_t. As already mentioned, a fundamental difficulty with this assumption is that for any type of noise, when ϵ_0 is small (which is the case when nearest neighbors are used), λ_1 can always be positive, leading to misinterpreting noise as chaos. The reason is simple: ϵ_t will rapidly converge to the most probable distance between the constructed vectors, and thus will almost be surely larger than ϵ_0. However, when we define SDLE using Equation (103), we have not made any assumptions, except Δt being small (usually taken to be the sampling time interval). As we will see, chaos is characterized by a constant $\lambda(\epsilon)$ over a range of ϵ.

In the computation of SDLE, we first examine which embedding vectors defined by Equation (32) fall within the series of shells defined by Equation (40). Then, the evolution of those vector pairs (V_i, V_j) can be monitored, and their average behavior of divergence (not necessarily exponential) can be computed. So far as exponential or power law divergence are concerned, we can exchange the order of taking the logarithm and averaging. Then, Equation (104) becomes

$$\lambda(\epsilon_t) = \frac{\left\langle \ln \| V_{i+t+\Delta t} - V_{j+t+\Delta t} \| - \ln \| V_{i+t} - V_{j+t} \| \right\rangle}{\Delta t} \tag{106}$$

where t and Δt are measured in terms of the sampling time, and the average, denoted by the angle brackets, is over all indices i, j with their corresponding vectors satisfying Equation (40).

The program for computing SDLE is explained in detail in [202], and can be obtained from the authors. The major scaling laws of SDLE that are most relevant for analyzing complex data are summarized below [189]:

- For deterministic chaos,

$$\lambda(\epsilon) \sim \text{constant}, \tag{107}$$

Amazingly, this property can even be observed in finite high-dimensional data, including the Lorenz'96 system, which has dimensions close to 30 [193], and in turbulent isotropic fluid with an integral scale Reynolds number reaching 6200 [203]. In such systems, estimation of dimensions is infeasible.

- As observational data are always contaminated by noise, it is important to have a scaling law for noisy chaos and noise-induced chaos [82,118]. The law reads

$$\lambda(\epsilon) \sim -\gamma \ln \epsilon, \tag{108}$$

  The law pertains to small scales, and $\gamma > 0$ controls the speed of information loss.
- For $1/f^{2H+1}$ processes,

$$\lambda(\epsilon) \sim H\epsilon^{-1/H}. \tag{109}$$

- For α-stable Levy processes,

$$\lambda(\epsilon) \sim \frac{1}{\alpha}\epsilon^{-\alpha}. \tag{110}$$

- For stochastic oscillations, both scaling laws $\lambda(\epsilon) \sim -\gamma \ln \epsilon$ and $\lambda(\epsilon) \sim H\epsilon^{-1/H}$ can be observed when different embedding parameters are used.
- When the dynamics of a system are very complicated, one or more of the above scaling laws may manifest themselves on different ϵ ranges.

It is now clear that with the help of these scaling laws, distinguishing chaos from noise can be readily solved. More importantly, we can now understand very well the nature of each type of behavior of the data by obtaining the defining parameters for that behavior.

To illustrate how SDLE characterizes chaotic features and the effect of noise, let us briefly discuss Boolean chaos in a ring oscillator. Boolean chaos normally refers to the continuous time dynamics of a system of interconnected digital gates whose output updates are not regulated by an external clock. Recently, an alternative Boolean architecture for generating chaotic oscillations was proposed by Blakely et al. [204]. See Figure 30. Three typical kinds of waveforms for the variable v_3 are shown in Figure 31. The chaotic behaviors of the oscillations can be aptly characterized by SDLE, as shown in Figure 32—the Figure actually has shown more than chaos: the chaotic behavior is best defined for the variable v_1, and the effect of noise is most clearly visible for the variable v_3. The reason is straightforward: in this series circuit, the noise at the third gate is the largest.

Among the many properties of SDLE, two make it unique. One is its skill of dealing with nonstationarity, including detecting intermittent chaos from models as well as observational data [155,195]. To understand intermittency, it is useful to consider the evolution of river flow dynamics over 1 year. With some thinking, one can readily realize that the time period may be divided into two periods, wet and dry, where the wet season may be associated with frequent rain and snow melting, and the dry sea may be largely associated with no or little rain, and constant evaporation. The river flow dynamics must be very different in these two periods. Since standard methods for detecting chaos assume the existence of a single chaotic attractor, those methods are ill positioned to unambiguously determine whether river flow dynamics are chaotic or not. To illustrate how intermittent chaos can be detected by SDLE, Figure 33 shows an example of the Umpgua river in Oregon. The exponential divergence is evidently shown by the linear $\ln \epsilon(t)$ vs. t curve for t going from about 20 days to about 100–150 days. Consequentially, there are well-defined plateaus of SDLE, i.e., a constant SDLE, shown in Figure 33a2 (the blue curves). It is also interesting to note the scaling law of Equation (108) on small scales. This is caused by the faster-than-exponential growth of small distances in the initial period (less than 20 days), and it is mainly due to stochasticity, i.e., randomly driven by snow melting, rain, etc., besides measurement noise. The chaotic and the noisy dynamics depicted in Figure 33 can be improved by using the adaptive algorithm discussed earlier. The results using the filtered data are shown in Figure 33 as the red curves.

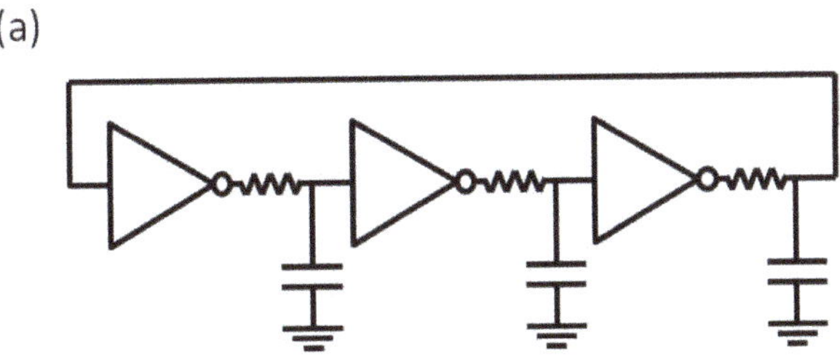

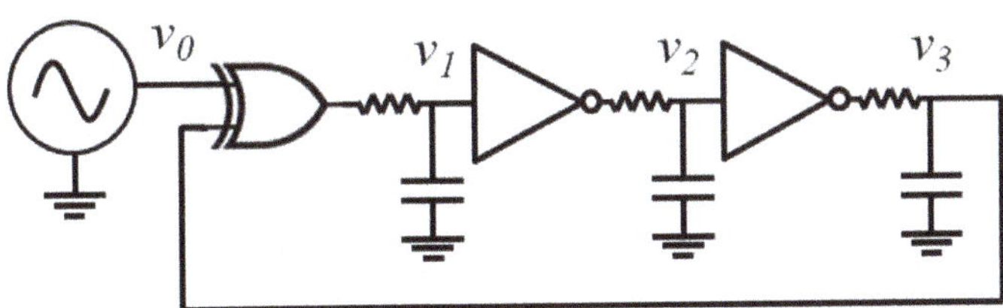

Figure 30. (**a**) A three inverter ring oscillator. (**b**) A ring oscillator driven by an external periodic signal. The resistor-capacitor stages may represent either discrete components or the finite bandwidth of non-ideal inverters.

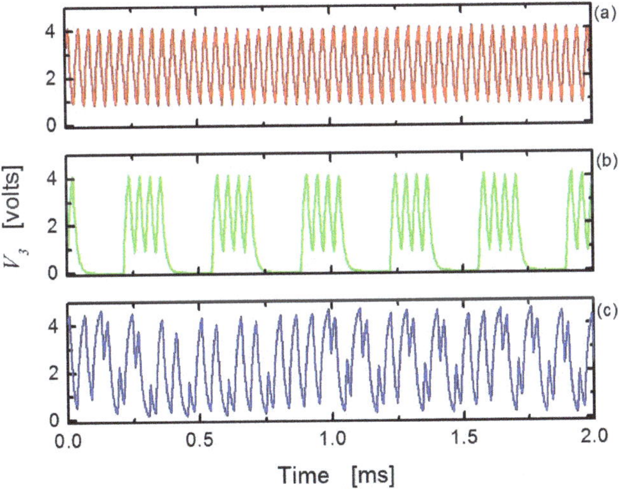

Figure 31. Typical oscillations displayed by an experimentally implemented ring oscillator. (**a**) Self oscillations occur with the input held constant above the threshold. (**b**) Slow driving produces periodic bursts of self oscillation. (**c**) Faster driving produces an irregular oscillation.

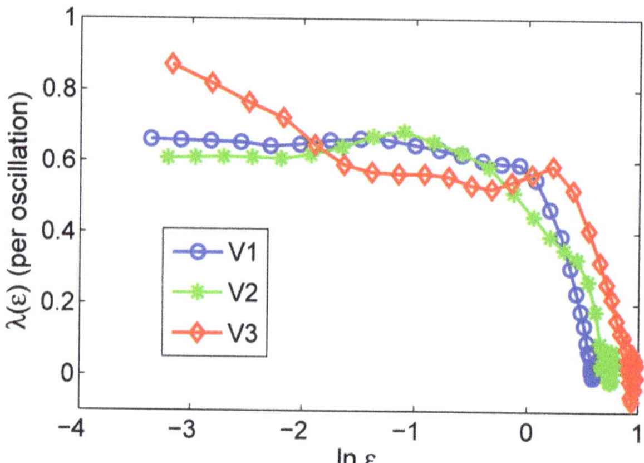

Figure 32. SDLE calculated from the experimental time series of v_1 (blue), v_2 (green), and v_3 (red).

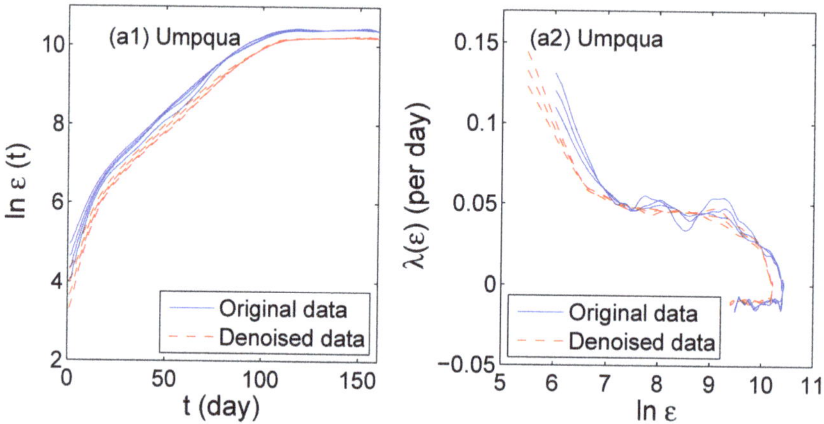

Figure 33. Intermittent chaos in the Umpqua river. Shown in (**a1,a2**) are the error growth curves and SDLE curves, respectively. The blue curves are for the original data, while the red curves are for the filtered data. The embedding parameters used in the computation are $m = 6$, $L = 3$. Three different shells specified by Equation (40) are used. These curves collapse on each other, except when t is small. This highlights that the computational results are essentially independent of the initial shells chosen.

The other unique property of SDLE is that it provides a unified framework to understand other complexity measures. Concretely, the values of other complexity measures can be inferred from the values of SDLE at specific scales. This statement is best appreciated by using signals with phase-transition-like changes (or regime changes). Because of this, let us use electroencephalography (EEG) data with epileptic seizures. A typical result is illustrated in Figure 34, where we observe that the temporal variations of the Lyapunov exponent, the correlation dimension, the correlation entropy, and the Hurst parameter are similar to the values of SDLE either on smaller or on larger scales. In fact, the list of the complexity measures can be expanded to include the permutation entropy, the LZ complexity, and the energy of the EEG waves such as $\alpha, \beta, \delta, \theta$. For the details, we refer to [38]. While here these connections are illustrated using EEG data, the issue is relevant to many other situations, including paleoclimatological data and fMRI data analysis. This property highly suggests that SDLE can serve as a basis for unifying commonly used complexity measures.

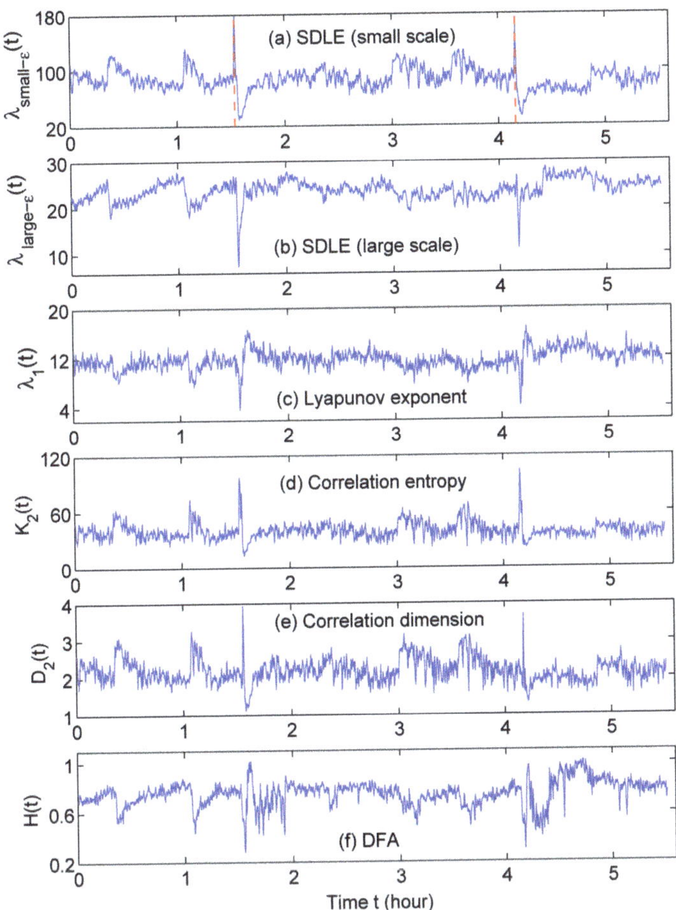

Figure 34. Epileptic seizure detection from continuous EEG data of a patient, illustrating that SDLE can serve as a basis to unify commonly used complexity measures. Shown in the figure are the temporal variations of (**a**) $\lambda_{\text{small}-\epsilon}$, (**b**) $\lambda_{\text{large}-\epsilon}$, (**c**) the LE, (**d**) the K_2 entropy, (**e**) the D_2, and (**f**) the Hurst parameter. Seizure occurrence times were determined by clinical experts and were indicated here as the vertical dashed lines.

5. Toward a Theory of Social Complexity

World civilization continues to progress. Yet, difficulties and suffering befall the world from time to time. While many difficulties and sufferings are from nature, some are inflicted by mankind itself. The major problems facing humanity are constantly changing over time. Modern problems that confound humans include: How can we avoid the chain collapse of the stock markets? How soon will the American politics, which was so divided during Trump's presidency, be back to "normal"? Will the COVID-19 virus completely disappear? Why do some terrorist organizations use suicide bombers, and others do not? While there are many more similarly important current issues, there are also fundamental problems of a different nature that span the long river of time: How have the major problems of each era evolved into these problems today? Are there similarities in major issues in different eras? Is there a unified theory to understand the evolution of history? With the Internet and social media generating unprecedented amounts of data related to individual and group behaviors, these and many other major issues can finally be hoped to be addressed by computational means.

Computational social science was born out of the big data of the Internet and social media [205] and will continue to be the biggest beneficiary of big data. Indeed, many fascinating studies on the detailed behaviors of individuals and their interactions have been published. Now it is time to seriously ponder how to develop a theory of social complexity with lasting value. Natural science has been making every effort to pushing its frontiers to the largest and the smallest scales. In social science, the smallest scale is individuals, and the largest are countries and regions consisting of a number of countries. To make social science truly a science, the country-wide scale has to be focused on. Therefore, a significant portion of the theory of social complexity has to be centered on the quantification of evolution of political processes of countries and international relations. Realizing this, one can be sure that complexity science will definitely play a fundamental role in social science that is not rivaled by black-box machine-learning based approaches, since machine-learning cannot be 100% correct, and the cost inflicted by any mistake in forming important policies could be enormous. This is completely different from e-commerce, as errors or mistakes there, although still costly, could be remedied. Here, we focus on the scaling law governing the complexity of world-wide political evolution.

Major data for demographic research include data from the web, social media, cell phone and credit card usage, digitized historical data, and massive media reports data, including printed newspapers. While all of them are useful for studying individual behavior and human interactions, the last, the massive media reports data, are most appropriate for the purpose of studying the complexity of world-wide political evolution since every aspect of social interactions has been more or less covered by news reports. Fortunately, such data are available now. It is called the Global Database of Events, Language, and Tone (GDELT). It is a new initiative based on terabytes of information to construct a catalog of all major human societal activity across all countries of the world, containing more than 650 million unique events across all countries, during the period from 1979 to the present. GDELT events are drawn from a wide array of news media, both in English and non-English, from across the world, ranging from international to local sources in nearly every country. Each event has a number of attributes, including two actors, such as USA and China, coordinates of geolocation, time of the event, average tone of the report, and most importantly, a value called Goldstein-scale intensity [206], which measures the degree of cooperation or conflicts between the two actors. Altogether, there are 20 classes of events, where each class also consists of a few to a few dozen independent events, yielding a total of 290 independent events. This strategy separates GDELT from all other keyword-based analyses, and mathematically speaking is more desirable, as working with independent events is fully consistent with the probability axiom system of Kolmogorov.

GDELT was produced by the TABARI automated coding software (http://eventdata. psu.edu/software.dir/tabari.html) using the CAMEO event and actor coding system [207]. TABARI works with dictionaries of a very large set of verb phrases (>15,000 phrases) and noun phrases (>40,000 phrases) in combination with shallow parsing of English language sentences to identify grammatical structures such as subject-verb-object, compound subjects and objects, and compound sentences. CAMEO is an update of earlier (1960s) event coding taxonomies, with changes introduced by automated coding and new behaviors, such as suicide bombings. CAMEO provides a detailed and systematic taxonomy for coding contemporary political actors, including international, supranational, transnational, and internal actors. An earlier version of this system recently was successfully employed in the DARPA ICEWS project [208] to code 25 gigabytes of Asian news reports involving more than 6.7 million stories, which provided the key input for forecasting models with accuracy, sensitivity, and specificity all exceeding DARPA's pre-set criteria. The data are updated every 15 min and are open access at http://gdelt.utdallas.edu; tools for working with the data are discussed both on that web site and at http://gdeltblog.wordpress.com.

Political processes have a number of important attributes, such as large momentum, lack of predictability, and apparently similar patterns across history. While the last attribute may entice one to model historical processes using periodic models (e.g., cliodynamics [209,210]), to accommodate all the attributes of political/historical processes, one has to go way beyond modeling by cyclic processes. We surmise that random fractal theory [38] may offer an interesting means to quantify political processes. Our finding based on Googlebook's Ngram data that social phenomena and human response to natural phenomena possess different kinds of long-range correlations [149] further motivates us to employ the key concept from random fractal theory [38], the Hurst parameter H, to determine whether political processes may possess long-range correlations and, if yes, to understand their consequences.

As we have mentioned, one of the most important attributes of the political events data is the Goldstein scale [206], which characterizes the degree of conflict or cooperation between the two actors of the event. As on each single day, for each country, there are many events. Therefore, one can readily compute the daily average of the Goldstein scale for the country. This daily average changes with time, i.e., it is a time series. Therefore, we can analyze this time series by computing the Hurst parameter using the most robust method, the adaptive fractal analysis introduced earlier. More concretely, we can partition the daily average Goldstein scale time series into small segments, compute H for each segment, and examine the variation of H with time. By overlapping adjacent segments by 1 month, the temporal resolution of the H curve is 1 month. Four examples of the variation of H with time are shown in Figure 35, for USA, China, Turkey, and Indonesia. In fact, in each subplot, two curves are plotted. The blue curve has a temporal resolution of 1 month, while the red one has a temporal resolution of 1 year. To better understand these curves, we focus on the red curves. First, we observe that all curves lie between 0.5 and 1, meaning that all political processes are characterized by long-range correlations. Second, we observe that the variation of $H(t)$ is different for different countries. In fact, this variation is dictated by the major political events that occurred in the respective countries. In the case of USA, for example, there are three large decreases in $H(t)$. The last two can be easily associated with the two Iraq wars. The most interesting is the first sharp drop in $H(t)$ that occurred around 1987. This suggests that the cold war between the USA and former Soviet Union also had greatly strained the US. In the case of China, local maxima and minima of the $H(t)$ curve correspond to changes of national leaders very well (concretely, one local maximum is at 1997, when DENG Xiaoping died, and JIANG Zemin took over the leadership; two local mimina are at 2002 and 2012, when HU Jintao took over the power from JIANG, and when XI Jinping took over the power from HU, respectively). This is also observed for many other countries. In general, $H(t)$ will increase when policies in a country are enhanced and will decrease when internal/external conditions change such that many policies of a country have to be modified or replaced by new ones. Therefore, the temporal variation of $H(t)$ parsimoniously and accurately summarizes the evolution of the political processes (and hence history) of a country.

There is an important implication of the above understanding to the overseas infrastructure investment. This is a key issue that has to be seriously considered by China in the implementation of the Belt and Road Initiative, and by any other countries who wish to make infrastructure investments overseas. The necessary condition for the smooth implementation of a project is that the duration of the construction of the project is shorter than half of the average cycle of policy changes in a targeted country. To understand this, consider construction of high-speed rail as an example. We can now understand why the Ankara-Istanbul line, even though constructed for 11 years, from 2003 to 2014, was successfully completed. It was in an increasing $H(t)$ episode. Such long episodes are rare among all the countries in the world though. In contrast, the $H(t)$ curve for Indonesia varies with a much higher frequency. Indeed, there is a strong anti-China sentiment in Indonesia partly induced by the construction of a high-speed rail there.

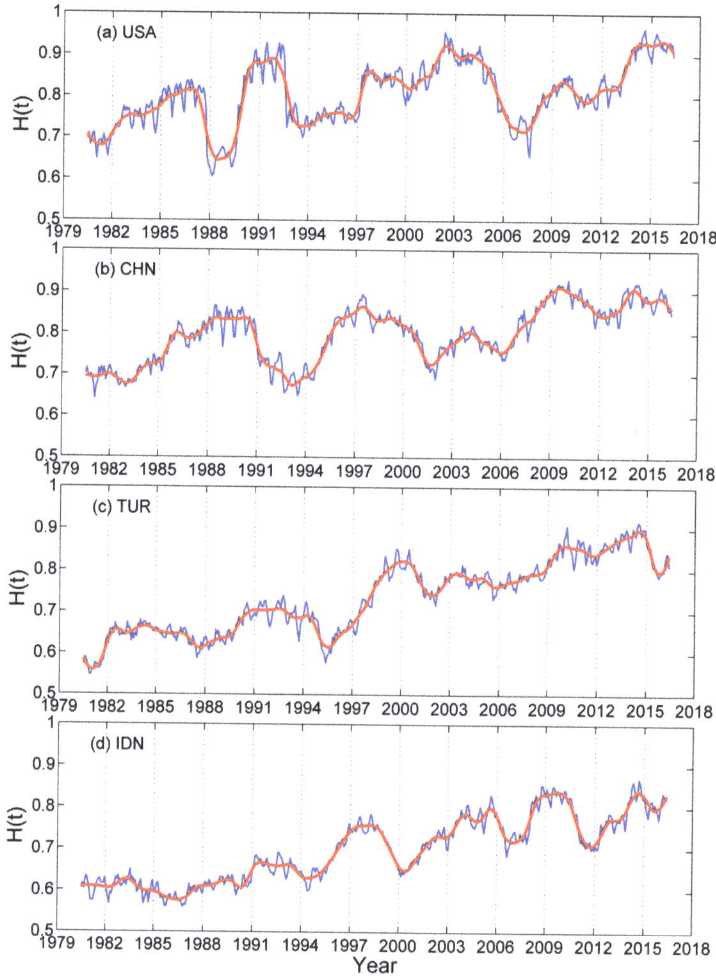

Figure 35. Long-range correlations (or inertia) of political processes in four countries: (**a**) USA, (**b**) China, (**c**) Turkey, and (**d**) Indonesia. The blue curve has a temporal resolution of 1 month, while the red one has a temporal resolution of 1 year.

6. Concluding Remarks and Future Directions

With the rapidly approaching 5G era, and 6G also on the horizon, the rapidly accumulating big data in science, engineering, and society will soon become enormously bigger. No one can afford not to grasp such an unprecedented opportunity. While computer scientists are diligently developing more powerful database management and machine-learning approaches, it is time to go to the next phase. This next phase has to start from deeply studying the dynamics of all the dynamical processes that have been captured by the big data and the mechanisms of how the human brain works. So far as data analysis is concerned, we can easily envision that mainstream machine-learning and complexity science based approaches will not only complement but also interact with each other increasingly tightly in future. To help accelerate this marriage, we advocate to synergistically use mainstream machine learning based approaches and multiscale approaches from complexity science. Concretely, we have discussed two multiscale approaches. One is based on adaptive filtering. It can accurately determine arbitrary trends from any kind of complex data, reduce noise from data, and estimate the Hurst parameter and multifractal spectrum for complex time series. The other originates from chaos theory and can unify the major complexity measures that have been used today. They are especially useful in obtaining key parameters characterizing a dynamical system and thus can be used to help design better unsupervised machine learning schemes. To help readers better understand these techniques, the article is written both as a tutorial and a survey. It can be used as a course material, including summer extensive training course—in fact, the material presented here has been shaped by a few summer extensive training courses conducted by one of the authors (J. Gao). When the material is used for teaching purposes, it will be beneficial to motivate students to have hands-on experiences with the many methods discussed in the paper. Instructors as well as readers interested in the computer programs (mostly in matlab) for the analysis are welcome to contact the corresponding author.

While various applications of the concepts and methods presented in the paper are discussed, to further stimulate readers to think and apply the methodology, we formulate a number of theoretically or practically important questions to end the paper:

- In Section 2.3.5, we find that citations to the original works on chaos synchronization decay exponentially. We also know that the general citation of scientific works decay as a power law. Can a model be developed that not only reconciles this marked difference but also finds a causal connection between them?
- We have observed in Figure 3 that the distribution of forest fires in USA and China is very different. It is known that casualties in fire fighting are much bigger in China than in the USA. Can the information in the distribution of forest fires be used to design better fire fighting strategies so that casualty and property loss can be both minimized?
- What is the fundamental difference between nation states with and without negative feedbacks?
- Which kinds of data are better in modeling the fundamental dynamics of cultural changes, the sparse data from poll/survey or massive real-time data streams acquired through sensors, mobile platforms, and the Internet?
- Will chaos theory in the strict mathematical sense be relevant to social emergent behaviors such as popular uprising? For this purpose, reading some fascinating descriptions from Victor Hugo's *Les Miserables* (Penguin Classics, Translated and with an introduction by Norman Denny) could be stimulating:
  "Nothing is more remarkable than the first stir of a popular uprising. Everything, everywhere happens at once. It was foreseen but is unprepared for; it springs up from pavements, falls from the clouds, looks in one place like an ordered campaign and in another like a spontaneous outburst. A chance-comer may place himself at the head of a section of a crowd and lead it where he chooses. This first phase is filled with terror mingled with a sort of terrible gaiety ..."

Author Contributions: Conceptualization, J.G. and B.X.; methodology, J.G.; software, J.G.; validation, J.G. and B.X.; formal analysis, J.G.; investigation, J.G.; resources, J.G.; data curation, J.G.; writing—original draft preparation, J.G.; writing—review and editing, J.G. and B.X.; visualization, J.G.; supervision, J.G.; project administration, J.G. and B.X.; funding acquisition, J.G. and B.X. Both authors have read and agreed to the published version of the manuscript.

Funding: This research was supported by the National Natural Science Foundation of China under Grant Nos. 71661002 and 41671532, the Fundamental Research Funds for the Central Universities, and the National Key Research and Development Program of China, grant number 2019AAA0103402.

Institutional Review Board Statement: Not applicable.

Informed Consent Statement: Not applicable.

Data Availability Statement: Not applicable, as all used data are publicly available.

Acknowledgments: The authors thank three anonymous reviewers for very helpful comments. One of the authors (JG) benefited tremendously from participating the long-term program on culture analytics organized by the Institute for Pure and Applied Mathematics (IPAM) at UCLA, which was supported by the National Science Foundation. The authors also thank Bin Liu for preparing Figure 3 and Zhenzhen Wang for preparing Figure 6.

Conflicts of Interest: The authors declare no conflict of interest.

References

1. Manyika, J.; Chui, M.; Brown, B.; Bughin, J.; Dobbs, R.; Roxburgh, C.; Byers, A.H. Big Data: The Next Frontier for Innovation, Competition, and Productivity. 2011. Available online: https://www.mckinsey.com/business-functions/mckinsey-digital/our-insights/big-data-the-next-frontier-for-innovation (accessed on 17 June 2021).
2. Boyd, D.; Crawford, K. *Six Provocations for Big Data*; The Center for Open Science: Charlottesville, VA, USA, 2017. [CrossRef]
3. Sakaki, T.; Okazaki, M.; Matsuo, Y. Earthquake shakes Twitter users: Real-time event detection by social sensors. In Proceedings of the 19th International Conference on World Wide Web, Raleigh, NC, USA, 26–30 April 2010.
4. Ginsberg, J.; Mohebbi, M.H.; Patel, R.S.; Brammer, L.; Smolinski, M.S.; Brilliant, L. Detecting influenza epidemics using search engine query data. *Nature* **2009**, *457*, 1012–1014. [CrossRef] [PubMed]
5. Butler, D. When Google got flu wrong. *Nature* **2013**, *494*, 155–156. [CrossRef] [PubMed]
6. Available online: http://baike.baidu.com/link?url=zP_UWpBFHUI5PYen8cvlzKsXUhprdWaw97tSQ3L7ffOjjUYCTfnq_NMnxZG6IsKS5t0y85b2vMuIPa02atZFjStLmWoJMAFEvlfGlfvJ7zK#f-comment (accessed on 17 June 2021).
7. Dash, S.; Shakyawar, S.K.; Sharma, M.; Kaushik, S. Big data in healthcare: Management, analysis and future prospects. *J. Big Data* **2019**, *6*, 1–25. [CrossRef]
8. Raghupathi, W.; Raghupathi, V. Big data analytics in healthcare: Promise and potential. *Health Inf. Sci. Syst.* **2014**, *2*, 1–10. [CrossRef] [PubMed]
9. Reichman, O.J.; Jones, M.B.; Schildhauer, M.P. Challenges and opportunities of open data in ecology. *Science* **2011**, *331*, 703–705. [CrossRef]
10. O'Donovan, P.; Leahy, K.; Bruton, K.; O'Sullivan, D.T.J. Big data in manufacturing: A systematic mapping study. *J. Big Data* **2015**, *2*, 1–22. [CrossRef]
11. Karakatsanis, L.P.; Pavlos, E.G.; Tsoulouhas, G.; Stamokostas, G.L.; Mosbruger, T.; Duke, J.L.; Pavlos, G.P.; Monos, D.S. Spatial constrains and information content of sub-genomic regions of the human genome. *Iscience* **2021**, *24*, 102048. [CrossRef]
12. Rosenhead, J.; Franco, L.A.; Grint, K.; Friedland, B. Complexity theory and leadership practice: A review, a critique, and some recommendations. *Leadersh. Q.* **2019**, *30*, 101304. [CrossRef]
13. Rusoja, E.; Haynie, D.; Sievers, J.; Mustafee, N.; Nelson, F.; Reynolds, M.; Sarriot, E.; Williams, B.; Swanson, R.C. Thinking about complexity in health: A systematic review of the key systems thinking and complexity ideas in health. *J. Eval. Clin. Pract.* **2018**, *24*, 600–606. [CrossRef]
14. Lecun, Y. How does the brain learn so much so quickly? In Proceedings of the Cognitive Computational Neuroscience (CCN), New York, NY, USA, 6–8 September 2017.
15. Shannon, C.E.; Weaver, W. *The Mathematical Theory of Communication*; The University of Illinois Press: Champaign, IL, USA, 1949.
16. Kolmogorov, A.N. Entropy per unit time as a metric invariant of automorphism. *Dokl. Russ. Acad. Sci.* **1959**, *124*, 754–755.
17. Sinai, Y.G. On the Notion of Entropy of a Dynamical System. *Dokl. Russ. Acad. Sci.* **1959**, *124*, 768–771.
18. Kolmogorov, A. On Tables of Random Numbers. *Sankhy Indian J. Stat. Ser. A* **1963**, *25*, 369–375. [CrossRef]
19. Kolmogorov, A. On Tables of Random Numbers. *Theor. Comput. Sci.* **1998**, *207*, 387–395. [CrossRef]
20. Chaitin, G.J. On the Simplicity and Speed of Programs for Computing Infinite Sets of Natural Numbers. *J. ACM* **1969**, *16*, 407–422. [CrossRef]
21. Ziv, J.; Lempel, A. Compression of individual sequences via variable-rate coding. *IEEE Trans. Inf. Theory* **1978**, *24*, 530–536. [CrossRef]
22. Kastens, K.A.; Manduca, C.A.; Cervato, C.; Frodeman, R.; Goodwin, C.; Liben, L.S.; Mogk, D.W.; Spangler, T.C.; Stillings, N.A.; Titus, S. How geoscientists think and learn. *Eos Trans. Am. Geophys. Union* **2009**, *90*, 265–272. [CrossRef]
23. Goldstein, J. Emergence as a Construct: History and Issues. *Emergence* **1999**, *1*, 49–72. [CrossRef]
24. Corning, P.A. The Re-Emergence of "Emergence": A Venerable Concept in Search of a Theory. *Complexity* **2002**, *7*, 18–30. [CrossRef]
25. Lin, C.C.; Shu, F.H. On the spiral structure of disk galaxies. *Astrophys. J.* **1964**, *140*, 646–655. [CrossRef]
26. Vasavada, A.R.; Showman, A. Jovian atmospheric dynamics: An update after Galileo and Cassini. *Rep. Prog. Phys.* **2005**, *68*, 1935–1996. [CrossRef]
27. Zhang, G.M.; Yu, L. Emergent phenomena in physics. *Physics* **2010**, *39*, 543. (In Chinese)
28. Hemelrijk, C.K.; Hildenbrandt, H. Some Causes of the Variable Shape of Flocks of Birds. *PLoS ONE* **2011**, *6*, e22479. [CrossRef]
29. Hildenbrandt, H.; Carere, C.; Hemelrijk, C.K. Self-organized aerial displays of thousands of starlings: A model. *Behav. Ecol.* **2010**, *21*, 1349–1359. [CrossRef]
30. Shaw, E. Schooling fishes. *Am. Sci.* **1978**, *66*, 166–175. [CrossRef]
31. Reynolds, C.W. Flocks, herds and schools: A distributed behavioral model. *Comput. Graph.* **1987**, *21*, 25–34. [CrossRef]

32. D'Orsogna, M.R.; Chuang, Y.L.; Bertozzi, A.L.; Chayes, L.S. Self-Propelled Particles with Soft-Core Interactions: Patterns, Stability, and Collapse. *Phys. Lett.* **2006**, *96*, 10. [CrossRef] [PubMed]
33. Hemelrijk, C.K.; Hildenbrandt, H. Self-Organized Shape and Frontal Density of Fish Schools. *Ethology* **2007**, *114*, 3. [CrossRef]
34. Kroy, K.; Sauermann, G.; Herrmann, H.J. Minimal model for sand dunes. *Phys. Rev. Lett.* **2002**, *88*, 054301. [CrossRef]
35. Manson, S.M. Simplifying complexity: A review of complexity theory. *Geoforum* **2001**, *32*, 405–414. [CrossRef]
36. Tang, L.; Lv, H.; Yang, F.; Yu, L. Complexity testing techniques for time series data: A comprehensive literature review. *Chaos Solitons Fractals* **2015**, *81*, 117–135. [CrossRef]
37. Newman, M.E.J. Power laws, Pareto distributions and Zipf's law. *Contemp. Phys.* **2005**, *46*, 323–351. [CrossRef]
38. Gao, J.B.; Cao, Y.H.; Tung, W.W.; Hu, J. *Multiscale Analysis of Complex Time Series—Integration of Chaos and Random Fractal Theory, and Beyond*; Wiley: Hoboken, NJ, USA, August 2007.
39. Pareto, V. La legge della domanda. In *Ecrits d'Economie Politique Pure*; Pareto, Ed.; Librairie Droz: Geneve, Switzerland, 1895; Chapter 11, pp. 295–304.
40. Benford, F. The Law of Anomalous Numbers. *Proc. Am. Philos. Soc.* **1938**, *78*, 551–572.
41. Pietronero, L.; Tosatti, E.; Tosatti, V.; Vespignani, A. Explaining the uneven distribution of numbers in nature: The laws of Benford and Zipf. *Phys. A* **2011**, *293*, 297–304. [CrossRef]
42. Varian, H. Benford's Law (Letters to the Editor). *Am. Stat.* **1972**, *26*, 65.
43. From Benford to Erdös. Radio Lab. Episode 2009-10-09. 30 September 2009. Available online: https://www.wnycstudios.org/podcasts/radiolab/segments/91699-from-benford-to-erdos (accessed on 19 June 2021).
44. Election forensics, The Economist (22 February 2007). Available online: https://www.economist.com/science-and-technology/2007/02/22/election-forensics (accessed on 19 June 2021).
45. Deckert, J.; Myagkov, M.; Ordeshook, P.C. Benford's Law and the Detection of Election Fraud. *Political Anal.* **2011**, *19*, 245–268. [CrossRef]
46. Mebane, W.R. Comment on Benford's Law and the Detection of Election Fraud. *Political Anal.* **2011**, *19*, 269–272. [CrossRef]
47. Goodman, W. The promises and pitfalls of Benford's law. *Significance R. Stat. Soc.* **2016**, *13*, 38–41. [CrossRef]
48. Sehity, T.; Hoelzl, E.; Kirchler, E. Price developments after a nominal shock: Benford's Law and psychological pricing after the euro introduction. *Int. J. Res. Mark.* **2005**, *22*, 471–480. [CrossRef]
49. Durant, W.; Durant, A. *The Story of Civilization, The Age of Louis XIV*; Simon & Schuster: New York, NY, USA, 1963; p. 720.
50. Durant, W.; Durant, A. *The Story of Civilization, Rousseau and Revolution*; Simon & Schuster: New York, NY, USA, 1967; p. 643.
51. Gao, J.B.; Hu, J.; Mao, X.; Zhou, M.; Gurbaxani, B.; Lin, J.W.-B. Entropies of negative incomes, Pareto-distributed loss, and financial crises. *PLoS ONE* **2011**, *6*, e25053. [CrossRef]
52. Fan, F.L.; Gao, J.B.; Liang, S.H. Crisis-like behavior in China's stock market and its interpretation. *PLoS ONE* **2015**, *10*, e0117209. [CrossRef]
53. Bowers, M.C.; Tung, W.W.; Gao, J.B. On the distributions of seasonal river flows: Lognormal or powerlaw? *Water Resour. Res.* **2012**, *48*, W05536. [CrossRef]
54. Deligne, N.; Coles, S.; Sparks, R. Recurrence rates of large explosive volcanic eruptions. *J. Geophys. Res.* **2010**, *115*, B06203. [CrossRef]
55. Tsallis, C. Possible generalization of Boltzmann-Gibbs statistics. *J. Stat. Phys.* **1988**, *52*, 479–487. [CrossRef]
56. Tsallis, C. *Introduction to Nonextensive Statistical Mechanics: Approaching a Complex World*; Springer: Berlin/Heidelberg, Germany, 2009.
57. Pavlos, G.P.; Karakatsanis, L.P.; Xenakis, M.N.; Pavlos, E.G.; Iliopoulos, A.C.; Sarafopoulos, D.V. Universality of non-extensive Tsallis statistics and time series analysis: Theory and applications. *Phys. A Stat. Mech. Appl.* **2014**, *395*, 58–95. [CrossRef]
58. Barabasi, A.L.; Albert, R. Emergence of scaling in random networks. *Science* **1999**, *286*, 509–512. [CrossRef] [PubMed]
59. Gao, J.B.; Han, Q.; Lu, X.L.; Yang, L.; Hu, J. Self organized hotspots and social tomography. *EAI Endorsed Trans. Complex Syst.* **2013**, *13*, e1. [CrossRef]
60. Jones, M. Phase space: Geography, relational thinking, and beyond. *Prog. Hum. Geogr.* **2009**, *33*, 487–506. [CrossRef]
61. Henon, M. A two-dimensional mapping with a strange attractor. *Commun. Math. Phys.* **1976**, *50*, 69–77. [CrossRef]
62. Shields, P. *The Theory of Bernoulli Shifts*; Univ. Chicago Press: Chicago, IL, USA, 1973.
63. Atmanspacher, H.; Scheingraber, H. A fundamental link between system theory and statistical mechanics. *Found. Phys.* **1987**, *17*, 939–963. [CrossRef]
64. Grassberger, P.; Procaccia, I. Estimation of the Kolmogorov entropy from a chaotic signal. *Phys. Rev. A* **1983**, *28*, 2591–2593. [CrossRef]
65. Feigenbaum, M.J. Universal behavior in nonlinear systems. *Phys. D* **1983**, *7*, 16–39. [CrossRef]
66. Ruelle, D.; Takens, F. On the nature of turbulence. *Commun. Math. Phys.* **1971**, *20*, 167. [CrossRef]
67. Pomeau, Y.; Manneville, P. Intermittent transition to turbulence in dissipative dynamical systems. *Commun. Math. Phys.* **1980**, *74*, 189–197. [CrossRef]
68. Gao, J.B.; Rao, N.S.V.; Hu, J.; Jing, A. Quasi-periodic route to chaos in the dynamics of Internet transport protocols. *Phys. Rev. Lett.* **2005**, *94*, 198702. [CrossRef] [PubMed]
69. Packard, N.H.; Crutchfield, J.P.; Farmer, J.D.; Shaw, R.S. Geometry from a time series. *Phys. Rev. Lett.* **1980**, *45*, 712–716. [CrossRef]

70. Takens, F. Detecting strange attractors in turbulence. In *Dynamical Systems and Turbulence, Lecture Notes in Mathematics*; Rand, D.A., Young, L.S., Eds.; Springer: Berlin/Heidelberg, Germany, 1981; Volume 898, p. 366.
71. Sauer, T.; Yorke, J.A.; Casdagli, M. Embedology. *J. Stat. Phys.* **1991**, *65*, 579–616. [CrossRef]
72. Abarbanel, H.D.I. *Analysis of Observed Chaotic Data*; Springer: Berlin/Heidelberg, Germany, 1996.
73. Gao, J.B.; Zheng, Z.M. Local exponential divergence plot and optimal embedding of a chaotic time series. *Phys. Lett. A* **1993**, *181*, 153–158. [CrossRef]
74. Gao, J.B.; Zheng, Z.M. Direct dynamical test for deterministic chaos and optimal embedding of a chaotic time series. *Phys. Rev. E* **1994**, *49*, 3807–3814. [CrossRef]
75. Wolf, A.; Swift, J.B.; Swinney, H.L.; Vastano, J.A. Determining Lyapunov exponents from a time series. *Phys. D* **1985**, *16*, 285–317. [CrossRef]
76. Gao, J.B.; Hu, J.; Tung, W.W.; Zheng, Y. Multiscale analysis of economic time series by scale-dependent Lyapunov exponent. *Quant. Financ.* **2013**, *13*, 265–274. [CrossRef]
77. Rosenstein, M.T.; Collins, J.J.; De Luca, C.J. Reconstruction expansion as a geometry-based framework for choosing proper delay times. *Phys. D* **1994**, *73*, 82–98. [CrossRef]
78. Kantz, H. A robust method to estimate the maximal Lyapunov exponent of a time series. *Phys. Lett. A* **1994**, *185*, 77–87. [CrossRef]
79. Gao, J.B.; Zheng, Z.M. Direct dynamical test for deterministic chaos. *Europhys. Lett.* **1994**, *25*, 485–490. [CrossRef]
80. Gao, J.B.; Tung, W.W. Pathological tremors as diffusional processes. *Biol. Cybern.* **2002**, *86*, 263–270. [CrossRef]
81. Gao, J.B. Recognizing randomness in a time series. *Phys. D* **1997**, *106*, 49–56. [CrossRef]
82. Gao, J.B.; Chen, C.C.; Hwang, S.K.; Liu, J.M. Noise-induced chaos. *Int. J. Mod. Phys. B* **1999**, *13*, 3283–3305. [CrossRef]
83. Hu, S.Q.; Raman, A. Chaos in Atomic Force Microscopy. *Phys. Rev. Lett.* **2006**, *96*, 036107. [CrossRef]
84. Grassberger, P.; Procaccia, I. Characterization of strange attractors. *Phys. Rev. Lett.* **1983**, *50*, 346–349. [CrossRef]
85. Theiler, J. Spurious dimension from correlation algorithms applied to limited time-series data. *Phys. Rev. A* **1986**, *34*, 2427–2432. [CrossRef]
86. Gao, J.B.; Hu, J.; Liu, F.Y.; Cao, Y.H. Multiscale entropy analysis of biological signals: A fundamental bi-scaling law. *Front. Comput. Neurosci.* **2015**, *9*, 64. [CrossRef]
87. Theiler, J.; Eubank, S.; Longtin, A.; Galdrikian, B.; Farmer, J.D. Testing for nonlinearity in time series: The method of surrogate data. *Phys. D Nonlinear Phenom.* **1992**, *58*, 77–94. [CrossRef]
88. Lancaster, G.; Iatsenko, D.; Pidde, A.; Ticcinelli, V.; Stefanovska, A. Surrogate data for hypothesis testing of physical systems. *Phys. Rep.* **2018**, *748*, 1–60. [CrossRef]
89. Cuomo, K.M.; Oppenheim, A.V. Circuit implementation of synchronized chaos with applications to communications. *Phys. Rev. Lett.* **1993**, *71*, 65–68. [CrossRef] [PubMed]
90. Uchida, A.; Amano, K.; Inoue, M.; Hirano, K.; Naito, S.; Someya, H.; Oowada, I.; Kurashige, T.; Shiki, M.; Yoshimori, S.; et al. Fast physical random bit generation with chaotic semiconductor lasers. *Nat. Photon.* **2008**, *2*, 728. [CrossRef]
91. Sciamanna, M.; Shore, K. Physics and applications of laser diode chaos. *Nat. Photon.* **2015**, *9*, 151. [CrossRef]
92. Soriano, M.C.; Garcia-Ojalvo, J.; Mirasso, C.R.; Fischer, I. Complex photonics: Dynamics and applications of delay-coupled semiconductors lasers. *Rev. Modern Phys.* **2013**, *85*, 421. [CrossRef]
93. Harayama, T.; Sunada, S.; Yoshimura, K.; Muramatsu, J.; Arai, K.-i.; Uchida, A.; Davis, P. Theory of fast nondeterministic physical random-bit generation with chaotic lasers. *Phys. Rev. E* **2012**, *85*, 046215. [CrossRef]
94. Mikami, T.; Kanno, K.; Aoyama, K.; Uchida, A.; Ikeguchi, T.; Harayama, T.; Sunada, S.; Arai, K.-i.; Yoshimura, K.; Davis, P. Estimation of entropy rate in a fast physical random-bit generator using a chaotic semiconductor laser with intrinsic noise. *Phys. Rev. E* **2012**, *85*, 016211. [CrossRef]
95. Sunada, S.; Harayama, T.; Davis, P.; Tsuzuki, K.; Arai, K.; Yoshimura, K.; Uchida, A. Noise amplification by chaotic dynamics in a delayed feedback laser system and its application to nondeterministic random bit generation. *Chaos* **2012**, *22*, 047513. [CrossRef]
96. Durt, T.; Belmonte, C.; Lamoureux, L.P.; Panajotov, K.; Van den Berghe, F.; Thienpont, H. Fast quantum-optical random-number generators. *Phys. Rev. A* **2013**, *87*, 022339. [CrossRef]
97. Yoshimura, K.; Muramatsu, J.; Davis, P.; Harayama, T.; Okumura, H.; Morikatsu, S.; Aida, H.; Uchida, A. Secure Key Distribution Using Correlated Randomness in Lasers Driven by Common Random Light. *Phys. Rev. Lett.* **2012**, *108*, 070602. [CrossRef] [PubMed]
98. Kanno, K.; Uchida, A. Consistency and complexity in coupled semiconductor lasers with time-delayed optical feedback. *Phys. Rev. E* **2012**, *86*, 066202. [CrossRef] [PubMed]
99. Li, X.-Z.; Zhuang, J.-P.; Li, S.-S.; Gao, J.B.; Chan, S.C. Randomness evaluation for an optically injected chaotic semiconductor laser by attractor reconstruction. *Phys. Rev. E* **2016**, *94*, 042214. [CrossRef]
100. Pecora, L.M.; Carroll, T.L. Synchronization in chaotic systems. *Phys. Rev. Lett.* **1990**, *64*, 821–824. [CrossRef]
101. Fujisaka, H.; Yamada, T. Stability theory of synchronized motion in coupled-oscillator systems. *Prog. Theor. Phys.* **1983**, *69*, 32. [CrossRef]
102. Carroll, T.L.; Pecora, L.M. Synchronizing chaotic circuits. *IEEE Trans. Circ. Syst.* **1991**, *38*, 453–456. [CrossRef]
103. Afraimovich, V.S.; Verichev, N.N.; Rabinovich, M.I. Stochastic synchronization of oscillations in dissipative systems. *Radiophys. Quantum Electron.* **1986**, *29*, 795. [CrossRef]

104. Yamada, T.; Fujisaka, H. Stability theory of synchronized motion in coupled-oscillator systems. II. *Prog. Theor. Phys.* **1983**, *70*, 1240. [CrossRef]
105. Afraimovich, V.S.; Verichev, N.N.; Rabinovich, M.I. Stochastic synchronization of oscillations in dissipative systems. *Izv. Vyssh. Uchebn. Zaved. Radiofiz.* **1986**, *29*, 1050. [CrossRef]
106. Yamada, T.; Fujisaka, H. Stability theory of synchronized motion in coupled-oscillator systems. III. *Prog. Theor. Phys.* **1984**, *72*, 885. [CrossRef]
107. Fujisaka, H.; Yamada, T. Stability theory of synchronized motion in coupled-oscillator systems. IV. *Prog. Theor. Phys.* **1985**, *74*, 918. [CrossRef]
108. Pikovskii, A.S. Synchronization and stochastization of array of selfexcited oscillators by external noise. *Radiophys. Quantum Electron.* **1984**, *27*, 390. [CrossRef]
109. Volkovskii, A.R.; Rulkov, N.F. Experimental study of bifurcations at the threshold for stochastic locking. *Sov. Tech. Phys. Lett.* **1989**, *15*, 249.
110. Aranson, I.S.; Rulkov, N.F. Nontrivial structure of synchronization zones in multidimensional systems. *Phys. Lett. A* **1989**, *139*, 375. [CrossRef]
111. Pikovskii, A. On the interaction of strange attractors. *Z. Phys. B* **1984**, *55*, 149. [CrossRef]
112. Locquet, A. Chaos-Based secure optical communications using semiconductor lasers. In *Handbook of Information and Communication Security*; Stavroulakis, P., Stamp, M., Eds.; Springer: Berlin/Heidelberg, Germany, 2010; pp. 451–478.
113. Argyris, A.; Syvridis, D.; Larger, L.; Annovazzi- Lodi, V.; Colet, P.; Fischer, I.; Garcia-Ojalvo, J.; Mirasso, C.R.; Pesquera, L.; Shore, K.A. Chaos-based communications at high bit rates using commercial fibre-optic links. *Nature* **2005**, *438*, 343–346. [CrossRef]
114. Crutchfield, J.P.; Huberman, B.A. Fluctuationa and the onset of chaos. *Phys. Lett.* **1980**, *74*, 407. [CrossRef]
115. Crutchfield, J.P.; Farmer, J.D.; Huberman, B.A. Fluctuations and simple chaotic dynamics. *Phys. Rep.* **1982**, *92*, 45–82. [CrossRef]
116. Kautz, R.L. Chaos and thermal noise in the RF-biased Josephson junction. *J. Appl. Phys.* **1985**, *58*, 424. [CrossRef]
117. Hwang, K.; Gao, J.B.; Liu, J.M. Noise-induced chaos in an optically injected semiconductor laser. *Phys. Rev. E* **2000**, *61*, 5162–5170. [CrossRef]
118. Gao, J.B.; Hwang, S.K.; Liu, J.M. When can noise induce chaos? *Phys. Rev. Lett.* **1999**, *82*, 1132. [CrossRef]
119. Alexandrov, D.V.; Bashkirtseva, I.A.; Ryashko, L.B. Noise-induced chaos in non-linear dynamics of El Ninos. *Phys. Lett. A* **2018**, *382*, 2922–2926. [CrossRef]
120. Lei, Y.M.; Hua, M.J.; Du, L. Onset of colored-noise-induced chaos in the generalized Duffing system. *Nonlinear Dyn.* **2017**, *89*, 1371–1383. [CrossRef]
121. Mandelbrot, B.B. *The Fractal Geometry of Nature*; Freeman: San Francisco, CA, USA, 1982.
122. Bassingthwaighte, J.B.; Liebovitch, L.S.; West, B.J. *Fractal Physiology*; Oxford University Press: Oxford, UK, 1994.
123. Pandey, A. *Practical Microstrip and Printed Antenna Design*; Artech House: Norwood, MA, USA, 2019; pp. 253–260, ISBN 9781630816681.
124. Gao, J.B.; Cao, Y.H.; Lee, J.M. Principal Component Analysis of 1/f Noise. *Phys. Lett. A* **2003**, *314*, 392–400. [CrossRef]
125. Li, W.; Kaneko, K. Long-range correlation and partial 1/f-alpha spectrum in a noncoding DNA-sequence. *Europhys. Lett.* **1992**, *17*, 655–660. [CrossRef]
126. Voss, R.F. Evolution of long-range fractal correlations and 1/f noise in DNA base sequences. *Phys. Rev. Lett.* **1992**, *68*, 3805. [CrossRef]
127. Peng, C.K.; Buldyrev, S.V.; Goldberger, A.L.; Havlin, S.; Sciortino, F.; Simons, M.; Stanley, H.E. Long-range correlations in nucleotide sequences. *Nature* **1992**, *356*, 168. [CrossRef] [PubMed]
128. Gao, J.; Qi, Y.; Cao, Y.; Tung, W.W. Protein coding sequence identification by simultaneously characterizing the periodic and random features of DNA sequences. *J. Biomed. Biotechnol.* **2005**, *2005*, 139–146. [CrossRef]
129. Hu, J.; Gao, J.B.; Cao, Y.H.; Bottinger, E.; Zhang, W.J. Exploiting noise in array CGH data to improve detection of DNA copy number change. *Nucleic Acids Res.* **2007**, *35*, e35. [CrossRef]
130. Gilden, D.L.; Thornton, T.; Mallon, M.W. 1/f noise in human cognition. *Science* **1995**, *267*, 1837–1839. [CrossRef] [PubMed]
131. Chen, Y.; Ding, M.; Kelso, J.A.S. Long Memory Processes ($1/f^\alpha$ type) in Human Coordination. *Phys. Rev. Lett.* **1997**, *79*, 4501. [CrossRef]
132. Collins, J.J.; Luca, C.J.D. Random Walking during Quiet Standing. *Phys. Rev. Lett.* **1994**, *73*, 764. [CrossRef]
133. Furstenau, N. A nonlinear dynamics model for simulating long range correlations of cognitive bistability. *Biol. Cybern.* **2010**, *103*, 175–198. [CrossRef] [PubMed]
134. Gao, J.B.; Billock, V.A.; Merk, I.; Tung, W.W.; White, K.D.; Harris, J.G.; Roychowdhury, V.P. Inertia and memory in ambiguous visual perception. *Cogn. Process.* **2006**, *7*, 105–112. [CrossRef]
135. Ivanov, P.C.; Rosenblum, M.G.; Peng, C.K.; Mietus, J.; Havlin, S.; Stanley, H.E.; Goldberger, A.L. Scaling behaviour of heartbeat intervals obtained by wavelet-based time-series analysis. *Nature* **1996**, *383*, 323. [CrossRef]
136. Amaral, L.A.N.; Goldberger, A.L.; Ivanov, P.C.; Stanley, H.E. Scale-independent measures and pathologic cardiac dynamics. *Phys. Rev. Lett.* **1998**, *81*, 2388. [CrossRef] [PubMed]
137. Ivanov, P.C.; Rosenblum, M.G.; Amaral, L.A.N.; Struzik, Z.R.; Havlin, S.; Goldberger, A.L.; Stanley, H.E. Multifractality in human heartbeat dynamics. *Nature* **1999**, *399*, 461. [CrossRef]

138. Bernaola-Galvan, P.; Ivanov, P.C.; Amaral, L.A.N.; Stanley, H.E. Scale invariance in the nonstationarity of human heart rate. *Phys. Rev. Lett.* **2001**, *87*, 168105. [CrossRef]

139. Gao, J.B. Analysis of Amplitude and Frequency Variations of Essential and Parkinsonian Tremors. *Med. Biol. Eng. Comput.* **2004**, *52*, 345–349. [CrossRef]

140. Kuznetsov, N.; Bonnette, S.; Gao, J.B.; Riley, M.A. Adaptive fractal analysis reveals limits to fractal scaling in center of pressure trajectories. *Ann. Biomed. Eng.* **2012**, *41*, 1646–1660. [CrossRef]

141. Gao, J.B.; Hu, J.; Buckley, T.; White, K.; Hass, C. Shannon and Renyi Entropies To Classify Effects of Mild Traumatic Brain Injury on Postural Sway. *PLoS ONE* **2011**, *6*, e24446. [CrossRef]

142. Gao, J.B.; Gurbaxani, B.M.; Hu, J.; Heilman, K.J.; Emauele, V.A.; Lewis, G.F.; Davila, M.; Unger, E.R.; Lin, J.S. Multiscale analysis of heart rate variability in nonstationary environments. *Front. Comput. Physiol. Med.* **2013**, *4*, 119.

143. Gao, J.B.; Hu, J.; Tung, W.W. Complexity measures of brain wave dynamics. *Cogn. Neurodynamics* **2011**, *5*, 171–182. [CrossRef] [PubMed]

144. Zheng, Y.; Gao, J.B.; Sanchez, J.C.; Principe, J.C.; Okun, M.S. Multiplicative multifractal modeling and discrimination of human neuronal activity. *Phys. Lett. A* **2005**, *344*, 253–264. [CrossRef]

145. Hu, J.; Zheng, Y.; Gao, J.B. Long-range temporal correlations, multifractality, and the causal relation between neural inputs and movements. *Front. Neurol.* **2013**, *4*, 158. [CrossRef] [PubMed]

146. Zhu, H.B.; Gao, J.B. Fractal behavior in the headway fluctuation simulated by the NaSch model. *Phys. A* **2014**, *398*, 187–193. [CrossRef]

147. Bowers, M.; Gao, J.B.; Tung, W.W. Long-Range Correlations in Tree Ring Chronologies of the USA: Variation within and Across Species. *Geophys. Res. Lett.* **2013**, *40*, 568–572. [CrossRef]

148. Gao, J.B.; Fang, P.; Liu, F.Y. Empirical scaling law connecting persistence and severity of global terrorism. *Phys. A* **2017**, *482*, 74–86. [CrossRef]

149. Gao, J.B.; Hu, J.; Mao, X.; Perc, M. Culturomics meets random fractal theory: Insights into long-range correlations of social and natural phenomena over the past two centuries. *J. R. Soc. Interface* **2012**, *9*, 1956–1964. [CrossRef] [PubMed]

150. Wolf, M. 1/f noise in the distribution of prime numbers. *Phys. A* **1997**, *241*, 493. [CrossRef]

151. Gao, J.; Hu, J.; Tung, W.W.; Cao, Y.; Sarshar, N.; Roychowdhury, V.P. Assessment of long range correlation in time series: How to avoid pitfalls. *Phys. Rev. E* **2006**, *73*, 016117. [CrossRef] [PubMed]

152. Peng, C.K.; Buldyrev, S.V.; Havlin, S.; Simons, M.; Stanley, H.E.; Goldberger, A.L. Mosaic organization of dna nucleotides. *Phys. Rev. E* **1994**, *49*, 1685–1689. [CrossRef] [PubMed]

153. Arneodo, A.; Bacry, E.; Muzy, J.F. The thermodynamics of fractals revisited with wavelets. *Phys. A* **1995**, *213*, 232–275.

154. Gao, J.B.; Hu, J.; Tung, W.W. Facilitating joint chaos and fractal analysis of biosignals through nonlinear adaptive filtering. *PLoS ONE* **2011**, *6*, e24331. [CrossRef]

155. Tung, W.W.; Gao, J.B.; Hu, J.; Yang, L. Recovering chaotic signals in heavy noise environments. *Phys. Rev. E* **2011**, *83*, 046210. [CrossRef] [PubMed]

156. Gao, J.B.; Sultan, H.; Hu, J.; Tung, W.W. Denoising nonlinear time series by adaptive filtering and wavelet shrinkage: A comparison. *IEEE Signal Process. Lett.* **2010**, *17*, 237–240.

157. Riley, M.A.; Kuznetsov, N.; Bonnette, S.; Wallot, S.; Gao, J.B. A Tutorial Introduction to Adaptive Fractal Analysis. *Front. Fractal Physiol.* **2012**, *3*, 371. [CrossRef]

158. Frisch, U. *Turbulence—The Legacy of A.N. Kolmogorov*; Cambridge University Press: Cambridge, UK, 1995

159. Gouyet, J.F. *Physics and Fractal Structures*; Springer: Berlin/Heidelberg, Germany, 1995.

160. Frederiksen, R.D.; Dahm, W.J.A.; Dowling, D.R. Experimental assessment of fractal scale similarity in turbulent flows—Multifractal scaling. *J. Fluid Mech.* **1997**, *338*, 127–155. [CrossRef]

161. Mandelbrot, B.B. Intermittent turbulence in self-similar cascades: Divergence of high moments and dimension of carrier. *J. Fluid Mech.* **1974**, *62*, 331–358. [CrossRef]

162. Parisi, G.; Frisch, U. On the singularity structure of fully developed turbulence. In *Turbulence and Predictability in Geophysical Fluid Dynamics and Climate Dynamics*; Ghil, M., Benzi, R., Parisi, G., Eds.; North-Holland: Amsterdam, The Netherlands, 1985; pp. 71–84.

163. Gao, J.B.; Rubin, I. Multifractal modeling of counting processes of long-range-dependent network traffic. *Comput. Commun.* **2001**, *24*, 1400–1410. [CrossRef]

164. Gao, J.B.; Rubin, I. Multiplicative multifractal modeling of long-range-dependent network traffic. *Int. J. Commun. Syst.* **2001**, *14*, 783–801. [CrossRef]

165. Tung, W.W.; Moncrief, M.W.; Gao, J.B. A systemic view of the multiscale tropical deep convective variability over the tropical western Pacific warm pool. *J. Clim.* **2004**, *17*, 2736–2751. [CrossRef]

166. Hu, J.; Tung, W.W.; Gao, J.B. Detection of low observable targets within sea clutter by structure function based multifractal analysis. *IEEE Trans. Antennas Propag.* **2006**, *54*, 135–143. [CrossRef]

167. Osborne, A.R.; Provenzale, A. Finite correlation dimension for stochastic-systems with power-law spectra. *Phys. D* **1989**, *35*, 357–381. [CrossRef]

168. Provenzale, A.; Osborne, A.R.; Soj, R. Convergence of the K2 entropy for random noises with power law spectra. *Phys. D* **1991**, *47*, 361–372. [CrossRef]

169. Hu, J.; Gao, J.B.; Principe, J.C. Analysis of biomedical signals by the Lempel-Ziv complexity: The effect of finite data size. *IEEE Trans. Biomed. Eng.* **2006**, *53*, 2606–2609.
170. Galatolo, S.; Hoyrup, M.; Rojas, C. Effective symbolic dynamics, random points, statistical behavior, complexity and entropy. *Inf. Comput.* **2010**, *208*, 23–41. [CrossRef]
171. Gao, J.B.; Hu, J.; Tung, W.W. Entropy measures for biological signal analysis. *Nonlinear Dyn.* **2012**, *68*, 431–444. [CrossRef]
172. Bandt, C.; Pompe, B. Permutation entropy: A natural complexity measure for time series. *Phys. Rev. Lett.* **2002**, *88*, 174102. [CrossRef]
173. Cao, Y.H.; Tung, W.W.; Gao, J.B.; Protopopescu, V.A.; Hively, L.M. Detecting dynamical changes in time series using the permutation entropy. *Phys. Rev. E* **2004**, *70*, 1539–3755. [CrossRef] [PubMed]
174. Wang, J.; Wei, H.; Ye, C.; Ding, Y. Fractal behavior of traffic volume on urban expressway through adaptive fractal analysis. *Phys. A* **2016**, *443*, 518–525.
175. Shen, S.; Ye, S.J.; Cheng, C.X.; Song, C.Q.; Gao, J.B.; Yang, J.; Ning, L.X.; Su, K.; Zhang, T. Persistence and Corresponding Time Scales of Soil Moisture Dynamics During Summer in the Babao River Basin, Northwest China. *J. Geophys. Res. Atmos.* **2018**, *123*, 8936–8948. [CrossRef]
176. Zhang, T.; Shen, S.; Cheng, C.; Song, C.Q.; Ye, S.J. Long range correlation analysis of soil temperature and moisture on A'rou hillsides, Babao River basin. *J. Geophys. Res. Atmos.* **2018**, *123*, 12606–12620. [CrossRef]
177. Yang, J.; Su, K.; Ye, S. Stability and long-range correlation of air temperature in the Heihe River Basin. *J. Geogr. Sci.* **2019**, *29*, 1462–1474. [CrossRef]
178. Gao, J.B.; Fang, P.; Yuan, L.H. Analyses of geographical observations in the HeiheRiver Basin: Perspectives from complexity theory. *J. Geogr. Sci.* **2019**, *29*, 1441–1461. [CrossRef]
179. Jiang, A.; Gao, J. Fractal analysis of complex power load variations through adaptive multiscale filtering. In Proceedings of the International Conference on Behavioral, Economic and Socio-Cultural Computing (BESC—2016), Durham, NC, USA, 11–13 November 2016.
180. Li, Q.; Gao, J.B.; Zhang, Z.W.; Huang, Q.; Wu, Y.; Xu, B. Distinguishing Epileptiform Discharges from normal Electroencephalograms Using Adaptive Fractal and Network Analysis: A Clinical Perspective. *Front. Physiol.* **2020**, *11*, 828. [CrossRef] [PubMed]
181. Zheng, F.; Chen, L.; Gao, J.; Zhao, Y. Fully Quantum Modeling of Exciton Diffusion in Mesoscale Light Harvesting Systems. *Materials* **2021**, *14*, 3291. [CrossRef]
182. Gao, J.B.; Jockers, M.L.; Laudun, J.; Tangherlini, T. A multiscale theory for the dynamical evolution of sentiment in novels. In Proceedings of the International Conference on Behavioral, Economic and Socio-Cultural Computing (BESC—2016), Durham, NC, USA, 11–13 November 2016.
183. Hu, Q.Y.; Liu, B.; Thomsen, M.R.; Gao, J.B.; Nielbo, K.L. Dynamic evolution of sentiments in Never Let Me Go: Insights from multifractal theory and its implications for literary analysis. *Digit. Scholarsh. Humanit.* **2020**. [CrossRef]
184. Wever, M.; Nielbo, K.L.; Gao, J.B. Tracking the Consumption Junction: Temporal Dependencies in Dutch Newspaper Articles and Advertisements. *Digit. Humanit. Q.* **2020**, *14*, 2. Available online: http://www.digitalhumanities.org/dhq/vol/14/2/000445/000445.html (accessed on 19 June 2021).
185. Nielbo, K.L.; Baunvig, K.F.; Liu, B.; Gao, J.B. A curious case of entropic decay: Persistent complexity in textual cultural heritage. *Digit. Scholarsh. Humanit.* **2018**. [CrossRef]
186. Hu, J.; Gao, J.B.; Wang, X.S. Multifractal analysis of sunspot time series: The effects of the 11-year cycle and Fourier truncation. *J. Stat. Mech.* **2009**, *2009*, P02066. [CrossRef]
187. Wu, Z.H.; Huang, N.E.; Long, S.R.; Peng, C.K. On the trend, detrending, and variability of nonlinear and nonstationary time series. *Proc. Natl. Acad. Sci. USA* **2007**, *104*, 14889–14894. [CrossRef]
188. Podobnik, B.; Stanley, H.E. Detrended cross-correlation analysis: A new method for analyzing two non-stationary time series. *Phys. Rev. Lett.* **2008**, *100*, 84–102. [CrossRef] [PubMed]
189. Gao, J.B.; Hu, J.; Tung, W.W.; Cao, Y.H. Distinguishing chaos from noise by scale-dependent Lyapunov exponent. *Phys. Rev. E* **2006**, *74*, 066204. [CrossRef]
190. Torcini, A.; Grassberger, P.; Politi, A. Error Propagation in Extended Chaotic Systems. *J. Phys. A Math. Gen.* **1995**, *28*, 4533. [CrossRef]
191. Aurell, E.; Boffetta, G.; Crisanti, A.; Paladin, G.; Vulpiani, A. Growth of non-infinitesimal perturbations in turbulence. *Phys. Rev. Lett.* **1996**, *77*, 1262. [CrossRef] [PubMed]
192. Aurell, E.; Boffetta, G.; Crisanti, A.; Paladin, G.; Vulpiani, A. Predictability in the large: An extension of the concept of Lyapunov exponent. *J. Phys. A* **1997**, *30*, 1–26. [CrossRef]
193. Gao, J.B.; Tung, W.W.; Hu, J. Quantifying dynamical predictability: The pseudo-ensemble approach (in honor of Professor Andrew Majda's 60th birthday). *Chin. Ann. Math. Ser. B* **2009**, *30*, 569–588. [CrossRef]
194. Gao, J.B.; Hu, J.; Mao, X.; Tung, W.W. Detecting low-dimensional chaos by the "noise titration" technique: Possible problems and remedies. *Chaos Solitons Fractals* **2012**, *45*, 213–223. [CrossRef]
195. Hu, J.; Gao, J.B.; Tung, W.W.; Cao, Y.H. Multiscale analysis of heart rate variability: A comparison of different complexity measures. *Ann. Biomed. Eng.* **2010**, *38*, 854–864. [CrossRef]

196. Hu, J.; Gao, J.B.; Tung, W.W. Characterizing heart rate variability by scale-dependent Lyapunov exponent. *Chaos Interdiscip. J. Nonlinear Sci.* **2009**, *19*, 028506. [CrossRef]
197. Ryan, D.A.; Sarson, G.R. The geodynamo as a low-dimensional deterministic system at the edge of chaos. *EPL* **2008**, *83*, 49001. [CrossRef]
198. Fan, Q.B.; Wang, Y.X.; Zhu, L. Complexity analysis of spatial—Ctemporal precipitation system by PCA and SDLE. *Appl. Math. Model.* **2013**, *37*, 4059–4066. [CrossRef]
199. Hu, J.; Gao, J.B. Multiscale characterization of sea clutter by scale-dependent Lyapunov exponent. *Math. Probl. Eng.* **2013**, *2013*, 584252. [CrossRef]
200. Blasch, E.; Gao, J.B.; Tung, W.W. Chaos-based Image Assessment for THz Imagery. In Proceedings of the 2012 11th International Conference on Information Science, Signal Processing and Their Applications, Montreal, QC, Canada, 3–5 July 2012.
201. Li, Q.; Gao, J.B.; Huang, Q.; Wu, Y.; Xu, B. Distinguishing Epileptiform Discharges from Normal Electroencephalograms Using Scale-Dependent Lyapunov Exponent. *Front. Bioeng. Biotechnol.* **2020**, *8*, 1006. [CrossRef]
202. Gao, J.B.; Hu, J.; Tung, W.W.; Blasch, E. Multiscale analysis of physiological data by scale-dependent Lyapunov exponent. *Front. Fractal Physiol.* **2012**, *2*, 110. [CrossRef]
203. Berera, A.; Ho, R.D.J.G. Chaotic Properties of a Turbulent Isotropic Fluid. *Phys. Rev. Lett.* **2018**, *120*, 024101. [CrossRef]
204. Blakely, J.N.; Corron, N.J.; Pethel, S.D.; Stahl, M.T.; Gao, J.B. Non-autonomous Boolean chaos in a driven ring oscillator. In *New Research Trends in Nonlinear circuits—Design, Chaotic Phenomena and Applications*; Kyprianidis, I., Stouboulos, I., Volos, C., Eds.; Nova Publishers: New York, NY, USA, 2014; Chapter 8, pp. 153–168.
205. Lazer, D.; Pentland, A.; Adamic, L.; Aral, S.; Barabasi, A.L.; Brewer, D.; Christakis, N.; Contractor, N.; Fowler, J.; Van Alstyne, M.; et al. Computational social science. *Science* **2009**, *323*, 721–723. [CrossRef] [PubMed]
206. Goldstein, J.S. A Conflict-Cooperation Scale for WEIS Events Data. *J. Confl. Resolut.* **1992**, *36*, 369–385. [CrossRef]
207. Schrodt, P.A.; Gerner, D.J.; Ömür, G. Conflict and Mediation Event Observations (CAMEO): An Event Data Framework for a Post Cold War World. In *International Conflict Mediation: New Approaches and Findings*; Bercovitch, J., Gartner, S., Eds.; Routledge: New York, NY, USA, 2009.
208. O'Brien, S.P. Crisis Early Warning and Decision Support: Contemporary Approaches and Thoughts on Future Research. *Int. Stud. Rev.* **2010**, *12*, 87–104. [CrossRef]
209. Turchin, P. *Historical Dynamics: Why States Rise and Fall*; Princeton University Press: Princeton, NJ, USA, 2003.
210. Turchin, P. Arise 'cliodynamics'. *Nature* **2008**, *454*, 34–35. [CrossRef] [PubMed]

MDPI

St. Alban-Anlage 66

4052 Basel

Switzerland

www.mdpi.com

Applied Sciences Editorial Office

E-mail: applsci@mdpi.com

www.mdpi.com/journal/applsci